Birds of Two Worlds

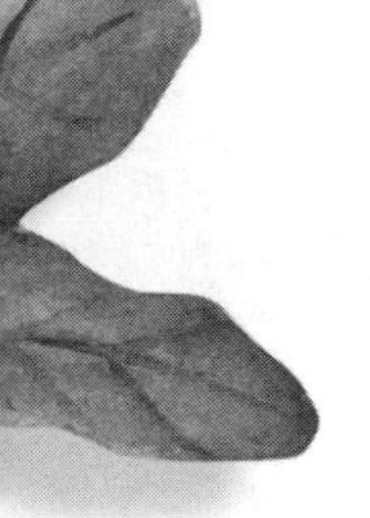

THE ECOLOGY AND EVOLUTION OF MIGRATION

Birds of Two Worlds

Edited by RUSSELL GREENBERG
and PETER P. MARRA
Smithsonian Institution, Washington, D.C.

THE JOHNS HOPKINS UNIVERSITY PRESS
Baltimore and London

Printed in the United States of America on acid-free paper

9 8 7 6 5 4 3 2 1

The Johns Hopkins University Press
2715 North Charles Street
Baltimore, Maryland 21218-4363
www.press.jhu.edu

Library of Congress Cataloging-in-Publication Data

Birds of two worlds: the ecology and evolution of migration / edited by Russell Greenberg and Peter P. Marra.
p. cm.
Includes bibliographical references and index.
ISBN 0-8018-8107-2 (hardcover : alk. paper)
1. Birds—Migration. I. Greenberg, Russell. II. Marra, Peter P.
QL698.9.B575 2004
598.156′8—dc22 2004019611

A catalog record for this book is available from the British Library.

Illustrations on title page and part openers by Dan Lane

Contents

Contributors

Allan J. Baker, Centre for Biodiversity and Conservation Biology, Royal Ontario Museum, and Department of Zoology, University of Toronto, Canada

Christopher Paul Bell, Conservation Programmes, Zoological Society of London, England

Carroll G. Belser, Department of Biological Sciences, Clemson University, United States

Eldredge Bermingham, Smithsonian Tropical Research Institute, Panama

Keith L. Bildstein, Acopian Center, Hawk Mountain Sanctuary, United States

Katrin Böhning-Gaese, Institut für Zoologie, Abteilung V, Johannes Gutenberg–Universität Mainz, Germany

Luke K. Butler, Burke Museum and Department of Zoology, University of Washington, United States

R. Terry Chesser, Australian National Wildlife Collection, CSIRO Sustainable Ecosystems, Australia

Sonya M. Clegg, Department of Biological Sciences, Imperial College London, England

William W. Cochran, Department of Electrical Engineering, University of Illinois at Champaign-Urbana, United States

Inês de Lima Serrano do Nascimento, Australasian Wader Studies Group, Australia

Alfred M. Dufty Jr., Department of Biology, Boise State University, United States

John Faaborg, Division of Biological Sciences, University of Missouri–Columbia, United States

Sylvia M. Fallon, Department of Biology, University of Missouri–St. Louis, United States
Robert C. Fleischer, Genetics Program, National Museum of Natural History, Smithsonian Institution, United States
Jukka T. Forsman, Department of Biology, University of Oulu, Finland
Daniel R. Froehlich, Burke Museum and Department of Zoology, University of Washington, United States
Sidney A. Gauthreaux Jr., Department of Biological Sciences, Clemson University, United States
Patricia M. González, Fundacion Inalafquen, Argentina
Russell Greenberg, Smithsonian Migratory Bird Center, National Zoological Park, United States
Sean Gross, Department of Ecology and Evolution, University of Chicago, United States
Keith A. Hobson, Prairie and Northern Wildlife Research Center, Canadian Wildlife Service, Canada
Philip A. R. Hockey, Percy FitzPatrick Institute of African Ornithology, University of Cape Town, South Africa
Rebecca L. Holberton, Department of Biological Sciences, University of Maine, United States
Richard T. Holmes, Department of Biological Sciences, Dartmouth College, United States
Darren E. Irwin, Department of Animal Ecology, Lund University, Sweden
Jessica H. Irwin, Department of Animal Ecology, Lund University, Sweden
Matthew D. Johnson, Department of Wildlife, Humboldt State University, United States
Leo Joseph, Department of Ornithology, Philadelphia Academy of Natural Sciences, United States
William H. Karasov, Department of Wildlife Ecology, University of Wisconsin, Madison, United States
Mari Kimura, Center for Tropical Research and Department of Biology, San Francisco State University, United States
Steven C. Latta, Point Reyes Bird Observatory, United States
Bernd Leisler, Max Planck Research Center for Ornithology, Vogelwarte Radolfzell, Germany
Douglas J. Levey, Department of Zoology, University of Florida, United States
Åke Lindström, Department of Animal Ecology, Lund University, Sweden
Irby J. Lovette, Evolutionary Biology Program, Cornell Laboratory of Ornithology, United States
Peter P. Marra, Smithsonian Environmental Research Center, United States
Scott R. McWilliams, Department of Natural Resources Science, University of Rhode Island, United States
Claudia Mettke-Hofmann, Max Planck Research Center for Ornithology, Department of Biological Rhythms and Behavior, Germany
Jenny E. Michi, Department of Biological Sciences, Clemson University, United States
Borja Milá, Department of Ecology and Evolutionary Biology and Center for Tropical Research, Institute of the Environment, University of California–Los Angeles, United States
Clive D. T. Minton, Centro de Estudos de Migraçoes de Aves (CEMAVE-IBAMA), Brazil
Anders Pape Møller, Laboratoire de Parasitologie Evolutive, Université Pierre et Marie Curie, France
Mikko Mönkkönen, Department of Biology, University of Oulu, Finland
Frank R. Moore, Department of Biological Sciences, University of Southern Mississippi, United States
Eugene S. Morton, Conservation and Research Center, Smithsonian Institution, United States
Lawrence J. Niles, Endangered and Nongame Species Program, New Jersey Division of Fish and Wildlife, United States
Theunis Piersma, Animal Ecology Group, Center for Ecological and Evolutionary Studies, University of Groningen, The Netherlands
Trevor E. Pitcher, Department of Zoology, University of Toronto, Canada
Trevor Price, Department of Ecology and Evolution, University of Chicago, United States
Robert E. Ricklefs, Department of Biology, University of Missouri–St. Louis, United States
Christopher M. Rogers, Department of Biological Sciences, Wichita State University, United States
Danny I. Rogers, School of Environmental and Information Sciences, Charles Stuart University, Australia
Sievert Rohwer, Burke Museum and Department of Zoology, University of Washington, United States
Kristen C. Ruegg, Center for Tropical Research and Department of Biology, San Francisco State University, United States
Michael C. Runge, United States Geological Survey, Patuxent Wildlife Research Center, United States
Volker Salewski, Forschungsstelle für Ornithologie der Max-Planck-Gesellschaft, Germany
Roland Sandberg, Department of Animal Ecology, Lund University, Sweden
Thomas W. Sherry, Department of Ecology and Evolutionary Biology, Tulane University, United States
T. Scott Sillett, Smithsonian Migratory Bird Center, National Zoological Park, United States
Robert J. Smith, Department of Biology, University of Scranton, United States
Thomas B. Smith, Department of Ecology and Evolutionary Biology and Center for Tropical Research, Institute of the Environment, University of California–Los Angeles, United States
David W. Steadman, Florida Museum of Natural History, University of Florida, United States
Allan M. Strong, School of Natural Resources, University of Vermont, United States
Bridget J. M. Stutchbury, Department of Biology, York University, Canada

Bethany L. Swanson, Department of Biology, University of Missouri–St. Louis, United States

Tibor Szép, Department of Environmental Sciences, College of Nyíregyháza, Hungary

Michael S. Webster, School of Biological Sciences, Washington State University, United States

Martin Wikelski, Department of Ecology and Evolutionary Biology, Princeton University, United States

David W. Winkler, Department of Ecology and Evolutionary Biology, Cornell University, United States

Hans Winkler, Konrad Lorenz Institute of Comparative Ethology, Austria

Jorje I. Zalles, Acopian Center, Hawk Mountain Sanctuary, United States

Leo Zwarts, Rijksinstituut voor Integraal Zoetwater-beheer en Afvalwaterbehandeling, The Netherlands

Preface

MOVEMENT IS THE HALLMARK of avian life. The freedom with which most birds fly through their environment shapes every aspect of their lives—from the food items they consume and their response to predators, to the size of their territories, and their dependence upon long-distance vocal signals to communicate. But mobility of birds plays its most dramatic role in migration. Migration is the regular movement from a breeding site to a location (or locations) to survive the rest of the year. We may never know when birds first started migrating, but over the tens of millions of years of these annual pilgrimages, migration has proliferated and lengthened, to the point that thousands of species migrate some distance and hundreds of species move thousands of miles between temperate, arctic, and tropical environments.

The attention of ornithologists is riveted on the very long temperate-tropical migration systems partly because of the spectacular feat of moving vast distances, often at high altitude and over some of the most formidable barriers. Behaviorists fascinated with how information is stored and processed have long been challenged by how birds find and remember their route between very precisely defined locations in different continents. Physiologists and morphologists probe the adaptations that allow birds to complete these Olympian journeys. Still other researchers wrestle with how genes and the environment interact to shape the precise physiological and behavioral programs that migration demands.

As fascinating and important as these inquiries into the machinery of migration are, the ultimate challenge is understanding how and why these systems evolved and how the act of migration affects a species overall life history. Answering these broader questions requires the more reductionistic study of how birds accomplish these feats, but these pursuits need to be folded into a more holistic approach to the geography, history, ecology, and demographic trade-offs of the migratory strategy.

This holistic view of temperate-tropical migration systems has been evolving much the way that biological evolution proceeds: changes in perspective give rise to periods of rapid change followed by years of steady empirical progress. The evolution of thought on these migration systems can be traced through a series of major symposia that have punctuated the last 25 or so years.

In 1977, a group of about 50 bird ecologists held a meeting at the Smithsonian's Conservation and Research Center (CRC) in Front Royal, Virginia, to discuss the important findings and the most important questions remaining in migration ecology. The excited chatter of the conference focused on newly obtained information on how birds spent their time in the Tropics. The meeting was a small effort to redress the years of imbalanced concentration on the ecology of migratory birds on the temperate breeding grounds and culminated in a symposium volume entitled *Migrant Birds in the Neotropics: Ecology, Behavior, Distribution, and Conservation* (Keast and Morton, eds., 1980). That volume became one of the most heavily cited bird books ever produced from a scientific meeting and an excellent complement to Moreau's classic volumes on the Palearctic-African migration system. Twelve years later, another meeting was held at the Manomet Bird Observatory, Massachusetts, and also resulted in a landmark volume, *Ecology and Conservation of Neotropical Migrant Landbirds* (Hagan and Johnston, eds., 1992). The Manomet symposium rode on the crest of a wave of renewed interest in the challenges facing the conservation of migratory species dependent upon habitat along their long journeys through the (predominantly Western) Hemisphere. Both conferences were influential in driving research agendas on New World migratory birds (primarily songbirds) at the end of the twentieth century.

It was only fitting that the Smithsonian Institution hosted the next conference, 25 years after the first, to synthesize the cutting-edge findings about the basic ecology and evolution of migratory birds of all worlds. "Birds of Two Worlds: Advances in the Ecology and Evolution of Temperate-Tropical Migration Systems" took place in March 2002 at the National Conservation and Training Center in Shepardstown, West Virginia, and lasted 3½ days. The conference included a series of invited talks organized into symposia, round table discussions, and a contributed poster session.

The present volume, resulting from that conference, emphasizes basic evolutionary and ecological questions that surround the phenomenon of long-distance migration in birds and includes both New and the Old World examples. Rather than focusing on the mechanisms and adaptations for migration itself, we concentrate on how migration systems evolve and how the challenges of moving between vastly different environments can influence the life history, morphology, and behavior of the species involved. What properties distinguish long-distance migrants from resident birds and how migratory birds cope with the transitions during the annual cycle are major threads woven through all the chapters. Finally, we explore the factors that underlie the distribution of migratory species across ecosystems and continents. Many of the questions addressed in this volume have haunted ornithologists for decades, but the recent advent of conceptual models and the technological tools for probing these difficult questions have allowed ornithologists to stretch the boundaries of our understanding.

The volume is organized into seven thematic sections that establish a narrative trail. When this path is traversed the reader will see the breadth of biology required to understand the evolution and ecology of these migratory systems. Each section begins with a brief introduction by a leading ornithologist that is designed to provide a broad but cogent synthesis of the chapters that follow. To provide an overall roadmap, below we describe the parts of the volume and the big questions and research themes that illuminate the research described in the individual contributions.

Evolution of Migration Systems

Temperate-tropical migration systems represent the end point of evolutionary processes that select for shorter seasonal movements. For decades, ornithologists have grappled with the theoretical questions of what factors drive the evolution of seasonal movements. What are the demographic trade-offs for different migration strategies? Ornithologists have also speculated upon the historical aspects of the development of migration systems: How rapidly do they evolve? Did tropical resident populations give rise to temperate-tropical migrants? How often does long-distance migration evolve within particular taxa? Chapters in this section address these shared questions from distinctly different perspectives and methodologies.

Adaptations for Two Worlds

Migratory birds must often function efficiently and competitively in two highly divergent habitats. Nowhere is this more apparent than in the temperate-tropical migration systems. The contributors approach the problem of adaptive compromise facing migratory birds at the level of adaptations that characterize species and populations and the challenges that face individuals. Do these adaptive compromises have a tangible effect on highly migratory species when compared with birds that remain resident in a region throughout the year? How do birds obtain and use information about the environment at the ends and pathways of migration? And how do individual migrants balance the obvious need to display a high degree of ecological flexibility with the contravening demands for specialization, and what behavioral mechanisms maintain this balance?

Biogeography

Biogeographers search for patterns in the distribution of organisms across large spatial scales and develop hypotheses about what factors account for the patterns. Most biogeographers deal with a single geographic range for the species they study. What makes the study of the geography of migratory birds so interesting is that at least two ranges are involved, as well as the path between them. As migratory birds crisscross the globe, they often share their ranges and routes to varying degrees with many species (such as those that migrate from the Western Palearctic to sub-Saharan Africa), thus forming discernible migration systems. Authors in this section explore factors that account for the large-scale patterns of distribution of migratory birds and incorporate ecological, historical, and biophysical processes into their hypotheses.

Connectivity

Up until about 10 years ago, we knew almost nothing about dispersal into breeding populations or the geographic relationship between specific breeding and nonbreeding grounds of birds. A recent explosion of technological advances, from the application of molecular techniques to the use of stable isotopes, now allows scientists to address these issues and others. The chapters in this section describe the technology and, more importantly, explore the ecological and evolutionary implications of understanding population connectivity.

Migration Itself

Long-distance migration is one of the most physically demanding events in the animal kingdom. The adaptations and preparations for the journey have ramifications for their entire life history and give rise to many big questions: What factors influence departure and stopover schedules? How are migratory routes shaped by natural selection and how rapidly can they be modified? In keeping with the theme of the volume, the chapters here explore and emphasize how these adaptations to the extreme sport of migration influence the performance of birds at other times of year.

Behavioral Ecology

The chapters in this section explore the mating systems, use of space, and flocking behavior of migratory birds during the breeding and, particularly, the nonbreeding season. The ultimate aim of this section is to improve our understanding of how cross-seasonal interactions influence the behavior and ecology of birds at both ends of their migratory trips.

Population Ecology

The dynamics of breeding populations of migratory birds vary considerably: many populations show long periods of stability, others display long-term cycles, and still others show less predictable annual fluctuations. Understanding why some population numbers change and others do not is a central issue in ecology and crucial to understanding the anthropogenic causes of population change. Are migrants most limited during the breeding season, the nonbreeding season, or migration? When are density-dependent factors, such as food supply, depredation, and habitat availability, important, and when do major climatic occurrences, such as an ill-timed hurricane or El Niño event, determine population size? What is the role of social factors, such as dominance and territoriality? And finally, can we improve our ability to predict population changes over time?

This is an exciting time in migratory bird research and we believe that this volume captures that enthusiasm and goes one step further to explicitly point aspiring students toward the research gaps still in need of attention. Our hope is that this volume will inspire students and researchers alike and continue to push the envelope of our understanding of migratory bird biology.

Acknowledgments

FIRST AND FOREMOST, WE THANK our wives, Judy Gradwohl and Anne Perrault, and our children, Jeremy, Natalie and Aline. Their patience, support, and humor throughout this entire project have been vital to its completion. We also greatly appreciate the Smithsonian Institution (the National Zoological Park and the Smithsonian Environmental Research Center, in particular) for providing the freedom and support to bring this volume to completion.

This book evolved from a conference at the National Conservation Training Center in Shepardstown, West Virginia. The scientific program, which formed the backbone for this volume, was developed with the help of a Board of Advisors that included Richard Holmes, Eugene Morton, Doug Levey, Dave Winkler, Ellen Ketterson, and Robert Fleischer. We thank them for all of their creative ideas and hard work. Mary Deinlein, Bill DeLuca, Greg Gough, Robert Reitsma, Robert Rice, Karen Roux, Colin Studds, and Scott Sillett all contributed in different ways to the making of a successful conference.

We are particularly indebted to Natasha Atkins for painstakingly copyediting almost every last word between these covers. We are also grateful to Lynn Cassell and Sara Campbell for their work in helping organize the different sections of the book. Each manuscript received a thorough reading by two peer reviewers. Because many were anonymous to the authors, we will not name the reviewers here. Nonetheless, we thank them all for their constructive suggestions. The line drawings for the section title pages were superbly rendered by Dan Lane. All of the aforementioned struggled under our often challenging deadlines.

Financial support for the conference and aspects of manuscript preparation were provided by the Smithsonian Migratory Bird Center, Friends of the National Zoo, and the Office of Fellowships of the Smithsonian Institution.

The previous two symposium volumes have set a high standard. We commend the editors of those volumes, Allan Keast and Eugene Morton and John Hagan and David Johnston for setting us out on the important course of understanding tropical migratory birds. Their work inspired us and we hope (and expect) that in the years to come others will carry on with the work to understand the mysteries of bird migration.

Birds of Two Worlds

DFL '03

Part 1 / EVOLUTION OF MIGRATION SYSTEMS

RUSSELL GREENBERG

Overview

THE PHENOMENON OF MIGRATION is best developed in the class Aves, and because of the number of species involved, the number of ecosystems traversed, and the profound adaptations required to make these journeys, avian migration has captured the imagination of evolutionary biologists for a long time. Migration is so inexorably woven into the adaptive fabric of the species involved that it would be fair to say that all of the chapters in this volume and the research themes they communicate could be placed in a section on the evolution of migration. What distinguishes the chapters in Part 1 is their greater focus on the following questions: How did this global exchange of birds evolve? What are the small selective steps involved in the evolution of migratory patterns within individual species? And what are the major environmental factors that account for the development of the major migration systems? These questions are not new. What is new is the advent of conceptual and technical tools that have become available to migratory bird biologists in the recent decades. These tools have allowed the examination of time-honored hypotheses as well as the development of new theory regarding the evolution of migration.

Migration as it is manifested on Earth today is a slice in time in a unique story associated with the myriad populations and species involved. As scientists we seek to describe and understand this historical narrative as we search for underlying principles. Both the story and the principles become foci of the study of the evolution of migration. The term "evolution of migration," therefore, conjures up a number of different visions within the ornithological world. The chapters in this section have approached the evolution of migration from several different vantage points, all of which can be marked by the banner of evolutionary studies.

One approach is historical, focusing on the conditions under which migration first developed in the global or in regional avifaunas. This perspective, which is pursued in geologic time by David Steadman and examined through inferences based on current distributional patterns by Philip Hockey. Certainly the relationship between migration and climate is not in and of itself an epiphany, but the increasing sophistication of our knowledge of conditions through geologic time and current climatic patterns allows for increasingly quantitative assessments. Steadman ably steers us through the types of traditional paleoecological evidence that are available to hypothesize when the broad features of avian migration systems might have developed. While presenting an eloquent synopsis of what we can infer from paleontological studies, his chapter makes it clear that the details of the history of most migration systems are difficult to decipher from the relatively scanty fossil record in most migratory groups. However, it is very clear that what has been learned in recent decades about the long-term history of climate and vegetation, particularly in the Tropics, will greatly inform our understanding of the evolution of migration systems. In our zeal to answer the question "Where did migration develop (north or south . . .)?" Steadman's chapter is a good reminder that the geography of vegetation and climate has changed dramatically and repeatedly over recent geological time. A greater integration of paleoecological and ecological approaches is surely one of the next great steps in the study of the evolution of migration. Hockey both reviews and develops inferences by taking a global view of present-day environmental conditions. He shows that concepts of seasonality and climatic buffering go a long way toward explaining large-scale patterns of the distribution of migratory and resident taxa; more importantly, he describes the limits of this approach in terms of predicting which species in an avifauna will be the migratory species. Another approach is to focus on the ecological antecedents of migration. Changes in global climate through evolutionary time and global variation at present help provide a broad-scale explanation for when and where avifaunas as a whole may have become more or less migratory. However, these approaches are less successful in explaining which species in an avifauna will develop migration and where they will go when they migrate. Perhaps a more interesting question is what attributes found in birds displaying incipient migration patterns might have led in the past to the evolution of full-fledged migration patterns. Hockey develops this theme as he examines the ecological attributes of both long-distance and short-distance migrants in tropical areas. Clearly our knowledge of the distribution and natural history of migratory and resident birds has grown immensely, particularly in Africa, Australia, and the Neotropics, although this type of information remains difficult to access for the Asian Tropics.

A third approach is to study the mode and timing of the evolution of migration within and between taxa. The recent explosion of molecular-based phylogenies of various taxa of birds has led to the mapping of migration as a character trait in particular clades. This work—represented in this volume in the chapter by Joseph—has clearly shown that even though migratory behavior probably has deep roots in many groups, migratory tendencies are surprisingly labile. Joseph further surmises that in the groups analyzed thus far, acquisition of a higher-latitude breeding ground appears to be the derived state. Finally, he demonstrates that breeding range expansions of temperate-breeding migrants are often almost explosive in their speed, resulting in minimal genetic structuring within populations. As more such analyses are completed, our ability to compare the patterns of the evolution of migration across taxa will provide a powerful tool to developing a more integrated picture of how entire migration systems evolved. This will be all the more powerful as it is integrated with detailed knowledge of changes in climate and vegetation in the regions where migratory taxa were likely to have evolved. Finally, combining phylogenetic histories with analysis of ecological attributes of related resident and migratory taxa will add depth to the studies of ecological correlates of the evolution of migration.

The above approaches have all focused on the particular story of the evolution of migratory taxa and migration systems. Evolutionary biology has grown immensely in recent decades with the development of theory that is based on the underlying costs and benefits trade-offs that lead to the evolution of life history traits. Christopher Bell has taken the lead in developing models that pinpoint particular trade-offs that account for when and where birds migrate. In his chapter he reviews and proposes models that help us understand leapfrog and other patterns of migration, phenomena that bridge the gap between the questions, Why do birds move? and Why do they move as far as they do? The value of this type of modeling will only increase as the tools to establish connections among populations improve along with our ability to estimate population parameters—such as season-specific and overall survivorship—and other correlates of fitness.

Finally, it has become increasingly apparent that migration and migratory patterns themselves have an impact on critical evolutionary processes. This impact can further influence the evolution of migratory patterns. The relationship between migration and evolution is a complex interaction. Darren and Jessica Irwin explore the role of migratory patterns on the speciation process itself. Migratory divides may contribute to the divergence of populations, sending populations on either side of the divide down their own evolutionary pathways. When a number of species are involved, such as in central Asia, this process facilitates the rapid divergence between entire migration systems. These sorts of higher-order interactions provide a promising avenue for further research.

From this brief overview of a handful of chapters in this volume, we can see that the "evolution-of-migration question" is really a number of different, interrelated questions. The field stands ready for young investigators to pursue research that integrates the various approaches and develops more crosscutting theories and hypotheses.

1

DAVID W. STEADMAN

The Paleoecology and Fossil History of Migratory Landbirds

One of the challenges to the study of the evolution of migration is determining when birds first made these journeys. The fossil evidence shows that some birds were capable of migratory flights at least 100 million years ago (Ma). It seems unlikely, however, that long-distance migration in landbirds was initiated or at least became widespread until the Cenozoic, no earlier than the episode of global cooling in the late Eocene and Oligocene, ca. 40 to 25 Ma. It was then that most modern orders and families of non-passerine birds are recognized as fossils, although passerines are unknown in the Northern Hemisphere until ca. 25 Ma. Following a somewhat warmer interval from ca. 25 to 15 Ma, global cooling (and increased seasonality at mid and high latitudes) began in the middle Miocene and has continued ever since. Especially because the diversification of passerines probably accelerated at this time, it could even be that long-distance migration did not become taxonomically or geographically widespread until the middle Miocene. The rich fossil record of birds over the past 15 million years, especially the last 5 million years, is dominated by living genera, whether passerine or non-passerine. The difficulty of identifying passerine bones has led to their poor representation in the published fossil record. This problem can be partially offset in the future by analyses of ancient DNA (late Quaternary fossils only) and a larger investment of time in passerine comparative osteology, although the fossil record of passerines probably never will be as comprehensive as that of non-passerines.

The 22 glacial-interglacial cycles of the past 1.8 million years represent the coolest time in the history of living genera and species of birds. In North America and probably elsewhere, long-distance migration may have involved more species of birds, traveling over longer distances, during the warmer intervals (interglacials, such as today) than during times of expanded continental ice sheets. Glacial-interglacial shifts in climate rearrange the species composition of plant communities. The sets of species that make up any modern plant community, whether temperate or tropical, may be unique in geological time, particularly in light of the massive extinction of large mammals only about 13,000 years ago. Therefore, over millennial or longer time scales, it is highly adaptive for the breeding, migratory, and wintering distribution and habitat preference of any species of temperate-tropical migrant landbird to be flexible.

INTRODUCTION

To help set the stage for other chapters in this volume, I briefly review the paleontological and paleoecological evidence for avian migration, with a focus on landbirds rather than seabirds. I note some of the major contributions that paleoecology, paleoclimatology, and avian paleontology have made to understanding the history of migration. I also point out some poorly understood topics, with the hope that my speculations and queries will help to stimulate new research. A geographic bias toward the Holarctic region is unavoidable because the fossil record of birds is more comprehensive here than in any other major biogeographical region. This is appropriate, however, because the northern continents, especially North America, sustain many more species of long-distance migrants than anywhere else.

MESOZOIC

The first unequivocal birds were the osteologically "primitive" late Jurassic *Archaeopteryx lithographica* and its close relatives (Elzanowski 2002). The feather impressions of *Archaeopteryx,* as well as those from unnamed early Cretaceous birds from Mongolia, include asymmetrical primaries with interlocking barbs, evidence of powered flight, even if not for long distances (Kurochkin 1985:fig. 2; Feduccia 1996; Kellner 2002). The vast majority of described Mesozoic birds are from other primitive groups, such as the Enantiornithes, that did not survive beyond the Cretaceous (Feduccia 1996; Chiappe 2002). Bones of the wing and pectoral girdle from some other Mesozoic birds, however, suggest sustained flight and perhaps, therefore, long-distance migration. For example, *Ambiortus dementjevi* (early Cretaceous of Mongolia; only ca. 10–15 million years younger than *Archaeopteryx*) was fully carinate (with a keeled sternum) and had numerous features in its shoulder and wing bones indicative of a good ability to fly (Kurochkin 1985, 1999). Nevertheless, given the evidence for warm, aseasonal global climates during the Cretaceous (A. Graham 1999:142), there would seem to have been little if any climatic stimulus for long-distance migration during this time.

CENOZOIC (TERTIARY)

Paleocene through Oligocene

To generalize from the rich early Tertiary fossil record in North America and Europe, most living orders of birds are recognizable by some time in the late Paleocene through Eocene, about 60–34 Ma (Olson 1989; Feduccia 2003). The diverse Eocene avifaunas of the Holarctic region feature many extinct families of small "perching birds" that belong in or are related to the Apodiformes, Coraciiformes, Piciformes, and other non-passerine orders (Mayr 1998a, 1998b, 2000; Mayr and Mourer-Chauviré 2000; Feduccia 2003). Although it cannot be determined with certainty, these non-passerine landbirds probably were residents rather than long-distance migrants, given the tropical-subtropical conditions that characterized most latitudes through the early to middle Eocene (A. Graham 1999:176) and the fact that most modern members of these orders are resident rather than migratory. In Eocene and Oligocene (55–24 Ma) fossil deposits of the mid-latitude Northern Hemisphere (e.g., Green River, Wyoming; London Clay, England; Messel, Germany; Quercy, France) are representatives of families of generally nonmigratory birds, such as oilbirds, mousebirds, and todies, that are confined today to the Tropics. For example, of all the non-passerine families currently endemic to continental Africa, only the Scopidae (Hamerkop) and Balaenicipitidae (Shoebill) have not been found as Tertiary fossils in Europe, Asia, or North America, especially the former (Blondel and Mourer-Chauviré 1998). Some of the tropical element in Tertiary mid-latitude avifaunas persisted into the Miocene.

Long-distance migration in landbirds may have been initiated, or at least may have become widespread, during the episode of global cooling (and drying?) that began in the late Eocene and continued through the Oligocene, ca. 40–24 Ma. During this time interval, North American vegetation types gradually became less tropical (A. Graham 1999:231–233), and many more modern families of non-passerine birds can be distinguished in the fossil record (Olson 1985, 1989).

Passerine birds dominate most modern avifaunas and include many more species of long-distance migrants than any other order. Zoogeographic, paleontological, morphological, and molecular evidence suggests that passerines originated in the Southern Hemisphere (Feduccia and Olson 1982; Edwards and Boles 2002; Ericson et al. 2002). The fossil record is of little help here because small avian fossils of any sort are rare or absent from early Tertiary sites on all four southern continents. By far the earliest fossils assigned to Passeriformes are from the early Eocene (ca. 55 Ma) of Queensland, Australia, with two undetermined taxa represented (Boles 1995, 1997). The next earliest passerine fossil from the Southern Hemisphere is a humeral fragment of a probable suboscine from the early-middle Miocene of Argentina (Noriega and Chiappe 1993). Passerines are unknown in the Northern Hemisphere until the late Oligocene (25 Ma) of France (Mourer-Chauviré et al. 1989, 1996).

These French fossils represent an oscine rather than suboscine, although family-level assignment is not possible.

Miocene and Pliocene

By the very early Miocene (ca. 20 Ma), several forms of oscines were present in Europe although only a lark (Alaudidae) has been identified reliably to family (Ballmann 1972). In Florida, the Thomas Farm Site (early Miocene, Hemingfordian land mammal age, ca. 18 Ma [MacFadden 2001]) has yielded more than 100 fossils of small and medium-sized oscine passerines, which I have begun to study (September 2004). That the earliest evidence for the radiation of New World passerines has gone uninvestigated until now reflects the difficulty of interpreting the osteology of passerines as well as how little original research is being done in Cenozoic avian paleontology.

After the warm interval from ca. 25 to 15 Ma, strong global cooling (with increased seasonality at mid and high latitudes) began in the middle Miocene, intensified in the Pliocene, and has continued ever since (Williams et al. 1997; Willis et al. 1999, Zachos et al. 2001) (fig. 1.1), with associated development of temperate and boreal vegetation types in North America (A. Graham 1999:267). The diversification of passerines, including migratory forms, probably accelerated during the Miocene (Olson 2001). The North American fossil record of birds during the last 15 million years is dominated by living genera, whether passerine or non-passerine (table 1.1). Compatible with late Miocene through Quaternary increased cooling and seasonality is

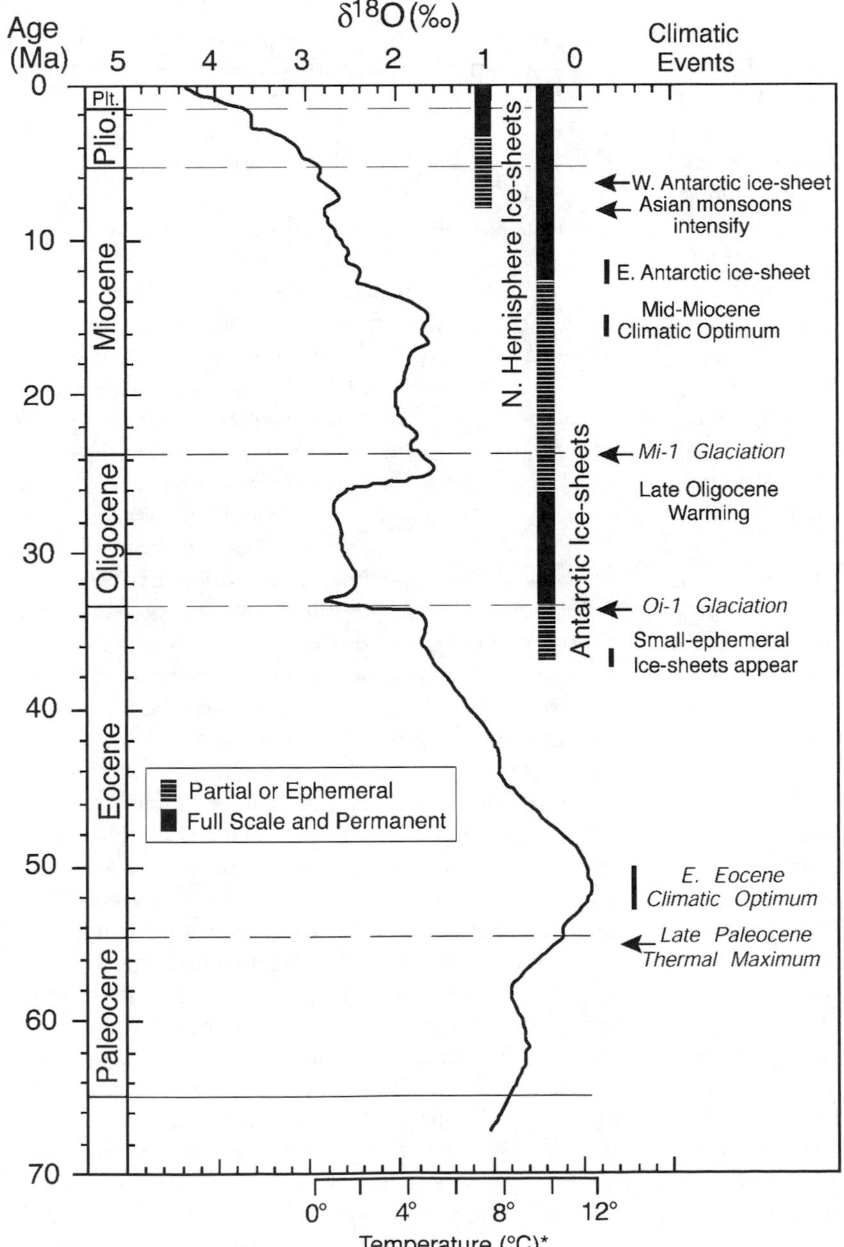

Fig. 1.1. Major climate trends during the Cenozoic. Modified from Zachos et al. (2001).

Table 1.1 Miocene through Pleistocene records of genera of North American landbirds in some orders or families with a number of extant long-distance migrants

	Late Arikareean + Hemingfordian		Barstovian + Clarendonian		Hemphillian		Blancan		Irvingtonian		Rancholabrean	
Family / Order	N	% Living	N	% Living	N	% Living	N	% Living	N	% Living	N	% Living
Anatidae	3	33	8	50	10	70	16	88	11	73	15	87
Accipitriformes	9	22	5	40	11	82	8	88	8	75	15	73
Rallidae	—	—	2	100	2	100	5	100	4	100	7	100
Charadriidae + Scolopacidae	—	—	4	100	6	100	9	100	5	100	13	100
Passeriformes	—	—	1	0	3	100	6	83	22	100	67	96
Total	12	25	20	60	32	84	44	91	50	90	117	92

Note: Column headings are North American land mammal ages. Late Arikareean + Hemingfordian = 24–16 Ma; Barstovian + Clarendonian = 16–9 Ma; Hemphillian = 9–4.8 Ma; Blancan = 4.8–1.6 Ma; Irvingtonian = 1.6–0.3 Ma; Rancholabrean = 0.3–0.01 Ma. Accipitriformes = Accipitridae + Pandionidae + Falconidae. Data compiled from Howard (1963, 1983), Brodkorb (1964, 1967, 1978), Lundelius et al. (1983), Steadman (1984), Becker (1987), Bickart (1990), Chandler (1990), Emslie (1998), and Olson and Rasmussen (2001).

the phylogeny-based proposal by Outlaw et al. (2003) that permanent residency is ancestral within the lineage of *Catharus* thrushes, with migratory behavior evolving three times among the five Nearctic/Neotropical migrant species. Similarly, mtDNA-based phylogenies of *Dendroica* warblers, pipits (*Anthus*), and wagtails (*Motacilla*) suggest multiple switches between residency and migration (Lovette and Bermingham 1999; Voelker 1999, 2002).

Much of the interest in modern long-distance migration from the Nearctic to the Neotropics involves nine-primaried oscines (e.g., warblers, tanagers, buntings, sparrows, and orioles; Emberizidae s.l.). Unfortunately, this huge assemblage of songbirds is very difficult to characterize at the species level with postcranial osteology. Some species are difficult or impossible to identify even with an entire rostrum or mandible, especially paruline warblers. The only two published Tertiary records of New World nine-primaried oscines involve living genera. Whether they represent resident or migrants is uncertain. The first is a rostrum from the late Miocene (late Clarendonian–early Hemphillian, ca. 9 Ma) Long Island local fauna of north-central Kansas that represents the extinct *Ammodramus hatcheri,* which is barely distinguishable from the living, widespread Grasshopper Sparrow (*A. savannarum*) (Steadman 1981).

The second is an extinct species of intermediate-sized *Passerina* from an early Pliocene (late Hemphillian) site at Yepómera, west-central Chihuahua (Steadman and McKitrick 1982). This unnamed bunting is represented by three humeri that are smaller than in *P. caerulea* (Blue Grosbeak) but larger than in other species of *Passerina* s.s., several of which breed primarily or exclusively in temperate areas and winter in the Subtropics/Tropics (Payne 1992; Lowther et al. 1999). The presence of this extinct species in northern Mexico at ca. 5 Ma is congruent with estimated divergence times in *Passerina* based on mtDNA (Klicka et al. 2001), where *P. caerulea* and the smaller *P. amoena* (Lazuli Bunting) have been proposed as sister taxa that diverged at ca. 3.7–2.4 Ma. The fossil bunting from Yepómera is intermediate between these two living species both qualitatively and quantitatively. Combining paleontological data with phylogenies of living species is a fertile area for future research.

CENOZOIC (QUATERNARY)

Climate and Vegetation

The roughly 22 glacial-interglacial cycles that began at 1.8 Ma (the Quaternary) represent the overall coolest time interval in the history of living orders, families, genera, and species of birds, if not in the entire history of the Class Aves. When studying the Quaternary evolution and biogeography of birds, it is important to keep in mind that conditions as warm and ice-free as those of the past several thousand years probably characterize less than 10% of the Quaternary (see Williams et al. 1997, Fischer et al. 1999, and Rutherford and D'Hondt 2000). By contrast, even today's climate is warm only relative to glacial intervals; oxygen isotope data indicate that most or all of the Tertiary was much warmer (and less seasonal) than at any time in the Quaternary.

In North America and probably in other regions of mid to high latitude (such as Europe [see Blondel and Mourer-Chauviré 1998]), long-distance migration may have involved more species of birds, traveling over longer distances, during warm (interglacial) intervals than during times of expanded continental ice sheets. Because most areas of North America north of 40° N were covered with ice or were periglacial during the last glacial maximum (22 to 18 thousand years ago [ka]), the latitudinal extent of long-distance migration probably was substantially shorter than today for the various migratory waterfowl, raptors, shorebirds, and passerines that currently nest primarily or only north of 35° N. This is especially so because glacial cooling and perhaps drying of Neotropical lowlands, although substantial at ca. 2° to 6°C mean annual temperature cooler than modern (Leyden 1984; van der Hammen and Absy 1994; Colinvaux et al. 1996), was generally not enough to preclude frost-free conditions and the persistence of a tropical flora between 20° N and 20° S in the lowlands. Considering its great range of latitude, longitude, and elevation, I believe that Mexico during glacial times would have sustained breeding populations of many species of birds that now nest only to its north.

Because North America is blessed with hundreds of late Quaternary sites that preserve pollen, spores, and/or macrofossils of plants (Betancourt et al. 1990; Jackson et al. 1997; A. Graham 1999), much information is available on how vegetation changed during the transition from the late Pleistocene (the last glacial interval) to the Holocene (the past ca. 12,000 years, an interglacial interval). Changing plant communities may stimulate migratory behavior because new local plant associations may not satisfy a species of bird's annual need for food, nest sites, and so on. Here I restrict my discussion to the New World; see Greenberg et al. (1999) for a historical approach to the relationship between long-term vegetation changes and the ecology of Old World foliage-gleaning insectivorous birds, most of which are migratory.

In the northeastern United States, a decline in spruce (*Picea*) and an increase in oak (*Quercus*) are two of the most striking changes (fig. 1.2). With local variation, this represents the replacement of a largely coniferous woody vegetation (spruce as well as fir [*Abies*], larch [*Larix*], and jackpine [*Pinus banksiana*]) with a mixture of hardwoods (e.g., oak [*Quercus*], beech [*Fagus*], and maple [*Acer*]) and conifers, the latter consisting mainly of other pines (e.g., *Pinus strobus*) or hemlock (*Tsuga*) rather than spruce. Some of the warmer floral elements, such as hickories (*Carya*) and black gum (*Nyssa*), did not reach New England until several thousand years later than the oaks (Shuman et al. 2001). The late Pleistocene fossil plant data from eastern North America also suggest that open wooded habitats, variously called woodlands, parklands, or savannas, were more widespread than the closed-canopy forests that, in the absence of logging, characterize much of the region today. A likely factor here is the loss of most of North America's large mammals near the close of the Pleistocene (see the next section).

The much greater availability of water in the currently arid western United States filled many basins with major lakes during the glacial maxima (Bachhuber 1992; Reheis 1999; Broughton et al. 2000). These lakes and associated aquatic vegetation undoubtedly had a major influence on the distribution of migratory waterfowl and shorebirds, among others. Across the West, high-elevation conifers lived at lower elevations during the cooler and wetter glacial times than today (Betancourt 1990; Grigg and Whitlock 1998). In the Southwest, late Pleistocene pinyon-juniper woodlands were replaced in the Holocene by desertscrub. From the Big Bend region of southwestern Texas, for example, plant macrofossils from packrat middens portray major changes during the glacial-interglacial transition from ca. 12 to 10 ka (Van Devender 1990) (fig. 1.3). Pinyon-juniper woodland, dominated by papershell pinyon (*Pinus remota*), juniper (*Juniperus* sp.), and Hinckley oak (*Quercus hinckleyi*) was replaced by desertscrub with honey mesquite (*Prosopis glandulosa*), creosote bush (*Larrea divaricata*), and blind prickly pear (*Opuntia rufida*). Several species persisted through the transition, such as Mormon tea (*Ephedra* sp.), variable prickly pear (*O. phaeacantha*), and lechugilla (*Agave lechugilla*).

Extinct Megafauna

Some 34 genera and more than 40 species of large North American mammals, including armadillos, glyptodonts, ground sloths, giant beavers, capybaras, lions, sabertooths,

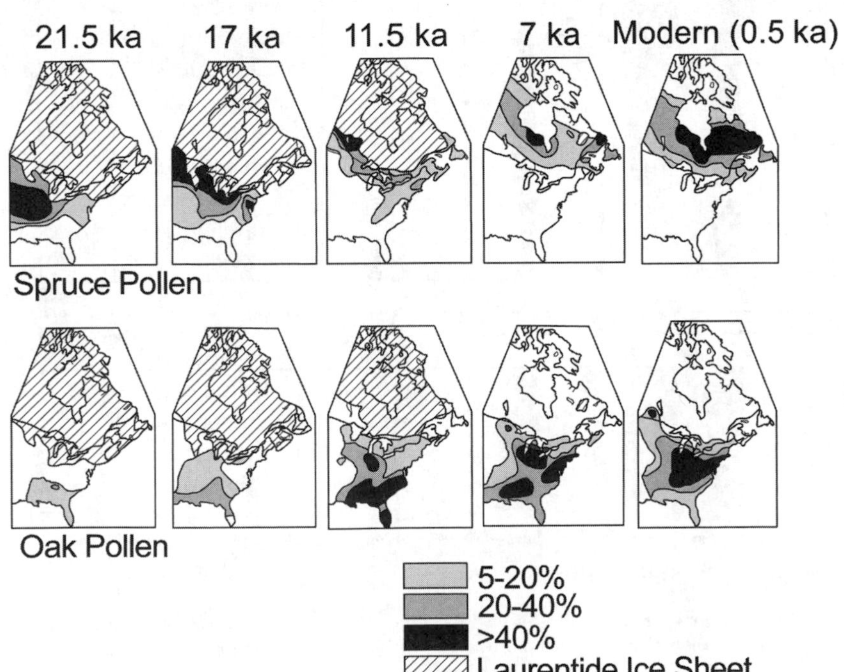

Fig. 1.2. Distribution of spruce (*Picea* spp.) and oak (*Quercus* spp.) pollen in central and eastern North America over the past 21,500 years. From Davis and Shaw (2001:fig. 1).

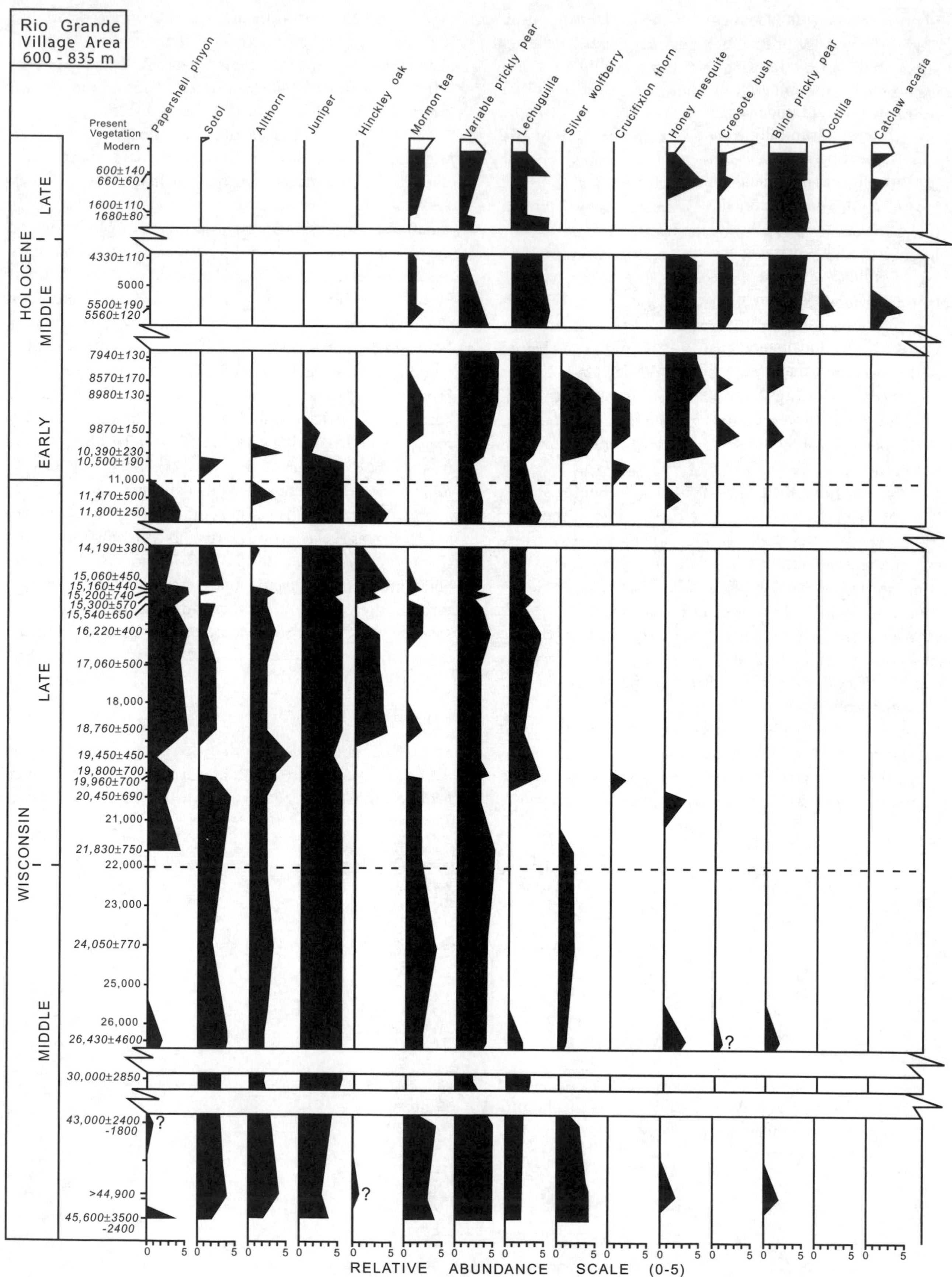

Fig. 1.3. Chronological summary of 15 important plant macrofossils from packrat middens of the Rio Grande Village area (29° 11–16′ N, 102° 58′–103° 01′ W; elev. 600–835 m), Big Bend National Park, Brewster County, Texas. Relative abundance scale: 0, absent; 1, rare; 2, uncommon; 3, common; 4, very common; 5, abundant. From Van Devender (1990:fig. 7.7).

peccaries, camels, pronghorns, musk-oxen, mountain-goats, horses, tapirs, mammoths, and mastodons, became extinct at the end of the Pleistocene, along with associated commensals such as scavenging birds (Steadman and Martin 1984; Graham and Lundelius 1994; Alroy 1999; Martin and Steadman 1999). Ironically, North American plants suffered little if any extinction at this time, in spite of major distributional changes as continental and alpine glaciers retreated. This rapid episode of extinction changed North America (and the rest of the New World) from something that must have resembled an African game park to a megafaunally impoverished region where large indigenous mammals play less of a role in ecosystem functioning than they have since mammals first attained a large body size 45 Ma. For an extremely brief period of geological time (only the past 13,000 years, which is less than 1% of time even within the Quaternary), most North American ecosystems have been operating with cervids (deer, elk, moose, caribou), bison, bears, wolves, pumas, and humans as the dominant large mammals, a dramatic substitution for 40-plus extinct species that spanned many orders, families, and feeding guilds.

From a landscape/habitat perspective, proboscideans probably are the New World's most drastic mammalian loss. Mammoths (*Mammuthus* spp.) and mastodons (*Mammut americanum*) ranged across most of North America into Mexico and Central America, whereas gomphotheres (*Cuvieronius, Haplomastodon, Notiomastodon*) occupied much of Central and South America (Webb and Rancy 1996). In Africa today, elephants (*Loxodonta*) keep large swaths of wooded habitat in early stages of succession (Owen-Smith 1987). The currently thick riparian forests of cottonwood, willow, ash, and walnut in the western United States probably were riddled with wide, well-manured elephant trails until just 13,000 years ago (Martin, in press). Woody legumes, which provide excellent habitat for wintering migratory passerines, probably were more common and widespread in southern Mexico and Central America in the late Pleistocene than during the megafaunally impoverished Holocene (Janzen and Martin 1982; Greenberg et al. 1997).

Range Changes and Identifying Fossils

Given the major changes in climate, sea level, and plant communities that took place throughout the Quaternary, not to mention the arrival of people and loss of most large mammals at the close of the last ice age, the migratory species of birds that survive are a testimony to adaptation. It is clear that, over millennial or longer time scales, the breeding, migratory, and wintering distribution of any species of temperate-tropical migrant landbird must be flexible. This concept is supported by molecular phylogenies that suggest independent gains and losses of migration in various lineages of parulid warblers and within three species of thrush, warbler, and sparrow (Smith et al., Chap. 18, this volume). Such proposals also are compatible with the minor osteological differences that sometimes distinguish between resident and migratory congeneric species (Calmaestra and Moreno 2000).

Fossil evidence for major Quaternary changes in distribution is extensive for plants (see above), insects (especially Coleoptera [Elias 1994]), mammals of all sizes (the small ones mostly extant; the large ones mostly extinct [see R. W. Graham and Lundelius 1994 and R. W. Graham et al. 1996]), and for resident species of birds such as certain hawks, grouse, columbids, and corvids (Lundelius et al. 1983; Emslie 1998). For migratory landbirds, the evidence for glacial-interglacial distributional changes is much more meager because of several complications. First, to establish local nesting through fossils, one must find bones of nestlings or recently fledged birds (which can be difficult or impossible to identify) or evidence of medullary deposits (fat- and calcium-rich deposits in the femur or tibiotarsus indicative of laying females). Paleontologists often do not note such features, if detectable, in the fossils they study.

Another limitation in determining range shifts is the oft-mentioned difficulty of using isolated skeletal elements to make reliable, species-level identifications in most groups of passerine birds, especially those with many small species, such as New World nine-primaried oscines or Old World sylviid warblers. The North American fossil record is heavily biased toward non-passerines for this reason, and also because very small fossils often were not recovered from sites until the use of fine-mesh sieves became widespread in the 1950s and 1960s. The list that I compiled (in Lundelius et al. 1983) for 84 late Pleistocene (ca. 25–10 ka) avian fossil sites across North America features 211 species of non-passerines, 77 species of probable resident passerines, and only 14 species of probable Nearctic/Neotropical migrant passerines. Of these 302 species, as many as 46 are extinct, dominated by storks, condors, anatids, hawks, eagles, accipitrid vultures, flamingos, turkeys, owls, and icterids.

From the Plio-Pleistocene of Florida (Blancan through Rancholabrean land mammal ages), Emslie (1998) listed 93 species of non-passerines, 24 species of probable resident passerines, and only five species of probable Nearctic/Neotropical migrant passerines, the latter consisting of *Catharus* sp., cf. *Hylocichla mustelina, Dumetella carolinensis, Vermivora* cf. *V. celata,* and *Passerina* sp. Of these 122 species, the 21 extinct ones consist of a cormorant, an anhinga, a stork, a teratorn, three condors, a duck, three eagles, an accipitrid vulture, a cracid, a turkey, two rails, a phorusracid, a woodcock, a pigeon, and two owls. The very extensive set of birds reported for the entire Pleistocene of the Palearctic region (Tyrberg 1998) features an impressive 198 species of passerines, including many long-distance migrants, although I challenge Palearctic paleontologists to provide the osteological evidence to verify, for example, identification of the 11 species of *Sylvia* and *Phylloscopus* listed by Tyrberg (1998:570–571).

Although the problem of identifying passerine fossils could be partially offset by a larger investment of time in comparative osteology, there is little reason for optimism in species-rich yet osteologically similar groups such as *Empidonax* flycatchers or paruline warblers, especially *Dendroica*. This situation is exacerbated by the fact that very few students are willing to invest the years of hands-on training required to become skilled in comparative osteology. Those who do typically are swamped with important (even if often less challenging) research on non-passerines.

Ancient DNA also is limited in its potential to unravel a chronometrically accurate and precise history of migratory passerines. Although it would be wonderful if we could obtain long sequences of both mitochondrial and nuclear DNA from Pleistocene fossils of warblers to compare with modern data sets such as in Smith et al. (Chap. 18, this volume), DNA in the protein fraction (collagen) of bird bone degrades rapidly (<10 ka) under most sedimentary conditions. From various sites on tropical islands in the Pacific and Caribbean, I have found through radiocarbon dating that collagen is absent or highly degraded in about one-half of the 50-plus reptile, bird, and mammal bones tested, none of which exceeds 10 ka and most of which are less than 5 ka. Dry and/or very cold conditions are the best for long-term preservation of DNA, as indicated by a 390-base-pair (bp) sequence of mitochondrial HVRI being obtained from 66% of Adélie Penguin (*Pygoscelis adeliae*) bones up to 7,000 years old from Antarctica (Lambert et al. 2002), and 135- and 60-bp fragments of control region mtDNA from bones up to 60,000 years old of Brown Bears (*Ursus arctos*) from Beringia (Barnes et al. 2002).

The ability to use stable isotopes in the collagen of bird bones to track latitudinal signals over long time scales probably suffers from more or less the same preservational limitations as those for ancient DNA. Although credible data on stable isotopes have been generated by using tooth enamel in fossil mammals dating to 5 Ma or older (MacFadden et al. 1999; Rogers and Wang 2002), studies of stable isotopes in the presumably less stable inorganic fraction of bird bones have not been conducted. However, isotopic ratios of pre-Pleistocene bird bones are likely to have been modified by post-depositional chemical processes. This alteration likely complicates the potential for stable isotopes to track latitudinal or longitudinal signals in fossils, as has been done for the Black-throated Blue Warbler (*Dendroica caerulescens*) using modern feathers (Rubenstein et al. 2002).

ISLANDS

Although species of birds that breed in the Nearctic are common as migrants or winter residents in the West Indies, very few of their fossils have been identified in West Indian fossil sites. For example, of the 54 species of landbirds recorded prehistorically in the Lesser Antilles (Pregill et al. 1994), only six—the duck *Aythya collaris*, osprey *Pandion haliaetus*, the rail *Rallus limicola*, and shorebirds *Calidris melanotos*, *Calidris* small sp., and *Limnodromus griseus*—are likely to be migrants rather than residents. For New Providence Island in the Bahamas, only two (the rail *Porzana carolina* and warbler *Dendroica* sp.) of the 32 species of landbirds reported from the late Pleistocene and Holocene sites by Olson and Hilgartner (1982) are probable migrants. Similarly, of the 34 species of landbirds that I have identified thus far from the late Pleistocene and Holocene site of Indian Cave on Middle Caicos (Turks and Caicos Islands), the only currently migratory species is the sapsucker *Sphyrapicus varius*.

The area and isolation of islands, whether continental (land-bridge) or oceanic (never connected to a major land mass), vary with fluctuations in sea level. Among the most drastic of these effects are in the Bahamas, where the current 29 islands and 661 cays would have been consolidated into five major islands and several smaller ones during the last glacial maximum, with an overall increase in land area from 11,406 km^2 to ca. 124,716 km^2 (Olson and Hilgartner 1982; Buden 1987). Conversely, very little land at all existed in the Bahamas during the high-sea-level stand of the last interglacial at ca. 125 ka (Hearty and Neumann 2002). The winter range of Kirtland's Warbler (*Dendroica kirtlandii* [see Sykes and Clench 1998]) and other long-distance migrants in the Bahamas has had to be flexible through time, although given the extreme osteological challenge of identifying species of *Dendroica* (see above), I doubt that fossils will ever reveal changes in the winter or breeding range of *D. kirtlandii*.

Several species of shorebirds that breed at high northern latitudes reside extensively or exclusively on remote Pacific islands outside of the breeding season. These include the Pacific Golden Plover (*Pluvialis [dominica] fulva*), Wandering Tattler (*Heteroscelus incanus*), and Bristle-thighed Curlew (*Numenius tahitiensis*). They and other less common species of shorebirds occur regularly in bone deposits in the Hawaiian Islands (Olson and Hilgartner 1982; Olson and James 1991). In the case of *P. fulva* and *N. tahitiensis*, fossils from more than 120 ka have been found (James 1987), which now are believed to date to more than 500 ka (S. L. Olson, pers. comm. 2002). In Tonga I have found *P. fulva* in a late Pleistocene context. Assuming that such continental tundra-breeders never nested on remote, tropical islands, these shorebirds probably have wintered on oceanic islands for at least hundreds of millennia, regardless of changes in their breeding distribution because of expansion and contraction of continental glaciers.

Bones of migratory charadriid and scolopacid shorebirds also are found regularly in late Holocene (<3 ka) archaeological (cultural) sites in all regions of Oceania. As is the case today, shorebirds are generally more common and more diverse in prehistoric sites from western Oceania (Micronesia, Melanesia [see Steadman and Intoh 1994, Steadman 1999, and Steadman and Kirch 1998]) than in Polynesian sites farther east in the Pacific (Steadman and Justice 1998; Steadman et al. 2002). Only in the case of *Numenius tahitiensis*, which has a limited breeding distribution and a drastic molt schedule on its wintering grounds that may render it flightless for several weeks (Marks 1993), did prehistoric predation by people possibly have much effect on populations.

DISCUSSION AND CONCLUSIONS

The last 5 million years, which include the entire tenure of most or all extant species, have been the coolest (and therefore probably the most seasonal) period in the history of birds. This cool interval has culminated in 22 glacial-interglacial cycles over the past 1.8 million years. Global fluctuations in temperature, precipitation, and plant distributions generally increase in magnitude with increasing latitude. Combined with global changes in sea level, the Quaternary glacial-interglacial cycles provided strong stimuli for long-distance migration in birds, including distributions and habitat preferences that had to be flexible over long time scales to ensure survival. This is not to say that long-distance migration in birds developed only within the Quaternary; these tendencies undoubtedly have been developing independently in many groups of birds for the past 15 million years or longer. Another area where migratory birds have had to be adaptable involves the Earth's magnetic field. Polar wandering, global magnetic reversals, and other geomagnetic fluctuations have occurred through geological time (Bloxham et al. 2002; Tarduno et al. 2002), yet we have no idea how this might influence the many species that use magnetism for directional orientation during migration. The last geomagnetic reversal occurred ca. 780,000 years ago; 19 other reversals have taken place over the past 5 million years (Berggren et al. 1995).

A relatively recent biotic change that probably has had a major influence on the distribution and relative abundance of landbirds was the late Pleistocene collapse of large mammal communities on all continents except Africa. In the Pleistocene presence of diverse large mammalian browsers and grazers (see distributional and taxonomic summaries in R. W. Graham and Lundelius 1994), savannas and patchier forests may have covered many of the areas in both temperate and tropical America destined to become vastly forested in the Holocene. Thus a very positive legacy of the extinct megafauna, which survived until only 13,000 years ago, may be the ability of so many species of birds, whether resident or migratory, to occupy patchy forests, successional forests, or simply nonforested habitats. This is beneficial to modern migratory birds on both their breeding and wintering grounds. Although we are justifiably concerned with modern deforestation in the Neotropics, we should not overlook the fact that extensive parts of the Mayan region (and elsewhere) were deforested in prehistory, with much of the forest recovery coming only after the tragic decimation of Amerindian populations through European contact (Deevey et al. 1979; Binford et al. 1987; Rosenmeier et al. 2002; Heckenberger et al. 2003). Palynological evidence argues for tropical/subtropical pine-juniper-oak woodland or savannah as the dominant vegetation in Petén during glacial periods (Leyden 1984); thus vegetation similar to the seasonally moist tropical forests that occur there now probably has been dominant for only a minority of time during the past 2 million years.

In North America, other sources of uncertainty in relating ancient to modern data from plant and animal communities are the recent (past century or two) severe range contractions of top carnivores—the wolverine (*Gulo gulo*), gray wolf (*Canis lupus*), red wolf (*Canis niger*), puma (*Felis concolor*), grizzly bear (*Ursus horribilis*)—a browser (elk, *Cervus canadensis*), and two important trees (American chestnut [*Castanea dentata*]; American elm [*Ulmus americana*]). A related consideration is the human-caused extinction a century ago of the Passenger Pigeon (*Ectopistes migratorius*). Bony evidence from prehistoric sites in New York, Pennsylvania, and adjacent states (Steadman 1998, 2001, 2003) suggests that written records of this pigeon's extreme abundance in the sixteenth through nineteenth centuries were not exaggerated, and that it was probably abundant throughout the Holocene. The immense migratory flocks and nesting concentrations of Passenger Pigeons must have profoundly influenced the forest ecosystems in eastern North America. Although the Passenger Pigeon was primarily a seed predator, its huge numbers and tendency to travel would mean that even a very low level of mortality (with or without predation) would disperse viable seeds that had not yet passed from its crop to its gizzard. We can only speculate whether living without Passenger Pigeons and with so few large mammals might be beneficial to some species of migratory birds and not to others.

In spite of current global warming and its hypothesized effects on long-term climate (Berger and Loutre 2002) and on the health and distribution of both resident and migrant birds (Harvell et al. 2002; Peterson et al. 2002; Malakoff 2003), the next glacial interval still may begin before too long in a geological sense. The coastlines will expand rather than contract as more of the Earth's water is tied up in ice. Most of Canada will be uninhabitable. North America will experience its first glacial interval uninfluenced by mammoths, mastodons, or sabretooths. If oaks decline in range as in the last glacial maximum (fig. 1.3), then species that typically breed in oak forests, such as the Black-and-White Warbler (*Mniotilta varia*), might decline. The breeding ranges of spruce-loving species such as Cape May and Bay-breasted Warblers (*Dendroica tigrina, D. castanea*) will shift to the south. Tennessee Warblers (*Vermivora peregrina*) may finally have a chance to breed in Tennessee. If it survives, Kirtland's Warbler may become more numerous as pioneer stands of jackpine expand south and east of Michigan (see Little 1971:maps 46-N, 46-E for modern distribution of *Pinus banksiana*) and enlarged Bahamian islands provide more winter range. Bicknell's Thrush (*Catharus bicknelli*) might breed in spruce-fir forests across the central and southern Appalachians and winter in the Turks and Caicos Islands as well as Hispaniola. On millennial or longer time scales, and whether the climate gets warmer or cooler, migratory landbirds face a future that is guaranteed to be different from the past.

ACKNOWLEDGMENTS

I thank Pete Marra and Russ Greenberg for supporting my participation in their fascinating conference about migratory birds. For comments that improved the chapter, I thank Steve Emslie, Russ Greenberg, Jeremy Kirchman, Andy Kratter, Bruce MacFadden, Pete Marra, and Storrs Olson. Jeremy Kirchman and Sarah Schaak assisted in preparing the manuscript. Margaret Davis, Thomas Van Devender, and James Zachos kindly granted permission to use their published figures. This research was funded in part by National Science Foundation grants EAR-9714819 and DEB-0228682.

LITERATURE CITED

Alroy, J. 1999. Putting North America's end-Pleistocene megafaunal extinction in context: large-scale analyses of spatial patterns, extinction rates, and size distribution. Pages 105–143 *in* Extinctions in Near Time: Causes, Contexts and Consequences (R. D. MacPhee, ed.). Kluwer Academic/Plenum Publishers, New York.

Bachhuber, F. W. 1992. A pre-late Wisconsin paleolimnologic record from the Estancia Valley, central New Mexico. Geological Society of America Special Paper 270:289–307.

Ballmann, P. 1972. Les oiseaux Miocènes de La Grive-Saint-Alban (Isère). Geobios 2:157–204.

Barnes, I., P. Matheus, B. Shapiro, D. Jensen, and A. Cooper. 2002. Dynamics of Pleistocene population extinctions in Beringian brown bears. Science 295:2267–2270.

Becker, J. J. 1987. Neogene Avian Localities of North America. Smithsonian Institution Press, Washington, D.C.

Berger, A., and M. F. Loutre. 2002. An exceptionally long interglacial ahead? Science 297:1287–1288.

Berggren, W. A., D. V. Kent, M.-P. Aubrey, and J. Hardenbol (eds.). 1995. Geochronology, time scales and global stratigraphic correlations. Society for Sedimentary Geology, Special Publication 54.

Betancourt, J. L. 1990. Late Quaternary biogeography of the Colorado Plateau. Pages 259–293 *in* Packrat Middens: The Last 40,000 Years of Biotic Change (J. L. Betancourt, T. R. Van Devender, and P. S. Martin, eds.). University of Arizona Press, Tucson.

Betancourt, J. L., T. R. Van Devender, and P. S. Martin (eds.). 1990. Packrat Middens: The Last 40,000 Years of Biotic Change. University of Arizona Press, Tucson.

Bickart, K. J. 1990. Recent advances in the study of Neogene fossil birds: Pt. 1. The birds of the late Miocene-early Pliocene Big Sandy Formation, Mohave County, Arizona. Ornithological Monographs no. 44. American Ornithologists' Union, Washington, D.C.

Binford M. W., M. Brenner, T. J. Whitmore, A. Higuera-Gundy, E. S. Deevey, and B. Leyden. 1987. Ecosystems, paleoecology and human disturbance in subtropical and tropical America. Quaternary Science Reviews 6:115–128.

Blondel, J., and C. Mourer-Chauviré. 1998. Evolution and history of the western Palaearctic avifauna. Trends in Ecology and Evolution 13:488–492.

Bloxham, J., S. Zatman, and M. Dumberry. 2002. The origin of geomagnetic jerks. Nature 420:65–68.

Boles, W. E. 1995. The world's oldest songbird. Nature 374:21–22.

Boles, W. E. 1997. Fossil songbirds (Passeriformes) from the early Eocene of Australia. Emu 97:43–50.

Brodkorb, P. 1964. Catalogue of fossil birds: Pt. 2. Anseriformes through Galliformes. Bulletin of the Florida State Museum, Biological Sciences 8:195–335.

Brodkorb, P. 1967. Catalogue of fossil birds: Pt. 3. Ralliformes, Ichthyornithiformes, Charadriiformes. Bulletin of the Florida State Museum, Biological Sciences 11:99–220.

Brodkorb, P. 1978. Catalogue of fossil birds: Pt. 5. Passeriformes. Bulletin of the Florida State Museum, Biological Sciences 23:139–228.

Broughton, J. M., D. B. Madsen, and J. Quade. 2000. Fish remains from Homestead Cave and lake levels of the past 13,000 years in the Bonneville Basin. Quaternary Research 53:392–401.

Buden, D. W. 1987. The birds of southern Bahamas: an annotated check-list. British Ornithologists' Union Check-list no. 8.

Calmaestra, R. G., and E. Moreno. 2000. Ecomorphological patterns related to migration: a comparative osteological study with passerines. Journal of the Zoological Society of London 252:495–501.

Chandler, R. M. 1990. Recent advances in the study of Neogene fossil birds: Pt. 2. Fossil birds of the San Diego Formation, late Pliocene, Blancan, San Diego County, California. Ornithological Monographs no. 44. American Ornithologists' Union, Washington, D.C.

Chiappe, L. M. 2002. Basal bird phylogeny: problems and solutions. Pages 448–472 *in* Mesozoic Birds (L. M. Chiappe and L. M. Witmer, eds.). University of California Press, Berkeley.

Colvinaux, P. A., P. E. De Oliveira, J. E. Moreno, M. C. Miller, and M. B. Bush. 1996. A long pollen record from lowland Amazonia: forest and cooling in glacial times. Science 274:85–88.

Davis, M. B., and R. G. Shaw. 2001. Range shifts and adaptive responses to Quaternary climate change. Science 292:673–679.

Deevey, E. S., D. S. Rice, P. M. Rice, H. H. Vaughan, M. Brenner, and M. S. Flannery. 1979. Mayan urbanism: impact on a tropical karst environment. Science 206:298–306.

Edwards, S. V., and W. E. Boles. 2002. Out of Gondwana: the origin of passerine birds. Trends in Ecology and Evolution 17:347–349.

Elias, S. A. 1994. Quaternary Insects and Their Environments. Smithsonian Institution Press, Washington, D.C.

Elzanowski, A. 2002. Archaeopterygidae (Upper Jurassic of Germany). Pages 129–159 *in* Mesozoic Birds (L. M. Chiappe and L. M. Witmer, eds.). University of California Press, Berkeley.

Emslie, S. D. 1998. Avian community, climate, and sea-level changes in the Plio-Pleistocene of the Florida peninsula. Ornithological Monographs no. 50. American Ornithologists' Union, Washington, D.C.

Ericson, P. G. P., L. Christidis, A. Cooper, M. Irestedt, J. Jackson, U. S. Johansson, and J. A. Norman. 2002. A Gondwanan origin of passerine birds supported by DNA sequences of the endemic New Zealand wrens. Proceedings of the Royal Society, London, Series B, Biological Sciences 269:235–241.

Feduccia, A. 1996. The Origin and Evolution of Birds. Yale University Press, New Haven.

Feduccia, A. 2003. "Big bang" for Tertiary birds. Trends in Ecology and Evolution 18:172–176.

Feduccia, A., and S. L. Olson. 1982. Morphological similarities between the Menurae and Rhinocryptidae, relict passerine birds of the Southern Hemisphere. Smithsonian Contributions to Zoology 366:1–22.

Fischer, H., M. Wahlen, J. Smith, D. Mastroianni, and B. Deck. 1999. Ice core records of atmospheric CO_2 around the last three glacial terminations. Science 283:1712–1714.

Graham, A. 1999. Late Cretaceous and Cenozoic History of North American Vegetation, North of Mexico. Oxford University Press, Oxford.

Graham, R. W., and E. L. Lundelius, Jr. 1994. FAUNMAP: A database documenting late Quaternary distributions of mammal species in the United States. Illinois State Museum Scientific Papers, Vol.25, no. 1.

Graham, R. W., E. L. Lundelius, Jr., et al. [18 other authors]. 1996. Spatial response of mammals to late Quaternary environmental fluctuations. Science 272:1601–1606.

Greenberg, R., P. Bichier, and J. Sterling. 1997. Acacia, cattle and migratory birds in southeastern Mexico. Biological Conservation 80:235–247.

Greenberg, R., V. Pravosudov, J. Sterling, A. Kozlenko, and V. Kontorschikov. 1999. Divergence in foraging behavior of foliage-gleaning birds of Canadian and Russian boreal forests. Oecologia 120:451–462.

Grigg, L. D., and C. Whitlock. 1998. Late-glacial vegetation and climate change in western Oregon. Quaternary Research 49:287–298.

Harvell, C. D., C. E. Mitchell, J. R. Ward, S. Altizer, A. P. Dobson, R. S. Ostfeld, and M. D. Samuel. 2002. Climate warming and disease risks for terrestrial and marine biota. Science 296:2158–2162.

Hearty, P. J., and A. C. Neumann. 2002. Rapid sea level and climate change at the close of the last interglaciation (MIS 5e): evidence from the Bahama Islands. Quaternary Science Reviews 20:1881–1895.

Heckenberger, M. J., A. Kuikuro, J. C. Russell, M. Schmidt, C. Fausto, and B. Franchetto. 2003. Amazonia 1492: pristine forest of cultural parkland? Science 301:1710–1714.

Howard, H. 1963. Fossil birds from the Anza-Borrego Desert. Los Angeles County Museum Contributions in Science 73:1–33.

Howard, H. 1983. A list of the extinct fossil birds of California. Bulletin of the Southern California Academy of Sciences 82:1–11.

Jackson, S. T., J. T. Overpeck, T. Webb III, S. E. Keattch, and K. H. Anderson. 1997. Mapped plant-microfossil and pollen records of late Quaternary vegetation change in eastern North America. Quaternary Science Reviews 16:1–70.

James, H. F. 1987. A Late Pleistocene avifauna from the island of Oahu, Hawaiian Islands. Documents des Laboratoires de Géologie de la Faculté de Science de Lyon 99:221–230.

Janzen, D. H., and P. S. Martin. 1982. Neotropical anachronisms: the fruit that gomphotheres ate. Science 215:19–27.

Kellner, A. W. A. 2002. A review of avian Mesozoic fossil feathers. Pages 389–404 *in* Mesozoic Birds (L. M. Chiappe and L. M. Witmer, eds.). University of California Press, Berkeley.

Klicka, J., A. J. Fry, R. M. Zink, and C. W. Thompson. 2001. A cytochrome-*b* perspective on *Passerina* bunting relationships. Auk 118:611–623.

Kurochkin, E. N. 1985. A true carinate bird from Lower Cretaceous deposits in Mongolia and other evidence of Early Cretaceous birds in Asia. Cretaceous Research 6:271–278.

Kurochkin, E. N. 1999. The relationships of the early Cretaceous *Ambiortus* and *Otogornis*. Smithsonian Contributions to Paleobiology 89:275–284.

Lambert, D. M., P. A. Ritchie, C. D. Millar, B. Holland, A. J. Drummond, and C. Baroni. 2002. Rates of evolution in ancient DNA from Adélie Penguins. Science 295:2270–2273.

Leyden, B. W. 1984. Guatemalan forest synthesis after Pleistocene aridity. Proceedings of the National Academy of Sciences USA 81:4856–4859.

Little, E. L., Jr. 1971. Atlas of United States Trees, Vol. 1, Conifers and Important Hardwoods. U.S. Department of Agriculture Miscellaneous Publication no. 1146.

Lovette, I. J., and E. Bermingham. 1999. Explosive speciation in the New World *Dendroica* warblers. Proceedings of the Royal Society, London, Series B, Biological Sciences 266:1629–1636.

Lowther, P. E., S. M. Lanyon, and C. W. Thompson. 1999. Painted Bunting. The Birds of North America, no. 398 (A. Poole and F. Gill, eds.). The Birds of North America, Inc., Philadelphia.

Lundelius, E. L., R. W. Graham, E. Anderson, J. Guilday, J. A. Holman, D. W. Steadman, and S. D. Webb. 1983. Terrestrial vertebrate faunas. Pages 311–353 *in* Late-Quaternary Environments of the United States, Vol. 1, The Late Pleistocene (S. C. Porter, ed.). University of Minnesota Press, Minneapolis.

MacFadden, B. J. 2001. Three-toed browsing horse *Anchitherium clarencei* from the Early Miocene (Hemingfordian) Thomas Farm, Florida. Bulletin of the Florida Museum of Natural History 43:79–109.

MacFadden, B. J., N. Solounias, and T. E. Cerling. 1999. Ancient diets, ecology, and extinction of 5-million-year-old horses from Florida. Science 283:824–827.

Malakoff, D. 2003. Science agencies get most of what they want, finally. Science 299:1160–1161.

Marks, J. S. 1993. Molt of Bristle-thighed Curlews in the Northwestern Hawaiian Islands. Auk 110:573–587.

Martin, P. S. In press. Overkill: Mystery in Near Time. University of California Press, Berkeley.

Martin, P. S., and D. W. Steadman. 1999. Prehistoric extinctions on islands and continents. Pages 17–55 *in* Extinctions in Near Time: Causes, Contexts and Consequences (R. D. MacPhee, ed.). Kluwer Academic / Plenum Publishers, New York.

Mayr, G. 1998a. "Coraciiforme" und "piciforme" Kleivögel aus dem Mittel-Eozän der Grube Messel (Hessen, Deutschland). Courier Forschungsinstitut Senckenberg 205:1–101.

Mayr, G. 1998b. A new family of Eocene zygodactyl birds. Senckenbergiana Lethaea 78:199–209.

Mayr, G. 2000. Tiny hoopoe-like birds from the Middle Eocene of Messel (Germany). Auk 117:964–970.

Mayr, G., and C. Mourer-Chauviré. 2000. Rollers (Aves: Coraciiformes s.s.) from the Middle Eocene of Messel (Germany) and the Upper Eocene of the Quercy (France). Journal of Vertebrate Paleontology 20:533–546.

Mourer-Chauviré, C., M. Hugueney, and P. Jonet. 1989. Découverte de Passeriformes dans l'Oligocène supérieur de France. Comptes Rendus de l'Académie des Sciences de Paris, Série II, 309:843–849.

Mourer-Chauviré, C., M. Hugueney, and P. Jonet. 1996. Paleogene avian localities of France. Pages 567–598 *in* Tertiary avian localities of Europe (J. Mlikovsky, ed.). Acta Universitatis Caroliniae Geologica 39:519–846.

Noriega, J. I., and L. M. Chiappe. 1993. An early Miocene passeriform from Argentina. Auk 110:936–938.

Olson, S. L. 1985. The fossil record of birds. Avian Biology 8:79–252.

Olson, S. L. 1989. Aspects of global avifaunal dynamics during the Cenozoic. Pages 2023–2029 *in* Proceedings of the 19th International Ornithological Congress, Vol. 2.

Olson, S. L. 2001. Why so many kinds of passerine birds? Bioscience 51:268–269.

Olson, S. L., and W. B. Hilgartner. 1982. Fossil and subfossil birds from the Bahamas. Pages 22–56 *in* Fossil Vertebrates from the Bahamas (S. L. Olson, ed.). Smithsonian Contributions to Paleobiology no. 48.

Olson, S. L., and H. F. James. 1982. Prodromus of the fossil avifauna of the Hawaiian Islands. Smithsonian Contributions to Zoology no. 365.

Olson, S. L., and H. F. James. 1991. Descriptions of thirty-two new species of birds from the Hawaiian Islands: Pt 1. Nonpasserines. Ornithological Monographs no. 45. American Ornithologists' Union, Washington, D.C.

Olson, S. L., and P. C. Rasmussen. 2001. Miocene and Pliocene birds from the Lee Creek Mine, North Carolina. Pages 233–365 *in* Geology and Paleontology of the Lee Creek Mine, North Carolina, III (C. E. Ray and D. J. Bohaska, eds.). Smithsonian Contributions to Paleobiology no. 90. Smithsonian Institution Press, Washington, D.C.

Outlaw, D. C., G. Voelker, B. Milá, and D. Girman. 2003. The evolution of long-distance migration and historical biogeography of the *Catharus* thrushes: a molecular phylogenetic approach. Auk 120:299–310.

Owen-Smith, N. 1987. Pleistocene extinctions: the pivotal role of megaherbivores. Paleobiology 13:351–362.

Payne, R. B. 1992. Indigo Bunting. The Birds of North America, no. 4 (A. Poole and F. Gill, eds.). The Birds of North America, Inc., Philadelphia.

Peterson, A. T., M. A. Ortega-Huerta, J. Bartley, V. Sánchez-Cordero, J. Soberón, R. H. Buddemeier, and D. R. B. Stockwell. 2002. Future projections for Mexican faunas under global climate change scenarios. Nature 416:626–629.

Pregill, G. K., D. W. Steadman, and D. R. Watters. 1994. Late Quaternary vertebrate faunas of the Lesser Antilles: historical components of Caribbean biogeography. Bulletin of Carnegie Museum of Natural History 30:1–44.

Reheis, M. 1999. Highest pluvial-lake shorelines and Pleistocene climate of the western Great Basin. Quaternary Research 52:195–205.

Rogers, K. L., and Y. Wang. 2002. Stable isotopes in pocket gopher teeth as evidence of a late Matuyama climate shift in the southern Rocky Mountains. Quaternary Research 57:200–207.

Rosenmeier, M. F., D. A. Hodell, M. Brenner, and J. H. Curtis. 2002. A 4000-year lacustrine record of environmental change in the Southern Maya lowlands, Petén, Guatemala. Quaternary Research 57:183–190.

Rubenstein, D. R, C. P. Chamberlain, R. T. Holmes, M. P. Ayers, J. R. Waldbauer, G. R. Graves, and N. C. Tuross. 2002. Linking breeding and wintering ranges of a migratory songbird using stable isotopes. Science 295:1062–1065.

Rutherford, S., and S. D'Hondt. 2000. Early onset and tropical forcing of 100,000-year Pleistocene glacial cycles. Nature 408:72–75.

Shuman, B., J. Bravo, J. Kaye, J. A. Lynch, P. Newby, and T. Webb III. 2001. Late Quaternary water-level variations and vegetation history at Crooked Pond, southeastern Massachusetts. Quaternary Research 56:401–410.

Steadman, D. W. 1981. A re-examination of *Palaeostruthis hatcheri* (Shufeldt), a late Miocene sparrow from Kansas. Journal of Vertebrate Paleontology 1:171–173.

Steadman, D. W. 1984. A middle Pleistocene (late Irvingtonian) avifauna from Payne Creek, central Florida. Special Publication of the Carnegie Museum no. 8:47–52.

Steadman, D. W. 1998. From glaciers to global warming: the long-term ecology of birds in New York State. Pages 56–71 *in* Birds of New York State (E. Levine, ed.). Cornell University Press, Ithaca.

Steadman, D. W. 1999. The prehistory of vertebrates, especially birds, on Tinian, Aguiguan, and Rota, Northern Mariana Islands. Micronesica 31:59–85.

Steadman, D. W. 2001. A long-term history of terrestrial birds and mammals in the Chesapeake/Susquehanna drainage. Pages 83–108 *in* Discovering the Chesapeake: The History of an Ecosystem (P. D. Curtin, G. S. Brush, and G. W. Fisher, eds.). Johns Hopkins University Press, Baltimore.

Steadman, D. W. 2003. Long-term change and continuity in the Holocene bird community of western New York State. Bulletin of the Buffalo Society of Natural Sciences 37:121–132.

Steadman, D. W., and M. Intoh. 1994. Biogeography and prehistoric exploitation of birds from Fais Island, Yap State, Federated States of Micronesia. Pacific Science 48:116–135.

Steadman, D. W., and L. J. Justice. 1998. Prehistoric exploitation of birds on Mangareva, Gambier Islands, French Polynesia. Man and Culture in Oceania 14:81–98.

Steadman, D. W., and P. V. Kirch. 1998. Biogeography and prehistoric exploitation of birds in the Mussau Islands, Papua New Guinea. Emu 98:13–22.

Steadman, D. W., and P. S. Martin. 1984. Extinction of birds in the late Pleistocene of North America. Pages 466–477 *in* Quaternary Extinctions: A Prehistoric Revolution (P. S. Martin and R. G. Klein, eds.). University of Arizona Press, Tucson.

Steadman, D. W., and M. C. McKitrick. 1982. A Pliocene bunting from Chihuahua, Mexico. Condor 84:240–241.

Steadman, D. W., A. Plourde, and D. V. Burley. 2002. Prehistoric butchery and consumption of birds in the Kingdom of Tonga, South Pacific. Journal of Archaeological Science 29:571–584.

Sykes, P. W., and M. H. Clench. 1998. Winter habitat of Kirtland's Warbler: an endangered Nearctic/Neotropical migrant. Wilson Bulletin 110:244–261.

Tarduno, J. A., R. D. Cottrell, and A. V. Smirnov. 2002. The Cretaceous superchron geodynamo: observations near the tangent cylinder. Proceedings of the National Academy of Sciences USA 99:14021–14025.

Tyrberg, T. 1998. Pleistocene birds on the Palearctic: a catalogue. Publications of the Nuttall Ornithological Club no. 27.

van der Hammen, T., and M. L. Absy. 1994. Amazonia during the last glacial. Palaeogeography, Palaeoclimatology, Palaeoecology 109:247–261.

Van Devender, T. R. 1990. Late Quaternary vegetation and climate of the Chihuahuan Desert, United States and Mexico. Pages 104–133 *in* Packrat Middens: The Last 40,000 Years of Biotic Change (J. L. Betancourt, T. R. Van Devender, and P. S. Martin, eds.). University of Arizona Press, Tucson.

Voelker, G. 1999. Molecular evolutionary relationships in the avian genus *Anthus* (Pipits: Motacillidae). Molecular Phylogenetics and Evolution 11:84–94.

Voelker, G. 2002. Systematics and historical biogeography of wagtails: dispersal versus vicariance revisited. Condor 104:725–739.

Webb, S. D., and A. Rancy. 1996. Late Cenozoic evolution of the Neotropical mammal fauna. Pages 335–358 *in* Evolution and Environment in Tropical America (J. B. C. Jackson, A. F. Budd, and A. G. Coates, eds.). University of Chicago Press, Chicago.

Williams, D. F., J. Peck, E. B. Karabanov, A. A. Prokopenko, V. Kravchinsky, J. King, and M. I. Kuzmin. 1997. Lake Baikal record of continental climate response to orbital insolation during the past 5 million years. Science 278:1114–1117.

Willis, K. J., A. Kleczkowski, and S. J. Crowhurst. 1999. 124,000-year periodicity in terrestrial vegetation change during the late Pliocene epoch. Nature 397:685–688.

Zachos, J., M. Pagani, L. Sloan, E. Thomas, and K. Billups. 2001. Trends, rhythms, and aberrations in global climate 65 Ma to Present. Science 292:686–693.

2

LEO JOSEPH

Molecular Approaches to the Evolution and Ecology of Migration

THIS CHAPTER IS A REVIEW of new and emerging DNA-based methods to study the evolution and ecology of migration. It provides a hierarchical conceptual framework, with examples from specific studies of migratory species, to understand how these techniques illuminate aspects of the evolution of migration. A comparative phylogenetic approach can help determine the attributes of species in which migratory behavior is most likely to evolve. Ancestral ranges of migrants can be estimated, and distributional shifts that accompanied the evolution of migration can be traced by using both phylogenetic and historical biogeographic analyses of closely related migratory species. From these types of analyses applied to several unrelated groups of birds, the hypothesis that breeding range often shifts as migration evolves has found fresh support. Other issues, which have been clarified through population genetics and phylogeography of individual species, include the evolution of migration routes and the possible histories of the often extremely close relationship of migrants and their nonmigratory conspecific relatives. Such studies also address the role, geographical source, and timing of recent range expansions of migrants and allow researchers to begin to explore the role of historical diversity in the evolution of individual migrations within a migration system.

MOLECULAR APPROACHES TO THE EVOLUTION OF MIGRATION

Increasingly, molecular tools are being used to address the phylogeny, historical biogeography, and population genetics of migratory species. These new tools are bringing new insights to the evolution of migration. Molecular data sets lend themselves to being applied hierarchically, from the study of entire migration systems, to the evolution of migration within clades of species, and finally down to processes acting within a single migratory species. In this chapter I illustrate, through example, how molecular and analytical tools provide a broad range of insights into the study of bird migration. Throughout, new questions and perspectives are offered on how past and present processes, whether biotic or abiotic, have interacted to shape the evolution of migration as we see it today. The scope of the review covers "when" and "where" (historical biogeography) and "who" (which species are likely to become migratory) of migration's evolution, not its "why" (ecology) or "how" (navigation mechanisms). Although I have confined myself to avian examples, molecular approaches are being used to explore similar questions in other migratory animals, particularly mammals such as bats and whales (Baker et al. 1994; Petit and Mayer 2000).

Comparative Phylogenetic Approach: Migration Systems

Chesser and Levey (1998) used a comparative phylogenetic approach to help define the pool of species from which migrants are most likely to evolve. They tested Levey and Stiles's (1992) hypothesis that migration most likely evolves in lineages with at least partially frugivorous or nectarivorous species. In addition to these dietary characteristics, Chesser and Levey (1998) focused on the climatic buffering characteristics of the most important habitats used by different species. They used the term "non-buffered" to describe edge, canopy, and open habitats and reserved the classification of "buffered" for forest interiors. They found that temperate-tropical migration in all New World passerines is more prevalent in frugivorous lineages and in lineages characterized by species of "non-buffered" habitats. However, only the latter habitat-based relationship is significant. They concluded that although migration appears to be constrained from evolving in insectivorous species of forest interiors, release from those constraints has not necessarily led to its evolution. Chesser and Levey (1998) recognized that more detailed phylogenies of Neotropical birds at lower taxonomic levels would be necessary to fully explore their approach. Their method, however, opened up new ways of thinking about the ecological patterns and processes underlying the evolution of an entire migration system.

Phylogeny and Historical Biogeography in the Evolution of Migration

With a phylogenetic analysis of a clade of migrant and nonmigrant species in hand, the simplest questions to ask are whether the migratory species are each other's closest relatives and whether single or multiple gains and losses of migration need be invoked (Burns 1998; Joseph et al. 1999, 2004; Klein and Brown 1994; Outlaw et al. 2003). If the migratory species are each others' closest relatives, that is, they form a clade themselves, then postulating a single origin of migration is sufficient. Should a nonmigratory species be nested within an otherwise migratory clade, a loss of migration is evident and an example of this is in the *Piranga* tanagers (Burns 1998).

Detailed studies of phylogenies have uncovered apparently complex patterns of origins and losses of migration. The five migratory and seven resident species of Nearctic-Neotropical *Catharus* thrushes (excluding *Hylocichla mustelina*) (Outlaw et al. 2003) provide an excellent example of this complexity and what we can learn from it. Recognizing that migration and its directions can evolve rapidly (Berthold and Querner 1982; Berthold et al. 1992; Klein and Brown 1994), Outlaw et al. (2003) applied maximum likelihood-based reconstructions of migratory status (Pagel 2000). Maximum-likelihood methods allow one to reconstruct character evolution while using a particular model of evolution. It is hoped that this approach is more realistic than simply asking what is the most parsimonious reconstruction of a character's evolution on a particular tree. This permitted more rigorous testing of alternative hypotheses than in other similar studies (e.g., *Charadrius* plovers and their relatives [Joseph et al. 1999]). Outlaw et al. concluded that migration most likely arose at least three times in *Catharus*.

A molecular phylogeny can enable students of migration to ask when and where vicariance (appearance of a physical barrier in the geographical range of a taxon thus splitting it in two) and dispersal need be invoked to explain present-day, disjunct breeding and nonbreeding distributions of migrants and whether the ancestral range of each migratory species was its present-day breeding or nonbreeding distribution, that is, which distribution has been derived through a range shift (Joseph et al. 1999, 2003; Outlaw et al. 2003). The hypothesis that current ranges of migratory birds reflect the return to an "ancestral home" (Gauthreaux 1982) can be cast in a broader historical biogeographic context.

Because they often have at least two disjunct ranges, migrants pose a particular challenge to traditional ways of determining ancestral areas and areas of origin (Bremer 1992). *Charadrius* plovers and their relatives (e.g., *Oreopholus, Vanellus* [Joseph et al. 1999]) and the *Catharus* thrushes (Outlaw et al. 2003) have been test cases for using molecular data in developing novel approaches to this challenge. In the *Charadrius* study, breeding and nonbreeding ranges were treated as separate characters. Different character states were different continental regions, for example, South America, North America, eastern Asia, western Asia. A species such as the Semipalmated Plover (*Charadrius semipalmatus*), with a breeding range in North America and a nonbreeding range in South America, was assigned a breeding range character state of North America and a nonbreeding range character state of South America. The Tawny-throated Dotterel (*Oreopholus ruficollis*), on the other hand, which breeds and migrates entirely within southern South America, was assigned a character state of South

America for both breeding and nonbreeding ranges. One can of course refine the geographical scale at which range character states are assigned. The coarse level in the examples just given, however, was appropriate to the questions being asked in the Joseph et al. (1999) study.

These range characters were mapped onto a phylogeny derived independently from mitochondrial DNA (mtDNA). Shifts in character states of breeding and nonbreeding ranges were reconstructed by using parsimony. If, for example, nonbreeding grounds have shifted during the evolution of migration, then when breeding and nonbreeding grounds are mapped on a phylogeny, the breeding grounds would not be expected to change, whereas the nonbreeding grounds would. In this example, as one traces for a given species the relevant branches of the phylogenetic tree on which the character evolution is being mapped, there would be little or no change in the character state of breeding range along those branches, whereas there would be substantial change in the trace of nonbreeding range. The reciprocal would apply if breeding grounds had shifted.

Results for *Charadrius* plovers and their relatives suggest that they first evolved in the Southern Hemisphere, arguably in South America. Current migrations from North to South America, for example, then evolved from a shift in breeding ranges from south to north, as inferred from the distribution of two species (*Charadrius semipalmatus, C. wilsonia*).

Outlaw et al. (2003) applied dispersal-vicariance analysis (DIVA [Ronquist 1997]) to reconstruct the ancestral area for each node of the *Catharus* tree. DIVA's primary objective is to measure the "cost" of assigning either vicariance or dispersal at different nodes of a tree in order to derive the present-day distributions of the taxa involved. A secondary function is reconstruction of ancestral areas. They concluded that a Neotropical origin of *Catharus* was most likely, and migrant species probably arose from common ancestors shared with resident species of the highlands of Mexico and Central America. This supports Cox's (1985) hypothesis, which was not based on any phylogenetic information, that the seasonal Subtropics have been a staging area for the evolution of long-distance migration in the Americas.

In sum, the combined phylogenetic and biogeographic approach to the evolution of migration has supported a working hypothesis that emerged in the 1970s (e.g., papers in Keast and Morton 1980): that migratory birds generally evolve from ancestors in the present-day nonbreeding range through shifts or displacements of the breeding range. It also allows us to begin a synthesis of the diverse biotic and abiotic factors and processes that led to the evolution of migration, at least for temperate-tropical migrants breeding in North America. That is, the maximization of productivity arising from bursts of summer food availability in higher boreal latitudes was likely a key, ultimate selective pressure that initiated evolution of now-familiar temperate-tropical migrations. Probably linked to this selective pressure are other physiological, ecological, and environmental factors. For example, Levey and Stiles (1992) discussed frugivorous species that move widely in search of patchily distributed food resources. They suggested that such species were predisposed to undertake geographically longer movements of temperate-tropical migration by which they maximize their productivity from summer flushes of insect resources. The role of prevailing climatic forces about the Gulf of Mexico in facilitating these long-distance movements was discussed by Rappole and Ramos (1994). Thus, all of these biotic and abiotic factors likely came into play gradually and at different times in different species as habitats and selective pressures changed. In the Northern Hemisphere at least, the receding of ice sheets was also critical. Finally, selective pressure to return to the ancestral home would have come from maximizing survival by avoiding the harsh winters at progressively higher latitudes. Viewed from these various phylogenetic, ecological, and physical perspectives, the temperate-tropical migrant birds we are so familiar with today are flying north for summer as much as—if not more than—they are flying south for winter. Parenthetically, it is worth noting that the ecological and physical components of the synthesis just suggested arise largely from the literature on boreally breeding temperate-tropical migrants in North America and the northern Neotropics; we know virtually nothing of whether similar pressures led to temperate-tropical migration within South America or Australia in species such as Swainson's Flycatcher (*Myiarchus swainsoni*) and the White-browed Woodswallow (*Artamus superciliosus*), respectively, for example.

POPULATION GENETICS AND PHYLOGEOGRAPHY IN THE EVOLUTION OF MIGRATION

The tools of molecular population genetics permit descriptions and measurements of how much genetic diversity is present within a species and how much of it is distributed within and among populations. Molecular phylogeography describes the present-day geographical distribution of genetic diversity within a species and the phylogenetic relationships among the various populations. From these analyses, the past processes that generated the present-day structure can be inferred. Despite the explosive increase in published studies of the population genetics and phylogeography of birds, we are only just beginning to use this kind of knowledge to refine our understanding of how migration evolved. Where these studies have been done in species with one or more migratory populations, they have clear implications for our understanding of the evolution of migration. Examples cover species as diverse as shorebirds and seabirds with circumpolar breeding ranges, especially *Calidris* sandpipers (Wenink et al. 1993, 1996; Tiedemann 1999; Wennerberg et al. 1999, 2002; Wennerberg 2001), *Uria* murres (Friesen et al. 1996a, 1996b), *Thalassarche* albatrosses of the southern oceans (Abbott and Double 2003), and Eurasian passerines such as *Phylloscopus* warblers (Bensch et al. 1999; Irwin et al. 2001), *Motacilla* wagtails (Odeen and Bjorklund 2003; Pavlova et al. 2003), *Anthus* pipits (Voelker

1999a, 1999b), and *Luscinia* bluethroats (Zink et al. 2003). To date, however, relatively few molecular studies of birds have been designed explicitly to address issues in the evolution of migration (but see Bermingham et al. 1992, Klein and Brown 1994, Cicero and Johnson 1998, Buerkle 1999, Milot et al. 2000, Ruegg and Smith 2002, and Joseph et al. 2003). Nonetheless, some clear trends, which likely reflect environmental history as much as the evolution of migration itself, have emerged from this work and are the focus of the next part of this review.

Example 1: Close Relationship of Migrants and Nonmigrants within a Species

Buerkle (1999) examined the Prairie Warbler (*Dendroica discolor*), which has a migratory subspecies, *D. d. discolor,* a relative habitat generalist breeding in the eastern United States and wintering in Florida and the West Indies, and a nonmigratory subspecies, *D. d. paludicola,* which is endemic to the mangroves of subtropical Florida. A "tree" and "network diagram" depicting relationships of the various mtDNA sequences found within a species were constructed (see example in fig. 2.1). Analysis of Molecular Variance (AMOVA) addressed how much of the species' present-day genetic diversity is distributed within and between the two subspecies, and between regions occupied by different populations of the subspecies. AMOVA uses DNA-level analogues of Wright's F statistics (Wright 1951), which are based on allele frequencies and have been the traditional statistical approach to studying geographic genetic structure. Finally, the number of individuals exchanged between populations was estimated from the familiar population genetics variable *Nm,* where *N* is the population size and *m* is the rate of migration. *Nm* is equal to the *n,* number of individuals migrating per generation, and is an easier variable to work with than *n* itself.

The largest share of the variance (27–42%) is explained by comparisons between samples of the nonmigratory and migratory subspecies. An additional one-fifth of the total genetic variance is explained by between-region variation, whereas at most only 9% of the variance is explained by pairwise comparisons within the migratory subspecies. Significant differences in population structure of this sort were unexpected given the proximity of the two subspecies in the southeastern United States and their obvious capacity for long-distance movement. Individuals are exchanged between the two subspecies at a rate of less than one per generation. Phylogenetic analysis, however, found no evidence for any deep, historical separation between the two subspecies. Buerkle (1999) reconciled the significant difference between present-day population structure of the two subspecies and the lack of phylogenetic divergence between them by concluding that their isolation and concomitant morphological and behavioral differentiation (most likely the loss of migration in *D. d. paludicola*) must have arisen relatively recently in their history.

Example 2: Linking Phylogeny and Population Genetics in *Myiarchus* Tyrant-Flycatchers: A Temporal Framework

Analysis of Swainson's Flycatcher (*Myiarchus swainsoni*) of South America revealed a situation parallel to but more complex than that in the Prairie Warbler. The same popu-

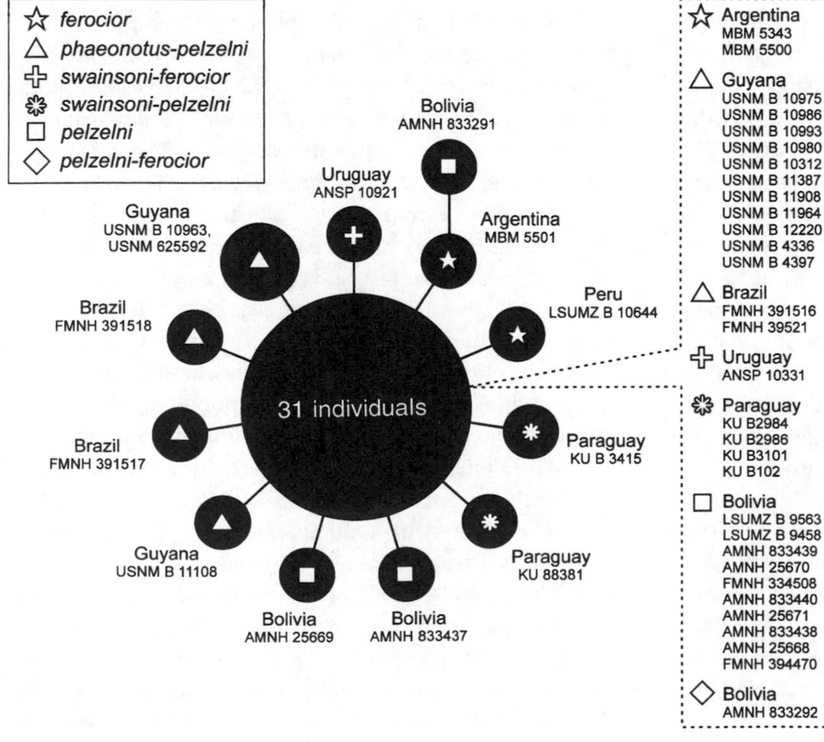

Fig. 2.1. Statistical parsimony network of mtDNA from 44 individuals of migratory and nonmigratory species for 842 base pairs of the mitochondrial ATPase 8 and 6 genes. Each circle represents a unique haplotype for the ATPase genes, and its size reflects the numbers of individuals found to have that sequence. Museum registration numbers of each individual are indicated (AMNH: American Museum of Natural History; ANSP: Academy of Natural Sciences; FMNH: Field Museum of Natural History; KU; University of Kansas; LSUMZ: Louisiana State University Museum of Zoology; MBM: Marjorie Barrick Museum of Natural History; USNM: United States National Museum). The length of each line connecting the circles reflects the number of differences between the connected haplotypes. The network's essentially starlike shape is expected in a population that has undergone a recent expansion. Reproduced by permission from *Journal of Biogeography.*

lation genetic and phylogeographic approaches were used as in Buerkle's (1999) study as well as some additional ones (Joseph et al. 2003) that estimated relevant temporal parameters. These were coupled with phylogenetic perspectives provided by an analysis of relationships among *Myiarchus* species generally (Joseph et al. 2004).

Swainson's Flycatcher as currently construed (Lanyon 1978) comprises two migratory subspecies, *M. s. swainsoni* and *M. s. ferocior,* and two nonmigratory ones, *M. s. pelzelni* and *M. s. phaeonotus.* Migratory *M. s. swainsoni* is an evolutionarily old member of the South American radiation of *Myiarchus.* Its closest relative could not be discerned; certainly it is not at all closely related to any other population of the *M. swainsoni* complex that was studied. Migratory *M. s. ferocior,* in contrast, is very closely related to each of the following: nonmigratory *M. s. pelzelni* and *M. s. pelzelni / M. s. phaeonotus* "intergrades" (individuals from populations that are geographically and phenotypically intermediate between breeding populations of two or more subspecies), and migratory *M. s. ferocior / M. s. swainsoni* and *M. s. pelzelni / M. s. swainsoni* "intergrades." Over 80% of the total variance was explained by within-population variation (AMOVA) in a sample of 44 individuals collected across 4,000 km of South America from the latter five populations. A corollary of this was zero net sequence divergence across that entire range.

In addition to a lack of interpopulation divergence, the genetic evidence suggests a recent and rapid population expansion. Mismatch analysis (the observed numbers of differences is plotted against the frequency of those numbers [Rogers 1995]), based on a pairwise comparison of the DNA sequences in the same clade of 44 individuals, showed the unimodal distribution characteristic of a recently expanded population. Historically stable populations are expected to show a multimodal, ragged distribution. The starlike shape of the network of relationships among the mtDNA sequences themselves is typical of a recently expanded population (fig. 2.1). Mismatch analysis also estimated that the time since the expansion is likely well within the last 100,000 years. So, for this clade of 44 migratory and nonmigratory individuals, the findings parallel those in *D. discolor:* relatively rapid morphological and behavioral (i.e., migratory status) differentiation against a backdrop of no historical separation is again inferred. The divergent position of nominate *M. s. swainsoni,* which likely warrants taxonomic recognition as a separate species, signals that evolution of temperate-tropical migration in South America has still more layers of spatio-temporal complexity. That is, the biogeographical history of migratory *M. s. swainsoni* on the South American landscape almost certainly played out at different times than that of *M. s. ferocior,* and the two likely evolved migration independently. The basic tools used to derive this picture, however, were levels of nucleotide diversity and their spatial distribution analyzed with AMOVA, a statistical test for population expansion versus stability (Fu's *Fs* [see Fu 1997]), mismatch analysis, and the shape of the network diagram of mtDNA relationships.

Example 3: Further Linking Phylogenetic and Population Genetics: Unraveling Evolution of Migration in Swainson's Thrush (*Catharus ustulatus*)

In their broader analysis of *Catharus* thrushes, Outlaw et al. (2003) estimated that Swainson's Thrush (*Catharus ustulatus*), a long-distance migrant, diverged from its closest relative as long as 4 Ma, but they did not speculate about when migration originated in this species. Ruegg and Smith (2002), however, coupled the genetic approaches we have seen so far (levels of nucleotide diversity, mismatch analyses, mtDNA networks) with banding data to probe at least some of the evolutionary history of migration in *C. ustulatus* more deeply.

Ruegg and Smith (2002) focused on the evolution of the two distinct migratory pathways used by Swainson's Thrush. One migration route is used by populations that breed along North America's Pacific Coast. It is a more or less direct north-south route between those breeding grounds and Mexican and Costa Rican nonbreeding grounds. Alternatively, populations that breed in Alaska first fly east across northern North America to the eastern United States and Canada. Then, they continue their migration south. All of these breeding populations from eastern North America and Alaska overwinter primarily in Panama and northern South America (Brewer et al. 2000).

MtDNA in *C. ustulatus* falls into two groups that are concordant with the two migratory routes. These groupings indicate that populations breeding in Alaska are more closely related to those breeding in the eastern United States and Canada than to the geographically closer populations of the Pacific coastal group. Both mtDNA groups had low nucleotide diversity, starlike mtDNA networks, and unimodal mismatch distributions, all typical of recently expanded populations, as we have seen. The level of divergence between the two mtDNA groups, while low at 0.69%, is nonetheless greater than within either group. Tentative application of conventional estimates of how much divergence accumulates in mtDNA sequences with time, that is, the rate of mtDNA evolution (see Lovette 2004), suggests that the two mtDNA groups separated as recently as 10,000 years before present (B.P.) and certainly within the late Pleistocene. Ruegg and Smith (2002) concluded that the species' migration routes are effectively retracing historical routes of expansion rather than following the most direct migration route. Differentiation in morphology among the various subspecies involved is again inferred to be recent. Here we should recall that migration often evolves through displacement of the breeding distribution. Ruegg and Smith's (2002) analysis coupled with Outlaw et al.'s (2003) inference of probable Neotropical origins for *Catharus* generally then raises the question of what caused recent differentiation between the two groups of populations in the Neotropics. Differentiation among breeding populations within the two groups, as recognized in present-day subspecies, presumably came later, as similarly inferred in the Prairie Warbler.

Examples 4 and 5: Further Insight into Temperate-Tropical Migration in the Americas

Milá et al. (2002) found similar evidence for a Pleistocene range expansion in the MacGillivray's Warbler (*Oporornis tolmiei*), another North American temperate-tropical migrant, based on the pattern of generally low nucleotide diversities, unimodal mismatch distributions, significant Fu's *Fs*, and star-shaped mtDNA networks. It was most clear, however, in the populations that breed in the United States, not those that breed in Mexico, though all are migratory. Milá et al. (2002) estimated the time of expansion at approximately 12,500 years B.P. This coincides reasonably well with the recession of ice sheets after the Last Glacial Maximum (LGM) at approximately 18,000 years B.P. Such rapid expansions, which have also been found in nontropical migrants such as Red-winged Blackbirds (*Agelaius phoeniceus* [Ball et al. 1988]), Song Sparrows (*Melospiza melodia* [Fry and Zink 1998]), Swamp Sparrows (*M. georgiana* [Greenberg et al. 1998]), and Chipping Sparrows (*Spizella passerina* [Zink and Dittman 1993]), suggest that the ecology to predispose an essentially tropical bird to evolve migration to temperate breeding grounds was in place long before the ice sheets receded; and once they did recede, selective pressure to evolve migration came into play.

Migratory populations of the Yellow Warbler (*Dendroica petechia*) across Canada and Alaska fall into eastern and western groups as defined by mtDNA (Milot et al. 2000). Interestingly, sequences from two of the 46 eastern birds are more closely related to a clade of 13 western birds' sequences than to other eastern sequences. This implies that the entire western population was derived as an offshoot of just part of the eastern population. This parallels, at least potentially, the findings in *C. ustulatus*, where Alaskan populations are closely related to eastern North American ones. To examine this further, the authors used a coalescent analysis implemented in GENETREE (Griffiths 1998). Coalescence traces present-day genetic diversity back along the branches of a tree of relationships to a common ancestral gene while simultaneously modeling the mutation process that generated the tree. The analysis showed with a probability of greater than 95% that western haplotypes originated within the eastern population and that the earliest age of the most recent common ancestor for the whole complex is again 18,500 years B.P., around the height of the LGM. It is not surprising that the LGM and subsequent retraction of ice sheets have been factors in shaping present-day genetic architectures in individual species. However, this example shows molecular data being able to provide temporal frameworks for *geographical* resolution of relatively fine-scale events in the evolutionary histories of migratory species.

Example 6: Molecules and Morphology in the Evolution of Migration

In the European Willow Warbler (*Phylloscopus trochilus*), Bensch et al. (1999) studied molecular and morphological variation across a 350-km-wide zone where two populations meet and presumably hybridize. This intergrade zone has been dubbed a migratory divide because the two morphologically different populations also show widely divergent migration routes even from the area of contact. Analyses based on rapidly evolving microsatellite loci that sampled the nuclear genome were integrated with extensive morphological analyses of variation across the migratory divide. The morphological analyses strongly suggested that the migratory divide is indeed a hybrid zone resulting from two populations that evolved in allopatry but which are now in secondary contact. Morphological clines in the zone of contact are concordant with recolonization of the Scandinavian peninsula after the LGM. In contrast to the morphological findings, however, and adding complexity to the evolution of migration in the birds, is that exactly as in the case of *D. discolor* and *M. swainsoni,* the two populations of Willow Warblers are panmictic for mtDNA, as well as microsatellites, and show no phylogenetic differentiation at these same loci. Bensch et al. (1999) ascribed this either to the morphological and behavioral differences being maintained by strong selection on the genes that control these phenotypes with high levels of current gene flow or to a very recent divergence in morphology and behavior. Buerkle (1999) and Joseph et al. (2003) clearly favored the latter explanation in their respective studies. The former explanation would predict that older, and therefore more divergent, sequences should still be detectable, and this would be reflected in higher nucleotide diversity, a parameter that measures frequency of individual sequences and their divergence from each other.

CONCLUSIONS

Molecular studies of migratory birds and their nonmigratory relatives bring insights to the evolution of migration at numerous levels. Most broadly, we have seen that a comparative phylogenetic approach applied across different families of birds has been used to suggest broad ecological characteristics that have been precursors to the evolution of migration. At the level of individual clades of closely related migratory and nonmigratory species, a robust phylogenetic framework derived from molecular data can be used to infer how often migration has been gained and lost in the clade's history. Further, we can at least tentatively infer times of origin of various migratory and nonmigratory lineages. Thus we can begin to derive a temporal framework with which to reconstruct the environmental history associated with the evolution of migration. Inference of that history can then be refined. Various biogeographical methods attempt to reconstruct the history of distributional shifts that have accompanied the evolution of migration. These methods try to determine the ancestral range of migratory species and thus illuminate the broad macrogeographical shifts involved in evolution of widely disjunct breeding and nonbreeding distributions. Finally, the tools of molecular population genetics yield the finer details of population structure within a species. Then the processes generating the present-day structure can be inferred and a time frame in which they operated can be established—again, however, tentatively. Thus we can tease apart how different migration routes have evolved within a species, and we can probe the relationships among migratory and nonmigratory populations of a species. This kind of work has shown that Pleistocene glaciations have had a profound effect in shaping the present-day genetic architecture of avian species, at least for those with Northern Hemisphere breeding ranges. Although the role of glaciation has been emerging in avian molecular studies more generally for some years, not just those of migratory species (e.g., Avise and Walker 1998), it is particularly relevant to our ability to recover the history of migration from the levels of diversity left in the DNA record. The studies are also salient reminders that inferences from broader phylogenetic analyses (e.g., those cited for *Catharus, Anthus,* and *Charadrius*) about the number of times migration has evolved within a clade are tentative. They also exemplify the methodology offered by molecular tools and the complementary, historical dimensions that they can bring to present-day ecological diversity. Finally, studies of *Myiarchus, Phylloscopus, Catharus,* and *Dendroica* emphasize how the hierarchical framework offered by molecular studies can help develop a full understanding of the evolution and ecology of migration. In each of these, phylogenetic analyses of most or all member species of each clade offer deeper evolutionary perspective to detailed population genetic and phylogeographical analyses of particular migratory species. The scope for further extending this approach is as broad as the number of migratory species that can be studied. Particularly striking at the moment is the dearth of examples from Australia-Asia and Africa–western Europe. The potential breadth of future molecular study of migrants is suggested by Zink's (2002) caution that migratory status is likely not a single character with two character states, present or absent. As he stresses, it is more a shorthand for the genetic machinery underlying migratory physiology and orientation and so likely comprises several different adaptive systems. One looks forward to an exciting period of interdisciplinary collaboration and benefit.

ACKNOWLEDGMENTS

I thank Gary Voelker for discussion, Russ Greenberg and Pete Marra for their support and much-valued advice about the content of this chapter, Natasha Atkins for her skilled editing, and the authors of all works reviewed here for their contributions. Three anonymous reviewers contributed valuable comments. I should like to thank my principal collaborators in my own work on migration: Deryn Alpers, David Stockwell, and Thomas Wilke. Thomas Wilke also helped with preparation of the figure.

LITERATURE CITED

Abbott, C. L., and M. C. Double. 2003. Phylogeography of shy and white-capped albatrosses inferred from mitochondrial DNA sequences: implications for population history and taxonomy. Molecular Ecology 12:2747–2758.

Avise, J. C., and D. Walker. 1998. Pleistocene phylogeographic effects on avian populations and the speciation process. Proceedings of the Royal Society of London, Series B, Biological Sciences 265:457–463.

Baker, C. S., R. W. Slade, J. L. Bannister, and eight other authors. 1994. Hierarchical structure of mitochondrial DNA gene flow among humpback whales *Megaptera novaeangliae,* worldwide. Molecular Ecology 3:313–327.

Ball, R. M., F. C. Freeman, E. Bermingham, and J. C. Avise. 1988. Phylogeographic population structure of red-winged blackbirds assessed by mitochondrial DNA. Proceedings of the National Academy of Sciences USA 85:1558–1562.

Bermingham, E., S. Rohwer, S. Freeman, and C. Wood. 1992. Vicariance biogeography in the Pleistocene and speciation in North American wood warblers: a test of Mengel's model. Proceedings of the National Academy of Sciences USA 89:6624–6628.

Berthold, P. 1994. Bird Migration: A General Survey. Oxford University Press, Oxford.

Berthold, P., and U. Querner. 1982. Partial migration in birds: experimental proof of polymorphism as a controlling system. Experientia 38:805–806.

Berthold, P., A. J. Helbig, G. Mohr, and U. Querner. 1992. Rapid microevolution of migratory behaviour in a wild bird species. Nature 360:668–669.

Bremer, K. 1992. Ancestral areas: a cladistic reinterpretation of the center of origin concept. Systematic Biology 41:436–445.

Brewer, D., A. Diamond, E. J. Woodsworth, B. T. Collins, and E. H. Dunn. 2000. Canadian Atlas of Bird Banding. Canadian Wildlife Service, Puslinch, Ontario.

Buerkle, C. A. 1999. The historical pattern of gene flow among migratory and nonmigratory populations of Prairie Warblers (Aves: Parulinae). Evolution 53:1915–1924.

Burns, K. J. 1998. Molecular phylogenetics of the genus *Piranga:* implications for biogeography and the evolution of morphology and behavior. Auk 115:621–634.

Chesser, R. T., and D. J. Levey. 1998. Austral migrants and the evolution of migration in New World birds: diet, habitat, and migration revisited. American Naturalist 152:311–319.

Cicero, C., and N. Johnson. 1998. Molecular phylogeny and ecological diversification in a clade of New World songbirds (genus *Vireo*). Molecular Ecology 10:1359–1370.

Cox, G. W. 1985. The evolution of avian migration systems between temperate and tropical migration regions of the new world. American Naturalist 126:451–474.

Fleischer R., C. McIntosh, and C. Tarr. 1998. Evolution on a volcanic conveyor belt, using phylogeographic reconstructions and K-Ar based ages of the Hawaiian Islands to estimate molecular evolutionary rates. Molecular Ecology 7:533–545.

Fretwell, S. 1980. Evolution of migration in relation to factors regulating bird numbers. Pages 517–528 *in* Migrant Birds in the Neotropics: Ecology, Behavior, Distribution, and Conservation (J. A. Keast and E. S. Morton, eds.). Smithsonian Institution Press, Washington, D.C.

Friesen, V. L., W. A. Montevecchi, A. J. Gaston, R. T. Barrett, and W. S. Davidson. 1996a. Molecular evidence for kin groups in the absence of large-scale genetic differentiation in a migratory bird. Evolution 50:924–930.

Friesen, V. L., W. A. Montevecchi, A. J. Baker, R. T. Barrett, and W. S. Davidson. 1996b. Population differentiation and evolution in the common guillemot *Uria aalge*. Molecular Ecology 5:793–805

Fry, A. J., and R. M. Zink. 1998. Geographic analysis of nucleotide diversity and song sparrow (Aves: Emberizidae) population history. Molecular Ecology 7:1303–1313.

Fu, Y.-X. 1997. Statistical tests of neutrality of mutations against population growth, hitchhiking and background selection. Genetics 147:915–925.

Gauthreaux, S. A. 1982. The ecology and evolution of avian migration systems. Pages 93–168 *in* Avian Biology, Vol. 6 (D. S. Farner, J. R. King, and K. C. Parkes, eds.). Academic Press, New York.

Greenberg, R., P. J. Cordero, S. Droege, and R. C. Fleischer. 1998. Morphological adaptation with no mitochondrial DNA differentiation in the coastal plain swamp sparrow. Auk 115:706–712.

Griffiths, R. C. 1998. GENETREE. Available from www.maths.monash.edu.au/~mbahlo/mpg/gtree.html.

Hagan, J. W., and D. W. Johnston. 1992. Ecology and Conservation of Neotropical Migrant Landbirds. Smithsonian Institution Press, Washington, D.C.

Irwin, D., S. Bensch, and T. D. Price. 2001. Speciation in a ring. Nature 409:333–337.

Joseph, L., E. Lessa, and L. Christidis. 1999. Phylogeny and biogeography in the evolution of migration: shorebirds of the *Charadrius* complex. Journal of Biogeography 26:329–342.

Joseph, L., T. Wilke, and D. Alpers. 2003. Independent evolution of migration on the South American landscape in a long-distance temperate-tropical migratory bird, Swainson's Flycatcher *Myiarchus swainsoni*. Journal of Biogeography 30:925–937.

Joseph, L., T. Wilke, E. Bermingham, D. Alpers, and R. Ricklefs. 2004. Towards a phylogenetic framework for the evolution of shakes, rattles and rolls in *Myiarchus* tyrant-flycatchers (Aves: Passeriformes: Tyrannidae). Molecular Phylogenetics and Evolution 31:139–152.

Keast, A., and E. S. Morton. 1980. Migrant Birds in the Neotropics: Ecology, Behavior, Distribution, and Conservation. Smithsonian Institution Press, Washington, D.C.

Klein, N. K., and W. M. Brown. 1994. Intraspecific molecular phylogeny in the Yellow Warbler (*Dendroica petechia*), and implications for avian biogeography in the West Indies. Evolution 48:1914–1932.

Lanyon, W. E. 1978. Revision of the *Myiarchus* flycatchers of South America. Bulletin of the American Museum of Natural History 161:429–627.

Levey, D. J., and F. G. Stiles. 1992. Evolutionary precursors of long-distance migration: resource availability and movement patterns in Neotropical landbirds. American Naturalist 140:467–491.

Lovette, I. J. 2004. Mitochondrial dating and mixed support for the "2%" rule in birds. Auk 121:1–6.

Maddison, W. P., and D. R. Maddison. 2004. Mesquite—a modular system for evolutionary analysis, version 1.01. Available at http://mesquiteproject.org/

Milá, N., D. J. Girman, M. Kimura, and T. Smith. 2000. Genetic evidence for the effect of a postglacial population expansion on the phylogeography of a North American songbird. Proceedings of the Royal Society of London, Series B, Biological Sciences 267:1033–1040.

Milot, E., H. L. Gibbs, and K. A. Hobson. 2000. Phylogeography and genetic structure of northern populations of the yellow warbler (*Dendroica petechia*). Molecular Ecology 9:667–681.

Odeen, A., and M. Bjorklund. 2003. Dynamics in the evolution of sexual traits: losses and gains, radiation and convergence in yellow wagtails (*Motacilla flava*). Molecular Ecology 12:2113–2130.

Outlaw, D. C., G. Voelker, B. Milá, and D. J. Girman. 2003. Evolution of migration in and historical biogeography of *Catharus* thrushes: a molecular phylogenetic approach. Auk 120:299–310.

Pagel, M. 1999. The maximum likelihood approach to reconstructing ancestral character states of discrete characters in phylogenies. Systematic Biology 48:612–622.

Pavlova, A., R. M. Zink, S. V. Drovetski, Y. Red'kin, and S. Rohwer. 2003. Phylogeographic patterns in *Motacilla flava* and *M. citreola:* species limits and population history. Auk 120:744–758.

Petit, E., and F. Mayer. 2000. A population genetic analysis of migration: the case of the noctule bat (*Nyctalus noctula*). Molecular Ecology 9:683–690.

Rappole, J. H., and M. A. Ramos. 1995. Factors affecting migratory bird routes over the Gulf of Mexico. Bird Conservation International 4:251–262.

Rogers, A. R. 1995. Genetic evidence for a Pleistocene population explosion. Evolution 49:608–615.

Ronquist, F. 1997. Dispersal-vicariance analysis: a new approach to the quantification of historical biogeography. Systematic Biology 46:195–203.

Ruegg, K. C., and T. B. Smith. 2002. Not as the crow flies: a historical explanation for circuitous migration in Swainson's Thrush (*Catharus ustulatus*). Proceedings of the Royal Society of London, Series B, Biological Sciences 269:1375–1381.

Tiedemann, R. 1999. Seasonal changes in the breeding origin of migrating dunlins (*Calidris alpina*) as revealed by mitochondrial DNA sequencing. Journal fur Ornithologie 140:319–323.

Voelker, G. 1999a. Molecular evolutionary relationships in the avian genus *Anthus* (Pipits: Motacillidae). Molecular Phylogenetics and Evolution 11:84–94.

Voelker, G. 1999b. Dispersal, vicariance, and clocks: historical biogeography and speciation in a cosmopolitan passerine genus (*Anthus:* Motacillidae). Evolution 53:1536–1552.

Wenink, P. W., A. J. Baker, and M. G. J. Tilanus. 1993. Hypervariable control region sequences reveal global population structuring in a long-distance migrant shorebird, the Dunlin (*Calidris alpina*). Proceedings of the National Academy of Sciences USA 90:94–98.

Wenink, P. W., A. J. Baker, H-U. Rosner, and M. G. J. Tilanus. 1996. Global mitochondrial DNA phylogeography of Holarctic breeding dunlins (*Calidris alpina*). Evolution 50:318–330.

Wennerberg, L. 2001. Breeding origin and migration pattern of Dunlin (*Calidris alpina*) revealed by mitochondrial DNA analysis. Molecular Ecology 10:1111–1120.

Wennerberg, L., N. M. A. Holmgren, P. E. Jonsson, and T. von Schantz. 1999. Genetic and morphological variation in Dunlin (*Calidris alpina*) breeding in the Palearctic tundra. Ibis 141:391–398.

Wennerberg, L., M. Klaassen, and A. Lindstrom. 2002. Geographical variation and population structure in the White-rumped Sandpiper (*Calidris fuscicollis*) as shown by morphology, mitochondrial DNA and carbon isotope ratios. Oecologia 131:380–390.

Wright, S. 1951. The genetical structure of populations. Annals of Eugenics 15:323–354.

Zink, R. M. 2002. Towards a framework for understanding the evolution of migration. Journal of Avian Biology 33:433–436.

Zink, R. M., and D. Dittman. 1993. Population structure and gene flow in the chipping sparrow and a hypothesis for evolution in the genus *Spizella*. Wilson Bulletin 105:399–413.

Zink, R. M., S. V. Drovetski, S. Questiau, I. V. Fadeev, E. V. Nesterov, M. C. Westberg, and S. Rohwer. 2003. Recent evolutionary history of the bluethroat (*Luscinia svecica*) across Eurasia. Molecular Ecology 12:3069–3075.

3

DARREN E. IRWIN
AND JESSICA H. IRWIN

Siberian Migratory Divides

The Role of Seasonal Migration in Speciation

MIGRATORY BEHAVIOR MAY PLAY an important role in the evolution of new species. Differences in migratory behavior could promote reproductive isolation among related groups in several ways. First, they can lead directly to premating isolation, for example, if two groups have different arrival times on the breeding grounds. Second, postmating isolation may occur if hybrids have intermediate but suboptimal migratory behavior, for instance, if two groups migrate by different routes around a geographic barrier and hybrids migrate across the barrier. Third, selection against hybrids can promote premating isolating mechanisms. We did a survey of migratory routes of all passerine species breeding in Siberia, and the results suggest that the Tibetan Plateau is a major barrier to migration. Of 97 long-distance migrants in Siberia, most (85%) use only one route around Tibet (42 through Kazakhstan, 40 through eastern China). Of the 15 species that use both routes, seven of these are known to have migratory divides between western and eastern subspecies. In at least one group, the Greenish Warblers (*Phylloscopus trochiloides*), the western and eastern Siberian forms are reproductively isolated in Siberia, although there is a chain of populations connecting them around the Tibetan Plateau to the south. In four additional cases, migratory divides occur between western and eastern sister species. These patterns suggest that two very different migratory programs can seldom coexist in a single gene pool, and that migration may play a strong role in speciation in Siberia. The need to migrate can also hinder range expansion, thus preventing colonization of new regions and limiting opportunities for speciation.

INTRODUCTION

It is well known that migratory behavior can have strong influences on the evolution of ecological, behavioral, and physiological characteristics of individual species (Alerstam 1990; Dingle 1996). Less studied is the possibility that migratory behavior may influence the process by which one species splits into two or more species. In this chapter we explore how the process of speciation might be promoted or hindered by migratory behavior in different biogeographical situations. By influencing when and how new species originate, migratory behavior might influence the numbers, distributions, and relationships of birds around the world.

To explore the possible role that migration has in speciation in passerine birds, we focus on biogeographic patterns in central Siberia. By considering migratory routes one can gain much insight regarding biogeographic history, distributions of species and subspecies, and the possible role of divergent migratory behaviors in causing reproductive isolation. We conclude that the Tibetan Plateau, a high-altitude desert in central Asia, is a major barrier to migration and has had a major influence in shaping biogeographic patterns in northern Asia. Central Siberia contains many examples of "migratory divides" in which divergent but geographically adjacent groups of a single species use strongly differing routes to migrate to their winter grounds. Migration could be playing a significant role in generating reproductive isolation and hence promoting speciation in these cases.

The arguments made in this chapter rely on five basic assumptions, each of which has empirical and theoretical support. First, the routes used by migratory birds are genetically influenced. Second, birds are under selection to use optimal migration routes (which are influenced by various factors such as distance, opportunities for refueling, predation risk, winds, and elevation changes), and some of these routes include significant detours around regions that are difficult to migrate across. Third, when two groups with different migratory behaviors hybridize, the hybrids might instinctively choose a migratory route that is intermediate to the two parental routes. Fourth, hybrids may be selected against if their intermediate migratory behavior leads them across unsuitable regions. Fifth, selection against hybrids (i.e., post-zygotic isolation) can cause selection for behaviors that prevent the two parental groups from interbreeding (i.e., premating isolation). We review the evidence for these assumptions below.

Genetics of Migration

Successful migration is usually the result of a complex and finely tuned set of morphological, physiological, cognitive, and behavioral traits (Dingle 1991; Berthold 1999b). Although learning has a role, many of these migratory traits have been shown to have a genetic basis, and it appears that most are influenced by multiple genes (Berthold 1999b). Hybrids between two groups that differ in migratory traits often have intermediate characteristics. For example, instinctive migratory direction has been studied extensively in Blackcaps (*Sylvia atricapilla* [Helbig 1991, 1996; Berthold et al. 1992]). In central Europe during the autumn, Blackcaps from western populations orient toward the southwest, whereas those from eastern populations orient toward the southeast. Hybrids that were raised in the lab tended to orient directly south, in an intermediate direction with respect to the parental forms (Helbig 1991). Such intermediacy of hybrids has been shown with respect to the time of the onset of migration, the length of migration, rate of fat deposition, and wing shape (Berthold and Querner 1981; Berthold 1999a). The many traits involved in migration appear to be fine-tuned for each species, the result of many genes that have evolved together.

Ecological Barriers and Migratory Divides

One of the difficulties that migratory birds often face is that breeding and wintering ranges are usually separated by regions of unsuitable habitat (Alerstam 1990). The Gulf of Mexico and the deserts of northern Mexico and the southwestern United States separate many breeding ranges in North America from wintering ranges in Central and South America. European breeding ranges are separated from winter ranges in central Africa by the Mediterranean Sea and the Sahara Desert. And north Asian breeding ranges are separated from wintering areas in southern Asia by a variety of deserts and mountains in central Asia. These ecological barriers offer little opportunity for feeding or resting, presenting a major challenge because migrating birds need stopover sites to refuel (Moore et al., Chap. 20, this volume). In fact, the great majority of time during migration is spent refueling rather than flying (Hedenström and Alerstam 1997). In response to the challenge presented by ecologically poor areas between breeding and wintering areas, migratory birds have developed two basic strategies: they either can build up huge fat reserves for nonstop flights directly across the barrier or they can make a series of shorter flights around the barrier using the best available detour. Of course, birds may also use a combination of these approaches.

Some species that fly around the barriers have more than one optimal route, and this can contribute to the formation of migratory divides, in which two adjacent breeding populations take different routes to their wintering grounds. For example, several European species that migrate to sub-Saharan Africa, such as Blackcaps and Willow Warblers (*Phylloscopus trochilus*), have a migratory divide between two forms, one of which migrates across the western side of the Mediterranean and Sahara, the other of which migrates across the eastern side of these areas (fig. 3.1A) (Hedenström and Pettersson 1987; Helbig 1991; Bensch et al. 1999). By using these routes, the birds avoid the central Mediterranean and Sahara, where those areas are widest. In northwestern North America, Swainson's Thrushes (*Catharus ustulatus*) have a migratory divide between a form that migrates down the west coast to Central America and another form that migrates to eastern North America before flying

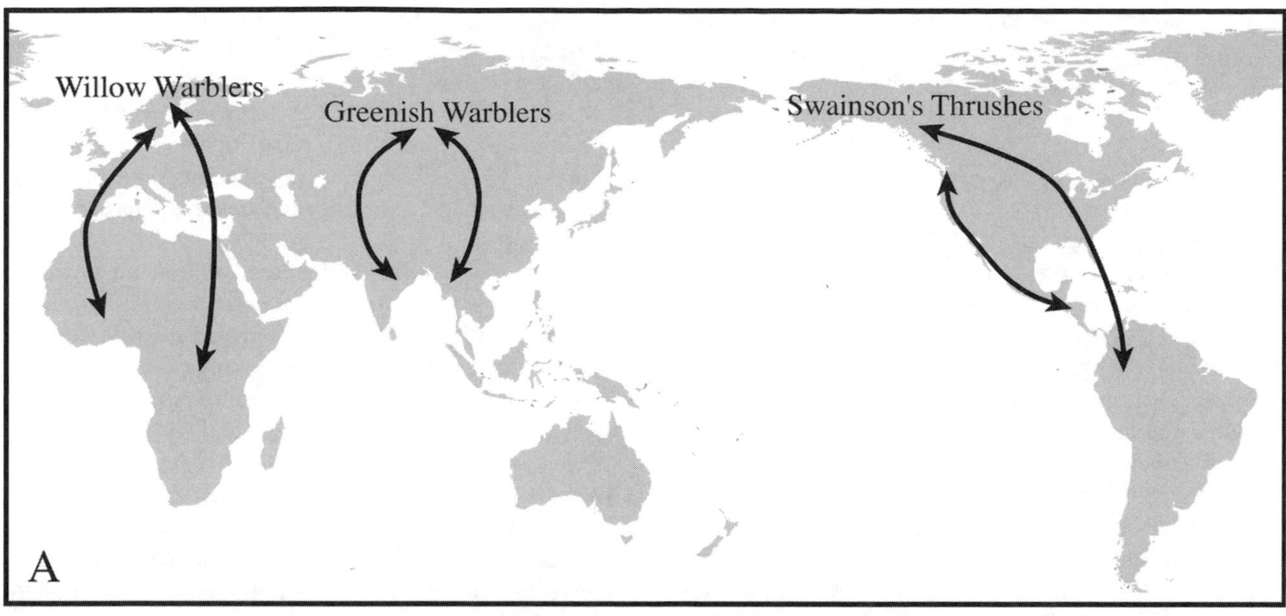

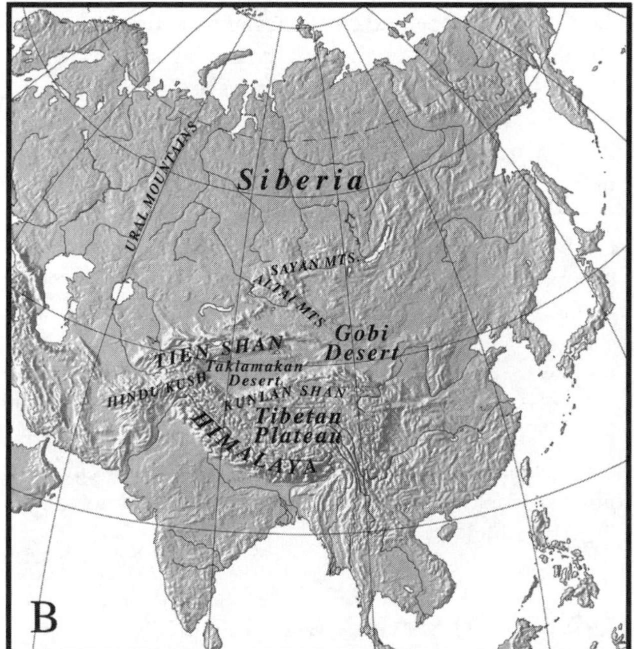

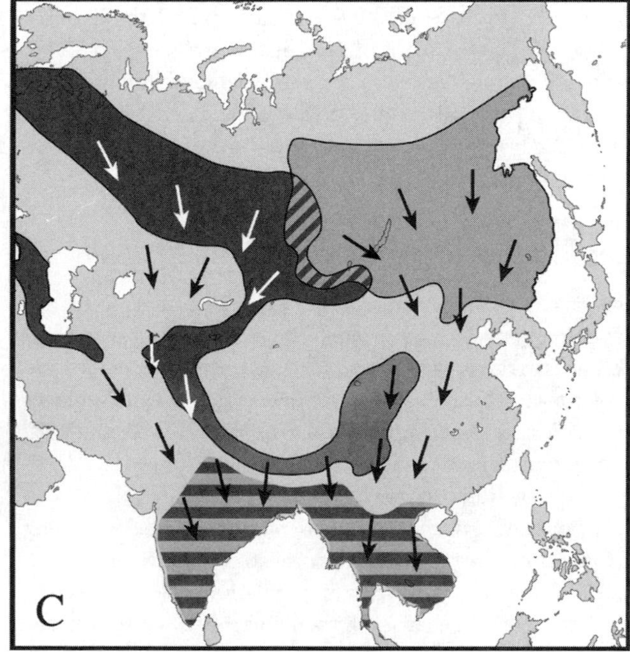

Fig. 3.1. Maps of the world and Asia, showing (A) migratory routes of three species with migratory divides, (B) geographical features of Asia, and (C) breeding range, migratory routes, and wintering range (horizontal gray stripes) of Greenish Warblers (*Phylloscopus trochiloides*). Greenish Warblers breed in forests that occur across middle and southern Siberia as well as in a ring of mountains encircling a region of deserts (the Tibetan Plateau and the Taklamakan and Gobi deserts). Two distinct forms of Greenish Warbler meet in central Siberia (*P. t. viridanus* in the west, *P. t. plumbeitarsus* in the east) but migrate along different western and eastern routes to their wintering ranges.

south across the Caribbean into South America (fig. 3.1A) (Ruegg and Smith 2002). It is possible that birds use these routes in part to avoid deserts and mountains in the southwestern United States and northern Mexico.

Another migratory divide is located in Asia, where two forms of Greenish Warblers (*Phylloscopus trochiloides*) coexist in the forests of central Siberia but migrate by different routes to their wintering areas (fig. 3.1A). The west Siberian form, *P. t. viridanus,* migrates through west-central Asia to winter in forests within India, whereas the east Siberian form, *P. t. plumbeitarsus,* migrates through eastern China to forests in Southeast Asia (fig. 3.1B,C) (Ticehurst 1938). By taking these routes, the birds travel either to the west or the east of a large region of deserts that includes the Tibetan Plateau as well as the low-altitude deserts of the Gobi and the Taklamakan. This region also contains some of the tallest mountain ranges in the world, including the Himalayas, the Tian Shan, and the Kunlun Shan (fig. 3.1B). It is likely that forest-dependent insectivorous birds such as warblers would have an extremely difficult time finding food in this area, especially during early spring and late fall. Populations of Greenish Warblers also occur to the west, south, and east of the Tibetan plateau, and all of these populations migrate roughly southward to their wintering areas (fig. 3.1C).

To fully understand patterns of migration, we must consider historical factors as well as current ecological and geographical ones. The migratory divides mentioned above appear to be, in part, the result of secondary contact of two expanding range fronts. The migratory divide in the Swainson's Thrushes coincides with sudden changes in genetics and morphology, suggesting secondary contact between forms that were isolated in western and eastern refugia during the last glacial maximum (Ruegg and Smith 2002). Populations of Swainson's Thrushes in western Canada and Alaska, on the northeastern side of the migratory divide, are thought to have originated from range expansion from a glacial refugium in southeastern North America. Their migration route seems to retrace this route of expansion, with birds migrating to eastern North America before turning south toward South America (Ruegg and Smith 2002). Likewise, the two forms of Greenish Warblers in Siberia resulted from two divergent groups expanding into the same area from populations to the west and east of Tibet (Irwin et al. 2001b). The different migratory routes reflect these ancestral origins.

Speciation and Migration

Speciation can be defined as the evolution of reproductive isolation between two groups (Mayr 1942; Price 1998). Differences in migratory behavior could promote reproductive isolation in three basic ways. First, migratory differences could directly lead to premating isolation, preventing mating between individuals from the two groups. Such premating isolation could result from different arrival times on the breeding grounds or from pairing on the wintering grounds or during migration. Second, migratory differences might cause postmating isolation due to selection against hybrids because of their inferior migratory traits. Third, selection against hybrids may lead to selection for premating isolation, in a process known as reinforcement (Dobzhansky 1940; Howard 1993; Liou and Price 1994). In this way differences in migratory behavior might have caused the evolution of greater differences in traits involved in mate choice, such as song.

Migratory divides could be maintained by each of these processes or a combination of them. Helbig (1991) argued that the migratory divide in Blackcaps is maintained by assortative mating (premating isolation), which could result from different arrival times on the breeding grounds, or by selection against hybrids (postmating isolation), which could result from inferior migratory behavior in hybrids, as suggested by orientation experiments in which hybrids oriented in an intermediate direction toward the south. Helbig (1991) suggested that hybrids in the wild would also orient southward, taking them directly across the Alps, the Mediterranean Sea, and the Sahara Desert where they are widest. In their analysis of variation in Willow Warblers, Bensch et al. (1999) concluded that the migratory divide in central Sweden must be maintained by assortative mating and/or selection against hybrids, perhaps due to intermediate instinctive migratory direction; the zone of contact between the two groups is too narrow to be explained by secondary contact alone.

The few migratory divides that have been studied in detail, most notably those of the Blackcaps and Willow Warblers, illustrate the possible importance of migratory behavior in maintaining or promoting reproductive isolation and speciation. But how often migration plays a role in speciation is not yet known. Are the Blackcaps, Willow Warblers, Swainson's Thrushes, and Greenish Warblers unusual cases? Or could migration play a major role in speciation in passerines, structuring the biogeographical patterns in an entire region? To address this question in a systematic way we examined migration routes in Asia, a region that has received relatively little attention from migration researchers compared with Europe and North America. Our research on Greenish Warblers (Irwin 2000; Irwin et al. 2001b; Irwin et al. 2001c) led us to postulate that many species that breed in Siberia and spend their winters in southern Asia might avoid flying across the Tibetan Plateau and the deserts to the north. If so, individual migrants must either migrate to the west or the east of Tibet. Because these routes differ so dramatically, and because migratory behavior is probably genetically based, we postulated that the use of both routes would rarely occur in a single gene pool (i.e., in a single species). If so, species that breed in Siberia should migrate only west of Tibet or only east of Tibet. Another possibility is that, as in the Greenish Warblers, a species has two or more distinct Siberian forms (i.e., subspecies) with a migratory divide between them. Such migratory divides indicate possible situations in which migratory behavior is promoting speciation.

MATERIALS AND METHODS

We surveyed the literature to uncover information on migratory routes of all long-distance passerine migrants breeding in north-central Eurasia. We focused on passerines because their juveniles usually migrate separately from adults, suggesting that genetic programs for migratory behavior play a dominant role; many non-passerines migrate in family groups, in which the offspring learn migratory routes from older individuals (Sutherland 1998). Because we were primarily interested in the role that the Tibetan Plateau and nearby deserts may have had in structuring patterns of migration and speciation, we limited our survey to those species that breed somewhere in western and central Siberia, an area that is roughly north of the Tibetan Plateau and Gobi Desert. We defined this area as being north of Russia's southern border (49–56° N latitude), east of the Ural Mountains (60° E longitude), west of 120° E longitude,

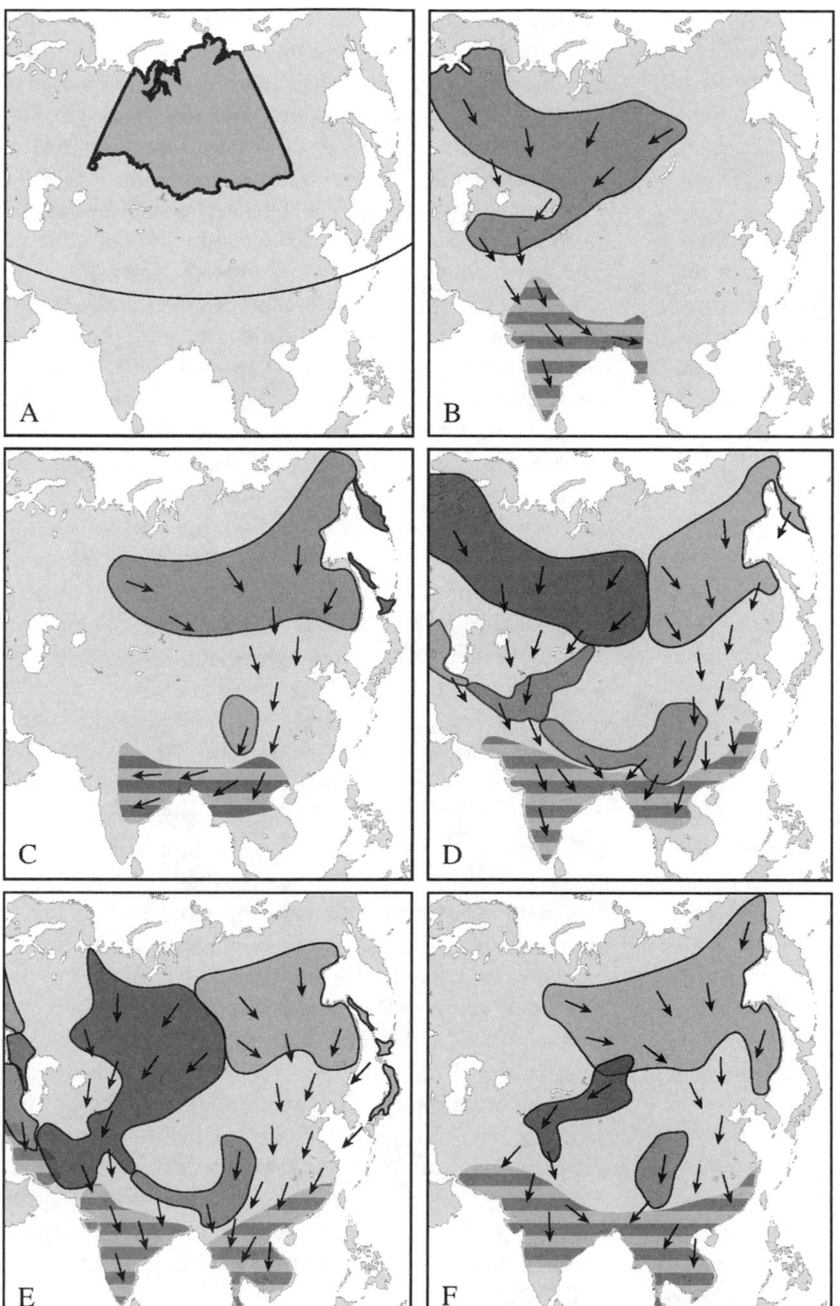

Fig. 3.2. Maps of Asia showing (A) the area in which species must breed to be included in our survey (western and central Siberia) and the 30° N latitude south of which a species must winter to be classified as a long-distance migrant; and (B–F) ranges and migratory routes for five groups of taxa. Breeding ranges are shown by gray regions, and wintering areas are shown by horizontal stripes. Different subspecies or species are indicated by different shades of gray. (B) Blyth's Reed Warbler; (C) Siberian Rubythroat; (D) Common Rosefinch; (E) Siberian Stonechat; (F) Hume's Leaf Warbler and Yellow-browed Warbler.

and south of the Arctic Ocean (fig. 3.2A). We chose the eastern boundary to deliberately exclude many species that breed in the Russian Far East but not farther west. We also excluded several central Asian species with ranges that extend only slightly into steppe or mountain habitats of extreme southern Russia (e.g., Red-headed Bunting, *Emberiza bruniceps*).

To determine which species of passerines breed in this area, we consulted Dementev and Gladkov (1954), Flint et al. (1984), Sibley and Monroe (1990), and Rogacheva (1992). We then used those references and the following to compile information regarding wintering areas and migration routes: McClure (1974) provided broad information on migration in Asia; MacKinnon and Phillips (2000) described ranges in China, with some information on migration; Ali and Ripley (1987–1999), Grimmett et al. (1999), and Kazmierczak and van Perlo (2000) described ranges and migration routes in India and Pakistan; Robson (2000) detailed birds wintering in Southeast Asia; Keith et al. (1992) provided information on birds that winter in Africa; and Cramp (1988–1994) provided useful migration data on species that extend into the western Palearctic. We also consulted tables in two books (Dolnick 1985, 1987) that summarize the species identity of 12,994 passerine birds that were caught during spring and fall migration at a wide variety of research sites in central Asia (i.e., Kazakhstan, Kyrgyzstan,

Uzbekistan, Tajikistan, and Turkmenistan). For more information on specific groups of passerines, we consulted specialty books on swallows and martins (Turner and Rose 1989), pipits and wagtails (Alström and Mild 2003), stonechats (Urquhart 2002), warblers (Baker 1997; Shirihai et al. 2001), thrushes (Clement and Hathway 2000), finches and sparrows (Clement et al. 1993), buntings (Byers et al. 1995), and crows and jays (Madge and Burn 1994). Using these sources, we were able to gain a good understanding of migration routes and wintering ranges of the large majority of species breeding in western and central Siberia.

The Tibetan Plateau would be expected to play a role in shaping migration routes primarily in those species that actually migrate far enough south to potentially encounter it during migration. We therefore focused our survey on long-distance migrants, which we defined as those with a winter range that is primarily south of 30° N latitude (fig. 3.2A). This is approximately the latitude of Lhasa, in south-central Tibet. Thus our list of long-distance migrants includes those wintering in India, Southeast Asia, and Africa, but not those that winter primarily in central Asia or northern China. We examined all available range maps and used the consensus to determine whether the majority of the wintering range of each species was south of 30° N latitude. We then recorded whether each species migrated to the west of Tibet, to the east of Tibet, or over Tibet. Whenever a species consisted of multiple subspecies, we recorded information specific to each subspecies, when available. Our goal was to determine whether different subspecies had different migratory behaviors. Some of the species or subspecies are given different names and taxonomic treatments by different authors; we generally followed Beaman's (1994) list of Palearctic birds or a more recent authority.

RESULTS

We counted 171 species of passerine birds breeding in western and central Siberia. Of these, 97 are long-distance migrants, with their wintering range primarily south of 30° N latitude (fig. 3.2A). Roughly one-third of these winter in Africa, one-third in India, and one-third in Southeast Asia, although some species winter in more than one of those regions.

Most of the long-distance migrants appear to avoid migrating across the Tibetan Plateau and the deserts of northwest China. Rather, all of the species have been found in large numbers during migration in at least one of two major flyways: on the west through the central Asian countries of Kazakhstan, Uzbekistan, Kyrgyzstan, Tajikistan, and Turkmenistan (Dolnik 1990) and on the east through eastern China (McClure 1974; MacKinnon and Phillips 2000).

Most species use only one of these migration routes. Of the 97 species of long-distance migrants (see the Appendix to this chapter), 42 apparently migrate only to the west of Tibet and 40 migrate only to the east of Tibet. Of the western migrants, about half spend the winter in India. An example is the Blyth's Reed Warbler (*Acrocephalus dumetorum*) (fig. 3.2B), which apparently migrates only west of Tibet even though its breeding range extends to eastern Siberia and its wintering range extends east to Bangladesh and Myanmar. Many of the western migrants winter in Africa, for example, the Willow Warbler (*Phylloscopus trochilus acredula* and *P. t. yakutensis*), which has a breeding range extending to far-eastern Siberia. The eastern migrants, on the other hand, winter primarily in southern China and Southeast Asia, although some have wintering ranges that extend west into India. For example, the Siberian Rubythroat (*Luscinia calliope*) (fig. 3.2C) breeds as far west as the Ural Mountains and winters as far west as central India, but does not migrate through central Asia in large numbers, instead flying during autumn migration to east Siberia, turning south across eastern China, and finally turning west.

This leaves 15 species that use both the western and the eastern migratory routes (table 3.1). However, seven of these species have a migratory divide between recognized subspecies. In most of these cases, two subspecies breed in adjacent western and eastern ranges in Siberia, with the western form migrating along the western route to India and the eastern form migrating along the eastern route to Southeast Asia. Examples include the Greenish Warbler (fig. 3.1C), the Common Rosefinch (*Carpodacus erythrinus*) (fig. 3.2D), and the Siberian Stonechat (*Saxicola maura*) (fig. 3.2E). Such migratory divides between subspecies are found in a diverse group of passerine species (table 3.1). Migratory divides also sometimes occur between taxa that are classified as separate species; there are at least four cases in which two closely related species migrate along different sides of Tibet and have breeding ranges that meet in a narrow region of Siberia (table 3.1). One example consists of the Hume's Leaf Warbler (*Phylloscopus humei humei*), which migrates through central Asia and winters in India, and the Yellow-browed Warbler (*Phylloscopus inornatus*), which migrates through eastern China and winters in Southeast Asia (fig. 3.2F). Because of their morphological similarity, only recently have these two taxa been recognized as separate species (Svensson 1987; Formozov and Marova 1991; Irwin et al. 2001a).

Out of the 97 long-distance migrants breeding in western and central Siberia, eight have at least one subspecies that apparently uses both the western and eastern migratory routes (table 3.1). Six of these taxa are pipits and wagtails, which are known to migrate in mixed subspecies flocks, suggesting an important role for cultural, rather than genetic, transmission of migratory behavior (Rogacheva 1992). Wagtails also commonly inhabit open areas, suggesting that Tibet and nearby areas may not serve as a strong migratory barrier in these taxa. Two of the taxa, the Red-throated Pipit (*Anthus cervinus*) and the Citrine Wagtail (*Motacilla citreola*), have been described by some authors as having a migratory divide between two subspecies, but later treatments have classified them as monotypic across Siberia, where there is clinal west-to-east variation (see footnotes in table 3.1). One of the remaining species, the Dark-

Table 3.1 Long-distance migrating passerine taxa (species or pairs of sister species) that breed in central Siberia and migrate along both sides of Tibet to their wintering areas

Taxa	West of Tibet	East of Tibet	Taxa	West of Tibet	East of Tibet
Sand Martin			Bluethroat		
Riparia riparia riparia	X		*Luscinia svecica svecica*	X	X
R. r. ijimae		X	*L. s. pallidogularis*	X	
Barn Swallow			Siberian Stonechat		
Hirundo rustica rustica	X		*Saxicola maura maura*	X	
H. r. tytleri		X	*S. m. stejnegeri*		X
Common House Martin			Dark-throated Thrush		
Delichon urbica urbica	X		*Turdus ruficollis ruficollis*	X	X
D. u. lagopoda		X	*T. r. atrogularis*	X	X
Red-throated Pipit[1]			Great Reed Warbler / Oriental Great		
Anthus cervinus	X	X	Reed Warbler		
Tree Pipit / Olive-backed Pipit[2]			*Acrocephalus arundinaceus zarudnyi*	X	
Anthus trivialis trivialis	X		*A. orientalis*		X
A. hodgsoni yunnanensis		X	Hume's Leaf Warbler / Yellow-browed		
Water Pipit			Warbler		
Anthus spinoletta blakistoni	X	X	*Phylloscopus humei*	X	
Yellow Wagtail			*P. inornatus*		X
Motacilla flava thunbergi	X	X	Greenish Warbler		
M. f. beema	X		*Phylloscopus trochiloides viridanus*	X	
M. f. tschutschensis		X	*P. t. plumbeitarsus*		X
Citrine Wagtail[3]			Red-breasted Flycatcher		
Motacilla citreola citreola	X	X	*Ficedula parva parva*	X	
Grey Wagtail			*F. p. albicilla*	Few	X
Motacilla cinerea cinerea	X	X	Red-backed Shrike / Brown Shrike		
White Wagtail			*Lanius collario*	X	
Motacilla alba alba	X		*L. cristatus*		X
M. a. personata	X		Common Rosefinch		
M. a. baicalensis	X	X	*Carpodacus erythrinus erythrinus*	X	
M. a. ocularis		X	*C. e. grebnitzkii*		X

Note: In each case, we list subspecies or species that breed in central Siberia and indicate whether they migrate on the western or eastern side of Tibet. Migratory divides between subspecies or sister species occur in 11 cases.

[1]Although recent authors have treated the Red-throated Pipit as monotypic (Cramp 1988-1994; Alström and Mild 2003), Dementev and Gladkov (1954) described two subspecies, *A. c. rufogularis* in the west and *A. c. cervinus* in the east, with a migratory divide between them.

[2]These species overlap substantially over a broad region of Siberia, yet they take different routes to their wintering areas.

[3]Most authors (e.g., Dementev and Gladkov 1954; Cramp 1988–1994) have treated Citrine Wagtails in western Siberia as two subspecies, *M. c. werae* in the west and *A. c. citreola* in the east, with a migratory divide between them. Here we follow the latest authority, Alström and Mild (2003), in treating the Siberian populations as a single subspecies.

throated Thrush (*Turdus ruficollis*), is apparently not an obligate migrant; in most years birds winter in southern Asia, but in years with a good berry crop many birds spend the winter in Siberia (Cramp 1988:vol. 5). The last remaining case, the Bluethroat (*Luscinia svecica*), is an obligate migrant, but little is known about its migratory behaviors in Siberia. It is possible in all of these cases that there are narrow migratory divides and that the morphology across these divides is so similar that different subspecies have not been recognized.

DISCUSSION

One caveat regarding these results is that relatively little ornithological work has been done in Tibet and northwest China, so we cannot presently rule out the possibility that some species do migrate across those regions. In fact, a few passerine species breed on the Tibetan Plateau and necessarily migrate across it, but most of these are alpine, desert, or steppe specialists that do not have ranges that extend far into Siberia. For example, Black Redstarts (*Phoenicurus ochruros*), which breed in Tibet and the Altai Mountains of southern Siberia, migrate both across Tibet and around it on the west side (MacKinnon and Phillips 2000). However, several lines of evidence suggest that the great majority of passerine species breeding in Siberia avoid Tibet. First, studies on the west side (Dolnik 1990) have documented that many birds take a circuitous route through that region; during spring migration, birds tend to fly toward the northwest first and then turn to the northeast, staying in the lowland deserts rather than crossing the snow-covered mountain ranges. Second, most taxa have been observed migrating in large numbers to either the west or east of Tibet, suggesting that crossing Tibet is not the predominant strategy for any of them. Third, Dolnick (1990), who summarized decades of research on birds migrating across central Asia (i.e., the western route around Tibet), concluded that most species stop to refuel often while crossing that region; birds usually leave stopover sites with only a small amount of fat, enough to fly for less than 4.5 h. Such a strategy would probably be deadly if used during migration across Tibet or the Gobi Desert, which offer little opportunity for feeding compared with the milder deserts along the western fly route.

Migratory behavior is closely associated with subspecific variation in a variety of passerine species. This pattern is consistent with a role for different migratory behaviors in preventing gene flow between subspecies and promoting speciation. However, differences between western and eastern subspecies in Siberia, both in migration routes and in other traits, are also the result of biogeographic history. During Pleistocene glaciations, most of Siberia was unsuitable for many of the species that are there now, especially for the species that inhabit forests (Grichuk 1984; Ukraintseva 1993; Adams and Faure 1997; Price et al. 1997). According to Adams and Faure (1997), who reconstructed vegetation cover from many types of evidence, just after the last glacial maximum at about 17,000 to 15,000 years ago (^{14}C years) "conditions all across northern Eurasia appear to have been dry and treeless, dominated by polar desert or semi-desertic steppe-tundra." Glaciers were covering northwest Siberia, and all of central and eastern Siberia consisted of polar desert, steppe-tundra, or extreme desert. Forests were confined to central China, Japan, the Himalayas, southern India, and Southeast Asia. It is also likely that some of the mountains of central Asia, such as the Tian Shan range, harbored forests (Grichuk 1984). Following a warming and moistening of the climate at about 13,000 ^{14}C years ago, open woodlands of birch and boreal conifers gradually expanded into Siberia, but strong cooling between 11,200 and 10,200 ^{14}C years ago then caused the trees to recede. Tree cover again expanded into Siberia following a rapid warming of the climate about 10,000 ^{14}C years ago, and by about 8,000 ^{14}C years ago most areas of northern Eurasia were even more forested than they would be today under natural conditions (Ukraintseva 1993; Adams and Faure 1997). Although the amount of forest in Siberia has fluctuated some since then, it has never fully receded (Khotinsky 1984). Given this inferred history, contact zones between divergent forest-dependent taxa in central Siberia must have formed within the last 10,000 years or so.

During the last glaciation, extreme desert conditions prevailed over the Tibetan, Taklamakan, and Gobi desert region (Grichuk 1984; Ukraintseva 1993; Adams and Faure 1997), forming a large ecological barrier in the center of Asia. When forest moved into Siberia, many species may have expanded into Siberia along two pathways on either side of this ecological barrier. For example, in the case of the Greenish Warblers, molecular evidence has supported the hypothesis (Ticehurst 1938) that an ancestral species in southern Asia expanded northward along two pathways on either side of Tibet, resulting in distinct forms with differing migration routes when the two expanding fronts met in central Siberia (Irwin et al. 2001b). The many subspecies and species boundaries in central Siberia suggest that many taxa have similar histories of expansion into Siberia along two pathways. Ancestral migration routes appear to have been conserved during these expansions, such that species or subspecies that expanded from central Asia into Siberia still migrate through central Asia to India, whereas those that expanded from eastern China into Siberia still migrate through eastern China to Southeast Asia.

Although the different migratory behaviors are partly a result of history, the future of a species may be influenced by migratory behavior. When two divergent forms meet, at least two outcomes are possible. First, they may interbreed freely, exchanging genes and blending the differences between the forms. Second, they may have some amount of reproductive isolation, perhaps allowing or causing them to diverge further into separate species. In central Siberia, migratory behavior may make the second scenario much more likely to occur. As discussed in the Introduction, migratory differences can cause premating or postmating reproductive isolation, and postmating isolation due to migratory behavior might lead to reinforcement (i.e., selection for premating isolation).

Some of the pairs of subspecies that have a migratory divide between them are sometimes treated as separate species. Four other migratory divides (see Results) consist of taxa that are generally considered separate species. In all of these cases, speciation has been completed or has proceeded almost to completion. These may be cases in which divergent forms met in Siberia, and different migratory behavior, along with differences in other traits, promoted the development of stronger reproductive isolation. For example, the two Siberian forms of the Greenish Warbler appear to be separate species, differing distinctly in molecular markers, songs, song recognition, plumage patterns, and migration routes. These traits change gradually through the chain of populations that encircle the Tibetan Plateau to the south (Ticehurst 1938; Irwin 2000; Irwin et al. 2001b), making the Greenish Warblers an example of a "ring species," in which a ring of populations has only a single species boundary (Mayr 1942; Irwin et al. 2001c). Geographical variation within a ring species can illustrate the evolutionary changes that can occur during divergence of two species from their common ancestor. Irwin (2000) and Irwin et al. (2001b) emphasized the likely role of divergence in male song in causing premating reproductive isolation between the Siberian forms. But speciation is most likely to occur when both premating and postmating reproductive isolation are present, and migratory behavior could have caused both. If *viridanus* and *plumbeitarsus* did ever hybridize, the hybrids probably had inferior migratory behavior. By causing selection against hybrids in this way, migratory behavior may have reduced gene flow between *viridanus* and *plumbeitarsus,* promoting speciation (Irwin et al. 2001c). The overall patterns of reproductive isolation in the Greenish Warblers are consistent with the hypothesized role of migratory divides in speciation: in central Siberia, on the northern side of the migratory barrier (where there is a migratory divide), there is a species boundary, whereas in the Himalayas, along the southern side of the migratory barrier (i.e., presumably where migratory divides are not an important factor), there are no known species boundaries (Irwin et al. 2001b, 2001c).

Two other species in Asia have distributions and migratory patterns that are strikingly similar to those of the Greenish Warblers. These are the Common Rosefinch (fig. 3.2D) and Siberian Stonechat (fig. 3.2E). Both consist of a ring of populations encircling the Tibetan Plateau, and both have a migratory divide between distinct subspecies in Siberia. Perhaps the biogeographic histories of these species are similar to that of the Greenish Warblers, and perhaps they are ring species as well.

Although we have emphasized how migratory behavior might directly cause reproductive isolation, migration might also promote speciation indirectly by promoting the divergence of sexually selected traits. Researchers have proposed that migrants experience more intersexual selection on song than nonmigrants (Catchpole 1980, 1982; Morton 1996) because females must choose mates quickly after arriving on the breeding grounds, and song is a signal that can be assessed quickly. In resident species, songs may be more important in male-male territorial interactions. This hypothesis was supported by Read and Weary (1992), who conducted a broad survey of passerine species and found that migrants on average had larger song repertoires, something that females usually prefer (Catchpole and Slater 1995; Searcy and Yasukawa 1996). Researchers have also proposed that intersexual selection can lead to rapid divergence of mating signals (West-Eberhard 1983; Iwasa and Pomiankowski 1995; Price 1998). By combining these ideas, we postulate that increased migratory behavior may lead to increased divergence in songs and song preferences between geographically distant populations, thereby promoting the evolution of reproductive isolation and speciation. The patterns of variation in the Greenish Warblers are consistent with this hypothesis: songs in the north, where birds migrate farther and spend less time on the breeding grounds, are much longer and complex than those in the south (Irwin 2000). Apparently, complex songs have evolved from simple ones during both northward expansions, but the form of complexity differs between the two Siberian forms, probably contributing to premating reproductive isolation between them (Irwin et al. 2001b, 2001c).

In addition to revealing a number of migratory divides in Siberia, our survey revealed that migratory routes in Siberia appear to be highly conserved. Few species migrate along both western and eastern sides of Tibet without having a migratory divide between subspecies, and some populations of many species seem to have suboptimal migratory routes. Clearly, both the western route through central Asia and the eastern route through eastern China are suitable routes for a wide variety of species. Why, then, do some species, such as the Blyth's Reed Warbler (fig. 3.2B), only use the western route, although the eastern one would be shorter for birds breeding in the eastern side of the range? Likewise, why do some species, such as the Siberian Rubythroat (fig. 3.2C), use only the eastern route, although the western one would be shorter for some birds? We conclude from these patterns that large and sudden changes in migration routes seldom occur during evolution and that two strongly divergent migratory routes are seldom used by a single subspecies.

The concept of fitness landscapes (Fear and Price 1998) is a useful aid in understanding both the formation of migratory divides and the conservation of ancestral migratory routes. A fitness landscape illustrates graphically how fitness depends on the traits of an individual (note that we are referring to individual fitness landscapes, not "adaptive surfaces," which show the mean fitness of a population). In fig. 3.3, we show how fitness landscapes might change during northward expansion of a species along two pathways around a barrier such as Tibet. Each graph shows how the fitness of a bird breeding in a certain location depends on two traits, in this case two instinctive migratory directions. In this hypothetical example, the birds migrate according to a simple rule; they have an initial instinctive migratory direction, and after some time change to a second instinctive direction. Orientation experiments have demonstrated that Garden Warblers (*Sylvia borin*) migrate in this way, enabling them to avoid crossing the widest parts of the Mediterranean and Sahara (Gwinner and Wiltschko 1978). In our example (fig. 3.3), birds in the south have optimal behavior if they migrate directly south to their wintering area. As the species expands north, the optimal behavior gradually changes, such that birds on the northwest side of Tibet should first migrate southwest and then turn southeast, and birds on the northeast side of Tibet should first migrate southeast and then turn southwest. When the two populations meet directly north of Tibet, there are two optimal migratory behaviors (i.e., two peaks on the fitness landscape), but hybrids with intermediate migratory behaviors are at a fitness disadvantage (i.e., a valley on the fitness landscape). Speciation can be thought of as the evolution of a species from a single fitness peak onto two fitness peaks separated by a valley. Gradual change in migratory behavior during expansion around a barrier provides a way for a species to evolve onto two peaks gradually without ever crossing a fitness valley. Although we have illustrated this concept by using traits for migratory direction, many other traits (e.g., migratory distance, refueling frequency, physiology, molt) might evolve along similar fitness landscapes and thus promote speciation.

The fitness landscapes in fig. 3.3 also illustrate how suboptimal migratory routes may be conserved. A species that expands into Siberia from only a single refugium, on either the west or east side of Tibet, might be "trapped" on a single adaptive peak in Siberia. Another, higher fitness peak may exist somewhere in genotypic and phenotypic space, but if the two peaks are far enough apart with a deep enough valley between them, the necessary mutations may not occur in a single individual. Furthermore, even if one individ-

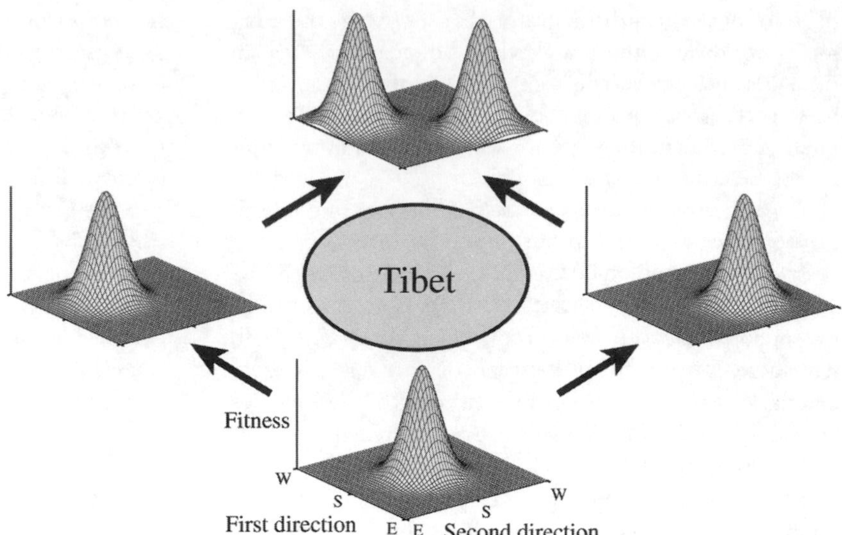

Fig. 3.3. A depiction of hypothetical fitness landscapes at four locations around a barrier to migration such as Tibet. Each graph shows how fitness depends on two traits, in this case a first and second instinctive migratory direction. In the south, there is a single optimal behavior (i.e., a single fitness peak), but these diverge to the north, on the west and east sides of the barrier. On the north side of the barrier, there are two optimal migratory behaviors (i.e., two fitness peaks), but intermediate behaviors are selected against (i.e., a valley on the fitness surface). Starting on the southern side of the barrier, a species could expand northward along two pathways, diverging and eventually ending up on different fitness peaks (i.e., speciating), without ever crossing a fitness valley.

ual did somehow "jump" to the other peak (i.e., the other migratory behavior), it might then breed with an individual on the first peak, perhaps leaving their offspring in the fitness valley. This may explain why populations of some species, such as the Blyth's Reed Warbler (fig. 3.2B) and Siberian Rubythroat (fig. 3.2C) have not evolved more direct migratory routes to their wintering ranges. Furthermore, the conserved migratory routes of these species may be constraining their ability to expand their breeding ranges farther west or east. As a species spreads farther and farther from its wintering range, the height of the fitness peak that it is on may decrease because of the challenge of migrating farther.

Although we have emphasized constraints on evolving new migratory programs, there is abundant evidence for rapid evolutionary change in migratory behavior in many species. Phylogenetic reconstructions of both Old World Warblers (Helbig 2003) and New World Warblers (I. J. Lovette and T. B. Smith, unpubl. data) show that migratory species often evolve from resident species and vice versa. Sutherland (1998) listed 43 cases in which birds have changed migration routes in historical times. A particularly well-studied example is the Blackcap: before 1960, Blackcaps breeding in Germany generally wintered in Spain or southwest France, but now many use a migration route that brings them to Britain for the winter (Berthold et al. 1992). However, this west-northwest migratory direction probably falls within the normal range of genetic variation (Berthold et al. 1992), and Blackcaps often winter along the west coast of France. Sutherland (1998) observed that changes in migration routes tend to be rather small, and thus concluded that migratory behaviors can change rapidly in some situations but are highly constrained in others, leading to many cases of suboptimal migratory routes. Interestingly, Sutherland (1998) noted that all species with seemingly suboptimal migratory routes (14 cases) have short periods of parental care, suggesting that genetically determined migratory behaviors are more constrained than culturally determined migratory behaviors. A change between a migratory route along the west of Tibet to one along the east of Tibet, or vice versa, would be a much larger change than those that have been observed to date. Furthermore, many of the observed changes can occur by a series of small steps; intermediate forms might not suffer a loss in fitness. Both flexibility and constraints in the evolution of migratory behavior (Sutherland 1998) can be understood by thinking in terms of fitness surfaces (e.g., fig. 3.3). If there is no fitness valley between the presently used route and the optimal route, gradual evolution can lead to the population's using the new route. But if there is a deep valley between the present route and the optimal route, the population may not be able to evolve to use the optimal route.

In this chapter we have emphasized the role that migratory behavior may have in promoting speciation. However, the possibility that migratory behavior constrains the ability of a species to expand into uninhabited regions brings up the possibility that migratory behavior can also hinder speciation. If a species is prevented from colonizing new regions, it may be prevented from differentiating in response to novel selection pressures or genetic drift in combination with geographic isolation. It has often been observed that migratory species usually have smaller range sizes than nonmigratory species (e.g., Edward Blyth quoted by Darwin 1838, Notebook C [in Barrett et al. 1987:p. 36]; Mayr 1976; Böhning-Gaese et al. 1998; Bensch 1999). For example, Böhning-Gaese et al. (1998) showed that migratory species are less likely to occur in both North America and Europe, and Bensch (1999) showed that migratory species are less likely to have a range that includes both Scandinavia and eastern Siberia. These results are notable given the intuitive prediction that migratory species should have more potential than resident species to disperse and colonize new regions. Both Böhning-Gaese et al. (1998) and Bensch (1999) argued that migrants are more constrained than residents by the need to migrate to distant wintering grounds and to evolve a new migratory program as they expand into new regions. Many of the remarkable differences in the avian fauna of Asia and North America (e.g., the Old World warblers [Silviidae] vs. the New World warblers [Parulidae], the Old World flycatchers [Muscicapidae] vs. the New World flycatchers [Tyrannidae]) may be a result of the inability of these highly migratory groups to expand from one continent to the other because of their inability to evolve new migratory programs. In contrast, families dominated by resident species (e.g., Corvidae, Paridae, Emberizidae) are usually widespread on both continents.

The few migratory species that have successfully colonized North America from Asia or vice versa illustrate the constraints imposed by migration. The Bluethroat (*Luscinia svecica svecica*), the Arctic Warbler (*Phylloscopus borealis kennicotti*), and the Yellow Wagtail (*Motacilla flava tschutschensis*) have colonized Alaska from Siberia, but each continues to winter in Southeast Asia. The Grey-cheeked Thrush (*Catharus minimus*) has gone in the opposite direction, expanding from Alaska into eastern Siberia, while still wintering in South America. The most dramatic example is the Northern Wheatear (*Oenanthe oenanthe*), which has colonized far into North America along two fronts, one from eastern Siberia into Alaska and the other from Europe into Greenland and northeast Canada. All of these populations continue to migrate back to sub-Saharan Africa for the winter. An intriguing question is, what would happen if these expanding fronts meet in central Canada? The two populations would have different instinctive migratory directions, the western one orienting westward and the eastern one orienting eastward, creating a dramatic migratory divide. Hybrids would almost certainly have inferior migratory behavior, perhaps leading to a species boundary forming in Canada. If so, the Wheatear would become a circumpolar ring species.

In this chapter we have documented strong associations between migratory routes and subspecific and specific variation in northern Asia. Although these patterns are consistent with a role for migration in causing reproductive isolation and speciation, they are certainly not proof. Generally, migratory divides correspond to places where two groups have come into secondary contact after diverging in allopatry. The two groups may differ in many traits, some of which could play an important role in generating reproductive isolation (e.g., song, plumage, habitat preferences). We have argued that differences in instinctive migratory routes may be particularly powerful in causing selection against hybrids, but it is also possible that other factors play more important roles in speciation of central Siberian migratory birds. In the Greenish Warblers, for instance, song differences between the two expanding fronts may have been sufficient to cause complete reproductive isolation when the two fronts met (Irwin et al. 2001b).

FUTURE DIRECTIONS

We suggest several approaches that researchers could take to more directly examine whether reproductive isolation in migratory divides is in fact caused by divergent migratory behaviors. One approach is to compare contact zones of migratory and resident taxa that have otherwise similar biogeographic histories. If migration has played an important role in generating reproductive isolation, we would expect reproductive isolation to be greater in contact zones between migratory taxa than in contact zones between resident taxa. We are aware of only three contact zones in passerine birds of central Siberia that have been studied in detail. Of these three, the two migratory cases (the Greenish Warblers, and the Hume's Leaf / Yellow-browed Warblers) show complete reproductive isolation (Ticehurst 1938; Irwin et al. 2001a, 2001b), whereas the one resident case (Yellowhammer [*Emberiza citrinella*] / Pine Bunting [*Emberiza leucocephalos*]) shows extensive interbreeding between the taxa (Panov et al. 2003). This sort of comparison could extend beyond birds; it is interesting to note that the Siberian Larch (*Larix sibirica*), in west Siberia, and the Daurian Larch (*Larix dahurica*), in east Siberia, hybridize across a contact zone in central Siberia (Knystautus 1987). A difficulty with this approach is that migratory and resident taxa differ in many ways aside from migratory behavior itself; some other factor that is correlated with migratory behavior might influence reproductive isolation. A more direct approach is to conduct intensive population studies of the divergent taxa that meet at migratory divides. Most informative would be a study in the contact zone between taxa that hybridize occasionally. Orientation experiments might reveal that the two taxa have different instinctive migratory orientation (e.g., to the southwest and to the southeast) and that hybrids have an intermediate orientation (e.g., to the south). By banding nestlings and recording their return to the study area the following year, researchers might observe that hybrid offspring have a lower rate of return than pure offspring. These results would strongly indicate that differences in migratory orientation are causing post-zygotic isolation. By studying patterns of variation in traits used in mate choice along with the fitness consequences of mating decisions (i.e., offspring return rates), researchers might be able to determine the strength of selection for premating isolation due to differences in migratory behavior. Eventually, researchers may even be able to affix passerine birds with satellite-tracking devices, perhaps revealing in detail how the two parental groups avoid a barrier to migration, while the hybrids attempt, unsuccessfully, to fly directly across.

APPENDIX

Long-Distance Migrants That Breed in Western or Central Siberia and Migrate Only West of Tibet

Tree Pipit (*Anthus trivialis*)
Tawny Pipit (*Anthus campestris*)
Black-throated Accentor (*Prunella atrogularis*)
Altai Accentor (*Prunella himalayana*)
European Robin (*Erithacus rubecula*)

Thrush Nightingale (*Luscinia luscinia*)
Eversmann's Redstart (*Phoenicurus erythronota*)
Black Redstart (*Phoenicurus ochruros*) (also goes over Tibet)
Common Redstart (*Phoenicurus phoenicurus*)
Whinchat (*Saxicola rubetra*)
Isabelline Wheatear (*Oenanthe isabellina*)
Northern Wheatear (*Oenanthe oenanthe*)
Pied Wheatear (*Oenanthe pleschanka*)
Rufous-tailed Rock Thrush (*Monticola saxatilis*)
Song Thrush (*Turdus philomelos*)
Redwing (*Turdus iliacus*)
Common Grasshopper Warbler (*Locustella naevia*)
River Warbler (*Locustella fluviatilis*)
Savi's Warbler (*Locustella luscinioides*)
Aquatic Warbler (*Acrocephalus paludicola*)
Sedge Warbler (*Acrocephalus schoenobaenus*)
Paddyfield Warbler (*Acrocephalus agricola*)
Blyth's Reed Warbler (*Acrocephalus dumetorum*)
Great Reed Warbler (*Acrocephalus arundinaceus*)
Marsh Warbler (*Acrocephalus palustris*)
Booted Warbler (*Hippolais caligata*)
Icterine Warbler (*Hippolais icterina*)
Barred Warbler (*Sylvia nisoria*)
Lesser Whitethroat (*Sylvia curruca*)
Common Whitethroat (*Sylvia communis*)
Garden Warbler (*Sylvia borin*)
Blackcap (*Sylvia atricapilla*)
Hume's Leaf Warbler (*Phylloscopus humei*)
Chiffchaff (*Phylloscopus collybita*)
Willow Warbler (*Phylloscopus trochilus*)
Spotted Flycatcher (*Muscicapa striata*)
European Pied Flycatcher (*Ficedula hypoleuca*)
Eurasian Golden Oriole (*Oriolus oriolus*)
Red-backed Shrike (*Lanius collurio*)
Lesser Grey Shrike (*Lanius minor*)
Rose-coloured Starling (*Sturnus roseus*)
Ortolan Bunting (*Emberiza hortulana*)

Long-Distance Migrants That Breed in Western or Central Siberia and Migrate Only East of Tibet

Red-rumped Swallow (*Hirundo daurica*)
Olive-backed Pipit (*Anthus hodgsoni*)
Pechora Pipit (*Anthus gustavi*)
Buff-bellied Pipit (*Anthus rubescens*)
Richard's Pipit (*Anthus richardi*)
Blyth's Pipit (*Anthus godlewskii*)
Rufous-tailed Robin (*Luscinia sibilans*)
Siberian Rubythroat (*Luscinia calliope*)
Siberian Blue Robin (*Luscinia cyane*)
Red-flanked Bluetail (*Tarsiger cyanurus*)
Daurian Redstart (*Phoenicurus auroreus*)
White-throated Rock Thrush (*Monticola gularis*)
Scaly Thrush (*Zoothera dauma*)
Siberian Thrush (*Zoothera sibirica*)
Grey-backed Thrush (*Turdus hortulorum*)
Eyebrowed Thrush (*Turdus obscurus*)
Dusky Thrush (*Turdus naumanni*)
Spotted Bush Warbler (*Bradypterus thoracicus*)
Chinese Bush Warbler (*Bradypterus tacsanowskius*)
Pallas's Grasshopper Warbler (*Locustella certhiola*)
Lanceolated Warbler (*Locustella lanceolata*)
Gray's Grasshopper Warbler (*Locustella fasciolata*)
Oriental Great Reed Warbler (*Acrocephalus orientalis*)
Thick-billed Warbler (*Acrocephalus aedon*)
Arctic Warbler (*Phylloscopus borealis*)
Pallas's Leaf Warbler (*Phylloscopus proregulus*)
Yellow-browed Warbler (*Phylloscopus inornatus*)
Radde's Warbler (*Phylloscopus schwarzi*)
Dusky Warbler (*Phylloscopus fuscatus*)
Dark-sided Flycatcher (*Muscicapa sibirica*)
Asian Brown Flycatcher (*Muscicapa daurica*)
Mugimaki Flycatcher (*Ficedula mugimaki*)
Brown Shrike (*Lanius cristatus*)
Purple-backed Starling (*Sturnus sturninus*)
White-cheeked Starling (*Sturnus cineraceus*)
Black-faced Bunting (*Emberiza spodocephala*)
Yellow-browed Bunting (*Emberiza chrysophrys*)
Little Bunting (*Emberiza pusilla*)
Chestnut Bunting (*Emberiza rutila*)
Yellow-breasted Bunting (*Emberiza aureola*)

ACKNOWLEDGMENTS

We are especially grateful to Thomas Alerstam, Staffan Bensch, and Susanne Åkesson, as well as to the entire migration research group at Lund University for provoking our interest in migration. In addition, we thank Thomas Alerstam, Chris Bell, Staffan Bensch, Anders Hedenström, Åke Lindström, Trevor Price, Tom Smith, and Kasper Thorup for helpful discussion and comments on the manuscript. We are grateful to Peter Marra and Russell Greenberg for inviting us to participate in the Birds of Two Worlds Symposium and for suggesting improvements to the manuscript. For financial support we thank the International Research Fellowship Program of the National Science Foundation.

LITERATURE CITED

Adams, J. M., and H. Faure (eds). 1997. Review and Atlas of Palaeovegetation: Preliminary Land Ecosystem Map of the World since the Last Glacial Maximum. Oak Ridge National Laboratory, Tenn. [www.esd.ornl.gov/projects/qen/adams1.html]

Alerstam, T. 1990. Bird Migration. Cambridge University Press, Cambridge.

Ali, S., and S. D. Ripley. 1987–1999. Handbook of the Birds of India and Pakistan, Together with those of Bangladesh, Nepal, Bhutan and Sri Lanka, Vols. 5–10 (second ed.). Oxford University Press, New Delhi.

Alström, P., and K. Mild. 2003. Pipits and Wagtails. Princeton University Press, Princeton.

Baker, K. 1997. Warblers of Europe, Asia and North Africa. Princeton University Press, Princeton.

Barrett, P. H., P. J. Gautrey, S. Herbert, D. Kohn, and S. Smith. 1987. Charles Darwin's Notebooks, 1836–1844. Cambridge University Press, Cambridge.

Beaman, M. 1994. Palearctic Birds: A Checklist of the Birds of Europe, North Africa and Asia North of the Foothills of the Himalayas. Harrier Publications, Stonyhurst, UK.

Bensch, S. 1999. Is the range size of migratory birds constrained by their migratory program? Journal of Biogeography 26:1225–1235.

Bensch, S., T. Andersson, and S. Åkesson. 1999. Morphological and molecular variation across a migratory divide in Willow Warblers, *Phylloscopus trochilus.* Evolution 53:1925–1935.

Berthold, P. 1999a. Geographic variation and the microevolution of avian migratory behavior. Pages 164–179 *in* Geographic Variation in Behavior (S. A. Foster and J. A. Endler, eds.). Oxford University Press, Oxford.

Berthold, P. 1999b. A comprehensive theory for the evolution, control and adaptability of avian migration. Ostrich 70:1–11.

Berthold, P., A. J. Helbig, G. Mohr, and U. Querner. 1992. Rapid microevolution of migratory behaviour in a wild bird species. Nature 360:668–670.

Berthold, P., and U. Querner. 1981. Genetic basis of migratory behavior in European warblers. Science 212:77–79.

Böhning-Gaese, K., L. I. González-Gusmán, and J. H. Brown. 1998. Constraints on dispersal and the evolution of the avifauna of the Northern Hemisphere. Evolutionary Ecology 12:767–783.

Byers, C., U. Olsson, and J. Curson. 1995. Buntings and Sparrows: A Guide to the Buntings and North American Sparrows. Pica Press, Sussex.

Catchpole, C. K. 1980. Sexual selection and the evolution of complex songs among European warblers of the genus *Acrocephalus.* Behaviour 74:149–166.

Catchpole, C. K. 1982. The evolution of bird sounds in relation to mating and spacing behavior. Pages 297–319 *in* Acoustic Communication in Birds, Vol. 1 (D. E. Kroodsma and E. H. Miller, eds.). Academic Press, New York.

Catchpole, C. K., and P. J. B. Slater. 1995. Bird Song: Biological Themes and Variations. Cambridge University Press, Cambridge.

Clement, P., A. Harris, and J. Davis. 1993. Finches and Sparrows: An Identification Guide. Princeton University Press, Princeton.

Clement, P., and R. Hathway. 2000. Thrushes. Christopher Helm, London.

Cramp, S. (ed.). 1988–1994. Handbook of the Birds of Europe, the Middle East and North Africa. Oxford University Press, Oxford.

Dementev, G. P., and N. A. Gladkov (eds.). 1954. Birds of the Soviet Union, Vols. 5/6 (English translation 1968–1970). Israel Program for Scientific Translations, Jerusalem.

Dingle, H. 1991. Evolutionary genetics of animal migration. American Zoologist 31:253–264.

Dingle, H. 1996. Migration: The Biology of Life on the Move. Oxford University Press, Oxford.

Dobzhansky, T. 1940. Speciation as a stage in evolutionary divergence. American Naturalist 74:312–321.

Dolnik, V. R. (ed.). 1985. Energy resources of the birds migrating across arid and mountain regions of middle Asia and Kazakhstan. Trudy Zoologicheskogo Instituta, Akademiya Nauk SSSR, Leningrad, 137. [In Russian with English summaries.]

Dolnik, V. R. (ed.). 1987. Study of bird migration in arid and mountainous regions of Middle Asia and Kazakhstan. Trudy Zoologicheskogo Instituta, Akademiya Nauk SSSR, Leningrad, 173. [In Russian with English summaries.]

Dolnik, V. R. 1990. Bird migration across arid and mountainous regions of middle Asia and Kasakhstan. Pages 368–386 *in* Bird Migration: Physiology and Ecophysiology (E. Gwinner, ed.). Springer-Verlag, Berlin.

Fear, K. K., and T. Price. 1998. The adaptive surface in ecology. Oikos 82:440–448.

Flint, V. E., R. L. Boehme, Y. V. Kostin, and A. A. Kuznetsov. 1984. A Field Guide to the Birds of the USSR. Princeton University Press, Princeton.

Formozov, N. A., and I. M. Marova. 1991. Secondary contact zones of leaf warbler taxa in southern Tuva (based on bioacoustic data). Ornitologia 25:29–30. [In Russian.]

Grichuk, V. P. 1984. Late Pleistocene vegetation history. Pages 155–178 *in* Late Quaternary Environments of the Soviet Union (A. A. Velichko, ed.). University of Minnesota Press, Minneapolis.

Grimmett, R., C. Inskipp, and T. Inskipp. 1999. Birds of India, Pakistan, Nepal, Bangladesh, Bhutan, Sri Lanka, and the Maldives. Princeton University Press, Princeton.

Gwinner, E., and W. Wiltschko. 1978. Endogenously controlled changes in migratory direction of the Garden Warbler *Sylvia borin.* Journal of Comparative Physiology 125:267–273.

Hedenström, A., and T. Alerstam. 1997. Optimum fuel loads in migratory birds: distinguishing between time and energy minimization. Journal of Theoretical Biology 189:227–234.

Hedenström, A., and J. Pettersson. 1987. Migration routes and wintering areas of willow warblers *Phylloscopus trochilus* (L.) ringed in Fennoscandia. Ornis Fennica 64:137–143.

Helbig, A. J. 1991. Inheritance of migratory direction in a bird species: a cross-breeding experiment with SE- and SW-migrating blackcaps (*Sylvia atricapilla*). Behavioral Ecology and Sociobiology 28:9–12.

Helbig, A. J. 1996. Genetic basis, mode of inheritance and evolutionary changes of migratory directions in Palearctic warblers (Aves: Sylviidae). Journal of Experimental Biology 199:49–55.

Helbig, A.J. 2003. Evolution of bird migration: A phylogenetic and biogeographic perspective. Pages 3–20 *in* Avian Migration (P. Berthold, E. Gwinner, and E. Sonnenschein, eds). Springer-Verlag, Berlin.

Howard, D. J. 1993. Reinforcement: origin, dynamics, and fate of an evolutionary hypothesis. Pages 46–69 *in* Hybrid Zones and the Evolutionary Process (R. G. Harrison, ed.). Oxford University Press, Oxford.

Irwin, D. E. 2000. Song variation in an avian ring species. Evolution 54:998–1010.

Irwin, D. E., P. Alström, U. Olsson, and Z. M. Benowitz-Fredericks. 2001a. Cryptic species in the genus *Phylloscopus* (Old World leaf warblers). Ibis 143:233–247.

Irwin, D. E., S. Bensch, and T. D. Price. 2001b. Speciation in a ring. Nature 409:333–337.

Irwin, D. E., J. H. Irwin, and T. D. Price. 2001c. Ring species as bridges between microevolution and speciation. Genetica 112/113:223–243.

Iwasa, Y., and A. Pomiankowski. 1995. Continual change in mate preferences. Nature 377:420–422.

Kazmierczak, K., and B. van Perlo. 2000. A Field Guide to the Birds of the Indian Subcontinent. Pica Press, Sussex.

Keith, S., E. K. Urban, and C. H. Fry (eds.). 1992. The Birds of Africa, Vol. 4. Academic Press, London.

Khotinsky, N. A. 1984. Holocene vegetation history. Pages 179–200 *in* Late Quaternary Environments of the Soviet Union (A. A. Velichko, ed.). University of Minnesota Press, Minneapolis.

Knystautus, A. 1987. The Natural History of the USSR. McGraw-Hill, New York.

Liou, L. W., and T. D. Price. 1994. Speciation by reinforcement of premating isolation. Evolution 48:1451–1459.

MacKinnon, J., and K. Phillips. 2000. A Field Guide to the Birds of China. Oxford University Press, Oxford.

Madge, S., and H. Burn. 1994. Crows and Jays: A Guide to the Crows, Jays and Magpies of the World. Christopher Helm, London.

Mayr, E. 1942. Systematics and the Origin of Species. Dover Publications, New York.

Mayr, E. (ed.). 1976. History of the North American Bird Fauna: Evolution and the Diversity of Life. Selected Essays. Belknap-Harvard University Press, Cambridge, Mass.

McClure, H. E. 1974. Migration and Survival of the Birds of Asia. Applied Scientific Research Corporation of Thailand, Bangkok.

Morton, E. S. 1996. A comparison of vocal behavior among tropical and temperate passerine birds. Pages 258–268 *in* Ecology and Evolution of Acoustic Communication in Birds (D. E. Kroodsma and E. H. Miller, eds.). Cornell University Press, Ithaca.

Panov, E. N., A. S. Roubtsov, and D. M. Monzikov. 2003. Hybridization between Yellowhammer and Pine Bunting in Russia. Dutch Birding 25:17–31.

Price, T. 1998. Sexual selection and natural selection in bird speciation. Philosophical Transactions of the Royal Society of London, Series B, Biological Sciences 353:251–260.

Price, T. D., A. J. Helbig, and A. D. Richman. 1997. Evolution of breeding distributions in the Old World Leaf Warblers (genus *Phylloscopus*). Evolution 51:552–561.

Read, A. F., and D. M. Weary. 1992. The evolution of bird song: comparative analyses. Philosophical Transactions of the Royal Society of London, Series B, Biological Sciences 338:165–187.

Robson, C. 2000. A Guide to the Birds of Southeast Asia. Princeton University Press, Princeton.

Rogacheva, H. 1992. The Birds of Central Siberia. Husum Druck- und Verlagsgesellschaft, Husum, Germany.

Ruegg, K. C., and T. B. Smith. 2002. Not as the crow flies: a historical explanation for circuitous migration in the Swainson's Thrush (*Catharus ustulatus*). Proceedings of the Royal Society of London, Series B, Biological Sciences 269:1375–1381.

Searcy, W. A., and K. Yasukawa. 1996. Song and female choice. Pages 454–473 *in* Ecology and Evolution of Acoustic Communication in Birds (D. E. Kroodsma and E. H. Miller, eds.). Cornell University Press, Ithaca.

Shirihai, H., G. Gargallo, A. J. Helbig, A. Harris, and D. Cottridge. 2001. Sylvia Warblers: Identification, Taxonomy and Phylogeny of the Genus *Sylvia*. Christopher Helm, London.

Sibley, C. G., and B. L. Monroe Jr. 1990. Distribution and Taxonomy of Birds of the World. Yale University Press, New Haven.

Sutherland, W. J. 1998. Evidence for flexibility and constraint in migration systems. Journal of Avian Biology 29:441–446.

Svensson, L. 1987. More about *Phylloscopus* taxonomy. British Birds 80:580–581.

Ticehurst, C. B. 1938. A Systematic Review of the Genus *Phylloscopus*. British Museum, New York.

Turner, A., and C. Rose. 1989. A Handbook to the Swallows and Martins of the World. Christopher Helm, London.

Ukraintseva, V. V. 1993. Vegetation Cover and Environment of the "Mammoth Epoch" in Siberia. The Mammoth Site of Hot Springs, Hot Springs, S.D.

Urquhart, E. 2002. Stonechats: A Guide to the Genus *Saxicola*. Yale University Press, New Haven.

West-Eberhard, M. J. 1983. Sexual selection, social competition, and speciation. Quarterly Review of Biology 58:155–183.

CHRISTOPHER PAUL BELL

Inter- and Intrapopulation Migration Patterns

Ideas, Evidence, and Research Priorities

A BROADLY SIMILAR SET of causal factors has been proposed to account for interpopulation migration patterns such as leap-frog and chain migration and for intrapopulation patterns of differential migration among sex and age classes. These include asymmetric competition for resources in wintering areas, variation in the intensity of competition for breeding resources, and factors independent of competition, such as variation in body size or the allocation of time between breeding and wintering areas. A variety of multifactorial models have also been proposed, which attribute migration patterns to competition at both seasons operating in combination with variation in habitat quality and migratory ability.

An alternative approach attributes leap-frog and chain patterns to variation among populations in the optimal latitude for premigratory fattening as a function of the variable timing of spring migration. Here this approach is applied for the first time to differential migration.

Progress in applying and testing migration pattern models has been hindered by the lack of adequate data on the movements and demographics of migratory birds. However, recent advances in field and analytical techniques have the potential to furnish the empirical base required to finally deduce the causes of migration patterns. Priorities for data collection and model development necessary to bring this about are discussed.

PATTERNS OF MIGRATION

Migratory movements have a directional and a distance component. The direction of migration, although generally governed by the direction of improvement in climate outside the breeding season, is also determined by constraints connected with evolutionary history and with the geographical distribution of habitats and land masses (Bildstein and Zalles, Chap. 13, this volume). Where different routes are taken it is not surprising that migration distances vary, as costs and benefits may differ radically. It is less clear, however, why migration distance should vary when individuals face ostensibly similar costs and constraints, such as when sharing a migration corridor. Nor is it obvious why broad-front migrations frequently give rise to a consistent pattern of winter distribution across a wide longitudinal range.

Variation in average length of migration can give rise to both inter- or intrapopulation patterns. The former frequently take the form of "leap-frog" migration, in which the winter distribution of breeding populations forms a mirror image of their position in the breeding season, or "chain" migration, where the populations are shunted along the axis of migration and retain the same spatial relationship as in the breeding season. Intrapopulation patterns form part of the syndrome of "differential" migration, in which different sex and age classes within a population differ in average timing and distance of migration. This chapter reviews the development of ecological and evolutionary thinking on the origin of these migration patterns, followed by a more detailed account of recent attempts at a synthesis, and concludes with a discussion of possible directions for future research.

The ultimate factors governing migratory behavior are the same as apply to other aspects of behavior and life history. Apart from the climatic and historical factors already alluded to, these include net primary productivity and/or food availability (Butler et al. 2001), temporal and spatial distribution of resources (Levey and Stiles 1992; Bell 2000), the action of pathogens and parasites (Piersma 1997), interspecific competition (Cox 1985), and predation (Fretwell 1980). Theories about migration patterns attempt to explain why the sex-age classes or populations of a migrant species respond differently to these underlying ecological factors, and generally focus either on the fundamental differences in biology that might cause independently varying responses or on the action of intraspecific competition within and between classes and populations.

Although ideas about inter- and intrapopulation patterns have influenced each other, they have largely focused on different species and have therefore remained rather separate, if parallel, lines of thought. For this reason it is convenient to begin with separate accounts of the development of theories about the two kinds of migration pattern.

Intrapopulation Patterns

Ideas about differential migration have been heavily influenced by thinking on the causes of partial migration, in which some members of a breeding population migrate, whereas others remain resident. Lack (1944a) reported a preponderance of females and juveniles in the migrant fraction of a range of partial migrants and speculated that individuals of low status, who are therefore less likely to survive competition for food, are more likely to accept whatever costs are attached to migration to avoid this competition when food is limiting (Lack 1954). Another partial migration theory proposes that migrants avoid the mortality risk faced by residents enduring harsh winters and so have greater overall survival, but lose out to residents in competition for the best breeding territories. Members of the territory-holding sex, generally males, are therefore more likely to sacrifice survival for enhanced breeding success by remaining resident (von Haartman 1968).

Ketterson and Nolan (1976) noted that both ideas could equally apply to fully migratory populations if low-status individuals migrate farther to avoid competition (the "dominance hypothesis") or if a shorter migration ensures more timely arrival at the destination, thus providing access to the best territories and resources (the "arrival time hypothesis"). They also suggested that size might underlie migratory strategy independently of its effect on social status if larger birds can better endure food shortage through superior fasting ability (the "body size hypothesis").

Though Ketterson and Nolan emphasized that the dominance, arrival time, and body size theories are not mutually exclusive, many attempts have been made to formally test the three ideas as competing hypotheses. Myers (1981) examined a sample of shorebirds with larger-bodied females but found that males still had the shorter migration, contrary to both the dominance and body size hypotheses. Males also arrived first on the breeding grounds except in two species without differential migration, and Myers concluded that knowledge of arrival time is "necessary and sufficient" to predict length of migration.

Ketterson and Nolan (1982, 1983) presented the results of an intensive field study of differential migration in the Dark-eyed Junco (*Junco hyemalis*) and concluded that none of the main hypotheses could account for the observed patterns, proposing instead a multifactorial hypothesis (see below). Nonetheless, work continued toward understanding the role of dominance in differential migration. Terrill (1987) showed that subordinate members of food-restricted captive pairs of Dark-eyed Juncos showed higher levels of nocturnal hopping (*zugunruhe*) at the end of the fall migratory period and well into midwinter, suggesting that subordinates might tend to migrate south facultatively if food became short in the wild. However, aviary experiments with pairs of juncos captured at different wintering latitudes found no difference in dominance status between northern and southern birds (Rogers et al. 1989; Cristol and Evers 1992). Moreover, adult males from southern wintering sites retained their higher status with respect to northern wintering young males (Rogers et al. 1989). Cristol et al. (1990) showed that earlier residency could reverse interclass dominance hierarchies in juncos, but this is unlikely to determine wintering latitude because juncos passing the winter at northern sites arrive no earlier, on average, than transients passing through to winter farther south (Nolan and Ketterson 1990).

Arnold (1991) expanded Myers's comparative analysis of reverse dimorphic migrants and found several examples contrary to the latter's assertion that arrival time predicts migration pattern, including differential migration of the sexes with simultaneous arrival on the breeding grounds and longer migration by the sex arriving first. As with previous comparative analyses, Arnold's study made no attempt to discriminate between body size and dominance hypotheses, instead simply assuming that larger birds are more dominant and so predicting the same contingency of size and migration patterns. Belthoff and Gauthreaux (1991), meanwhile, found evidence favoring the body size hypothesis against the dominance hypothesis when they examined migration in the House Finch (*Carpodacus mexicanus*) in which smaller females dominate males and discovered that females migrate farther.

The widest comparative analysis yet was undertaken by Cristol et al. (1999), who surveyed published evidence on over 150 mainly Holarctic-breeding migrants. On finding only a handful of species with no evidence of differential patterns, they concluded that such patterns are probably the norm rather than the exception, but like Arnold they found numerous examples contrary to predictions arising from the three main hypotheses.

Interpopulation Patterns

As with differential migration, patterns involving whole populations from different latitudes within the breeding range have long been linked to competition occurring on the winter range (Mayr and Meise 1930). However, whereas ideas about differential migration derive mainly from behavioral studies on the effects of social dominance, the discussion of interpopulation patterns has been influenced by ideas about competitive exclusion among species. Lack (1944b) noted that migrant species with overlapping breeding ranges often winter allopatrically, and he speculated that the ecological mechanisms that reduce interspecific competition and allow coexistence of closely related species within breeding habitats might be disrupted on the wintering grounds. If so, the resulting competition might be resolved by geographic isolation, an idea later invoked to explain interspecific leap-frog and chain migration patterns by Cox (1968). Salomonsen (1955) used the same hypothesis in explanation of interpopulation migration patterns within species, particularly leap-frog patterns. This was subsequently endorsed by Lack (1968), though neither he nor Salomonsen considered the mechanism of competitive exclusion or whether it involved interference (i.e., dominance related) or exploitation competition.

Alerstam and Högstedt (1980) noted the incompleteness of the "competitive exclusion" theory and suggested an additional factor to explain leap-frog patterns. Noting that such patterns commonly involve Arctic breeding populations leap-frogging those breeding at temperate latitudes, they speculated that the former are unable to winter close enough to breeding areas to detect variations in the onset of spring, and so move to low latitudes to avoid competition and harsh winters. Temperate breeders endure these factors to remain near their breeding areas, where they can detect the variable onset of spring and reap the benefits of timely arrival on the breeding grounds. The "spring predictability" hypothesis is therefore a restatement of the "arrival time" theory applied to differential migration, with low-latitude breeders instead of "territory holders" shortening migration to compete for breeding resources.

Pienkowski et al. (1985) questioned the assumptions of the arrival time hypothesis, arguing that arrival time is generally not critical to breeding success but that migration is so costly that extending it usually results in lower net survival for individuals choosing not to winter "as close to the breeding grounds as possible." The evidence presented in support of this position by Pienkowski et al. is largely anecdotal, but they also showed that within shorebird species wintering along the eastern Atlantic coastline, individuals wintering farther away from northern breeding areas are generally smaller. On this basis Pienkowski et al. proposed size-related dominance as the critical agent generating leap-frog patterns through interference competition occurring on the wintering area and on migration stopover sites, an idea essentially identical to the "dominance" hypothesis applied to differential migration.

In a rejoinder, Alerstam and Högstedt (1985) noted that the patterns presented by Pienkowski et al. are equally explicable under the "body size" hypothesis proposed in the context of differential migration, and they listed contrary examples in which larger-bodied birds undergo longer migratory journeys. They concluded with an assertion that leap-frog patterns are explicable in terms of a combination of asymmetric competition for breeding resources among birds using different wintering areas and asymmetric competition for winter resources among birds from different breeding areas, noting the need for a quantitative model to describe how this occurs. Lundberg and Alerstam (1986) published such a model the following year (see below).

A different perspective on interpopulation patterns was offered by Greenberg (1980). The "time allocation" hypothesis is really two ideas that stand or fall independently. The first is the concept that there is a survival cost to residence in the breeding area and that migrant birds remain there only as long as reproductive benefits outweigh this cost. If the resulting period of breeding residence is short, owing perhaps to a narrow seasonal resource peak, the survival cost is paid for only a short period, so annual survival (all other things being equal) is higher than where breeding residence is lengthy.

The second idea is that where survival per unit time on a wintering site increases with distance from the breeding area, survival outside the breeding season (both migrations plus wintering period) may be maximized by migrating to more distant wintering sites as the period of breeding residence shortens. Because breeding seasons get shorter at higher latitudes, leap-frog migration may result.

If the length of breeding season residence is determined as proposed in the first idea and choice of wintering site as in the second, the result could be a positive correlation between migration distance and annual survival. Attempts to test the time allocation hypothesis have tended to view this as a critical prediction and have generally found evidence contrary to it (Pienkowski and Evans 1985; Hestbeck et al. 1992; Sandercock and Jaramillo 2002). However, the explanation of leap-frog patterns relies only on the fact of contraction of breeding seasons at high latitudes, and it remains valid even if survival is unaffected by length of breeding residence, in which case there is no necessary correlation between survival and length of migration. Other tests have found patterns of winter residence and demography different from those predicted by the time allocation hypothesis (Myers et al. 1985; Hestbeck et al. 1992), but only in species that do not undergo leap-frog migration, so they provide little insight into the applicability of the hypothesis to such patterns.

Summary of Ideas on Migration Patterns

The hypotheses put forward to explain migration patterns can be classified as outlined in table 4.1, which reflects the convergence of ideas attributing inter- and intraspecific patterns either to competition on the breeding or wintering grounds or to factors independent of competition. Of the latter, the body size hypothesis has more immediate relevance to temperate-wintering differential migrants than to species showing interpopulation patterns, which often winter in the tropics. The applicability of the time allocation hypothesis to differential patterns is discussed below.

The many apparent exceptions to predictions arising from all the main hypotheses proposed to explain migration patterns have led to the development of a number of multifactor hypotheses, in which the different factors proposed act in combination.

MULTIFACTOR MODELS

Ketterson and Nolan (1983) proposed a composite model to explain differential migration in the Dark-eyed Junco, based on the conceptual framework of Baker (1978). Baker's approach aimed to provide a context for the evolutionary interpretation of all animal movement, but as a result it is so generalized that it provides little insight into a specific syndrome of movement such as differential migration. However, Ketterson and Nolan showed that when the generalized parameters of Baker's model are reduced to specific demographic variables, it translates readily to an orthodox optimality problem, with migration length as the decision variable. Ketterson and Nolan constructed such a model, based on demographic data from their long-term studies of juncos, in which age classes differ in rate of mortality as a function of migration distance, and the sex classes differ in suitability of wintering latitudes because of a greater advantage of proximity to the breeding area experienced by males.[1] The model predicts a latitudinal sequence similar to the one observed, with young males farthest north, adult females farthest south, and adult males and young females in the middle.

Lundberg and Alerstam (1986) also adopted the multifactor approach to explain interpopulation patterns, this time using game theory. They simulated population trends in a linear sequence of five sites, giving birds the option of breeding in sites 1–3 and wintering in sites 3–5, in which breeding success and survival were density-dependent. Stable outcomes were obtained under varying assumptions relating to underlying (density-independent) resource conditions, migration mortality, and asymmetric competition arising from proximity advantage at both migrations (i.e., breeding near wintering sites and wintering near breeding sites). Leap-frog, chain, and mixing patterns were all common outcomes, though the most pronounced leap-frog patterns occurred where migration costs were combined with proximity advantage at both seasons, and chain patterns with migration mortality and no proximity advantage.

Holmgren and Lundberg (1993) extended the modeling approach by investigating the consequences of despotic competition, where dominant individuals are able to preempt resources. Under such conditions, the relative quality of breeding and wintering sites is preserved as they are occupied by populations of migrants, unlike the situation explored by Lundberg and Alerstam, where density dependence equalizes site quality. Leap-frog and chain migration are again possible outcomes under varying assumptions relating to migration cost and asymmetric competition related to proximity advantage.

Cristol et al. (1999) revisited the multifactor hypothesis of Ketterson and Nolan (1983), this time in the form of a simulation model with five wintering sites in latitudinal sequence and varying in associated migration costs (in terms of survival and subsequent breeding success), density-independent survival, and a population size threshold for density dependence. The model was applied to the Dark-eyed Junco pattern under conditions equivalent to those assumed by Ketterson and Nolan (1983) in their optimality model, producing a similar result. The model was then developed by incorporating an effect of the observed dominance hierarchy among sex-age classes, with density-dependent survival affected only by members of the same class or one of higher status. Variation of parameter values enabled the model to reproduce some of the finer detail of the junco's pattern of winter distribution.

[1]Migration models generally incorporate the "arrival time" factor by assuming greater breeding success (or winter survival) among individuals using more proximate wintering (or breeding sites). In this context therefore, it is more accurate to refer to "proximity advantage" than to "arrival time advantage."

Table 4.1 Hypotheses on the origin of migration patterns

Causal factor	Intrapopulation (differential) patterns	Interpopulation (leap-frog, chain, etc?) patterns
Competition on wintering grounds	Dominance hypothesis (Lack 1954; Ketterson and Nolan 1976)	Ecological isolation hypothesis (Lack 1944b, 1968; Salomonsen 1955; Cox 1968) Dominance hypothesis (Pienkowski et al. 1985)
Competition on breeding grounds	Arrival time hypothesis (Von Haartman 1968; Ketterson and Nolan 1976)	Spring predictability hypothesis (Alerstam and Hogstedt 1980)
Noncompetition factors	Body size hypothesis (Ketterson and Nolan 1976)	Time allocation hypothesis (Greenberg 1980)

Evaluation of Multifactor Models

Though differing in structure, the factors built into all the composite models are basically the same, comprising one or more migration cost functions that vary among classes or populations, habitat "suitability" functions that equate to variation in density-independent mortality and/or breeding success, and varying degrees of density dependence that act asymmetrically among classes or populations.

The obvious criticism invited by these constructs is that in each case there is little reference to any empirical basis for the values ascribed to model functions. This lack of data therefore allows parameters of the various models to be varied at will to "explain" any conceivable pattern of migration. Thus, although a model may be able to reproduce the characteristics of real migration patterns there is little guarantee that it reflects the actual processes involved. This, together with a dearth of testable predictions generated by the models, might suggest that the approach offers little to the effort to understand migration patterns.

The fact remains, however, that such patterns may indeed be caused by a complex interaction of independent factors, in which case they can only be described by models of this kind. A substantive empirical basis for model components must therefore be established by measuring the relationship between fitness indices (i.e., survival and breeding success) and model parameters. Equally important, however, is critical scrutiny of the veracity of the demographic mechanisms themselves, as they too lack a firm basis because of the scarcity of accurate data on demographic patterns, however caused. Options for further development of the multifactor approach are discussed below.

TIME ALLOCATION AND SEASONALITY MODEL

Despite the increasing complexity of models developed to explain migration patterns, there remains the possibility that a relatively simple factor underlies variation in the distance birds migrate. One possibility is the concept, embodied in the time allocation hypothesis, that variation in the trade-off between costs and benefits of migration is governed primarily by the length of time spent on the wintering grounds. In developing this idea, Greenberg (1980) presented a relatively simple model with a choice of wintering ranges characterized by survivorship that is temporally constant but inversely related to migration costs involved in getting there.

Pienkowski and Evans (1985) argued to the effect that the choices faced by real migrants were unlikely to conform sufficiently to these assumptions for the model to retain its explanatory power. I explored the implications of this critique, by developing a time allocation model that relaxed the constraints written into Greenberg's original presentation (Bell 1996, 1997). The approach estimates survival as a function of temporal and latitudinal variation in food availability within a migrant's wintering range, incorporating variation among sections of the population in length of winter residence (hence the "time allocation *and* seasonality" [TAS] hypothesis). Estimates of food availability are derived from monthly actual evapotranspiration (AE) indices (Ricklefs 1980) and calculated as the amount of evapotranspiration for a given time interval minus the annual minimum for a given latitude (excess AE).

Variation in length of winter residence implies variation in timing of migration, which may affect the cost per unit distance covered. In particular, early spring migrants might experience higher costs per unit distance traveled because of lower food availability en route (Pienkowski and Evans 1985). I incorporated temporal variation in migration costs per unit distance traveled in the TAS hypothesis, but I calculated this variation as a function of food availability on the wintering grounds immediately prior to migration. One consequence of this is that overall migration costs are actually reduced by extending migration if this brings a migrant into an area where food is abundant prior to prenuptial migration.

Formally, the TAS model is constructed as follows: Survival at a given wintering latitude during a time period i (generally a half-month) is modeled as a sigmoid function of excess AE during the same period (V_i), and survival over the whole wintering period (S_w) is then the product of survival in each period:

$$S_w = \prod_{i=1}^{n} \frac{\exp(a + bV_i)}{1 + \exp(a + bV_i)}$$

(a and b are constants defining the shape of the sigmoid function)

Cost of migration is incorporated by defining a quantity ($S_m^\star$) equal to unity minus "cost." $S_m^\star$ is modeled as a function of the difference between breeding and wintering latitudes ($\text{lat}_s - \text{lat}_w$) and of excess AE at the chosen wintering latitude during the period prior to departure on spring migration (V_n). The quantity is denoted $S^{\star}_m$ to emphasize that it is not simply a survival function like S_w, because costs include both mortality on migration and "arrival time"–related reduction in breeding output as a consequence of migrating away from the breeding area (Bell 1996, 1997):

$$S^{\star}_m = \exp((-C + dV_n)(\text{lat}_s - \text{lat}_w))$$

[C and d are constants defining the cost per unit latitude traversed on migration (C) and the degree to which each unit V_n mitigates this cost (d)]

Optimal wintering latitude, in the absence of density-dependent resource depression, is that which maximizes $S^{\star}_m.S_w$.

Because the TAS model has so far been applied only to leap-frog patterns, without any quantitative treatment of density dependence, it is apposite to extend the approach to a treatment of differential migration, in the form of a simulation model incorporating density-dependent mortality.

TAS Model Applied to the Dark-Eyed Junco

For the TAS model to explain differential migration among sex-age classes, there must be variation in timing of departure from the wintering area. Ketterson and Nolan (1983) were unable to find significant differences in departure dates among sex-age classes in the junco, although the median departure date was earlier among males. Other lines of evidence suggesting earlier male departure include an earlier peak of males among spring passage birds (Ketterson and Nolan 1983) and earlier commencement of migratory restlessness among males in a captive sample in spring (Ketterson and Nolan 1985). Rabenold and Rabenold (1985) also found an earlier return to breeding territories by males among Dark-eyed Juncos of the subspecies *J. h. carolinensis,* breeding in the southern Appalachians, so it is probably valid to include a slightly earlier male migration schedule among model assumptions.

As with previous applications of the TAS model, it is assumed that "excess AE" provides an index of food availability for the subject species. Figure 4.1 shows a profile of variation in excess AE in continental eastern North America during the period of the junco's winter residence and into the spring migration period. It is notable that during the peak of the junco's spring exodus in early–late March, excess AE is increasing rapidly in the main part of the wintering range.

Figure 4.2a shows latitudinal variation in the migration function ($S^{\star}_m$) and overwinter survival (S_w) functions derived from this excess AE profile. The trend in overwinter survival is the same for all members of the population, but the four sex-age classes face differing migration costs. Juveniles are assumed to have higher unit distance migration costs (Ketterson and Nolan 1983), so have lower $S^{\star}_m$ values at the northern extremity of the wintering range. South of this point the cost of migration begins to be compensated for by improved food availability prior to the return journey in spring, but this effect is greater in females because of their later migration. Consequently, juvenile females have higher $S^{\star}_m$ values (indicating lower migration costs) than adult males south of 39° N. The product of density-independent

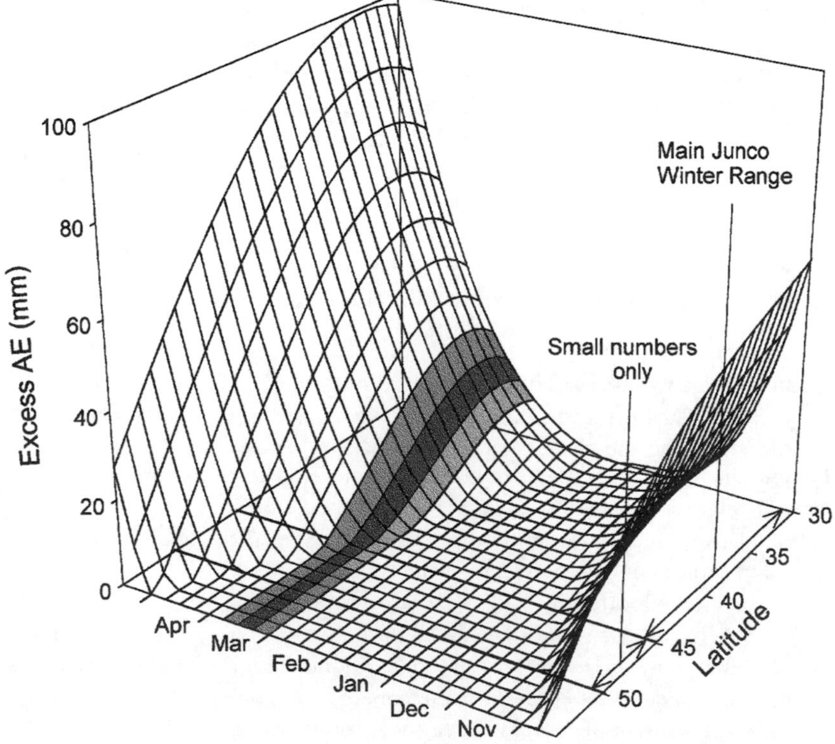

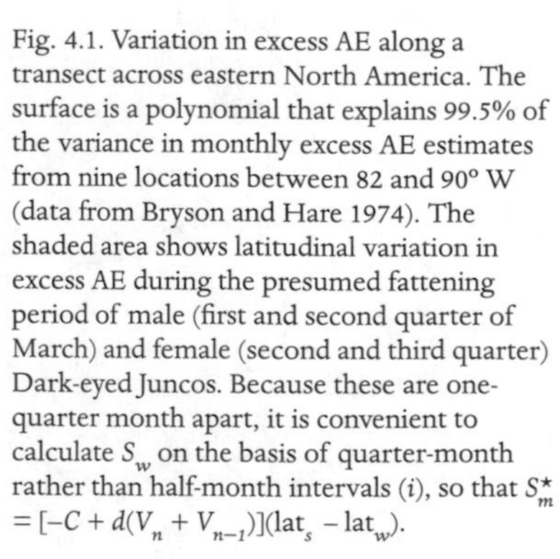

Fig. 4.1. Variation in excess AE along a transect across eastern North America. The surface is a polynomial that explains 99.5% of the variance in monthly excess AE estimates from nine locations between 82 and 90° W (data from Bryson and Hare 1974). The shaded area shows latitudinal variation in excess AE during the presumed fattening period of male (first and second quarter of March) and female (second and third quarter) Dark-eyed Juncos. Because these are one-quarter month apart, it is convenient to calculate S_w on the basis of quarter-month rather than half-month intervals (i), so that $S^{\star}_m = [-C + d(V_n + V_{n-1})](\text{lat}_s - \text{lat}_w)$.

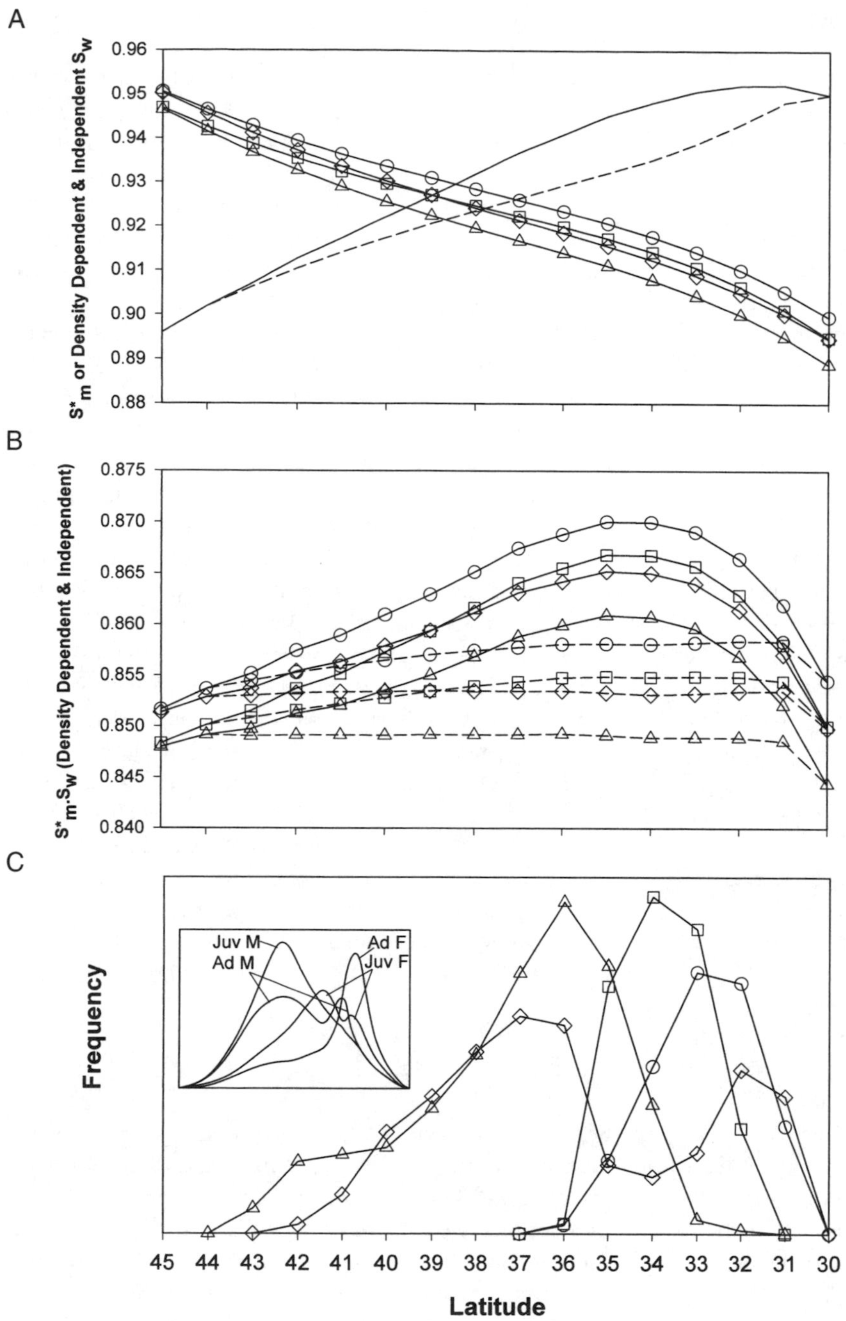

Fig. 4.2. Predicted latitudinal trends. Symbols: adult females (○), juvenile females (□), adult males (◇), juvenile males (△). In figs. 4.2A and B, unbroken lines show density-independent values, and broken lines show density-dependent values resulting from the predicted settling pattern shown in fig. 4.2C. (A) Migration function ($S^{\star}_{m}$) for each sex-age class (lines with symbols), and overwinter survival (S_w; lines lack symbols since values are equal across all sex-age classes). (B) Product of $S^{\star}_{m}$ and S_w (density-dependent and -independent) values shown in 4.2A. (C) Pattern generated by iterative simulation in which sex-age classes settle at optima of class-specific density-dependent $S^{\star}_{m}.S_w$ curves. The effect of population size on overwinter survival at a given latitude is generated by assuming that parameter *a* in the equation defining S_w declines linearly as population size increases. Note that the density-dependent effect of each individual is identical, regardless of sex-age class; that is, there is no asymmetric competition built into the model. The actual pattern derived from field observations (Ketterson and Nolan 1983) is inset.

cost and benefit trends ($S^{\star}_{m}.S_w$) indicates that all four classes have an optimal wintering latitude at around 34–35° N (fig. 4.2B). However, the shape of the optimality curve varies somewhat among the classes, which is significant for the density-dependent settling pattern.

The latter is simulated by iterating a sequence in which a cohort of each sex-age class settles at its optimal latitude, followed by recalculation of S_w to reflect population size. As the simulation proceeds, populations increase at initially optimal latitudes, where expected survival consequently declines. Other latitudes then become optimal, resulting in latitudinal expansion of the wintering population. The outcome of this process produces a distribution sharing many of the features documented for the winter distribution of the junco, with a roughly coincident northerly peak for adult and juvenile males and a minor southern peak for the former, an intermediate peak for juvenile females, and a southerly peak for adult females (fig. 4.2C).

The outcome of this settling pattern in terms of density-dependent S_w and $S^{\star}_{m}.S_w$ indicates that the model corresponds well to demographic as well as geographic patterns among wintering juncos (fig. 4.2A,B). Ketterson and Nolan (1982) found that overwinter survival is similar across sex-age classes at a given latitude but decreases toward the north, as reflected by the density-dependent S_w curve in fig. 4.2A. Field data also suggest that populations wintering at different latitudes have equal annual survival, and density-dependent $S^{\star}_{m}.S_{w}$ curves are flat within the latitudinal range occupied by each class (fig. 4.2B). If $S^{\star}_{m}$ primarily reflects survival rather than nonsurvival related "arrival time" costs, the product $S^{\star}_{m}.S_{w}$ should express wintering-latitude-dependent variation in annual survival, in which case the model again accurately reflects the patterns found in the field. Values around 0.85 for density-dependent $S^{\star}_{m}.S_{w}$ are substantially higher than annual survival estimates of 0.53–0.54 given by Ketterson and Nolan (1982), but this does not invalidate the model because important causes of mortality, such as predation and disease, may be independent of wintering latitude.

Evaluation of the TAS Model

Two requirements need to be fulfilled for this theory to contribute to an explanation of migration patterns: (i) Populations or sex-age classes migrating different distances must depart the wintering area on prenuptial migration at different times; and (ii) the population or class wintering farther from the breeding ground must accrue compensating benefit that is unavailable to that wintering closer by virtue of the difference in spring migration timing. The second requirement can only be properly tested by appropriate field studies of wintering migrants, options for which are discussed in the final section of this chapter.

The first requirement will almost always be fulfilled for leap-frog and chain migration patterns, because populations breeding at higher latitudes breed later in the year and so depart the wintering area later. The situation is less clear-cut for differential migration patterns among classes sharing the same breeding areas. Cristol et al. (1999) showed that most documented differential migrants show differences in spring arrival schedule, but they listed several exceptions in which classes differing in migration length arrive simultaneously. However, much of the evidence in the literature reviewed by Cristol et al. is anecdotal at best, emphasizing the need for focused field studies to quantitatively define the arrival schedules of sex-age classes in breeding populations of differential migrants.

It could be argued that the TAS model is simply another multifactorial model, prone to the same criticisms outlined earlier. However, the model is more parsimonious because it omits asymmetric competition and integrates proximity advantage and migration mortality in a single parameter. It is more empirically based, relying on a measure of habitat suitability derived from real environmental data and also generates a number of testable predictions, as is discussed below.

General Implications of the TAS Model for Temperate-Tropical Migration

The approach outlined in the TAS hypothesis has implications for an understanding of tropical and long-distance migration in general. Consider, as a brief example, the migration of Red Knot (*Calidris canutus*) populations on the East Atlantic seaboard.

Nearctic-breeding *C. c. islandica* winter in northwest Europe and commence breeding slightly earlier in June than the Siberian-breeding *C. c. canutus* populations that winter in West Africa. The latter fatten relatively slowly before leaving for western Europe in early May, at the same time as *C. c. islandica* populations are leaving western Europe for staging areas in Iceland and northern Norway (Piersma and Davidson 1992).

If low food availability in West Africa means that knots are unable to depart before early May, any *C. c. islandica* choosing to winter there would inevitably lag behind on spring migration. However, the later phenology of Siberian breeding areas means that *C. c. canutus* can take advantage of this mild West African wintering area and still arrive on the breeding grounds at the optimum time.

It may be, therefore, that food availability in the period immediately prior to spring migration could be the key to a more general explanation of why some temperate and Arctic breeding species and populations migrate to the Tropics, whereas others pass the winter at high latitudes much nearer the breeding area. The existence of resource-rich tropical wintering grounds offering potentially high levels of survival, even taking into account migration mortality, is of little consequence unless the phenology of such wintering grounds affords the opportunity to fatten and migrate back to high-latitude breeding areas in timely fashion. If not, a short migration may always prove the superior strategy, despite high overwinter mortality, because it affords timely arrival at the breeding site and therefore high breeding success.

FUTURE DIRECTIONS

Some clear priorities for the next phase of research on migration patterns have emerged from this discussion, with three areas of work requiring particular attention: (i) development of the structure of multifactor models to include more realistic demographic mechanisms, (ii) improvement of the empirical base available to "constrain" model parameter values, and (iii) field tests of the predictions derived from the TAS hypothesis. The rationale for each of these priority areas is discussed below, along with some suggestions for graduate-postdoctoral research projects that could help to fill in current gaps in knowledge.

Theoretical Development of Multifactor Models

Some of the assumptions of multifactor models would appear to be unquestionable: for example, that there is a mortality "cost" associated with migratory movement; that habitat varies in suitability; and that there is dominance-related asymmetric competition between sex-age classes and populations. Some other assumptions that are crucial to the ability of multifactor models to "explain" migration patterns are less secure, not least the way in which the demographic effects of dominance are assumed to underlie migration patterns.

One of the few testable predictions arising from the multifactor approach is that subordinate classes should have lower overwinter survival than dominants in the same wintering population (Cristol et al. 1999). Lack of evidence for this among wintering Dark-eyed Juncos was one of the main factors leading Ketterson and Nolan (1982, 1983) to question the role of dominance in generating migration patterns. However, more recent studies suggest that the action of dominance hierarchies may be more subtle than simply an immediate impact on survival.

Subordinate female American Redstarts (*Setophaga ruticilla*) are pushed into inferior habitats by males (Marra 2000), but they appear to survive the winter as well as males, despite showing evidence of poorer body condition (Marra and Holmes 2001). However, there is evidence that poor conditions experienced overwinter by subordinates may reduce breeding success by delaying arrival on the breeding grounds and also reduce survival outside the period of winter residence (Marra et al. 1998; Marra and Holmes 2001). Similar cross-seasonal effects have previously been noted in the Barnacle Goose (*Branta leucopsis*), where higher population density on the breeding grounds leads to higher mortality on autumn migration (Owen and Black 1991). In the light of such evidence, a more fruitful approach to modeling migration patterns as an outcome of dominance hierarchies might be to link social status on the wintering grounds to subsequent migration mortality or breeding success, instead of to overwinter survival.

The possible role of dominance also needs to be investigated from an evolutionary perspective. There is reason to believe that much of the variation in migratory behavior underlying migration patterns is heritable (Terrill and Berthold 1989; Holberton 1993), and multifactor models assume that the effects of dominance are a significant factor selecting for such variation. However, given that status is condition-dependent with a large degree of overlap among classes and populations, it might be questioned whether dominance-related selection can favor a longer migration for a particular class. If, for instance, the sexes have the same mean heritable migration length and males are dominant on average to females, it may be that average females can improve fitness by lengthening migration but that high-status females cannot. Because average females cannot exceed the fitness of high-status females, they will still experience negative selection, even after extending migration. If so, dominance is more likely to be a factor during the facultative phase of migration because selection might favor a tendency to extend migration given low realized status (Terrill 1987). In any event, a game-theoretic model of status-dependent migration length is required to clarify issues.

The idea that proximity to a migratory destination promotes enhanced breeding success through "timely" arrival remains untested. However, some support for the hypothesis is provided by a model of optimal arrival developed by Kokko (1999), which confirms the intuitive notion that the arrival time optimum is both earlier and narrower when the cost of arriving late is high. A population or class with a narrower optimum may find that the benefits of a more distant wintering site are outweighed by greater costs of mistimed arrival. Modeling might therefore be used to explore the plausibility of any "proximity" effect by relating breeding success to: (i) migration-length-dependent ability to adjust to annual variation in phenology, and (ii) the shape of class- or population-specific arrival time optima.

Improvement of the Empirical Base

For the improvements in theoretical focus outlined above to yield genuine insight into migration patterns, a complementary program of field studies is required to provide reliable evidence about the demography of migrant bird species and populations, and thereby "constrain" the values of model parameters. Measurement of migration "costs" is particularly crucial, as without a way of quantifying the cost of migrating a given distance, it remains impossible even to say whether the putative trade-off with improved overwinter conditions is steep or shallow.

Mortality en route is the most obvious cost attached to migration, but quantifying it is far from straightforward. Resighting studies of marked individuals of site-tenacious species on the wintering grounds have, at best, enabled the partitioning of survival estimates between the wintering period and the rest of the year (Evans and Pienkowski 1984; Owen and Black 1991). However, a recent study spanning both breeding and wintering grounds has yielded indirect estimates of migration mortality (Sillett and Holmes 2002). Studies of this kind on populations that vary in distance migrated could provide species-specific estimates of distance-dependent migration mortality.

For species that are not site tenacious, it may only be possible to obtain annual survivorship estimates from recovery data. The same applies even for site-tenacious species if they have characteristics such as a skulking habit, large home ranges, or early departure following nest failure, which reduce the probability of resightings. Populations with differing migration strategies are likely to vary in survival rates throughout the year, so annual survival can provide only limited insight into the contribution to mortality of migration itself (Nichols 1996). However, the limited data available suggest that migration mortality may be negligible in groups such as waders and waterfowl but highly significant in long-distance migrant songbirds (Ens et al. 1994), so for the latter, comparisons of annual mortality between populations may provide a stronger "signal" concerning the effect of migration.

For long-distance migrants there is little prospect of deriving accurate survival estimates from ringed birds because, with the exception of hunted species, recoveries are rare. However, a new technique for aging animals by using blood samples (Haussmann and Vleck 2002) has the potential to greatly simplify the gathering of data on age distributions, so estimates of annual survival may become much easier to obtain for a wide range of migratory species and populations. At the very least, therefore, calibration studies for the aging technique should be carried out for a range of migratory species. Routine taking of blood samples for analysis during trapping studies involving migratory birds at their breeding grounds would then provide an invaluable database of demographic data.

The technique of stable isotope analysis (Hobson, Chap. 19, this volume) not only offers the prospect of revealing migration patterns in unprecedented detail, but also creates new opportunities to quantify model parameters. Studies of differential migrants that deduce the wintering area of individuals whose arrival date on the breeding grounds is known could be used to quantify any relationship between migration distance and arrival time or arrival time variance. Use of several isotopes in combination may be necessary to narrow down a wintering area sufficiently (Webster et al. 2002), but if arrival time optima can also be defined for sex-age classes, it may then be possible to estimate the relative magnitude of any "proximity advantage."

Despite the paucity of critical predictions arising from multifactorial models, the approach outlined above may provide an effective means of "testing" model structures, because the models will be viable only if they continue to reproduce observed migration patterns when their parameters are confined within empirically determined limits. It may then ultimately be possible to derive models that are generalizable across groups of closely related species and reproduce species-specific migration patterns when primed with data derived from comparative studies of social structure and life history (Cristol et al. 1999).

Tests of the TAS Hypothesis

I have previously outlined the program of work required to test the TAS hypothesis as applied to leap-frog and chain migration (Bell 1996, 1997), and the approach described can easily be generalized to include differential migration. For patterns where the earlier-departing class or population winters closest to the breeding grounds, the model predicts that food abundance, and therefore fattening rate, should be relatively low during the departure period. Within the same time frame, a similar level of food abundance should be experienced by the class or population wintering farther from the breeding ground, but this should increase markedly by the time of the latter's premigratory fattening period, and the resultant rate of fattening should be significantly higher than that achieved by the earlier-departing population or class. Under time-minimization (Weber and Houston 1997), the improved rate of fat deposition should at least compensate for the time required to lay down the additional fat required for the longer journey, unless there is additional compensation from improved overwinter survival.

Feeding and fattening conditions for populations of migrants that have overwintered at different latitudes should therefore be investigated throughout the period of spring emigration for all the populations concerned. Fattening rates prior to migration need to be measured directly via trapping programs, which can be backed up by indirect methods of measuring food availability, such as recording foraging rates of birds and the overall proportion of the time budget devoted to foraging. Direct measurement of food abundance should also be carried out, using appropriate methods such as arthropod trapping.

CONCLUDING REMARKS

Two decades ago, Myers et al. (1985:p. 526) described the study of migration patterns as "poor in data and rife with imprecise theory," and they bemoaned the lack of evidence linking underlying ecological factors with the demographic mechanisms thought to drive migratory behavior. Though the situation today remains much as Myers found it, the prospect for progress is much brighter.

Improved methodologies, including new molecular techniques, have the potential to close important gaps in knowledge. As a result, it may be possible for the first time to develop robust, empirically based models that capture the main demographic and behavioral processes underlying the rich variety of avian migration patterns. Equally, the development of such models can provide a strategic framework for data collection and ensure that scarce resources are used to maximum benefit.

Implementation of the program outlined here will provide an essential empirical and theoretical basis for the further development of models that use individual-based optimization criteria to predict the consequences of habitat change for migratory bird populations (e.g., Sutherland and Dolman 1994; Goss-Custard et al. 1995). Arguably the successful extension of the individual-based approach to species with widely dispersed wintering populations (e.g., Pettifor et al. 2000) will depend on a thorough understanding of factors regulating migratory patterns. At the same time, progress in modeling migration patterns will rely on continued progress toward an understanding of factors limiting populations of migratory birds (Sherry et al., Chap. 31, this volume), hopefully leading to quantitative estimates of density dependence (e.g., Baillie et al. 2000).

The study of migration patterns is in transition from being a largely speculative field, of interest mainly to academics and theoreticians, to one that makes a major contribution to the understanding of migratory (i.e., essentially all [Bell 2000]) bird populations. Provision of adequate resources is essential for this contribution to be realized, but the prospect of creating a practical toolkit for the conservation and management of migratory birds is one that conservationists can ill afford to ignore.

LITERATURE CITED

Alerstam, T., and G. Högstedt. 1980. Spring predictability and the evolution of leap-frog migration. Ornis Scandinavica 11:196–200.

Alerstam, T., and G. Högstedt. 1985. Leap-frog arguments: reply to Pienkowski, Evans and Townshend. Ornis Scandinavica 16:71–74

Arnold, T. W. 1991. Geographic variation in sex ratios of wintering American Kestrels *Falco sparverius*. Ornis Scandinavica 22:20–26.

Baillie, S. R., W. J. Sutherland, S. N. Freeman, R. D. Gregory, and E. Paradis. 2000. Consequences of large-scale processes for the conservation of bird populations. Journal of Applied Ecology 37 (Supplement 1):88–102.

Baker, R. R. 1978. The Evolutionary Ecology of Animal Migration. Hodder and Stoughton, London.

Bell, C. P. 1996. Seasonality and time allocation as causes of leapfrog migration in the Yellow Wagtail, *Motacilla flava*. Journal of Avian Biology 27:334–342.

Bell, C. P. 1997. Leap-frog migration in the Fox Sparrow: minimizing the cost of spring migration. Condor 99:470–477.

Bell, C. P. 2000. Process in the evolution of bird migration and pattern in avian ecogeography. Journal of Avian Biology 31:258–265.

Belthoff, J. R., and S. A. Gauthreaux Jr. 1991. Partial migration and differential winter distribution of house finches in the eastern United States. Condor 93:374–382.

Bryson, R. A., and F. K. Hare. 1974. Climates of North America. World Survey of Climatology, Vol. 1. Elsevier, Amsterdam.

Butler, R. W., N. C. Davidson, and R. I. G. Morrison. 2001. Global scale shorebird distribution in relation to productivity of near-shore ocean waters. Waterbirds 24:224–232.

Cox, G. W. 1968. The role of competition in the evolution of migration. Evolution 22:180–192.

Cox, G. W. 1985. The evolution of avian migration systems between temperate and tropical regions of the New World. American Naturalist 126:451–474.

Cristol, D. A., and D. C. Evers. 1992. Dominance status and latitude are unrelated in wintering Dark-eyed Juncos. Condor 94:539–542.

Cristol, D. A., V. Nolan Jr., and E. D. Ketterson. 1990. Effect of prior residence of dominance status of Dark-eyed Juncos, *Junco hyemalis*. Animal Behaviour 40:580–586.

Cristol, D. A., M. B. Baker, and C. Carbone. 1999. Differential migration revisited Latitudinal Segregation by Age and Sex Class. Current Ornithology 15:33–88.

Ens, B. J., T. Piersma, and J. M. Tinbergen. 1994. Towards predictive models of bird migration schedules: theoretical and empirical bottlenecks. NIOZ-Rapport 1994–5. Netherlands Institute for Sea Research, Texel.

Evans, P. R., and M. W. Pienkowski. 1984. Population dynamics of shorebirds. Pages 83–123 *in* Shorebirds: Breeding Biology and Populations (J. Burger and B. M. Olla, eds.). Behaviour of Marine Animals, Vol. 5. Plenum Press, New York.

Fretwell, S. 1980. Evolution of Migration in Relation to Factors Regulating Bird Numbers. Pages 517–527 *in* Migrant Birds in the Neotropics: Ecology, Behavior, Distribution, and Conservation (A. Keast and E. S. Morton, eds.). Smithsonian Institution Press, Washington, D.C.

Goss-Custard, J. D., R. T. Clarke, S. E. A. le V. dit Durell, R. W. G. Caldow, and B. J. Ens. 1995. Population consequences of winter habitat loss in a migratory shorebird: Pt. 2. Model predictions. Journal of Applied Ecology 32:337–351.

Greenberg, R. 1980. Demographic aspects of long-distance migration. Pages 493–504 *in* Migrant Birds in the Neotropics: Ecology, Behavior, Distribution, and Conservation (A. Keast and E. S. Morton, eds.). Smithsonian Institution Press, Washington, D.C.

Haussmann, M. F., and C. M. Vleck. 2002. Telomere length provides a new technique for aging animals. Oecologia 130:325–328

Hestbeck, J. B., J. D. Nichols, and J. E. Hines. 1992. The relationship between annual survival rate and migration distance in mallards: an examination of the time-allocation hypothesis for the evolution of migration. Canadian Journal of Zoology 70:2021–2027.

Holberton, R. L. 1993. An endogenous basis for differential migration in the Dark-eyed Junco. Condor 95:580–587.

Holmgren, N., and S. Lundberg. 1993. Despotic behaviour and the evolution of migration patterns in birds. Ornis Scandinavica 24:103–109.

Ketterson, E. D., and V. Nolan Jr. 1976. Geographic variation and its climatic correlates in the sex ratio of eastern-wintering Dark-eyed Juncos (*Junco hyemalis hyemalis*). Ecology 57:679–693.

Ketterson, E. D., and V. Nolan Jr. 1982. The role of migration and winter mortality in the life history of a temperate-zone migrant, the Dark-eyed Junco, as determined from demographic analyses of winter populations. Auk 99:243–259.

Ketterson, E. D., and V. Nolan Jr. 1983. The evolution of differential bird migration. Current Ornithology 1:357–402.

Ketterson, E. D., and V. Nolan Jr. 1985. Intraspecific variation in avian migration: evolutionary and regulatory aspects. Pages 553–579 *in* Migration: Mechanisms and Adaptive Significance (M. A. Rankin, ed.). Contributions in Marine Science, Vol. 27 (Supplement). University of Texas Marine Sciences Institute, Port Aransas, Tex.

Kokko, H. 1999. Competition for early arrival in migratory birds. Journal of Animal Ecology 68:940–950.

Lack, D. 1944a. The problem of partial migration. British Birds 37:122–130, 143–150.

Lack, D. 1944b. Ecological aspects of species-formation in passerine birds. Ibis 86:260–286.

Lack, D. 1954. The Natural Regulation of Animal Numbers. Clarendon Press, Oxford.

Lack, D. 1968. Bird migration and natural selection. Oikos 19:1–9.

Levey, D. J., and F. G. Stiles. 1992. Evolutionary precursors of long-distance migration: resource availability and movement patterns in neotropical landbirds. American Naturalist 140:447–476

Lundberg, S., and T. Alerstam. 1986. Bird migration patterns: conditions for stable geographical population segregation. Journal of Theoretical Biology 123:403–414.

Marra, P. P. 2000. The role of behavioural dominance in structuring patterns of habitat occupancy in a migrant bird during the non-breeding season. Behavioural Ecology 11:299–308.

Marra, P. P., and R. T. Holmes. 2001. Consequences of dominance-mediated habitat segregation in American Redstarts during the nonbreeding season. Auk 118:92–104.

Marra, P. P., K. A. Hobson, and R. T. Holmes. 1998. Linking winter and summer events in a migratory bird by using stable-carbon isotopes. Science 282:1884–1886.

Mayr, E., and W. Meise. 1930. Theoretisches zur Geschichte des Vogelzuges. Der Vogelzug 1:149–172.

Myers, J. P. 1981. A test of three hypotheses for latitudinal segregation in the sexes of wintering birds. Canadian Journal of Zoology 59:1527–1534.

Myers, J. P., J. L. Maron, and M. Sallaberry. 1985. Going to extremes: why do sanderlings migrate to the Neotropics? Ornithological Monographs 36:520–535.

Nichols, J. D. 1996. Sources of variation in migratory movements of animal population: statistical inference and a selective review of empirical results for birds. Pages 147–197 *in* Population Dynamics in Ecological Space and Time (O. E. Rhodes Jr., R. K. Chesser, and M. H. Smith, eds.). University of Chicago Press, Chicago.

Nolan, V., Jr., and E. D. Ketterson. 1990. Timing of autumn migration and its relation to winter distribution in Dark-eyed Juncos. Ecology 71:1267–1278.

Owen, M., and J. M. Black. 1991. The importance of migration mortality in non-passerine birds. Pages 361–372 *in* Bird Population Studies: Relevance to Conservation and Management (C. M. Perrins, J.-D. Lebreton, and G. J. M. Hirons, eds.). Oxford University Press, Oxford.

Pettifor, R. A., R. W. G. Caldow, J. M. Rowcliffe, J. D. Goss-Custard, J. M. Black, K. H. Hodder, A. I. Houston, A. Lang, and J. Webb. 2000. Spatially explicit, individual-based, behavioural models of the annual cycle of two migratory goose populations. Journal of Applied Ecology 37 (Supplement 1):103–135.

Pienkowski, M. W., and P. R. Evans. 1985. The role of migration in the population dynamics of birds. Pages 331–352 *in* Behavioural Ecology: Ecological Consequences of Adaptive Behaviour (R. M. Sibly and R. H. Smith, eds.). 25th Symposium of the British Ecological Society, Reading, 1985. Blackwell Scientific, Oxford.

Pienkowski, M. W., P. R. Evans, and D. J. Townshend. 1985. Leap-frog and other migration patterns of waders: a critique of the Alerstam and Högstedt hypothesis and some alternatives. Ornis Scandinavica 16:61–70.

Piersma, T. 1997. Do global patterns of habitat use and migration strategies co-evolve with relative investments in immunocompetence due to spatial variation in parasite pressure? Oikos 80:623–631.

Piersma, T., and N. C. Davidson. 1992. The migrations and annual cycles of five subspecies of Knots in perspective. Wader Study Group Bulletin 64 (Supplement): 187–197.

Rabenold, K. N., and P. P. Rabenold. 1985. Variation in altitudinal migration and site tenacity in two subspecies of Dark-eyed Junco in the southern Appalachians. Auk 102:805–819.

Ricklefs, R. E. 1980. Geographical variation in clutch size among passerine birds: Ashmole's hypothesis. Auk 97:38–49.

Rogers, C. M., T. L. Thiemer, V. Nolan Jr., and E. D. Ketterson. 1989. Does dominance determine how far Dark-eyed Juncos, *Junco hyemalis,* migrate into their winter range? Animal Behaviour 37:498–506.

Salomonsen, F. 1955. The evolutionary significance of bird-migration. Det Kongelige Danske Videnskabernes Selskab Biologiske Meddelelser 22:1–62.

Sandercock, B. K., and A. Jaramillo. 2002. Annual survival rates of wintering sparrows: assessing demographic consequences of migration. Auk 119:149–165.

Sillett, T. S., and R. T. Holmes. 2002. Variation in survivorship of a migratory songbird throughout its annual cycle. Journal of Animal Ecology 71:296–308.

Sutherland, W. J., and P. M. Dolman. 1994. Combining behaviour and population dynamics with applications for predicting the consequences of habitat loss. Proceedings of the Royal Society of London, Series B, Biological Sciences 255:133–138.

Terrill, S. B. 1987. Social dominance and migratory restlessness in the Dark-eyed Junco (*Junco hyemalis*). Behavioral Ecology and Sociobiology 21:1–11.

Terrill, S. B., and P. Berthold. 1989. Experimental evidence for endogenously programmed differential migration in the Blackcap (*Sylvia atricapilla*). Experientia 45:207–209.

von Haartman, L. 1968. The evolution of resident versus migratory habit in birds: some considerations. Ornis Fennica 45:1–7.

Weber, T. P., and A. I. Houston. 1997. A general model for time-minimising avian migration. Journal of Theoretical Biology 185: 447–458.

Webster, M. S., P. P. Marra, S. M. Haig, S. Bensch, and R. T. Holmes. 2002. Links between worlds: unravelling migratory connectivity. Trends in Ecology and Evolution 17:76–83.

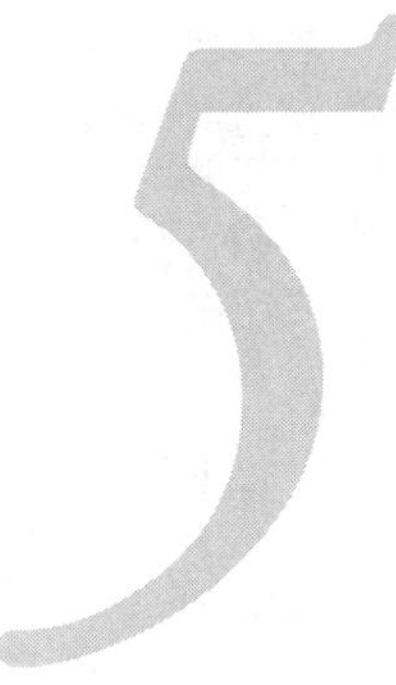

PHILIP A. R. HOCKEY

Predicting Migratory Behavior in Landbirds

THIS CHAPTER DEVELOPS A HOLISTIC OVERVIEW of environmental conditions and avian ecological attributes that influence migration. Environmentally, these parameters range from temperature and rainfall on a macro-scale to within-habitat thermal buffering on a micro-scale. Ecological attributes include diet and foraging mode. At all latitudes, mean midwinter temperature is a powerful predictor of the proportion of breeding species that will be migratory: once this exceeds 20°C, 90% or more of species that breed at a particular locality are likely to be resident. Migration away from high latitudes in winter is primarily a response to food availability rather than to thermal stress. Climate alone, however, cannot predict which species will be migratory. At the micro-scale, there is evidence that thermal buffering, most pronounced in structurally complex and well-stratified habitats, further influences the probability of migration. However, no study has explicitly disentangled the effects of complexity and buffering. Habitat occupancy by migrants on their nonbreeding grounds differs across flyways: although it may be possible to develop paradigms linking habitat features and migration on the regional scale, there is no evidence for a global paradigm. Strong links exist among migration, diet, and foraging mode, but not all of these links exist across all flyways. Despite large differences in habitat availability in the Afrotropics and Asia, migrants on both flyways are typically insectivorous, nomads are granivorous, and residents are frugivorous or predators of terrestrial vertebrates. In the Neotropics, where the receiving environment is most analogous to Asia, frugivory characterizes a greater proportion of migrants. Prey capture techniques are linked to diet, and migratory tendency decreases along a continuum from aerial foraging to ground gleaning. Overall, the majority of factors that influence migration (e.g., temperature and food availability) have probably been identified. However, in many cases there is a degree of autocorrelation among these variables. Only through rigorous dissection of these autocorrelations will our predictive capacity be enhanced. The search for global, cross-flyway ecological paradigms, however, may remain elusive until analyses are controlled for phylogeny.

INTRODUCTION

Several factors might force or favor the evolution of migration. At the organismal level, it could be driven by demographics or population size, with competition (density dependence) driving at least partial migration (e.g., Cox 1985); this has been proposed as a likely trigger for the evolution of migration in relatively stable, tropical regions (Rappole 1995). Migration could be driven environmentally by the fitness advantages of tracking predictable spatio-temporal variations in resource availability, especially food. Should such cycles of resource availability be predictable, but their amplitude vary, this may lead to facultative migration, whereby the proportion of a population that migrates varies from year to year. A combination of spatial and temporal unpredictability in resource levels, such as occurs in arid regions in response to rainfall, is more likely to promote nomadism (e.g., Dean 1997, 2004).

Two broad fields of theory seeking an explanation of the origins of migratory behavior have emphasized the roles of density-dependent demographics (Cox 1985; Rappole 1995; Bell 2000) and inheritance from pre-avian ancestors (Berthold 1999a). Both schools of thought promote an intrinsic, as distinct from an environmental, driving mechanism and, respectively, do and do not require a tropical origin of the behavior.

Experimental evidence suggests that migratory behavior at the population, rather than individual, level is genetically labile, at least among some higher taxa, and that marked changes in migration patterns can (and do) evolve within a few generations (Berthold et al. 1992; Berthold 1999b). Intuitively, however, it may be "easier" to suppress migration (and then re-evolve it) than to evolve migration de novo. An analysis of present-day migration patterns led Levey and Stiles (1992) to propose that migration is a response to a combination of habitat and diet. This hypothesis was retested and accepted, with a strong emphasis on habitat, by Chesser and Levey (1998). However, there is no a priori reason to assume that the ultimate factor driving the evolution of migration in a particular bird species is necessarily the proximate influence that maintains it (e.g., Zink 2002). If, for example, a tropical resident evolved migration in response to a density-dependent reduction in reproductive fitness (e.g., Rappole 1995), the behavior could be subsequently maintained by seasonal variations in resource availability acting selectively on survivorship (e.g., Greenberg 1980). The processes of the present are thus not necessarily a window on the pressures of the past, although some migration patterns appear to retain historical components that could be construed as suboptimal (e.g., Williams and Webb 1996; Sutherland 1998). It is important, therefore, when examining correlates of modern migratory behavior, to appreciate that the patterns of today could mostly represent behavioral responses to current ecological conditions (assuming that the behavior is adaptive). They may be long divorced from their evolutionary origins, the two being coupled only in as much as the transition from one to another must have been possible.

Attempts to explain present-day migration patterns date back more than 100 years (e.g., Gätke 1895). Close correlations have been proposed with climate (Newton and Dale 1996a, 1996b; Chesser 1998), habitat (MacArthur 1959; Chesser and Levey 1998), diet (Levey and Stiles 1992; Hockey 2000), and foraging mode (Hockey 2000). The need to explain the driving forces behind migration patterns has intensified in recent years: there are clear indications, especially from boreal latitudes, that migration patterns are already changing in response to climate change (reviewed by Burton 1995).

This chapter attempts to formulate a generalized description of the environmental-ecological conditions to which migration is, or is not, a likely response—a need highlighted by Chesser and Levey (1998). The study tries to develop a predictive framework for identifying the links among environment, foraging ecology, and migration. The approach is necessarily probabilistic and only semiquantitative. I have adopted a hierarchical approach, starting with macro-scale environmental correlates of migration behavior and progressing through considerations of habitat and microenvironment to diet and foraging behavior.

MACRO-SCALE CORRELATES OF MIGRATORY BEHAVIOR

Several authors have undertaken analyses of migratory behavior on large geographical scales, using as the dependent variable either the proportion of breeding taxa that are migratory (e.g., Newton and Dale 1996a, 1996b; Chesser 1998) or the proportion that are resident (Hockey 2000). In terms of predicting the likelihood or otherwise of migratory behavior, the proportion of breeding species within a community that is resident may be the most appropriate parameter, assuming residency to be the ancestral condition and migration to be derived (Rappole 1995). These analyses were stimulated by MacArthur's (1959) study of long-distance Nearctic-Neotropical migrants (short-distance migrants and nomads were treated as residents). In broad terms he found that (a) the proportion of migrant taxa within breeding assemblages in the Nearctic decreased from north to south, but (b) this pattern was strongly modified by habitat. For example, migrants accounted for a greater proportion of species breeding in deciduous forests of the east relative to the damp, western coniferous forests. He interpreted this pattern as reflecting the relative amplitude of seasonal variation in food resources (being much greater in the east) and proposed that the habitat-linked and latitude-linked patterns he observed were largely decoupled. This interpretation was challenged by Willson (1976), who hypothesized (but did not test) that climate should ultimately shape migratory patterns and that, as a necessary corollary, this response should be detectable within habitats across climatic gradients. This was tested by Herrera (1978), who found that (a) the average temperature of the coldest month (T_{CM}) was a robust predictor of the proportion of migrant birds in a community, with the proportion of migrants increasing as T_{CM} decreased; and (b) this pattern was repeated across habitat types. Subsequently, Newton and Dale (1996a, 1996b) reconfirmed this finding along latitudinal gradients in both the western Palearctic and the eastern Nearctic. Chesser (1998) examined the incidence of migration among Neotropical tyrannid flycatchers, and found that latitude explained a high proportion of the variance ($r^2 = 0.82$): this pattern was repeated across all species and within habitats. Latitude itself is not a "variable" to which birds are likely to respond, being no more than a surrogate measure for ecological factors that might influence migration (or other life history parameters [e.g., Martin 2000]). Migratory tendency among tyrannids was linked with several temperature-related variables, including mean temperatures of both the hottest (T_{HM}) and coldest (T_{CM}) months, and the relative temperature range [$(T_{HM} - T_{CM})/T_{HM}$)]. Temperature-linked variables overrode the effects of others, including habitat complexity (cf. MacArthur 1959), and led Chesser (1998) to conclude that the relationship between temperature variables and migratory tendency was "globally significant." In support of this conclusion, Newton and Dale's (1996a, 1996b) studies indicated that a greater proportion of species was migratory in the eastern Nearctic than at equivalent latitudes in the western Palearctic, winter temperatures being lower in the former.

The only gradient study that has spanned equatorial latitudes is that of Hockey (2000), who, incorporating Newton and Dale's (1996a) data, compared the proportion of breeding species that were resident (P_{BR}) with T_{CM} at latitudes along the Palearctic-Afrotropical flyway between 80° N and 35° S. The relationship was linear, with T_{CM} explaining most of the variance in P_{BR} ($r^2 = 0.96$, $p << 0.0001$). Once T_{CM} exceeded about 20°C, 90% or more of species breeding at a particular site were likely to be resident. This analysis excluded all predominantly coastal and marine species to obviate bias between localities with and without a coastline. The analysis also explained a greater proportion of the variance in migratory propensity than any previous study, probably because of its wider latitudinal range. Comparing equivalent T_{CM} north and south of the equator, there was a slight tendency for southern communities to have a smaller proportion of migratory taxa (an interesting parallel to the more "tropical" life histories of Southern Hemisphere birds [Martin 2000]). However, the overall relationship between P_{BR} and latitude was highly symmetrical about the equator.

Together, the above studies clearly demonstrate a strong link between low winter temperatures and migration: however, it has been questioned whether this effect is direct or indirect. Perhaps not surprisingly, ecologists (e.g., MacArthur 1959; Herrera 1978; Newton and Dale 1996a, 1996b) have interpreted temperature-linked responses as indirect effects, in particular to food availability, whereas physiologists have favored stress as an explanation. Root (1988a, 1988b) argued that the temperate limits to species' nonbreeding ranges were constrained by their physiological tolerance of low temperatures, a view challenged by Repasky (1991) on the basis that birds spanning the full range of body sizes are found in the coldest places. While acknowledging that food availability and physiological stress are unlikely to be entirely decoupled, Chesser (1998) correctly pointed out that Root's (1988a, 1988b) findings could equally well be explained by food limitation. He presented three lines of evidence in support of this, two correlative and one quasi-experimental. At the correlative level, species' thermoregulatory patterns are adapted to climate (Weathers 1979) and winter ranges are influenced by diet (Newton 1995). More convincingly, winter ranges of several North American species have spread northward in response to supplementary food at bird feeders (Tramer 1974). Given that the prey/food items of birds are almost invariably smaller than the birds themselves, it could be predicted for species eating animal prey that the physiological stresses associated with low temperatures would affect the behavior and availability of the prey before that of the predator. For species eating seeds and fruit, this is unlikely to be the case.

It is possible that the range limits of some migratory taxa are indeed set by thermal stress: there is, for example, a tendency for small migratory species to be concentrated at tropical latitudes in the nonbreeding season (J. V. Hamblin and P. A. R. Hockey, unpubl. data). However, many taxa are migratory within the Tropics, where temperature alone is highly unlikely to constrain their distribution (e.g., Hockey 2000). At the global level, therefore, the link between temperature and migration cannot be explained directly by temperature stress.

Although apparently universally applicable, the recognition (at least for landbirds) of a "temperature-migration paradigm" does little more than provide a backdrop to migratory behavior. It tells us that a greater proportion of species will be migratory in environments that experience intense seasonal cold than in those that are warm throughout the year. However, it provides no insight into *which* species will be migratory.

HABITAT AND THE MICROENVIRONMENT

If low temperatures either directly or indirectly promote migratory behavior, then it can be predicted as a corollary that environmental features that buffer the effects of temperature extremes themselves influence migration. For example, the litter layer of a Mediterranean shrubland is likely to be less buffered from temperature extremes than the litter layer of a closed-canopy forest.

The concept of buffering is not new (MacArthur 1959; Fager 1968) and its influence on migratory behavior at the species level was neatly illustrated by a comparison of the woodpecker assemblages of Minnesota, Maryland, and Guatemala (Askins 1983). The T_{CM} of these sites varied from –11°C (Minnesota) to 21°C (Guatemala). Woodpeckers eat mostly wood-boring insects, obtaining their food from a "buffered" microenvironment below the bark of trees: they also roost at night in tree cavities, themselves 5°–6°C above ambient temperature (Askins 1981). In the coldest site (Minnesota), only 12 woodland bird species were resident, and four of these were woodpeckers. Of these four species of temperate woodpeckers, only female Downy Woodpeckers (*Dendrocopos villosus*) regularly migrate south in winter. Downy Woodpecker is the smallest of the four temperate woodpeckers and is one of the least reliant on excavation to obtain its food. However, females are at least as heavy as males (Short 1982), suggesting that foraging constraints rather than physiological ones drive the former to migrate.

Despite the intuitive importance quantifying thermal buffering in microenvironments, I am unaware of any such study within even a single habitat in one climatic zone, even though some studies rely on such buffering as an explanation for observed patterns (e.g., Askins 1983; Chesser and Levey 1998). In general, increased buffering is associated with increasing habitat complexity. However, without data that link complexity to thermal buffering, separating the direct influences of habitat complexity (e.g., influence on diversity of food resources) from influences of the thermal consequences of this complexity (e.g., influence on seasonal abundance of different food resources) is problematic.

At the regional level, the proportion of Neotropical passerines that is migratory increases, although not monotonically, along a gradient of decreasing habitat complexity from forest and woodland to scrub and open country (Chesser 1994). Similarly, if birds of the southern Afrotropics are compared across a simple, four-stage gradient of increasing habitat complexity (from grassland and semidesert habitats at one extreme to forests at the other), the same pattern emerges strongly (fig. 5.1). Globally, patterns of migration among the Cuculidae (cuckoos and coucals) are also comparable, with taxa of primary forest far less likely to be migratory than taxa occurring in open woodland and scrub (Hockey 2000). In an analysis of 12 Neotropical families and subfamilies, Chesser and Levey (1998) found a close correlation (the only exception being the antbirds Thamnophilidae) between occupancy of "unbuffered" habitats (forest edge and canopy, secondary and open habitats) and migration, and between occupancy of buffered forest interiors and residency. The same pattern is reported from New Guinea (Beehler et al. 1986). This pattern traditionally has been interpreted as a consequence of competition, with residents of the "stable" forest interior being competitively dominant and forcing nonbreeding migrants to occupy habitats characterized by ephemeral food resources. This implies that during their nonbreeding residency period, migrants, being competitive subdominants (e.g., Willis 1966; Leck 1972), are at a fitness disadvantage relative to residents. Levey and Stiles (1992) and Chesser and Levey (1998) propose an inverse explanation whereby the habitat occupancy of migrants is not so much a consequence of their migratory behavior as a cause of it. If this is the case, there is no need to invoke a fitness differential between residents and migrants to explain differences in habitat use. Indeed, it has been proposed more than once that competition among migrants is greater than competition between residents and migrants (e.g., Lack 1976; Willis 1980).

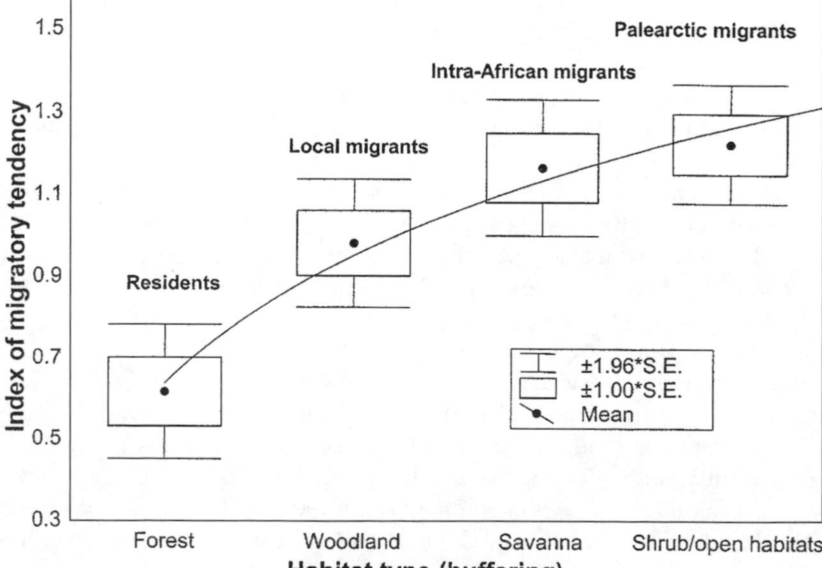

Fig. 5.1. Relationship between migratory tendency and habitat occupancy derived from 742 landbird species of the southern Afrotropics. Labels within the figure refer to the dominant type of migrant within each habitat. The index of migratory tendency is the average for all species (ranked from 0 = sedentary to 4 = intercontinental migrant) occurring within each habitat. (M. S. L. Mills and P. A. R. Hockey, unpubl. data.)

The buffered/unbuffered habitat dichotomy between residents and migrants, respectively, is not necessarily evident within specific habitats. A comparison of resident and migrant tropical forest birds in peninsular Malaysia (331 species, 15% migratory) found that a slightly smaller proportion of migrants than residents foraged in the lower strata of the forest. The reverse was true for the upper canopy and emergent strata (G. Balme and P. A. R. Hockey, unpubl. data) (fig. 5.2). These differences were not significant, however, implying that residents do not occupy more buffered forest microhabitats than migrants. This casts further doubt on the importance of competition in determining habitat use and accords with Wells' (1990) observation from the Sunda region that migrants were not precluded by residents from occupying primary forest habitats, including the forest floor. Indeed, one of the most abundant forest floor birds in Malaysia, and one that displays both territoriality and philopatry, is the migratory Siberian Blue Robin (*Luscinia cyanae* [Wells 1990]). Other strongly forest-dependent migrants that forage in the lower strata of Malaysian forests include Slaty-legged Crake (*Rallina euryzonoides*), Hooded and Blue-winged Pittas (*Pitta sordida / moluccensis*), Asian Paradise-Flycatcher (*Terpsiphone paradisi*), Brown-chested Jungle-Flycatcher (*Phinomyias brunneata*), Blue-throated Flycatcher (*Cyornis rubeculoides*), and Pale-legged Leaf-Warbler (*Phylloscopus tenellipes*). For forest-dwelling birds, therefore, it is not possible to distinguish migrants from residents in tropical Asia on the basis of microhabitat occupancy, although this might change if analyses were controlled for phylogeny. Despite the lack of difference between habitat occupancy by migrants and residents in Malaysian forests, there were considerable differences between them in diet and foraging mode (fig. 5.2). Residents primarily gleaned nonvolant invertebrates and fruit, whereas migrants ate mostly volant insects, caught primarily by perch-and-sally hunting. Diet is certainly one axis along which competition can be either reduced or obviated. Similar analyses of spatial and dietary differentiation (or lack of) between migrants and residents in a range of habitats within a region would help to shed some light on whether competition does play a significant role in determining nonbreeding habitat choice by migrants. Evidence to date, however, suggests that such competition, if it exists, is generally weak (e.g., Leck 1972; Lack 1976).

Only if habitat use is not a response to competition does habitat occupancy becomes a potentially useful predictor of migratory propensity. The problem is how to describe habitats in a way that has predictive power. This problem is exacerbated by whether it is the breeding or nonbreeding habitat (or both) that should be considered, because super-

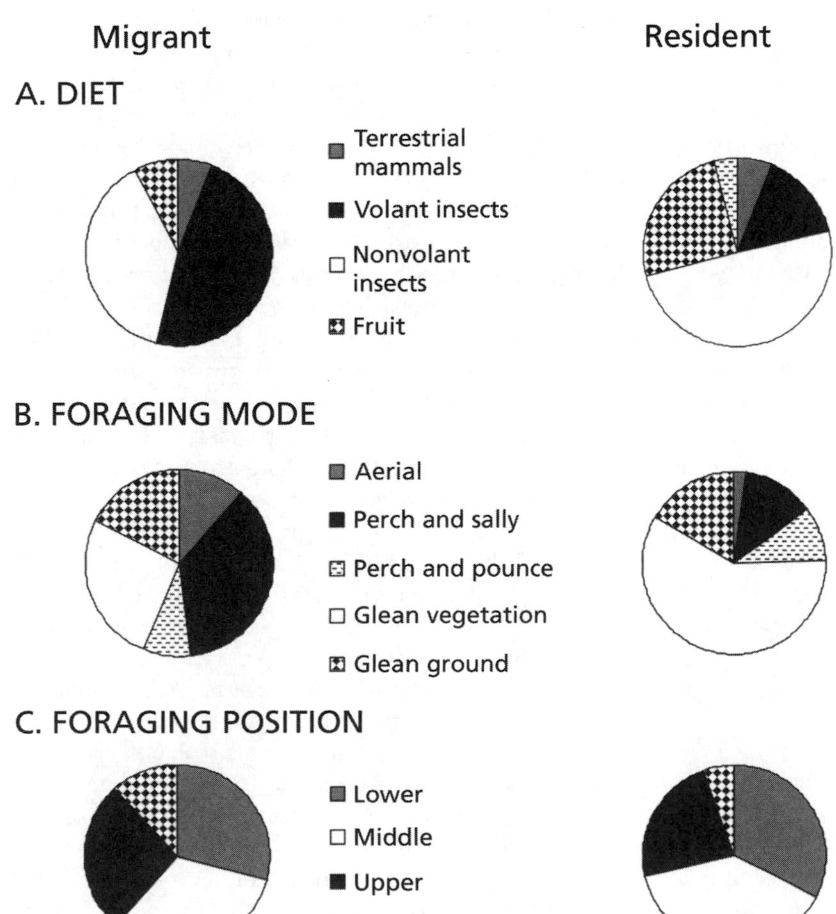

Fig. 5.2. Comparison of (A) diets, (B) foraging modes, and (C) foraging strata of migrant and resident landbirds in the forests of peninsular Malaysia. (G. Balme and P. A. R. Hockey, unpubl. data.)

imposed on the one-dimensional axis of habitat choice is a second axis of habitat availability. For example, most Austral-Neotropical passerine migrants breed in open and scrub habitats, whereas the majority of species migrating south to the Neotropics from breeding grounds in the Nearctic breed in woodland and forest (Chesser 1994). In this case, the effect of habitat availability on habitat occupancy appears to be of overriding importance (although habitat availability was not quantified in this study).

A comparison of habitat availability with habitat occupancy by migratory landbirds across the three major flyways (Nearctic-Neotropical, Western Palearctic–Afrotropical, and Eastern Palearctic–Oriental (excluding Australia) suggests that evolutionary history may have had a strong influence on present-day habitat use. Classifying habitats as either well-buffered (forest, closed woodlands), moderately buffered (savanna, shrublands), or unbuffered (open habitats, e.g., grassland, deserts, and semideserts), the two most similar flyways in terms of relative habitat availability on both breeding and nonbreeding grounds are those of the Nearctic and eastern Palearctic. Patterns of habitat occupancy (proportions of migrant taxa occupying different habitats) are, however, very different. In the New World, moderately buffered habitats are favored on the breeding grounds and habitat occupancy on the Neotropical nonbreeding grounds very closely reflects habitat availability. Species breeding in the eastern Palearctic also favor woodlands, but on the nonbreeding grounds these species avoid dense forests, strongly favoring unbuffered habitats. In the western Palearctic, breeders also favor woodlands and avoid forests on the Afrotropical nonbreeding grounds. Contrary to habitat-based expectations, therefore, patterns of nonbreeding habitat occupancy in Southeast Asia are closer to those of the Afrotropics than the Neotropics (J. V. Hamblin and P. A. R. Hockey, unpubl. data). Interestingly, of those Palearctic-breeding taxa migrating to Southeast Asia, those that also migrate to Africa show the strongest avoidance of well-buffered habitats in Asia (J. V. Hamblin and P. A. R. Hockey, unpubl. data).

The avoidance of well-buffered habitats by Palearctic-breeding birds on both the breeding and nonbreeding grounds provides indirect support for MacArthur's (1959) hypothesis of migration being a response to the amplitude of seasonal resource fluctuations—a hypothesis developed using data from the Nearctic. Indeed, of the six possible combinations of flyway and hemisphere, the "odd man out" seems to be the (random) habitat choice of migrants in the Neotropics; in all other instances, habitat choice is strongly nonrandom. In Asia, the avoidance of forests by migrants could be interpreted as resulting from competition, yet those migrants that do occupy forests do not differ from residents in microhabitat selection, although they do differ in diet, targeting volant insects (which may be the most seasonally ephemeral of the forests' food resources).

Within the Afrotropics, species with different types of migrations favor different habitats. Palearctic-breeding and partial migrants favor open habitats, intra-African migrants typify savannas, and residents characterize woodlands and forests (fig. 5.1). Bell (2000) linked the use of open habitats by trans-Saharan migrants to extensive discontinuities in the availability of forested stopover sites between the Palearctic and the Afrotropics. Along the Nearctic-Neotropical and Eastern Palearctic–Oriental flyways, forests are present "at all latitudes," and Bell (2000) considers these migrations to be "analogues" of one another. However, the analogy is inappropriate given the avoidance of forests by (long-distance) migrants in Southeast Asia.

If habitat structure / buffering is important in influencing migration, its impacts are likely to be manifest in one of two processes. Firstly, and perhaps rarely, is a purely physiological question of when or whether habitat structure can moderate the thermal environment to allow residency in species that would otherwise be forced to migrate. Secondly is how habitat buffering affects seasonal extremes in food availability and thus induces a fitness playoff between residency and migration. The latter is almost certainly the more important of the two. Negative relationships between habitat buffering and resource seasonality have been shown for flowers, fruits, and insects (reviewed by Levey and Stiles 1992).

The interplay between climate extremes and habitat thermal buffering is only one way in which climate and habitat structure might interact to influence migration. The ways in which climate drives migration differ between high and low latitudes. At high latitudes, temperature is undoubtedly the key. At low latitudes, at least in the Afrotropics, movements are mediated ultimately by rainfall independently of temperature (Hockey 2000). Even in the Tropics, rainfall seasonality can have a very strong influence on resource availability (e.g., Janzen 1973). There is a mass movement of intra-African migrants from nonbreeding equatorial latitudes north to the Sahelian savannas during the boreal wet season (May to September) and south to the southern savannas during the austral wet season (October to March). In many cases the same species (but different populations) undertakes migrations in both directions (Hockey 2000).

The link between habitat structure and migratory behavior is far less clear than the link between temperature or rainfall and migration. Chesser and Levey (1998) argued a strong case for the overriding importance of habitat in influencing migration patterns of Neotropical birds, yet the Neotropics is the one area where habitat selection (by migrants from the Nearctic) appears to be random with respect to habitat availability (J. V. Hamblin and P. A. R. Hockey, unpubl. data). If a global, habitat-linked paradigm exists, equivalent to the "temperature-migration paradigm," research to date has not identified it. The best correlate we can hope for with habitat structure as a single explanatory variable is the proportion of migratory species in an assemblage; it is very unlikely that habitat parameters alone will allow us to predict which species will be migratory. This is supported by the fact that Chesser and Levey's (1998) "model for migrants," developed in the Neotropics (viz. frugivores and nectarivores of dry and unbuffered habitats), would rapidly be falsified if it were used as the basis for predicting migrant taxa in the Afrotropics or the Orient (Hockey 2000). Both areas conform to Chesser and Levey's (1998) habitat predictions (indeed, selection for unbuffered habitats is stronger away from the Neotropics) but not (and not anywhere close) to dietary predictions. Phylogenetically rigorous, worldwide comparisons of habitat use versus migratory tendency are needed to resolve the issue of whether habitat structure is useful in predicting *which* taxa will be migratory.

FOOD AVAILABILITY AND DIET

The lack of convincing evidence for thermal stress limiting the poleward distribution of birds in the nonbreeding season begs the conclusion that migrations are driven proximally by variations in food supply, as proposed by MacArthur (1959). This is hardly a novel idea, but surprisingly few studies have addressed the significance of how diet might affect migratory behavior.

Diverse attributes of a species' prey base are likely to influence its absolute and seasonal availability to birds. For example, the tropical forests of South America and Asia are rich in fruiting species of the Lauraceae and Palmae families, whereas these are poorly represented in Africa (Snow 1980). Theoretically, a low diversity of fruiting trees could constrain the diversity of both resident and migrant frugivores in African forests and could also promote migration among African frugivores (relative to those of other tropical regions) by virtue of a reduced aggregate fruiting season. African forests are indeed species-poor relative to their Neotropical and Asian counterparts, but they are typified by a lack of migratory frugivores (Hockey 2000). This pattern may be a legacy of dramatic contraction of the Congo Basin Forest ca. 65,000 years ago (Lövei 1989), leading to an extinction of forest specialists and persistence of opportunists able to switch diet seasonally and thus obviate the need to migrate (Hockey 2000).

Granivory is widespread among birds, especially small birds, many of which are flock foragers. In tropical and subtropical areas, abundant seed crops frequently are localized responses to episodic and unpredictable rain events, but the crops themselves may persist for extended periods, especially in the absence of further rain to stimulate germination. In the southern and eastern Afrotropics, there is little evidence of migration among granivores, but many species are nomadic (Lack 1986; Dean 1997; Hockey 2000). Nomadic granivores characterize arid and semiarid habitats worldwide (Dean 2004).

Invertebrates make up most or all of the diet of the majority of migratory landbirds. Some, such as small, volant insects with short generation times, undergo rapid fluctuations in population size in response to temperature changes. For example, the dense swarms of midges that plague Arctic researchers during the boreal summer disappear in winter, along with any predators dependent on this resource. Invertebrates buried in sand, mud, or wood, by comparison, do not show such seasonal extremes in abundance.

From an analysis of 73 southern Afrotropical bird families, Hockey (2000) concluded that migration was strongly associated with insectivory, nomadism with granivory, and residency with frugivory. A more detailed dietary analysis (742 southern Afrotropical species) confirmed this pattern and further identified that aquatic vertebrate feeders characteristically undertake local movements and that terrestrial vertebrate feeders are mostly resident (M. S. L. Mills and P. A. R. Hockey, unpubl. data). The lack of migration among frugivores in sub-Saharan Africa is pronounced: of 83 Holarctic-breeding landbirds that migrate to southern Africa, not one is a frugivore (Hockey 2000). This is in strong contrast to the Mediterranean, where the majority of overwintering passerines are frugivorous (Lövei 1989). On all three major flyways, the proportion of migrants that is insectivorous increases with distance from the breeding grounds, with a concomitant decrease in frugivory and granivory (J. V. Hamblin and P. A. R. Hockey, unpubl. data). Among Holarctic-breeding migrants whose journeys take them south of at least 30° N, at least 78% of Nearctic breeders are at least partially insectivorous and 24% at least partially frugivorous/nectarivorous on the nonbreeding grounds. Equivalent figures for the western Palearctic are 88 and 7%, and for the eastern Palearctic are 84 and 7%, respectively (J. V. Hamblin and P. A. R. Hockey, unpubl. data). This highlights a further similarity between the two Old World flyways and, once again, places the New World, where frugivory is proportionally 3.4 times more common, as the "odd man out." Of 58 Neotropical migrants that include some fruit/nectar in their diet, 24% are obligate frugivores/nectarivores—there are no obligately frugivorous/nectarivorous migrants in the Afrotropics or the Orient (J. V. Hamblin and P. A. R. Hockey, unpubl. data). The greater prevalence of frugivory in the Neotropics may be linked to the fact that New World migrants do not avoid forests on the nonbreeding grounds.

Although there are some broad correlations between migration and diet, these can be further refined by considering the ways in which birds catch their prey. Among African insectivores, for example, the probability of migration among species decreases in the following order: those catching volant insects above the canopy, perch-and-sally hunters, perch-and-pounce hunters, ground gleaners (Hockey 2000). This progression could be interpreted as a response to habitat buffering, with these categories of predators catching their prey progressively closer to the ground. However, establishing an empirical link between migration and food availability does require measurement of the latter, which is notoriously difficult. Nonetheless, if food availability is the proximate stimulus to which migratory birds respond (or have evolved an anticipatory response), its quantification becomes of primary importance. As a corollary, it could be predicted that within a given dietary guild of birds, residents will have more catholic diets and more versatile foraging repertoires than migrants and are likely to exhibit prey-switching at the time of year when migrants are absent. Among migrants wintering south of 30° N, only 13% (both

Old World flyways) to 19% (New World) switch diet between the breeding and nonbreeding seasons (J. V. Hamblin and P. A. R. Hockey, unpubl. data).

OVERVIEW

For two reasons, it is difficult to develop a conceptual model that encompasses all environmental and ecological attributes believed to contribute to maintaining migratory behavior in birds. Firstly, the spatial scales at which these factors operate range from the macroclimatic to the microenvironmental. Secondly, some patterns are not consistently replicated across the major flyways (e.g., habitat selection and diet). Some of this variability may be attributable to "accidents of history," but below I provide a broad overview of plausible generalizations and identify some of the apparent exceptions.

- At high latitudes, seasonal fluctuations in temperature, and especially the temperature at the coldest time of year, strongly influence migratory behavior independently of habitat type and, to some degree, independently of food type. At lower latitudes, seasonal movements are strongly dictated by rainfall and diet (at least in Africa), but the point at which rainfall supersedes temperature as the driving variable is unknown. Evidence suggests that both temperature and rainfall influence movements indirectly through their effect on food supplies.
- Habitat features that buffer the effects of temperature (and presumably of rainfall) also influence migration. Overall, when controlling for latitude, a greater proportion of species is resident in complex than in simple habitats. Although the buffering effect has frequently been invoked to explain this difference, it has apparently never been measured; consequently, the influences of habitat buffering and habitat complexity have not been satisfactorily separated.
- Predicting seasonal patterns of habitat use is problematic because habitat availability has mostly been ignored as an explanatory variable. Tropical African and Asian forests, for example, are avoided by Palearctic-breeding migrants (which aggregate in less buffered habitats), but Neotropical forests are not avoided by Nearctic migrants. Similarly, among those taxa that do occupy forests, patterns of small-scale habitat partitioning between residents and migrants in the Neotropics are not obviously mirrored in Southeast Asia.
- Type and seasonal abundance patterns of prey strongly influence migration: these vary latitudinally, longitudinally, and within habitats. For example, although there are many migratory frugivores in western Eurasia and the Neotropics, frugivores in sub-Saharan Africa are almost exclusively resident. The way in which food type and foraging mode influence migratory behavior is best illustrated by the invertebrate-feeders. Migration is most prevalent among aerial insectivores and least pronounced among terrestrial gleaners foraging on nonvolant forms. Arranged within and along the same gradient are perch-and-sally and perch-and-pounce hunters. This gradient of foraging techniques links back to one of apparent habitat buffering and its mediating effect on invertebrate life histories.

Conceptually, we are closer to understanding the conditions associated with migration than was the case a decade ago. However, our understanding of the links among habitat availability, occupancy, and tolerance is weak. Many of the broad-scale patterns reported here are based on analyses of migrant species richness and not on the relative abundances of species. Although they may contribute toward a framework of ideas, many of these ideas cannot be taken forward until appropriate field studies are carried out.

FUTURE DIRECTIONS

Of the three major flyways, the Nearctic-Neotropical is the best studied and the Eastern Palearctic–Oriental the least well known. In seeking generalizations about migratory behavior, a major limitation of most studies to date is that they have not, with some exceptions (e.g., Mönkkönen et al. 1992; Bell 2000; Hockey 2000) made interflyway comparisons; this may have led to spurious debate. Those that have made such comparisons have not controlled for phylogeny.

From the few interflyway comparisons that have been made, perhaps the only truly global paradigm about migratory patterns is their link to winter temperatures (even this has not been tested empirically on the Asian flyway). There is also a general tendency for migrants to avoid forested habitats on both breeding and nonbreeding grounds, with the Neotropics being an exception (J. V. Hamblin and P. A. R. Hockey, unpubl. data). The similarities in habitat availability between the Neotropics and Southeast Asia beg an explanation of why behavior of migrants in Asia most closely approximates that of migrants in the Afrotropics.

Also unexplored to a large degree is the role of large-scale physical geography, especially tropical geography, in influencing dispersion patterns. The wide-open, unfragmented landmass of Africa, the narrow (and recent) land bridge of the Isthmus of Panama, and the island mosaic of Indonesia are hardly physically comparable—do the numerous sea crossings of the latter explain the almost total lack of Palearctic-breeding landbirds migrating to Australia? There are many long-distance migrants at equivalent latitudes in South America and Africa (and no shortage of local migrants within Australia and between Australia and New Guinea). In the same vein, the ratio of breeding to wintering land areas on the eastern and western Old World flyways is dramatically reversed—what are the potential consequences of this difference?

In the Neotropics, forest-dwelling migrants appear to be disproportionately concentrated at forest fringes and in the canopy, relative to residents. This dichotomy appears not to exist in tropical Asia but there has been no field study of this apparent difference—is it real or imagined?

Perhaps, however, the greatest stumbling block of all in the search for global truths lies in phylogeny. The greatest evolutionary similarities exist between birds of the two Old World flyways. On the winter receiving areas, however, the greatest environmental similarities exist between the Neotropics and Southeast Asia. The few analyses available to date suggest that the behavior of Oriental migrants most closely resembles that of their Afrotropical counterparts, implying that history is placing a clear footprint on modern migration patterns. If this is indeed the case, there is a pressing need for large-scale, phylogenetically controlled analyses to identify the nature and extent of this footprint. In their absence, the purely ecological search for paradigms that hold true across flyways may be doomed to failure.

LITERATURE CITED

Askins, R. A. 1981. Survival in winter: the importance of roost holes to resident birds. Loon 53:179–184.

Askins, R. A. 1983. Foraging ecology of temperate-zone and tropical woodpeckers. Ecology 64:945–956.

Beehler, B. M., T. K. Pratt, and D. A. Zimmerman. 1986. Birds of New Guinea. Princeton University Press, Princeton.

Bell, C. P. 2000. Process in the evolution of bird migration and pattern in avian ecography. Journal of Avian Biology 31:258–265.

Berthold, P. 1999a. A comprehensive theory for the evolution, control and adaptability of avian migration. *In* Proceedings of the 22nd International Ornithological Congress, Durban (N. J. Adams and R. H. Slotow, eds.). Ostrich 70:1–11.

Berthold, P. 1999b. Geographical variation and the microevolution of avian migratory behaviour. Pages 164–179 *in* Geographic Variation in Behaviour: Perspectives on Evolutionary Mechanisms (S. A. Foster and J. A. Endler, eds.). Oxford University Press, New York.

Berthold, P., A. Helbig, G. Mohr, and U. Querner. 1992. Rapid microevolution of migratory behaviour in a wild bird species. Nature 360:668–669.

Burton, J. F. 1995. Birds and Climate Change. Christopher Helm, London.

Chesser, R. T. 1994. Migration in South America: an overview of the austral system. Bird Conservation International 4:91–107.

Chesser, R. T. 1998. Further perspective on the breeding distribution of migratory birds: South American austral migrant flycatchers. Journal of Animal Ecology 67:69–77.

Chesser, R. T., and D. J. Levey. 1998. Austral migrants and the evolution of migration in New World birds: diet, habitat and migration revisited. American Naturalist 152:311–319.

Cox, G. W. 1985. The evolution of avian migration systems between temperate and tropical regions of the New World. American Naturalist 126:451–474.

Dean, W. R. J. 1997. The distribution and biology of nomadic birds in the Karoo, South Africa. Journal of Biogeography 24:769–779.

Dean, W. R. J. 2004. Nomadic Desert Birds. Springer-Verlag, Berlin.

Fager, E. W. 1968. The community of invertebrates in decayed oak wood. Journal of Animal Ecology 37:121–142.

Gätke, H. 1895. Heligoland as an Ornithological Observatory. David Douglas, Edinburgh.

Greenberg, R. 1980. Demographic aspects of long-distance migration. Pages 493–504 *in* Migrant Birds in the Neotropics (A. Keast and E. S. Morton, eds.). Smithsonian Institution Press, Washington, D.C.

Herrera, C. M. 1978. On the breeding distribution pattern of European migrant birds: MacArthur's theme re-examined. Auk 95:496–509.

Hockey, P. A. R. 2000. Patterns and correlates of bird migrations in sub-Saharan Africa. Emu 100:410–417.

Janzen, D. 1973. Sweep samples of tropical foliage insects: effects of seasons, vegetation types, elevation, time of day, and insularity. Ecology 54:687–708.

Lack, D. 1976. Island Biology Illustrated by the Land Birds of Jamaica. Blackwell Scientific Publications, Oxford.

Lack, P. C. 1986. Ecological correlates of migrants and residents in a tropical African savanna. Ardea 74:111–119.

Leck, C. F. 1972. The impact of some North American migrants at fruiting trees in Panama. Auk 89:842–850.

Levey, D. J., and F. G. Stiles. 1992. Evolutionary precursors of long-distance migration: resource availability and movement patterns in Neotropical land birds. American Naturalist 140:447–476.

Lövei, G. L. 1989. Passerine migration between the Palearctic and Africa. Current Ornithology 6:143–174.

MacArthur, R. H. 1959. On the breeding distribution patterns of North American migrant birds. Auk 76:318–325.

Martin, T. E. 2000. Parental care and clutch sizes in North and South American birds. Science 287:1482–1485.

Mönkkönen, M., P. Helle, and D. Welsh. 1992. Perspectives on Palearctic and Nearctic bird migration; comparisons and overview of life-history and ecology of migrant passerines. Ibis 134 (Supplement 1):7–13.

Newton, I. 1995. The contribution of some recent research on birds to ecological understanding. Journal of Animal Ecology 65:137–146.

Newton, I., and L. Dale. 1996a. Relationship between migration and latitude among west European birds. Journal of Animal Ecology 65:137–146.

Newton, I., and L. C. Dale. 1996b. Bird migration at different latitudes in eastern North America. Auk 113:626–635.

Rappole, J. H. 1995. The Ecology of Migrant Birds: A Neotropical Perspective. Smithsonian Institution Press, Washington, D.C., and London.

Repasky, R. R. 1991. Temperature and the northern distributions of wintering birds. Ecology 72:2274–2285.

Root, T. A. 1988a. Environmental factors associated with avian distributional boundaries. Journal of Biogeography 15:489–505.

Root, T. L. 1988b. Energy constraints on avian distributions and abundances. Ecology 69:330–339.

Short, L. L. 1982. Woodpeckers of the World. Delaware Museum of Natural History Monograph Series no. 4. Greenville, Del.

Snow, D. W. 1980. Regional differences between tropical floras and the evolution of frugivory. Pages 1192–1198 *in* Proceedings of the 17th International Ornithological Congress (R. Nöhring, ed.). University of Ottawa Press, Ottawa.

Sutherland, W. J. 1998. Evidence for flexibility and constraint in migration systems. Journal of Avian Biology 29:441–446.

Tramer, E. J. 1974. On latitudinal gradients in avian diversity. Condor 76:123–130.

Weathers, W. W. 1979. Climatic adaptation in avian standard metabolic rate. Oecologia 42:81–89.

Wells, D. R. 1990. Migratory birds and tropical forest in the Sunda region. Pages 357–369 *in* Biogeography and Ecology of Forest Bird Communities (A. Keast, ed.). SPB Academic Publishing, The Hague.

Williams, T. C., and T. Webb III. 1996. Neotropical bird migration during the ice ages: orientation and ecology. Auk 113:105–111.

Willis, E. O. 1966. The role of migrant birds at swarms of army ants. Living Bird 5:187–231.

Willis, E. O. 1980. Ecological roles of migratory and resident birds on Barro Colorado Island, Panama. Pages 205–225 *in* Migrant Birds in the Neotropics (A. Keast and E. S. Morton, eds.). Smithsonian Institution Press, Washington, D.C.

Willson, M. F. 1976. The breeding distribution of North American migrant birds: a critique of MacArthur (1959). Wilson Bulletin 88:582–587.

Zink, R. M. 2002. Towards a framework for understanding the evolution of avian migration. Journal of Avian Biology 33:433–436.

Part 2 / ADAPTATIONS FOR TWO WORLDS

DOUGLAS J. LEVEY

Overview

THE DEFINING CHARACTERISTIC of a migratory bird is its seasonal occupation of breeding and wintering grounds that are far removed from each other, thereby requiring the ability to make long journeys approximately every 6 months. Those aerial voyages are obvious, having captured the attention of naturalists for centuries. Not so obvious are the ecological consequences of living in two worlds. The breeding and wintering worlds are likely to differ in vegetation structure, resource availability, climate, competitors, and predators. Selective pressures are sure to differ as well. On the breeding grounds, the name of the game is reproductive success, whereas on the wintering grounds the primary goal is to stay alive and garner enough resources for a quick and early return to the temperate zone. The theme of this section is how migratory birds have responded to the disparate challenges imposed by leading two lives in two places.

From the outset, it is important to keep in mind that the metaphor of living in "two worlds" is potentially misleading. Like most popular metaphors, it offers a useful yet sometimes seductive framework. Do migratory birds encounter only two worlds? What about the world(s) encountered during migration itself? What about migratory species that are nomadic on their wintering grounds? What about migratory species that experience dramatic seasonal shifts in climate, resources, and competitors while in the Tropics? Adaptations to such shifts are perhaps no less important than, and are certainly linked to, adaptations to the "two worlds" that we typically envision for migratory species.

The point is not that migration biologists should study all worlds but rather that the two-world perspective may hinder our understanding of how migratory birds are unique. For example, it overlooks the fact that most nonmigratory birds in the Temperate Zone essentially change worlds from summer to winter. Likewise, as pointed out herein by McWilliams and Karasov, some migratory birds may be so skilled at tracking resources that their world stays relatively constant, despite their travels (e.g., albatrosses). Thus, any comparison between ecological or evolutionary strategies of migratory birds needs to be carefully framed. In particular, it should not be based on a straight comparison of nonmigratory and migratory species but on a comparison of migratory species that encounter dramatic seasonal shifts in their environment and nonmigratory species that do not. A parallel comparison of migratory and nonmigratory species that all encounter such seasonal shifts would reveal whether migratory birds are truly unique in their abilities to adapt to different environments.

Migratory birds can respond in two fundamental ways to the ecological and evolutionary challenges of living in different places. The first four chapters in this section focus on morphological or anatomical correlates of migration, whereas the last two discuss behavioral strategies of migratory birds. A common theme is that morphological and behavioral attributes associated with migration may both help and hinder a bird's ability to cope with the disparate demands of the breeding and nonbreeding seasons.

Winkler and Leisler examine morphological correlations of migration within and among species in four passerine lineages with well-resolved phylogenies. They find that migrant species tend to have higher muscle mass associated with flight and lower muscle mass associated with terrestrial locomotion. More surprising, migratory sylviid and parulid warblers have smaller forebrains than nonmigratory conspecifics. They suggest that this difference may reflect the relatively short time that migratory species have for growth and development. More generally, they conclude that additive genetic variance for morphological traits is high, thereby allowing rapid evolution of features beneficial to migration. This conclusion echoes a consensus throughout this book—that migration has evolved quickly and independently in many lineages.

Ecomorphological studies typically focus on traits that are relatively static over an individual's life (e.g., wing length). These traits change over evolutionary time. In contrast are traits that can change over much shorter time spans. McWilliams and Karasov draw attention to the avian gut as a prime example of the latter. They explore the conditions under which digestive constraints may influence the pace of migration, examine the degree of spare capacity in the gut, and discuss potential trade-offs between acclimation of the digestive system to migratory flights and the rate at which energy can be extracted from ingested food during stopover periods. They conclude that the nutritional challenges faced during migration are largely met through spare capacity and phenotypic plasticity of the birds' digestive system.

Rohwer et al. examine how patterns of molt can be tied to geography and migratory behavior. They begin by pointing out that migrant passerines in eastern North America typically molt on their breeding ground before migration, whereas those in the West tend to initiate migration before molt. This difference is explained by exceedingly dry conditions in the West at the end of summer, coupled with a flush of vegetation caused by Mexican monsoon rains in New Mexico, eastern Arizona, and northern Mexico. In effect, Rohwer et al. suggest that many species of western passerines treat this area as a "third world," one in which they can undergo molt before pushing farther south to their wintering grounds. Other differences in the spatial and temporal pattern of molt are pursued in a similar manner, highlighting several adaptive hypotheses.

Rogers draws attention to the behavioral ecology of migrants on their wintering grounds. He explains how fluctuations in components of body mass can reveal potential trade-offs between caloric intake and predation risk. As birds prepare for migration to the breeding grounds or as predator abundance changes, such trade-offs can shift in predictable ways, thereby allowing one to tease apart alternative hypotheses of fat regulation. The Food Limitation Hypothesis states that body mass is directly limited by food supply, whereas the Predator-Food Hypothesis states that body mass is regulated by the combined risks of starvation and predation, not by food abundance alone. Distinguishing between these hypotheses is important. If, for example, the Predator-Food Hypothesis is supported, the common assumption that body fat can be used as a surrogate of habitat quality becomes tenuous.

Finally, Mettke-Hofmann and Greenberg take the theme of migrant behavior one step further. Because migrant birds face unique challenges, the authors propose that migrants and nonmigrants differ in cognitive attributes. They review a handful of such attributes, including exploration behavior, response to novelty, spatial awareness, and long-term memory. Although the data collection on such traits is still in its infancy, migrants and nonmigrants often appear to differ in expected directions. The implication is that migrants and nonmigrants not only experience different worlds but also respond to them in fundamentally different ways. This field of "cognitive ecology" is poorly explored but holds great promise. As with most other authors in this section, Mettke-Hofmann and Greenberg conclude by laying out questions and approaches.

SCOTT R. MCWILLIAMS
AND WILLIAM H. KARASOV

Migration Takes Guts

Digestive Physiology of Migratory Birds and Its Ecological Significance

PHENOTYPIC FLEXIBILITY IN THE digestive system of migratory birds is critically important in allowing birds to successfully overcome the physiological challenges of migration. However, phenotypic flexibility in the digestive system of birds has limits that can influence the pace of migration. For example, partial atrophy of the gut after one to two days without feeding limits utilization of ingested food energy and nutrients and thereby slows the pace of migration. Lack of certain digestive enzymes can directly limit diet choice and utilization of foods that require such enzymes for digestion. Regular switching of diets also may reduce utilization of ingested food energy and nutrients of a given diet. Finally, maximal food intake of migratory birds may ultimately be limited by associated increases in gut size that negatively affect flight performance. Determining when rates of digestion constrain diet choice or re-fattening rates in migratory birds requires understanding the magnitude of spare volumetric or biochemical capacity relative to the magnitude of change in food quantity or quality. We discuss an approach for studying the importance of spare capacity in limiting performance of migratory birds and review the few such studies that have been conducted on migratory songbirds. We conclude that digestive constraints are likely to influence the pace of migration in birds when birds must refuel after one to two days without feeding, when birds lack certain digestive enzymes, when birds regularly switch diets, and when birds are hyperphagic and must also fly.

INTRODUCTION

Birds during migration are like Olympic marathon athletes in that to be successful they must satisfy the physiological demands associated with formidable feats of athletic endurance. Just as exercised muscles increase in size in well-trained human athletes, flight muscles increase in size and capacity when birds migrate (Marsh 1984; Driedzic et al. 1993; Bishop et al. 1996). More recent studies demonstrate that lean mass (including flight muscles) of birds is quite dynamic over even short time scales (e.g., hours and days) and that birds store and use both lean and fat mass during migration (Piersma 1990; Piersma and Jukema 1990; Lindstrom and Piersma 1993; Bauchinger and Biebach 2001). The increase in muscle capacity in migratory birds is necessary to satisfy the physiological demands of long-distance flight (reviewed in Butler and Bishop 2000). Furthermore, these changes in muscle size and capacity are reversible for both birds and mammals as demonstrated by the seasonal hypertrophy and atrophy of pectoralis and cardiac muscle in migratory birds (e.g., Gaunt et al. 1990; Jehl 1997).

Although these changes in skeletal muscle are relatively obvious in birds (and mammals), there are also coincident changes in internal, vital organs that are less apparent but just as important. For example, digestive organs such as the small intestine, gizzard, and liver increase in size and capacity with increased food intake and energy demands such as those associated with exercise (McWilliams and Karasov 2001). The increase in digestive organ capacity converts more food energy into usable metabolic energy to fuel the increased energetic costs of exercising enlarged skeletal muscle. These same energy-supplying organs are some of the most metabolically active and energetically expensive organs in vertebrates (Martin and Fuhrman 1955; Alexander 1999), and this may explain why they atrophy when energy demands are reduced (e.g., once migration is complete [Piersma 2002]).

The fundamental issue addressed by these examples is that the capacity of many physiological systems, including that of the digestive system, is matched to the prevailing demand but can be modulated in response to changes in demand (Diamond and Hammond 1992; Hammond and Diamond 1997). In other words, as demands on the physiological system increase or decrease there is a coincident increase or decrease in the capacity of key organs. Rapid reversible changes in physiological systems such as the digestive system provide examples of flexible norms of reaction (Stearns 1989; Travis 1994) or phenotypic flexibility in that they enable individuals to respond flexibly to changes in the environment (Piersma and Lindstrom 1997; Piersma 2002; Piersma and Drent 2003). Phenotypic flexibility in physiological traits may itself be a critical component of the adaptive repertoire of animals that may influence diet diversity, niche width, feeding rate, and thus the acquisition of energy and essential nutrients (Karasov 1996; Kersten and Visser 1996; Pigliucci 1996; McWilliams et al. 1997; Piersma and Lindstrom 1997; Piersma 2002).

A related fundamental issue is that at any given time the key organs of a physiological system are not exactly matched to the prevailing demand, but instead they provide some limited excess capacity (Diamond and Hammond 1992; Diamond 1998). This "spare capacity" is measured as the excess in capacity of the system over the load on the system (Toloza et al. 1991). The level of spare capacity is ecologically important because it defines the limits of short-term response in animals. For the digestive system in particular, the amount of spare capacity determines, for example, how much an animal can change its feeding rate or diet before digestive efficiency is reduced (Toloza et al. 1991). Because changes in feeding rate and diet are common in migratory birds, understanding the extent of spare capacity in their digestive systems provides insights into when digestion may constrain diet choice and feeding rate.

These concepts of phenotypic flexibility and spare capacity are illustrated in fig. 6.1, using the digestive system of migratory birds as a model physiological system. Two points are worth highlighting in fig. 6.1: (1) at any given time, a migratory bird has some limited spare capacity (called "immediate spare capacity"), but this decreases in extent as the digestive system reaches its ultimate capacity; and (2) phenotypic flexibility of the digestive organs is primarily re-

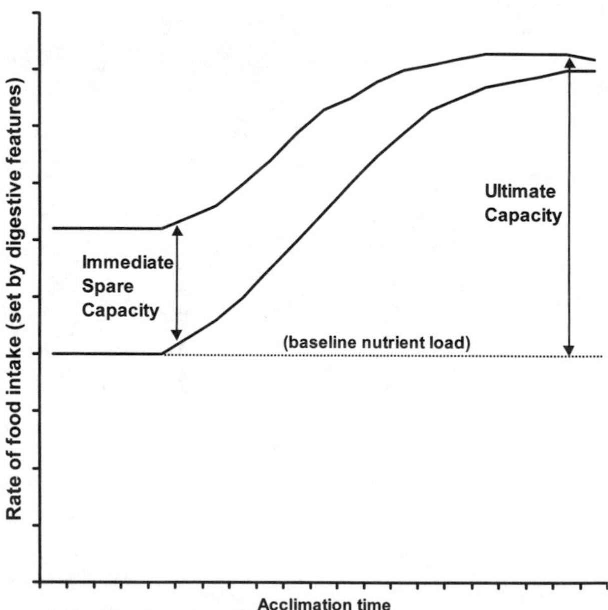

Fig. 6.1. Immediate spare capacity and ultimate capacity (phenotypic flexibility plus immediate spare capacity) for a hypothetical bird exposed to increasing energy demands (e.g., during migration or during cold weather). The solid lower line represents the nutrient load from feeding. Its baseline corresponds to the bird's routine energy demands (e.g., not during migration or at thermoneutral temperatures). The solid upper line represents the capacity of the gut for processing that nutrient load. Capacity on the *y*-axis could be volumetric intake rate (as shown), nutrient uptake capacity, rate of digestive enzyme activity, or some other performance measure of the bird. The *x*-axis is time since the start of an increase in energy demand or change in diet quality. When energy demands are near maximum and the bird has been given time to fully acclimate to these elevated energy demands, then phenotypic flexibility in the digestive system of the bird enables increased energy intake.

sponsible for a bird's ability to change food intake and diet (i.e., it represents the majority of the "ultimate capacity"); however, such phenotypic flexibility requires acclimation time.

A central theme of this chapter is that birds during migration face considerable nutritional challenges and that phenotypic flexibility and spare capacity of their digestive system play key roles in how they respond to these challenges. We focus here on the digestive system of migratory birds because recent research has shown that: (1) features of the gut (e.g., size, nutrient uptake rates, digestive enzyme activity) are modulated in response to changes in the quality and quantity of the diet (see McWilliams and Karasov 2001 and Piersma 2002); and (2) these digestive adjustments are likely important for permitting the high feeding rate of migratory birds and conceivably could constrain the rate of energy intake and the diets of birds during migration (Piersma 2002). Because there are other recent, comprehensive reviews of phenotypic flexibility in the digestive systems of birds and mammals (Karasov 1996; Piersma and Lindstrom 1997; Starck 1999b; McWilliams and Karasov 2001; Karasov and McWilliams 2005), we concentrate in this chapter on providing examples of this phenomenon and its ecological importance for migratory passerine birds generally, and particularly in Old and New World warblers when possible. In the next section, we use the phenomenon of hyperphagia in migratory birds to illustrate how spare capacity of the digestive system can be measured and to demonstrate the importance of phenotypic flexibility in digestive organs for migratory birds.

PREPARING FOR MIGRATION

In preparation for migration, birds increase their food intake (i.e., become hyperphagic) and store the energy and nutrient reserves necessary to fuel the costs of subsequent migratory flight(s) (Alerstam and Lindstrom 1990; Blem 1990; Biebach 1996). What changes in the digestive system might facilitate hyperphagia in migratory birds? Studies of free-living migratory birds (mostly geese and ducks) have reported significantly larger gut size during migratory periods compared with nonmigratory periods (reviewed in Starck 1999b and in McWilliams and Karasov 2001). However, results from such studies of wild birds are not definitive evidence for increases in gut size with elevated food intake because most wild birds change their diets during migratory periods, and this also affects their digestive system (see section below).

Direct evidence for modulation of digestive features in response to changes in food intake comes from work with captive birds. Studies of passerine birds report increased surface area and volume of the gut with long-term increases in food intake (Dykstra and Karasov 1992; Karasov 1996; Piersma and Lindstrom 1997; McWilliams et al. 1999). Only Dykstra and Karasov (1992) and McWilliams et al. (1999), however, have simultaneously measured adjustments in gut anatomy, retention time of digesta, digestive biochemistry (i.e., enzyme hydrolysis rates and/or nutrient absorption rates), and digestive efficiency in response to increased food intake. They found that the rate of digestive enzyme activity and nutrient uptake per unit of small intestine in House Wrens (*Troglodytes aedon*) and Cedar Waxwings (*Bombycilla cedrorum*) did not change with fourfold higher food intake. Instead, the main digestive adjustment to increased food intake was an increase in gut length, mass, and volume, which largely compensated for increased digesta flow at high intake rates so that digestive efficiency remained constant as food intake increased.

If changes in food intake occur faster than the time scale required for digestive adjustment, rather than gradually and slowly, then increased food intake may have quite different effects on digestive performance. We know little about how short-term changes in food intake affect digestive performance in wild birds. When their food was taken away for 2- to 3-h intervals throughout the day, Cedar Waxwings and Yellow-rumped Warblers (*Dendroica coronata*) increased their short-term (hourly) food intake 25 and 50%, respectively, compared with ad libitum conditions (McWilliams and Karasov 1998a, 1998b). These short-term increases in food intake did not result in changes in digestive efficiency or retention time, suggesting some spare digestive capacity when food intake increases by as much as 50%.

We designed a comprehensive study with White-throated Sparrows (*Zonotrichia albicollis*) to determine their response to both rapid and gradual increases in energy demand so that we could estimate the level of spare capacity and phenotypic flexibility in their digestive systems in response to changes in feeding rate. The experiment involved manipulating ambient temperature, which causes changes in the metabolic rate of sparrows (i.e., increased metabolic rate with lower ambient temperature) and thus induces changes in their food intake as they maintain a constant body temperature. By random assignment, sparrows were either held continuously at +21°C, switched rapidly from +21° to −20°C, or gradually acclimated to −20°C over 50 days. For all sparrows in these three treatment groups, we measured daily food intake, digestive efficiency and retention time of starch (the primary nutrient in their semisynthetic diet), ingesta-free mass of the digestive tract (gizzard, small intestine, large intestine), and mass of the liver and pancreas (see McWilliams et al. 1999 for specific methods). The primary prediction was that sparrows switched rapidly from warm to cold temperatures would maintain constant digestive efficiency only if some safety margin of nutrient absorption capacity over nutrient intake existed before the temperature switch.

White-throated Sparrows at −20°C required 83% more food than birds at +21°C, as indicated by their greater feeding rates while maintaining body mass (fig. 6.2). When birds were switched rapidly from +21° to −20°C they increased feeding rate only 45% and lost body mass (fig. 6.2). Interestingly, birds in all three treatment groups had similar digestive efficiency and retention times (fig. 6.2), as measured using a

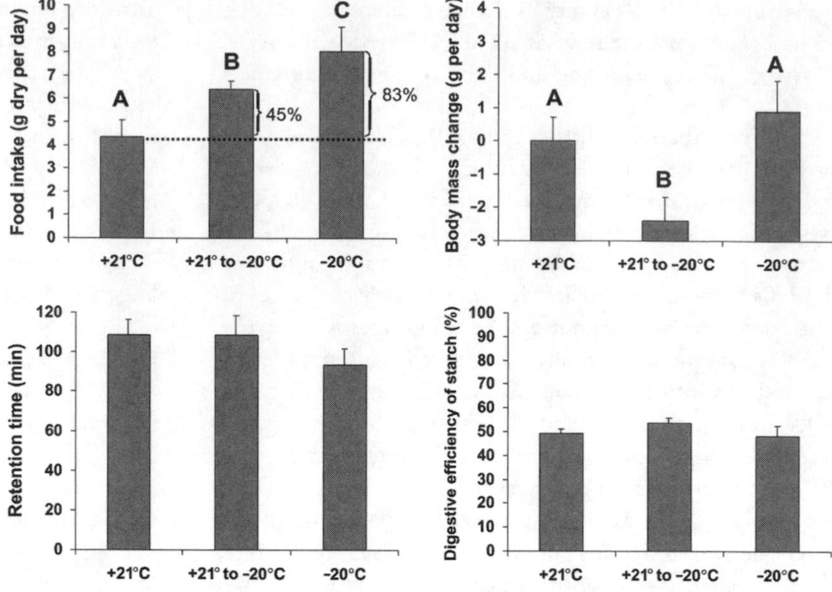

Fig. 6.2. Food intake, body mass change, retention time of digesta, and digestive efficiency of White-throated Sparrows (*Zonotrichia albicollis*) that were either acclimated to +21° or –20°C or switched immediately from +21° to –20°C. Sparrows acclimated to +21°C have a limited spare capacity of about 45% as indicated by an increase in food intake of this magnitude for birds switched rapidly to colder temperatures. These results and those in fig. 6.3 suggest that phenotypic flexibility in digestive features is necessary for sparrows to achieve their ultimate capacity.

radiolabeled inert marker and starch (see McWilliams et al. 1999 for specific methods). Thus, sparrows have some spare capacity (of about 45%) but this was not enough to satisfy the energy demands imposed by a rapid switch from +21° to –20°C. If given enough time for acclimation to the cold, however, sparrows can satisfy the elevated energy demands associated with living in the cold as evidenced by their ability to maintain body mass after 50 days of acclimation at –20°C.

The digestive adjustments to increased feeding rate that occurred during acclimation to the cold included an increase in mass of small intestine (fig. 6.3), large intestine, and liver but not gizzard and pancreas. We are currently completing analyses of digestive enzyme activity and nutrient uptake rates to determine if adjustments in these digestive features are involved along with changes in gut size. Note that the 57% increase in small intestine was enough to accommodate the 83% higher feeding rate in birds acclimated at –20°C. This is apparent because mean retention time, efficiency digesting starch, and body mass did not decline significantly with cold acclimation (fig. 6.2). If one considers that sparrows acclimated to +21°C had a spare capacity of 45% to start with, adding an increase in gut size of 57% to that can more than account for the 83% increased ability to process food. The two measures together imply that sparrows acclimated to –20°C probably still had some spare capacity, perhaps 22% (calculated from the ratio [45 + 57]/83). This makes sense, because it is known that captive White-throated Sparrows can tolerate temperatures down to –29°C, where feeding rates are 2.26 (126%) times higher than at +21°C (Kontogiannis 1968). Thus, the results from the experiment with White-throated Sparrows, along with those by Kontogiannis (1968), conform nicely to the model presented in fig. 6.1 and imply that immediate spare capacity is around 45% but that after long-term acclimation the ultimate capacity is around 126% above "baseline."

Given the lack of change in rates of digestive enzyme activity and nutrient uptake per unit of small intestine in House Wrens and Cedar Waxwings exposed to lowered ambient temperature (Dykstra and Karasov 1992; McWilliams et al. 1999), we suspect that the primary digestive adjustment in White-throated Sparrows (as well as most other passerine birds) to increased food intake is an increase in gut length, mass, and volume. Theoretically, there must be some limit to an animal's ability to enhance gut size, increase food intake, and sustain elevated metabolic rates (see, e.g., Ricklefs 1996, Hammond and Diamond 1997, and Piersma 2002 for recent reviews). For migratory birds that must fly, gut size increases with energy expenditure, but the increase in gut

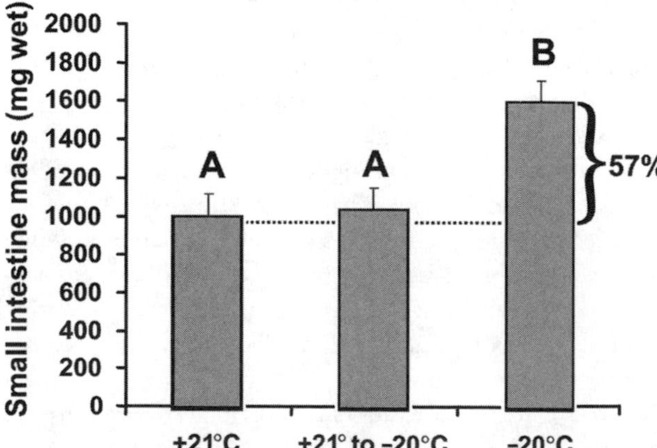

Fig. 6.3. Small intestine mass (mg) of White-throated Sparrows (*Zonotrichia albicollis*) that were either acclimated to +21° or –20°C or switched immediately from +21° to 20°C. See the text for a discussion of how these increases in gut size along with the results shown in fig. 6.2 can be used to estimate the immediate spare capacity and ultimate capacity of White-throated Sparrows (depicted hypothetically in fig. 6.1 and actually in fig. 6.7).

size may be limited by other physiological and morphological constraints associated with flying.

ATROPHY OF BIRD GUTS DURING MIGRATION

For most migratory songbirds, migration itself involves many flights interspersed with layovers at "stopover" sites where energy and nutrient reserves are rebuilt. Thus, birds during migration alternate between periods of high feeding rate at migratory stopover sites and periods without feeding as they travel between stopover sites. These intervals without food may be relatively short (e.g., less than 8 h) for birds migrating short distances at a given time, or they may last for days for birds migrating over oceans or other large ecological barriers (e.g., deserts, mountains).

Ecological field studies of passerine birds have revealed that recovery of body condition after arrival at stopover sites is typically slow for 1 to 2 days and then much more rapid despite apparently abundant food resources (Davis 1962; Nisbet et al. 1963; Muller and Berger 1966; Langslow 1976; Rappole and Warner 1976; Biebach et al. 1986; Moore and Kerlinger 1987; Hume and Biebach 1996; Gannes 1999). Although ecological conditions influence the rate of recovery (Rappole and Warner 1976; Moore and Kerlinger 1987; Hansson and Pettersson 1989; Kuenzi et al. 1991), birds exhibit the two-step recovery after fasting even when provided food ad libitum in the laboratory (Ketterson and King 1977; Klaassen and Biebach 1994; Hume and Biebach 1996; Gannes 1999; Karasov and Pinshow 2000).

Physiological mechanisms to explain the initially slow recovery of body mass after arrival at a stopover site are largely unexplored (Berthold 1996; Biebach 1996). The gut-limitation hypothesis (see McWilliams and Karasov 2001 for alternative hypotheses) suggests that the initially slow rate of mass gain at stopover sites occurs because birds lose digestive tract tissue and hence digestive function during fasting, and rebuilding of the gut takes time and resources and itself restricts the supply of energy and nutrients from food. Shorebirds studied by Piersma (Piersma 1998, 2002; Piersma and Gill 1998) had reduced digestive organs just before migratory departure, presumably to reduce the energetic costs of carrying larger guts during migratory flight. Other studies of shorebirds as well as passerine birds documented reductions in digestive organs during migratory flights (Biebach 1998; Battley et al. 2000, 2001). Migrant Blackcap Warblers (*Sylvia atricapilla*) at a desert oasis stopover site had reduced digestive organs that increased in size when they were provided food ad libitum (Karasov and Pinshow 1998), and Yellow-rumped Warblers killed by colliding into a radio tower in central Wisconsin during their nocturnal migration had small intestines that were smaller than captive birds (fig. 6.4) (D. A. Afik, pers. comm.).

What are the consequences for migratory birds of having smaller guts after a migratory flight? The effect of reduced digestive organs on food intake and digestive efficiency has been studied in wild-caught Garden Warblers (*Sylvia borin*), Blackcap Warblers, Thrush Nightingale (*Luscinia luscinia*), Yellow-rumped Warblers, and White-throated Sparrows (Klaassen and Biebach 1994; Hume and Biebach 1996; Klaassen et al. 1997; Karasov and Pinshow 2000; Lee et al. 2002; Pierce and McWilliams 2004), and may differ depending on whether gut size was reduced by fasting or food restriction (i.e., reduced feeding). When digestive organs were reduced by fasting, digestible dry matter intake of Blackcap (fig. 6.4) and Garden Warblers were also reduced even though the birds were provided ad libitum food after the fast. In contrast, when Blackcap (fig. 6.4) and Yellow-rumped Warblers were food restricted (ca. 50% of ad libitum daily intake), digestive organs such as the small intestine were reduced by 20% but food intake and digestive

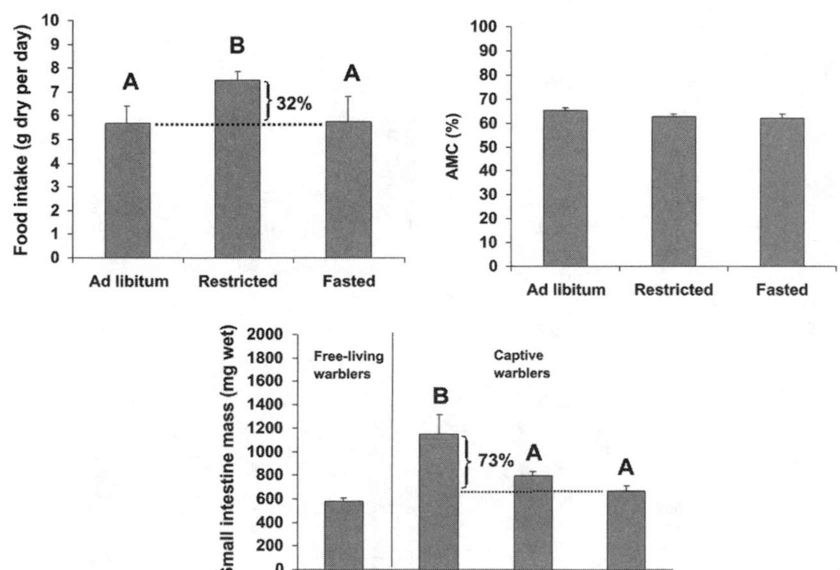

Fig. 6.4. Food intake, apparent metabolizability coefficient (AMC), and small intestine mass of Blackcap Warblers (*Sylvia atricapilla*) on the day the birds were returned to ad libitum feeding after being fasted for 2 days ("fasted"; $n = 7$), fed ad libitum ($n = 6$) or fed for 3 days at 45% ad libitum ("restricted"; $n = 6$). Small intestines of blackcaps were reduced by fasting and food restriction although blackcaps caught just after completing a migration had even smaller intestines ("migrants" [from Karasov and Pinshow 1998]). However, digestible food intake, the product of intake and digestive efficiency, was lower in fasted blackcaps than in food-restricted blackcaps.

efficiency did not change relative to unrestricted birds. Only one published study of a species of wild bird has investigated the effects of fasting or food restriction on biochemical aspects of digestion (Lee et al. 2002). Food restriction reduced digestive enzyme hydrolysis rates by 37–48% in Yellow-rumped Warblers (depending on the type of enzyme [fig. 6.5]).

These studies of captive songbirds demonstrate that reductions in digestive organs occur in both fasted and food-restricted migratory birds, but the reductions in organs of digestion seem to limit refueling rates only in fasted birds. This would suggest that digestion is most likely to constrain refueling rates in long-distance migrants or in other migrants that go without feeding for at least a day at a time. In fact, short-distance migrants, which rarely face extended periods without food, may exhibit phenotypic flexibility in their digestive system like that observed in long-distance migrants, but the spare capacity of their digestive system may usually allow these birds to avoid digestive constraints. For example, Lee et al. (2002) showed that food-restricted warblers had a spare biochemical capacity of at least 37–48% as indicated by their ability to maintain constant food intake and digestive efficiency despite a reduction in digestive organ size and enzyme hydrolysis rates. However, refueling rates of short-distance migrants such as Yellow-rumped Warblers may still be constrained by digestion if their spare capacity is depleted. For example, the free-living Yellow-rumped Warblers that were killed at night by colliding with the radio tower had guts that were more atrophied even than fasted warblers in captivity. If these free-living warblers had survived migration that night, they probably would have had to rebuild their guts before resuming high feeding rates and efficient digestion.

BIRDS SWITCH THEIR DIETS DURING MIGRATION

Migratory birds often switch their diets seasonally. For example, many insectivorous songbirds switch to feeding primarily on fruits during migration (Evans 1966; Herrera 1984; Izhaki and Safriel 1989; Bairlein 1990, 1991; Bairlein and Gwinner 1994; Biebach 1996; Parrish 1997, 2000). In addition, birds on migration may frequently switch their diet because of changes in food availability. For example, migrating birds may one day encounter preferred fruits that are ubiquitous, whereas the next day they may encounter few fruits but insects are ubiquitous. In general, dramatic changes in dietary substrate, for example, from protein- and fat-rich insects to carbohydrate-rich fruits, offer significant physiological challenges for birds (Afik et al. 1995; Afik and Karasov 1995; Karasov 1996; Levey and Martinez del Rio 2001; McWilliams and Karasov 2001). Whether such dietary changes are constrained or facilitated by digestive processes is a central issue in foraging ecology and digestive physiology (Karasov 1990, 1996).

Digestive efficiency for a particular diet eaten by a bird depends in part on the nutrient composition of the diet (Karasov 1990). For example, birds digest nectar almost completely (>95%), and they can assimilate most (ca. 75%) of the energy in seeds, whole vertebrates, insects, and fruits, but birds digest plant foliage relatively poorly (<50%). Similar relative differences in digestive efficiency were observed intraspecifically in omnivorous Yellow-rumped Warblers acclimated to seed, insect, and fruit diets (Afik and Karasov 1995). However, Yellow-rumped Warblers switched from low-fat to high-fat diets and American Robins (*Turdus migratorius*) and European Starlings (*Sturnus vulgaris*) switched

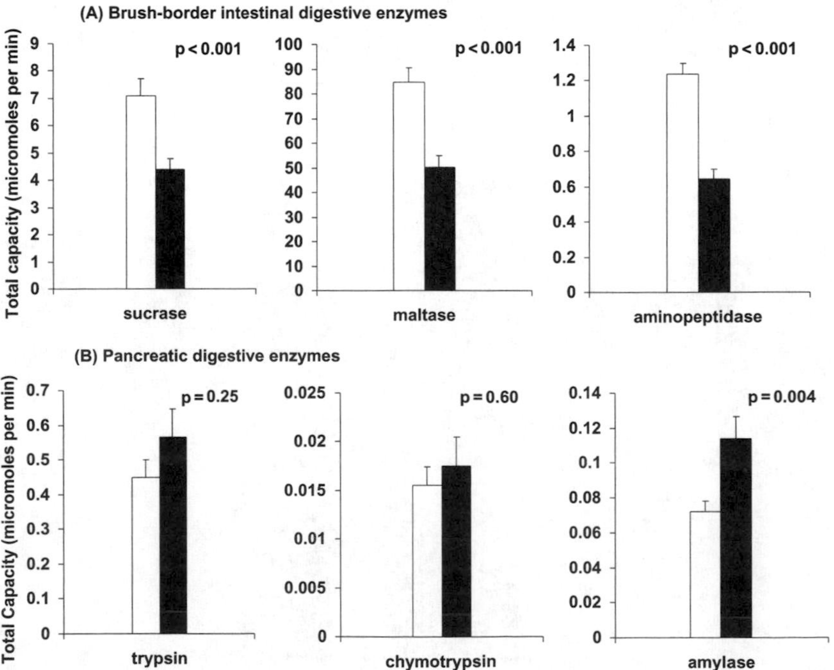

Fig. 6.5. Activity rates of key intestinal (A) and pancreatic (B) digestive enzymes in Yellow-rumped Warblers (*Dendroica coronata*) fed ad libitum (white bars) or fed at 45% ad libitum ("restricted"; black bars) (Lee et al. 2002). Enzyme hydrolysis rates ("total capacity" in micromoles per min) were calculated for the entire small intestine and pancreas given measured mass-specific hydrolysis rates and the mass of the digestive organs. Despite the 37–48% reductions in activity rates of intestinal digestive enzymes and reductions of ca. 20% in small intestine mass, food intake and digestive efficiency were similar in food-restricted and ad libitum fed birds on the day after the birds were returned to ad libitum feeding (see Lee et al. 2002). Thus, these warblers have some spare digestive capacity to compensate for the reductions in gut size and enzyme hydrolysis rates.

from fruit to insect diets had reduced digestive efficiency the day after a diet switch and achieved diet-specific efficiency levels only after 2 to 3 days (Levey and Karasov 1989; Afik and Karasov 1995 [fig. 6.6]). Thus, a migratory bird that switches its diet more frequently than every few days will likely digest a given diet less efficiently than a bird that does not switch diets.

This reduction in digestive efficiency associated with frequent diet switching occurs in part because it takes days for complete modulation of gut size, digestive enzymes, and nutrient uptake transporters (Biviano et al. 1993; Afik et al. 1995; Karasov and Hume 1997; Levey et al. 1999). For example, changes in gut size have been reported in many birds in relation to seasonal changes in diet composition (Pendergast and Boag 1973; Moss 1974, 1983; Ankney 1977; Dubowy 1985; Al-Dabbagh et al. 1987; Walsberg and Thompson 1990; Moorman et al. 1992; Piersma et al. 1993). Such changes in digestive organ size may occur within 1 or 2 days, and certainly within a week, as indicated by measurements of organ size change over time, the rate of cell proliferation, and turnover time of intestine (Piersma et al. 1999; Starck 1999a, 1999b; Dekinga et al. 2001). However, these estimates are based largely on studies of nonpasserine birds (e.g., shorebirds, ducks, quail, and chickens) and only a few notable recent studies of passerine birds (reviewed in Starck 1999b).

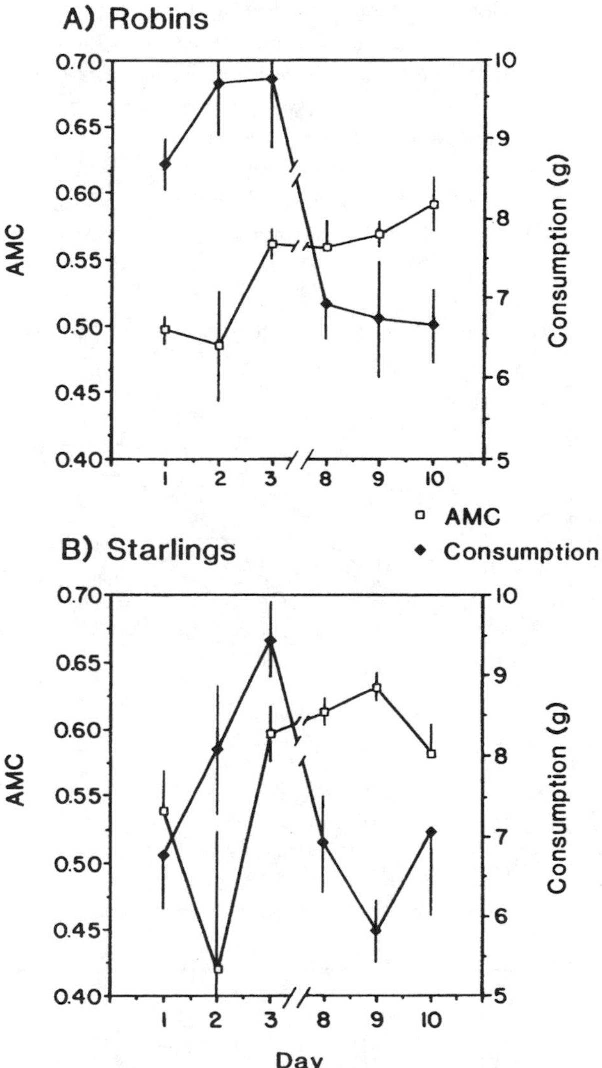

Fig. 6.6. Consumption (g) and apparent metabolizability coefficient (AMC) of crickets (*Acheta domestica*) eaten by American Robins (*Turdus migratorius*) and European Starlings (*Sturnus vulgaris*) as a function of time since switching the birds from a fruit mash diet (Denslow et al. 1987) to crickets. AMC reached diet-specific efficiency levels only after 3 days of feeding on crickets. Figure from Levey and Karasov (1989) with permission of The American Ornithologists' Union and Allen Press, Inc.

Moreover, it is not always true that after a diet switch the initially reduced digestive efficiency improves and eventually reaches levels similar to those in birds regularly eating that diet. Digestive features of some birds are relatively fixed and essentially define the bird's diet. For example, passerine birds in the Sturnidae-Muscicapidae taxon lack sucrase and behaviorally avoid diets with sucrose (Martinez del Rio 1990). Similarly, all vertebrates lack cellulase, the enzyme that can digest plant cellulose. Accordingly, the majority of birds species (97%) do not eat plant leaves or stems, and the few bird species that primarily eat leaves are relatively large birds that use fermentation by indigenous gut microbes to digest cellulose (McWilliams 1999). In the remainder of this section, we discuss digestive constraints associated with switching between certain types of diets.

Among vertebrates in general, nutrient absorption rates and activity of digestive enzymes correlate with the amount of dietary substrate (e.g., protein-digesting enzymes increase with their respective dietary protein substrate, carbohydrate-digesting enzymes increase with their respective dietary carbohydrate substrate [Ferraris and Diamond 1989; Stevens and Hume 1995; Karasov and Hume 1997]). Although there are relatively few studies of digestive enzymes in wild birds, comparisons of bird species with different feeding habits (e.g., insectivores, frugivores, granivores) have generally supported this general pattern (e.g., Afik et al. 1995; Witmer and Martinez del Rio 2001).

In contrast, intraspecific studies of modulation of digestive enzymes and nutrient uptake rate in response to changes in diet composition have not consistently found that enzyme activity or uptake rate changes in proportion to the amount of dietary substrate. For example, activity of digestive enzymes in Pine Warblers (*Dendroica pinus*) changes in proportion to dietary substrate (Levey et al. 1999), but the few other wild passerine birds studied to date exhibit somewhat different patterns of modulation in digestive enzymes. Specifically, wild birds fed diets with higher carbohydrate concentrations did not increase their digestive disaccharidases, whereas birds fed diets with higher protein concentrations increased their aminopeptidase-N activity (Afik and Karasov 1995; Martinez del Rio et al. 1995; Sabat et al. 1998; Caviedes-Vidal et al. 2000). Similarly, none of the four species of omnivorous birds studied to date showed modulation of mediated glucose transport activity, and amino acid uptake increased with dietary protein in only two of the four species studied (Caviedes-Vidal and Karasov 1996; Karasov 1996; Afik et al. 1997a; Chediack et al. 2003). The absence of modulation of mediated glucose transport in birds may occur because birds rely less on active transport for absorption of glucose and more on passive absorption of glucose (Karasov and Cork 1994; Levey and Cipollini 1996; Afik et al. 1997b).

The pattern of modulation of digestive peptidases but not disaccharidases holds across bird species that are dietary generalists and specialists (Sabat et al. 1998), although more comparative studies are needed to determine if dietary flexibility is generally unrelated to this type of digestive plasticity (e.g., see Levey et al. 1999). The lack of modulation of carbohydrate-digesting enzymes suggests that birds switching from insects (mostly protein and fat with little carbohydrate) to fruits (mostly carbohydrate or fat with little protein) may digest the carbohydrates in fruits less efficiently than fruit specialists. Although there are few studies of digestive enzymes in wild frugivorous birds, available evidence provides some support for this hypothesis. For example, omnivorous Pine Warblers up-regulated carbohydrase activity when switched to fruit diets, yet they could not maintain body mass on pure fruit diets and they had relatively low rates of carbohydrase activity compared with the predominantly frugivorous Cedar Waxwing (Levey et al. 1999; Witmer and Martinez del Rio 2001). In contrast, omnivorous Yellow-rumped Warblers did not modulate carbohydrase activity in response to changes in diet (Afik et al. 1995), yet they could maintain body mass on pure fruit diets and their carbohydrase activity was much higher than that of Pine Warblers (Levey et al. 1999) but it was still lower than in Cedar Waxwings (Witmer and Martinez del Rio 2001). Clearly, more research is needed before general patterns of digestive constraints and diet breadth are revealed.

THE PACE OF MIGRATION IN RELATION TO THE PACE OF DIGESTIVE CHANGE

If energy and nutrient demands cannot be satisfied because of inadequate rates of digestion or capacities of the digestive system, then digestive features can constrain choice of diet and food intake. For most birds, the maximum size of the digestive tract is likely limited by constraints associated with flying. For birds that lack certain digestive enzymes, digestive features clearly constrain diet choice. Determining when rates of digestion constrain diet choice or re-fattening rates in migratory birds requires understanding the magnitude of spare volumetric or biochemical capacity relative to the magnitude of change in food quantity or quality.

Studies of short-term changes in food intake suggest that digestion does not appreciably constrain the animal as long as increases in food intake are less than 50% above ad libitum levels (McWilliams and Karasov 1998a, 1998b). Doubling of food intake occurs commonly in birds preparing for migration (Berthold 1975; Blem 1980; Karasov 1996) and in birds at cold temperatures (Dawson et al. 1983; Dykstra and Karasov 1992; McWilliams et al. 1999). Recent studies have shown that increases in food intake of two- to fourfold are possible without measurable effects on digestive efficiency as long as birds have had sufficient time to acclimate (see "Preparing for Migration" above and fig. 6.7). Presumably, if a fourfold increase in food intake occurred before compensatory changes in these digestive features (i.e., over less than 3 days), then digestive efficiency would decrease.

Whereas the primary digestive adjustment to changes in food quantity is change in the amount of gut and not in the absorption rate of tissue-specific enzymes or nutrients, changes in diet quality cause a suite of digestive adjustments including modulation of digestive enzymes and nutrient uptake rates, as well as gut size (Karasov 1996; Starck 1999a; Dekinga et al. 2001; McWilliams and Karasov 2001). The few studies of short-term changes in food quality suggest that certain changes in diet are possible without measurable effects on digestive efficiency as long as birds have had sufficient time to acclimate (see "Birds Switch Their Diets during Migration" above). However, certain changes in diet (e.g., insects to leaves) are impossible for most birds because of constraints associated with digestion.

How long does it take for the digestive system to become acclimated to such changes in food intake and diet quality? Fasted songbirds (e.g., Blackcaps and White-throated Sparrows) progressively increased their absorption rates to a maximum over 3 days (Karasov and Pinshow 2000). Intestinal turnover time is 2 to 3 days for small birds compared to

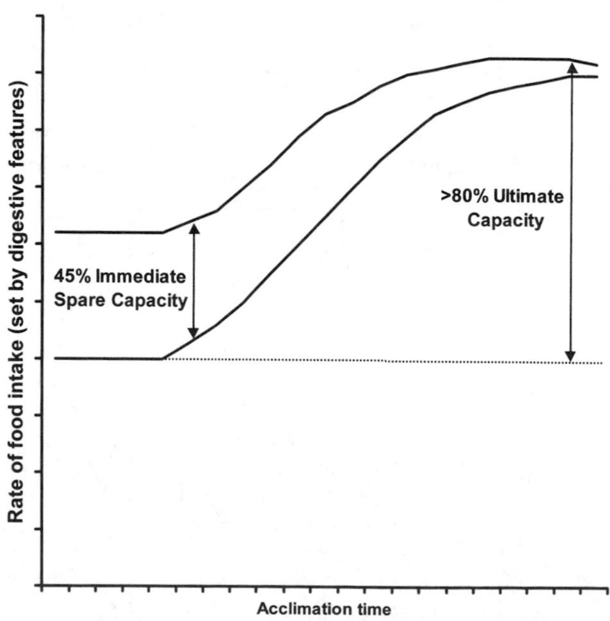

Fig. 6.7. Immediate spare capacity and ultimate capacity (phenotypic flexibility plus immediate spare capacity) for White-throated Sparrows (*Zonotrichia albicollis*) exposed to increasing energy demands associated with cold ambient temperatures. White-throated Sparrows acclimated for at least 50 days at −20°C required 83% more food than birds acclimated at +21°C. When birds were switched rapidly from 21° to −20°C they increased feeding rate only 45% and this was not sufficient to satisfy the extra energy demands given that these birds lost body mass. We estimated that sparrows acclimated to −20°C probably still had some spare excess capacity, perhaps 22% (see text).

8 to 12 days for larger birds (Starck 1999b). Activity of digestive enzymes and nutrient transporters increased within 2 days (Karasov and Hume 1997). Digestive organs of birds increased in size within 1 to 6 days after switching diet or increasing food intake (Dekinga et al. 2001; McWilliams and Karasov 2001). Thus, digestive adjustments in response to certain changes in diet quantity and quality appear to require at least a few days and perhaps as much as a week or more, depending on the type of digestive adjustment (see fig. 6.7). For an actively migrating bird, this pace of digestive change may be too slow and so digestive constraints may directly retard the pace of bird migration.

CONCLUSIONS AND FUTURE DIRECTIONS

If the biological cost associated with maintaining large guts were minimal or the consumable resources used by migratory birds were relatively constant, then migratory birds would likely maintain a relatively constant and enlarged digestive system with significant excess capacity. However, phenotypic flexibility in the digestive system of migratory birds is pervasive and we have shown here that birds maintain significant but limited spare capacity. For migratory birds, phenotypic flexibility and spare capacity in the digestive system have likely evolved both to reduce the biological costs associated with maintaining large guts and to solve the difficult physiological problems associated with a changing environment.

In general, maintaining large guts is costly, in part because digestive organs are some of the most metabolically active tissues in vertebrates. Quantitative measurements of the costs associated with maintaining extensive spare capacity and phenotypic flexibility in the digestive system of migratory birds have not yet been attempted. These costs would include not only the energetic and nutritional costs of maintaining more gut than immediately necessary, but also the trade-offs associated with using the finite space within the animal for digestive organs and tissues instead of other necessary physiological systems (Diamond 1998). For migratory birds in particular, carrying the extra mass of enlarged digestive organs imposes an additional energetic cost during flight. Thus, although empirical data are lacking, the biological costs associated with maintaining large guts are likely to be significant for migratory birds.

Phenotypic flexibility and spare capacity are adaptive for animals that live in a changing environment. Certainly, the migratory habit of some birds has evolved to take advantage of seasonal environments that are quite dynamic over time. We have argued that most migratory birds change both what they eat and the amount they eat as a normal part of migration; thus, food resources used by migratory birds are dynamic and, in this sense, too, migratory birds live in a changing environment. What remains to be determined is whether phenotypic flexibility and spare capacity are greater in migratory birds than in nonmigratory birds.

Theoretically, animals in highly variable environments or those that are mobile and so encounter quite different environments during their annual cycle (e.g., migratory birds) are more likely to have evolved phenotypic flexibility. We have provided examples of how phenotypic flexibility and spare capacity of the digestive system are critically important to the success of migratory birds. However, there is little evidence that these features of the digestive system are unique to migratory birds. This is, in part, because too few studies have focused on this aspect of birds as a group. No study has compared the extent of phenotypic flexibility and spare capacity in migratory and nonmigratory birds with appropriate controls for phylogeny. We caution that such a study would have to be carefully planned. Some migratory birds (e.g., albatrosses) travel great distances as they track the same food resource(s), for example, so their food resources are relatively unchanging despite their extensive travels. As well, relatively sedentary birds that live in northern latitudes, such as Ruffed Grouse (*Bonasa umbellus*), must successfully overcome dramatic seasonal changes in their environment. Thus, migration alone is unlikely to select for phenotypic flexibility in the digestive system of birds.

Migratory birds that vary in the extent to which they change diets and the amount they eat would be the most useful subjects for comparative studies of phenotypic flexibility and spare capacity. We predict that phenotypic flexibility in the digestive system will be most extensive in migratory birds that regularly switch their diet or that dramatically alter their food intake as a regular part of their migration. Studies cited here provide examples of how to measure phenotypic flexibility and spare capacity of the digestive system in birds. What remains to be discovered is how the flexibility and capacity of the avian digestive system vary across taxa and how they are related to the phylogeny, ecology, and life history of birds.

ACKNOWLEDGMENTS

Supported by the National Science Foundation (IBN-9318675 and IBN-9723793 to WHK and IBN-9984920 to SRM).

LITERATURE CITED

Afik, D., E. Caviedes-Vidal, C. Martinez del Rio, and W. H. Karasov. 1995. Dietary modulation of intestinal hydrolytic enzymes in yellow-rumped warblers. American Journal of Physiology 269:R413–R420.

Afik, D., B. W. Darken, and W. H. Karasov. 1997a. Is diet shifting facilitated by modulation of intestinal nutrient uptake? Test of an adaptational hypothesis in Yellow-rumped Warblers. Physiological Zoology 70:213–221.

Afik, D., and W. H. Karasov. 1995. The trade-offs between digestion rate and efficiency in warblers and their ecological implications. Ecology 76:2247–2257.

Afik, D., S. R. McWilliams, and W. H. Karasov. 1997b. Test for passive absorption of glucose in the yellow-rumped warbler and its ecological implications. Physiological Zoology 70:370–377.

Al-Dabbagh, K. Y., J. H. Jiad, and I. N. Waheed, 1987. The influence of diet on the intestine length of the white-cheeked bulbul. Ornis Scandinavica 18:150–152.

Alerstam, T., and A. Lindstrom. 1990. Optimal bird migration: the relative importance of time, energy and safety. Pages 331–351 *in* Bird migration: physiology and ecophysiology (E. Gwinner, ed.). Springer-Verlag, New York.

Alexander, R. M. 1999. Energy of animal life. Oxford University Press, Oxford.

Ankney, C. D. 1977. Feeding and digestive organ size in breeding Lesser Snow Geese. Auk 94:275–282.

Bairlein, F. 1990. Nutrition and food selection in migratory birds. Pages 198–213 *in* Bird migration: physiology and ecophysiology (E. Gwinner, ed.). Springer-Verlag, Berlin.

Bairlein, F. 1991. Nutritional adaptations to fat deposition in the long-distance migratory Garden Warbler (*Sylvia borin*). Pages 2149–2158 *in* Proceedings of the 20th International Ornithological Congress, Christchurch.

Bairlein, F., and E. Gwinner. 1994. Nutritional mechanisms and temporal control of migratory energy accumulation in birds. Annual Review of Nutrition 14:187–215.

Battley, P. F., M. W. Dietz, T. Piersma, A. Dekinga, S. Tang, and K. Hulsman. 2001. Is long-distance bird flight equivalent to a high-energy fast? Body composition changes in freely migrating and captive fasting Great Knots. Physiological and Biochemical Zoology 74:435–449.

Battley, P. F., T. Piersma, M. W. Dietz, S. Tang, A. Dekinga, and K. Hulsman. 2000. Empirical evidence for differential organ reductions during trans-oceanic bird flight. Proceeding of the Royal Society of London, Series B, Biological Sciences 267:191–195.

Bauchinger, U., and H. Biebach. 2001. Differential catabolism of muscle protein in Garden Warblers (*Sylvia borin*): flight and leg muscle act as a protein source during long-distance migration. Journal of Comparative Physiology, Series B, Biochemical, Systematic, and Environmental Physiology 171:293–301.

Berthold, P. 1975. Migration: control and metabolic physiology. Pages 77–128 *in* Avian Biology (D. S. Farner, and J. R. King, eds.). Academic Press, New York.

Berthold, P. 1996. Control of Bird Migration. Chapman and Hall, New York.

Biebach, H. 1996. Energetics of winter and migratory fattening. Pages 280–323 *in* Avian Energetics and Nutritional Ecology (C. Carey, ed.). Chapman and Hall, New York.

Biebach, H. 1998. Phenotypic organ flexibility in Garden Warblers *Sylvia borin* during long-distance migration. Journal of Avian Biology 29:529–535.

Biebach, H., W. Friedrich, and G. Heine. 1986. Interaction of body mass, fat, foraging and stopover period in trans-Sahara migrating passerine birds. Oecologia 69:370–379.

Bishop, C. M., P. J. Butler, A. J. E. Haj, S. Egginton, and M. J. J. E. Loonen. 1996. The morphological development of the locomotor and cardiac muscles of the migratory barnacle goose (*Branta leucopsis*). Journal of Zoology, London 239:1–15.

Biviano, A. B., C. Martinez del Rio, and D. L. Phillips. 1993. Ontogenesis of intestine morphology and intestinal disaccharidases in chickens (*Gallus gallus*) fed contrasting purified diets. Journal of Comparative Physiology, Series B, Biochemical, Systematic, and Environmental Physiology 163:508–518.

Blem, C. R. 1980. The energetics of migration. Pages 175–224 *in* Animal Migration, Orientation, and Navigation (S. A. Gauthreaux, Jr., ed.). Academic Press, New York.

Blem, C. R. 1990. Avian energy storage. Pages 59–113 *in* Current Ornithology (M. Power, ed.). Plenum Press, New York.

Butler, P. J., and C. M. Bishop. 2000. Flight. Pages 391–435 *in* Sturkie's Avian Physiology (G. C. Whittow, ed.) (fifth ed.). Academic Press, New York.

Caviedes-Vidal, E., D. Afik, C. Martinez del Rio, and W. H. Karasov. 2000. Dietary modulation of intestinal enzymes of the house sparrow (*Passer domesticus*): testing an adaptive hypothesis. Comparative Biochemistry and Physiology A 125:11–24.

Caviedes-Vidal, E., and W. H. Karasov. 1996. Glucose and amino acid absorption in house sparrow intestine and its dietary modulation. American Journal of Physiology 40:R561–R568.

Chediack, J. G., E. Caviedes-Vidal, V. Fasulo, L. J. Yamin, and W. H. Karasov. 2003. Intestinal passive absorption of water-soluble compounds by sparrows: effect of molecular size and luminal nutrients. Journal of Comparative Physiology, Series B, Biochemical, Systematic, and Environmental Physiology 173:187–197.

Davis, P. 1962. Robin recaptures on Fair Isle. British Birds 55:225–229.

Dawson, W. R., R. L. Marsh,., and M. E. Yacoe. 1983. Metabolic adjustments of small passerine birds for migration and cold. American Journal of Physiology 245:R755–R767.

Dekinga, A., M. W. Dietz, A. Koolhaas, and T. Piersma. 2001. Time course and reversibility of changes in the gizzards of red knots alternately eating hard and soft food. Journal of Experimental Biology 204:2167–2173.

Denslow, J. S., D. J. Levey, T. C. Moermond, and B. C. Wentworth. 1987. A synthetic diet for fruit-eating birds. Wilson Bulletin 99:131–134.

Diamond, J. 1998. Evolution of biological safety factors: a cost/benefit analysis. Pages 21–27 *in* Principles of Animal Design: The Optimization and Symmorphosis Debate (E. Weibel, C. R. Taylor, and L. Bolis, eds.). Cambridge University Press, New York.

Diamond, J., and K. Hammond. 1992. The matches, achieved by natural selection, between biological capacities and their natural loads. Experientia 48:551–557.

Driedzic, W. R., H.L. Crowe, P. W. Hicklin, and D. H. Sephton. 1993. Adaptations in pectoralis muscle, heart mass, and energy metabolism during premigratory fattening in semipalmated sandpipers (*Calidris pusilla*). Canadian Journal of Zoology 71:1602–1608.

Dubowy, P. J. 1985. Seasonal organ dynamics in post-breeding male Blue-winged Teal and Northern Shovelers. Comparative Biochemistry and Physiology A 82:899–906.

Dykstra, C. R., and W. H. Karasov. 1992. Changes in gut structure and function of House Wrens (*Troglodytes aedon*) in response to increased energy demands. Physiological Zoology 65:422–442.

Evans, P. R. 1966. Migration and orientation of passerine night-migrants in northeast England. Journal of Zoology 150:319–369.

Ferraris, R. P., and J. M. Diamond. 1989. Specific regulation of intestinal nutrient transporters by their dietary substrates. Annual Review of Physiology 51:125–141.

Gannes, L. Z. 1999. Flying, fasting and feeding: the physiology of bird migration in old world Sylvid and Turdid thrushes. Ph.D. dissertation. Princeton University, Princeton, N.J.

Gaunt, A. S., R. S. Hikida, J. R. Jehl Jr., and L. Fenbert. 1990. Rapid atrophy and hypertrophy of an avian flight muscle. Auk 107:649–659.

Hammond, K. A., and J. M. Diamond. 1997. Maximal sustained energy budgets in humans and animals. Nature 386:457–462.

Hansson, M., and J. Pettersson. 1989. Competition and fat deposition in Goldcrests (*Regulus regulus*) at a migration stop-over site. Vogelwarte 35:21–31.

Herrera, C. M. 1984. Adaptation to frugivory of Mediterranean avian seed dispersers. Ecology 65:609–617.

Hume, I., and H. Biebach. 1996. Digestive tract function in the long-distance migratory garden warbler, *Sylvia borin*. Journal of Comparative Physiology, Series B, Biochemical, Systematic, and Environmental Physiology 166:388–395.

Izhaki, I., and U. N. Safriel. 1989. Why are there so few exclusively frugivorous birds? Experiments on fruit digestibility. Oikos 54:23–32.

Jehl, J. R. J. 1997. Cyclical changes in body composition in the annual cycle and migration of the Eared Grebe, *Podiceps nigricollis*. Journal of Avian Biology 28:132–142.

Karasov, W. H. 1990. Digestion in birds: chemical and physiological determinants and ecological implications. Studies in Avian Biology 13:391–415.

Karasov, W. H. 1996. Digestive plasticity in avian energetics and feeding ecology. Pages 61–84 *in* Avian Energetics and Nutritional Ecology (C. Carey, ed.). Chapman and Hall, New York.

Karasov, W. H., and S. J. Cork. 1994. Glucose absorption by a nectarivorous bird: the passive pathway is paramount. American Journal of Physiology 267:G18–G26.

Karasov, W. H., and I. D. Hume. 1997. Vertebrate gastrointestinal system. Pages 409–480 *in* Comparative physiology, Vol. 1 (W. H. Dantzler, ed.). Handbook of Physiology, Section 13. Oxford University Press, New York.

Karasov, W. H., and S. R. McWilliams. 2005. Digestive constraints in mammalian and avian ecology. *In* Physiological and Ecological Adaptations to Feeding in Vertebrates (J. M. Starck, and T. Wang, eds.). Science Publishers, Enfield, N.H. (in press).

Karasov, W. H., and B. Pinshow. 1998. Changes in lean mass and in organs of nutrient assimilation in a long-distance migrant at a springtime stopover site. Physiological Zoology 71:435–448.

Karasov, W. H., and B. Pinshow. 2000. Test for physiological limitation to nutrient assimilation in a long-distance passerine migrant at a springtime stopover site. Physiological and Biochemical Zoology 73:335–343.

Kersten, M., and W. Visser. 1996. The rate of food processing in the oystercatcher: intake and energy expenditure constrained by a digestive bottleneck. Functional Ecology 10:440–448.

Ketterson, E. D., and J. R. King 1977. Metabolic and behavioral responses to fasting in the White-crowned Sparrow (*Zonotrichia leucophrys gambelii*). Physiological Zoology 50:115–129.

Klaassen, M., and H. Biebach. 1994. Energetics of fattening and starvation in the long-distance migratory garden warbler, *Sylvia borin,* during the migratory phase. Journal of Comparative Physiology, Series B, Biochemical, Systematic, and Environmental Physiology 164:362–371.

Klaassen, M., A. Lindstrom, and R. Zijlstra. 1997. Composition of fuel stores and digestive limitations to fuel deposition rate in the long-distance migratory Thrush Nightingale, *Luscinia luscinia*. Physiological Zoology 70:125–133.

Kontogiannis, J. E. 1968. Effect of temperature and exercise on energy intake and body weight of the White-throated Sparrow, *Zonotrichia albicollis*. Physiological Zoology 41:54–64.

Kuenzi, A. J., F. R. Moore, and T. R. Simmons, T. R. 1991. Stopover of neotropical landbird migrants on East Ship Island following trans-Gulf migration. Condor 93:869–883.

Langslow, D. R. 1976. Weights of blackcap on migration. Ringing and Migration 1:78–91.

Lee, K. A., W. H. Karasov, and E. Caviedes-Vidal. 2002. Digestive response to restricted feeding in migratory yellow-rumped warblers. Physiological and Biochemical Zoology 5:314–323.

Levey, D. J., and M. L. Cipollini,. 1996. Is most glucose absorbed passively in northern bobwhite? Comparative Biochemistry and Physiology 113A:225–231.

Levey, D. J., and W. H. Karasov. 1989. Digestive responses of temperate birds switched to fruit or insect diets. Auk 106:675–686.

Levey, D. J., and C. Martinez del Rio. 2001. It takes guts (and more) to eat fruit: lessons from avian nutritional ecology. Auk 118:819–831.

Levey, D. J., A. R. Place, P. J. Rey, and C. Martinez del Rio. 1999. An experimental test of dietary enzyme modulation in pine warblers *Dendroica pinus*. Physiological and Biochemical Zoology 72:576–587.

Lindstrom, A., and T. Piersma. 1993. Mass changes in migrating birds: the evidence for fat and protein storage re-examined. Ibis 135:70–78.

Marsh, R. L. 1984. Adaptations of the Gray Catbird *Dumetella carolinensis* to long-distance migration: flight muscle hypertrophy associated with elevated body mass. Physiological Zoology 57:105–117.

Martin, A. W., and F. A. Fuhrman. 1955. The relationship between summated tissue respiration and metabolic rate in the mouse and dog. Physiological Zoology 28:18–34.

Martinez del Rio, C. 1990. Dietary, phylogenetic, and ecological correlates of intestinal sucrase and maltase activity in birds. Physiological Zoology 63:987–1011.

Martinez del Rio, C., K. Brugger, M. Witmer, J. Rios, and E. Vergara. 1995. An experimental and comparative study of dietary modulation of intestinal enzymes in European starlings (*Sturnus vulgaris*). Physiological Zoology 68:490–511.

McWilliams, S. R. 1999. Digestive strategies of avian herbivores. Pages 2198–2207 *in* Proceedings of the 22nd International Ornithological Congress, Durban (N. Adams and R. Slotow, eds.).

McWilliams, S. R., D. Afik, and S. Secor. 1997. Patterns and processes in the vertebrate digestive system: implications for the study of ecology and evolution. Trends in Ecology and Evolution 12:420–422.

McWilliams, S. R., E. Caviedes-Vidal, and W. H. Karasov. 1999. Digestive adjustments in cedar waxwings to high feeding rates. Journal of Experimental Zoology 283:394–407.

McWilliams, S. R., and W. H. Karasov. 1998a. Test of a digestion optimization model: effects of costs of feeding on digestive parameters. Physiological Zoology 71:168–178.

McWilliams, S. R., and W. H. Karasov. 1998b. Test of a digestion optimization model: effects of variable-reward feeding schedules on digestive performance of a migratory bird. Oecologia 114:160–169.

McWilliams, S. R., and W. H. Karasov. 2001. Phenotypic flexibility in digestive system structure and function in migratory birds and its ecological significance. Comparative Biochemistry and Physiology Part A 128:579–593.

Moore, F., and P. Kerlinger. 1987. Stopover and fat deposition by North American wood-warblers (Parulinae) following spring migration over the Gulf of Mexico. Oecologia 74:47–54.

Moorman, T. E., G. A. Baldassarre, and D. M. Richard. 1992. Carcass mass, composition, and gut morphology dynamics of Mottled Ducks in fall and winter in Louisiana. Condor 94:407–417.

Moss, R. 1974. Winter diet, gut length, and interspecific competition in Alaskan ptarmigan. Auk 91:737–746.

Moss, R. 1983. Gut size, body weight, and digestion of winter food by grouse and ptarmigan. Condor 85:185–193.

Muller, H. C., and D. D. Berger. 1966. Analyses of weight and fat variations in transient Swainson's Thrushes. Bird-banding 37:83–111.

Nisbet, I. C. T., W. H. Drury, and J. Baird. 1963. Weight loss during migration, Part 1, Deposition and consumption of fat by the blackpoll warbler. Bird-banding 34:107–138.

Parrish, J. D. 1997. Patterns of frugivory and energetic condition in Nearctic landbirds during autumn migration. Condor 99:681–697.

Parrish, J. D. 2000. Behavioral, energetic, and conservation implications of foraging plasticity during migration. Studies in Avian Biology 20:53–70.

Pendergast, B. A., and D. A. Boag. 1973. Seasonal changes in the internal anatomy of Spruce Grouse in Alberta. Auk 90:307–317.

Pierce, B., and S. R. McWilliams. 2004. Diet quality and food limitation affect the dynamics of body composition and digestive organs in a migratory songbird (*Zonotrichia albicollis*). Physiological and Biochemical Zoology 77:471-483.

Piersma, T. 1990. Pre-migratory "fattening" usually involves more than the deposition of fat alone. Ringing & Migration 11:113–115.

Piersma, T. 1998. Phenotypic flexibility during migration: optimization of organ size contingent on the risks and rewards of fueling and flight. Journal of Avian Biology 29:511–520.

Piersma, T. 2002. Energetic bottlenecks and other design constraints in avian annual cycles. Integrative and Comparative Biology 42:51–67.

Piersma, T., M. W. Dietz, A. Dekinga, S. Nebel, J. van Gils, P. F. Battley, and B. Spaans. 1999. Reversible size-changes in stomachs of shorebirds: when, to what extent, and why? Acta Ornithologica 34:175–181.

Piersma, T., and J. Drent. 2003. Phenotypic flexibility and the evolution of organismal design. Trends in Ecology and Evolution 18:228–233.

Piersma, T., and R. E. J. Gill. 1998. Guts don't fly: small digestive organs in obese Bar-tailed Godwits. Auk 115:196–203.

Piersma, T., and J. Jukema. 1990. Budgeting the flight of a long-distance migrant: changes in nutrient reserve levels of Bar-tailed Godwits at successive spring staging sites. Ardea 78:315–337.

Piersma, T., A. Koolhaas, and A. Dekinga. 1993. Interactions between stomach structure and diet choice in shorebirds. Auk 110:552–564.

Piersma, T., and A. Lindstrom. 1997. Rapid reversible changes in organ size as a component of adaptive behaviour. Trends in Ecology and Evolution 12:134–138.

Pigliucci, M. 1996. How organisms respond to environmental changes: from phenotypes to molecules (and vice versa). Trends in Ecology and Evolution 11:168–173.

Rappole, J. H., and D. W. Warner. 1976. Relationships between behavior, physiology and weather in avian transients at a migration stopover site. Oecologia 26:193–212.

Ricklefs, R. E. 1996. Avian energetics, ecology, and evolution. Pages 1–30 *in* Avian Energetics and Nutritional Ecology (C. Carey, ed.). Chapman and Hall, New York.

Sabat, P., F. Novoa, F. Bozinovic, and C. Martinez del Rio. 1998. Dietary flexibility and intestinal plasticity in birds: a field and laboratory study. Physiological Zoology 71: 226–236.

Starck, J. M. 1999a. Phenotypic flexibility of the avian gizzard: rapid, reversible and repeated changes of organ size in response to changes in dietary fibre content. Journal of Experimental Biology 202:3171–3179.

Starck, J. M. 1999b. Structural flexibility of the gastro-intestinal tract of vertebrates: implications for evolutionary morphology. Zoologischer Anzeiger 238:87–101.

Stearns, S. C. 1989. The evolutionary significance of phenotypic plasticity. Bioscience 39:436–445.

Stevens, C. E., and I. D. Hume. 1995. Comparative physiology of the vertebrate digestive system. Cambridge University Press, Cambridge.

Toloza, E. M., M. Lam and J. M. Diamond, J. 1991. Nutrient extraction by cold-exposed mice: a test of digestive safety margins. American Journal of Physiology 261:G608–G620.

Travis, J. 1994. Evaluating the adaptive role of morphological plasticity. Pages 99–122 *in* Ecological Morphology: Integrative Organismal Biology (P. C. Wainwright and S. M. Reilly, eds.). University of Chicago Press, Chicago.

Walsberg, G. E., and C. W. Thompson. 1990. Annual changes in gizzard size and function in a frugivorous bird. Condor 92:794–795.

Witmer, M. C., and C. Martinez del Rio. 2001. The membrane-bound intestinal enzymes of waxwings and thrushes: adaptive and functional implications of patterns of enzyme activity. Physiological and Biochemical Zoology 74:584–593.

HANS WINKLER AND BERND LEISLER

To Be a Migrant

Ecomorphological Burdens and Chances

PREVIOUS WORK SHOWS THAT, in general, long-distance migrants have more pointed wings, shorter tails, and relatively less mass in the hind limb than nonmigrants. Here we focus on the constraints that may result from these and other characteristics of migrants and from covariances among traits. With Common Principal Components we compare intra- and interspecific variation in a migratory and a nonmigratory species of wheatear. The analysis finds axes of covariation and shows that, assuming that the underlying genetic covariance pattern is stable, selection for the migratory habit would not have opposed a preexisting component of covariation. Then we show in mimids, parulids, and sylviids that migration affects the hind limb as well. However, how the hind limb is modified depends on the prevalent locomotion style in a lineage. Migrant sylviid and parulid warblers and some other songbird migrants have relatively smaller, flatter skulls and smaller forebrains. This we interpret as first evidence for the hypothesis that in addition to the requirements of efficient long-distance flight, other factors such as temporal limitations on growth also may impinge on the morphology of migrants.

INTRODUCTION

Migration affects almost all aspects of the internal and external morphology of birds. Morphological differences between migrants and sedentary birds may be related to flight performance or to other ecological factors that act differently in these groups.

Several studies have shown that adaptations for long-distance migration account for significant portions of interspecific morphological variation (Leisler and Winkler 1985, 2003; Winkler and Leisler 1985, 1992). For example, nonpasserine migrants possess larger hearts than residents, and migrants allocate less mass to their hind limbs (Winkler and Leisler 1992). Migrant songbirds have longer sterna and deeper keels than closely related sedentary species, and there are corresponding differences in the muscular system (Calmaestra and Moreno 1998, 2000). Distal skeletal elements of the wing are longer in long-distance migrants than in sedentary birds, and differences are also reflected in external features of the wing. Although there is no consistent difference in wing-loading between migrants and residents, in most families migrants have higher-aspect-ratio wings, and wing tips are frequently more pointed, particularly in songbirds (Leisler and Winkler 1985, 2003; Winkler and Leisler 1985, 1992; Mönkkönen 1995; Lockwood et al. 1998; Fiedler 1998; Voelker 2001). The tail too is subjected to change when long-distance migration evolves. However, the role of the tail for flight performance has been studied in detail only recently (e.g., Thomas 1996; Maybury and Rayner 2001). Comparative studies in songbirds, at least, provide evidence that migrants tend to have squarer and shorter tails as opposed to the often long and graduated tails found in nonmigrants (Fitzpatrick 1999; Voelker 2001; Leisler and Winkler 2003). Other morphological consequences of migration not directly related to efficient flight over long distances are virtually unstudied. Because migrants and residents differ in their foraging strategies, food requirements, and habitat use (Leisler 1990), they also should develop corresponding morphologies.

Migration physiology and the tight seasonal schedule that migrants have to follow also may constrain the development of certain morphological features and thus contribute to consistent differences between migrants and residents.

Species change more or less easily in their characteristics. The degree of resistance to change has profound effects upon the rate of adaptive responses. Low additive genetic variance in the traits involved and genetic covariances that oppose the needed complex changes, for instance, would slow down evolution toward an optimal configuration (Lande 1979; Björklund and Merilä 1993; Schluter 2000). Migratory behavior as such seems to evolve rather quickly. House Finches introduced to eastern North America evolved partial migratory behavior within a few decades (Able and Belthoff 1998). And studies in sylviid songbirds subjected to artificial selection have shown that important physiological and behavioral traits associated with long-distance migration, such as fat deposition and migratory restlessness, can be completely changed in either direction within six generations (Berthold 1996, 1999). If the main axis of intraspecific covariation among traits points in the same direction as the changes required by selective pressures for long-distance migration, a migrant morphology could also evolve rapidly and without great selective costs.

One may argue that some lineages live in habitats that select for efficient forward flight anyway and that in such cases migration does not impose additional morphological changes (Eck and Bub 1992). Such habitats include open areas (and woodlands) typical, for example, for plovers and waterbirds, and also wheatears and pipits. The upper forest canopy is often, especially in the tropics, used by nomadic species and birds that track localized and highly dispersed resources (e.g., fruiting trees). Other behavior, for instance, certain display flights, may precondition for a migrant morphology or reinforce the effects of migration (Voelker 2001). In all these cases migration may not have created additional adaptations, or the necessary changes are small and thus easily achieved.

An important question in the ecomorphology of migration is whether adaptations for long-distance movements compromise other essential functions such as habitat use or flight displays (Leisler and Winkler 1985; Winkler and Leisler 1985, 1992; Hedenström and Møller 1992; Voelker 2001). One position is that the flight style most commonly used in daily life determines the morphology of the flight apparatus, such that long-distance migration accounts for some intraspecific variation, at best (Rayner 1988). Most studies, however, support the other view that adaptations for long-distance migration override the demands of possible other flight types, thus creating the already-mentioned conflicts. These conflicts are specifically expressed in traits that are associated with more than one function at the same time, and were called constraining features by Winkler and Leisler (1992). Thus, adaptations in the flight apparatus for long-distance flight are expected to compromise the ability of migrants to utilize all available habitats for aerial locomotion, especially in the context of foraging or escape. Selective pressures on long-distance flight are related to mass allocation, energetics, and drag. They should act particularly on the hind limbs because they are relatively large and do not actively contribute to aerial propulsion. We therefore expect to find many constraining features not only in the flight apparatus, but also in the hind limb.

One consequence of strong selection for long-distance migration is that we find much convergence and thus morphological similarities among migrants in different groups of birds. From this it also follows that the degree of intraspecific polymorphism, especially sexual dimorphism, should be lower in migrants, provided both sexes migrate. Significant dimorphism in wing shape as found in manakins, in which females are adapted for high mobility in large home ranges, whereas males' wings are adapted for particular display movements (Théry 1997), is not likely to occur in migrants. The prominent sexual dimorphism in wing and tail characters in most falcons is, for instance, absent in migrant species (Kemp and Crowe 1993). Willson et al. (1975) found that even dimorphism in bill shape is lower in migratory songbirds. Cooch et al. (1996) showed that temporal constraints on development may reduce sexual dimorphism in late broods of Snow Geese. This type of constraint has so far played no role in the discussion on the ecomorphology of migrants and certainly deserves more attention.

In this chapter we focus on the various constraints that adaptations for migration may pose for birds. Specifically, we test whether correlations among traits may pose a problem for the evolution of migrant morphologies, whether preadaptations for migration exist, and whether morphological adaptations for migration conflict with other ecological demands. We also provide first evidence for the hypothesis that in addition to the requirements of efficient long-distance flight, other factors such as temporal limitations on growth may also impinge on the morphology of migrants.

MATERIAL AND METHODS

Species Studied

Our comparative data came from the following four different species-rich lineages of passerine birds with well-resolved molecular phylogenies (partly published in Leisler and Winkler 2003): Old World *Sylvia;* acrocephaline warblers, genera *Acrocephalus* and *Hippolais;* pipits, genus *Anthus;* and New World parulid warblers, with particular emphasis on species of the genus *Dendroica.* Data on wheatears (*Oenanthe*) were taken from Leisler (1990, unpubl. material). We generally compared species, but in the pipits we also included subspecies with marked differences in migratory behavior. Another data set pertains to 27 species of songbirds (Portmann 1947). Their phylogeny was taken from Sibley and Ahlquist (1990), with further data from Sheldon and Winkler (1993), Arnaiz-Villena et al. (1998), and Groth (1998); we also calculated distances and topology of the corvids in the set from published GenBank cytochrome *b* sequences. We also included ecomorphological data from a Neotropical rain forest bird community in southern Venezuela (Winkler and Preleuthner 1999, 2001) comprising 33 species of the forest interior and 33 living in the forest canopy. The phylogenetic information for the 66 species involved was also basically derived from Sibley and Ahlquist (1990), with additional information gleaned from Lanyon (1985), Hackett and Rosenberg (1990), Prum (1990), Burns (1997), Hackett and Lehn (1997), Espinosa de los Monteros (1998), Lanyon and Omland (1999), Klicka et al. (2000), and Johnson et al. (2001).

Morphometric Analysis

External morphological measurements were taken from study skins and internal ones on skeletons from the collections at Tring, Tervuren, Vienna, Bonn, Berlin, Munich, Stuttgart, Washington, New York, Pretoria, Cape Town, Paris, and Radolfzell. The 17 external and 17 internal morphological features of three functional complexes (feeding, flight apparatus, and hind limb) were measured as defined in Leisler and Winkler (1985, 2003). We also included measurements of the neurocranium of sylviids and parulids.

In *Sylvia* and acrocephaline warblers we examined both external and internal characters, that is, 34 traits. Böhning-Gaese et al. (2003) allowed us to use external measurements of *Sylvia* species taken from skins (for sample sizes see Leisler and Winkler 2003). Parulid warblers were studied by use of internal characters only whereas the analysis of pipits was based on external measurements (Leisler and Winkler 2003). Almost all measurements were taken by one person (K.-H. Siebenrock). We used species means, which were corrected for size by dividing by the cube-root of body mass and then log-transformed (with the exception of tail graduation, which can attain zero or negative values) because we intended to analyze aspects of allometric shape rather than size. Information about body mass was taken from the handbooks (given in Leisler and Winkler 2003), Dunning (1992), and labels of the study skins. For the mimid data, taken from Engels (1940), a different approach was used; here, the humerus was taken as a size-related trait. In the analyses of a rain forest avifauna we avoided confounding effects of size by using the ratio of the extent of the primary projection to wing length (Kipp's ratio). Data for the brain sizes of passerines were taken from Portmann (1947). For migration distances we simply measured the distances between midpoint breeding range to midpoint wintering range of the (sub)species studied. Information was taken from the handbooks (stated in Leisler and Winkler 2003). Vegetation height data of *Sylvia* were taken from Leisler and Winkler (1992) and Shirihai et al. (2001).

Data Analysis

The method suitable for comparing within-species covariance structures among species is the analysis of Common Principal Components (Flury 1988; Steppan et al. 2002). We assumed that because of random variation, including measuring errors, not all components would be shared between species, so we used an algorithm that produces satisfactory approximations (Flury 1988) to calculate Partial Common Principal Components.

All multiple regressions are the result of exhaustive searches through all possible combinations of a fixed number of variables. The combination with the highest R^2 was selected, with no restriction on interpredictor correlations. We felt that the sample sizes (number of species) in the regression analyses justified the use of only three predictors, although the Akaike and the Bayesian Information Criteria (Burnham and Anderson 1998) in all cases suggested a higher number of independent variables. We then analyzed this set of variables in a Generalized Linear Model (Dobson 1990) that included phylogenetic information as suggested by Martins and Hansen (1997) and Pagel (1997) to account for correlations among species values. For the analysis of the mimid data, and to show the association of phylogenetic changes in frugivory and migration distance in *Sylvia* warblers, we also used independent contrasts (Felsenstein 1985).

RESULTS AND DISCUSSION

Phylogenetic Burdens and Preadaptations

We compared two species of the rather uniform genus *Oenanthe, O. lugubris* representing an African resident and *O. oenanthe* a long-distance migrant. We used Partial Common Principal Components to analyze the relationship between the within-species correlations and the between-species differences. We found two components common to the intraspecific variation in both species. The second most important one was strongly related to wing pointedness (primary projection). The individual scores of the two species overlapped greatly along the first axis, but not at all along the second one. The changes toward a migrant morphology were not related to the major axis of intraspecific variation. However, they followed the second main axis of covariation. The fact that this axis is orthogonal to the other axes implies, therefore, that the transition to migration is not burdened by strong intracorrelations opposing this change, even when assuming that the underlying genetic covariance pattern is stable. The implication is that the additive genetic variability within the residential population of *O. lugubris* is sufficient to allow the development of migratory habits rather easily. That also suggests that also in other genera in which migration evolved several times and migrants evolved from residents and residents from migrants, like *Vireo, Sylvia, Acrocephalus,* and *Anthus* (Cicero and Johnson 1998; Voelker 2001; Leisler and Winkler 2003), selective costs due to morphological reshaping were low.

Pipits of the genus *Anthus* live in open habitats and should be preadapted to migration. If this habit were to produce the same adaptations as migration, one would expect to find no relationship between morphology and migration distance. The regression analysis (33 species, 17 external characters) yielded a significant correlation between migration distance and longer primary projection, decreasing wing slotting, and less graduated tails ($R^2 = 0.643$, $p < 0.0001$; $R^2 = -0.637$, $p < 0.0001$, after inclusion of phylogeny). Although the relationship was weaker than in our other examples, the expectation was not met. The most likely explanation is that in this group open-country life did produce some adaptations similar to those needed for migration, but that the requirements of migration produced additional demands and adaptations. Voelker (2001) also found that wing shape and tail length were, to a limited extent, affected by migration in this group. In pipits and other species migratory behavior can be correlated with the possession of particular display flights (Voelker 2001; Hedenström and Møller 1992).

Levey and Stiles (1992) reasoned that dependency on spatially and temporally variable resources necessitates high mobility for those birds that track these resources, even in the Tropics. Canopy fruits represent such a resource, and these authors found that frugivores and nectarivores from canopy and edge habitats were overrepresented among altitudinal migrants. If seasonal dependency on plant reproduction in the Tropics was driving the evolution of long-distance migration (Levey and Stiles 1992) tropical frugivorous canopy birds should show morphological similarities with migrants. In the Neotropical bird community we analyzed, canopy birds were more frugivorous than those of the interior (Winkler and Preleuthner 2001). Canopy birds also had significantly more pointed wings (*t*-test, $p < 0.0002$), confirming the ideas of Levey and Stiles (1992). However, because of the strong association between stratum use and phylogenetic origin (a problem associated with the specific history of the Neotropical bird fauna [Ricklefs 2002]) this relation could not be corroborated in the phylogenetic analysis. Although Levey and Stiles (1992) listed the Parulinae among the counterexamples, there is an association between frugivory and long-distance migration at least in *Dendroica.* All the frugivorous species (Levey and Stiles 1992:appendix D) migrate more than 3,500 km (Fisher's exact test, $p < 0.02$). Seasonal frugivory is common among temperate long-distance migrants. In *Sylvia,* for example, evolutionary change to longer migration was associated with a change to more frugivory ($r = 0.708$, $p < 0.002$ for the independent contrasts [see Leisler and Winkler 2003]).

Ecomorphological Constraints

To examine a further test case for the proposed relationship between hind limb and flight, and hence migration, we took the skeletal measurements of nine species of mockingbirds and thrashers published by Engels (1940) and analyzed the relationship between peripheral wing elements (ulna, metacarp) and leg elements (femur, tarsus). The sample includes the migratory Catbird (*Dumetella carolinensis*) and Sage Thrasher (*Oreoscoptes montanus*). Humerus length was used to correct for size as described in the methods. It is strongly size correlated, much more than any of the more peripheral skeletal elements (H. Winkler, unpubl. results), whereas the distal wing elements are the most strongly related to long-distance flight performance (Winkler and Leisler 1992; Leisler and Winkler 2003). Canonical correlation analysis shows the predicted relationship (fig. 7.1). This strong relationship ($r = 0.934$, $X^2 = 13.33$, 4 d.f., $p < 0.01$) results from contrasting correlations of the wing (–0.999 for the ulna, and –0.919 for the metacarp) and hind limb elements (0.932 for the femur, 0.134 for the tarsus) with this factor. These results were also confirmed by analyzing the independent contrasts. The highest correlation between hind limb and forelimb traits in the dependent data was the negative correlation between femur and ulna $r = -0.869$ ($p < 0.001$), and $r = 0.877$ ($p < 0.01$) between the corresponding contrasts. In other words, an evolutionary change to a better development of the distal flight apparatus leads to a reduction in the femur, a proximal leg element.

We also analyzed morphometric data of Old World

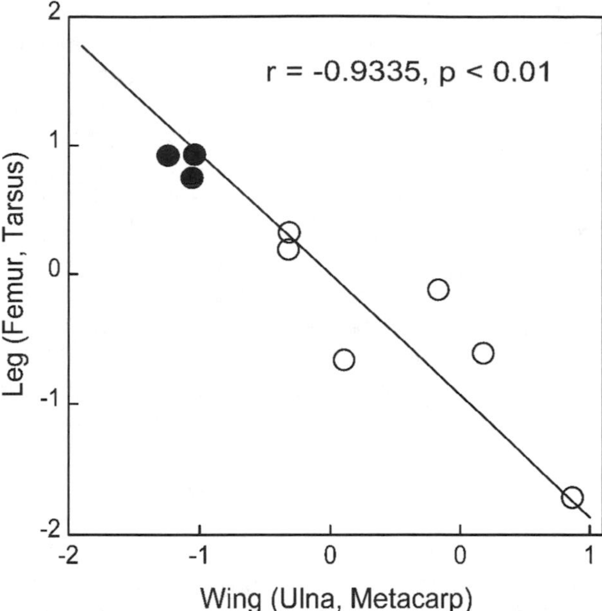

Fig. 7.1. Canonical correlation between leg elements (femur, tarsus) and distal-wing elements (ulna, metacarp) in thrashers (data from Engels 1940; see text for details). Species marked with filled circles represent ground-digging specialists (*Toxostoma lecontei, T. redivivum, T. dorsale*).

warblers (13 *Acrocephalus* and *Hippolais* species) and 49 parulid species. Both groups contain long-distance migrants as well as residents. We searched for the subset of three variables that explain migration distance of these species best. The best predictors explained 92.3% of the variance in the migration distance in acrocephaline warblers ($F_{3,9} = 36.20$, $p < 0.00005$) and 77.1% in parulid warblers ($F_{3,45} = 50.634$, $p < 0.00001$). The longer the migration distance was in acrocephaline species, the shorter was the notch of the inner web of the primaries, and the less were tail graduation and tarsal diameter. Femur length and the length of the carpometacarp were positively related, and skull height entered with a negative coefficient in parulid warblers (Leisler and Winkler 2003).

Although the analysis of the parulid data shows that the hind limb is related to migration, it seems to contradict the conclusions we reached in the analysis of the mimid data and the suggestion by Eaton et al. (1963) that lengthening of the forelimb goes along with a compensatory shortening of the hind limb. The significant partial relationship between lengthening of the femur and migration distance ($t = 10.727$, $p < 0.0001$) can be consolidated, however, with the results of earlier studies. Palmgren (1937) and Rüggeberg (1960) proposed that in long-legged perching birds, tarsus and femur should be of equal length, which implies a stronger relative increase in femur length. Commisso (1988) analyzed hind limb myology and bipedal locomotion in wood warblers and found that in long-legged perching species musculature was weakly developed. This suggests that length data alone may be misleading. The important variables are muscle mass and mode of locomotion (Leisler and Winkler 2003).

In earlier studies we found that there is a general trade-off between flight muscle mass and hind limb muscle mass (Winkler and Leisler 1992). In the acrocephaline warblers, measurements of tarsal diameters also indicate that migrants possess low-mass hind limbs.

To discuss the consequences of these complex relations for habitat use, we turn to the genus *Sylvia*, which includes long-distance and short-distance migrants, as well as permanent residents. In these species, too, just three morphological characters (metacarpal length and bill width correlate positively, skull height negatively) suffice to explain 92.5% of the variation in migration distance ($p < 0.0001$, $N = 17$). But many more traits are related to migration. We therefore applied a PCA on the morphological data to reduce their dimensionality (table 7.1). The first component (explaining 31.4% of the variance) was related to flight, with prominent positive contributions by the primary projection, metacarp, and keel height, and negative ones by tail graduation, tail length, and skull height and width; the second component (20.31% of the variance) related to all functional complexes; and the third (11.25%) represented variation with respect to a curved and narrow bill. A bootstrap analysis (10,000 samples) confirmed that all the eigenvalues associated with these components were significantly different from one at the $p < 0.0001$ level.

In the Introduction, we defined a trait as being constraining if it is involved in more than one function. In our example we can demonstrate this with migration and vegetation height. The first PC correlated well with both of these variables ($r = 0.878, p < 0.001; r = 0.596, p < 0.05$). The constraining component is obviously the first one and relates to the weak positive correlation between vegetation height and migration distance ($r = 0.459, p < 0.1$). The physical cause lies in the fact that in our example low vegetation also means dense vegetation and that the requirements for high maneuverability in dense vegetation are at variance with those for long-distance flight (Rayner 1988).

Table 7.1 Results of a PCA on 34 morphological variables in *Sylvia* Warblers

PC1		PC2	
Primary projection	0.930	Middle claw	0.745
Tail graduation	−0.887	Ulna	0.728
Metacarp	0.879	Hind claw	0.648
Keel height	0.861	Sternum width	0.647
Tail length	−0.822	Pelvis width	0.645
Distalwing tip	0.794	Humerus	0.626
Wing length	0.736	Notch length	0.583
Skull height	−0.731	Wing length	0.574
Coracoid	0.725	Skull length	0.541
Tarsus	−0.693	Middle toe	0.537
Bill width	−0.692	Bill length	0.535
% Variance explained	31.35		20.31

Note: Only those loadings are shown that substantially contribute to the interpretation of the components. See text for details.

The relationships between the hind limb and the flight apparatus are easy to interpret. The apparent association between skull dimensions and migration is more difficult to understand. It was revealed both in the regression analysis as well as in the PCA (first component) of the *Sylvia* warblers. The tendency of migrants to have flatter skulls was also found in parulids. Smaller and specifically flatter skulls in migrants could then mean that migration constrains forebrain size. We could test this idea only with another data set, completely independent of the previous ones. We used the data published by Portmann (1947), and to keep comparability within reasonable limits, we restricted the analysis to the 27 species of songbirds contained in that set. The hemisphere indices of Portmann (1947), the ratio of forebrain mass to the mass of the remaining brain, were first regressed on bird body mass. The residuals from this regression were used to analyze the relationship with migration (fig. 7.2). The species were classified as residents, short-distance migrants, and long-distance migrants. The results showed that a significant negative relationship between migration habit and forebrain size exists. The more migratory a species, the smaller its relative forebrain size ($r = -0.748$, $p < 0.0001$; $r = -0.727$, $p < 0.0001$, after inclusion of phylogeny).

CONCLUSIONS

Morphological integration or genetic correlations (Björklund and Merilä 1993; Schluter 2000) seem not to restrict the evolution of migratory behavior. It is therefore not surprising that the results of analyses with and without phylogenetic corrections differ only slightly (see also Leisler and Winkler [2003]). Furthermore, mapping migratory behavior on the phylogenies reveals that it may show up independently in different lineages and evolve quickly. Birds seem to be very adaptable even in morphological characters and we suggest that this flexibility, rather than migratory behavior as such (Winker 2000), constitutes an important avian key innovation. Our analyses indicate that birds of open habitats and tropical canopies, especially those that rely on ephemeral food sources, already have some prerequisites that increase their chances of becoming migrants. However, migration has its own strong requirements, which are rarely fully concordant with other demands. It is interesting to note that on the other end of the scale, as it were, lifestyles for which short wings are advantageous predispose for flightlessness (McCall et al. 1998).

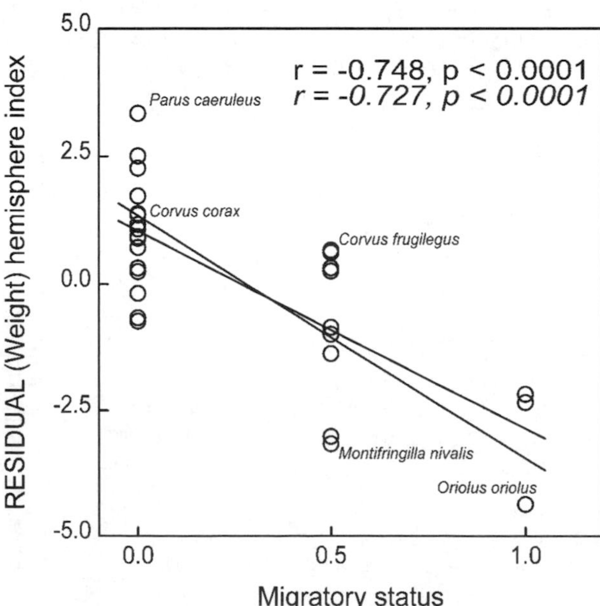

Fig. 7.2. Relationship between relative forebrain size (residuals of Portmann's 1947 hemispheric indices regressed on body mass) and migratory habits in 27 species of songbirds (0 = resident, 0.5 = short-distance migrant, 1 = long-distance migrant). Italics and the steeper regression line refer to an analysis with phylogeny included.

The morphological correlations and convergences in the flight apparatus related to migration are well established (Rayner 1988; Winkler and Leisler 1992; Leisler and Winkler 2003). Furthermore, the conflicting demands of different flight styles (Rayner 1988) are well understood. Thus we focus our discussion on the less obvious ecomorphological consequences of efficient long-distance travel. Migration constrains bipedal locomotion because of the problem of mass allocation. This may not be reflected in simple measurements of linear dimensions. The decisive variable is muscle mass. Running and hop-perching can be performed with relatively weak muscles. Certain clinging techniques, such as those used by *Acrocephalus* warblers (Leisler et al. 1989), may also involve little muscle mass, whereas those of parids and antbirds, for example, may be more demanding in this respect. There are obviously several solutions to this conflict and each lineage has to be studied carefully with respect to prevalent locomotion styles and morphological configuration. Looking at single measurements alone could otherwise yield contradictory results, and broad-scale comparisons like those by Calmaestra and Moreno (2000) may not be able to detect the effects of migration on hind limb morphology.

The relationship between skull shape and size and migratory behavior described here is little understood. It seems to have escaped attention by other workers and we only can speculate about its significance. Its emergence in two phylogenetically distinct lineages, the sylviids and the parulids, indicates that a common general principle is involved. Most intriguing is the finding that migrants develop smaller forebrains. This may impair certain cognitive functions including memory, behavioral flexibility, and social competence. Possible explanations fall into two categories, those involving flight performance and those related to other characteristics of migrants. Large brains add to the payload and consume much energy and may thus be too costly for migrants. The aerodynamic effects of reduced skull heights are probably too small to be important. Adaptations in the jaw apparatus associated with certain types of food (fruits) or foraging modes could pose a problem, as discussed in Leisler and Winkler (2003). After having checked this, our conclusion is that the effects of migration on skull and brain measurements are not mediated by foraging habits. Developmental constraints on growth due to the tight seasonal schedule of migrants may play a role, however.

The data show that migration can develop in almost any group of birds. In most aspects other than those closely linked to long-distance flight, migrants appear not to have very specialized morphologies. They may be, however, developmentally constrained because of their tight seasonal schedule. This possibility, namely that the limited time window open for postnatal growth has significant morphological consequences, was a new aspect emerging from our analyses, a result that inspires further study.

Migrants are opportunists that depend on easily accessible resources that they track by wandering between two worlds, and their behavior and ecology is well embodied in their morphology.

ACKNOWLEDGMENTS

We thank M. Adams, E. Bauernfeind, J. Blondel, T. Cassidy, A. Craig, R. van den Elzen, A. Kemp, M. Louette, E. Pasquet, R. Prys-Jones, J. Reichholf, H. Schifter, B. Stephan, M. P. Walters, and F. Woog for access to and loan of specimens. Our special thank goes to K.-H. Siebenrock, who meticulously measured all the specimens. We are grateful to A. Helbig, M. Schuda, and K. Böhning-Gaese, who provided us with molecular and morphometric data on *Sylvia* species, and to M. Preleuthner, who collected the morphological data on Neotropical birds. This latter work was funded by the Fonds zur Förderung der wissenschaftlichen Forschung, Austria, project P11563-BIO granted to H. Winkler.

LITERATURE CITED

Able, K. P., and J. R. Belthoff. 1998. Rapid "evolution" of migratory behaviour in the introduced house finch of eastern North America. Proceedings of the Royal Society of London, Series B, Biological Sciences 265:2063–2071.

Arnaiz-Villena, A., M. Álvarez-Tejado, C. Ruíz-del-Valle, C. García de-la-Torre, P. Varela, M. J. Recio, S. Ferre, and J. Martínez-Laso. 1998. Phylogeny and rapid Northern and Southern Hemisphere speciation of goldfinches during the Miocene and Pliocene epochs. Cellular and Molecular Life Sciences 54:1031–1041.

Berthold, P. 1996 Control of Bird Migration. Chapman and Hall, London.

Berthold, P. 1999. A comprehensive theory for the evolution, control and adaptability of avian migration. *In* Proceedings of the 22nd International Ornithological Congress, Durban (N. J. Adams and R. H. Slotow, eds.). Ostrich 70(1):1–11.

Björklund, M., and J. Merilä. 1993. Morphological differentiation in *Carduelis* finches: adaptive vs. constraint models. Journal of Evolutionary Biology 6:359–373.

Böhning-Gaese, K., M. D. Schuda, and A. J. Helbig. 2003. Weak phylogenetic effects on ecological niches os Sylvia warblers. Journal of Evolutionary Biology 16:956–965.

Burnham, K. P., and D. R. Anderson. 1998. Model Selection and Inference: A Practical Information-Theoretic Approach. Springer-Verlag, New York.

Burns, K. J. 1997. Molecular systematics of tanagers (Thraupinae): evolution and biogeography of a diverse radiation of Neotropical birds. Molecular Phylogenetics and Evolution 8:334–348.

Calmaestra, R. G., and E. Moreno. 1998. Ecomorphological patterns related to migration: a myological study with passerines. *In* Proceedings of the 22nd International Ornithological Congress, Durban (N. J. Adams and R. H. Slotow, eds.). Ostrich 69:466. [Abstract.]

Calmaestra, R. G., and E. Moreno. 2000. Ecomorphological patterns related to migration: a comparative osteological study with passerines. Journal of Zoology 252:495–501.

Cicero, C., and N. K. Johnson. 1998. Molecular phylogeny and ecological diversification in a clade of New World songbirds (genus *Vireo*). Molecular Ecology 7:1359–1370.

Commisso, F. W. 1988. Ecomorphology of locomotion of the parulid hind limb. Pages 2246–2264 *in* Proceedings of the 19th International Ornithological Congress.

Cooch, E. G., D. B. Lank, and F. Cooke 1996. Intraseasonal variation in the development of sexual size dimorphism in a precocial bird: evidence from the lesser snow goose. Journal of Animal Ecology 65:439–450.

Dobson, A. J. 1990. An Introduction to Generalized Linear Models. Chapman and Hall, London.

Dunning, J. B. 1992. CRC Handbook of Avian Body Masses. CRC Press, Boca Raton, Fla.

Eaton, S. W., P. D. O'Connor, M. B. Osterhaus, and B. Z. Anicete. 1963. Some osteological adaptations in Parulidae. Pages 71–83 *in* Proceedings of the 13th International Ornithological Congress.

Eck, S., and H. Bub. 1992. Die "Flügelspitze," ein wichtiges Maβ am Vogelflügel (mit speziellen Bemerkungen über die palaearktischen *Fringilla*- und *Anthus*-Arten). Anzeiger des Vereins Thüringer Ornithologen 1:79–84.

Engels, W. L. 1940. Structural adaptations in thrashers (Mimidae: genus *Toxostoma*) with comments on interspecific relationships. University of California Publications in Zoology 42:341–400.

Espinosa de los Monteros, A. 1998. Phylogenetic relationships among the trogons. Auk 115:937–954.

Felsenstein, J. 1985. Phylogenies and the comparative method. American Naturalist 125:1–15.

Fiedler, W. 1998. Der Flugapparat der Mönchsgrasmücke (*Sylvia atricapilla*): Erfassungsmethoden, intraspezifische Variabilität, phäno- und gentypische Varianz, Selektionsgründe und ökophysiologische Bedeutung. Ph.D. dissertation, Universität Tübingen, Tübingen, Germany.

Fitzpatrick, S. 1999. Tail length in birds in relation to tail shape, general flight ecology and sexual selection. Journal of Evolutionary Biology 12:49–60.

Flury, B. H. 1988. Common Principal Components and Related Multivariate Models. John Wiley and Sons, New York.

Groth, J. G. 1998. Molecular phylogenetics of finches and sparrows: consequences of character state removal in cytochrome *b* sequences. Molecular Phylogenetics and Evolution 10:377–390.

Hackett, S. J., and C. A. Lehn. 1997. Lack of genetic divergence in a genus (*Pteroglossus*) of Neotropical birds: the connection between life-history characteristics and levels of genetic divergence. Ornithological Monographs 48:267–279.

Hackett, S. J., and K. V. Rosenberg. 1990. Comparison of phenotypic and genetic differentiation in South American antwrens (Formicariidae). Auk 107:473–489.

Hedenström, A., and A. P. Møller. 1992. Morphological adaptations to song flight in passerine birds: a comparative study. Proceedings of the Royal Society of London, Series B, Biological Sciences 247:183–137.

Johnson, K .P., S. de Kort, K. Dinwoodey, A. C. Mateman, C. ten Cate, C. M. Lessells, and D. H. Clayton. 2001. A molecular phylogeny of the dove genera *Streptopelia* and *Columba*. Auk 118:874–887.

Kemp, A., and T. Crowe. 1993. A morphometric analysis of *Falco* species. Pages 223–232 *in* Biology and Conservation of Small Falcons (M. K. Nicholls and R. Clarke, eds.). Hawk and Owl Trust, London.

Klicka, J., K. P. Johnson, and S. M. Lanyon. 2000. New World nine-primaried oscine relationships: constructing a mitochondrial DNA framework. Auk 117:321–336.

Lande, R. 1979. Quantitative genetic analysis of multivariate evolution, applied to brain: body size allometry. Evolution 33:402–416.

Lanyon, S. M. 1985. Molecular perspective on higher-level relationships in the Tyrannoidea (Aves). Systematic Zoology 34:404–418.

Lanyon, S. M., and K.E. Omland. 1999. A molecular phylogeny of the blackbirds (Icteridae): five lineages revealed by cytochrome-*b* sequence data. Auk 116:629–639.

Leisler, B. 1990. Selection and use of habitat of wintering migrants. Pages 156–174 *in* Bird Migration (E. Gwinner, ed.). Springer-Verlag, Berlin.

Leisler, B., H. W. Ley, and H. Winkler. 1989. Habitat, behaviour and morphology of *Acrocephalus* warblers: an integrated analysis. Ornis Scandinavica 20:181–186.

Leisler, B., and H. Winkler. 1985. Ecomorphology. Current Ornithology 2:155–186.

Leisler, B., and H. Winkler. 2003. Morphological consequences of migration in passerines. Pages 175–186 *in* Avian Migration (P. Berthold, E. Gwinner, and E. Sonnenschein, eds.). Springer-Verlag, Berlin.

Levey, D. J., and F. G. Stiles. 1992. Evolutionary precursors of long-distance migration: resource availability and movement patterns in Neotropical landbirds. American Naturalist 140:447–476.

Lockwood, R., J. P. Swaddle, and J. M. V. Rayner. 1998. Avian wingtip shape reconsidered: wingtip shape indices and morphological adaptations to migration. Journal of Avian Biology 29:273–292.

Martins, E. P., and T. F. Hansen. 1997. Phylogenies and the comparative method: a general approach to incorporating phylogenetic information into the analysis of interspecific data. American Naturalist 149:646–667.

Maybury, W. J., and J. M. V. Rayner. 2001. The avian tail reduces body parasite drag by controlling flow separation and vortex shedding. Proceedings of the Royal Society of London, Series B, Biological Sciences 268:1405–1410.

McCall, R. A., S. Nee, and P. H. Harvey. 1998. The role of wing length in the evolution of avian flightlessness. Evolutionary Ecology 12:569–580.

Mönkkönen, M. 1995. Do migrant birds have more pointed wings? A comparative study. Evolutionary Ecology 9:520–528.

Pagel, M. 1997. Inferring evolutionary processes from phylogenies. Zoologica Scripta 26:331–348.

Palmgren, P. 1937. Beiträge zur biologischen Anatomie der hinteren Extremitäten der Vögel. Acta Societatis Fauna Flora Fennica 60:137–161.

Portmann, A. 1947. Études sur la cérébralisation chez les oiseaux: Pt. 2. Les indices intra-cérébraux. Alauda 15:1–15.

Prum, R. O. 1990. Phylogenetic analysis of the evolution of display behavior in the neotropical manakins (Aves: Pipridae). Ethology 84:202–231.

Rayner, J. M. V. 1988. Form and function in avian flight. Current Ornithology 5:1–66.

Ricklefs, R. E. 2002. Splendid isolation: historical ecology of the South American passerine fauna. Journal of Avian Biology 33:207–211.

Rüggeberg, T. 1960. Zur funktionellen Anatomie der hinteren Extremität einiger mitteleuropäischer Singvogelarten. Zeitschrift für wissenschaftliche Zoologie 164:1–106.

Schluter, D. 2000. The Ecology of Adaptive Radiation. Oxford University Press, Oxford and New York.

Sheldon, F. H., and D. W. Winkler. 1993. Intergeneric phylogenetic relationships of swallows estimated by DNA-DNA hybridization. Auk 110:798–824.

Shirihai, H. G. Gargallo, and A. J. Helbig. 2001. *Sylvia* Warblers: Identification, Taxonomy and Phylogeny of the Genus *Sylvia*. Helm, London.

Sibley, C. G., and J. E. Ahlquist. 1990. Phylogeny and classification of birds. Yale University Press, New Haven.

Steppan, S. J., P. C. Philips, and D. Houle. 2002. Comparative quantitative genetics: evolution of the G matrix. Trends in Ecology and Evolution 17:320–327.

Théry, M. 1997. Wing-shape variation in relation to ecology and sexual selection in five sympatric lekking manakins (Passeriformes: Pipridae). Ecotropica 3:9–19.

Thomas, A. L. R. 1996. Why do birds have tails? The tail as a drag reducing flap, and trim control. Journal of theoretical Biology 183:247–253.

Voelker, G. 2001. Morphological correlates of migratory distance and flight display in the avian genus *Anthus*. Biological Journal of the Linnean Society 73:425–435.

Willson, M. F., J. R. Karr, and R. R. Roth. 1975. Ecological aspects of avian bill-size variation. Wilson Bulletin 87:32–44.

Winker, K. 2000. Migration and speciation. Nature 404:36.

Winkler, H., and B. Leisler. 1985. Morphological aspects of habitat selection in birds. Pages 415–433 *in* Habitat Selection in Birds (M. L. Cody, ed.). Academic Press, Orlando, Fla.

Winkler, H., and B. Leisler. 1992. On the ecomorphology of migrants. Ibis 134 (Supplement): S21–S28.

Winkler, H., and M. Preleuthner. 1999. The ecomorphology of Neotropical frugivores. Acta Ornithologica 34:141–148.

Winkler, H., and M. Preleuthner. 2001. Behaviour and ecology of birds in tropical rain forest canopies. Plant Ecology 153:193–202.

SIEVERT ROHWER, LUKE K. BUTLER,
AND DANIEL R. FROEHLICH

Ecology and Demography of East-West Differences in Molt Scheduling of Neotropical Migrant Passerines

AMONG NEOTROPICAL MIGRANT PASSERINES that breed in North America, adults of just five of 55 eastern breeders migrate in the fall before molting, whereas adults of at least 13 of 26 western breeders undertake all or part of their fall migration before molting, a difference that remains significant when limited to phylogenetically independent contrasts. Several recently studied species that begin migration before molting (Bullock's Oriole, Western Warbling Vireo, Lazuli Bunting, Western Tanager) have been shown to move to the region of the Mexican Monsoon, where they pause to molt before continuing on to their wintering areas. Two complementary environmental forces favor this scheduling of molt and migration in western passerines. Most of the lowland West is exceedingly dry and unproductive in late summer, but a late-summer food flush favorable for molting is generated by the Mexican Monsoon rains of New Mexico, eastern Arizona, and northern Mexico. Some authors have suggested that migrating before the fall molt is favored for defending winter territories, but we found no comparative support for this hypothesis. Just 18% of winter territorial species migrate before molting, whereas 33% of species that are not territorial in winter migrate before molting. Tyrannid flycatchers vary greatly in the scheduling of molt and migration, with several eastern and transcontinental species migrating before their fall molt. This variation is poorly understood, but two factors may be especially important for larger tyrannids: breeding-season time constraints (associated with prolonged parental care) and the need for slow flight-feather replacement to maintain maneuverability for capturing large insects. In some North American passerines juveniles molt before migrating but adults molt after; in other species both juveniles and adults migrate before molting. Part of this variation may be explained by interspecific differences in the quality of the juvenile plumage. Species in which both juveniles and adults migrate before molting have better-quality juvenile plumage than species in which juveniles, but not adults, molt before migrating. Finally, passerines of the western Palearctic-African migration system differ from Neotropical migrants in that none (except swallows) delay molt of their body plumage until after migration. In contrast, many Palearctic-African migrant passerines delay replacement of their flight feathers until late in winter, and we argue that they do so because their African wintering habitats are more abrasive than the habitats used by North American passerines wintering in the Neotropics.

INTRODUCTION

Our goal in this chapter is to provide a synthetic overview of molt-migration scheduling in Neotropical migrant passerines that breed in North America. The impetus for this study is the recent discovery of an east-west contrast in the scheduling of the prebasic molt relative to the fall migration in a number of closely related east-west species pairs (Rohwer and Manning 1990; Young 1991; Voelker and Rohwer 1998). Surprisingly, the special scheduling of the fall molt of passerines breeding in western North America seems without precedent in European passerines that winter in sub-Saharan Africa.

Breeding, migrating, and molting place special energetic demands on birds and are often segregated in time. Given that environmental constraints appear to prevent winter breeding in all Palearctic and Neotropical migrants (hence, fixing the timing of migrations and breeding), only molt can vary with respect to its scheduling in the annual cycle. The complete annual molt in adults may occur on the breeding grounds before the fall migration or on the wintering grounds after the fall migration, or parts of it may occur in different places. To our knowledge, all Neotropical migrant passerine species that replace different components of their plumage in different regions replace body plumage on the breeding grounds but delay replacement of most flight feathers until after the fall migration (Jenni and Winkler 1994; Pyle 1997), although the reverse would be more difficult to document. In some species the molt is scheduled such that the fall migration is broken into two phases; the first phase takes these species to special stopover sites where they molt and the second phase takes them on to their final wintering area (Rohwer and Manning 1990; Young 1991; Jones 1995).

In the following analyses we use ecological and demographic variables to explain differences in the scheduling of the fall molt. We found it helpful to conceptualize these variables as "pushes" and "pulls." Pushes are ecological or demographic characteristics that disfavor molting in that region. Thus pushes favor birds leaving a region to molt elsewhere. Pulls are ecological or demographic characteristics of a region or population that favor individuals moving to that region before they begin a molt.

Large differences in precipitation and primary productivity in mid and late summer seem to drive differences in molt scheduling between eastern and western species. Remarkably, east-west differences between adults often do not apply to juveniles, and we suggest that species differences in the quality of the juvenile plumage, which may be a consequence of different rates of nest predation, may help explain these contrasts between age classes. We also propose that structural contrasts between the winter environments used by Palearctic and Nearctic migrants may explain continental contrasts in their molt scheduling better than previous hypotheses do. Finally, we make a special effort to identify weaknesses in the comparative data set for North American migrants and to identify species whose ecology or relationships make them particularly worthy of detailed study.

Scientific names of North American birds that are included in the Appendix to this chapter are omitted from the text.

EAST-WEST COMPARISONS

Categorizing Species

We base this analysis on the 115 migratory passerines (listed in the Appendix to this chapter) that breed in North America and mostly winter considerably south of the United States. We treated species as migratory if most individuals winter south of the United States (e.g., Lesser Goldfinch, Townsend's Warbler, Painted Bunting). Two groups were difficult to categorize with regard to migration—those species with only small parts of their breeding range in the United States and those few western breeders whose winter ranges scarcely extend south of Mexico. We excluded most migrant emberizids because they winter primarily north of Mexico. We further divided the 115 migratory species into four categories according to their breeding distributions: eastern, breeding largely east of the Rocky Mountains; western, breeding largely west of the Rocky Mountains; northern, having transcontinental ranges north of the arid intermontane region of western North America (hereafter the Great Basin); and continental, having more southern transcontinental ranges that include the Great Basin (Appendix to this chapter).

As is often the case in broad comparative studies, this one suffers from unevenness in the quality of the data. For some species the literature is conflicted over the timing and the location of the fall molt (e.g., Dickcissel [cf. Mengel 1965 and Pyle 1997]). For others there are simply no data, and differences among other close relatives caution against presumption. We used Pyle (1997) as our primary reference but provided additional citations (Appendix to this chapter) when we found better or updated references.

The fundamental question in this study was whether western passerines show multiple independent origins of molting after initiating the fall migration. We scored all Neotropical migrants as molting body feathers *before* or *after* a substantial southward movement. We used "substantial southward movement" because several western species interrupt their southward migration to molt before continuing on to their wintering grounds. For characters that have few states and that are highly labile over evolutionary history, the reliability of node reconstruction diminishes as one moves deeper into a phylogenetic tree (Omland and Lanyon 2000). Because the East has more than twice as many Neotropical migrants as the West, our test of east-west differences using phylogenetically independent contrasts was limited mostly by the number of western species with close relatives in the East. Some western species had no eastern relatives, and two, Scott's Oriole and Green-tailed Towhee, had such distant eastern relatives that we omitted them because of the uncertainty of correctly reconstructing internal nodes. Further, we excluded comparisons involving Least Flycatcher and Orchard Oriole because they had no close relatives in the West. With these restrictions we had 17 independent contrasts in molt-migration scheduling between eastern and western clades; all of these contrasts were comparisons of species in the same genus.

East-West Environmental Contrasts

Homogeneity and contrast characterize the late-summer molting environments encountered by eastern and western species, respectively. Annual precipitation is high in the East and low in most of the West. In the East, precipitation is evenly distributed throughout the year and primary productivity remains high in late summer. In contrast, most precipitation that results in primary productivity in the West comes as winter rain or snow (see rainfall maps in Espenshade 1991). Thus, throughout most of the West late-summer conditions are exceedingly dry. Variability in western environments is further augmented by the great topological relief of the region. Winter snow packs in the Pacific coastal ranges, the Cascades and, to a lesser extent, the Rockies and the Sierra Nevada Mountains, provide sufficient runoff for montane regions to remain productive into the late-summer molting season.

Late-summer conditions also vary substantially with latitude in the American West. To the south of the Great Basin, over a large arid region including New Mexico, eastern Arizona, and northern and western Mexico, most of the limited annual precipitation occurs in July, August, and September (hereafter the Mexican Monsoon, following Comrie and Glenn 1998). The flush in productivity that follows these rains has long been known to stimulate late-summer breeding in several southwest species including the Curve-billed Thrasher (*Toxostoma curvirostre* [Short 1974]), Rufous-winged Sparrow (*Aimophila carpalis* [Phillips 1951]), and Cassin's Sparrow (*Aimophila cassinii* [Ohmart 1966]), and several other insectivorous animals (Reichert 1979).

Given the energy and protein demands of molting, aridity seems to create a strong "push" for insectivorous species of the lowland West to depart their breeding grounds before molting. Many of these same species also seem to be "pulled" to more productive regions for molting. For example, various authors have suggested that the flush of productivity associated with the Mexican Monsoon constitutes a strong "pull" to the southwest for molting by various lowland western insectivorous birds (Rohwer and Manning 1990; Young 1991; Voelker and Rohwer 1998; Butler et al. 2002). In addition, species that breed at low elevations may be "pulled" to molt in more productive high-elevation habitats (Greenberg et al. 1974; Butler et al. 2002), whereas others may move directly to their Neotropical wintering grounds before molting (Johnson 1963).

EAST-WEST CONTRASTS IN MOLT SCHEDULING

Adult Prebasic Molts

Most eastern species molt in their breeding range. Just five of 55 eastern breeders (Yellow-bellied Flycatcher, Gray Kingbird, Eastern Wood-Pewee, Orchard Oriole, and central populations of Painted Bunting) are thought to leave the breeding range for the adult prebasic molt of body feathers (table 8.1A, Appendix to this chapter), and the categorizations of some of these five could change with more study. Orchard Orioles certainly leave without molting. East coast populations of Painted Bunting molt on the breeding grounds but central populations, which we considered eastern for consistency, move to the American Southwest before molting (Thompson 1991).

By contrast, adults of at least half of the 26 western breeders depart their breeding grounds before undergoing the prebasic molt (table 8.1A, Appendix to this chapter). It

Table 8.1 Breeding range and location of the adult prebasic molt of body feathers

A. Species counts without correction for relatedness ($p < 0.0001$ for East/West × off/on)

	Off	On	Totals
East	5	50	55
West	13	13	26
North	1	5	6
Continental	8	11	19
Southwestern	1	3	4
Totals	28	82	

B. Phylogenetically independent contrasts (one-tailed $p = 0.026$)

	Off	On	Totals
East	2	15	17
West	9	8	17

is well established that several of these western species molt in the Mexican Monsoon region: Bullock's Oriole (Rohwer and Manning 1990) , Lazuli Bunting (Young 1991), Western Warbling Vireo (Voelker and Rohwer 1998), and Western Tanager (Butler et al. 2002). Adults of several other species are known to depart the breeding grounds before molting, but whether they molt in the Mexican Monsoon region or on the wintering grounds requires further study (e.g., Western Kingbird, Western Wood-Pewee, Ash-throated Flycatcher, and Black-headed Grosbeak).

Independent Contrasts

The preceding summary of east-west contrasts included several groups of closely related species with similar scheduling of molt and migration. Thus, we used phylogenetically independent contrasts to explore whether the east-west differences in the full species list included enough independent changes to suggest adaptive change unequivocally. There are far fewer western than eastern species, so we matched each western species or each western clade having similar molt scheduling with its most closely related eastern species or clade. We then scored the molt scheduling of every species or clade as molting on or off the breeding grounds. For these comparisons we used the phylogenies of Klicka et al. (2001) for *Passerina,* Murray et al. (1994) and Cicero and Johnson (1998) for *Vireo,* Zink and Johnson (1984) for *Empidonax* and *Contopus,* Omland et al. (1999) for *Icterus,* Lovette and Bermingham (1999) and Lovette et al. (1999) for *Dendroica,* Burns (1998) for *Piranga,* and Zink et al. (2000) for *Vermivora.*

Phylogenetically independent contrasts revealed the same trend observed in the full data set, even though the number of species or clades that could be contrasted fell to 34 (Fisher's one-tailed $p = 0.026$ [table 8.1B]). We consider this analysis overly conservative because molt scheduling is sufficiently labile that close relatives with the same molt scheduling, which generate just one contrast, may represent replicates of species that could have changed but did not do so because their current molt scheduling is being maintained by stabilizing selection (Reeve and Sherman 1993; Hansen 1997). Several closely related east-west pairs show differences in molt scheduling (e.g., eastern and western Warbling Vireos, Black-headed and Rose-breasted Grosbeaks), and Painted Buntings even show population differences in molt scheduling between East Coast and Midwestern breeding populations (Thompson 1991). Furthermore, Western Tanagers feature variation among adults in molt scheduling, suggesting the presence of variation sufficient for rapid evolutionary changes in molt scheduling (Butler et al. 2002).

We may also explore whether phylogenetic constraints explain the exceptions to the general pattern described above. Specifically, are the few eastern species that depart their breeding ranges before molting closely related to western species that also migrate before molting? And are western species that molt on their breeding ranges closely related to eastern species that also molt before migrating? This further exploration of phylogenetic constraints is especially important for western insectivores that are not high-elevation breeders. Thus we exclude from this analysis western species that breed in the mountains because phylogeny and environment could be pulling them toward the same molt scheduling.

Of the eight species that oppose the rule, east "on" and west "off," five had sister clades "pulling" them to be opposed to the rule and three did not. Thus, there is weak but nonsignificant evidence of phylogenetic constraint explaining these exceptions ($p = 0.36$; one-tailed binomial test). Note, however, that this comparison may be strengthened by asking how much potential constraint is seen in those western species that molt "off." There were eight such western clades, all of which had an eastern relative molting "on," suggesting no phylogenetic constraint ($p = 0.004$; one-tailed binomial test).

WHERE DO WESTERN SPECIES MOLT?

Some lowland western insectivores are apparently "pushed" away from their breeding range for molting by late-summer aridity, and some are also "pulled" to the Mexican Monsoon region. A recent series of papers treating Bullock's Oriole (Rohwer and Manning 1990), Lazuli Bunting (Young 1991), Western Warbling Vireo (Voelker and Rohwer 1998), and Western Tanager (Butler et al. 2002) shows that adults of these species mostly molt in the southwest monsoon region. Unfortunately, the early literature on molt-migration scheduling did not assess stopover molting in the Mexican Monsoon region (Johnson 1963, 1974). Johnson (1963) implies that Gray Flycatchers molt on their winter range, but he provides no data on the number of molting specimens examined or on their dates and localities of collection.

The Mexican Monsoon region is clearly an important area for the conservation of the several species that are known to molt there, but little is known about their ecology and behavior during their fall molts. Furthermore, several other western species need to be examined for the possibility of stopover molting in the region. These include Dusky Flycatcher, Western Wood-Pewee, Western Kingbird, Ash-throated Flycatcher, Cordilleran Flycatcher, Pacific Slope Flycatcher, Lesser Goldfinch, Black-headed Grosbeak, western populations of Summer Tanager, northern populations of Hepatic Tanager, and Hooded Oriole. Several of these species arrive at Southeast Farallon Island in July or early August (Pyle and Henderson 1991), suggesting departure from their breeding grounds before molting. Despite the paucity of late-summer specimens from the region of the Mexican Monsoon, museum studies will likely provide the best first approach to examining the possibility of stopover molting in this region for these species.

Western topography may explain why several western breeders molt on their summer range. Most western mountain breeders that forage in conifers molt in their breeding range (Townsend's, Hermit, and Black-throated Gray Warblers, Hammond's Flycatcher, Audubon's Warbler). Other western species that breed in lower-elevation coniferous and riparian habitats and molt in their breeding range could easily move upslope in late summer for molting. For example, Black-throated Gray and Audubon's Warblers have populations that breed in relatively low-elevation coniferous forests of the Great Basin, and we suspect that they move upslope to molt in more productive habitats. Unfortunately, we know of few data addressing this issue of regional upslope movements for molting, though upslope movements have been documented for Nashville and Orange-crowned Warblers (Greenberg et al. 1974) and are implied for juvenile Western Tanagers (Butler et al. 2002).

Molt-migration patterns of other western species that molt on their breeding grounds can be explained by dietary differences, or are poorly studied. Like other migrant emberizids, Green-tailed Towhees molt on their arid western breeding grounds, where seeds are abundant in late summer. The remaining group of western species that apparently molt in their breeding range before moving south are poorly studied. At least two species, Orange-crowned and Nashville Warblers, are known to move upslope to more productive late-summer habitats for molting (Greenberg et al. 1974; Steele and McCormick 1995), and similar movements seem likely for additional species. Further, additional species are likely to be found to migrate to the region of the Mexican Monsoon before molting. Possibilities include Scott's Oriole, Hooded Oriole, Cassin's Vireo, Plumbeous Vireo, western populations of Summer Tanager, and northern populations of Hepatic Tanager.

WINTER TERRITORIALITY AND MOLT SCHEDULING

Several authors have suggested that winter territoriality "pulls" individuals to migrate before molting so they arrive early on their wintering grounds and benefit from site dominance (Alerstam and Högstedt 1982; Sealy and Biermann 1983; Lindstrom et al. 1993). This argument seems flawed on two counts. First, it ignores the cryptic behavior characteristic of molting birds because of the energetic and aerodynamic costs of molting and because of the risks of damaging soft, growing quills (Vega Riveria et al. 1998). Second, it is questionable that prior-residency advantages would be sufficient to enable early migrants molting on their winter territories to defend these territories from later migrants that could molt before migrating and, thus, be fully competitive upon arrival. Late-arriving birds in fresh plumage, unencumbered by energetic costs of molt and risks to growing feathers, should easily displace molting territory holders. Ironically, this use of the prior-residency argument is never applied to the breeding season, when it would be just as logical. That no passerine delays the spring molt until arrival on the breeding grounds strongly argues against territoriality as a pull favoring molting after migrating.

The prior-residency hypothesis predicts that Neotropical migrants that are territorial during winter should be more likely to molt after migrating compared with species that do not hold winter territories. We found data on winter territoriality and molt scheduling for 112 Neotropical migrants (Appendix to this chapter). In contrast to the prior-residency hypothesis, 33.3% of the 48 non-territorial species molt on the wintering grounds, but only 18.5% of the 54 territorial species molt on the wintering grounds (Fisher's exact p = 0.11, but opposite the predicted direction).

BREEDING-RANGE TIME CONSTRAINTS AND MOLT SCHEDULING

As aerial foragers, swallows are forced to molt slowly, making it difficult to complete the molt on the breeding grounds. Regardless of their distribution in North America, swallows molt and migrate simultaneously during the overland component of their fall migration (Niles 1972; Stutchbury and Rohwer 1990; Pyle 1997; Yuri and Rohwer 1997). Swallows wintering far into the Southern Hemisphere replace few flight feathers before arriving on their wintering grounds and arrest their molt before crossing the Gulf of Mexico. Those with more northern wintering distributions replace their flight feathers in late summer and fall, and eastern populations of Northern Rough-winged Swallows pause at the coast to complete flight-feather molt before crossing the Gulf of Mexico (Yuri and Rohwer 1997). Because swallows should molt slowly, time constraints likely favor most of the fall molt occurring after migration in those that are transequatorial migrants.

For North American passerines there are three general rules of molt-migration scheduling: eastern species molt before migration; western arid species molt after migration; and aerial foragers molt during overland migration. Tyrannids stand out as generating the most exceptions to these rules. Two eastern tyrannids molt their body and flight feathers after arriving on the winter range (Eastern Wood-Pewee and Gray Kingbird) and additional eastern tyrannids (Least Flycatcher, Great Crested Flycatcher, and, possibly, Yellow-bellied Flycatcher [Hussell 1982]) molt their body feathers (or at least begin this molt) before migrating, but delay replacement of their flight feathers until arrival on their winter grounds. A suite of transcontinental and western tyrannids also leaves its breeding range before molting (Eastern Kingbird, Willow Flycatcher, Alder Flycatcher, Olive-sided Flycatcher, Western Kingbird, Scissor-tailed Flycatcher).

As with swallows, something seems to be special about flycatchers. Both groups catch prey in the air, and they likely replace flight feathers as slowly as possible to maintain aerial efficiency (Johnson 1963). Because swallows forage while flying and migrate during the day, there should be little cost to molting during overland migrations. In contrast, tyrannids sally to forage and mostly migrate at night, making molting while migrating unlikely. Johnson (1963) observed that molts occurring after the fall migration tend to be protracted, citing data for tyrannids and swallows, but this idea has not been well tested in a comparative context. Two predictions might be examined. First, tyrannids should replace flight feathers much more slowly than gleaners and, possibly, even more slowly than swallows because they take larger prey that require greater maneuverability in flight. For this test it will be important to eliminate the confounding influence of day length by comparing tyrannids with other passerines molting after the fall migration. For temperate molters, decreasing day length in the fall is well known to accelerate the rate of molt (see Dawson 1994). Second, time constraints, driven by how early the eggs are laid in spring and in the length of post-fledging care, might explain additional variation within tyrannids. For example, post-fledging care is prolonged in Eastern Kingbirds, with young being fed several times an hour for nearly a month after fledging (Morehouse and Brewer 1968). This suggests that tyrannids may have longer periods of fledgling dependency than foliage-gleaning insectivores, presumably because acrobatic aerial foraging requires much time to learn. Temperate phoebes (*Sayornis*) molt on their breeding range, unlike other medium-sized tyrannids. Perhaps they have time to molt because they nest relatively early, and because they are smaller than kingbirds.

AGE CLASS CONCORDANCE

This section treats the scheduling of the first prebasic molt of body feathers by considering concordance between adults and yearlings. For 110 species the scheduling of both the adult (or definitive prebasic) molt and the first (post-juvenile) prebasic molt is known. Of these, 103 species feature age class concordance, and only seven do not (table 8.2). Moreover, every case of nonconcordance is a result of adults migrating *before* they molt and young migrating *after* they molt; thus, there are no "on-off" species featuring adults molting before migration and juveniles migrating southward in juvenile plumage before molting (table 8.2). The "off-on" group (adults off, juveniles on) includes two eastern species, four western species, and one transcontinental species. The "off-off" group (adults off, juveniles off) includes three eastern species, nine western species, and nine transcontinental species (Appendix to this chapter).

This summary of age class concordance suggests that there may be important reasons for the contrast between "off-off" species (in which both adults and young depart the breeding range before molting) and "off-on" species (in which adults molt after departing the breeding range, while young remain behind to molt on the breeding range before migrating). Why do the young of some species migrate southward in juvenile plumage and then molt *after* migrating (like adults), whereas juveniles of other species remain in their breeding range and molt *before* migrating (*unlike* adults)? In the following section we develop a hypothesis based on variation in quality of juvenile plumages to explain this age class discordance in molt scheduling.

Table 8.2 Concordance between adults and yearlings on the location of the prebasic molt of body feathers

		Yearlings		
		Off	On	Totals
Adults	Off	21	7	28
	On	0	82	82
	Totals	21	89	110

FEATHER QUALITY AND ADULT-YEARLING DISCORDANCE

Ornithologists have long known that birds in juvenile plumage can often be recognized by the loose texture of their body plumage (Göhringer 1951; Dwight 1900). Juvenile feathers often have fewer barbs per unit of rachis and reduced connectivity between adjacent barbs, even on the distal, most vanelike, portion of their feathers (Rohwer and Manning 1990; Speidel and Rohwer, unpubl.). The functional significance of this difference is largely unexplored, but contrasts between species are striking (Butler et al., unpubl.). Juveniles with highly "decomposed" barbs give the impression of having grown cheap feathers that facilitate rapid fledging, perhaps because of intense nest predation.

Thus, small passerines may have two counts against undertaking the fall migration in juvenile plumage. First, aerodynamic drag is due in part to frictional forces between the air and the surface of the body (skin friction). The relative contribution of skin friction to overall aerodynamic drag increases as body size decreases. For species with equally decomposed juvenile plumage, therefore, smaller species should suffer more from frictional aerodynamic drag than larger species. Second, as body size decreases, the risk of nest predation likely goes up, because larger birds are more capable of nest defense. Thus, small species not only experience more of their total drag as skin friction, but they also may be more likely to grow cheap "decomposed" juvenile feathers to facilitate rapid fledging. By this logic nestling predation drives the quality of the juvenile plumage, which entrains things like how long this plumage is worn and whether adults feed fledglings while they undergo their first prebasic molt of body feathers.

The preceding speculation about juvenile plumage quality suggests that "decomposed" juvenile plumages "hold back" juveniles from migrating with adults. To explore this question we compared feather quality in species that were off-off concordant with those that were off-on discordant (having adults that migrate before molting and juveniles after molting [table 8.2]). To assess feather quality we counted barbs on flank feathers of adults and juveniles. For each species, we measured three to five juvenile specimens and three to five adult specimens, with similar proportions of each sex in each age class. For each specimen, we counted the total number of barbs (medial and lateral side of each feather) on the distal 1 cm of three flank feathers, starting within 2–3 mm from the tip of the feather, at the point where barbs began being regularly spaced. Mean barb counts were computed for each specimen and used to compute age class means for each species. All barb measurements were made by LKB using a lighted, 10× magnifying glass.

To correct for species differences in mean barb number we compared $(C_a - Cj)/C_a$, where C_a is the adult barb count and C_j is the juvenile barb count. Values above 0 characterize species for which adults have higher-quality feathers than juveniles, and values below 0 would represent species for which adults have lower-quality feathers than juveniles (fig. 8.1).

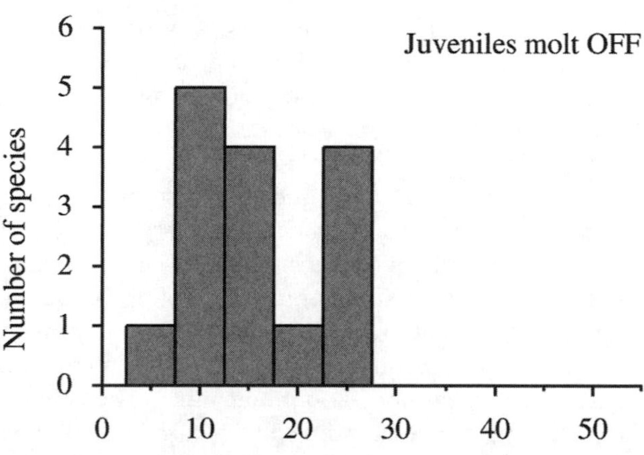

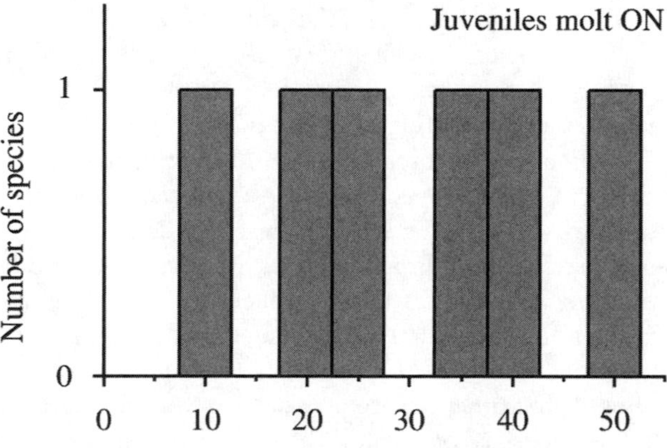

Fig. 8.1. Quality contrasts between the adult definitive basic and the juvenile body plumage for species in which adults migrate before they molt. The top graph contrasts these plumages for species in which juveniles molt after they have migrated in juvenile plumage; the bottom graph contrasts species in which juveniles remain behind to replace their body plumage on the breeding grounds and migrate in first basic plumage.

Unfortunately, the list of comparisons we were able to make is limited because there are few off-on species (table 8.2), so phylogenetically independent contrasts could not be attempted. Nonetheless, those species characterized by discordance (adults molting after migration, juveniles molting before migration) have barb count differences significantly greater than species in which juveniles migrate with adults before molting ($p = 0.016$, Mann-Whitney U-test [fig. 8.1]). Further exploration of this comparative contrast will require incorporating migrants from other parts of the world with age contrasts in molt scheduling (if any exist!), because there are so few Neotropical migrants with age class discordance. Nevertheless, the success of this comparison suggests that feather quality merits further exploration: Will drag forces be higher in species with more decomposed juvenile plumages? Does a highly decomposed juvenile plumage reflect the demography of nesting success, or is it related to something entirely different, such as thermoregulation?

COMPARISONS WITH PALEARCTIC MIGRANTS WINTERING IN AFRICA

Does the Push of Aridity Apply to Trans-Saharan Migrants?

The contrast in molt scheduling between eastern and western Neotropical migrant passerines prompted us to ask how general might be the "push" of late-summer aridity. Thus, we asked if the subset of trans-Saharan migrants having predominantly Mediterranean breeding distributions in Europe and western Asia tended to molt after their southward migration. We examined the 51 species listed in table 8.2 of Jenni and Winkler (1994) for this comparison. Other than swallows, all of the Palearctic migrants included in their table renew body plumage before migrating. Thus, the only comparison relevant to the Nearctic migrants of western North America was the scheduling of flight feather replacement.

To our surprise, species that breed in the Mediterranean region were no more likely to molt after migrating than species that are more widely distributed in the eastern Palearctic. For the 16 species with breeding ranges restricted to the Mediterranean region, 43.8% molted their flight (and body) feathers after the fall migration. For the 35 species with larger breeding ranges in the western Palearctic, 42.9% molted after the fall migration. Although late-summer aridity is far more severe and affects a much larger region of western North America than is true of the Mediterranean region of Europe, this result weakly suggests that the "pull" of the Mexican Monsoon may be as important as the "push" to depart a drying breeding range for passerines of the American West.

Molt Scheduling in Trans-Saharan Migrants

Trans-Saharan migrants of the western Palearctic generally molt body plumage in the breeding range, but they are remarkably variable in the timing of their primary molt relative to the fall migration (Moreau 1972; Alerstam and Högstedt 1982; Jenni and Winkler 1994; Jones 1995; Svensson and Hedenstrom 1999). In contrast, Neotropical migrants typically molt their body and flight feathers together in the fall.

Using *Phylloscopus* warblers, Svensson and Hedenström (1999) showed seven to ten independent origins of body and flight feather molt being separated by migration. Evolutionary lability in molt scheduling is strongly supported, but there is little consensus about ecological and demographic causes. As has been suggested for North American species (Sealy and Biermann 1983), winter territoriality is sometimes suggested as favoring departure from the breeding range before molting (Alerstam and Högstedt 1982; Lindstrom et al. 1993), but we are unaware of comparative tests of this hypothesis for Palearctic migrants. Time constraints are suggested by two associations. First, delaying molt until winter is significantly associated with longer migrations in *Phylloscopus* (Svensson and Hedenström 1999). Second, in some Eurasian species, multibrooding is associated with a delay in replacement of flight feathers until arrival on their African wintering grounds, where the birds spend most of the year (Alerstam and Högstedt 1982; Jones 1995).

Insights from Nearctic-Palearctic Comparisons

The hypothesis that winter territoriality favors molting after migration is based on unlikely assumptions and, as shown above, it fails empirically for Neotropical migrants. Time constraints must be important for molt scheduling, particularly for the flight feathers of large birds (Pietiäinen et al. 1984; Langston and Rohwer 1995, 1996; Rohwer 1999). Given the low variability in the rate of flight feather growth in birds of all sizes (Rohwer 1999), time constraints on molt scheduling should be minimal for most small passerines because the summed length of their primaries is small. However, in Arctic breeding passerines the molt is often so accelerated that birds in heavy molt are nearly flightless (see Jenni and Winkler 1994). Time constraints are important for aerial foragers that must molt slowly to minimize foraging costs. Indeed, time constraints may be serious enough for Chuck-will's-widows (*Caprimulgus carolinensis*) that they cannot forage during very intense late stages of their molt (Sutton 1969; Rohwer 1971; Mengel 1976). Among Nearctic and Palearctic migrants, time constraints seem particularly likely to apply to larger tyrannids and bee-eaters, which replace their flight feathers after arriving on the winter range.

Although distance of migration is associated with multiple origins of winter flight feather molt in Palearctic migrants (Svensson and Hedenström 1999), we doubt that migratory distance is a primary cause. Among Nearctic migrants, eastern breeders generally migrate much farther than western breeders, yet western species often delay their molt until after the fall migration. That nearly all eastern Nearctic migrants molt before migration seriously challenges migration distance as a causal variable for winter flight feather replacement in Palearctic migrants.

Winter Habitat and Molt Scheduling

Is there a general hypothesis that can explain the differences in molt scheduling between the long-distance migrants of North America and Eurasia? We propose that differences in exposure to sun and in the abrasiveness of winter habitats may be the primary cause of contrasts between Nearctic and western Palearctic migrants. In both of these migration systems, species inhabiting grasslands or reed beds have two complete or near-complete molts each year because of excessive abrasion of their plumage by siliceous deposits in grasses and sedges (e.g., sedge wrens [*Cistothorus*] and *Ammodramus* sparrows of North America [Woolfenden 1956], and some *Locustella* and *Acrocephalus* warblers of Eurasia [Svensson and Hedenström 1999]). These molts seem to be driven solely by plumage abrasion because these species do not change color seasonally and several migrate only short distances.

Most trans-Saharan Palearctic migrants live in harsher, more exposed, and more abrasive winter habitats than do Neotropical migrants. Eurasian migrants wintering in sub-Saharan Africa are strongly concentrated in thorn scrub and acacia savannas of the Sahel, Sudan, East Africa, and southern Africa (Moreau 1972; Jones 1995). Such habitats are more likely to abrade plumages than Neotropical forests, and prolonged exposure to UV light in African savannahs likely renders feathers brittle and easily damaged (Bergman 1982). Almost no western Palearctic migrants winter in the tropical forests of Central and West Africa, presumably because of intensive competition from forest residents (Moreau 1972). In contrast, temperate forest breeders of the New World typically winter in tropical forests with more shade and softer foliage. Interestingly, except for Bobolinks, no passerine migrant to the Neotropics replaces its primaries twice a year, something that the trans-Saharan migrant *Phylloscopus trochilus* does, as do three other open-country Palearctic migrants that winter in Africa or Asia, *Lanius cristatus, L. tigrinus,* and *Pericrocotus divaricatus.* Environmental abrasion seems minimal during winter for forest-dwelling Neotropical migrants, because these species show minimal wear of their flight feathers when they arrive in North America in the spring, despite having carried these feathers throughout the winter.

We propose that contrasts in vegetation structure and exposure to sun between the winter habitats of western Palearctic and Nearctic migrants result in dramatic differences between these groups in abrasion of their flight feathers during winter. If, as we suggest, flight feathers deteriorate rapidly in the harsh and sunny winter habitat occupied by Palearctic migrants to Africa, this feather wear may favor late-winter flight feather replacement in trans-Saharan migrants. This hypothesis is consistent with the following observations:

- Several long-distance *Phylloscopus* migrants that winter in Asian forests (*P. inornatus, P. proregulus, P. fuscatus*) replace their flight feathers before, rather than after, migrating (Svensson and Hedenström 1999).

- Trans-Saharan migrants appear to delay replacement of the flight feathers until as late in winter as foraging conditions allow. Many move to the Sahel and Sudan, where food conditions are good for 1 to 2 months after their arrival. Those that remain in these northern regions for the winter molt their flight feathers early because these areas become dry and unproductive later in winter. However, many other species stop here for a month or so and then, instead of molting, fatten and perform a second migration to meet the rains in eastern and southern Africa. Those species that proceed farther southward in this second movement do not molt until late winter (Jones 1995).

Comparative tests of habitat-related differences in the rates of feather wear would be informative. An obvious test would be to look for differences in rates of feather damage caused by abrasion and exposure to sunlight between migrants to the Neotropics and Palearctic migrants to sub-Saharan Africa.

METHODOLOGICAL RECOMMENDATIONS

Our experience attempting to document molt-migration scheduling from museum specimens prompts us to offer several suggestions for improving such studies. On the one hand, museum specimens offer comprehensive and directly comparable data from throughout breeding, migratory, and winter ranges. This is a huge advantage. On the other hand, collecting effort varied enormously throughout the year with the effect that, although birds may be abundant in particular times and places, museum specimens from those times and places may be sparse. This is strikingly true of North American migrants in the Neotropics from September to December (see Rohwer 1986 for Indigo Buntings). Such collecting biases are also apparent for July, August, and September from the Mexican Monsoon region. Furthermore, throughout much of its history, most collecting in North America targeted breeding birds for the description of geographic variation. Collectors also avoided molting birds, and the extreme heat and humidity during the Mexican Monsoon has resulted in little fieldwork in this region at exactly the time when it is used as a stopover site for molting.

Shortages of specimens taken during the late summer and fall in the American Southwest and the Neotropics are further exacerbated by the relative abundance of specimens from late summer from areas to the north, where more late-summer and fall collecting and salvage take place. For example, although most adult Western Tanagers depart the North to molt in the Southwest, nearly as many adults in primary feather molt come from the breeding range as from the American Southwest (Butler et al. 2002). In this case establishing that most adults move before molting was easy for two reasons. First, although molting adults collected on the breeding grounds in late summer are relatively uncommon, there are hundreds of molting juveniles in collections from this region. These young birds, as well as many additional nonmolting adults from the north, assure us that molting adults are not missing from this region simply because collectors were not active. Second, if a period when most adults are in molt is defined (16 July–31 October for the Western Tanager), it then becomes possible to compare the proportion of adults taken in this period from the breeding range and from the Southwest that are in molt. Although few adults have been collected in the Southwest during this period, most are in molt. The converse is true for areas to the north, suggesting that nonmolting late-summer birds taken in the North are mostly late breeders that would still have left before molting (Butler et al. 2002).

Because of systematic collecting biases, researchers should use molt rates (e.g., percent specimens in molt by region and time period) or measures of collecting effort (e.g., [adults]/[adults + young]) to reliably evaluate whether most birds leave their breeding area to molt. More general measures of collecting effort should now be possible with data from increasing numbers of collections being available online. For example, the total number of specimens of other species collected can be used as an index of collecting effort for a selected region and time period. Incorporating measures of collecting effort should greatly increase the likelihood of detecting subtle molt-related movements.

The common practice of targeting only molting-season specimens minimizes the possibility of distinguishing short-distance molt-related movements. For example, Western Tanager juveniles hatched in the Great Basin appear to disperse to nearby high-elevation forests to molt. The departure of juveniles out of the Great Basin could not have been established without comparing the geographical distribution of molting juveniles with the distribution of "hatchling" juveniles, which was measured by the distribution of breeding adults (Butler et al. 2002). Thus, studies of molt-related movements should consider the distribution of breeding-season specimens, which are normally excluded to reduce the size of specimen loans.

Finally, we note that the importance of the Mexican Monsoon region as a molt stopover site for other species could be evaluated by using stable isotopes from the huge numbers of specimens available from the breeding grounds. However, this methodology depends on whether the Mexican Monsoon region has a unique isotope signature (Marra et al. 1998; Hobson 1999).

FUTURE DIRECTIONS

Molt is among the most important activities driving the evolution of avian life histories. Despite its high energetic costs (Murphy and King 1992), implications for survival (Nilsson and Svensson 1996; Dawson et al. 2000), and enormous time requirements (Rohwer 1999), few avian biologists have explored the relevance of molting to avian life history evolution. Every section of this chapter reveals serious shortcomings in our understanding of the scheduling of molt and migration in migrant passerines. Only recently have we discovered the importance of the region of the Mexican Monsoon as a molting ground for Neotropical migrant passerines. Future collection-based studies of molt scheduling in individual species should raise the bar for data quality by using on-line data bases to generate powerful interspecific indices of collecting effort. With such indices we expect that additional species with breeding ranges that include the region of the Mexican Monsoon, but that also extend well north of this region, will be added to the list of species that rely on this region for their fall molt. Indeed, this could even include Lark Sparrows (*Chondestes grammacus*), which we recently encountered in this region in August and September in numbers far beyond breeding densities.

With our growing awareness of the importance of this region for molting, we badly need to understand the ecology of species that move there to molt (Leu and Thompson 2002). How many are using riparian corridors, which have been seriously degraded by agriculture and the invasion of salt cedar (*Tamarix*)? Our recent (and inconclusive) survey suggested that molt migrants were mostly using lowland acacia flats, a habitat that seemed less degraded than riparian corridors, but is heavily affected by grazing and efforts to make pastures more productive. Exactly which habitats in this region so many North American birds are using to molt remain unknown, mainly because no one has looked for them.

Similarly, we know little of the ecology of molting for those western species (or age classes) that breed in arid regions and that do not move to the monsoon region to molt. In an extensive search of the literature we found just two references (Greenberg et al. 1974; Steele and McCormick 1995) documenting upslope movements for molting. Surely there are good data residing in banding files for numerous MAPS stations that could be used to explore such movements for additional species.

Finally, for all the importance of feathers in the lives of birds, the avian literature is remarkable for its absence of papers interpreting the significance of structural variation among contour feathers. As we have shown here, lax juvenile plumages may prevent juveniles in some species from migrating before they molt in the fall. Studies of the significance of variation in feather quality have the potential of becoming career research programs. Similarly, studies of how rates of wear vary across habitats and, thus, how differences in wear could affect the scheduling of molt between western Palearctic-African migrants and Neotropical migrants will likely be elucidating.

ACKNOWLEDGMENTS

Thanks to Jed Burtt, Chris Filardi, Russ Greenberg, and Keith Hobson for help with references, and to Peter Pyle and Russ Greenberg for helpful reviews. Specimens examined for this study are housed at the University of Washington Burke Museum and at the National Museum of Natural History. Support for this project came from Burke Museum Eddy Fellowships to LKB and DRF, a National Science Foundation fellowship to LKB, and from the Endowment for Ornithology at the Burke Museum.

LITERATURE CITED

Alerstam, T., and G. Högstedt. 1982. Bird migration and reproduction in relation to habitats for survival and breeding. Ornis Scandinavica 13:25–37.

Bergman, G. 1982. Why are the wings of *Larus fuscus fuscus* so dark? Ornis Fennica 59:77–83.

Burns, K. J. 1998. Molecular phylogenetics of the genus *Piranga:* implications for biogeography and the evolution of morphology and behavior. Auk 115:621–634.

Butler, L. K., M. G. Donahue, and S. Rohwer. 2002. Molt-migration in Western Tanagers (*Piranga ludoviciana*): age effects, aerodynamics, and conservation implications. Auk 119:1010–1023.

Cicero, C., and N. K. Johnson. 1998. Molecular phylogeny and ecological diversification in a clade of New World songbirds (genus *Vireo*). Molecular Ecology 7:1359–1370.

Comrie, A. C., and E. C. Glenn. 1998. Principal components-based regionalization of precipitation regimes across the southwest United States and northern Mexico, with an application to monsoon precipitation variability. Climate Research 10:201–215.

Dawson, A. 1994. The effects of daylength and testosterone on the initiation and rate of moult in Starlings *Sternus vulgaris.* Ibis 136:335–340.

Dawson, A., S. A. Hinsley, P. N. Ferns, R. H. C. Bonser, and L. Eccleston. 2000. Rate of moult affects feather quality: a mechanism linking current reproductive effort to future survival. Proceedings of the Royal Society of London, Series B, Biological Sciences 267:2093–2098.

Dwight, J., Jr. 1900. The sequence of plumages and moults of the passerine birds of New York. Annals of the New York Academy of Sciences 13:73–360.

Espenshade, Jr., E. B. (ed.). 1991. Goode's World Atlas (eighteenth ed.). Rand McNally, Chicago.

Göhringer, R. 1951. Vergleichende Untersuchungen über das Juvenil- und Adultkleid bei der Amsel (*Turdus merula* L.) und beim Star (*Sturnus vulgaris* L.). Revue Suisse de Zoologie 58:279–358.

Greenberg, R., T. Keller-Wolf, and V. Keeler-Wolf. 1974. Wood warbler populations in the Yolla Bolly Mountains of California. Western Birds 5:81–90.

Hansen, T. E. 1997. Stabilizing selection and the comparative analysis of adaptation. Evolution 51:1341–1351.

Hobson, K. A. 1999. Tracing origins and migration of wildlife using stable isotopes: a review. Oecologia 120:314–326.

Hussell, D. J. T. 1982. The timing of the fall migration in Yellow-bellied Flycatchers. Journal of Field Ornithology 533:1–6.

Jenni, L., and R. Winkler. 1994. Moult and aging of European passerines. Academic Press, London.

Johnson, N. K. 1963. Comparative molt cycles in the tyrannid genus *Empidonax.* Pages 870–883 *in* Proceedings of the 13th International Ornithological Congress.

Johnson, N. K. 1974. Molt and age determination in Western and Yellowish Flycatchers. Auk 91:111–131.

Jones, P. J. 1995. Migration strategies of Palearctic passerines in Africa. Israel Journal of Zoology 41:393–406.

Klicka, J., A. J. Fry, R. M. Zink, and C. W. Thompson. 2001. A cytochrome-b perspective on *Passerina* bunting relationships. Auk 118:611–623.

Langston, N. E., and S. Rohwer. 1995. Unusual patterns of incomplete primary molt in Laysan and Black-footed Albatrosses. Condor 97:1–19.

Langston, N. E., and S. Rohwer. 1996. Molt-breeding tradeoffs in albatrosses: life history implications for big birds. Oikos 76:498–510.

Leu, M., and C. W. Thompson. 2002. The potential importance of migratory stopover sites as flight feather molt staging areas: a review for neotropical migrants. Biological Conservation 106:45–56.

Lindström, Å., D. J. Pearson, D. Hasselquist, A. Hedenström, S. Bensch, and S. Åkesson. 1993. The moult of Barred Warblers *Sylvia nisoria* in Kenya: evidence for a split wing-moult pattern initiated during the birds' first winter. Ibis 135:403–409.

Lovette, I. J., and E. Bermingham. 1999. Explosive speciation in the New World *Dendroica* Warblers. Proceedings of the Royal Society of London, Series B, Biological Sciences 266:1629–1636.

Lovette, I. J., E. Birmingham, S. Rohwer, and C. Wood. 1999. Mitochondrial restriction fragment length polymorphism (RFLP) and sequence variation among closely related avian species and the genetic characterization of hybrid *Dendroica* warblers. Molecular Ecology 8:1431–1441.

Marra, P. P., K. A. Hobson, and R. T. Holmes. 1998. Linking winter and summer events in a migratory bird using stable carbon isotopes. Science 282:1884–1886.

Mengel, R. M. 1976. Rapid tail molt and temporarily impaired flight in the Chuck-will's-widow. Wilson Bulletin 88:351–353.

Mengel, R. M. 1965. The Birds of Kentucky. Ornithological Monographs 3:1–581.

Moreau, R. E. 1972. The Palearctic–African Bird Migration Systems. Academic Press, London.

Morehouse, E. L., and R. Brewer. 1968. Feeding of nestling and fledgling Eastern Kingbirds. Auk 85:44–54.

Murphy, M. E., and J. R. King. 1992. Energy and nutrition use during molt by white-crowned sparrows *Zonotrichia leucophrys gambelii*. Ornis Scandinavica 23:304–313.

Murray, B. et al. 1994. The use of cytochrome *B* sequence variation in estimation of phylogeny in the Vireonidae. Condor 96:1037–1054.

Niles, D. M. 1972. Molt cycles of purple martins (*Progne subis*). Condor 74:61–71.

Nilsson, J.-A., and E. Svensson. 1996. The cost of reproduction: a new link between current reproductive effort and future reproductive success. Proceedings of the Royal Society of London, Series B, Biological Sciences 236:711–714.

Ohmart, R. D. 1966. Breeding record of the Cassin's sparrow (*Aimophila cassinii*) in Arizona. Condor 68:400.

Omland, K. E., and S. M. Lanyon. 2000. Reconstructing plumage evolution in orioles (*Icterus*): repeated convergence and reversal in patterns. Evolution 54:2119–2133.

Omland, K. E., S. M. Lanyon, and S. J. Fritz. 1999. A molecular phylogeny of the New World orioles (*Icterus*): the importance of dense taxon sampling. Molecular Phylogenetics and Evolution 12: 224–239.

Pietiäinen, H, P. Saurola, and H. Kolunen. 1984. The reproductive constraints on moult in the Ural Owl *Strix uralensis*. Annales Zoologici Fennici 21:277–281.

Phillips, A. R. 1951. The molts of the Rufous-winged Sparrow. Wilson Bulletin 63:323–326.

Pyle, P. 1997. Identification guide to North American birds: Pt. 1. Slate Creek Press, Bolinas, Calif.

Pyle, P., and R. P. Henderson. 1991. The birds of Southeast Farallon Island: occurrence and seasonal distribution of migratory species. Western Birds 22:41–84.

Reeve, H. K., and P. W. Sherman. 1993. Adaptation and the goals of evolutionary research. Quarterly Review of Biology 68:1–32.

Reichert, S. E. 1979. Development and reproduction in desert animals. Pages 797–822 *in* Arid-land Ecosystems, Vol. 1 (D. W. Goodall and R. A. Perry, eds.). Cambridge University Press, Cambridge.

Rohwer, S. A. 1971. Molt and the annual cycle of the Chuckwill's-widow, *Caprimulgus carolinensis*. Auk 88:485–519.

Rohwer, S. 1986. A previously unknown plumage of 1st-year Indigo Buntings and theories of delayed plumage maturation. Auk 103:281–292.

Rohwer, S. 1999. Time constraints and moult-breeding tradeoffs in large birds. Pages 568–581 *in* Proceedings of the 22nd International Ornithological Congress, Durban (N. J. Adams and R. H. Slotow, eds.).

Rohwer, S., and J. Manning. 1990. Differences in timing and number of molts for Baltimore and Bullocks Orioles: implications to hybrid fitness and theories of delayed plumage maturation. Condor 92:125–140.

Sealey, S. G., and G. C. Biermann. 1983. Timing of breeding and migrations in a population of Least Flycatchers in Manitoba. Journal of Field Ornithology. 54:113–224.

Short, L. L. 1974. Nesting of southern Sonoran birds during the summer rainy season. Condor 76:21–32.

Steele, J., and J. McCormick. 1995. Partitioning of the summer grounds by Orange-crowned Warblers into breeding grounds, adult molting grounds and juvenile staging areas. North American Bird Bander 20:152.

Stutchbury, B. J., and S. Rohwer. 1990. Molt patterns in the Tree Swallow (*Tachycineta bicolor*). Canadian Journal of Zoology-Revue Canadienne de Zoologie 68:1468–1472.

Sutton, G. M. 1969. A Chuck-Will's-Widow in postnuptial molt. Oklahoma Ornithological Society 2:9–11.

Svensson, E., and A. Hedenstrom. 1999. A phylogenetic analysis of the evolution of moult strategies in Western Palearctic warblers (Aves : Sylviidae). Biological Journal of the Linnean Society 67:263–276.

Thompson, C. W. 1991. The sequence of molts and plumages in Painted Buntings and implications for theories of delayed plumage maturation. Condor 93:209–235.

Vega Riveria, J. H., W. J. McShea, J. A. Rappole, and C. A. Haas. 1998. Pattern and chronology of prebasic molt for the Wood Thrush and its relation to reproduction and migration departure. Wilson Bulletin 110:384–392.

Voelker, G., and S. Rohwer. 1998. Contrasts in scheduling of molt and migration in Eastern and Western Warbling Vireos. Auk 115:142–155.

Woolfenden, G. E. 1956. Comparative breeding behavior of *Ammospiza caudacuta* and *A. maritima*. University of Kansas Publications of the Museum of Natural History 10:45–75.

Young, B. E. 1991. Annual molts and interruption of the fall migration for molting in Lazuli Buntings. Condor 93: 236–250.

Yuri, T., and S. Rohwer. 1997. Molt and migration in the Northern Rough-winged Swallow. Auk 114:249–262.

Zink, R. M., R. C. Blackwell-Rago, and F. Ronquist. 2000. The shifting roles of dispersal and vicariance in biogeography. Proceedings of the Royal Society of London, Series B, Biological Sciences 267:497–503.

Zink, R. M., and N. K. Johnson. 1984. Evolutionary genetics of Flycatchers: Pt. 1. Sibling species in the genera Empidonax and Contopus. Systematic Zoology 33:205–216.

APPENDIX: CATEGORIZATIONS OF SPECIES USED IN THE COMPARATIVE ANALYSES OF ROHWER, BUTLER, AND FROEHLICH AND FROEHLICH, ROHWER, AND STUTCHBURY

This table provides a list of how we categorized species with respect to the location of their breeding grounds, molts winter territoriality. The data on molt mostly came from Pyle (1997); better or updated references that change Pyle's categorizations are listed in the text of this chapter, but are not flagged in this appendix. Because there are few detailed studies of Neotropical migrants on their wintering grounds, the classification of species with regard to winter territoriality is difficult. For this reason we have provided a list of the references we consulted in making these decisions. Many of these references were also used in the analyses by Froehlich et al. (Chap. 25, this volume). To avoid duplication, we have presented the complete list of references on winter territoriality here. Scientific names follow the AOU Checklist (1998), except that we treat eastern and western populations of Warbling Vireo as separate species.

Common name	Species	Breeding grounds	Adult prebasic molt	First-year prebasic molt	Winter territorial	Winter territoriality references
Olive-sided Flycatcher	*Contopus cooperi*	Continental	Winter	Winter	Yes	Meyer de Schauensee et al. 1978; Ridgely & Tudor 1994; Stiles & Skutch 1989; Nickell 1968
Western Wood-Pewee	*Contopus sordidulus*	Western	Winter	Winter	Yes	Bemis & Rising 1999
Eastern Wood-Pewee	*Contopus virens*	Eastern	Winter	Winter	Yes	Fitzpatrick 1980; Willis 1966
Yellow-bellied Flycatcher	*Empidonax flaviventris*	Eastern	Winter	Breeding	Yes	Rappole & Warner 1980; Stiles & Skutch 1989; Dickey & van Rossem 1938
Acadian Flycatcher	*Empidonax virescens*	Eastern	Breeding	Breeding	Yes	Willis 1966; Morton 1980; Stiles & Skutch 1989; Ridgely & Gwynne 1989
Alder Flycatcher	*Empidonax alnorum*	Northern	Winter	Winter	Yes	Ridgely & Tudor 1994; Lowther 1999;
Willow Flycatcher	*Empidonax traillii*	Continental	Winter	Winter	Yes	Gorski 1969; Ridgely & Tudor 1994
Least Flycatcher	*Empidonax minimus*	Eastern	Breeding	Breeding	Yes	Ely 1973; Rappole & Warner 1980
Hammond's Flycatcher	*Empidonax hammondii*	Western	Breeding	Breeding		
Gray Flycatcher	*Empidonax wrightii*	Western	Winter	Winter	Yes	Sterling 1999
Dusky Flycatcher	*Empidonax oberholseri*	Western	Winter	Winter		
Pacific-Slope Flycatcher	*Empidonax difficilis*	Western	Winter	Winter	Yes	Hutto 1987; Lowther 2000;
Cordilleran Flycatcher	*Empidonax occidentalis*	Western	Winter	Winter	Yes	Hutto 1987; Lowther 2000;
Eastern Phoebe	*Sayornis phoebe*	Eastern	Breeding	Breeding		
Say's Phoebe	*Sayornis saya*	Western	Breeding	Breeding		
Ash-throated Flycatcher	*Myiarchus cinerascens*	Western	Stopover/ winter	Stopover/ winter		
Great-crested Flycatcher	*Myiarchus crinitus*	Eastern	Breeding	Breeding	Yes	Willis 1966; Morton 1980
Cassin's Kingbird	*Tyrannus vociferans*	Western	?	?		
Western Kingbird	*Tyrannus verticalis*	Western	Winter	Winter	No	Stiles & Skutch 1989; Gamble & Bergin 1996; Dickey & van Rossem 1938
Eastern Kingbird	*Tyrannus tyrannus*	Continental	Winter	Winter	No	Fitzpatrick 1980; Morton 1980; Ridgely & Tudor 1994
Gray Kingbird	*Tyrannus dominicensis*	Eastern	Winter	Winter	No	Meyer de Schauensee et al. 1978
Scissor-tailed Flycatcher	*Tyrannus forficatus*	Eastern	?	?	No	Stiles & Skutch 1989; Regosin 1998
Purple Martin	*Progne subis*	Continental	Winter	Winter	No	Ridgely & Tudor 1989; Brown 1997;
Tree Swallow	*Tachycineta bicolor*	Continental	Migration	Migration	No	Ridgely & Gwynne 1989; Howell & Webb 1995;
Violet-green Swallow	*Tachycineta thalassina*	Western	Migration	Migration	No	Dickey & van Rossem 1938; Bent 1942
N. Rough-winged Swallow	*Stelgidopteryx serripennis*	Continental	Migration	Migration	No	Dejong 1996
Bank Swallow	*Riparia riparia*	Continental	Winter	Winter	No	Garrison 1999
Cliff Swallow	*Hirundo pyrrhonota*	Continental	Winter	Winter	No	Ridgely & Tudor 1989

Common name	Species	Breeding grounds	Adult prebasic molt	First-year prebasic molt	Winter territorial	Winter territoriality references
Barn Swallow	*Hirundo rustica*	Continental	Winter	Winter	No	Brown & Brown 1999
Gray Catbird	*Dumetella carolinensis*	Eastern	Breeding	Breeding	Yes	Nickell 1968; Rappole & Warner 1980
Veery	*Catharus fuscescens*	Eastern	Breeding	Breeding	No	Willis 1966
Gray-cheeked Thrush	*Catharus minimus*	Northern	Breeding	Breeding	No	Willis 1966
Bicknell's Thrush	*Catharus bicknelli*	Eastern	Breeding	Breeding	Yes	Rimmer, pers. comm.
Swainson's Thrush	*Catharus ustulatus*	Continental	Breeding	Breeding	No	Willis 1966
Hermit Thrush	*Catharus guttatus*	Continental	Breeding	Breeding	Yes	Brown et al. 2000
Wood Thrush	*Hylocichla mustelina*	Eastern	Breeding	Breeding	Yes	Morton 1980; Rappole & Warner 1980; Rappole et al. 1992; Winker et al. 1990; Willis 1966
Blue-gray Gnatcatcher	*Polioptila caerulea*	Continental	Breeding	Breeding	No	Hutto 1994
White-eyed Vireo	*Vireo griseus*	Eastern	Breeding	Breeding	Yes	Barlow 1980; Rappole & Warner 1980
Bell's Vireo	*Vireo bellii*	Continental	Breeding	Breeding	Yes	Barlow 1980
Black-capped Vireo	*Vireo atricapillus*	Eastern	Breeding	Breeding	Yes	Barlow 1980
Gray Vireo	*Vireo vicinior*	Western	Breeding	Breeding	Yes	Barlow 1980; Bates 1992; Barlow et al. 1999
Yellow-throated Vireo	*Vireo flavifrons*	Eastern	Breeding	Breeding	Yes	Barlow 1980; Morton 1980; Stiles & Skutch 1989
Plumbeous Vireo	*Vireo plumbeus*	Western	Breeding	Breeding	Yes	Barlow 1980
Cassin's Vireo	*Vireo cassinii*	Western	Breeding	Breeding	Yes	Barlow 1980
Blue-headed Vireo	*Vireo solitarius*	Eastern	Breeding	Breeding	Yes	Barlow 1980; James 1998
Eastern Warbling Vireo	*Vireo gilvus*	Eastern	Breeding	Breeding	No	Dickey & van Rossem 1938; Meyer de Schauensee et al. 1978; but see Barlow 1980
Western Warbling Vireo	*Vireo swainsonii*	Western	Stopover	Breeding	No	Hutto 1994
Philadelphia Vireo	*Vireo philadelphicus*	Eastern	Breeding	Breeding	No	Barlow 1980; Dickey & van Rossem 1938; Wetmore et al. 1984; Moskoff & Robinson 1996; Tramer & Kemp 1980; Stiles & Skutch 1989; but see Loftin 1977
Red-eyed Vireo	*Vireo olivaceus*	Eastern	Breeding	Breeding	No	Barlow 1980
Black-whiskered Vireo	*Vireo altiloquus*	Eastern			No	Barlow 1980
Lesser Goldfinch	*Carduelis psaltria*	Western	Stopover	Stopover	No	Howell & Webb 1995
Blue-winged Warbler	*Vermivora pinus*	Eastern	Breeding	Breeding	Yes	Morton 1980
Golden-winged Warbler	*Vermivora chrysoptera*	Eastern	Breeding	Breeding	Yes	Morton 1980; Tramer & Kemp 1980; Stiles & Skutch 1989
Tennessee Warbler	*Vermivora peregrina*	Northern	Breeding	Breeding	No	Morton 1980; Tramer & Kemp 1980; Stiles & Skutch 1989; Ridgely & Gwynne 1989; Dickey & van Rossem 1938; Meyer de Schauensee et al. 1978; but see Thurber & Villeda 1976
Orange-crowned Warbler	*Vermivora celata*	Continental	Breeding*	Breeding*	Yes	Rappole & Warner 1980
Nashville Warbler	*Vermivora ruficapilla*	Continental	Breeding*	Breeding*	No	Hutto 1994; Beebe in Bent 1953
Virginia's Warbler	*Vermivora virginiae*	Western	Breeding	Breeding	No	Curson et al. 1994; Olson & Martin 1999; but see Howell & Webb 1995
Lucy's Warbler	*Vermivora luciae*	Southwestern	Breeding	Breeding	No	Howell & Webb 1995; Johnson et al. 1997
Northern Parula	*Parula americana*	Eastern	Breeding	Breeding	No	Faaborg & Arendt 1984; Staicer 1992; but see Diamond & Smith 1973
Yellow Warbler	*Dendroica petechia*	Continental	Breeding	Breeding	Yes	Morton 1980; Rappole & Warner 1980; Greenberg & Salgado-Ortiz 1994; Warkentin & Hernandez 1996; Hutto 1981a
Chestnut-sided Warbler	*Dendroica pensylvanica*	Eastern	Breeding	Breeding	Yes	Gradwohl & Greenberg 1980; Morton 1980; Greenberg 1984
Magnolia Warbler	*Dendroica magnolia*	Eastern	Breeding	Breeding	Yes	Rappole & Warner 1980; Morton 1980; Greenberg & Salgado-Ortiz 1994; Greenberg 1979
Cape May Warbler	*Dendroica tigrina*	Eastern	Breeding	Breeding	Yes	Eaton 1953; Staicer 1992; Latta & Wunderle 1996
Black-thr. Blue Warbler	*Dendroica caerulescens*	Eastern	Breeding	Breeding	Yes	Holmes et al. 1989; Wunderle 1995; Wunderle & Latta 2000

continued

Common name	Species	Breeding grounds	Adult prebasic molt	First-year prebasic molt	Winter territorial	Winter territoriality references
Audubon's Warbler	*Dendroica coronata*	Western	Breeding	Breeding	No	Rappole & Warner 1980; Latta & Wunderle 1996; Howell 1972; Greenberg 1979
Myrtle Warbler	*Dendroica coronata*	Northern	Breeding	Breeding	No	Rappole & Warner 1980; Latta & Wunderle 1996; Howell 1972; Greenberg 1979
Black-thr. Gray Warbler	*Dendroica nigrescens*	Western	Breeding	Breeding	No	Curson et al. 1994
Golden-cheeked Warbler	*Dendroica chrysoparia*	Eastern	Breeding	Breeding	Yes?	Vidal, R. M. et al. 1994; King & Rappole 2000; Ladd & Gass 1999
Black-thr. Green Warbler	*Dendroica virens*	Eastern	Breeding	Breeding	Yes	Skutch in Bent 1953; Rappole & Warner 1980; Latta & Wunderle 1996; Tramer & Kemp 1980; Greenberg 1979; but see Dickey & van Rossem 1938
Townsend's Warbler	*Dendroica townsendi*	Western	Breeding	Breeding	No	Froehlich, unpubl. data; Curson et al. 1994; Dunn & Garrett 1997; but see Tramer & Kemp 1980 and Thurber & Villeda 1976
Hermit Warbler	*Dendroica occidentalis*	Western	Breeding	Breeding	No	Curson et al. 1994
Blackburnian Warbler	*Dendroica fusca*	Eastern	Breeding	Breeding	No	Chipley 1980; Ridgely & Tudor 1989; but see Stiles & Skutch 1989
Yellow-throated Warbler	*Dendroica dominica*	Eastern	Breeding	Breeding	Yes	Howell 1972; Curson et al. 1994; Howell & Webb 1995; but see Hall 1996
Grace's Warbler	*Dendroica graciae*	Southwestern	Breeding	Breeding	No	Curson et al. 1994; Howell & Webb 1995
Kirtland's Warbler	*Dendroica kirtlandii*	Eastern	Breeding	Breeding	Yes	Mayfield 1992
Prairie Warbler	*Dendroica discolor*	Eastern	Breeding	Breeding	Yes	Staicer 1992; Latta & Wunderle 1996; Diamond & Smith 1973
Palm Warbler	*Dendroica palmarum*	Eastern	Breeding	Breeding	No	Eaton 1953; Ridgely & Gwynne 1989; but see Howell 1972
Bay-breasted Warbler	*Dendroica castanea*	Eastern	Breeding	Breeding	No	Greenberg 1984
Blackpoll Warbler	*Dendroica striata*	Northern	Breeding	Breeding	No	Meyer de Schauensee et al. 1978; but see Sick 1971
Cerulean Warbler	*Dendroica cerulea*	Eastern	Breeding	Breeding	No	Meyer de Schauensee et al. 1978; Robbins et al. 1992;
Black-and-White Warbler	*Mniotilta varia*	Eastern	Breeding	Breeding	Yes	Morton 1980; Rappole & Warner 1980; Rappole et al. 1992; Latta & Wunderle 1996; Wunderle & Latta 2000; Ornat & Greenberg 1990
American Redstart	*Setophaga ruticilla*	Eastern	Breeding	Breeding	Yes	Rappole & Warner 1980; Latta & Wunderle 1996; Holmes et al. 1989; Ornat & Greenberg 1990; Marra et al. 1993; Warkentin & Hernandez 1996; Wunderle & Latta 2000
Prothonotary Warbler	*Protonotaria citrea*	Eastern	Breeding	Breeding	No	Morton 1980; Lefebvre et al. 1994; Warkentin & Hernandez 1996
Worm-eating Warbler	*Helmitheros vermivorus*	Eastern	Breeding	Breeding	Yes	Rappole & Warner 1980; Rappole et al. 1992; Diamond & Smith 1973
Swainson's Warbler	*Limnothlypis swainsonii*	Eastern	Breeding	Breeding	Yes	Eaton 1953; Diamond & Smith 1973; Brown & Dickson 1994
Ovenbird	*Seiurus aurocapillus*	Eastern	Breeding	Breeding	Yes	Rappole & Warner 1980; Rappole et al. 1992; Faaborg & Arendt 1984; Eaton 1953; Warkentin & Hernandez 1996
Northern Waterthrush	*Seiurus noveboracensis*	Northern	Breeding	Breeding	Yes	Rappole & Warner 1980; Morton 1980; Warkentin & Hernandez 1996; Schwartz 1964
Louisiana Waterthrush	*Seiurus motacilla*	Eastern	Breeding	Breeding	Yes	Eaton 1953; Rappole & Warner 1980; Robinson 1995; Eaton 1953; Morton 1980

Common name	Species	Breeding grounds	Adult prebasic molt	First-year prebasic molt	Winter territorial	Winter territoriality references
Kentucky Warbler	*Oporornis formosus*	Eastern	Breeding	Breeding	Yes	Morton 1980; Rappole & Warner 1980; Mabey & Morton 1992; Rappole et al. 1992; Willis 1966
Connecticut Warbler	*Oporornis agilis*	Eastern	Breeding	Breeding	Yes	Meyer de Schauensee et al. 1978; Pitocchelli et al. 1997; Curson et al. 1994
Mourning Warbler	*Oporornis philadelphia*	Eastern	Breeding	Breeding	Yes	Meyer de Schauensee et al. 1978; Morton 1980; Dunn & Garrett 1997; Stiles & Skutch 1989
Macgillivray's Warbler	*Oporornis tolmiei*	Western	Breeding	Breeding	Yes	Hutto 1981a; Stiles & Skutch 1989; Dunn & Garrett 1997
Common Yellowthroat	*Geothlypis trichas*	Continental	Breeding	Breeding	Yes	Hutto 1981a; Ornat & Greenberg 1990; Rappole & Warner 1980
Hooded Warbler	*Wilsonia citrina*	Eastern	Breeding	Breeding	Yes	Rappole & Warner 1980; Rappole et al. 1992; Ornat & Greenberg 1990
Wilson's Warbler	*Wilsonia pusilla*	Continental	Breeding	Breeding	Yes	Skutch in Bent 1953; Hutto 1987; Tramer & Kemp 1980; Hutto 1981a; Loftin 1977; Rappole & Warner 1980; Rappole et al. 1992
Canada Warbler	*Wilsonia canadensis*	Eastern	Breeding	Breeding	Yes	Greenberg & Gradwohl 1980; but see Willis 1966
Yellow-breasted Chat	*Icteria virens*	Continental	Breeding	Breeding	Yes	Diamond & Smith 1973; Stiles & Skutch 1989; Rappole & Warner 1980
Hepatic Tanager	*Piranga flava*	Southwestern	Breeding	Breeding	No	Meyer de Schauensee et al. 1978; Howell & Webb 1995; Stiles & Skutch 1989
Scarlet Tanager	*Piranga olivacea*	Eastern	Breeding	Breeding	Yes	Ridgely & Tudor 1989; Mowbray 1999; Howell & Webb 1995; but see Stiles & Skutch 1989
Summer Tanager	*Piranga rubra*	Continental	Breeding	Breeding	Yes	Morton 1980; Rappole & Warner 1980; Stiles & Skutch 1989; Howell & Webb 1995
Western Tanager	*Piranga ludoviciana*	Western	Stopover	Breeding[a]	Yes	Stiles & Skutch 1989; Howell & Webb 1995
Green-tailed Towhee	*Pipilo chlorurus*	Western	Breeding	Breeding	No	Dobbs et al. 1998
Rose-breasted Grosbeak	*Pheucticus ludovicianus*	Eastern	Breeding	Breeding	No	Loftin 1977; Stiles & Skutch 1989; Howell & Webb 1995
Black-headed Grosbeak	*Pheucticus melanocephalus*	Western	Stopover	Breeding	No	Howell & Webb 1995
Blue Grosbeak	*Passerina caerulea*	Continental	Winter	Breeding	No	Howell 1972; Stiles & Skutch 1989; Howell & Webb 1995
Lazuli Bunting	*Passerina amoena*	Western	Stopover	Breeding	No	Howell & Webb 1995; Greene et al. 1996
Indigo Bunting	*Passerina cyanea*	Eastern	Breeding	Breeding	No	Nickell 1968; Rappole & Warner 1980; Payne 1992; Stiles & Skutch 1989; Howell & Webb 1995
Painted Bunting (E. coast)	*Passerina ciris*	Eastern	Breeding	Breeding	Yes	Froehlich, pers. obs.
Painted Bunting (Central)	*Passerina ciris*	Eastern	Stopover	Stopover	No	Rappole & Warner 1980; Howell & Webb 1995
Dickcissel	*Spiza americana*	Eastern	Breeding	?	No	Meyer de Schauensee et al. 1978; Ridgely & Gwynne 1989; Basili 1998
Bobolink	*Dolichonyx oryzivorus*	Eastern	Breeding	Breeding	No	Ridgely & Tudor 1989
Hooded Oriole	*Icterus cucullatus*	Southwestern	Winter	Winter	No	Howell & Webb 1995
Baltimore Oriole	*Icterus galbula*	Eastern	Breeding	Breeding	No	Morton 1980; Stiles & Skutch 1989; Howell & Webb 1995
Bullock's Oriole	*Icterus bullockii*	Western	Stopover	Stopover	No	Howell & Webb 1995
Orchard Oriole	*Icterus spurius*	Eastern	Winter	Winter	No	Greenberg & Salgado Ortiz 1994; Stiles & Skutch 1989; Dickey & Van Rossem 1938; Nickell 1968; Morton 1980; Howell & Webb 1995
Scott's Oriole	*Icterus parisorum*	Western	Breeding	Breeding	No	Howell & Webb 1995

[a]Individuals of these species generally move upslope to higher elevations in the intermontane West after breeding or fledging.

REFERENCES FOR THE APPENDIX

Note: All *Birds of North America* references are edited by A. Poole and F. Gill, published by The Birds of North America, Inc., Philadelphia.

American Ornithologists' Union. 1983. Check-list of North American Birds (seventh ed.). American Ornithologists' Union, Washington, D.C.

Barlow, J. C. 1980. Patterns of ecological interactions among migrant and resident vireos on the wintering grounds. Pages 79–107 *in* Migrant Birds in the Neotropics: Ecology, Behavior, Distribution, and Conservation (A. Keast and E. S. Morton, eds.). Smithsonian Institution Press, Washington, D.C.

Barlow J. C., S. N. Leckie, and C. T. Barill. 1999. Gray Vireo (*Vireo vicinior*). The Birds of North America, no. 447.

Basili G. 1998. What's in a name: Dickcissel, Pajaro Arrocero, *Spiza americana:* a revealing look at the names of a neotropical migrant. Bird Conservation 13.

Bates, J. M. 1992. Winter territorial behavior of Gray Vireos. Wilson Bulletin 104:425–433.

Bemis, C., and J. D. Rising. 1999. Western Wood-Pewee (Contopus sordidulus). The Birds of North America, no. 451.

Bent, A. C. 1942. Life histories of North American flycatchers, larks, swallows, and their allies. Smithsonian Institution Bulletin 179:555.

Bent, A. C. 1953. Life histories of North American wood warblers. Smithsonian Institution Bulletin 203:734.

Brown, C. R. 1997. Purple Martin (*Progne subis*). The Birds of North America, no. 287.

Brown, C. R., and M. B. Brown. 1999. Barn Swallow (*Hirundo rustica*). The Birds of North America, no. 452.

Brown, D. R., P. C. Stouffer, and C. M. Strong. 2000. Movement and territoriality of wintering Hermit Thrushes in southeastern Louisiana. Wilson Bulletin 112:347–353.

Brown, R. E., and J. G. Dickson. 1994. Swainson's Warbler (*Limnothlypis swainsonii*). The Birds of North America, no. 126.

Chapman, F. M. 1890. On the winter distribution of the Bobolink (*Dolichonyx oryzivorus*) with remarks on its routes of migration. Auk 7:39–45.

Chipley, R. M. 1980. Nonbreeding ecology of the Blackburnian Warbler. Pages 309–317 *in* Migrant Birds in the Neotropics: Ecology, Behavior, Distribution, and Conservation (A. Keast and E. S. Morton, eds.). Smithsonian Institution Press, Washington, D.C.

Cruz, A. 1975. Ecology and breeding biology of the Solitary Vireo. Journal of the Colorado-Wyoming Academy of Sciences 7:36–37.

Curson, J., D. Quinn, and D. Beadle. 1994. Warblers of the Americas: An Identification Guide. Houghton Mifflin, New York.

Dejong, M. J. 1996. Northern Rough-winged Swallow (*Stelgidopteryx serripennis*). The Birds of North America, no. 234.

Diamond, A. W., and R. W. Smith. 1973. Returns and survival of banded warblers wintering in Jamaica. Bird-Banding 44:221–224.

Dickey, S. R., and A. J. van Rossem. 1938. The Birds of El Salvador. Field Museum of Natural History Publications, Zoological Series 23:1–609.

Dobbs, R. C., P. R. Martin, and T. E. Martin. 1998. Green-tailed Towhee (*Pipilo chlorurus*). The Birds of North America, no. 368.

Dunn, J. L., and K. L. Garrett. 1997. A Field Guide to the Warblers of North America. Houghton Mifflin, Boston.

Eaton, S. W. 1953. Wood warblers wintering in Cuba. Wilson Bulletin 65:169–174.

Ely, C. A. 1973. Returns of North American birds to their wintering grounds in southern Mexico. Bird-Banding 44: 228–229.

Enstrom, D. A. 1992. Delayed plumage maturation in the Orchard Oriole (*Icterus spurius*): tests of winter adaptation hypotheses. Behavioral Ecology and Sociobiology 30:35–42.

Faaborg, J., and W. J. Arendt. 1984. Population sizes and philopatry of winter resident warblers in Puerto Rico. Journal of Field Ornithology 55:376–378.

Faaborg, J., and J. E. Winters. 1979. Winter resident returns and longevity and weights of Puerto Rican birds. Bird-Banding 50:216–223.

Fitzpatrick, J. W. 1980. Wintering of North American Tyrant flycatchers in the Neotropics. Pages 67–76 *in* Migrant Birds in the Neotropics: Ecology, Behavior, Distribution, and Conservation (A. Keast and E. S. Morton, eds.). Smithsonian Institution Press, Washington, D.C.

Fitzpatrick, J. W. 1982. Northern birds at home in the tropics. Natural History 91:40–47.

Gamble, L. R., and T. M. Bergin. 1996. Western Kingbird (*Tyrannus verticalis*). The Birds of North America, no. 227.

Garrison, B. A. 1999. Bank Swallow *Riparia riparia.* The Birds of North America, no. 414.

George, T. L. 1987. Behavior of territorial male and female Townsend's Solitaires (*Myadestes townsendi*) in winter. American Midland Naturalist 118:121–127.

Gorski, L. J. 1969. Traill's Flycatchers of the "fitz-bew" songform wintering in Panama. Auk 86:745–747.

Gradwohl, J., and R. Greenberg. 1980. The formation of antwren flocks on Barro Colorado Island, Panama. Auk 97:385–395.

Graves, G. R. 1996. Censusing wintering populations of Swainson's Warblers: surveys in the Blue Mountains of Jamaica. Wilson Bulletin 108:94–103.

Greenberg, R. 1979. Body size, breeding habitat, and winter exploitation systems in Dendroica. Auk 96:756–766.

Greenberg, R. 1984. The winter exploitation systems of Bay-breasted and Chestnut-sided Warblers in Panama. University of California Publications in Zoology 116:1–107.

Greenberg, R. 1986. Competition in migrant birds in the non-breeding season. Current Ornithology 3:281–307.

Greenberg, R. 1992. Forest migrants in non-forest habitats on the Yucatan Peninsula. Pages 273–286 *in* Ecology and Conservation of Neotropical Migrant Landbirds (J. M. Hagan III and D. W. Johnston, eds.). Smithsonian Institution Press, Washington, D.C.

Greenberg, R., D. K. Niven, S. Hopp, and C. Boone. 1993. Frugivory and coexistence in a resident and a migratory vireo on the Yucatan Peninsula. Condor 95:990–999.

Greenberg, R., and J. Salgado Ortiz. 1994. Interspecific defense of pasture trees by wintering Yellow Warblers. Auk 111:672–682.

Greenberg, R. S., and J. A. Gradwohl. 1980. Observations of paired Canada Warblers *Wilsonia canadensis* during migration in Panama. Ibis 122:509–512.

Greene, E., V. R. Muehter, and W. Davison. 1996. Lazuli Bunting (*Passerina amoena*). The Birds of North America, no. 232.

Hall, G. A. 1996. Yellow-throated Warbler (Dendroica dominica). The Birds of North America, no. 223.

Holmes, R. T., T. W. Sherry, and L. Reitsma. 1989. Population structure, territoriality and overwinter survival of two migrant warbler species in Jamaica. Condor 91:545–561.

Howell, S. N. G., and S. Webb. 1995. A Guide to the Birds of Mexico and Northern Central America. Oxford University Press, Oxford and New York.

Howell, T. R. 1972. Birds of the lowland pine savanna of northeastern Nicaragua. Condor 74:316–340.

Hutto, R. L. 1980. Winter habitat distribution of migratory land birds in western Mexico, with special reference to small foliage-gleaning insectivores. Pages 181–203 *in* Migrant Birds in the Neotropics: Ecology, Behavior, Distribution, and Conservation (A. Keast and E. S. Morton, eds.). Smithsonian Institution Press, Washington, D.C.

Hutto, R. L. 1981a. Seasonal variation in the foraging behavior of some migratory western wood warblers. Auk 98:765–777.

Hutto, R. L. 1981b. Temporal patterns of foraging activity in some wood warblers in relation to the availability of insect prey. Behavioral Ecology and Sociobiology 9:195–198.

Hutto, R. L. 1987. A description of mixed-species insectivorous bird flocks in western Mexico. Condor 89:282–292.

Hutto, R. L. 1992. Habitat distributions of migratory landbird species in western Mexico. Pages 221–239 *in* Ecology and Conservation of Neotropical Migrant Landbirds (J. M. Hagan III and D. W. Johnston, eds.). Smithsonian Institution Press, Washington, D.C.

Hutto, R. L. 1994. The composition and social organization of mixed-species flocks in a tropical deciduous forest in western Mexico. Condor 96:105–118.

Isler, M. L., and P. R. Isler. 1999. Tanagers. Christopher Helm, London.

James, R. D. 1998. Blue-headed Vireo (*Vireo solitarius*). The Birds of North America, no. 379.

Johnson, M. D., and T. W. Sherry. 2001. Effects of food availability on the distribution of migratory warblers among habitats in Jamaica. Journal of Animal Ecology 70:546–560.

Johnson, R. R., H. K. Yard, and B. T. Brown. 1997. Lucy's Warbler (*Vermivora luciae*). The Birds of North America, no. 318.

King, D. I., and J. H. Rappole. 2000. Winter flocking of insectivorous birds in montane pine-oak forests in Middle America. Condor 102:664–672.

Kricher, J. C., and W. E. Davis Jr. 1986. Returns and winter-site fidelity of North American migrants banded in Belize, Central America. Journal of Field Ornithology 57:48–52.

Ladd, C., and L. Gass. 1999. Golden-cheeked Warbler (*Dendroica chrysoparia*). The Birds of North America, no. 420.

Latta, S. C., and M. E. Baltz. 1997. Population limitation in Neotropical migratory birds: comments. Auk 114:754–762.

Latta, S. C., and J. Faaborg. 2001. Winter site fidelity of Prairie Warblers in the Dominican Republic. Condor 103:455–468.

Latta, S. C., and J. M. Wunderle Jr. 1996. The composition and foraging ecology of mixed-species flocks in pine forests of Hispaniola. Condor 98:595–607.

Lefebvre, G., and B. Poulin. 1996. Seasonal abundance of migrant birds and food resources in Panamanian mangrove forests. Wilson Bulletin 108:748–759.

Lefebvre, G., B. Poulin, and R. McNeil. 1994. Spatial and social behavior of Nearctic warblers wintering in Venezuelan mangroves. Canadian Journal of Zoology-Revue Canadienne de Zoologie 72:757–764.

Lerner, S. D. Z., and D. F. Stauffer. 1998. Habitat selection by Blackburnian Warblers wintering in Colombia. Journal of Field Ornithology 69:457–465.

Loftin, H. 1977. Returns and recoveries of banded North American birds in Panama and the tropics. Bird-Banding 48:253–258.

Lovette, I. J., and R. T. Holmes. 1995. Foraging behavior of American Redstarts in breeding and wintering habitats: implications for relative food availability. Condor 97:782–791.

Lowther, P. E. 1999. Alder Flycatcher (Empidonax alnorum). The Birds of North America, no. 446.

Lowther, P. E. 2000. Pacific-slope Flycatcher (Empidonax difficilis) and Cordilleran Flycatcher (Empidonax occidentalis). The Birds of North America, no. 556.

Lynch, J. F. 1989. Distribution of overwintering Nearctic migrants in the Yucatan Peninsula: Pt. 1. General patterns of occurrence. Condor 91:515–544.

Lynch, J. F., E. S. Morton, and M. E. van der Voort. 1985. Habitat segregation between the sexes of wintering Hooded Warblers (*Wilsonia citrina*). Auk 102:714–721.

Mabey, S. E., and E. S. Morton. 1992. Demography and territorial behavior of wintering Kentucky Warblers in Panama. Pages 329–336 *in* Ecology and Conservation of Neotropical Migrant Landbirds (J. M. Hagan III and D. W. Johnston, eds.). Smithsonian Institution Press, Washington, D.C.

Marra, P. P. 2000. The role of behavioral dominance in structuring patterns of habitat occupancy in a migrant bird during the nonbreeding season. Behavioral Ecology 11:299–308.

Marra, P. P., and R. T. Holmes. 2001. Consequences of dominance-mediated habitat segregation in American Redstarts during the nonbreeding season. Auk 118:92–104.

Marra, P. P., T. W. Sherry, and R. T. Holmes. 1993. Territorial exclusion by a long-distance migrant warbler in Jamaica: a removal experiment with American Redstarts (*Setophaga ruticilla*). Auk 110:565–572.

Mayfield, H. F. 1992. Kirtland's Warbler (*Dendroica kirtlandii*). The Birds of North America, no. 19.

McNeil, R. 1982. Winter resident repeats and returns of austral and boreal migrant birds banded in Venezuela. Journal of Field Ornithology 53:125–132.

Meyer de Schauensee, R., and A. L. Mack. 1982. A Guide to the Birds of South America. Pan American Section, International Council for Bird Preservation. Place of publication not known.

Meyer de Schauensee, R., W. H. Phelps Jr., G. Tudor, H. W. Trimm, J. Gwynne, and K. D. Phelps. 1978. A Guide to the Birds of Venezuela. Princeton University Press, Princeton.

Morton, E. S. 1976. The adaptive significance of dull coloration in Yellow Warblers. Condor 78:423.

Morton, E. S. 1979. A comparative survey of avian social systems in northern Venezuelan habitats. Pages 233–259 *in* Vertebrate Ecology in the Northern Neotropics (J. F. Eisenberg, ed.). Smithsonian Institution Press, Washington D.C.

Morton, E. S. 1980. Adaptations to seasonal changes by migrant land birds in the Panama Canal Zone. Pages 437–453 *in* Migrant Birds in the Neotropics: Ecology, Behavior, Distribution, and Conservation (A. Keast and E. S. Morton, eds.). Smithsonian Institution Press, Washington, D.C.

Morton, E. S., J. F. Lynch, K. Young, and P. Mehlhop. 1987. Do male Hooded Warblers exclude females from nonbreeding territories in tropical forest? Auk 104:133–135.

Moskoff, W., and S. K. Robinson. 1996. Philadelphia Vireo (*Vireo philadelphicus*). The Birds of North America, no. 214.

Mowbray, T. B. 1999. Scarlet Tanager. The Birds of North America, no. 479.

Neudorf, D. L., and S. A. Tarof. 1998. The role of chip calls in winter territoriality of Yellow Warblers. Journal of Field Ornithology 69:30–36.

Nickell, W. P. 1968. Returns of northern migrants to tropical winter quarters and banded birds recovered in the United States. Bird-Banding 39:107–116.

Olson, C. R., and T. E. Martin. 1999. Virginia's Warbler (*Vermivora virginiae*). The Birds of North America, no. 477.

Ornat, A. L., and R. Greenberg. 1990. Sexual segregation by habitat in migratory warblers in Quintana-Roo, Mexico. Auk 107:539–543.

Ornat, A. L., J. F. Lynch, and B. M. Demontes. 1989. New and noteworthy records of birds from the eastern Yucatan Peninsula. Wilson Bulletin 101:390–409.

Parrish, J. D., and T. W. Sherry. 1994. Sexual habitat segregation by American Redstarts wintering in Jamaica: importance of resource seasonality. Auk 111:38–49.

Payne, R. B. 1992. Indigo Bunting (*Passerina cyanea*). The Birds of North America, no. 4.

Pitocchelli, J., J. Bouchie, and D. Jones. 1997. Connecticut Warbler (*Oporornis agilis*). The Birds of North America, no. 320.

Price, T. 1981. The ecology of the Greenish Warbler *Phylloscopus trochiloides* in its winter quarters. Ibis 123:131–145.

Pyle, P. 1997. Identification Guide to North American Birds: Pt. 1. Slate Creek Press, Bolinas, Calif.

Ramos, M. A., and J. H. Rappole. 1994. Local movements and translocation experiments of resident and migratory birds in southern Veracruz, Mexico. Bird Conservation International 4:175–180.

Rappole, J. H. 1983. Analysis of plumage variation in the Canada Warbler. Journal of Field Ornithology 54:152–159.

Rappole, J. H. 1988. Intra- and intersexual competition in migratory passerine birds during the nonbreeding season. Pages 2308–2317 *in* Proceedings of the 19th International Ornithological Congress (H.Ouellet, ed.). University of Ottawa Press, Ottawa.

Rappole, J. H., D. I. King, and W. C. Barrow Jr. 1999. Winter ecology of the endangered Golden-cheeked Warbler. Condor 101:762–770.

Rappole, J. H., D. I. King, and P. Leimgruber. 2000. Winter habitat and distribution of the endangered Golden-cheeked Warbler (*Dendroica chrysoparia*). Animal Conservation 3:45–59.

Rappole, J. H., E. S. Morton, and M. A. Ramos. 1992. Density, philopatry, and population estimates for songbird migrants wintering in Veracruz. Pages 337–344 *in* Ecology and Conservation of Neotropical Migrant Landbirds (J. M. Hagan III and D. W. Johnston, eds.). Smithsonian Institution Press, Washington, D.C.

Rappole, J. H., M. A. Ramos and K. Winker. 1989. Wintering Wood Thrush movements and mortality in southern Veracruz. Auk 106:402–410.

Rappole, J. H., and D. W. Warner. 1980. Ecological aspects of migrant bird behavior in Veracruz, Mexico. Pages 353–393 *in* Migrant Birds in the Neotropics: Ecology, Behavior, Distribution, and Conservation (A. Keast and E. S. Morton, eds.). Smithsonian Institution Press, Washington, D.C.

Regosin, J. V. 1998. Scissor-tailed Flycatcher (*Tyrannus forficatus*). The Birds of North America, no. 342.

Ridgely, R. S., and J. A. Gwynne. 1989. A guide to the birds of Panama with Costa Rica, Nicaragua, and Honduras. Princeton University Press, Princeton and Oxford.

Ridgely, R. S., and G. Tudor. 1989. The Birds of South America, Vol. 1, The Oscine Passerines. University of Texas Press, Austin.

Ridgely, R. S., and G. Tudor. 1994. The Birds of South America, Vol. 2, The Suboscine Passerines: Ovenbirds and Woodcreepers, Typical and Ground Antbirds, Gnateaters and Tapaculos, Tyrant Flycatchers, Cotingas and Manakins. Oxford University Press, Oxford and Tokyo.

Rimmer, C., and K. McFarland. 1995. Investigations of high elevation bird communities in the northeastern United States and the Dominican Republic, Progress Report 1995. Vermont Institute of Natural Sciences, Woodstock, Vt.

Robbins, C. S., J. W. Fitzpatrick, and P. B. Hamel. 1992. A warbler in trouble: *Dendroica cerulea*. Pages 549–562 *in* Ecology and Conservation of Neotropical Migrant Landbirds (J. M. Hagan III and D. W. Johnston, eds.). Smithsonian Institution Press, Washington, D.C.

Robinson, S. K., J. Terborgh, and J. W. Fitzpatrick. 1988. Habitat selection and relative abundance of migrants in southeastern Peru. Pages 2298–2307 *in* Proceedings of the 19th International Ornithological Congress, Vol. 2 (H. Ouellet, ed.). University of Ottawa Press, Ottawa.

Robinson, W. D. 1995. Louisiana Waterthrush (*Seiurus motacilla*). The Birds of North America, no. 151.

Rogers, D. T., Jr., D. L. Hicks, E. W. Wischusen, and J. R. Parrish. 1982. Repeats, returns, and estimated flight ranges of some North American migrants in Guatemala. Journal of Field Ornithology 53:133–138.

Schwartz, P. 1964. The Northern Waterthrush in Venezuela. Living Bird 3:169–184.

Shane, T. G. 2000. Lark Bunting (*Calamospiza melanocorys*). The Birds of North America, no. 542.

Sherry, T. W., and R. T. Holmes. 1989. Age-specific social-dominance affects habitat use by breeding American Redstarts (*Setophaga ruticilla*): a removal experiment. Behavioral Ecology and Sociobiology 25:327–333.

Sherry, T. W., and R. T. Holmes. 1996. Winter habitat quality, population limitation, and conservation of Neotropical-Nearctic migrant birds. Ecology 77:36–48.

Sick, H. 1971. Blackpoll Warbler on winter quarters in Rio de Janeiro, Brazil. Wilson Bulletin 83:198–200.

Sick, H. 1993. Birds in Brazil: A Natural History. Princeton University Press, Princeton.

Slud, P. 1964. The birds of Costa Rica. Bulletin of the American Museum of Natural History 128:1–372.

Staicer, C. A. 1992. Social behavior of the Northern Parula, Cape May Warbler, and Prairie Warbler wintering in second-growth forest in southwestern Puerto Rico. Pages 308–321 *in* Ecology and Conservation of Neotropical Migrant Landbirds (J. M. Hagan III and D. W. Johnston, eds.). Smithsonian Institution Press, Washington, D.C.

Sterling, J. C. 1999. Gray Flycatcher (*Empidonax wrightii*). The Birds of North America, no. 458.

Stiles, F. G., and A. F. Skutch. 1989. A Guide to the Birds of Costa Rica. Comstock Publishing Associates, New York.

Stotz, D. F., R. O. Bierregaard, M. Cohn Haft, P. Peterman, J. Smith, A. Whittaker, and S. V. Wilson. 1992. The status of North American migrants in central Amazonian basin. Condor 94:608–621.

Strong, A. M., and T. W. Sherry. 2000. Habitat-specific effects of food abundance on the condition of ovenbirds wintering in Jamaica. Journal of Animal Ecology 69:883–895.

Stutchbury, B. J. 1994. Competition for winter territories in a Neotropical migrant: the role of age, sex and color. Auk 111:63–69.

Terborgh, J. 1989. Where Have All the Birds Gone? Essays on the

Biology and Conservation of Birds That Migrate to the American Tropics. Princeton University Press, Princeton.

Thurber, W. A., and A. Villeda. 1976. Band returns in El Salvador, 1973–74 and 1974–75 seasons. Bird-Banding 47:277–278.

Tramer, E. J., and T. R. Kemp. 1979. Diet-correlated variations in social behaviour of wintering Tennessee Warblers. Auk 96:186–187.

Tramer, E. J., and T. R. Kemp. 1980. Foraging ecology of migrant and resident warblers and vireos in the highlands of Costa Rica. Pages 285–296 *in* Migrant birds in the Neotropics: Ecology, Behavior, Distribution, and Conservation (A. Keast and E. S. Morton, eds.). Smithsonian Institution Press, Washington, D.C.

Twedt, D. J., and R. D. Crawford. 1995. Yellow-headed Blackbird (*Xanthocephalus xanthocephalus*). The Birds of North America, no. 192.

Vidal, R. M., C. Macias-Caballero, and C. D. Duncan. 1994. The occurrence and ecology of the Golden-cheeked Warbler in the highlands of Northern Chiapas, Mexico. Condor 96:684–691.

Wallace, G. E., H. G. Alonso, M. K. McNicholl, D. R. Batista, R. O. Prieto, A. L. Sosa, B. S. Oria, and E. A. H. Wallace. 1996. Winter surveys of forest-dwelling Neotropical migrant and resident birds in three regions of Cuba. Condor 98:745–768.

Warkentin, I. G., and D. Hernandez. 1996. The conservation implications of site fidelity: a case study involving Nearctic-Neotropical migrant songbirds wintering in a Costa Rican mangrove. Biological Conservation 77:143–150.

Wetmore, A., R. F. Pasquier, and S. L. Olson. 1984. The Birds of the Republic of Panama: Pt. 4. Smithsonian Institution Press, Washington, D.C.

Willis, E. O. 1966. The role of migrant birds at swarms of army ants. Living Bird 5:187–232.

Winker, K., and J. H. Rappole. 1992. The autumn passage of Yellow-bellied Flycatchers in south Texas. Condor 94:526–529.

Winker, K., J. H. Rappole, and M. A. Ramos. 1990. Population dynamics of the Wood Thrush in southern Veracruz, Mexico. Condor 92:444–460.

Wunderle, J. M. 1995. Population characteristics of Black-throated Blue Warblers wintering in three sites on Puerto Rico. Auk 112:931–946.

Wunderle, J. M., and S. C. Latta. 2000. Winter site fidelity of Nearctic migrants in shade coffee plantations of different sizes in the Dominican Republic. Auk 117:596–614.

CHRISTOPHER M. ROGERS

Food Limitation among Wintering Birds

A View from Adaptive Body Mass Models

STUDENTS OF THE ECOLOGY and population biology of tropical-wintering migratory birds often use measures of body mass components as surrogate measures of habitat quality. For example, low fat or nonfat mass is considered to reflect low habitat quality, that is, a habitat with low food supply and therefore low survival probability. The food limitation hypothesis (FLH) states that food supply directly limits one or more body mass components. However, the predation-food hypothesis (PFH) also explains observed trends in mass components among habitats of differing quality. This latter hypothesis, derived from models of adaptive regulation of avian body mass, suggests that food is inadequate to support high-mass components in poor habitats at the existing level of predation. The two hypotheses make a nonoverlapping set of predictions about lapses in foraging, effects on mass components of predator removal and food addition, and the extent of premigratory fat deposition in species populations that make initial long-distance migratory movements directly from the wintering grounds. During winter, body mass decisions influencing individual fitness are focused on choosing the mass strategy that minimizes the combined probabilities of starvation and predation. However, in early spring, fitness is suddenly related strongly to reaching the breeding grounds, as no breeding opportunities exist for migrants wintering in tropical regions. Only the PFH makes the prediction that with newly changed fitness priorities in spring, fat reserves increase suddenly as the bird prepares for a long-distance migratory flight, possibly over a large expanse of water. The FLH predicts that food supplies on the wintering grounds are inadequate to support any increases in either the fat or nonfat component of body mass. Distinguishing between hypotheses clarifies the role of simple food limitation versus the role of the food-predation interaction as factors affecting body mass regulation and mass-dependent migration and subsequent reproduction schedule.

INTRODUCTION

The theoretical development and empirical testing of models of adaptive body mass regulation have forged an increasingly sophisticated perspective of fitness maximization in small birds over the last 20 years (Lima 1986; Rogers 1987; McNamara and Houston 1990; Houston and McNamara 1993, 1999; Witter and Cuthill 1993; Bednekoff and Houston 1994; Houston and McNamara 1999; Cuthill et al. 2000; Pravosudov and Lucas 2001). Adaptive fat models view avian winter fat storage as a starvation-predation trade-off: by balancing costs and benefits of fat, winter survival probability is maximized. A main benefit of fat is assumed to be reduction of starvation risk, especially in environments of uncertain foraging gain; costs of fat potentially include increased exposure to predators while feeding to fatten (Ekman 1986; Witter and Cuthill 1993; Gosler et al. 1995; Cresswell 1998; Van Der Veen and Lindström 2000), and/or reduced flight performance at high fat levels (Witter et al. 1994; Metcalfe and Ure 1995; Küllberg et al. 1996, 2000; Lind et al. 1999; reviewed by Cuthill and Houston 1997). The exposure component of predation risk has recently been suggested to be more important than the reduced flight performance component (Dierschke 2003).

An increasing trend in studies of wintering migratory birds, especially Nearctic-Neotropical migrants, is to use various measures of body mass (mass/morphology or mass corrected for structural size) and components of body mass (visible subcutaneous fat class) as indirect measures of habitat quality, that is, food supply. The reasoning followed by most workers is that if body mass measures and/or visible fat is low in one habitat and high in another habitat, the latter habitat is of higher quality (e.g., has higher food supply, and may offer higher survival) than the former habitat. Furthermore, low-quality (low food) habitat therefore directly limits mass components. However, adaptive mass models also make clear predictions about how body mass changes when food supply changes. Specifically, one model (discussed in the next section) predicts reduced body fat at high food abundance as the optimal starvation-predation trade-off changes; note that this prediction is opposite to the prediction made by simple food limitation.

It is the purpose of the present chapter to address this gap between behavioral ecology and population biology, with emphasis on understanding the role of predation in food limitation during the wintering period. The goal is to use adaptive body mass models to understand how food and predation interact to affect, or limit, body mass components in the wintering season and to compare this perspective to that already established by workers considering simple food limitation. This chapter addresses the question not of *whether* but rather of *how* food is limiting in overwintering bird populations; the focus is on Nearctic-Neotropical migrants. Effects of individual condition on population processes and parameters in this group, for example, migration schedule and population output on the breeding grounds (Marra et al. 1998), can then be understood in terms of current multifactor body mass models that include predation.

PREDICTIONS OF ADAPTIVE BODY MASS MODELS

Similarities of Models: Predictions about Food Predictability/Variability

Models of adaptive body mass in non-caching birds (e.g., Lima 1986; Houston and McNamara 1993) each separately model two aspects of food supply: its predictability (variability) in space and time and its abundance. Both models predict that when food unpredictability or variability increases, resulting in increased starvation risk, body fat will increase; there is a large body of evidence to support this prediction (Rogers 1987; Ekman and Hake 1990; Ekman and Lilliendahl 1993; Rogers and Smith 1993; Bednekoff et al. 1994; Bednekoff and Krebs 1995; Rogers 1995; Witter et al. 1995; Smith and Metcalfe 1997; Witter and Swaddle 1997; Rogers and Reed 2003). Some findings have suggested that low social dominance status increases fat reserves of subordinates, owing to their unpredictable resource access, compared with dominants (Ekman and Lilliendahl 1993); in contrast, others suggest no effect of dominance status on fat reserves (Koivula et al. 1995) or higher fat reserves among dominants (Vehrlust and Hogstad 1996).

At the interspecific level, there is clear evidence for the predicted inverse relationship between food predictability and fat store in Neotropical bird species. In three avian foraging guilds wintering in Veracruz, Mexico, there was an inverse relationship between resource predictability and fat level (Rappole and Warner 1980; Rappole 1995). In an impressive data set, the percentage of individuals with moderate to heavy visible fat reserves increased from insectivores (high resource predictability), to frugivore/insectivores (intermediate predictability), to granivores (low resource predictability). Use of visible fat classes has been criticized, but this method remains the only effective way of noninvasively measuring fat reserves of small-bodied birds (Rogers 1991, 2003). In the only other comprehensive data set on fat reserves of wintering migrants (Greenberg 1992), visible fat reserves also showed substantial interspecific variation, although without knowledge of guild structure, it is difficult to draw conclusions concerning resource predictability and fat.

Differences between Models: Predictions about Food Abundance

The two adaptive mass models make opposing predictions about changes in food abundance. As McNamara and Houston (1990) summarize, differing predictions arise because the models differ in how they assess the temporal scale of this critical ecological factor. The original model (Lima 1986) includes between-day variability in foraging gain. This model predicts increased fat reserves when food availability is increased, because the bird usually does not perceive a reduction in starvation risk with an increase in food, and so it takes advantage of plentiful food by depositing fat to reduce starvation risk. Such a response indicates that the bird expects a variation in food intake between days, hence the value of body fat. Thus, in the Lima model, there is an association between changes in food predictability and changes in food abundance. In contrast, the Houston and McNamara model focuses on a within-day scale of environmental variability by including random interruptions in foraging within days. This model predicts a decrease in fat when food availability is increased because the bird perceives reduced (within-day) starvation risk with increased food, and so the optimal level of fat reserves is reduced. Predation probability is also lowered at higher gain in the latter model.

Which Prediction about Food Abundance Applies to the Tropics?

Owing to differences in the temporal scale of changing food supply, different models of adaptive fat regulation make opposing predictions concerning the effect of changing food supply on the avian fat reserve. In temperate environments like interior North America, foraging interruption is driven on a day-day time scale by inclement weather events, for example, snow and ice storms. Such critical environmental events would be survived with greater probability if reserves were increased. All studies to date that have varied food supply in temperate-dwelling bird species showed a slight increase in body fat at increased food supply in nature (Smith et al. 1980; Brittingham and Temple 1988; Rogers and Heath-Coss 2003; the last study in seven species).

In much warmer and snow-free environments such as tropical regions and regions with mild maritime climate, such severe events are rare, and starvation risk is more likely to be related to within-day interruptions possibly driven by density-dependent processes associated with higher population density of conspecifics and competitors. Therefore in the present analysis, I apply to tropical bird species the prediction made by the Houston and MacNamara model: the winter fat reserve will decrease with increasing food supply and increase with decreasing food supply. In doing so, I also consider the trade-off between maintaining the fat reserve and maintaining the nonfat reserve (the labile component probably consisting largely of the pectoralis musculature), as both are apparently labile in the short-term in tropical-wintering bird species (evidence reviewed below). Sacrifice of one component can be in support of another, a complication that must be considered for a complete perspective of overall body mass regulation.

PATTERNS OF FAT AND NONFAT BODY MASS COMPONENTS IN TROPICAL-WINTERING BIRDS

To proceed, I first describe temporal and spatial patterns of fat and nonfat body mass components and ask whether simple food limitation (Sherry et al., Chap. 31, this volume) or the hypothesis of adaptive fat regulation best explains these patterns.

There are surprisingly few detailed studies of body mass components and their relationship to varying food abundance in small migratory birds in the Tropics (both Neotropical and Paleotropical). In the Greenish Warbler (*Phylloscopus trochiloides*) wintering on the Indian subcontinent, food supply (invertebrate density) was high in years of higher winter precipitation, and body fat was low (Katti and Price 1999). In years of lower precipitation and consequent lower food abundance, body fat was higher. In contrast, the nonfat component showed the opposite pattern, being relatively low at low food and relatively high at high food. The pattern of nonfat variation was determined by inference: annual variation in visible fat score was not accompanied by annual variation in total body mass.

Ovenbirds (*Seiurus aurocapillus*) wintering in Jamaica, a large island in the Neotropics, showed exactly the same pattern in both body mass components (Strong and Sherry 2000). The main food of wintering Ovenbirds was terrestrial ants, with other major insect orders also represented in the diet. In this case, the pattern was within one winter; food supply declined toward late winter / early spring, and mass components were altered over time by individual birds. For purposes of evaluating the roles of food and predation in wintering migrants, the pattern shown by the former two species is here considered, as it was found in both Neotropical and Paleotropical migrants. As they accumulate, new studies of this nature can be evaluated with this theoretical framework.

ALTERNATIVE HYPOTHESES FOR MASS COMPONENT VARIATION IN TROPICAL-WINTERING BIRDS

Food Limitation Hypothesis

The patterns in body mass components in tropical birds described above (Greenish Warbler, Ovenbird) can be explained by two different mechanistic hypotheses. The first hypothesis is the food limitation hypothesis (FLH). This hypothesis is often cited to explain mass component variation among habitats, seasons, or years. Thus, for example, when a body mass component (fat, nonfat) is small, it is assumed outright that food supply in that habitat is low, and when a body component is high, the assumption is that food supply is high. Applying the FLH to the patterns observed in the Neotropics (Ovenbird) and Paleotropics (Greenish Warbler) in poor habitats, body fat may be high and nonfat low because food is simply inadequate to maintain both components at a level favored by the bird. Thus, fat is kept high as a hedge against future food shortages, but at the expense of a high nonfat component. The flight musculature is a major component of the avian nonfat mass, except in flightless bird species. Under the FLH, in poor habitats nonfat is kept at a minimum in support of elevated body fat. In good habitat, fat is low because starvation risk is low and high food supports a high nonfat component. In good habitats, no food limitation of any mass component occurs, so food limitation can be habitat-specific.

Marra et al. (1998) discuss population-level effects of individual mass decisions made by wintering Nearctic-Neotropical migrants. In poor habitats with low food supplies, especially the nonfat mass component can be low, with the effect of delayed vernal migration, leading to later arrival on the breeding grounds and lower reproductive output by the individual, hence the population. In habitats with greater food supplies, the nonfat component is high and departure time from the wintering grounds in spring is earlier, leading to earlier arrival on the breeding grounds, a longer reproductive period, and greater individual reproductive and population output. Simple food limitation can lead to all of these effects, in the absence of any role for predation.

Predation-Food Hypothesis

The second hypothesis explaining patterns in body mass components in tropical-wintering birds is the predation-food hypothesis (PFH). It states that food supply is inadequate to support both mass components at a preferred high level in poor habitats, seasons, or years at the existing background level of predation. In poor habitats with low food supplies, starvation risk is high, and the fat reserve is also high in response. However, the nonfat component is regulated at a low level, because to maintain a high value of this mass component would require more foraging beyond that needed to lower starvation risk, resulting in increased exposure to predators, which are common and diverse in many tropical regions. Maximizing daily survival probability, possibly leading to a lowered nonfat component and subsequent effects on spring reproduction, is more important than accepting an elevated winter predation risk during foraging that maintains both fat and nonfat components at high levels and minimizes delayed migration and negative effects on reproduction on the breeding grounds. In good habitats, fat is low because starvation risk is low; nonfat can be high with minimal increase in foraging rate and exposure to predators.

In Jamaica, site of the Ovenbird study, the avian predator community is limited to the American Kestrel (*Falco sparverius* [common]) and the Merlin (*F. columbarus* [rare]), both dwellers in open habitats and unlikely predators on small birds in dense habitats. However, a mongoose (*Herpestes auropunctatus*) was introduced into Jamaica in 1872, and the feral house cat (*Felis catus*) was introduced there hundreds of years before that (Henderson 1992). Both mammal species are noted predators on small birds (Nowak 1991) and could exert predation pressure on overwintering Nearctic-Neotropical migrants, although currently no data exist to test this possibility (Strong and Sherry 2000).

Differences between Hypotheses

The main difference between the above hypotheses lies in a role for predation as a factor limiting foraging and hence one, or perhaps more, body mass components. In more mechanistic terms, the FLH suggests that in a poor habitat, a foraging bird in winter simply cannot find enough food to keep both main body mass components at preferred levels. The PFH suggests that food supply may be abundant enough to support all preferred component levels, but predation risk is too high for the bird to maximize survival probability while exploiting food, so it makes a trade-off in poor habitats. In this fashion, food limitation is mediated by the ecological factor of predation risk, as foraging birds make individual decisions concerned with fitness maximization in the wintering period.

NONOVERLAPPING PREDICTIONS

The food limitation and predation-food hypotheses make a nonoverlapping set of predictions about behavior and mass component regulation. These are considered in turn, along with discussion of relevant observational and experimental approaches, and prioritized in the Future Directions section below.

Lapses in Foraging

The FLH maintains that there is inadequate food to support both high fat reserves and a high nonfat mass component. Thus, the hypothesis predicts that the wintering bird will forage the entire daylight period during each day of the wintering period. An implicit behavioral mechanism is that foraging birds acting under the hypothesis will not show sensitivity to predation risk.

In contrast, the PFH predicts a high probability of lapses in foraging. These should occur if predation risk is high enough to cause birds to feed intermittently, beyond any delay in foraging caused by digestive constraints. Lapses in foraging have rarely been investigated directly but have been observed in an insectivorous bird species (*Empidonax trailli*) during breeding, a time of high energy expenditure and high foraging rates (Ettinger and King 1980). It is a reasonable expectation that predation risk is high enough to cause foraging lapses in the Neotropics and Paleotropics because hawks of the genus *Accipiter* occur commonly in many areas of both biogeographical regions. In addition, two species in the hawk genus *Micrastur* are found in many regions of the Neotropics. All prey upon small birds in various habitats, including forest, an important habitat for wintering tropical migrants. Mammalian predators are also found throughout both tropical regions (see above), and could conceivably pose substantial predation risk to foraging birds, at least those that are primarily ground and shrub feeders.

Distinguishing between hypotheses can be accomplished by constructing daily time budgets of individually color-banded wintering birds. This might be most easily accomplished with territorial species, whose spatial distribution is easily predictable over time. Any foraging lapses would be easily detected, as would any conspicuous predator avoidance behavior, for example, scanning and other forms of vigilance, alarm notes, and facultative interspecific flocking in the assumed presence of a predator. Temporary interspecific flocking is a known feature of territorial tropical-wintering bird species (Rappole 1995). A caveat is that lapses can be confused with rest periods taken by the bird during the daily foraging routine. One possible approach is to measure the timing of foraging lapses in relation to the appearance of predators in the vicinity of feeding birds.

Predator Removal

Under the FLH, predator removal (live-trapping of predators and their temporary holding in captivity) is predicted to have no effect on any mass component. This outcome is expected because only food level is hypothesized to have an effect on any mass component, in any habitat (good or bad).

It is more difficult to establish a specific prediction for predator removal made by the PFH. However, limited progress can be made by considering habitat-specific responses. In low-quality habitat, fat is high and nonfat is low; thus if predators were removed the bird would gain the most future fitness by increasing nonfat, which would potentially facilitate earlier departure from the wintering grounds and more successful breeding (Marra et al. 1998). In high-quality habitat, fat is low because no benefit to fat is perceived by the bird; nonfat is high and already contributing to future fitness. An increase in fat level might occur here, as costs to fat would decrease with predator removal. Of concern, however, is the possibility that early predator removal could lead to more migrants choosing a favored predator-free area, leading to increased intraspecific competition for food. Predator removals may have to be conducted after migrants have terminated migration and settled on a wintering site.

Predator removal can be technically difficult to achieve because predators may prove difficult to capture and areas newly emptied of predators may be quickly filled by conspecifics seeking to exploit profitable empty habitat. Nevertheless, functional predator absence might be established long enough with intensive sustained effort. A prolonged predator removal campaign is not necessary because avian mass components are known to change rapidly with varying ecological conditions (Piersma and Lindström 1997). An alternative to predator removal is predator addition, which of necessity would be restricted to studies of ground- and shrub-foraging bird species potentially affected by terrestrial mammals. This predator group, as opposed to avian predators, could be introduced into territories of wintering birds by using humane caging methods. Finally, playback of recorded alarm calls is a useful possibility as an experimental manipulation.

General sensitivity to predators by wintering migrants was suggested by two studies in Belize. Petit et al. (1992) found that despite the presence of adequate habitat and food supply in deep forest interior, few migrants were observed there, the great majority choosing edge habitats, ostensibly without forest-interior avian predators (Barred Forest Falcon [*M. ruficollis*], Collared Forest Falcon [*M. semitorquatus*], and Bicolored Hawk [*A. bicolor*]). Migrants avoided forest interior and favored edge habitat, even when adequate food supply did not significantly vary between these two microhabitats (Petit et al. 1992). In a two-year study also conducted in Belize, C. M. Rogers (in prep.) documented the same pattern on census points variously located in riparian forest, parkland, deep forest interior, forest edge, young forest, milpa, and palm stands in offshore islands. All habitats except deep forest supported moderate to high numbers of wintering migrants, especially the Parulidae. These data are based largely on auditory detections of the birds, whose call notes were frequently detected within 50 m of the observer at the census point in all habitats.

However, at least two alternative hypotheses exist to explain edge avoidance of migrants on the wintering grounds. Although not completely independent of food supply within evolutionary and ecological time frames, competition with residents in forest interior may force migrants to less favorable edge environments; interspecific competition can be a significant ecological factor affecting avian populations in the Neotropics (Greenberg 1990). In addition, habitat selection behavior evolved in response to currently unknown selection regimes can be forwarded as a hypothesis explaining migrant avoidance of forest interior.

Alteration of Food Supply

If the FLH is correct, supplemental feeding of wintering birds should affect body mass components, particularly in low-quality habitat. Increasing food supply should effectively convert poor-quality habitat to better habitat, resulting in birds' showing the high-quality mass profile there (low fat, high nonfat). Assuming that the food supply in good habitat is already adequate, no effect of supplemental feeding is expected. Unfortunately, the FLH makes the same prediction: increased food would allow a greater food intake rate in low-quality habitat, reducing starvation risk, decreasing fat, and increasing nonfat. A second problem is that alteration of food supply is technically difficult to achieve, although increasing ant biomass with carbohydrate or other nutrient supplements might be feasible in Ovenbird territories (Strong and Sherry 2000).

Premigratory Fattening in Winter Habitat and Changed Fitness Priorities

Most species populations of Nearctic–Neotropical-wintering birds consist of long-distance migrants, which make lengthy migratory movements, often across great expanses of open water (e.g., trans-Gulf migrants) between the breeding and wintering grounds. As phenological conditions on the breeding grounds approach the capacity to support reproduction, individual priorities must change from maximizing fitness via winter survival to maximizing fitness by migrating in timely fashion to the breeding grounds (Changing priority model of nonbreeding mass regulation [fig. 9.1]). The success of such long-distance flights can depend on the deposition of extensive premigratory body fat. On the wintering grounds of the species concerned, food to support high fat depots is of course obtainable only from tropical habitats, initially including that at which the bird just spent the winter.

The FLH predicts that in poor habitats, food amount is inadequate to support elevated fat reserves because the bird is already foraging the entire day and scraping a living from the land that cannot be improved upon whatsoever. In contrast, under the PFH, avian body mass is regulated in close connection with current fitness priorities, especially the trade-off between avoiding starvation and avoiding predators. A broader application of the hypothesis (still reflecting a close relationship between body mass and fitness) predicts under changed individual fitness priorities in late winter, rapid and extensive vernal premigratory fat deposition in Nearctic-Neotropical migrants interested in reaching the breeding grounds on a favorable schedule of reproduction. Body components may be kept relatively low, in the interest of predator avoidance, but this fitness strategy becomes increasingly irrelevant to the soon-to-be migrant that has no option of breeding anywhere near its wintering site. Essentially, selection can be expected to favor a survival-oriented mass strategy after autumnal migration until spring, with body mass reflecting the starvation-predation trade-off. With the arrival of the vernal migratory phase of the annual physiological cycle, mass priorities quickly change as selection newly favors hyperphagia and hyperlipogenesis, and presumably also a high nonfat mass component, all of which are critical to successful long-distance migration.

In poor habitats, under the PFH, even with newly changed priorities in spring nonfat may show a delayed increase with increased food intake, and migration may still be delayed, with subsequent negative effects on reproduction on the breeding grounds. The vernal premigratory fat reserve may not show such latency, because fat deposits should be more quickly increased via hyperphagia.

A substantial literature already exists suggesting premigratory fat deposition on the wintering grounds among Nearctic-Neotropical migratory Parulidae. In the Northern Waterthrush (*Seiurus noveboracensis*) wintering in coastal mangrove in Venezuela, a gradual increase in body mass (with minor fluctuation) from September to April was followed by a sudden increase in body mass in May (Lefebvre et al. 1992). Among seven parulid species of six genera wintering in Jamaica, six species showed dramatic increases in both body mass and visible fat class in late winter or early spring; the seventh species showed only a late-season increase in visible fat (Diamond et al. 1977). Most species reached high levels of visible fat, falling into categories 8–12 on a 0–12 fat-class scale. One of these species, the American Redstart (*Setophaga ruticilla*), reached high species-specific body mass and high visible fat class in May, after showing low and declining values of these variables for most of the winter. Low body mass (and presumably also fat reserves) were observed in late winter for this species sampled in a variety of forested habitats (Holmes et al. 1989), and Marra and Holmes (2001) found a winter-long decline in size-corrected body mass in the Redstart in dry scrub, but not in mangrove; both studies were in Jamaica.

The above conceptual framework for clarifying the role of food supply in nonbreeding behavioral ecology of tropical-wintering migrants assumes that spring migration is not timed to be supported by invertebrate resource flushes on the wintering grounds in late winter, which would represent a food source not present even in good habitats in winter. The absence of late-winter resource flushes is supported by the well-known phenomenon of the late-winter dry season and its associated low invertebrate abundance observed in many regions of the Neotropics (e.g., Strong and Sherry 2000). If flush-related fattening does occur, the presence of extensive fat deposition in late winter (or early spring) among wintering migrants would not effectively distinguish

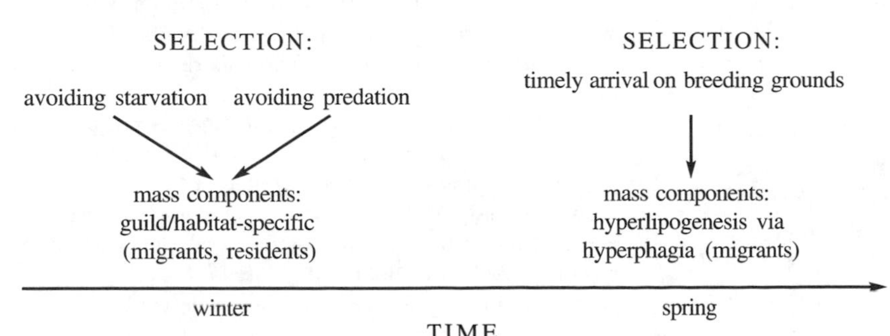

Fig. 9.1. Changing priority model of avian nonbreeding body mass regulation. Fitness maximizing behavior may change from maximizing survivorship via an optimal starvation-predation trade-off during winter to depositing extensive premigratory fat necessary for reaching distant breeding grounds on a schedule favorable for reproduction.

between the FLH and PFH: a sudden late-winter rise in body fat level could be explained by relaxed simple food limitation or by altered fitness priorities.

A more complex issue is the possibility of inexpensive, fat-independent short initial migratory movements to habitats with high food supplies capable of supporting rapid premigratory fat deposition. Such late-winter habitat shifts would suggest lack of food limitation in at least part of the wintering range of a given migrant species. On the other hand, if shifts are necessary for fat deposition, this would suggest real food limitation on the initial wintering grounds. Ultimately, reaching a preliminary conclusion on this matter depends on establishing a geographic scale of analysis.

In any event, as is the likely case for resource flushes, such "depositional" habitats seem unlikely in most places in the Neotropics during the widespread late-winter dry season, and even if they did exist, they could not be of such magnitude as to support many of the large numbers of Nearctic-Neotropical migrants moving northward in spring. Furthermore, any productive habitats can be expected to be occupied earlier by residents or by migrants establishing winter residency following the termination of fall migration from the breeding grounds.

FUTURE DIRECTIONS

This is appropriately a brief section, as the previous discussion considers at length research protocols that distinguish between the competing hypotheses concerned with food limitation on the wintering grounds of migrants. Only research priorities are suggested. It seems critical to test for predator-sensitivity of mass components, predicted by the PFH but not the FLH. Indeed, much if not all of the literature on the costs of maintaining a given mass component is based on studies of temperate, not tropical (including migrant), bird species. Determining sensitivity of mass components to experimentally altered food supply (with food predictability held constant) in the Tropics is obviously of high priority, although experiments may be limited to ant-eating species (above) or even granivorous or frugivorous species. In addition, it seems prudent to establish further that premigratory fattening, already demonstrated for a number of Nearctic-Neotropical migrant species, is supported by late-winter–early-spring resource flushes or newly occupied, highly productive late-winter habitats.

ACKNOWLEDGMENTS

Gratitude is expressed to Russ Greenberg and Pete Marra for organizing the Birds of Two Worlds Symposium, and to the Smithsonian Institution for facilitating the meeting. Belize fieldwork and manuscript preparation were supported by National Science Foundation grant 19229-KAN19230 and matching support from the State of Kansas.

LITERATURE CITED

Bednekoff, P.A., H. Biebach, and J.R. Krebs. 1994. Great tit fat reserves under unpredictable temperatures. Journal of Avian Biology 25:156–160.

Bednekoff, P. A., and A. I. Houston. 1994. Optimising fat reserves over the entire winter: a dynamic model. Oikos 71:408–415.

Bednekoff, P. A., and J. R. Krebs. 1995. Great Tit fat reserves: effects of changing and unpredictable feeding day length. Functional Ecology 9:457–462.

Brittingham, M. C., and S. A. Temple. 1988. Impacts of supplemental feeding on survival rates of Black-capped Chickadees. Ecology 69:581–589.

Cresswell, W. 1998. Diurnal and seasonal mass variation in blackbirds *Turdus merula:* consequences for mass-dependent predation risk. Journal of Animal Ecology 67:78–90.

Cuthill, I. C., and A. I. Houston. 1997. Managing time and energy. Pages 97–120 *in* Behavioural Ecology (J. R. Krebs and N. B. Davies, eds.). Oxford University Press, Oxford.

Cuthill, I. C., S. A. Maddocks, C. V. Weall, and E. K. M. Jones. 2000. Body mass regulation in response to changes in feeding predictability and overnight energy expenditure. Behavioral Ecology 11:189–195.

Diamond, A. W., P. Lack, and R. W. Smith. 1977. Weights and fat condition of some migrant warblers in Jamaica. Wilson Bulletin 89:456–465.

Dierschke, V. 2003. Predation hazard during migratory stopover: are light or heavy birds at risk? Journal of Avian Biology 34:24–29.

Ekman, J. 1986. Tree use and predator vulnerability of wintering passerines. Ornis Scandinavica 17:261–267.

Ekman, J., and A. Hake. 1990. Monitoring starvation risk: adjustments of body reserves in greenfinches (*Carduelis chloris* L.) during periods of unpredictable foraging success. Behavioral Ecology 1:62–67.

Ekman, J. B., and K. Lilliendahl. 1993. Using priority to food access: fattening strategies in dominance-structured willow tit (*Parus montanus*) flocks. Behavioral Ecology 4:232–238.

Ettinger, A. O., and J. R. King. 1980. Time and energy budgets of the Willow Flycatcher (*Empidonax traillii*) during the breeding season. Auk 97:533–546.

Gosler, A. G., J. J. D. Greenwood, and C. M. Perrins. 1995. Predation risk and the cost of being fat. Nature 377:621–623.

Greenberg, R. A. 1990. Competition among migrant birds in the nonbreeding season. Current Ornithology 7:281–307.

Greenberg, R. 1992. Forest migrants in nonforest habitats on the Yucatan Peninsula. Pages 273–286 *in* Ecology and Conservation of Neotropical Migrant Landbirds (J. W. Hagan and D. W. Johnston, eds.). Smithsonian Institution Press, Washington, D.C.

Henderson, R. W. 1992. Consequences of predator introductions and habitat destruction on amphibians and reptiles in the post-Columbus West Indies. Caribbean Journal of Science 28:1–10.

Holmes, R. T., T. W. Sherry, and L. Reitsma. 1989. Population structure, territoriality and overwinter survival of two migrant warbler species in Jamaica. Condor 91:545–561.

Houston, A. I., and J. M. McNamara. 1993. A theoretical investigation of the fat reserves and mortality levels of small birds in winter. Ornis Scandinavica 24:205–219.

Houston, A. I., and J. M. McNamara. 1999. Models of Adaptive Behaviour. Cambridge University Press, Cambridge.

Katti, M., and T. Price. 1999. Annual variation in fat storage by a migrant warbler overwintering in the Indian topics. Journal of Animal Ecology 68:815–823.

Koivula, K., M. Orell, S. Rytkönen, and K. Lahti. 1995. Fatness, sex and dominance: seasonal and daily body mass changes in Willow Tits. Journal of Avian Biology 26:209–216

Küllberg, C., T. Fransson, and S. Jacobsen. 1996. Impaired predator evasion in fat blackcaps (*Sylvia atricapilla*). Proceedings of the Royal Society of London, Series B, Biological Sciences 263:671–1675.

Küllberg, C. S, S. Jakobsson, and T. Fransson. 2000. High migratory fuel loads impair predator evasion in Sedge Warblers. Auk 117:1034–1038.

Lefebve, G., B. Poulin, and R. McNeil. 1992. Abundance, feeding behavior, and body condition of Nearctic warblers wintering in Venezuelan mangroves. Wilson Bulletin 104:400–412.

Lima, S. L. 1986. Predation risk and unpredictable feeding conditions: determinants of body mass in birds. Ecology 67:366–376.

Lind, J., T. Fransson, S. Jakobsson, and C. Kullberg. 1999. Reduced take-off ability in robins due to migratory fuel load. Behavioral Ecology and Sociobiology 46:65–70.

Marra, P.P., K.A. Hobson, and R.T. Holmes. 1998. Linking winter and summer events in a migratory bird by using stable-carbon isotopes. Science 282:1884–1886.

Marra, P. P., and R. T. Holmes. 2001. Consequences of dominance-mediated habitat segregation in American Redstarts during the nonbreeding season. Auk 118:92–104.

McNamara, J. M., and A. I. Houston. 1990. The value of fat reserves and the trade-off between starvation and predation. Acta Biotheoretica 38:37–61.

Metcalfe, N. B., and S. E. Ure. 1995. Diurnal variation in flight performance and hence potential predation risk in small birds. Proceedings of the Royal Society of London, Series B, Biological Sciences 261:395–400.

Nowak, R. M. 1991. Walker's Mammals of the World (fifth ed.). Johns Hopkins University Press, Baltimore.

Piersma, T., and Å Lindström. 1997. Rapid reversible changes in organ size as a component of adaptive behavior. Trends in Ecology and Evolution 12:134–138.

Petit, D. R., L. J. Petit, and K. G. Smith. 1992. Habitat associations of migratory birds overwintering in Belize, Central America. Pages 247–256 *in* Ecology and Conservation of Neotropical Migrant Landbirds (J. W. Hagan and D. W. Johnston, eds.). Smithsonian Institution Press, Washington, D.C.

Pravosudov, V. V., and J. R. Lucas. 2001. A dynamic model of short-term energy management in small food-caching and non-caching birds. Behavioral Ecology 12:207–218.

Rappole, J. H. 1995. The Ecology of Migrant Birds. Smithsonian Institution Press, Washington, D.C.

Rappole, J. H., and D. W. Warner. 1980. Ecological aspects of migrant bird behavior in Veracruz, Mexico. Pages 353–393 *in* Migrant Birds in the Neotropics: Ecology, Behavior, Distribution, and Conservation (A. Keast and E. S. Morton, eds.). Smithsonian Institution Press, Washington, D.C.

Rogers, C. M. 1987. Predation risk and fasting capacity: do wintering birds maintain optimal body mass? Ecology 68:1051–1061.

Rogers, C. M. 1991. An evaluation of the method of estimating body fat by quantifying visible subcutaneous fat. Journal of Field Ornithology 62:349–356.

Rogers, C. M. 1995. Experimental evidence for temperature-dependent winter lipid storage in the Dark-eyed Junco (*Junco hyemalis oreganus*) and the Song Sparrow (*Melospiza melodia morphna*). Physiological Zoology 68:277–289.

Rogers, C. M. 2003. New and continuing issues with using visible fat classes to estimate fat stores of birds. Journal of Avian Biology 34:129–133.

Rogers, C. M., and R. Heath-Coss. 2003. Effect of experimentally altered food abundance on fat reserves of wintering birds. Journal of Animal Ecology 72:822–830.

Rogers, C.M., and J.N.M. Smith. 1993. Predation risk and fasting capacity: Local trade-offs in body mass of wintering birds? Ecology 74:419–426.

Rogers, C. M., and A. K. Reed. 2003. Does avian winter fat storage integrate temperature and resource conditions? A long-term study. Journal of Avian Biology 34:112–118.

Smith, J. N. M., R. D. Mongomerie, M. J. Taitt, and Y. Yom-Tov. 1980. A winter feeding experiment on an island Song Sparrow population. Oecologia 47:164–170.

Smith, R. D., and N. B. Metcalfe. 1997. Diurnal, seasonal and altitudinal variation in energy reserves of wintering Snow Buntings. Journal of Avian Biology 28:216–222.

Strong, A. M., and T. W. Sherry. 2000. Habitat-specific effects of food abundance on the condition of ovenbirds wintering in Jamaica. Journal of Animal Ecology 69:883–895.

van der Veen, I., and K. M. Lindström. 2000. Escape flights of yellowhammers and greenfinches: more than just physics. Animal Behaviour 59:593–601.

Vehrlust, S., and O. Hogstad. 1996. Social dominance and energy reserves in flocks of Willow Tits. Journal of Avian Biology 27:203–208.

Witter, M. S., and I. C. Cuthill. 1993. The ecological costs of avian fat storage. Philosophical Transactions of the Royal Society of London 340:73–92.

Witter, M. S., I. C. Cuthill, and R. H. C. Bonser. 1994. Experimental investigations of mass-dependent predation risk in the European Starling, *Sturnus vulgaris*. Animal Behaviour 48:201–222.

Witter, M.S., and J.P. Swaddle. 1997. Mass regulation in juvenile Starlings: response to change in food availability depends on initial body mass. Functional Ecology 11:11–15.

Witter, M. S., J. P. Swaddle, and J. P. Cuthill. 1995. Periodic food availability and strategic regulation of body mass in the European starling, *Sturnus vulgaris*. Functional Ecology 9:568–574.

10

CLAUDIA METTKE-HOFMANN
AND RUSSELL GREENBERG

Behavioral and Cognitive Adaptations to Long-Distance Migration

OVER THE COURSE OF THE annual cycle, migratory birds must solve a variety of complex ecological problems related to habitat selection and resource use. We expect that aspects of exploration, response to novelty, and long-term memory reveal special adaptations for the challenges facing migrants. In this chapter we review the migratory bird year and suggest areas where migrants may show special cognitive attributes. In addition, we discuss the interplay between innate programming and learning and how this affects ecological plasticity throughout a migrant's year. Although studies are few, their results indicate that migrants may have evolved special cognitive abilities for a life on the move.

WHY STUDY COGNITION IN MIGRANTS?

As migrants travel over such great distances and visit a variety of habitats, they face many ecological problems that must be solved, either through natural selection shaping an innate behavioral program or through cognitive processes. Although many aspects of the migration program are partly genetically determined (e.g., Gwinner 1996), such as the approximate departure date, direction, and distance, every migratory bird embarks upon a trip that is in many aspects unique, and hence exploration, experience, and learning play a fundamental role in the many challenges that face a migratory bird. In this brief review, we outline the key decisions faced by migratory birds and discuss the role that cognitive processes may play. These key decisions include how birds gather information in natal areas to make an informed site selection when they return the following breeding season (site fidelity [Hund and Prinzinger 1979; Thompson and Hale 1989]); how tradition (Schüz et al. 1971) or individual decision making influences the selection of high-quality stopover sites; how the tension is resolved between adaptive specialization and plasticity in choosing habitats and foraging sites en route and at either end of the migratory trip; how the novelty of habitats and foods during the migratory cycle influence their selection; and how migrants retain a memory of high-quality sites selected during and between migrations. Throughout this chapter we examine the ways that migrants approach these decisions, distinguishing between what is known versus what is simply hypothesized. More importantly we focus on possible emergent cognitive abilities that distinguish migratory and resident species.

LEARNING THE NATAL AREA

The decision on what area to settle for the first breeding season is constrained in migrants by time and intraspecific competition. Selection favors birds that choose and establish territories rapidly and thus begin nesting quickly. One therefore might suppose that information gathered by young birds during their first post-breeding season assists in making the selection of possible settlement sites the following spring. A large body of literature shows that birds use information about future breeding territories collected during the post-breeding period to decide where to settle the next spring, a process known as "prospecting" (for a review see Reed et al. 1999). Decision making during prospecting is based on direct experience (personal information) and information about performance of conspecifics (public information [Doligez et al. 2002]). Because site fidelity is prevalent in returning adult birds, post-fledging exploration may be critical for life-long site selection (Cadiou 1999).

If information collected in a bird's first autumn is used to select suitable habitats or territories the following spring, this would represent an extreme example of "latent learning." Latent learning, a term introduced by Tolman in 1932, refers to learning in animals that may have no immediate benefit, yet may enhance problem solving at some future date. Although the precise mechanism is difficult to elucidate, migration provides an excellent example of where latent learning and long-term memory are of particular benefit.

As Winkler (Chap. 30, this volume) points out, first-year migrants generally return to the region, but not the specific site, of hatching. This sets up a dynamic between hard-wired navigational processes and individual learning and decision making. Post-fledging exploration, which is not restricted to the close vicinity of the site of hatching (see below), may well include the region of future settling. Although navigational processes are largely innate, the geographic position of the natal area is learned (Berndt and Winkel 1979). Löhrl (1959) investigated this learning process in more detail. Hand-raised Collared Flycatchers (*Ficedula albicollis*) were kept in outdoor aviaries with a good view of the landscape until 4 to 5 weeks after fledging and were then released at a novel place 90 km away. Next spring, all recoveries were at the site of displacement; no birds returned to the natal or rearing area (Löhrl 1959).

We do not know whether the lack of return to the rearing site was caused by the inability of the birds to assess habitat quality, and hence to establish site fidelity by viewing the environment from the cage, or by the inability to find this particular area on return because they were unfamiliar with the navigational target. Baker (1993) proposed that post-fledging exploration might be critical not only for learning about particular sites, but also for facilitating the orientation to the natal area itself (Baker 1993). If this were the case, post-fledging exploration in migrants would have an additional function beyond post-fledging exploration in residents, which only have to learn about specific sites but not about navigational targets. Baker (1993) expected that to meet the conditions required for navigation, post-fledging exploration in migrants would be more axial (i.e., bimodal) and north–south-oriented parallel to the migration axis, whereas in residents he expected no directional bias. The larger the area a bird explored prior to southward migration, the more likely it should be to encounter this area on return. An elongated familiar area around the future breeding site oriented parallel with the migration axis would enable a returning bird to find the familiar area even if the memory of latitude is not precise. Birds arriving at the appropriate latitude of their target area in spring might turn in a right angle and travel until they encounter the familiar target area. Ketterson and Nolan (1990) showed that at least in some species returning individuals recognize their breeding area even when encountered at an inappropriate time of the year, such as the beginning of spring migration.

Baker's (1993) expectation was supported by a comparison of post-fledging movements between two trans-Sahara migrants, the Eurasian Redstart (*Phoenicurus phoenicurus*) and the Northern Wheatear (*Oenanthe oenanthe*), and a resident, the European Robin (*Erithacus rubecula*). He analyzed recoveries of birds ringed within the previous 100 days as nestlings in Great Britain. Only recoveries to the north were used for the distance covered during post-fledging exploration. The two migrants explored a larger area to the north than the resident. Furthermore, the former oriented their exploration along a north-south axis parallel to the migration axis, whereas the robin showed no particular preference for any direction. Thus, it is likely that post-fledging exploration in migrants serves to create a navigational target to find the way back to the natal area in spring. However, much more comparative data are needed to verify these results and to clarify whether this is a general phenomenon in migrants.

SPECIALIZATION VERSUS PLASTICITY IN THE FACE OF HABITAT NOVELTY

Ecological Plasticity and the Annual Cycle

In all bird species there is a tension between the retention of species-typical resource use, where individuals select habitats, foraging sites, and food for which they are best adapted (ecological specialization), and the ability to respond to unpredictable resources. The ability to respond behaviorally to variable resources has been dubbed ecological plasticity (Morse 1980) and varies as an adaptation to the environment among species, among individuals within a species, and within individuals at different points in the annual cycle. In this discussion, we focus on the last point and distinguish between the transitory (migration itself) and the stationary (breeding, nonbreeding residency) periods of the annual cycle. This, of course, is an overgeneralization, because some migrants may continue to be nomadic throughout the breeding and nonbreeding periods. However, it is a simplification that allows the discussion to go forth.

We assume that the transitory period provides the greatest challenge to the retention of ecological specialization, because stochastic events during migration may play a large role in determining the habitat in which a migrant finds itself during stopovers, when birds need to restore their energy reserves (e.g., Biebach et al. 1986). Moreover, as distance from home increases, habitats and food become more and more dissimilar from those experienced in the natal area. It is easy to underestimate the importance of stopovers. When most people remember the great American novel *Huckleberry Finn,* they think of long periods when Huck and Jim are floating down the river on the raft. In fact, this idealized scene takes place in a handful of the hundreds of pages of this epic. The stopover adventures dominate the book. In much the same way, stopovers dominate bird migration; small warblers, for example, spend five to seven times more time at stopover sites than in flight during migration (Fransson 1995). The decision where to land is crucial for whether or not energy reserves can be built up. For instance, in nutritionally low-quality areas migrants have to change the habitat or even leave the area completely in search of a better resting place (Delingat and Dierschke 2000). This is time and energy consuming and increases the risk of predation. Evolutionary processes should favor mechanisms that ensure selection of the appropriate habitat.

We might suppose that, when possible, migrants select habitats that are most similar to their natal area; on the other hand, some relaxation of ecological specialization might be advantageous during migration when migrants may be forced to land on unfamiliar turf. Berthold and Terrill (1991) proposed that genetic programs facilitate appropriate habitat selection during migration. The innate preference for particular habitats would minimize search costs. Nevertheless, they expected a diversification of foraging behavior at stopover sites to meet increased metabolic demands.

At least some species possess something like an innate template of their habitat. Perhaps the best example of the experimental establishment of an innate habitat cue comes from resident tits. Naïve Coal Tits (*Parus ater*) preferred coniferous foliage and Blue Tits (*Parus caeruleus*) preferred deciduous foliage, their respective natural habitat types in double-choice experiments (Partridge 1974). Grünberger and Leisler (1993) also found such an innate preference in Coal Tits. However, their results reflected more of a bias, rather than a rigid preference. Even in this resident species that shows a clear habitat preference throughout the year, the innate preference could easily be changed by experience early in life. Coal Tits raised in artificial deciduous habitats preferred that habitat to artificial coniferous habitats at an age of 7 weeks (Grünberger and Leisler 1993). At least one long-distance migrant, the Arctic Warbler (*Phylloscopus borealis*) has demonstrated a clear innate preference for the appropriate breeding-habitat foliage; hand-raised individuals grew up to prefer coniferous versus broad-leafed foliage in accordance with the preference of the particular subspecies studied (Van Patten 1990).

How innate preferences are programmed has been a subject of speculation for ornithologists; however, how innate programs may change seasonally has received much less thought. In terms of what cues the birds respond to, James (1971) proposed that birds select habitats on the basis of overall *gestalt* or habitat structure. Lack (1933) hypothesized that under certain circumstances birds respond innately to specific habitat cues as a "sign stimulus" for habitat selection. Greenberg (1992) demonstrated, using naïve hand-raised Swamp (*Melospiza georgiana*) and Song Sparrows (*M. melodia*), that their rather precise habitat segregation during breeding season resulted from opposite responses to the presence of surface water. Even in this case, however, sparrows need to be responding to the presence of other habitat features and territorial conspecifics in deciding where to settle within the flooded or dry sites. Furthermore, the two species show broadly overlapping habitat preference during migration and in their winter quarters and often occur in the same flocks, suggesting that the strong response to particular habitat cues is relaxed outside of the breeding season.

With respect to a seasonal program for habitat use, an intriguing experiment in which juncos were presented with slide images in a Skinner box showed that the preference for breeding or nonbreeding habitat shifted seasonally (Roberts and Weigl 1984). Furthermore, manipulation of diurnal light cycles produced shifts consistent with the hypothesis that this habitat preference is under photoperiodic control. Under shorter photoperiods, preference for slides showing breeding-season habitat was correlated with body size. The authors suggested that larger and more dominant individuals are more likely to winter near the breeding habitat and arrive there earlier. As far as we know, this creative experimental approach to seasonal changes in habitat preferences has not been replicated.

Seasonal Foraging Specializations

Apart from the plasticity required of birds visiting a number of unfamiliar sites, many migratory birds show consistently different habitat use and foraging preferences during the stationary periods between migrations; in other words, species have two distinct patterns of ecological specialization. One of the classic examples of foraging specialization in tropical birds is the searching for insects in dead leaves that hang suspended in the vegetation of forest understories. In most Neotropical forest avifaunas the guild of dead-leaf specialists comprises one to seven species (Remsen and Parker 1984). A few species of migratory birds, particularly the Worm-eating Warbler (*Helmitheros vermivorus*), take up this specialization during the nonbreeding season as well, particularly on their tropical wintering grounds (Greenberg 1987). During the breeding season, Worm-eating Warblers forage for insects in live foliage; however, during the winter stationary period and throughout their Mesoamerican and Caribbean winter range, 70–80% of their foraging is focused on insects in dead curled leaves. Detailed observations and experiments with naïve hand-raised warblers demonstrated a complex interaction between innate preference, patterns of substrate exploration, and learning in the development of this seasonal specialization (Greenberg 1987). First, juveniles showed a strong innate preference for manipulating dead leaves over all other objects, including live foliage. This preference was weaker in birds during their first winter and the preference for live versus dead foliage could be increased and decreased with food reinforcement. However, Worm-eating Warblers show a strong propensity to probe and manipulate objects, and in particular, explore holes and surface irregularities. This pattern of exploration readily paves the way for acquiring the dead-leaf specialization. Dead leaves are more difficult to search than live foliage, but in tropical areas they support a far greater biomass of arthropod prey than live foliage.

Other species employ strikingly opportunistic foraging strategies during the nonbreeding season. Greenberg (1979) analyzed foraging opportunism among *Dendroica* warblers and hypothesized that conifer-breeding species that winter in tropical areas tend to employ the most opportunistic strategies in the genus. In winter, they are forced to occupy broadleafed habitats for which they are poorly adapted and, hence, have to use opportunistic strategies for locating, catching, and handling a diverse set of food types. Exploring and monitoring many resources instead of a few would seem more likely to engender other cognitive abilities than would retaining of a greater degree of foraging specialization.

Neophobia

Research on plasticity and innovation has generally focused on cognitive abilities, but the response to novelty itself may play a decisive role in determining the levels of ecological flexibility. Other potential mechanisms for altering the plasticity of ecological choices have been put forth. In 1990 Greenberg (1990b) proposed a hypothesis that stated that early experience can influence ecological plasticity later in life. Birds gather experience about food types and habitats at an early stage of life (post-fledging exploration). Preferences established during that period are protected from change by neophobic reactions, that is, birds avoid situations that they had not experienced early in life (neophobia threshold hypothesis). Because of differences in post-fledging exploration (e.g., see above) and complex interactions with their environment species differ in their early experiences. These differences in experience influence neophobic reactions (e.g., when reared under enriched or depauperate environments [e.g., Jones 1986]). Interspecific differences in neophobia, which seem to be the result of experiences gathered as a juvenile, have been shown in wood warblers (*Dendroica*). Greenberg (1983, 1984) experimentally tested feeding neophobia against food presented in novel microhabitats in feeding specialists and generalists of wood warblers during their first migration. The specialists hesitated longer to feed from a novel microhabitat than the generalists. Similar results were found in sparrows (Greenberg 1990a).

Preferences, innate or acquired during an early stage of life, can be a useful mechanism for initial habitat selection during migration. Unfortunately, we do not know whether any differences in this respect exist between residents and migrants because the experiments described above all were done with migrants. Although both residents and migrants might use the same mechanism to select suitable habitats—residents for reproduction, migrants for stopover, overwintering, and reproduction—it is conceivable that migrants are less neophobic, at least during migration, than residents. Seasonal variation in neophobia could be part of an endogenous migration program. Relaxation of ecological specialization during migration could thus be achieved by down-regulation of neophobia. Each place encountered on route is novel in the sense that this particular place was never visited before, even though the habitat type is familiar. Lower neophobia in migrants would be advantageous to reduce hesitancy to come into a stopover site. Residents might generally be more neophobic than migrants, but more specific differences might also be present. For example, habitat specialists that migrate might be less neophobic than habitat specialists that do not migrate. Similarly, migratory habitat generalists might show lower neophobia than resident habitat generalists.

SPATIAL EXPLORATION OF NEW SITES

After selection of a stopover site, familiarization with the unfamiliar environment is a prerequisite for survival. Knowing where the nearest cover is while feeding or which are the safest feeding patches and how they are spatially located in relation to other places reduces predation risks. Unfortunately, almost nothing is known of how migrants behave right after the arrival at a stopover site. Mettke-Hofmann (unpubl.) conducted a comparison of initial orientation in a novel environment among ten parrot species, which were either resident or nomadic. The birds could enter two unfamiliar aviaries adjacent to their familiar aviary. During the first minute, residents changed position very often, whereas nomads remained at one position (fig. 10.1). Later on both showed similar movement patterns. The differences in movement patterns might reflect differences in spatial orientation that would be expected in residents and migrants. Residents might use their adjacent familiar home range as a reference point (Eilam and Golani 1989), thus relating new features to a familiar spatial network (Burgess et al. 2000). Changing positions often might be advantageous to strengthen the network. For a migrant who is far away from its previous familiar area, it might be better to monitor an area from a fixed position to relate spatial relationships between different objects (e.g., trees, bushes, water holes) to a single reference point—where the bird sits (Mettke-Hofmann, unpubl.). This coincides with the orientation behavior in Eastern Chipmunks (*Tamias striatus*), which use the release point as a place of reference when in an unfamiliar environment (Thibault and Bovet 1999). Interpretation of the data must remain speculative, however, because in the experiment, the familiar aviary was directly adjacent to the novel rooms for both residents and nomads, and no comparable data for migrants are available. One might expect migrants to use strategies similar to those of nomads, because both encounter a great variety of habitats and are often confronted with unfamiliar areas. Future studies may bring some light into this exciting field of spatial orientation.

Once in a stopover site, a migrant assesses habitat quality and predation risk, and finally selects the habitat best suited for its demands (Ydenberg et al. 2002). During migration, however, birds are generally time constrained (Alerstam and Lindström 1990), and extensive exploration might be energetically costly. Furthermore, stopover durations are relatively short, and costs of exploration may rapidly exceed the benefits of increased information, especially as the bird needs to assess only temporary habitat suitability rather than long-term quality (Mettke-Hofmann and Gwinner 2004.). The trade-offs between the necessity to find food patches profitable enough to meet energetic requirements and the costs associated with such a search might mean that exploration would lead merely to a general overview about recent food resources rather than to a detailed knowledge of the habitat quality over the long term. That migrants seem to accept incomplete information about a stopover site is supported by a telemetry study of Summer Tanagers (*Piranga rubra*) during stopover (Aborn and Moore 1997). Birds that had newly arrived restricted their movements to a relatively small area. Nonetheless, they displayed movement patterns consistent with exploration (Aborn and Moore 1997).

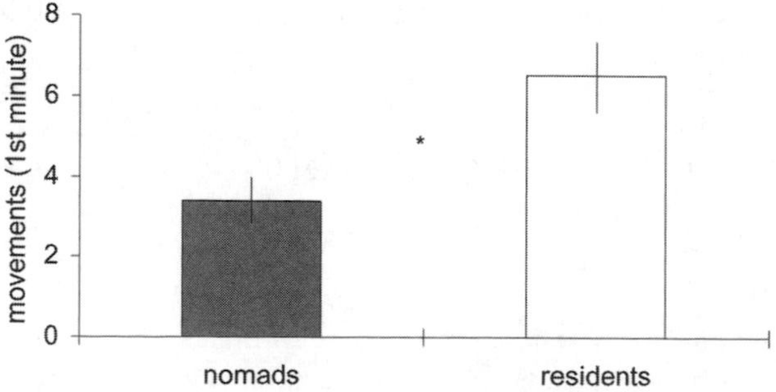

Fig. 10.1. Mean number of movements (± SE) by nomadic and resident species of parrots in a novel room within the first minute of exposure. $^*p < 0.05$.

On the other hand, when residents explore novel areas, for example, during dispersal, they need detailed information about current conditions of a territory, but also its long-term quality (Cadiou 1999; Reed et al. 1999). For example, a stable vegetation throughout the year (e.g., evergreen vegetation) is of great importance for residents (e.g., Airola and Barrett 1985). A detailed habitat assessment should be important for young birds not yet settled, but it is equally important for an already established territory owner that has decided to search for another territory. However, the decision to leave an established territory probably involves higher costs (the bird might lose its territory) than for a young bird.

Differences in assessment of environmental information between residents and migrants was investigated in two closely related and ecologically similar warbler species, the highly migratory Garden Warbler (*Sylvia borin*) and the resident Sardinian Warbler (*S. melanocephala momus*) (Mettke-Hofmann and Gwinner 2004). During autumn migration, individuals of either species could explore singly two novel rooms connected to their familiar cage. For analysis, the rooms were divided into similar-sized squares. Garden Warblers visited more squares per minute than Sardinian Warblers and hence explored in a way that would result in a rough overview, which is what we predicted for migrants. The Sardinian Warblers moved over short distances within the squares, a behavior consistent with more fine-toothed investigation (Mettke-Hofmann and Gwinner 2004). Further field and laboratory studies of migrant and resident species are needed to evaluate these results.

SELECTING A BREEDING TERRITORY

After a long journey of several months the migrant returns in many cases to its natal area in spring, where the most important task is the selection and acquisition of a breeding territory. Breeding habitat selection may be quite different between resident and migrant species in several ways. For one, residents have all year to assess habitat quality and they do so without a temporal break in their experience with the habitat. Second, territory turnover is often slow and competition for each vacancy is intense. For residents, territory acquisition has already begun in the post-breeding season, whereas for migrants, undefended habitat only becomes available over a rather short period of time after arrival in spring.

Breeding habitat selection is particularly challenging for the first-year returnee, because adults generally return to the same site (if not the same territory) they defended the previous year. Within a rather large range around the natal territory, young birds need to select a site and compete with older birds, which generally have site dominance. Habitat selection is constrained in two major ways. First, there is tremendous pressure for settling birds to make a decision rapidly (Alatalo and Lundberg 1984). Second, territory selection often occurs when conditions on the breeding grounds are quite different from what they will be later in the season. Direct assessment of habitat quality or food abundance is time consuming and potentially misleading. It is likely that birds use relatively simple cues—features of the habitat that correlate with overall quality during the breeding season—but the use of environmental cues during this brief period of exploration is poorly documented. Recently research has shifted to the importance of the presence or absence of birds themselves as being an important cue for settlement. Several migrant species (ducks, European and North American passerines), mainly habitat generalists, seem to use resident species as a cue for suitable habitats. A high density of residents attracts more migrants than a low density, presumably as an indication of high food abundance or low predation pressure (heterospecific attraction hypothesis [Mönkkönen et al. 1997]). That strategy might consume less time than direct investigations of habitats (Mönkkönen et al. 1999). Similar strategies are known from inexperienced resident birds; however, they use conspecifics as cues for profitable breeding sites (conspecific attraction hypothesis [Suryan and Irons 2001]). Whereas conspecific attraction is also proven in migrants for later-arriving and/or inexperienced colony-breeding individuals (Muller et al. 1997; Schjorring et al. 1999), heterospecific attraction, to our knowledge, is observed only in migrants. Possibly, heterospecific attraction evolved as a consequence of migration to reduce search time for breeding habitats. The question remains: How do the birds know which species are suitable as guidance? Heterospecific recognition either has to be learned, which requires high cognitive abilities (Mönkkönen et al. 1999), or must be innate. In the former case familiarization with sympatric species can occur immediately after fledging and during dispersal or after arrival at the breeding ground while in direct competition with other species. Mönkkönen et al. (1996) showed that at least some migrants recognize the song of resident species, indicating an early learning process.

The use of other species as cues for decision making seems not to be restricted to the breeding ground. During migration (Morse 1970) and on their winter quarters, many migrants commonly join mixed species foraging flocks, either with other migrants or around a nucleus of residents. The available interspecific associations should vary considerably along migration routes. Furthermore, on tropical winter ranges, the composition of mixed flocks, particularly the leader species, varies between and even within habitats. We can assume that selection has favored responses to some rather generalized aspects of, for example, vocalization, behavior, or morphology of the potential leader species. For example, Smith (1975) hypothesized that New World migrants show a strong response to the mobbing notes of nuclear species. These notes share acoustical features that are stylized in the "spishing" noise made by birders to attract birds.

How associations are selected and which behavioral adjustments are required to participate in these associations are poorly known for migrants, and the topic of mixed-species association is out of the scope of this chapter. However, in keeping with the theme that other birds can be substituted as a habitat cue, it should be noted that individuals of some species have been shown to adjust their winter-long territory to coincide with the territory of core flock members (Greenberg 1984). By using the location and boundaries of an existing resident flock of ecologically similar birds, the migrant can find a habitat that is predictable throughout the winter. Given that migrants often arrive during a rainy season and need to persist on a territory into the dry season (or in fewer cases, vice versa), such stable flock systems may be a valuable cue for long-term habitat quality.

Differences in residency may call for different strategies for monitoring long-term habitat change, be it seasonal, successional, or catastrophic. To discover changes in food abundance or distribution, for example, superior territories or new nest holes must be available so that residents can adjust movements and behaviors to the new situation. Seasonal changes might be secondary for migrants, which leave the breeding ground before conditions deteriorate. Differences in reaction to changes in familiar areas have been shown among four resident and four nomadic parrot species (Mettke-Hofmann 2000a). Fewer nomads made tactile contact with novel objects presented in the familiar aviary than residents. Furthermore, latencies between introduction of the objects and first tactile contact were longer in nomads (fig. 10.2). Unfortunately, phylogenetic relationships were not considered. At present, no data are available for a similar comparison between residents and mi-

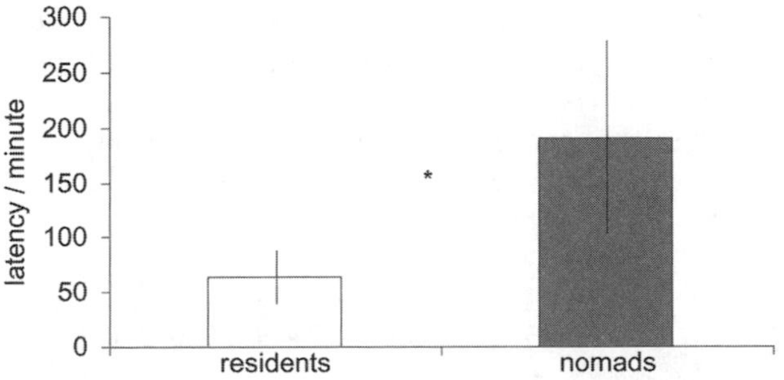

Fig. 10.2. Mean latencies (± SE) until first contact of a novel object for nomadic and resident parrot species. All birds were kept in conspecific pairs and the mean latency of the partners was used for analysis. Only results from individuals that contacted the objects during the experimental trials are included. *$p < 0.05$ (data from Mettke-Hofmann 2000a).

grants. If similar results are found, they would indicate that residents and migrants that occupy the same feeding and breeding niche extract different information from exactly the same environment.

LONG-TERM MEMORY IN MIGRATORY BIRDS

After breeding, migrants travel south again. How much are decisions for suitable stopover sites influenced by the experiences of previous migrations? Memory of such places would be a huge advantage for optimizing the migration route and increasing travel speed, thus hastening the return to winter quarters to claim or reclaim territories. Ringing data show that experienced birds are caught during a shorter period of time at a particular stopover site than inexperienced young birds, supporting a faster migration of the former (Jenni and Jenni-Eiermann 1987). Familiarity with the stopover site or better body condition (due to higher foraging efficiency), or a combination of both, may increase travel speed for adults as compared with juveniles. Many migrants use the same stopover sites or wintering areas in consecutive years (stopover and winter site fidelity [Nolan and Ketterson 1991; Cantos and Tellería 1994]).

Do long-distance migrants have special abilities related to long-term memory that assist in retracing previous migratory itineraries? Maintenance of memory should have some costs, although these costs might be negligible compared with the benefit gained. In contrast, maintenance costs in residents are expected to be high in relation to possible benefits. Residents stay in the same area year-round and can update their knowledge whenever necessary and do not require a long-lasting memory (Mettke-Hofmann and Gwinner 2003). Nonetheless, it is conceivable that they retain information about particular microhabitat characteristics from one year to the next. At present, only one study is available, which compared memory persistence with migratory behavior. Hand-raised birds of migratory Garden Warblers and resident Sardinian Warblers could explore two unfamiliar rooms, one with food, the other without food, from the familiar cage during the first autumn migration period of the Garden Warblers. Memory for food was tested after various delays. The Garden Warblers remembered the room with food for at least 12 months, whereas the Sardinian Warblers seemed to have a much shorter lasting memory (Mettke-Hofmann and Gwinner 2003). It remains to be tested whether this applies to other species as well.

SEASONAL CHANGES ON THE NEURAL SUBSTRATE OF MEMORY

Migrants are confronted with novel habitats and areas primarily during migration. The differences mentioned above between residents and migrants therefore might not exist throughout the entire year but emerge in a seasonal pattern corresponding to divergent challenges. For example, to process large amounts of spatial information during migration requires an efficient neural substrate. The hippocampal formation (HF) in the avian brain is known to be important for processing spatial information, and species that are highly dependent on spatial information (e.g., food-caching species [Basil et al. 1996]; brood parasitic species [Reboreda et al. 1996]) have a larger HF than species that do not rely so strongly on this type of information. Interestingly, it could be shown (Healy et al. 1996) that migrants and residents differ in the development of the HF. Comparison of the HF between experienced long-distance migratory Garden Warblers and those with no migration experience revealed a relatively larger HF in the experienced birds. In contrast, HF did not change in the same period of time in closely related but resident Sardinian Warblers. Thus, it is likely that migrants have special neural adaptations to process large amounts of spatial information while on migration. These capacities might not be needed during other parts of the season, for instance, during breeding.

Seasonal changes in the HF are known from food-caching species (Smulders et al. 1995), with the HF decreasing in size outside the caching season in fall. Clayton and Cristol (1996) showed seasonal variation in spatial memory in food-storing Marsh Tits (*Parus palustris*). Seasonal changes are also known in exploratory behavior in birds (Mettke-Hofmann 2000b) and mammals (Heth et al. 1987). Nothing is known about similar changes in the HF in migrants or whether such changes would have any implications on learning or memory formation. One might imagine that differences in memory formation between residents and migrants are present only when the migrant is in the migration stage.

The type of environmental information collected can also vary between species. For instance, food-caching species use primarily spatial cues to remember places, whereas noncaching species use both spatial and object-specific cues equally (Brodbeck and Shettleworth 1995). It remains to be tested whether comparable differences exist in migrants and residents with a preference for spatial cues in the former. If differences do exist, are they consistent throughout the year, or are there seasonal changes? For example, migrants might prefer spatial cues only during migration (because the HF might be reduced in the rest of the year).

EMERGENT ASPECTS OF COGNITION IN MIGRATORY BIRDS

In the preceding sections we followed a migrant bird in its first year and discussed the challenges with which it has to deal. Now, we switch from a detailed examination to a broader view. We have stressed the importance of attributes, such as low neophobia, high exploration, a high degree of generalization, and long-term memory. These features form a syndrome of characteristics that can be subsumed under the term "cognitive" abilities and do not necessarily reflect specific differences in feeding behavior or habitat preferences.

Combining the preliminary findings reveals the following picture of how factors necessary for a migrant or a resident might influence cognition:

Migrants in transit may relax their ecological specialization (possibly by lowered neophobia). They collect superficial information to get a snapshot overview about their environment and show little interest in monitoring long-term changes. The type of information they use should be primarily stable and, presumably, spatial (e.g., invariable habitat characteristics suitable to recognize stopover sites or the natal or breeding area). This stable information is stored for the long term.

Residents gather detailed information about their environment and react readily to changes to update their knowledge about, for example, food abundance or distribution in the annual cycle. Thus, the kind of information they are interested in is primarily of a variable nature and may include spatial as well as object-specific cues. Their memory is relatively short lived to prevent interference with new constellations of information (for interference, see Hampton et al. 1998). Whether or not these cognitive differences are actually present in a great variety of taxa remains to be seen. Modifications may arise, for instance in migratory species with low site fidelity.

When these differences hold true for residents and migrants, not only will birds with a migratory lifestyle pay attention to different stimuli and store different information than resident birds, but they will also react differently to the same challenge. These differences can have important implications for the survival of the species and conservation purposes. For example, residents might be better adapted to colonize new habitats or to open up new food resources because of their thorough exploration. Most bird species that live in urbanized areas are residents. Introduction experiments in New Zealand showed that residents are more likely than migrants to become successfully established (Veltman et al. 1996). Furthermore, successful invaders have a higher frequency of feeding innovations than unsuccessful ones (Sol and Lefebvre 2000), thus supporting our expectation. The short time that migrants spend in all of their areas, possibly in conjunction with their superficial exploration, might constrain their ability to adapt to environmental change. Such a limitation could be particularly important in stopover sites where habitats are altered by human activities. A few days' stopover is much too short to allow adaptation to altered conditions, and site fidelity will result in holding to familiar sites as long as possible without exploring other habitats. Here, the otherwise advantageous site fidelity turns into a disadvantage, especially for species in which stopover sites are transmitted from generation to generation. In contrast, heterospecific attraction should counteract the negative effects of site fidelity in a rapidly changing environment. Knowledge about these mechanisms can help to evaluate the consequences of environmental change in bird populations.

FUTURE DIRECTIONS

This overview describes what we believe is a promising and poorly explored field of migration research: cognitive ecology. All of the hypotheses require refinement and reformulation as we conduct investigations on a much broader range of species, ideally including a variety of taxa from the major migration systems. In the following, we give a summary of possible questions and hypotheses that should be addressed in the future.

- Are the differences found in post-fledging exploration between some residents and migrants a general phenomenon? Two approaches would be especially interesting. First, an analysis of recovery data from a wide range of species from different regions to verify the generality of the patterns of post-fledging distribution. Second, the use of radiotelemetry to follow post-fledging movements directly.
- Do migrants show lower neophobia toward entering a novel environment than residents? (a) Is there a general difference in neophobia between these two life-history traits? (b) Does neophobia decrease within a species during migration? (c) If differences exist, are they visible only within ecologically homogeneous groups, for example, within/among habitat specialists or among generalists?
- Do migrants and residents differ in their first orientation in a novel environment, as residents and nomads do?
- Is it a general phenomenon that migrants explore a novel environment using a different strategy than residents use? Do migrants explore more superficially, whereas residents explore more thoroughly?
- Is heterospecific attraction widespread or restricted to specific habitat types in migrants? (a) Does it also apply for habitat specialists? (b) When do migrants learn features of sympatric species? (c) Which features do they learn? (d) Does heterospecific attraction also function in selection of stopover sites?
- Do all migrants have a long-lasting memory or is this restricted to species that show site fidelity?
- Do residents and migrants react differently to changes in their environment? Both laboratory and field experiments can explore this question. For example, residents and migrants can be attracted to a feeder in the field (in the breeding area, at stopover sites or in the winter quarters). After habituation, novel objects can be presented near or at the feeder (see Greenberg 1989 for an appropriate test design). Latencies to approach the feeder with objects can easily be measured for residents and migrants. Additionally, reactions in the different areas (breeding, stopover, winter site) can be compared.
- Do residents and migrants differ in the types of information they gather? (a) Do migrants collect primarily spatial and/or stable information? (b) Do residents collect primarily variable information?
- Do learning and memory abilities vary with season?

Finally, the question arises as to whether such differences between migrants and residents occur throughout their life or primarily during periods of transit and transition. Because site fidelity to stopover sites, wintering areas, and natal or breeding area occurs in many migrants, it is primarily inexperienced young birds that have to cope with uncertainties and unfamiliar areas. Experienced birds, which are able to remember profitable or safe places visited in the past, will move from one familiar area to the next (Ketterson and Nolan 1990). Although small changes might occur from year to year, they might be negligible compared with the first year of migration. With increasing age do migrants behave progressively more like residents? This can have important implications for conservation. When the quality of the site has decreased or the site has been destroyed (by natural or human causes), it may be more difficult for experienced individuals than for inexperienced individuals to change the stopover site.

ACKNOWLEDGMENTS

CM-H thanks H. Fugger and E. Gwinner for helpful comments on the manuscript.

LITERATURE CITED

Aborn, D. A., and F. R. Moore. 1997. Pattern of movement by summer tanagers (*Piranga rubra*) during migratory stopover: a telemetry study. Behaviour 134:1077–1100.

Airola, D. A., and R. H. Barrett. 1985. Foraging and habitat relationships of insect-gleaning birds in a Sierra Nevada mixed-conifer forest. Condor 87:205–216.

Alatalo, R. V., and A. Lundberg. 1984. Density-dependence in breeding success of the Pied Flycatcher (*Ficedula hypoleuca*). Journal of Animal Ecology 53:969–977.

Alerstam, T., and A. Lindström. 1990. Optimal bird migration: the relative importance of time, energy, and safety. Pages 331–351 *in* Bird Migration: Physiology and Ecophysiology (E. Gwinner, ed.). Springer-Verlag, Berlin.

Baker, R. R. 1993. The function of post-fledging exploration: a pilot study of three species of passerines ringed in Britain. Ornis Scandinavica 24:71–79.

Basil, J. A., A. C. Kamil, R. P. Balda, and K. V. Fite. 1996. Differences in hippocampal volume among food storing corvids. Brain, Behavior and Evolution 47:156–164.

Berndt, R., and W. Winkel. 1979. Verfrachtungs-Experimente zur Frage der Geburtsortsprägung beim Trauerschnäpper (*Ficedula hypoleuca*). Journal für Ornithologie 120:41–53.

Berthold, P., and S. B. Terrill. 1991. Recent advances in studies of bird migration. Annual Review of Ecology and Systematics 22:357–378.

Biebach, H., W. Friedrich, and G. Heine. 1986. Interaction of bodymass, fat, foraging and stopover period in trans-Sahara migrating passerine birds. Oecologia 69:370–379.

Brodbeck, D. R., and S. J. Shettleworth. 1995. Matching location and color of a compound stimulus: comparison of a food-storing and a nonstoring bird species. Journal of Experimental Psychology: Animal Behavior Processes 21 (1):64–77.

Burgess, N., A. Jackson, T. Hartley, and J. O'Keefe. 2000. Predictions derived from modeling the hippocampal role in navigation. Biological Cybernetic 83:301–312.

Cadiou, B. 1999. Attendance of breeders and prospectors reflects the quality of colonies in the Kittiwake *Rissa tridactyla*. Ibis 141:321–326.

Cantos, F. J., and J. L. Tellería. 1994. Stopover site fidelity of four migrant warblers in the Iberian peninsula. Journal of Avian Biology 25:131–134.

Clayton, N. S., and D. A. Cristol. 1996. Effects of photoperiod on memory and food storing in captive Marsh Tits, *Parus palustris*. Animal Behaviour 52:715–726.

Delingat, J., and V. Dierschke. 2000. Habitat utilization by Northern Wheatears (*Oenanthe oenanthe*) stopping over on an offshore island during migration. Die Vogelwarte 40:271– 278.

Doligez, B., E. Danchin, and J. Clobert. 2002. Public information and breeding habitat selection in a wild bird population. Science 297:1168–1170.

Eilam, D., and I. Golani. 1989. Home base behavior of rats (*Rattus norvegicus*) exploring a novel environment. Behavioural Brain Research 34:199–211.

Fransson, T. 1995. Timing and speed of migration in North and West European populations of *Sylvia* warblers. Journal of Avian Biology 26:39–48.

Greenberg, R. 1979. Body size, breeding habitat, and winter exploitation systems in *Dendroica*. Auk 96:756–766.

Greenberg, R. 1983. The role of neophobia in determining the degree of foraging-specialization in some migrant warblers. American Naturalist 122:444–453.

Greenberg, R. 1984. Differences in feeding neophobia in the tropical migrant wood warblers *Dendroica castanea* and *D. pensylvanica*. Journal of Comparative Psychology 98:131–136.

Greenberg, R. 1987. Development of dead leaf foraging in a tropical migrant warbler. Ecology 68:130–141.

Greenberg, R. 1989. Neophobia, aversion to open space, and ecological plasticity in song and swamp sparrows. Canadian Journal of Zoology 67:1194–1199.

Greenberg, R. 1990a. Feeding neophobia and ecological plasticity: a test of the hypothesis with captive sparrows. Animal Behaviour 39:375–379.

Greenberg, R. 1990b. Ecological plasticity, neophobia, and resource use in birds. Studies in Avian Biology 13:431–437.

Greenberg, R. 1992. Responses to a simple habitat cue in naïve Swamp and Song Sparrows. Oecologia 92:299–300.

Grünberger, S., and B. Leisler. 1993. Die Ausbildung von Habitatpräferenzen bei der Tannenmeise (*Parus ater*): genetische Prädisposition und Einfluß der Jugenderfahrung. Journal für Ornithologie 134:355–358.

Gwinner, E. 1996. Circadian and circannual programmes in avian migration. Journal of Experimental Biology 199:39–48.

Hampton, R. R., S. J. Shettleworth, and R. P. Westwood. 1998. Proactive interference, recency, and associative strength: comparisons of Black-capped Chickadees and Dark-eyed Juncos. Animal Learning & Behavior 26 (4):475–485.

Healy, S. D., E. Gwinner, and J. R. Krebs. 1996. Hippocampal volume in migratory and non-migratory warblers: effects of age and experience. Behavioural Brain Research 81:61–68.

Heth, G., E. Nevo, and A. Beiles. 1987. Adaptive exploratory behaviour: differential patterns in species and sexes of subterranean mole rats. Mammalia 51:27–37.

Hund, K., and R. Prinzinger. 1979. Untersuchungen zur Ortstreue, Paartreue und Überlebensrate nestjunger Vögel bei der Mehlschwalbe *Delichon urbica* in Oberschwaben. Die Vogelwarte 30:107–117.

James, F. C. 1971. Ordination of habitat relationships among breeding birds. Wilson Bulletin 75:15–22.

Jenni, L., and S. Jenni-Eiermann. 1987. Der Herbstzug der Gartengrasmücke *Sylvia borin* in der Schweiz. Der Ornithologische Beobachter 84:173–206.

Jones, R. B. 1986. Responses of domestic chicks to novel food as a function of sex, strain and previous experience. Behavioural Processes 12:261–271.

Ketterson, E. D., and V. Nolan, Jr. 1990. Site attachment and site fidelity in migratory birds: experimental evidence from the field and analogies from neurobiology. Pages 117–129 *in* Bird Migration: Physiology and Ecophysiology (E. Gwinner, ed.). Springer-Verlag, Berlin.

Lack, D. 1933. Habitat selection in birds. Journal of Animal Ecology 2:239–269.

Löhrl, H. 1959. Zur Frage des Zeitpunktes einer Prägung auf die Heimatregion beim Halsbandschnäpper (*Ficedula albicollis*). Journal für Ornithologie 100:132–140.

Mettke-Hofmann, C. 2000a. Reactions of nomadic and resident parrot species, *Psittacidae,* to environmental enrichment at the Max-Planck Institut. International Zoo Yearbook 37:244–256.

Mettke-Hofmann, C. 2000b. Changes in exploration from courtship to the breeding state in Red-rumped parrots (*Psephotus haematonotus*). Behavioural Processes 49:139–148.

Mettke-Hofmann, C., and E. Gwinner. 2003. Long-term memory for a life on the move. Proceedings of the National Academy of Sciences USA 100:5863–5866.

Mettke-Hofmann, C., and E. Gwinner. 2004. Differential assessment of environmental information in a migratory and a non-migratory passerine. Animal Behaviour 68:1079–1086.

Mönkkönen, M., J. T. Forsman, and P. Helle. 1996. Mixed-species foraging aggregations and heterospecific attraction in boreal bird communities. Oikos 77:127–136.

Mönkkönen, M., R. Härdling, J. T. Forsman, and J. Tuomi. 1999. Evolution of heterospecific attraction: using other species as cues in habitat selection. Evolutionary Ecology 13:91–104.

Mönkkönen, M., P. Helle, G. J. Niemi, and K. Montgomery. 1997. Heterospecific attraction affects community structure and migrant abundances in northern breeding bird communities. Canadian Journal of Zoology 75:2077–2083.

Morse, D. H. 1970. Ecological aspects of some mixed-species foraging flocks of birds. Ecological Monograph 40:119–168.

Morse, D. H. 1980. Behavioral Mechanisms in Ecology. Harvard University Press, Cambridge.

Muller, K. L., J. A. Stamps, V. V. Krishnan, and N. H. Willits. 1997. The effects of conspecifics attraction and habitat quality on habitat selection in territorial birds (*Troglodytes aedon*). American Naturalist 150:650–661.

Nolan, V., Jr., and E. D. Ketterson. 1991. Experiments on winter-site attachment in young Dark-eyed Juncos. Ethology 87: 123–133.

Partridge, L. 1974. Habitat selection in titmice. Nature 247: 573–574.

Reboreda, J. C., N. S. Clayton, and A. Kacelnik. 1996. Species and sex differences in hippocampus size in parasitic and non-parasitic cowbirds. Neuroreport 7:505–508.

Reed, J. M., T. Boulinier, E. Danchin, and L. W. Oring. 1999. Informed dispersal: prospecting by birds for breeding sites. Current Ornithology 15:189–259.

Remsen, J. V., Jr., and T. A. Parker III. 1984. Arboreal dead-leaf-searching birds in the Neotropics. Condor 86:36–41.

Roberts, E. P., Jr., and P. D. Weigl. 1984. Habitat preference in the Dark-eyed Junco *Junco hyemalis:* the role of photoperiod and dominance. Animal Behaviour 32:709–714.

Schjorring, S., J. Gregersen, and T. Bregnballe. 1999. Prospecting enhances breeding success of first-time breeders in the great cormorant, *Phalacrocorax carbo sinensis.* Animal Behaviour 57:647–654.

Schüz, E., P. Berthold, E. Gwinner, and H. Oelke. 1971. Grundriß der Vogelzugkunde. Verlag Paul Parey, Berlin.

Smith, N. G. 1975. Spishing noise: biological significance of attraction and nonattraction by birds. Proceedings of the National Academy of Sciences USA 72:1411–1414.

Smulders, T. V., A. D. Sasson, and T. J. DeVoogd. 1995. Seasonal variation in hippocampal volume in a food-storing bird, the black-capped chickadee. Journal of Neurobiology 9:15–25.

Sol, D., and L. Lefebvre. 2000. Behavioural flexibility predicts invasion success in birds introduced to New Zealand. Oikos 90:599–605.

Suryan, R. M., and D. B. Irons. 2001. Colony and population dynamics of Black-legged Kittiwakes in a heterogeneous environment. Auk 118 (3):636–649.

Thibault, A., and J. Bovet. 1999. Homing strategy of the Eastern Chipmunk, *Tamias striatus* (Mammalia: Rodentia): validation of the critical distance model. Ethology 105:73–82.

Thompson, P. S., and W. G. Hale. 1989. Breeding site fidelity and natal philopatry in the Redshank *Tringa totanus.* Ibis 131:214–224.

Tolman, E. C. 1932. Purposive Behavior in Animals and Men. Appleton-Century-Crofts, New York.

Van Patten, K. C. 1990. Habitat choice in captive Arctic Warblers. Auk 107:434–437.

Veltman, C. J., S. Nee, and M. J. Crawley. 1996. Correlates of introduction success in exotic New Zealand birds. American Naturalist 147 (4):542–557.

Ydenberg, R. C., R. W. Butler, D. B. Lank, C. G. Guglielmo, M. Lemon, and N. Wolf. 2002. Trade-offs, condition dependence and stopover site selection by migrating sandpipers. Journal of Avian Biology 33:47–55.

DFL '03

Part 3 / BIOGEOGRAPHY

JOHN FAABORG

Overview

BIOGEOGRAPHY CAN MOST EASILY be defined as the study of distributions of organisms. It is an old science, with such famous practitioners as Wallace and Darwin in the past and MacArthur and Wilson more recently. Much of classical biogeography has been descriptive, starting with the mapping of major vegetation types and their associated fauna, adding patterns of diversity along such gradients as latitude, elevation, or island size, and ending with the designation of "biogeographic realms," those six categories into which the world can quite reasonably be divided given the distinct characteristics of flora and fauna in each. Many of these realms also coincide with the boundaries of continents, but sometimes major barriers such as mountains or deserts serve to split the realms in ways inconsistent with the structure of the continents themselves. History often plays a role in understanding the boundaries of realms, as with Wallace's Line, along which a continuous oceanic barrier clearly separates the Australian and Oriental realms. In other cases, as in the New World, the line between tropical and temperate realms is anything but clear-cut.

Migrant birds are not the classical biogeographer's taxon of choice if clear patterns for describing realms are the goal. Hundreds of species and billions of individuals move between continents on an annual basis, making a mockery of the lines drawn by scientists between the biogeographic realms. For example, work by my students and me in southern Tamaulipas, Mexico, found that about 40% of the species found there 7 to 8 months of the year breed in the United States and Canada. During this time, the migrants coexist with members of virtually all of the families used to characterize the Neotropical bird community. Many of these migrants seem to have specific roles within the diverse and highly structured mixed-species flocks that occur in tropical habitats, while others spend their nonbreeding season using specialized tropical resources such as army ant swarms. In some cases, as these winter residents move north to breed, they are replaced during the breeding season by species that spent the winter even farther south. As Stephen Fretwell once noted, "For most communities we must be genuinely puzzled by all the comings and goings."

As many of these long-distance migrants spend much more time on their nonbreeding grounds than on their breeding grounds, it is not unreasonable to ponder, "Whose birds are they? Where do they belong?" If they were originally tropical birds, what is it about them as a group that leads them to leave their tropical homes to breed in the north? How could they be sure that ecological space would be available for the energetically expensive act of breeding? If migrant species were really temperate in origin, how did they evolve ways to survive within tropical bird communities, which generally are as diverse as those found anywhere on Earth? Has the amount of either wintering or breeding habitat been a factor in determining the extent to which birds migrate? Has the ability to find the proper winds that allow migration over such barriers as oceans and mountains been a factor in determining how many birds migrate and where they go? Understanding migration involves examining a complex set of evolutionary trade-offs, including traits of survival and reproduction, and the costs of movements both within and between the two often very different biogeographic realms in which they spend their lives.

As scientists, our ability to examine what we tend to think of as a north-south or tropical-temperate phenomenon (a movement between two different climatic worlds) can be aided by the fact that hundreds of species have evolved migration on our planet in both the New and Old Worlds. The almost complete separation of individuals involved in migration in these two regions suggests that we can view them as separate evolutionary systems. Although one must always be careful, comparisons between Old World and New are the closest thing to experimental replication that a biogeographer can get. The value of this replication, though, is only as good as the consistency shown between these regions in habitat size and structure, history, and numerous other critical traits of both birds and regions.

Our survey of biogeographic trends begins with two papers that compare traits of migration in the Old World and New World. Mikko Mönkkönen and Jukka Forsman describe patterns of abundance of migrant and resident birds in forests of Europe and eastern North America. They show that migrant abundances peak at midlatitudes and suggest that migrant and resident abundances vary independently of each other at large geographic scales. Their examination of interactions between migrants and residents at local and regional scales, however, suggests that both positive and competitive interactions between these groups occur and affect patterns of abundance. They present a model of how these interactions may affect or be affected by abundance patterns but recognize that more work needs to be done.

Katrin Böhning-Gaese adopts a broader, macroecological approach to examine the influence of migrants on all temperate bird communities. She builds on recent work suggesting the tropical origins of most migrant species and shows that long-distance migrants have kept their original habitat preferences. This results in New World migration systems that involve mostly forest birds, whereas those of the Old World involve mostly birds of open habitats. These broad patterns mean that migrants tend to have lower abundances, smaller ranges, and more spatially clumped distributions than the resident species with which they coexist, although much more work is needed to reveal the mechanisms at work causing such patterns.

Keith Bildstein and Jorje Zalles continue with comparisons between Old and New Worlds, but they restrict their analyses to three genera of raptors, *Buteo, Falco,* and *Accipiter,* and they discuss the interplay between the differing flying abilities of these groups and the biogeographic patterns they show. Given that raptors are large and highly visible, use thermals, and usually avoid crossing water, it is relatively easy to chart the flyways they use during long-distance migration around the world. As a group, accipiters are poorer at soaring than buteos and not as efficient at powered flight as falcons, which may explain why fewer than 30% of *Accipiter* species show any tendency to migrate and why members of this group are common only on islands near continents. In contrast, both falcons and buteos occur regularly on more-isolated oceanic islands, presumably because they either are blown there (buteos) or regularly fly across water as part of normal migration (falcons). The authors suggest a process they call *migration dosing* to explain how migratory species may help generate endemic species.

Compared with the previous papers in this section, the contribution of Terry Chesser has a relatively narrow focus, dealing with austral migration in South America. Austral migrants tend to travel shorter distances than their temperate relatives and are dominated by members of the family Tyrannidae, although this group has radiated into a variety of ecological, behavioral, and morphological types in this region. By comparing patterns of distribution, habitat selection, and climate between austral migrants and residents, Chesser finds traits of migration that are comparable to those found in Europe and North America, despite the limited diversity of migrants taxonomically.

Our final contribution, by Sidney Gauthreaux, Jenny E. Michi, and Carroll Belser, adds descriptions of atmospheric circulation to the many other components of biogeography with the goal of understanding the annual difficulties (or lack thereof) associated with long-distance migration in the Gulf of Mexico and Caribbean Sea. By using wind data that have only recently been made available, they show that during most of spring migration, winds allow migrants to follow a "go-with-the-prevailing-flow" strategy that makes movement from the Tropics and even the Caribbean fairly simple and easy. In contrast, birds moving from northern breeding grounds to the Tropics must adopt a "sit-and-wait-for-favorable-winds" strategy in order to have efficient flying conditions in that season. Good autumnal flying conditions are often associated with cold fronts, which also may bring frost and a reduction in the food supply used by migratory birds. Thus, birds must balance waiting for good flying conditions with the risk of loss of food. Further examination of the details of annual variation in these wind patterns may also explain annual variation in migrant abundances, as we see how birds take the easiest path possible each spring. Expanding this analysis to other parts of the world will undoubtedly provide exciting insights into the difficulty or ease of migration elsewhere.

It is not surprising that in a symposium dealing with birds of two worlds, other sessions also include aspects of biogeography. Whether the focus be as narrow as hormonal patterns or as broad as evolutionary history, the position of continents and the habitats they contain serve as the template from which all aspects of migration develop.

11

MIKKO MÖNKKÖNEN
AND JUKKA T. FORSMAN

Ecological and Biogeographical Aspects of the Distribution of Migrants versus Residents in European and North American Forest Bird Communities

WE DESCRIBE THE PATTERNS of geographic variation in abundance of tropical migrants, short-distance migrants, and resident birds. Using a comprehensive collection of published breeding bird survey data, we contrast European and eastern North American forests. We also provide an overview of published studies about regional and local patterns in the abundance of resident and migrant birds. We discuss the alternative processes producing these patterns and particularly the role of local species interactions.

Resident abundance linearly decreased with latitude in both continents, but migrant abundance showed more complicated geographical trends. In both continents, tropical migrant abundance peaked at midlatitudes (55° N in Europe, 40° N in North America). Both resident and migrant abundances were consistently low in boreal forests but showed a wide range of variation farther south. We conclude that at large geographical scale resident and migrant abundances vary independently of each other and that interactions between resident and migrant birds are not the most likely explanation for these continental patterns.

Local-scale experimental work has provided evidence for both positive and competitive interactions between migrant and resident birds in breeding bird communities. We hypothesize that positive interactions prevail at low abundance of putative competitors (e.g., at northern latitudes), but with increasing competitor abundance the benefits from social aggregations become gradually outweighed by the negative effects of interspecific competition, and interaction turns into competitive. Our regional-scale analyses of the abundance associations between resident parids and migrant foliage gleaners provide support for this hypothesis. We found mainly positive associations between resident and migrant abundances in boreal forests, but a unimodal association farther south in temperate forests, where positive abundance relationships prevailed at low resident abundance but negative ones at high resident abundance. These associations were stronger in European forest bird communities than in North America.

INTRODUCTION

Species diversity patterns such as the latitudinal diversity gradient have been intensively studied ever since Wallace (1878). Geographic variation in abundance patterns has received far less attention even though it is likely that factors regulating species diversity and abundance are inherently the same. For example, productivity has often been implicated in the diversity patterns, but productivity primarily affects abundance and must be accompanied by other ecological or evolutionary processes to have an impact on species richness (Wilson 1992). Also, biotic interactions, which eventually modify species richness, must first have an effect on population abundance, and consequently, abundance patterns should reflect their effect before any diversity effects become discernible.

Breeding bird assemblages are to a varying degree composed of resident and migrant species. Migratory birds can further be divided into short-distance and tropical migrants. Many tropical migrants just briefly visit their breeding habitats for reproduction, but short-distance migrants may spend a large part of their yearly cycle on the breeding grounds. It is well known that the proportion of migrant species in regional faunas increases with latitude (Newton and Dale 1996a, 1996b). Likewise in local breeding bird assemblages, the relative proportion of the migratory species of the total bird abundance increases strongly toward higher latitudes (MacArthur 1959; Willson 1976; Herrera 1978; Helle and Fuller 1988), but a more precise and comprehensive description of the composition of northern breeding bird communities in terms of different migratory groups is still missing.

Theories and hypotheses to account for diversity and abundance patterns in local communities can roughly be classified into two categories according to the spatial scale. First, patterns in local communities can be considered predominantly reflections of processes on regional and continental scales, such as variation in productivity (e.g., species-energy theory [Wright et al. 1993]) or historical and geographical circumstances (Pianka 1966; Ricklefs 1987). By contrast, one may consider local-scale ecological processes, such as interspecific competition, predation, and parasitism, important for the coexistence of species and their relative abundance in their communities (see Wiens 1989, Krebs 1994, and Rosenzweig 1995 for reviews). Ecologists are becoming increasingly aware that these two approaches should not be mutually exclusive and that processes at both small and large spatio-temporal scales have their imprint in community patterns (Ricklefs 1987; McLaughlin and Roughgarden 1993; Travis 1996). The two scales may also be nested so that the importance and intensity of species interactions vary along with larger-scale environmental conditions. Darwin (1859) had already pointed out that interspecific competition may not be as intense in northern environments as in the south because of the harsh physical conditions in the north (see also Dobzhansky 1950; MacArthur 1972; Pielou 1979). There is some evidence for latitudinal variation in species interactions (Leonard 2000) but in general this hypothesis has not been rigorously examined.

The increasing proportion of migrant birds in regional fauna and local communities has been explained by increasing seasonality in food resources (MacArthur 1959). A more explicit mechanism was proposed by Herrera (1978), who suggested that in the north, resident bird numbers are limited by winter conditions well below summertime carrying capacity, which in turn enables more migrant birds to enter into breeding communities (see also Rabenold 1979). Herrera's (1978) hypothesis falls in line with the widely held view that communities are arenas of resource partitioning, where the underlying assumption is that resources are limiting. This hypothesis implicitly assumes that resident birds are competitively superior over migrant birds. Migrants are able to enter into a breeding community only if left-over resources are available after the prior occupation of the residents. In other words, he assumed diffuse competition between residents and migrants and density compensation between those groups (see also O'Connor 1981, 1990; Morse 1989). Implicit in Herrera's (1978) hypothesis is also the idea that the intensity of competition between resident and migrant birds varies geographically because of geographic variation in abiotic conditions (winter climate).

Although the relative contribution of different migratory groups and the geographical variation in it are well described for European and North American forest bird communities (MacArthur 1959; Willson 1976; Herrera 1978; Helle and Fuller 1988; Mönkkönen and Helle 1989), latitudinal variation in abundance has received far less attention (see Rabenold 1979). In this chapter we first describe the patterns of geographic variation in abundance of different migratory groups, particularly contrasting European and eastern North American forests. This study is based on a comprehensive collection of published breeding bird survey data from both continents. The composition of northern bird communities of different migratory habits raises the questions, How do these migratory groups of species interact locally? Can we see any reflections of these local interactions when viewing communities at larger biogeographical scales? We therefore provide an overview of published studies about regional and local patterns in the abundance of resident and migrant birds. We discuss the alternative processes producing these patterns and particularly the role of local species interactions. Comparison between hemispheres provides a possibility to see whether similar patterns emerge under ecologically similar conditions despite taxonomic differences. Migrants particularly tend to belong to different families of birds on these two continents, but residents are much more closely related between the two hemispheres (Haila and Järvinen 1990; Mönkkönen and Welsh 1994).

CONTINENTAL ABUNDANCE PATTERNS OF MIGRANTS AND RESIDENTS IN NORTH AMERICAN AND EUROPEAN FOREST BIRD COMMUNITIES

Material and Methods

We collected published breeding bird survey data from Europe ($N = 106$) and North America ($N = 98$). We selected only surveys conducted in mature forests (age >100 years and/or height >20 m) to standardize the data. In Europe, surveys originated from Spain (ca. 40° N) in the south to northern Finland (ca. 69° N) in the north, and from Great Britain (ca. 5° W) in the west to northeast European Russia (ca. 58° E) in the east. In North America, we included only Temperate Zone surveys collected east of 95° longitude because of clear faunal distinction between eastern and western parts of the continent (see, e.g., Mönkkönen 1992 and Mönkkönen and Viro 1997). To provide broad-enough latitudinal range, all boreal zone surveys available west to Yukon (129° W) were included. The southernmost and the northernmost data points were derived from the latitude of ca. 30° N and ca. 62° N, respectively. Therefore, both European and North American data sets represent a comparable amount of latitudinal (about 30°) and longitudinal (63°) variation, the largest longitudinal variation coming from the boreal forests. Because our one aim is to describe the large-geographic-scale patterns of variation in abundance of different migratory groups, this sort of scatter in data points is requisite. Note also that the data cover ecologically rather similar gradients of conditions, as the corresponding forest zones in North America lie about 10° farther south than in Europe.

From each survey we extracted bird abundance (pairs/10 ha) and geographic location. We also calculated the total abundance of forest birds to calculate the proportions of migratory groups. Here, forest birds include woodpeckers (Piciformes), cuckoos (Cuculiformes), and passerines (Passeriformes) except swallows (Hirundinidae), which were excluded as not being true forest birds. We focused on these taxa only because the standard census methods give reliable abundance estimates for these taxa only. The three included taxa comprise more than 90% of total abundance and more than 80% of all species in our European and North American survey data.

We used information in Marshall and Richmond (1992) and Snow and Perrins (1998) to categorize bird species into three migratory habit groups. Tropical migrants are species whose wintering areas are completely in the Tropics or Subtropics. Residents include only permanent resident species whose distributions show no difference between summer and winter. All other species were treated as short-distance migrants. Species such as the Yellow-rumped Warbler (*Dendroica coronata*) or the Chiffchaff (*Phylloscopus collybita*), which have both tropical and short-distance migrant populations, were considered short-distance migrants. We acknowledge that migratory habit is actually a gradient from truly permanent resident species to highly migratory species, but this categorization makes quantifiable comparisons between geographic locations possible.

We fitted general linear models to abundance and proportion of migratory groups. Modeling was started by entering latitude, longitude, their squared terms, and interaction between latitude and longitude as continuous covariates. This is a second-degree polynomial trend surface analysis (Legendre and Legendre 1998). In addition, method was entered as random factor and forest type as fixed factor. Surveys used four different census methods: the territory mapping, the line transect, the point count, and the single-visit study plot methods. Forest types were categorized as conifer, mixed conifer-deciduous, or deciduous forest, according to the information reported in the original publication.

We used backward elimination to end up with as simple a model as possible without dropping any significant ($p < 0.05$) term from the model. Terms were eliminated in order of their significance. We show only the final models including the significant terms (see tables 11.1 and 11.2). In no model did method explain a significant proportion of variation in abundance or proportion of any migratory group. Likewise, longitude and interaction between longitude and latitude could be dropped off from all models.

Results

Resident bird abundance decreases with latitude in both Europe and North America (table 11.1, fig. 11.1). On both continents, abundance varied widely at low latitudes but was invariably low at high latitudes. Resident abundance is on average about two times higher in Europe (13.6 pairs/10 ha) than in North America (7.8 pairs/10 ha). In North America, but not in Europe, forest type significantly explained resident abundance. Abundance was highest in deciduous and lowest in conifer forests. These abundance trends resulted in corresponding latitudinal trends in the proportion of resident birds in their breeding communities. Resident proportion decreased linearly in Europe from about 40% in the Mediterranean area to about 5% in northern Fennoscandia. In North America, the proportion of residents decreased, but not linearly, from about 30% in the south to 5% in the north. The effect of forest type was significant in North America; differences between forest types were similar to those shown for abundance (table 11.2, fig. 11.2).

No latitudinal trend in the abundance of short-distance migrants was found in North America but a significant nonlinear pattern emerged in Europe (table 11.1, fig. 11.1). In

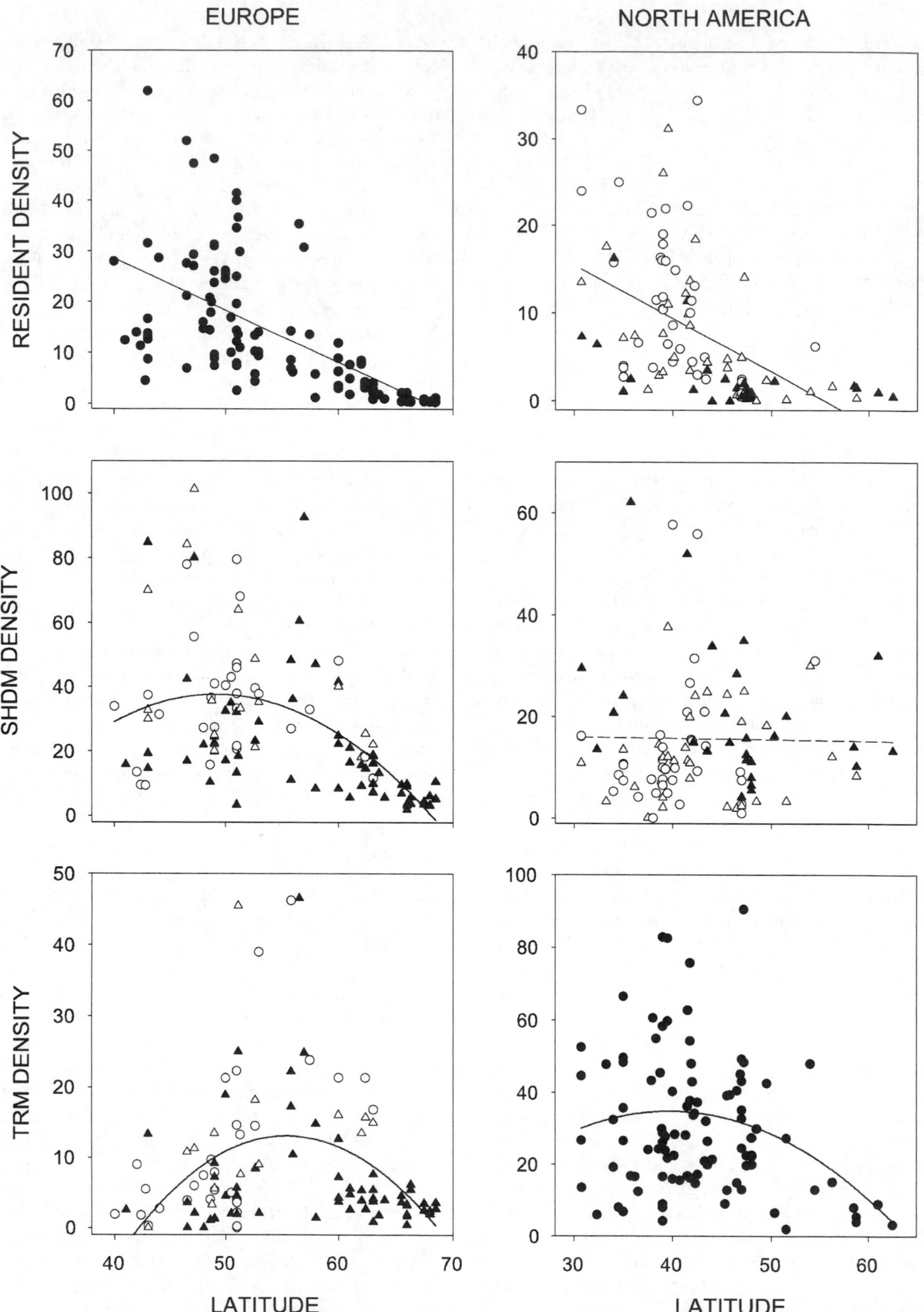

Fig. 11.1. Abundance (pairs/10 ha) of resident, short-distance migrant, and tropical migrant birds against latitude (°N) in European and eastern North American forest bird communities. Solid line indicates a significant (linear or quadratic, see table 11.1) relationship between abundance and latitude and dashed line nonsignificant trend. SHDM = short-distance migrants, TRM = tropical migrants. Open circles refer to deciduous forests, open triangles to mixed conifer-deciduous forests, and black triangles to coniferous forests. Symbols for forest types are shown only if the term was significant (see table 11.1). Black circles are used in cases where forest type was not significant.

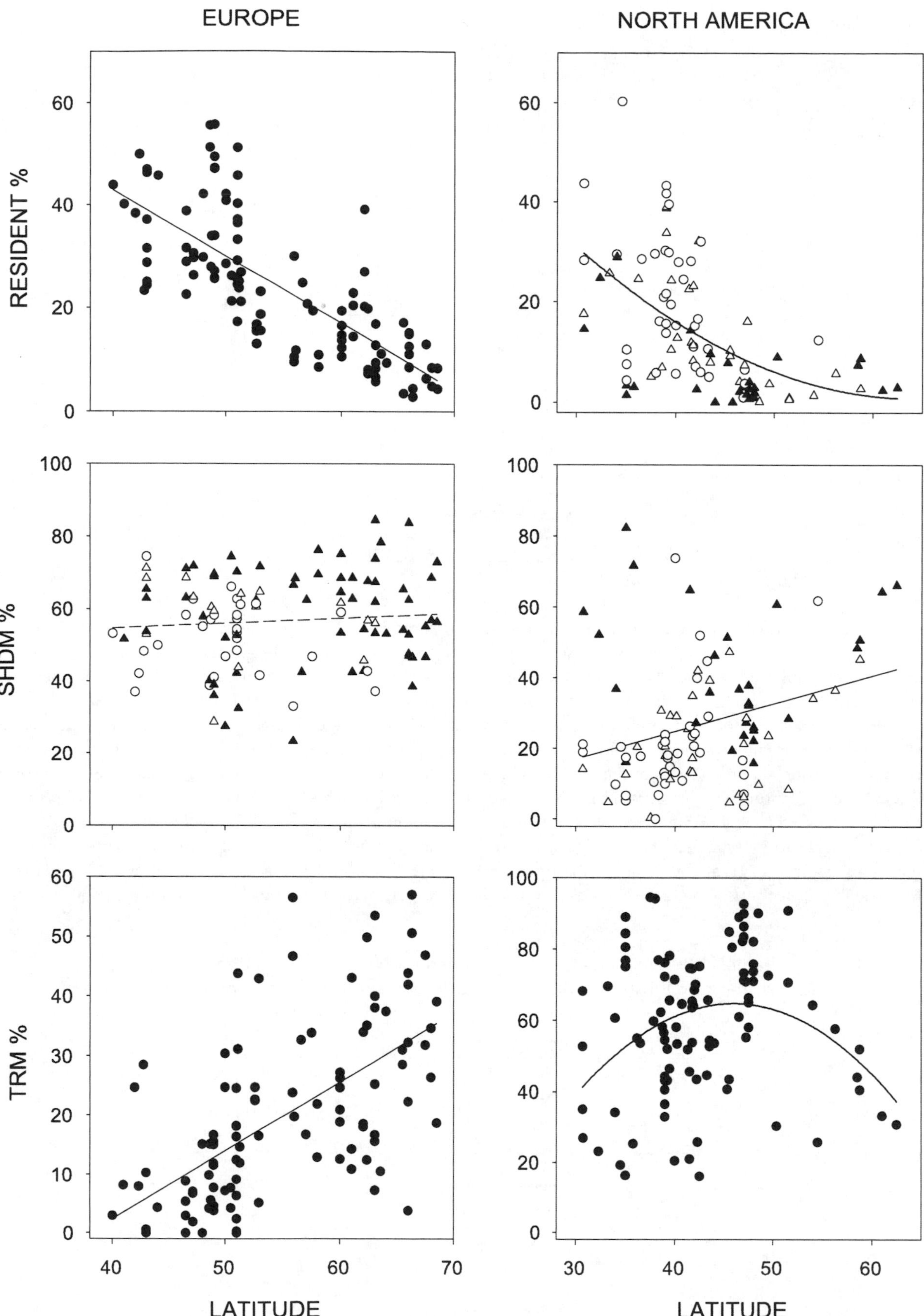

Fig. 11.2. Proportion of total pairs (%) of resident, short-distance migrant, and tropical migrant birds against latitude (°N) in European and eastern North American forest bird communities. Solid line indicates a significant (linear or quadratic) relationship between abundance and latitude and dashed line nonsignificant trend. SHDM = short-distance migrants, TRM = tropical migrants. For symbols see fig. 11.1.

Table 11.1 Final regression models for abundance variation in different migratory groups after backward elimination procedure

	Source of variation	F	d.f.	*p*
Europe				
Residents	Latitude, $r^2 = 0.391$	68.49	1	<0.001
Short-distance migrants	Latitude	6.97	1	0.010
	(Latitude)^2	8.68	1	0.004
	Forest type	3.17	2	0.046
	Model, $r^2 = 0.325$	13.61	4	<0.001
Tropical migrants	Latitude	21.15	1	<0.001
	(Latitude)^2	20.00	1	<0.001
	Forest type	4.33	2	0.016
	Model, $r^2 = 0.220$	8.26	4	<0.001
North America				
Residents	Latitude	19.39	1	<0.001
	Forest type	6.42	2	0.002
	Model, $r^2 = 0.220$	16.53	3	<0.001
Short-distance migrants	Forest type, $r^2 = 0.070$	4.67	2	0.012
Tropical migrants	Latitude	3.88	1	0.052
	(Latitude)^2	5.07	1	0.027
	Model, $r^2 = 0.088$	5.71	2	0.005

Note: Adjusted coefficient of determination (r^2) refers to a model with significant terms included.

Europe abundance peaked at about 49° N and decreased both south and north from that latitude. Abundance was on average twice as high in Europe (28.1 pairs/10 ha) as in North America (15.6 pairs/10 ha). Interestingly, these patterns were converted into totally different patterns in terms of short-distance migrant proportion (table 11.2, fig. 11.2). In Europe the proportion remained constant at about 57% irrespective of the latitude, but in North America, the proportion of short-distance migrants increased with latitude. In both continents, forest type significantly explained variation in short-distance migrant abundance and proportion. In Europe, abundance was higher in deciduous and mixed forests than in conifer forests, but in North America the pattern was opposite. Proportion of short-distance migrants was higher in conifer forests than in deciduous forests, and mixed forests were intermediate in both continents.

The average abundance of tropical migrants was more than three times higher in North American (30.4 pairs/10 ha) than in European (8.4 pairs/10 ha) forest bird communities, but in both continents latitudinal variation showed a unimodal pattern (table 11.1). In both continents, tropical migrant abundance peaked in northern parts of the temperate deciduous forest zone: in Europe at 55° N and in North America at 40° N (fig. 11.1). It is notable that tropical mi-

Table 11.2 Final linear models for variation in the proportion of different migratory groups after backward elimination procedure

	Source of variation	F	d.f.	*p*
Europe				
Residents	Latitude, $r^2 = 0.564$	136.80	1	<0.001
Short-distance migrants	No model			
Tropical migrants	Latitude, $r^2 = 0.379$	65.14	1	<0.001
North America				
Residents	Latitude	7.00	1	0.010
	(Latitude)^2	4.46	1	0.037
	Forest type	6.96	2	0.002
	Model, $r^2 = 0.220$	14.71	4	<0.001
Short-distance migrants	Latitude	4.12	1	0.045
	Forest type	6.38	2	0.003
	Model, $r^2 = 0.175$	7.85	3	<0.001
Tropical migrants	Latitude	5.10	1	0.026
	(Latitude)^2	5.39	1	0.022
	Model, $r^2 = 0.036$	2.81	2	0.065

Note: Adjusted coefficient of determination (r^2) refers to a model with significant terms included.

grant abundance in southern and northern Europe tended to vary relatively little but showed wide variation at midlatitudes. In North America the amount of variation in tropical migrant abundance decreased with latitude. In Europe, tropical migrant abundance was lower in conifer than in mixed or deciduous forests but in North America no forest-type effects were detected. When looking at the proportion of tropical migrants of the total pair numbers, the unimodal pattern remained for North America (peak at 45° N) but was transformed into a linearly increasing trend with latitude in Europe (table 11.2, fig. 11.2).

Convergent and Divergent Latitudinal Patterns

A major difference between Europe and North America turned out to be the overall composition of forest bird communities. In Europe, abundance and proportion of residents and short-distance migrants was higher than in North America, where tropical migrants were the dominant group. This may partially stem from differences in categorizing species into migratory habit groups. In the western Palearctic area, the distinction between short-distance migrants and tropical migrants is clear: short-distance migrants remain in the Palearctic region year-round, but most tropical migrants cross the Sahara Desert (others go to Southeast Asia). In North America, the line between short-distance and tropical migrants is less distinct because among Nearctic migrant birds there is a gradient of behaviors from truly long-distance migration to partial migration.

The intercontinental difference is not solely a question of definition, however, as shown by patterns in resident birds. Resident birds are taxonomically rather homogeneous groups; for example, more than 70% of resident pairs in both continents are titmice or chickadees (Paridae) or woodpeckers (Picidae). Resident birds were a markedly more dominant group, in terms of both absolute abundance and proportion of total pairs, in European than in North American forest bird communities. One possible explanation for the difference is the relatively more continental climate in eastern North America compared with western Europe, where most of our data come from. Forsman and Mönkkönen (2003) showed that winter climate is the strongest determinant of resident abundance in Europe. The relative harshness of the North American winter climates obviously favors long-distance migration over residency and short-distance migration. Earlier studies have shown a major difference in habitat association patterns of tropical migrants between eastern North America and Europe. In Europe, tropical migrants are particularly numerous in early and midsuccessional habitats, but they abound in mature forests in North America; these differences correspond to habitat availability patterns in the tropical wintering areas (Mönkkönen and Helle 1989).

Tropical migrant abundance reached the highest abundance at midlatitudes (40° N in North America, 55° N in Europe; see Fig 11.1); short-distance migrant abundance also showed a unimodal pattern in Europe. Given that winter climate is effectively explaining the geographic variation in abundance in resident birds, why might migrant abundance show the unimodal pattern? Herrera's (1978) hypothesis predicts that resident and migrant abundances are inversely related. Our results do not support this. In the south, abundances of residents and tropical migrants were inversely related to latitude, but north of the peak migrant abundance they varied in parallel. We therefore conclude that competition between resident and migrant birds is not the most likely explanation for these continental patterns (for discussion on interspecific interactions, see below). Likewise, a comparison between abundance patterns of short-distance and tropical migrants does not provide any support for competitive interactions being an important determinant of abundance of these species groups. However, in Europe (but not in North America) patterns of relative contribution of tropical migrant versus resident birds fit nicely with Herrera's idea (proportions were inversely related across the whole latitudinal extent of our data).

One possible explanation for the unimodal abundance pattern of migrants is that their abundance follows the geographic variation in productivity. Data on net primary productivity (Olson et al. 2001) for European forests show that productivity is highest at midlatitudes, approximately 50° N, and seems to peak at 35–40° N in North American forests. These figures correspond quite well to peak abundance of migrants (also short-distance migrants in Europe). We leave more detailed analyses of the relationship between productivity and migrant abundance as a challenge for future studies.

INTERACTIONS BETWEEN MIGRANT AND RESIDENT BIRDS

Local Patterns and a Hypothesis

Traditionally, interspecific competition is considered a major force in structuring animal communities. To test for numerical and fitness response of migrants to resident abundance in the breeding bird assemblages, we conducted a series of replicated field experiments where we manipulated resident titmice (*Parus* spp.) abundance. We then monitored numerical response and fitness consequences of these manipulations to migrant counterparts in the breeding bird communities. We hypothesized that if competition between residents and migrants is an important factor we should observe a decrease in numbers and fitness parameters of migrants as resident abundance increased.

Our results, encompassing two continents (Europe and North America) and a variety of conditions from northern (Lapland) to southern boreal forest zones (Minnesota), were very consistent (see Mönkkönen and Forsman 2002 for review). We did not find any support for interspecific competition but the general pattern turned out to be a positive response by migrants to increased titmice abundance. In all our experiments migrant species richness tended to be higher when titmice were present than when they were absent. Total migrant abundance (Mönkkönen et al. 1990; Forsman et al. 1998) and abundance of foliage gleaners (Mönkkönen et al. 1997) responded significantly and positively to increased titmice abundances. In each experiment one or two individual species showed positive—and none showed negative—response to titmice abundance.

These results led us to coin the term "heterospecific attraction" to refer to the preferential selection of habitat patches already occupied by individuals of another species. We suggest that migrant birds may use residents as cues to profitable breeding sites when direct and accurate assessing of the quality of available patches is difficult. Short available breeding time and large year-to-year variation in conditions, both characteristics of northern environments, render heterospecific attraction a profitable habitat selection strategy.

We have also shown that heterospecific attraction has a selective basis. Forsman et al. (2002) found significant fitness benefits for Pied Flycatchers (*Ficedula hypoleuca*) from nesting together with resident titmice. Flycatchers breeding in tight association with titmice initiated breeding earlier, had larger broods, and raised heavier fledglings than solitary breeding flycatchers. Migrant birds settling near resident titmice had an increase in brood size of about 10% (0.6 fledglings on average).

Results from local experimental work, however, do not lend themselves to far-reaching conclusions about the importance and generality of heterospecific attraction as a process. We therefore used analytical modeling to analyze ecological conditions that might lead to heterospecific attraction (Mönkkönen et al. 1999). The results of the model indicated that the ability to recognize heterospecific individuals is an efficient way to choose the best possible patch not only when benefits from social aggregation exceeded the effects of competition, but also when interspecific competition is strong enough that it is more beneficial to avoid patches occupied by residents. Heterospecific attraction of colonists would occur whenever the benefit from aggregating with residents exceeds the effects of competition. Some studies have shown that birds indeed can recognize vocalizations of other species (Mönkkönen et al. 1996; Forsman and Mönkkönen 2001).

To our knowledge the only other field experiment on forest birds that has provided evidence for positive interactions among bird species is Brawn et al. (1987). They manipulated abundances of breeding birds by putting up nest boxes in high-altitude pine forests in north-central Arizona. They found no evidence for interspecific competition; rather "patterns of density fluctuations indicated large positive covariances among species at both the community level and within guilds."

In heterospecific attraction, however, it is still unclear whether residents also benefit, or suffer, from the presence of migrants. For example, resident titmice and flycatchers have been shown to compete, flycatchers being the suffering party (von Haartman 1957; Gustafsson 1987, but see Forsman et al. 2002). The interactions between those species may, however, be more complicated than earlier thought. Anecdotal observations suggest that titmice nest box takeovers by flycatchers are common in some years but very rare in others (pers. obs.).

Other experimental studies have shown that interspecific competition during the breeding season can affect reproductive success (Högsted 1980; Minot 1981; Gustafsson 1987) or territory occupation (Reed 1982; Garcia 1983; Sæther 1983; Sasvári et al. 1987; Sherry and Holmes 1988). In general it seems that evidence for positive interactions comes from northern latitudes (our studies) or high altitudes (Brawn et al. 1987) and competitive interactions prevail at lower latitudes (or altitudes [see, e.g., Gustafsson 1987 and Sasvári et al. 1987]). An interesting study in this respect is that of Reed (1982). He demonstrated that Chaffinch (*Fringilla coelebs*) territories on the mainland of Scotland tended to co-occur with Great Tit (*Parus major*) territories more often than expected by chance alone. On islands, however, these two species competed with each other and occupied nonoverlapping territories. Reed (1982) concluded that the environment on the mainland is richer (more food) than on islands. These results suggest that interspecific interactions may vary considerably according to environmental conditions, from competition in situations where resources are limiting (islands in Reed's case) to positive interactions in others (mainland).

Geographic variation in conditions may also lead to changes in the quality, not just the intensity, of species interactions. Forsman et al. (2002) proposed a formal hypothesis suggesting that species interactions may shift from positive to competitive along environmental gradients. When environmental conditions vary, constructing a gradient of abundances of putative competitors, the outcome of their interactions may show a unimodal pattern. Accordingly, individual fitness or the direction of interspecific interactions, everything else being equal, is a nonlinear, unimodal function of the abundance of putative competitors (fig. 11.3). At low abundance (when negative effects of coexistence do not exceed the benefits) positive interactions prevail. With increasing competitor abundance, the benefits from social aggregations become gradually outweighed by the negative effects of interspecific competition, and interaction turns competitive. This scenario matches well with the theoretical examination of interspecific interactions (Mönkkönen et al. 1999), whereby attraction to heterospecifics is predicted to switch to avoidance if resource limitation becomes too strong.

Signs of Interactions at Biogeographic Scale

When applied to interactions between resident and migrant species, the hypothesis by Forsman et al. (2002) predicts that at low abundance levels of resident birds positive interactions prevail, whereas at high abundance competition is the more likely interaction. Two geographical hypotheses can be derived. First, in northern latitudes, where resident abundance is invariably low (see fig. 11.1), we should predomi-

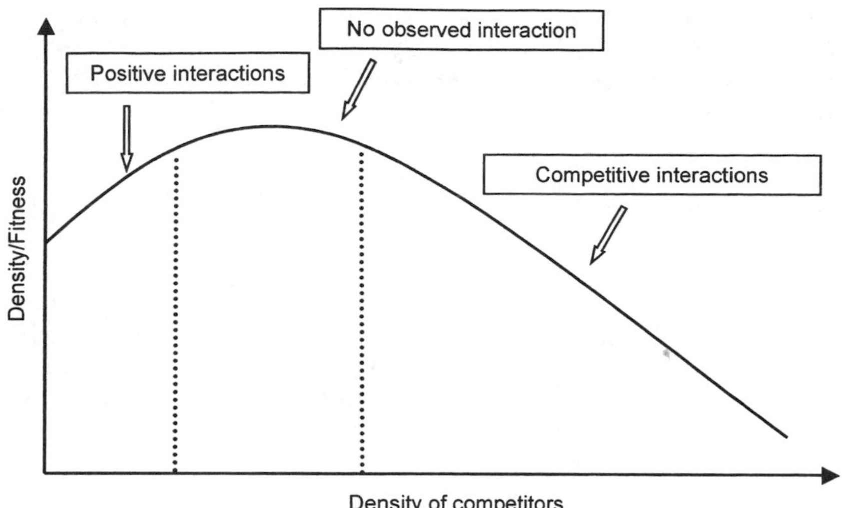

Fig. 11.3. A hypothesis suggesting a change in the intensity and direction of interactions along a gradient of abundance between potential competitors. Any increase in the abundance of a competitor is expected to result in positive effects (fitness or abundance) on the focal species (*y*-axis) at low abundance of a competitor (*x*-axis) and in negative effects at high abundance. The focal species reaches its maximum fitness and/or abundance at the intermediate abundance of a competitor, and any change in this region of abundance will have little or no effect on the focal species. Redrawn from Forsman et al. (2002). © The Royal Society Proceedings: Biological Sciences.

nantly find positive interactions. Second, farther south, where variation in resident abundance is more variable and, consequently, a long gradient of resident abundance can be found, a unimodal relationship between resident and migrant abundances should appear.

We tested these predictions with some of the data we used above for describing the abundance patterns of migratory habit groups (J. T. Forsman and M. Mönkkönen, unpubl.). We explored the hypothesis that interactions between resident and migrant birds in the local breeding bird communities vary geographically. We first divided the continental data into categories according to the main forest zones; we then analyzed the data separately in each zone to see if positive abundance relationships between residents and migrants appear in the north, and if unimodal abundance relationships can be found farther south, as predicted if interactions revert from positive to negative on a long gradient of resident abundance. In this analysis, we focused only on selected species groups in residents and migrants. Parids (genera *Parus* and *Poecile*) were selected to represent residents because these species are dominant resident members in breeding assemblages in both Europe and North America. We contrasted parid abundance with the abundance of migrant foliage gleaners and cavity nesters. Foliage gleaners are an ecologically defined group of species that potentially interact with parids because of common food resources (arboreal arthropods). Cavity nesters compete with parids over potential nest sites. Before contrasting resident and migrant abundances, we removed the latitudinal variation in abundances of both groups by using regression (residual abundance), because latitudinal variation alone potentially produces patterns in favor of our hypothesis (parallel abundance variation in the north and unimodal association farther south).

Our results showed that in boreal Europe, as predicted, migrants that belong to the same foraging guild as titmice had a strong and positive correlation with them, and titmice abundance explained a large proportion of variation in migrant abundance. In temperate forests of Europe, we also found the predicted unimodal relationship between resident and migrant abundances, with migrant abundance peaking at intermediate abundance of residents. In North America, abundance associations between titmice and migrants were weaker and we found only little support for the predicted latitudinal variation in abundance association. We found no indication in either continent of competition when only cavity nesters were analyzed.

The positive linear relationships between resident and migrant abundances in boreal Europe provide support for the role of positive interactions between these species groups. Although we cannot rule out the possibility that the positive relationship between resident and migrant abundance is due to parallel response to changes in environmental variation (e.g., productivity), our experimental studies in boreal forests certainly point out the role of direct positive interactions. Nonetheless, competition between migrants and residents appears to be unimportant in boreal regions.

Likewise, our correlative analysis does not unequivocally prove that in central European settings there is a shift from positive to negative interactions along the gradient of resident abundance. However, the unimodal pattern is difficult to explain without assuming a change in the quality and intensity of interactions. Another alternative is that the habitat requirements of residents and migrants are somewhat different (parallel preferences in some part of the environmental gradient but contrasting preference in some other part), but there is no evidence to support this explanation. Unfortunately, experimental evidence for positive interactions in temperate forests is also largely anecdotal (but see Reed 1982), although competitive interactions certainly exist (see above). One reason for this bias obviously is that experiments have been designed to test for the effects of interspecific competition and usually involved very high abundances of birds. For example, in Gustafsson's (1987) study, titmice abundance was manipulated to up to 30 pairs/10 ha, which is rather close to maximum titmice abundance in Europe. At high abundance, competitive interactions should prevail. In our study, the unimodal abundance relationship between resident and migrant birds remained even after latitudinal trends in abundance were removed, thus suggesting that the pattern does not stem from geographical variation in abundances.

At least two explanations are possible for the relatively weak abundance relationships between parids and migrants in North America. First, overall resident abundances are lower in North America than in Europe (fig. 11.1). Likewise, chickadees and titmice are a far less abundant group of forest birds in the Nearctic than in the western Palearctic. In our data, the titmice numbers were on average almost three times higher in Europe than in North America (8.7 vs. 3.1 pairs/10 ha). Moreover, local species richness of parids is usually one to two species in Nearctic forest bird communities but three to six in Europe. Parids therefore play a much smaller role in North America than in Europe, and accordingly, migrants in North America may not cue to titmice to the same extent as in European forests. However, our experiment in Minnesota (Mönkkönen et al. 1997) indicates that such cuing does happen in the Nearctic as well.

Second, Mönkkönen (1994) showed that there is much higher between-habitat diversity component, that is, higher dissimilarity in bird community structure between forest types, in Nearctic than in western Palearctic bird communities, suggesting that Nearctic species are relatively more specialized in their habitats. In other words, Nearctic forest birds are more tightly associated with structural or floristic features of their habitats than the more-generalist European birds and may not in general respond so readily to the presence or to the variation in abundance of other species.

Experimental evidence has shown a variety of types of interactions between residents and migrants, but we know rather little about the role of interactions in the distributions of different migrant birds. In our data, short-distance migrant birds were more abundant than tropical migrants in Europe but the reverse was true in North America. One possible explanation for this hemispheric difference is the competitive relationship between these two groups. More European birds are able to winter closer to the breeding grounds and therefore have earlier access to breeding sites at the expense of tropical migrants. Greenberg et al. (1999) showed a negative correlation between the proportion of warblerlike foliage gleaners (mainly tropical migrants) and foliage-gleaning finches (mainly short-distance migrants) in boreal forest bird communities. This correlation emerged from an intercontinental comparison that indicated a high-warbler / low-finch proportion in the New World (Ontario and MacKenzie) and a low-warbler / high-finch proportion in the Old World (European Russia and central Siberia). European study sites particularly stood out from all other areas in the overdominance of finches (the Chaffinch [*Fringilla coelebs*]) and in the low abundance of warblers. Greenberg et al. (1999) suggested the possibility that competitive interactions may mediate abundance of Chaffinches and warblers. Not only is there no evidence so far that Chaffinches are competitively dominant over warblers in access to good breeding sites, but there is some evidence that the reverse is true (Hogstad 1975).

CONCLUSIONS AND FUTURE DIRECTIONS

We conclude that it is important to realize that interspecific interactions have highly variable outcomes. Our results suggest that interspecific interactions may vary from positive to negative along environmental gradients (e.g., latitudinally) and there is a whole spectrum of potential interactions even within a limited set of species. The very same species pair may compete intensively under certain environmental conditions but may yet benefit from and be actively attracted to each other under other circumstances.

The latitudinal variation in resident and migrant abundances suggests that at large geographical scale abundances vary independently of each other. Resident abundance is tightly associated with, and therefore very likely regulated by, climatic conditions during winter (Forsman and Mönkkönen 2003). On the breeding grounds, migrant abundance seemed to follow the geographic variation in productivity because abundance peaked at the corresponding latitudes with net primary productivity. Future research should aim to reveal the extent to which migrant abundance follows the breeding season productivity of sites. If resident and migrant abundances generally reflect the variation in productivity in winter and during the breeding season, respectively, it would be possible to foresee the effects of climate change on the structure of forest bird communities at large geographic scale.

Results from regional-scale analysis and from local experiments by and large fall in line with the hypothesis that at low resident abundance positive interactions between migrants and residents prevail, whereas at high resident abundance competition is more important. So far, there is no single experimental study across a wide gradient of abundance of migrant and resident birds to test the hypothesis presented in fig. 11.3 unequivocally, and the evidence for it is circumstantial. Such a study would considerably improve our understanding of the variation in the quality and intensity of interactions and their role in community organization. Likewise, we certainly need more study to address the interactions between short-distance and long-distance migrants. These two groups on average comprise about 70–80% of breeding pairs.

ACKNOWLEDGMENTS

We are grateful to the Academy of Finland for financial support (to JTF). R. Greenberg, Y. Haila, P. Marra, and an anonymous reviewer provided constructive criticism and comments, which helped us to improve our chapter.

LITERATURE CITED

Brawn, J. D., W. J. Boecklen, and R. P. Balda. 1987. Investigations of density interactions among breeding birds in ponderosa pine forests: correlative and experimental evidence. Oecologia 72:348–357.

Darwin, C. 1859. On the Origin of Species by Means of Natural Selection or the Preservation of Favoured Races in the Struggle for Life. Facsimile copy. Culture et Civilisation, Brussels (1969).

Dobzhansky, T. 1950. Evolution in the tropics. American Scientist 38:208–221.

Forsman, J. T., and M. Mönkkönen. 2001. Responses by breeding birds to heterospecific song and mobbing call playback under varying predation risk. Animal Behaviour 62:1067–1073.

Forsman, J. T., and M. Mönkkönen. 2003. The role of climate in limiting European resident bird populations. Journal of Biogeography 30:55–70.

Forsman, J. T., M. Mönkkönen, P. Helle, and J. Inkeröinen. 1998. Heterospecific attraction and food resources in migrants' breeding patch selection in northern boreal forest. Oecologia 115:278–286.

Forsman, J. T., J. Seppänen, and M. Mönkkönen. 2002. Positive fitness consequences of interspecific interaction with a potential competitor. Proceedings of the Royal Society of London, Series B, Biological Sciences 269:1619–1623.

Garcia, E. F. J. 1983. An experimental test of competition for space between blackcaps *Sylvia atricapilla* and garden warblers *Sylvia borin* in the breeding season. Journal of Animal Ecology 52:795–805.

Greenberg, R., V. Pravosudov, J. Sterling, A. Kozlenko, and V. Kontorshchikov. 1999. Tits, warblers, and finches: foliage-gleaning birds of Nearctic and Palearctic boreal forests. Condor 101:299–310.

Gustafsson, L. 1987. Interspecific competition lowers fitness in Collared Flycatchers *Ficedula albicollis:* an experimental demonstration. Ecology 68:291–296.

Haila, Y., and O. Järvinen. 1990. Northern conifer forests and their breeding bird assemblages. Pages 61–85 *in* Biogeography and Ecology of Forest Bird Communities (A. Keast, ed.). SPB Academic Publishing, The Hague.

Helle, P., and R. J. Fuller. 1988. Migrant passerine birds in European forest succession in relation to vegetation height and geographical position. Journal of Animal Ecology 57:565–579.

Herrera, C. M. 1978. On the breeding distribution patterns of European migrant birds: MacArthur's theme reexamined. Auk 95:496–509.

Hogstad, O. 1975. Quantitative relations between hole-nesting and open-nesting species within a passerine community. Norwegian Journal of Zoology 23:261–267.

Högsted, G. 1980. Prediction and test of the effects of interspecific competition. Nature 283:64–66.

Krebs, C. J. 1994. Ecology (fourth ed.). HarperCollins College Publications, New York.

Legendre. P., and L. Legendre. 1998. Numerical Ecology (second English ed.). Elsevier, Amsterdam.

Leonard, G. H. 2000. Latitudinal variation in species interactions: a test in the New England rocky intertidal zone. Ecology 81:1015–1030.

MacArthur, R. H. 1959. On the breeding distribution pattern of North American migrant birds. Auk 76:318–325.

MacArthur, R. H. 1972. Geographical Ecology. Harper and Row, New York.

Marshall, R. M., and M. E. Richmond. 1992. The Northeastern Forest Database: 54 years of Breeding Bird Census Data. Office of Migratory Bird Management, Fish and Wildlife Service, U.S. Department of Interior, Washington, D.C.

McLaughlin, J. F., and J. Roughgarden. 1993. Species interactions in space. Pages 89–98 *in* Species Diversity in Ecological Communities: Historical and Geographical Perspectives (R. E. Ricklefs and D. Schluter, eds.). University of Chicago Press, Chicago.

Minot, E. O. 1981. Effect of interspecific competition for food on breeding Blue and Great Tits. Journal of Animal Ecology 50:375–386.

Mönkkönen, M. 1992. Life history traits of Palaearctic and Nearctic migrant passerines. Ornis Fennica 69:161–172.

Mönkkönen, M. 1994. Diversity patterns in Palaearctic and Nearctic forest bird assemblages. Journal of Biogeography 21:183–195.

Mönkkönen, M., and J. T. Forsman. 2002. Heterospecific attraction among forest birds: a review. Ornithological Science 1:41–51.

Mönkkönen, M., J. T. Forsman, and P. Helle. 1996. Mixed species foraging aggregations and heterospecific attraction in boreal bird communities. Oikos 77:127–137.

Mönkkönen, M., R. Härdling, J. T. Forsman, and J. Tuomi. 1999. Evolution of heterospecific attraction: using other species as cues in habitat selection. Evolutionary Ecology 13:91–104.

Mönkkönen, M., and P. Helle. 1989. Migratory habits of birds breeding in different stages of forest succession: a comparison between the Palaearctic and the Nearctic. Annales Zoologici Fennici 26:323–330.

Mönkkönen, M., P. Helle, G. J. Niemi, and K. Montgomery. 1997. Heterospecific attraction affects community structure and migrant abundances in northern breeding bird communities. Canadian Journal of Zoology 75:2077–2083.

Mönkkönen, M., P. Helle, and K. Soppela. 1990. Numerical and behavioral responses of migrant passerines to experimental manipulation of resident tits (*Parus* spp.): heterospecific attraction in northern breeding bird communities? Oecologia 85:218–225.

Mönkkönen, M., and P. Viro. 1997. Taxonomic diversity in avian and mammalian faunas of the northern hemisphere. Journal of Biogeography 24:603–612.

Morse, D. H. 1989. American Warblers. Harvard University Press, Cambridge.

Newton, I., and L. C. Dale. 1996a. Relationship between migration and latitude among west European birds. Journal of Animal Ecology 65:137–146.

Newton, I., and L. C. Dale. 1996b. Bird migration at different latitudes in eastern North America. Auk 113:626–635.

O'Connor, R. J. 1981. Comparison between migrant and non-migrant birds in Britain. Pages 167–195 *in* Animal Migration (D. J. Aidley, ed.). Society for Experimental Biology Seminar Series 13. Cambridge University Press, Cambridge.

O'Connor, R. J. 1990. Some ecological aspects of migrants and residents. Pages 175–182 *in* Bird Migration (E. Gwinner, ed.). Springer-Verlag, Berlin.

Olson, R. J., J. M. O. Scurlock, S. D. Prince, D. L. Zheng, and K. R. Johnson (eds.). 2001. NPP Multi-Biome: NPP and Driver Data for Ecosystem Model-Data Intercomparison. Available on-line [www.daac.ornl.gov/] from the Oak Ridge National Laboratory Distributed Active Archive Center, Oak Ridge, Tenn.

Pianka, E. R. 1966. Latitudinal gradients in species diversity: a review of concepts. American Naturalist 100:33–46.

Pielou, E. C. 1979. Biogeography. Wiley-Interscience, New York.

Rabenold, K. N. 1979. A reversed latitudinal diversity gradient in avian communities of eastern deciduous forests. American Naturalist 114:275–286.

Reed, T. M. 1982. Interspecific territoriality in the chaffinch and great tit on islands and the mainland of Scotland: playback and removal experiments. Animal Behaviour 30:171–181.

Ricklefs, R. E. 1987. Community diversity: relative roles of local and regional processes. Science 235:167–171.

Rosenzweig, M. L. 1995. Species Diversity in Space and Time. Cambridge University Press, Cambridge.

Sæther, B.-E. 1983. Mechanism of interspecific spacing out in a territorial system of the Chiffchaff *Phylloscopus collybita* and the Willow Warbler *P. trochilus.* Ornis Scandinavica 14:154–160.

Sasvári, L., J. Török, and L. Tóth. 1987. Density dependence effects between three competitive bird species. Oecologia 72:127–130.

Sherry, T. W., and R. T. Holmes. 1988. Habitat selection by breeding American Redstarts in response to a dominant competitor, the Least Flycatcher. Auk 105:350–364.

Snow, D. W., and C. M. Perrins. 1998. The Birds of the Western Palearctic (concise ed.). Oxford University Press, Oxford.

Travis, J. 1996. The significance of geographical variation in species interactions. American Naturalist 148 (Supplement): 1–8.

von Haartman, L. 1957. Adaptation in hole-nesting in birds. Evolution 11:339–347.

Wallace, A. R. 1878. Tropical Nature and Other Essays. Macmillan, New York.

Wiens, J. A. 1989. The Ecology of Bird Communities. Cambridge University Press, Cambridge.

Willson, M. F. 1976. The breeding distribution of North American migrant birds: a critique of MacArthur (1959). Wilson Bulletin 88:582–587.

Wilson, E. O. 1992. The Diversity of Life. Harvard University Press, Cambridge.

Wright, D. H., D. J. Currie, and B. A. Maurer. 1993. Energy supply and patterns of species richness on local and regional scales. Pages 66–76 *in* Species Diversity in Ecological Communities: Historical and Geographical Perspectives (R. E. Ricklefs and D. Schluter, eds.). University of Chicago Press, Chicago.

12

KATRIN BÖHNING-GAESE

Influence of Migrants on Temperate Bird Communities

A Macroecological Approach

LONG-DISTANCE MIGRANTS MOVE between temperate and tropical environments and are suspected to have considerable impact on both communities. Recent studies suggest that long-distance migratory landbirds are mostly derived from tropical ancestors. I used a macroecological approach to study the impact of the tropical origin of long-distance migrants on the structure of temperate bird communities. The results suggest that long-distance migrants kept their habitat preferences when shifting the breeding ranges from the Tropics to the temperate regions. This results in North American long-distance migrants preferring forest habitats and European long-distance migrants choosing open habitats. These preferences led to relatively high species richness in North American forests and European open landscapes. As a possible result of living in relatively species-rich habitats, long-distance migrants had lower abundances, smaller range sizes, and more spatially clumped distributions than residents. The patterns demonstrate that temperate bird communities are shaped not only by local processes operating presently within North America and Europe but also by long-term influences originating in the Tropics.

INTRODUCTION

One of the central patterns in biology is that the temperate regions and the Tropics exhibit major differences in the structure of ecological communities and in the traits of species. Tropical bird communities have, for example, higher species richness and evenness; bird species are on average less abundant and less widespread, and they have smaller body and clutch sizes than their temperate relatives (Stresemann 1927–1934; Allee et al. 1949; Cook 1969; Stevens 1989; Brown and Lomolino 1998).

For studying the differences between the temperate regions and the Tropics, long-distance migrants are an especially interesting group. Although they breed in the temperate regions, some of their traits might have evolved in response to selection pressures operating on their tropical wintering grounds. Thus, they differ from temperate and tropical bird species by living in and being shaped by both environments. And, vice versa—migrants themselves can be expected to have major impact on the bird assemblages in both environments, owing to the yearly influx of large numbers of species and individuals into temperate and tropical bird communities.

An important prerequisite for understanding what impact migrants might have on temperate or tropical bird communities is to understand the historical origin of migrants. Are long-distance migrants derived from tropical ancestors that shifted the breeding ranges to the north or are they derived from temperate species that moved the wintering ranges to the Tropics? Have they historically "invaded" the temperate or the tropical bird communities?

In this chapter, I present a framework under which a surprising number of patterns in temperate bird communities can be understood. I used a macroecological approach because by working with large numbers of species, a lot of idiosyncratic variation among bird species is averaged and patterns become apparent that cannot be seen by studying a single species or genus (Brown 1995; Gaston and Blackburn 2000). First, I use a macroecological analysis of biogeographic ranges to provide evidence that long-distance migratory landbirds are, in general, derived from tropical ancestors. Then, I demonstrate the potential consequences of the tropical origin of long-distance migrants for the ecology of these species and for the structure of temperate bird communities as a whole. I demonstrate that the tropical origin of long-distance migrants influences their habitat choice on the temperate breeding grounds. This habitat choice has consequences for the bird diversity of different habitat types in the temperate regions. The patterns in bird diversity can then be connected with differences in the abundance, range size, and spatial distribution of migrants and residents.

This chapter forms a consistent macroecological analysis of the biogeography and ecology of all North American and European landbird species and their Neotropical and Afrotropical relatives. The analyses about the biogeography, habitat choice, abundance, and range size of migrants and residents have already been published and are only summarized below. In addition, I present new data and analyses on the biodiversity of temperate habitats types and on the spatial distribution of migrants and residents. In the discussion, I compare the macroecological results with other results obtained by using different methods and working with fewer bird species, usually at smaller spatial scales. I also offer alternative explanations for the macroecological patterns.

WHAT ARE THE ANCESTORS OF LONG-DISTANCE MIGRANTS?

One approach to test whether long-distance migrants are derived from tropical or from temperate ancestors is to compare the biogeographic distributions of migrants and residents. This section is based on the work of Böhning-Gaese et al. (1998). We analyzed the biogeographic distribution of all 625 North American and European landbird species over the major biogeographic regions of the earth (Nearctic, Neotropical, Palearctic, Afrotropical, Oriental, Australian [Brown and Gibson 1983]). This and all following analyses were restricted to species that live in and obtain their resources from terrestrial habitats and excluded all species that use marine, coastal, freshwater, and wetland habitats, because both groups have distinctly different ecological requirements.

The results indicate that residents and short-distance migrants are readily shared between the Nearctic and Eurasia. In contrast, long-distance migrants are generally restricted to either the New or the Old World. The only long-distance migratory landbirds that show long-distance migration within both the New and the Old World are the Barn Swallow (*Hirundo rustica*) and the Bank Swallow (*Riparia riparia*).

This pattern disproves the hypothesis of the temperate origin of long-distance migrants. Forty-nine species of residents and short-distance migrants have distributions covering North America and Eurasia. If these taxa had easily been able to develop long-distance migration from the north to the south, then we would expect that more than two species (4%) would have done so. Additionally, we would expect at least some species with long-distance migration on one continent and residency or short-distance migration on the other continent. These species, however, do not exist.

The biogeographic pattern suggests that long-distance migrants are derived from tropical ancestors and shifted the breeding ranges from the Tropics to the temperate regions. Further range expansions between the New and the Old World, by crossing the Bering Strait, for example, are apparently very difficult. Range expansions seem to be constrained by genetically controlled physiological and behavioral adaptations that are connected with long-distance migrations (see Irwin and Irwin, Chap. 3, this volume). For example, the Northern Wheatear (*Oenanthe oenanthe*) was able to colonize the North American continent from the west, crossing the Bering Strait, and from the east, crossing the Atlantic. However, the species still migrates back to its former nonbreeding grounds in tropical Africa and did not evolve a new migratory route within the Americas.

The origin of temperate short-distance migrants appears to be twofold. Short-distance migrants found in genera with long-distance migration seem to be derived from long-distance migrants that successively moved their wintering ranges north from the Tropics, for example, to the Mediterranean (e.g., *Lanius, Sylvia, Catharus*). Short-distance migrants in genera without long-distance migration seem to be derived from temperate residents that moved to the south in particularly cold winters (e.g., *Bombycilla, Regulus*).

CONSEQUENCES FOR HABITAT CHOICE OF LONG-DISTANCE MIGRANTS

If long-distance migrants are derived from tropical ancestors, we would expect Nearctic long-distance migratory landbirds to be derived from Neotropical ancestors and European long-distance migrants mostly from Afrotropical ancestors. These differences in origin should lead to differences in habitat choice of Nearctic and Palearctic long-distance migrants. The following analyses are based on work by Böhning-Gaese and Oberrath (2003).

If the Neotropics have been covered by more forest than the Afrotropics and if long-distance migration evolved in proportion to the amount of available habitat, then we would expect long-distance migration to have evolved in the Neotropics in genera that are found in more-forested habitat types than in the Afrotropics. If habitat choice is phylogenetically conservative then long-distance migrants might have taken this habitat choice with them while successively moving the breeding range from the Tropics to the temperate regions. Thus, we would expect long-distance migrants in North America to be found in more-forested habitat types than in Europe. Furthermore, we expect short-distance migrants in genera with long-distance migration to show the same difference in habitat choice between North America and Europe as seen in long-distance migrants, because they appear to be derived from long-distance migrants.

To test these hypotheses we quantified percentage of forest cover for the Neotropical and Afrotropical biogeographic regions for the present situation (1995), for the reconstructed potential natural vegetation, and for the Pleistocene (Adams and Faure 1997; FAO 1999) defining "forest" in a consistent way for the different time periods. During the Pleistocene, climate and forest cover fluctuated dramatically between glacial and interglacial periods. Forest cover at the Last Glacial Maximum (18,000 ^{14}C y.a.) is probably typical for the peak of a Pleistocene glacial period, and forest cover of the potential natural vegetation typical for the peak of a Pleistocene interglacial (Adams and Faure 1997). Using a classification system consistent for all groups (see below), we classified habitat choice of all North American and European long-distance migratory landbirds on the breeding and nonbreeding grounds and of their Neotropical and Afrotropical relatives.

The results demonstrate that percentage of forest cover in the present vegetation, the potential natural vegetation,

Table 12.1 Percentage of forest cover of the ice-free land mass of Europe and the Nearctic, Afrotropical, and Neotropical biogeographic region for the present situation (1995), the reconstructed potential natural vegetation (PNV), and the Last Glacial Maximum (LGM, 18,000 ^{14}C y.a.)

Region	1995	PNV	LGM
Europe	30.3%	94.5%	2.4%
Nearctic	24.9%	59.2%	30.9%
Afrotropics	21.6%	24.8%	2.6%
Neotropics	47.4%	51.3%	18.0%

Note: Forest cover of LGM is probably typical for the peak of a Pleistocene glacial period, forest cover of PNV for the peak of a Pleistocene interglacial (Adams and Faure 1997).

and during Pleistocene glacial and interglacial periods has been at least twice as high in the Neotropics than in the Afrotropics (table 12.1). Besides, this difference appears to stretch back to at least the Pliocene (Behrensmeyer et al. 1992; Cox and Moore 1993). The forested habitat types chosen by Nearctic long-distance migrants on their Neotropical wintering grounds are not only the same as those chosen on their North American breeding grounds, but they are also the same as those chosen by their Neotropical relatives. In contrast, European long-distance migrants were found on the European breeding grounds in habitats that were slightly but significantly more open than the habitats they chose on their Afrotropical nonbreeding grounds. As with North American migrants, their choice of habitat on their nonbreeding grounds did not differ from that of their Afrotropical relatives. As a result, North American long-distance migrants were found in habitats that were significantly more forested than those chosen by European long-distance migrants. Furthermore, even short-distance migrants in genera with long-distance migration showed the same differences in habitat choice between North America and Europe that long-distance migrants exhibited.

The difference in habitat choice of North American and European long-distance migrants might have been caused by general differences in habitat availability and choice between North America and Europe. This hypothesis was first tested directly by quantifying for the Nearctic and for Europe the percentage of forest cover for the present vegetation, for the potential natural vegetation, and for the Pleistocene. Europe was defined as including the British Isles and continental Europe eastward to the 30° line of longitude. If habitat choice is related to vegetation cover in the Nearctic and Europe, then the preference of Nearctic long-distance migrants for more-forested habitats should be correlated with a higher percentage of forest cover in the Nearctic as compared to Europe. In contrast to this prediction, percentage of forest cover in the Nearctic is slightly lower than in Europe in the present vegetation and considerably lower than in Europe in the potential natural vegetation (table 12.1). In the Pleistocene, percentage of forest cover was higher in the Nearctic than in Europe during glacial periods and lower in the Nearctic than in Europe during interglacial periods (table 12.1). Thus, habitat distributions during the Pleistocene are inconclusive, because the habitat choice of long-distance migrants is consistent with percentage of forest cover during glacial periods and inconsistent with percentage of forest cover during interglacial periods.

We then tested the hypothesis indirectly by comparing the habitat choice of long-distance migrants with the habitat choice of residents and short-distance migrants in genera without long-distance migration. If habitat availability on the breeding grounds determines habitat choice, then residents and short-distance migrants should show the same difference in habitat choice between the Nearctic and Europe shown by long-distance migrants. We classified habitat choice of all North American and European residents and short-distance migrants by using the same classification system as for long-distance migrants (see next section). The results demonstrate that residents and short-distance migrants in genera without long-distance migration did not show differences in habitat choice between North America and Europe. Thus, general differences in habitat availability and choice between North America and Europe do not seem to explain the difference in habitat choice of long-distance migrants between the two regions.

CONSEQUENCES FOR THE BIODIVERSITY OF TEMPERATE HABITATS

Long-distance migrants prefer forested habitats in North America and open habitats in Europe, whereas residents and short-distance migrants do not show a difference in habitat choice between the two continents. At the same time, the amount of habitat area available to forest and open-country species in North America and Europe is, at present, very similar (table 12.1). If North American long-distance migrants "invade" forested habitats, this should lead to a profound influx of species into forests and a relatively high bird diversity of North American as compared with European forests. In contrast, if European long-distance migrants prefer open landscapes, this should result in a relatively high bird diversity in open landscapes in Europe. Thus, with the present analysis I wanted to test if the habitat preferences described above led to differences in the relative bird diversity of different habitat types in the temperate regions.

When comparing species richness between North America and Europe, I work with relative instead of absolute species numbers, for two reasons. First, North America and Europe have different land areas, which leads, per se, to different values of absolute species richness. Second, it is necessary to account for differences in habitat availability between the two regions (see table 12.1).

Methods

I first classified the migratory behavior of all North American and European landbird species. I defined long-distance migrants as species with the center of the wintering range south of the Tropic of Cancer for North American species and south of the Sahara for European species (National Geographic Society 1987; Heinzel et al. 1995). The 11 European long-distance migrants that migrate exclusively to the Oriental biogeographic region were removed from the data set because they seem to be derived from Oriental ancestors and might show different habitat preferences than those of the Afrotropical migrants and their ancestors. Short-distance migrants were defined as regularly wintering south of the breeding range, with the center of the wintering range north of the Tropic of Cancer for North American species and north of the Sahara for European species (National Geographic Society 1987; Heinzel et al. 1995). In addition, I distinguished between short-distance migrants in genera with and without long-distance migration (short/longs and short/resids, respectively). Resident species were classified as species not regularly wintering south of their breeding range.

Habitat choice was classified by transforming for each species the description of habitat choice in major handbooks (Ehrlich et al. 1988, 1994) into a gradient reflecting habitat structure with values of 1 (closed forest), 2 (open forest), 3 (forest edge), 4 (savanna, orchards, gardens), 5 (shrub land), 6 (open country with single trees or shrubs), and 7 (open country without trees or shrubs). Each species was assigned up to three different values that were then averaged. For example, the Blackbird (*Turdus merula*), described as breeding in "a wide range of areas with trees, bushes, from deep forests to inner cities" (Ehrlich et al. 1994:366) was given the values 1, 3, and 4, the average being 2.67. I defined species with a mean habitat value ≤ 3 as "forest species" and species with a mean habitat value >3 as "open-country species."

Results

The results confirm that in North America, forests were relatively rich in bird species whereas in Europe, the open landscape was comparatively species-rich (fig. 12.1). In North America, 46.1% (172) of the 373 landbird species were classified as forest species, 53.9% (201) as open-country species. In Europe, only 34.2% (82) of 240 species were defined as forest species, but 65.8% (158) as open-country species. The difference in relative species numbers is highly significant (fig. 12.1; χ^2 test: $\chi^2 = 8.6$, d.f. $= 1$, $p = 0.0034$).

This difference in species number can be attributed to differences in the habitat choice between long-distance migrants and short/longs. In North America, 58.9% (86) of the

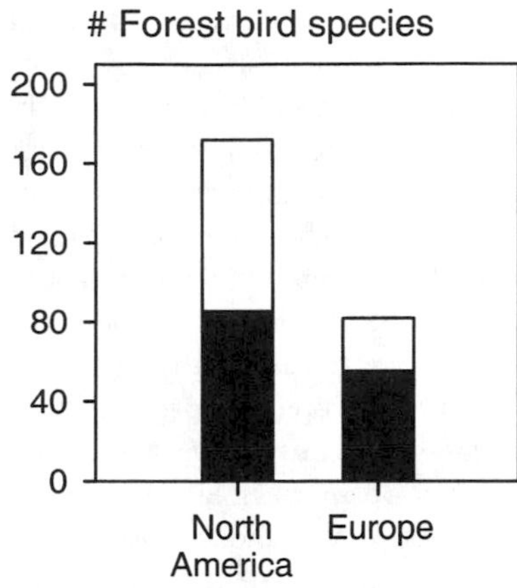

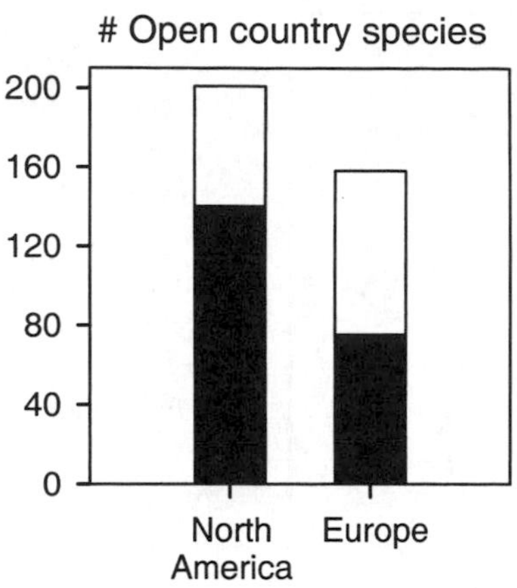

Fig. 12.1. Number of forest and open-country bird species in North America and Europe. White bars indicate long-distance migrants and short-distance migrants in genera with long-distance migration (referred to as "short/longs"). Black bars indicate residents and short-distance migrants in genera without long-distance migration (referred to as "short/resids"). In North America, most long-distance migrants are forest birds; in Europe, most are open-country birds. This results in relatively high species richness in North American forests and in European open landscapes.

146 long-distance migrants and short/longs were forest, and only 41.1% (60) were open-country birds. In Europe 24.1% (26) of the 108 long-distance migrants and short/longs were defined as forest, 75.9% (82) as open-country birds (fig. 12.1; χ^2 test: $\chi^2 = 30.6$, d.f. $= 1$, $p < 0.0001$).

In contrast, residents and short/resids did not differ in habitat choice between North America and Europe. In North America, 37.9% (86) of the 227 residents and short/resids were classified as forest, 62.1% (141) as open-country birds. Similarly, in Europe, 42.4% (56) of the 132 residents and short/resids were forest, 57.6% (76) open-country birds (fig. 12.1; χ^2 test: $\chi^2 = 0.7$, d.f. $= 1$, $p = 0.40$). Thus, as shown in fig. 12.1, the long-distance migrants and short/longs (white bars) are added on top of a proportionately similar number of residents and short/resids (black bars) in the different habitat types in North America and Europe.

These results indicate that in North America, the influx of long-distance migrants into forests contributes substantially to the relatively high bird species richness of this habitat type. In contrast, in Europe, the relative poverty of bird species in forests seems to be caused by long-distance migrants avoiding forest. In Europe, long-distance migrants prefer open habitat types, resulting in a relatively high bird species richness of these open landscapes.

CONSEQUENCES FOR THE ABUNDANCE AND RANGE SIZE OF LONG-DISTANCE MIGRANTS

Both in North America and in Europe, therefore, long-distance migrants live in relatively species-rich habitat types and thus under relatively high levels of interspecific competition. One consequence might be that long-distance migrants are less abundant and less widespread than short-distance migrants and residents. The following analyses are based on work by Böhning-Gaese and Oberrath (2001).

To test the hypothesis we compared the abundance and range size of long-distance migrants, short-distance migrants, and residents in the Lake Constance region in central Europe, using data from the breeding bird atlas *Lake Constance.* The atlas provides data collected in 1980–1981 and in 1990–1992 by the "Ornithologische Arbeitsgemeinschaft Bodensee" about the regional abundance, local abundance, and range size of 151 breeding bird species at the scale of 303 grid squares of 2 × 2 km (Schuster et al. 1983; Bauer and Heine 1992; Böhning-Gaese and Bauer 1996; Böhning-Gaese 1997). In this study, local abundance was defined as the mean abundance in occupied squares, and range size as the number of occupied squares. Regional abundance was defined as the product of local abundance and range size. To test whether differences in the ecology, life history, or phylogenetic relationship could account for differences in the abundance and range size among long-distance migrants, short-distance migrants, and residents, we controlled for many variables (habitat, diet, nest type, nest site, position of the Lake Constance region relative to the geographic range of the species, body mass, egg mass, clutch size, number of clutches per year, number of eggs per year, seasonal start of the breeding period, age at maturity, incubation time, fledging time, phylogenetic relatedness [Bezzel 1985, 1993]).

The results demonstrate that long-distance migrants had lower regional abundance than short-distance migrants, and short-distance migrants had lower abundance than residents (fig. 12.2). The difference among the three groups was significant even when we controlled for other ecological and life history traits as well as for phylogeny. The same pattern was found for local abundance and for range size. Long-distance migrants had lower local abundance and smaller range sizes than short-distance migrants and residents.

CONSEQUENCES FOR THE SPATIAL DISTRIBUTION OF LONG-DISTANCE MIGRANTS

In addition to its effects on abundance and range size, a high level of interspecific competition might also cause differ-

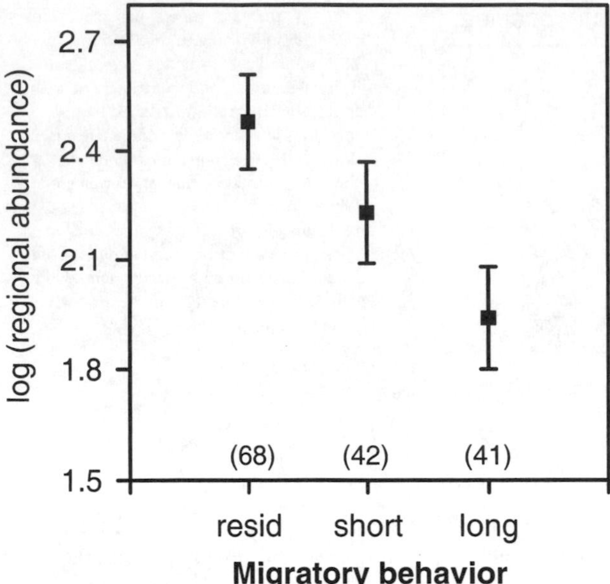

Fig. 12.2. Regional abundance of residents, short-distance migrants, and long-distance migrants in the Lake Constance region. Displayed are least square means (± 1 SE; JMP 1995) controlling for the significant effects of habitat, diet, nest site, position in the geographic range, and log (body mass) and for the nonsignificant effect of year. Values in parentheses are number of species per group.

ences in the spatial distribution among long-distance migrants, short-distance migrants, and residents within their habitat types.

To measure the degree of spatial clumping I used the data from the aforementioned breeding bird atlas. The degree of spatial clumping was determined by comparing the observed spatial distribution of a species with a "null distribution" expected by chance (Hurlbert 1979). As "null distribution" I used the Poisson distribution (Wright 1991). The expected number of occupied squares under the Poisson distribution (exp_range) is calculated with the equation exp_range $= p \times [1 - \exp(-N/p)]$, where N = regional abundance and p = total number of squares (Sokal and Rohlf 1995:86). To calculate how strongly a species deviated from its expected range size I used a clumping index. The index was calculated as log[(expected range size − observed range size) + 1]. A clumping index with the value 0 indicated no clumping, that is, a random spatial distribution. The higher the clumping index, the fewer squares a species occupied relative to the number expected from its regional abundance.

To test the influence of migratory status, breeding habitat, body mass, and diet on clumping, I assigned the species to categories following Böhning-Gaese and Oberrath (2001). Thus, I tested whether migrants and residents showed differences in spatial clumping controlling for the potentially confounding influences of habitat distribution, body mass, and diet.

When testing the relationship between the independent variables and the clumping index, not all species could be used because species with intermediate abundance deviated from expected values much more than species with very low and very high regional abundance. To allow for sufficient variation in clumping, I considered only species with "intermediate abundance." I decided arbitrarily that species should be allowed to deviate by at least 100 squares from their expected range size. The minimum regional abundance with this deviation was 197 breeding pairs, the maximum 3,547 breeding pairs. Thus, all species that had a regional abundance from 197 to 3,547 breeding pairs in either of the two census periods were considered as being of intermediate abundance (N = 58 species). However, even within this group of species, regional abundance influenced the variation in clumping. To control for those regional-abundance effects, I included a term for log(regional abundance) and for (log[regional abundance])^2 in the analysis.

In addition, the analysis contained a year term because the bird atlas data had been collected in two census periods, and thus the clumping index could be calculated for each period separately.

Migratory status had a significant influence on the degree of spatial clumping, with long-distance migrants and short-distance migrants being significantly more clumped than residents (fig. 12.3, table 12.2). That means that migrants were found in significantly fewer squares than expected by chance as compared to residents. The influence of migratory status was significant also when controlling for habitat, body mass, diet, regional abundance, and year effects (table 12.2). Thus, migrants had more spatially clumped distributions irrespective of the habitat they used.

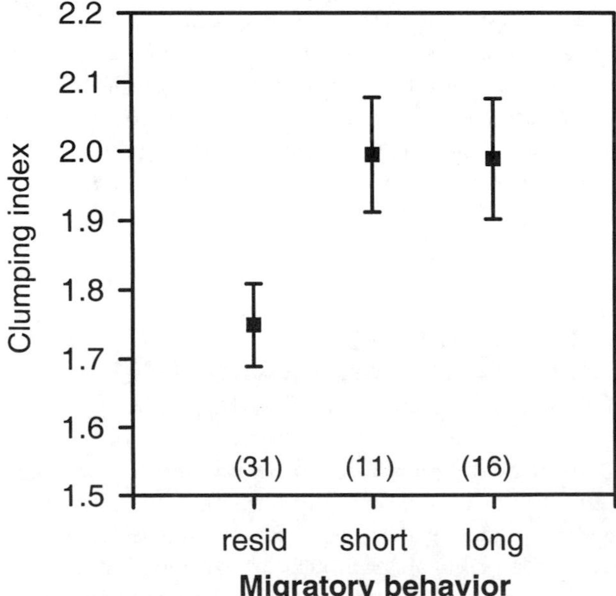

Fig. 12.3. Spatial clumping of residents, short-distance migrants, and long-distance migrants in the Lake Constance region. Displayed are least-square means (± 1 SE; JMP 1995) controlling for the significant effects of breeding habitat, log (body mass), diet, log (regional abundance), and log (regional abundance)^2, and for the nonsignificant effect of year. Values in parentheses are number of species per group.

Table 12.2 Results of a seven-factor ANCOVA analyzing the influence of migratory status, breeding habitat, body size, and diet on spatial clumping of 58 bird species with intermediate abundance in the Lake Constance region of central Europe

Variable	Model d.f.	Error d.f.	F	*p*	R^2
Model	11	104	6.45	<0.0001	40.6%
Migratory status	2		5.37	0.006	
Breeding habitat	3		8.04	<0.0001	
log(body mass)	1		5.80	0.018	
Diet	2		6.77	0.0017	
log(regional abundance)	1		9.51	0.0026	
[log(regional abundance)]^2	1		10.46	0.0016	
Year	1		0.08	0.78	

DISCUSSION

Ancestors of Long-Distance Migrants

The macroecological analysis strongly supports the tropical origin of long-distance migrants. In addition, the ancestral distribution of long-distance migrants has to be addressed by using an explicitly phylogenetic approach (Zink 2002). One way has been to use a molecular phylogeny of a number of taxa and to project the migratory behavior of the taxa into the phylogeny. All these analyses demonstrated that migratory behavior is phylogenetically very flexible, with gains and losses occurring rapidly (Böhning-Gaese and Oberrath 1999; Joseph et al. 1999; Lovette and Bermingham 1999; Voelker 1999, 2002; Helbig 2003). This approach is therefore of limited utility when trying to reconstruct ancestral distributions.

An alternative approach has been to project the biogeographic distribution of breeding and nonbreeding ranges into the molecular phylogeny and to reconstruct the most probable ancestral distributions. Using this approach, Joseph et al. (1999) demonstrated that the ancestral breeding and nonbreeding ranges of the genus *Charadrius* were Neotropical (South America). Outlaw et al. (2003) showed a Neotropical origin (Mexico / Central America) of the genus *Catharus* with migratory behavior evolving three times among the five Neotropical migrant species. However, more studies are needed to evaluate whether this pattern occurs more generally.

In addition, more indirect evidence exists that supports a tropical origin of long-distance migrants. A large proportion of long-distance migrants have close relatives in the Tropics. In North America 78% and in Europe 75% of the long-distance migrants have congeners that breed in the Tropics (Rappole 1995; Rappole and Jones 2002). Furthermore, the ecology and behavior of long-distance migrants are very similar to those of their tropical relatives (Jones 1995; Rappole 1995; Rappole and Jones 2002). Finally, species that have been introduced to new continents show range expansions mainly toward the north. The geographical ranges of these introduced species rarely extend to latitudes lower than those in their native ranges (Sax 2001).

Habitat Choice of Long-Distance Migrants and Biodiversity of Temperate Habitats

The macroecological analyses suggest that long-distance migrants kept their habitat preferences when shifting the breeding ranges from the Tropics to the temperate regions. Again, it would be important to approach the evolution of habitat choice by using an explicitly phylogenetic approach. To my knowledge, for birds this has been demonstrated only in the genus *Phylloscopus*, with species "tracking" their preferred habitat type during range expansions from the Himalayan region toward Siberia (Price et al. 1997).

The results of the present study are difficult to reconcile with the observation that single bird species did change their habitat choice over short periods of time (see examples in Cody 1985). However, patterns in the macroecological analyses are average effects over many species of birds. Thus, although it is possible that individual species can change their habitat choice (provided the genetic variance and selection pressure exists), species, on average, appear not to change habitat preferences during their evolutionary history. Furthermore, it is important to recognize that habitat choice is conservative only at certain phylogenetic levels. The present analysis suggests that, within genera, habitat choice is conservative and inherited from tropical ancestors. However, across genera, phylogenetic effects could not be detected (Böhning-Gaese and Oberrath 2003).

The macroecological analysis demonstrates that long-distance migrants were found in North America in significantly more-forested habitat types than in Europe. At the landscape scale, migrants in the eastern United States tended to be more abundant in landscapes with a greater proportion of forest habitats (Flather and Sauer 1996). Furthermore, studies at the habitat scale confirmed a preference of North American migrants for late and a preference of European migrants for early stages of forest succession (Whitcomb et al. 1981; Bilcke 1984; Helle and Fuller 1988; Mönkkönen and Helle 1989). Thus, the patterns in habitat choice seem to be invariant with regard to the spatial scale of the analysis.

Nevertheless, a variety of alternative hypotheses can be formulated to explain the difference in habitat choice of long-distance migrants between North America and Europe. In the present analysis, the reason why birds of more-forested habitats evolved long-distance migration in the Neotropics more than in the Afrotropics is reflected in the higher percentage of forest cover in the Neotropics (table 12.1); long-distance migrants might have evolved in proportion to the amount of available habitat. An alternative explanation is that the habitat the birds have to pass through on migration might have acted as a filter (Bell 2000;

Rappole and Jones 2002). In the New World, forests (among other habitat types) connect tropical and temperate regions, allowing range expansions of forest birds. In Africa, tropical and temperate regions are separated by deserts, semidesert, and other open habitats, thus promoting range expansions of open-county birds (Rappole and Jones 2002). Resolving the two hypotheses is difficult because they are not mutually exclusive. Within continents, vegetation cover on the tropical wintering grounds is often broadly similar to vegetation cover found on the migratory pathways.

Finally, the difference in habitat choice of long-distance migrants between North America and Europe could have been caused by habitat availability on the temperate breeding grounds. For example, differences in microhabitat choice and foraging behavior of foliage-gleaning birds between Canada and Russia can be explained by the Pleistocene vegetation history of North America and Eurasia (Greenberg et al. 1999). At a macroecological scale, however, comparisons of percentage of forest cover between North America and Europe for the present vegetation, for the potential natural vegetation, and for the Pleistocene either contradicted the "breeding ground" hypothesis or were inconclusive (table 12.1). Analysis of habitat choice of residents and short-distance migrants in genera without long-distance migration did not demonstrate differences between North America and Europe and, hence, also contradicted the "breeding ground" hypothesis.

Nonetheless, it is possible to develop scenarios that explain the present habitat choice of birds as a reflection of habitat availability on the breeding grounds. One scenario is that the habitat choice of long-distance migrants evolved at different time periods than that of residents. Alternatively, the habitat choice of North American birds evolved at different time periods than that of European birds. All these scenarios, however, require additional assumptions and are more complicated and less parsimonious than the "tropical origin" hypothesis. Eventually, only phylogenetic analyses of the biogeographic distribution and migratory behavior of related taxa, in connection with a carefully calibrated molecular clock and solid knowledge of the vegetation history of the breeding and wintering grounds, will be able to resolve these questions (see Steadman, Chap. 1, this volume; see Challenges for the Future, below).

Abundance, Range Size, and Spatial Distribution of Migrants

The present analysis demonstrated that long-distance migrants had, on average, lower abundance, smaller range sizes, and more clumped spatial distributions than residents. All the analyses were conducted at the scale of the Lake Constance region. However, when the analyses were repeated at the scale of Germany as a whole using data from Rheinwald (1993), the results were very similar (abundance and range size [Böhning-Gaese and Oberrath 2001]; spatial distribution [unpubl. data]). Furthermore, migrants are less abundant than residents in Britain (Cotgreave 1994). Finally, long-distance migrants have a more fragmented spatial distribution than residents within North America (Maurer and Heywood 1993). Thus, the patterns summarized in this study seem to hold true also at other spatial scales or, in the case of the spatial distribution, on other continents.

A proximate cause for the lower abundance, smaller range size, and more clumped spatial distribution of migrants might be stronger habitat specialization of migrants at a spatial scale not controlled for in this analysis. Thus, migrants might be more specialized with regard to microhabitat type or foraging behavior. This pattern might be especially pronounced for North American forest and European open-country species because they live in relatively species-rich habitats. Correspondingly, foliage gleaners in Nearctic forests were specialized on foliage and, in particular, conifer foliage, whereas Eurasian foliage gleaners were also frequently foraging on twigs, branches, or other substrates (Greenberg et al. 1999).

One possible mechanism causing narrow habitat niches of migrants is strong interspecific competition in relatively species-rich habitats. This might lead to migrants having narrower realized (or even fundamental) habitat niches than residents. A relationship between species richness of communities, the degree of interspecific competition, and niche breadth is a fundamental "rule" in ecology (Pianka 1994; Begon et al. 1996). However, this relationship has been very difficult to test and to prove (Wiens 1992).

A second mechanism causing narrow habitat niches of migrants might be that migrants might be able to track their "preferred" climatic conditions, habitat structures, and food resources through time. In contrast, residents, by wintering in the seasonal temperate regions, are selected for broader ecological niches. Thus, migrants might have narrower fundamental niches than residents. However, an ecomorphological analysis of 473 passerines comparing temperate residents, migrants, and tropical residents did not reveal a narrower "ecomorphological niche" of migrants in relation to residents (Ricklefs 1992). Similarly, an ecomorphological comparison of the same three groups within the genus *Sylvia* demonstrated that the "ecomorphological niche" of long-distance migrants was not narrower than that of residents. In contrast, the position of the "ecomorphological habitat niche" of migrants was intermediate between their temperate and tropical relatives, potentially allowing them to use both ends of a temperate-tropical habitat gradient (Böhning-Gaese et al. 2003).

A different explanation for the low abundance, small range size, and clumped spatial distribution of migrants might be the late arrival of long-distance migrants on the breeding grounds. Long-distance migrants start breeding on average 23 days later than short-distance migrants, short-distance migrants 9 days later than residents ($n = 150$ species [Böhning-Gaese, unpubl. data]). Late arrival might exclude migrants from good territories already occupied by residents or short-distance migrants, leading to low annual fecundity and low abundance. However, testing the influence of arrival date and annual fecundity on abundance and range size in multivariate analyses did not demonstrate a significant effect of arrival date or annual fecundity (Böhning-Gaese and Oberrath 2001).

CONCLUSIONS

Long-distance migrants, by "invading" temperate bird communities seem to "import" the structure of tropical communities. First, they considerably increase bird species richness in the temperate regions much beyond what would be sustainable at the respective latitudes. For many other groups of species, such as trees, amphibians, reptiles, and mammals, species richness is closely correlated with amount of available energy and drops continuously with latitude (Currie 1991). For birds, species richness in the temperate region is unusually invariant with latitude and drops only at very high latitudes (Currie 1991; Newton and Dale 1996; Hurlbert and Haskell 2003; Lemoine and Böhning-Gaese 2003).

The second way long-distance migrants "import" tropical community structure into temperate regions is through abundance and range size. The relatively high species richness of long-distance migrants in certain temperate habitats is associated with long-distance migrants having low abundance, small range sizes, and high degrees of spatial clumping. The same association between high species richness and small abundance and range size of the respective species is typically found in tropical bird communities (Stevens 1989; Bibby et al. 1992; Brown and Lomolino 1998).

To conclude, the tropical origin of long-distance migrants influences not only the ecology of long-distance migrants but also the structure of temperate bird communities as a whole. Thus, temperate bird communities are shaped not only by local processes operating presently within North America and Europe but also by long-term influences originating in the Tropics (Ricklefs and Schluter 1993). Long-distance migrants are therefore an interesting study system for understanding the fundamental patterns and processes operating in tropical and temperate ecological communities (Hurlbert and Haskell 2003).

CHALLENGES FOR THE FUTURE

The macroecological approach used in the present study has the advantage of revealing patterns in biogeographic distributions, habitat selection, abundance, and range size that could not have been detected by studying individual species or genera. The challenge, however, is to understand the mechanisms that cause these patterns, particularly the historical mechanisms that form today's communities. Current communities are shaped by an interplay of immigration, speciation, and extinction over millions of years in ever-changing environments. The macroecological approach has, so far, integrated little knowledge about the phylogenetic relationships of the species and the history of the environments. Thus, one of the challenges for future research is to combine macroecology, phylogenetic analyses, and historical biogeography (Webb et al. 2002).

A second challenge is understanding the current ecological mechanisms behind the macroecological patterns. Are the low abundance and small range size of long-distance migrants caused by interspecific competition and narrow realized niches, by selection for narrow fundamental niches, by late arrival and low annual fecundity, or by other factors? Here, the problem is that many of these processes can be tested experimentally only for a few species and at small spatial and temporal scales (if at all). However, the processes operating at small spatial and temporal scales are not necessarily the same as the ones that cause the macroecological patterns. Thus, a challenge for future research is to combine experimental tests with approaches that allow us to extrapolate results to other spatial and temporal scales, for example, computer simulations.

ACKNOWLEDGMENTS

I thank the chapter discussion group of the Abteilung Ökologie, R. Greenberg, and C. Both for very valuable comments on the manuscript. I also thank my principal collaborators in the work I have reviewed here: J. H. Brown, L. I. González-Guzmán, and R. Oberrath. Financial support was provided by the Deutsche Forschungsgemeinschaft (Bo 1221/3-1, Bo 1221/4-1).

LITERATURE CITED

Adams J. M., and H. Faure (eds.). 1997. Review and atlas of paleovegetation: preliminary land ecosystem maps of the world since the Last Glacial Maximum. Oak Ridge National Laboratory, Tennessee [www.esd.ornl.gov/projects/qen/adams4.html].

Allee, W. C., O. Park, A. E. Emerson, T. Park, and K. P. Schmidt. 1949. Principles of Animal Ecology. W. B. Saunders, Philadelphia.

Bauer, H.-G., and G. Heine. 1992. Die Entwicklung der Brutvogelbestände am Bodensee: Vergleich halbquantitativer Rasterkartierungen 1980/81 und 1990/91. Journal für Ornithologie 133:1–22.

Begon, M., J. L. Harper, and C. R. Townsend. 1996. Ecology: Individuals, Populations and Communities (third ed.). Blackwell Science, Oxford.

Behrensmeyer A. K., J. D. Damuth, W. A. DiMichele, R. Potts, H. D. Sues, and S. L. Wing. 1992. Terrestrial Ecosystems through Time: Evolutionary Paleoecology of Terrestrial Plants and Animals. University of Chicago Press, Chicago.

Bell, C. P. 2000. Process in the evolution of bird migration and pattern in avian ecogeography. Journal of Avian Biology 31:258–265.

Bezzel, E. 1985. Kompendium der Vögel Mitteleuropas: Nonpasseriformes—Nichtsingvögel. Aula Verlag, Wiesbaden, Germany.

Bezzel, E. 1993. Kompendium der Vögel Mitteleuropas: Passeres—Singvögel. Aula Verlag, Wiesbaden, Germany.

Bibby, C. J., N. J. Collar, M. J. Crosby, M. F. Heath, C. Imboden, T. H. Johnson, A. J. Long, A. J. Statterffield, and S. J. Thirgood. 1992. Putting Biodiversity on the Map: Priority Areas for Global Conservation. International Council for Bird Preservation, Cambridge, U.K.

Bilcke, G. 1984. Residence and non-residence in passerines: dependence on the vegetation structure. Ardea 72:223–227.

Böhning-Gaese, K. 1997. Determinants of avian species richness at different spatial scales. Journal of Biogeography 24:49–60.

Böhning-Gaese, K., and H.-G. Bauer. 1996. Changes of species abundance, distribution, and diversity in a central European bird community. Conservation Biology 10:175–187.

Böhning-Gaese, K., L. I. González-Guzmán, and J. H. Brown. 1998. Constraints on dispersal and the evolution of the avifauna of the Northern Hemisphere. Evolutionary Ecology 12:767–783.

Böhning-Gaese, K., and R. Oberrath. 1999. Phylogenetic effects on morphological, life-history, behavioural and ecological traits of birds. Evolutionary Ecology Research 1:347–364.

Böhning-Gaese, K., and R. Oberrath. 2001. Which factors influence the abundance and range size of Central European birds? Avian Science 1:43–54.

Böhning-Gaese, K., and R. Oberrath. 2003. Macroecology of habitat choice in long-distance migratory birds. Oecologia 137:296–303.

Böhning-Gaese, K., M. D. Schuda, and A. J. Helbig. 2003. Weak phylogenetic effects on ecological niches of *Sylvia* warblers. Journal of Evolutionary Biology 16:956–965.

Brown, J. H. 1995. Macroecology. University of Chicago Press, Chicago.

Brown, J. H., and A. C. Gibson. 1983. Biogeography. Mossby, St. Louis.

Brown, J. H., and M. V. Lomolino. 1998. Biogeography. Sinauer, Sunderland, Mass.

Cody, M. L. 1985. Habitat Selection in Birds. Academic Press, San Diego.

Cook, R. E. 1969. Variation in species density of North American birds. Systematic Zoology 18:63–84.

Cotgreave, P. 1994. Migration, body-size and abundance in bird communities. Ibis 136:493–495.

Cox, C. B., and P. D. Moore. 1993. Biogeography: An Ecological and Evolutionary Approach. Blackwell, Oxford.

Currie, D. V. 1991. Energy and large-scale patterns of animal and plant species richness. American Naturalist 137:27–49.

Ehrlich, P. R., D. S. Dobkin, and D. Wheye. 1988. The Birder's Handbook: A Field Guide to the Natural History of North American Birds. Simon and Schuster–Fireside, New York.

Ehrlich, P. R., D. S. Dobkin, D. Wheye, and S. L. Pimm. 1994. The Birdwatcher's Handbook: A Guide to the Natural History of the Birds of Britain and Europe. Oxford University Press, Oxford.

FAO. 1999. State of the World's Forests. Food and Agricultural Organization of the United Nations [www.fao.org/forestry/FO/SOFO/SOFO99/pdf/sofo_e/coper_en.pdf].

Flather, C. H., and J. R. Sauer. 1996. Using landscape ecology to test hypotheses about large-scale abundance patterns in migratory birds. Ecology 77:28–35.

Gaston, K. J., and T. M. Blackburn. 2000. Pattern and process in macroecology. Blackwell Science, Oxford.

Greenberg, R., V. Pravosudov, J. Sterling, A. Kozlenko, and V. Kontorschikov. 1999. Divergence in foraging behavior of foliage-gleaning birds of Canadian and Russian boreal forests. Oecologia 120:451–462.

Heinzel, H., R. Fitter, and J. Parslow. 1995. The Birds of Britain and Europe with North Africa and the Middle East. HarperCollins Publishers, London.

Helbig, A. J. 2003. Evolution of bird migration: a phylogenetic and biogeographic perspective. Pages 3–20 *in* Avian Migration (P. Berthold, E. Gwinner, and E. Sonnenschein, eds.). Springer, Berlin.

Helle, P., and R. J. Fuller. 1988. Migrant passerine birds in European forest successions in relation to vegetation height and geographical position. Journal of Animal Ecology 57:565–579.

Hurlbert, A. H., and J. P. Haskell. 2003. The effect of energy and seasonality of avian species richness and community composition. American Naturalist 161:83–97.

Hurlbert, S. H. 1979. Spatial distribution of the montane unicorn. Oikos 58:257–271.

JMP. 1995. JMP Statistics and Graphics Guide, Version 3.1. SAS Institute, Cary, N.C.

Jones, P. 1995. Migration strategies of Palearctic passerines in Africa. Israel Journal of Zoology 41:393–406.

Joseph, L., E. P. Less, and L. Christidis. 1999. Phylogeny and biogeography in the evolution of migration: shorebirds of the *Charadrius* complex. Journal of Biogeography 26:329–342.

Lemoine, N., and K. Böhning-Gaese. 2003. Potential impact of global climate change on species richness of long-distance migrants. Conservation Biology 17:577–586.

Lovette, I. J., and E. Bermingham. 1999. Explosive speciation in the New World *Dendroica* warblers. Proceedings of the Royal Society of London, Series B, Biological Sciences 266: 1629–1636.

Maurer, B. A., and S. G. Heywood. 1993. Geographic range fragmentation and abundance in neotropical migratory birds. Conservation Biology 7:501–509.

Mönkkönen, M., and P. Helle. 1989. Migratory habits of birds breeding in different stages of forest succession: a comparison between the Palearctic and the Nearctic. Annales Zoologici Fennici 26:323–330.

National Geographic Society. 1987. Field Guide to the Birds of North America. National Geographic Society, Washington, D.C.

Newton, I., and L. Dale. 1996. Relationship between migration and latitude among west European birds. Journal of Animal Ecology 65:137–146.

Outlaw, D. C., G. Voelker, B. Milá, and D. Girman. 2003. Evolution of long-distance migration in and historical biogeography of *Catharus* thrushes: a molecular phylogenetic approach. Auk 120:299–310.

Pianka, E. R. 1994. Evolutionary Ecology. HarperCollins, New York.

Price, T. D., A. J. Helbig, and A. D. Richman. 1997. Evolution of breeding distributions in the Old World leaf warblers (genus *Phylloscopus*). Evolution 51:552–561.

Rappole, J. H. 1995. The Ecology of Migrant Birds. Smithsonian Institution Press, Washington, D.C.

Rappole, J. H., and P. Jones. 2002. Evolution of Old and New World migration systems. Ardea 90:525–537.

Rheinwald, G. 1993. Atlas der Verbreitung und Häufigkeit der Brutvögel Deutschlands—Kartierung um 1985. Schriftenreihe des DDA 12, Bonn.

Ricklefs, R. E. 1992. The megapopulation: a model of demographic coupling between migrant and resident landbird populations. Pages 537–548 *in* Ecology and Conservation of Neotropical Migrant Landbirds (J. M. Hagan III and D. W. Johnston, eds.). Smithsonian Institution Press, Washington, D.C.

Ricklefs, R. E., and D. Schluter. 1993. Species Diversity in Ecological Communities: Historical and Geographical Perspectives. University of Chicago Press, Chicago.

Sax, D. F. 2001. Latitudinal gradients and geographic ranges of exotic species: implications for biogeography. Journal of Biogeography 28:139–150.

Schuster, S., V. Blum, H. Jacoby, G. Knötzsch, H. Leuzinger, M. Schneider, E. Seitz, and P. Willi. 1983. Die Vögel des Bodenseegebietes. Ornithologische Arbeitsgemeinschaft Bodensee, Konstanz.

Sokal, R. R., and F. J. Rohlf. 1995. Biometry (third ed.). W. H. Freeman, New York.

Stevens, G. C. 1989. The latitudinal gradients in geographic range: how so many species coexist in the tropics. American Naturalist 132:240–256.

Stresemann, E. 1927–34. Handbuch der Zoologie, Vol. 7 (2), Aves. Walter de Gruyter, Berlin.

Voelker, G. 1999. Molecular evolutionary relationships in the avian genus *Anthus* (Pipits: Motacillidae). Molecular Phylogenetics and Evolution 11:84–94.

Voelker, G. 2002. Systematics and historical biogeography of wagtails: dispersal versus vicariance revisited. Condor 104:725–739.

Webb, C. O., D. D. Ackerly, M. A. McPeek, and M. J. Donoghue. 2002. Phylogenies and community ecology. Annual Review of Ecology and Systematics 33:475–505.

Whitcomb, R. F., J. F. Lynch, M. K. Klimkiewicz, C. S. Robbins, B. L. Whitcomb, and D. Bystrak. 1981. Effects of forest fragmentation on avifauna of the eastern deciduous forest. Pages 125–205 *in* Forest Island Dynamics in Man-Dominated Landscapes (R. L. Burgess and D. M. Sharpe, eds.). Ecological Studies 41. Springer, New York.

Wiens, J. A. 1992. The Ecology of Bird Communities, Vol. 1, Foundations and Patterns. Cambridge University Press, Cambridge.

Wright, D. H. 1991. Correlations between incidence and abundance are expected by chance. Journal of Biogeography 18:463–466.

Zink, R. M. 2002. Towards a framework for understanding the evolution of avian migration. Journal of Avian Biology 33:433–436.

13

KEITH L. BILDSTEIN
AND JORJE I. ZALLES

Old World versus New World Long-Distance Migration in Accipiters, Buteos, and Falcons

The Interplay of Migration Ability and Global Biogeography

WE EXPLORE HOW MIGRATION ABILITY affects Old versus New World distributions and abundances of species of raptors by comparing biogeography and species richness in three falconiform genera, *Accipiter, Buteo,* and *Falco,* that differ considerably in their migration abilities. Together, these three genera (*Accipiter,* 50 species; *Buteo,* 28; *Falco,* 37) comprise 37% of all birds of prey and 38% of all migratory raptors. Overall, *Accipiter, Buteo,* and *Falco* differ considerably in wing structure and aerodynamics and typically use different flight mechanics while migrating. These differences, in turn, are reflected in differences in both migration tendencies and behavior. More than 80% of all buteos and falcons are complete, partial, or irregular migrants, whereas only 40% of all accipiters migrate. In addition, buteos and falcons are more likely to be transequatorial, long-distance migrants than are accipiters. Accipiters tend to be more island restricted, forest dependent, range restricted, and wholly tropical than buteos and falcons. Island-restricted accipiters tend to occur on fringing archipelago islands more than on truly isolated oceanic islands, whereas island-restricted buteos and falcons occur on both types of islands. Buteos probably occur on isolated oceanic islands because they tend to be wind drifted while migrating in flocks; falcons, probably because of their considerable overwater flight abilities. The hourglass configuration of continental landmasses and the north-south orientation of mountain ranges in the New World are associated with long-distance *Buteo* migration and speciation there. The vast open habitats of sub-Saharan Africa are associated with long-distance falcon migration and speciation there. The largely forested, fringing archipelagos of the South Pacific are associated with long-distance accipiter migration and speciation there. We describe a mechanism called *"migration dosing"* that we believe contributes to these relationships. Migration ability and continental geography are important emergent factors in determining Old versus New World patterns of raptor migration and species distribution.

INTRODUCTION

Migration abilities clearly affect the global distributions and abundances of birds. Several species of birds are naturally cosmopolitan, and many others have transcontinental distributions. Migration also has shaped the structures and relationships of regional avifaunas, including extensive large-scale, north-south relationships in both the Old and the New World (Darlington 1957; Snow 1978).

Although studies of avian migration and avian geography each have their own rich ornithological histories, investigations integrating these two important aspects of avian biology are generally lacking (see Brown and Lomolino 1998; Cox and Moore 2000; Mayr and Diamond 2001). To provide a framework for such studies, we explore how interactions between long-distance migration ability and global geography affect Old versus New World species distributions and abundances of the three most species-rich genera of diurnal birds of prey, *Accipiter, Buteo,* and *Falco.*

Raptors are ideal subjects for this study for several reasons. Falconiformes is one of the best-studied groups of birds, and its species geography and migration dynamics are reasonably well understood (Kerlinger 1989; Zalles and Bildstein 2000; Ferguson-Lees and Christie 2001). This group also exhibits a wide range of migration and dispersive behavior, including exceptional mobility that has produced some of the most capable and successful of all natural colonists (Stattersfield et al. 1998; Ferguson-Lees and Christie 2001).

METHODS

Taxonomy and Definitions

Taxonomy and common names follow del Hoyo et al. (1994). Migration is defined as directed, long-distance, recurring movement that alternates in direction and is temporally and spatially predictable. Complete migrants are species in which at least 90% of all individuals leave their breeding range during the nonbreeding season. Partial migrants are species in which fewer than 90% of all individuals migrate. Irruptive and irregular migrants are species whose migrations are less regular than those of complete and partial migrants (Kerlinger 1989). Migration geography largely follows Zalles and Bildstein (2000) and Ferguson-Lees and Christie (2001); flight strategies and mechanics follow Kerlinger (1989); ecology, distributions, and abundances largely follow Ferguson-Lees and Christie (2001). Ecoregions used to describe principal flyways follow Bailey (1989). Endemism follows Stattersfield et al. (1998) (range-restricted) and Bildstein et al. (1998) (country and regional).

Study Genera

ACCIPITER. Accipiters are largely forest-dependent, woodland, and woodland-savanna raptors found on six continents and on numerous oceanic islands. Most accipiters are small- to medium-bodied "true hawks" with relatively short, rounded wings and longish tails; they range in size from 62-g (male) Tiny Hawks (*Accipiter superciliosus fontanieri*) to 2,200-g (female) Northern Goshawks (*A. g. gentilis*). A few, including all three long-distance migrants, have relatively long pointed wings and medium-length tails. With at least 50 species, *Accipiter* represents 21% of Accipitridae and 16% of Falconiformes and is the most species-rich of all raptor genera (Brown and Amadon 1968; Wattel 1973; Kerlinger 1989; del Hoyo et al. 1994; Ferguson-Lees and Christie 2001).

BUTEO. Buteos are largely savanna and open-habitat, and sometimes woodland and forest, raptors found on five continents (not including Australia) and on many oceanic islands. Buteos are small- to mostly medium-bodied "buzzards" with broad wings and tails; they range in size from 200-g (male) Roadside Hawks (*Buteo magnirostris nattereri*) to 2,050-g (female) Upland Buzzards (*B. hemilasius*). With at least 28 species, *Buteo* represents 12% of Accipitridae and 9% of Falconiformes (Brown and Amadon 1968; Kerlinger 1989; del Hoyo et al. 1994; Ferguson-Lees and Christie 2001).

FALCO. Falcons are largely open-habitat raptors found on six continents and numerous oceanic islands. Falcons are small to medium-bodied raptors, with long pointed wings and moderately long tails, that range in size from 62-g (male) Seychelles Kestrels (*Falco araea*) to 2,100-g (female) Gyrfalcons (*F. rusticolus*). With at least 37 species, *Falco* represents 61% of Falconidae and 12% of Falconiformes (Brown and Amadon 1968; Cade 1982; Kerlinger 1989; del Hoyo et al. 1994; Ferguson-Lees and Christie 2001).

RESULTS

Characteristics and Geography of Raptor Migration

At least 183 of the world's 307 species of raptors migrate. Twenty species (7% of all raptors) are complete migrants (sensu Kerlinger 1989), 103 species (34%) are partial migrants, and 60 species (20%) are irruptive or local migrants (Zalles and Bildstein 2000). Most migration—particularly outside of the Tropics—is latitudinal (i.e., north-south). Raptors also migrate longitudinally (east-west) and altitudinally (Kerlinger 1989).

Raptors are lightly wing-loaded, and many species depend more heavily upon thermal- and slope-soaring to complete their migratory journeys than do many other species of birds (Brown and Amadon 1968; Kerlinger 1989; Berthold 2001). Most raptors migrate entirely by day, and most avoid long water crossings (Kerlinger 1989). A few species, including Levant Sparrowhawk (*Accipiter brevipes*), Amur Falcon (*Falco amurensis*), and Peregrine Falcon (*F. peregrinus*), migrate at night (Ferguson-Lees and Christie 2001). Although many raptors feed at least episodically when migrating, many larger species may forego eating for up to several weeks en route (Kerlinger 1989).

Both within and among species, migration tendencies increase with latitude, with high-latitude species that depend upon cold-blooded prey being particularly migratory. Overall, diurnal birds of prey are less migratory in the Southern than in the Northern Hemisphere (Kerlinger 1989); and, as is true for other groups of birds (Newton and Dale 1996), a lower percentage of raptors are migratory in Europe than in North America (Zalles and Bildstein 2000). Overall, 84 species migrate in Africa (61 excluding irruptive and local species); 75 (66) in mainland Asia, 45 (29) in South America, 44 (33) in North America, 39 (29) in the Pacific Islands, 38 (38) in Europe, and 22 (11) in Australia (Zalles and Bildstein 2000).

For at least portions of their migrations, most raptors migrate across broad fronts (Bednarz and Kerlinger 1989). Many species also engage in considerable small- or narrow-front migration (sensu Berthold 2001), usually along well-established migration corridors. Corridors typically occur along landscape features that include mountain ranges, river valleys, coastlines, and habitat discontinuities that act as "leading lines" (sensu Geyr von Schweppenburg 1963:192) to migration (Zalles and Bildstein 2000). Most corridors coalesce into one of five principal migration flyways, four of which are transequatorial (fig. 13.1). The main intercontinental flyways converge on narrow land bridges (such as Panama or Suez), or short water crossings (such as Gibraltar). Descriptions of these flyways, and the migrants that use them, follow.

TRANSAMERICAN FLYWAY. Each autumn, more than 5 million raptors travel along this 10,000-km overland system of corridors that stretches from central Canada to central Argentina (fig. 13.1). At least 32 species, including three accipiters, eight buteos, and five falcons, migrate along the flyway's central portion, the 4,000-km Mesoamerican Land Corridor, which connects North and South America. The bulk of the long-distance, transequatorial flight is dominated by world populations of Mississippi Kites (*Ictinia mississippiensis*), Broad-winged Hawks (*Buteo platypterus*), and Swainson's Hawks (*B. swainsoni*) and western North American breeding populations of Turkey Vultures (*Cathartes aura*) (Bildstein and Zalles 2000). Additional long-distance migrants include Rough-legged Hawks (*B. lagopus*) (northern third of the flyway only), Merlins (*Falco columbarius*), and Peregrine Falcons. No long-distance accipiter migrants use the flyway (fig. 13.2).

North to south, the Transamerican Flyway is tundra, mixed coniferous-deciduous forest, humid broad-leafed forest, steppe, dry steppe and shrub savanna, a mosaic of savanna, open woodland shrub, humid deciduous mixed forests, a mosaic of mixed rainforest and savanna, and steppe (Bailey 1989).

WESTERN EUROPEAN–WEST AFRICAN FLYWAY. Each autumn, approximately 200,000 raptors travel along the 5000-km overland system of corridors that stretches from Scandinavia to West Africa (fig. 13.1). At least 22 species, including two accipiters, two buteos, and five falcons, fly between western Europe and Africa along this flyway, which includes a short (<14-km) water crossing at the Strait of Gibraltar. The bulk of the long-distance flight is dominated by European breeding populations of Western Honey-buzzards (*Pernis apivorus*) and Black Kites (*Milvus migrans*). Additional long-distance migrants include Rough-legged

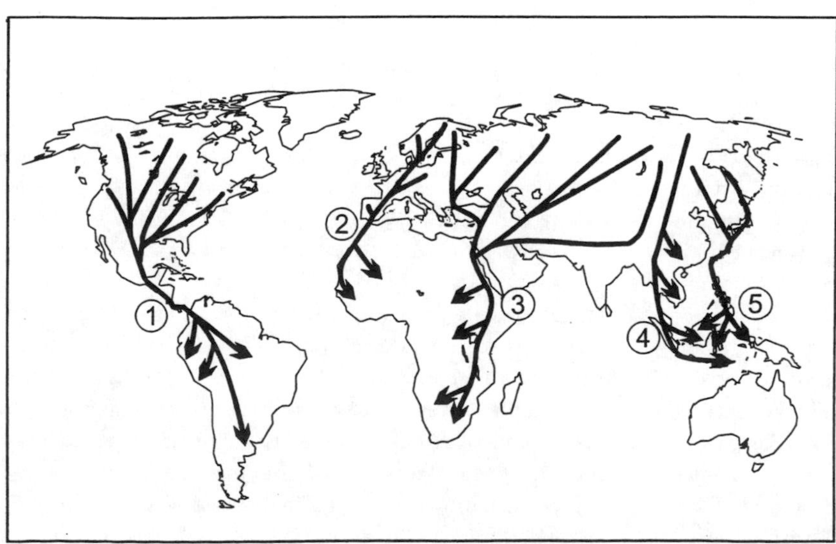

Fig. 13.1. Principal flyways used by accipiters, buteos, and falcons: (1) Transamerican Flyway, (2) Western European–West African Flyway, (3) Eurasian–East African Flyway, (4) East Asian Continental Flyway, and (5) East Asian Oceanic Flyway. (See text for details.)

Hawks (northern third of the flyway only), Merlins (northern half of the flyway only), Northern Hobbies (*Falco subbuteo*), and Peregrine Falcons. No long-distance accipiter migrants use the flyway (fig. 13.2).

North to south, the Western European–West African Flyway is tundra, coniferous or deciduous forest, Mediterranean scrub and desert, desert, arid scrub, and woodland savanna and rainforest (Bailey 1989).

EURASIAN–EAST AFRICAN FLYWAY. More than 1.5 million raptors travel along this 10,000-km system of largely overland corridors that stretches from eastern Scandinavia and western Siberia through the Middle East into southern Africa (fig. 13.1). At least 35 species, including three accipiters, three buteos, and nine falcons, use the flyway that follows the course of the Great Rift Valley and includes narrow water crossings at the Bosporus, Sinai Peninsula, and Babel-Mandeb Straits. Six Palearctic breeders, the Western Honey-buzzard, Black Kite, Levant Sparrowhawk, Eurasian (Steppe) Buzzard (*Buteo buteo vulpinus*), Lesser Spotted Eagle (*Aquila pomarina*), and Steppe Eagle (*A. nipalensis*), make up the bulk of the flight through the Middle Eastern part of the flyway. Additional long-distance migrants include Rough-legged Hawks (northern third of the flyway only), Lesser Kestrels (*Falco naumanni*), Western Red-footed Falcons (*F. vespertinus*), Merlins, Northern Hobbies, and Peregrine Falcons (fig. 13.2). The Amur Falcon may use portions of the flyway in spring.

North to south, the Eurasian–East African Flyway is tundra, coniferous forest, steppe, Mediterranean scrub and desert, woodland savanna, and a mosaic of open and closed woodland savanna (Bailey 1989).

EAST ASIAN CONTINENTAL FLYWAY. More than 1 million raptors travel along this 7,000-km mostly overland system of corridors that stretches from eastern Siberia to Southeast Asia and the Indonesian Archipelago, and that includes water crossings of 25–60 km at the Straits of Malacca, Sunda, Bali, and Lombok (fig. 13.1). At least 33 species, including four accipiters, three buteos, and seven falcons, migrate along portions of the flyway through eastern Asia. The bulk of the long-distance flight is dominated by Palearctic populations of Crested Honey-buzzards (*Pernis ptilorhynchus*), Grey-faced Buzzards (*Butastur indicus*), and Japanese Sparrowhawks (*Accipiter gularis*). Eurasian Buzzards (*B. b. japonicus*) migrate in the northern half of the flyway. Additional long-distance migrants include Rough-legged Hawks (northern third of the flyway only), Amur Falcons, Merlins, Northern Hobbies, and Peregrine Falcons (fig. 13.2).

North to south, this largely overland flyway is tundra, coniferous forest, steppe, desert and steppe or deciduous forest, and tropical rainforest with some savanna (Bailey 1989).

EAST ASIAN OCEANIC FLYWAY. As many as 150,000 to 500,000 raptors travel along this 5000-km, mostly overwater route that stretches from coastal eastern Siberia and Kamchatka to Japan, the Philippines, and into eastern Indonesia (fig. 13.1). At least 19 species, including four accipiters, two buteos, and four falcons, migrate along the flyway's main corridor, which extends from southern Japan through the Ryukyu Islands and Taiwan to the Philippines and eastern Indonesia. The bulk of the long-distance flight is dominated by Grey-faced Buzzards and Chinese Goshawks (*A. soloensis*). Additional long-distance migrants include Rough-legged Hawks (northern third of the flyway only), Eurasian Buzzards, Merlins, Northern Hobbies, and Peregrine Falcons (fig. 13.2). Long water crossings may restrict the numbers and variety of species that use this flyway.

North to south, the flyway is tundra, coniferous forest, deciduous forest, and tropical rainforest, interspersed throughout with short to long water crossings of up to 300 km (Bailey 1989).

Differences in the Migration Tendencies of *Accipiter, Buteo,* and *Falco*

Differences in the migration tendencies of accipiters, buteos, and falcons generally reflect anatomical differences among the three groups. Accipiters tend to have relatively short, rounded wings and long, narrow tails, whereas buteos have broad wings with considerable wing slotting, and relatively short, broad tails. Falcons have relatively long, high-aspect-ratio, pointed wings and longish tails (Kerlinger 1989). As a result, buteos are largely thermal- and slope-soaring migrants, whereas falcons are largely powered-flight migrants. Accipiters fall in between buteos and falcons in their use of soaring versus powered flight during migration.

Although thermal- and slope-soaring are difficult to achieve over water, direct, powered flight is particularly well suited for such travel, especially on tailwinds and crosswinds. As a result, buteos rarely undertake lengthy overwater crossings on migration, whereas falcons frequently do so.

Members of all three genera feed on migration. Accipiters and falcons tend to do so regularly, whereas buteos feed more episodically. All long-distance *Buteo* migrants and most, if not all, long-distance *Accipiter* migrants flock on migration, whereas only about half of the long-distance *Falco* migrants flock regularly (Kerlinger 1989; Zalles and Bildstein 2000; Ferguson-Lees and Christie 2001) (see table 13.1 and the Appendix to this chapter).

ACCIPITER. With only two complete migrants (4% of all accipiters) and 13 (26%) partial migrants, accipiters are one of the least migratory of all raptor genera (table 13.1). Most migrants in the genus, including the Western Hemisphere Sharp-shinned Hawk (*Accipiter striatus*), North American Cooper's Hawk (*A. cooperii*), Eurasian Shikra (*A. badius*), and Eurasian Sparrowhawk (*A. nisus*), are continentally based, short- and medium-distance, partial migrants that use undulating or flap-glide flight (sensu Kerlinger

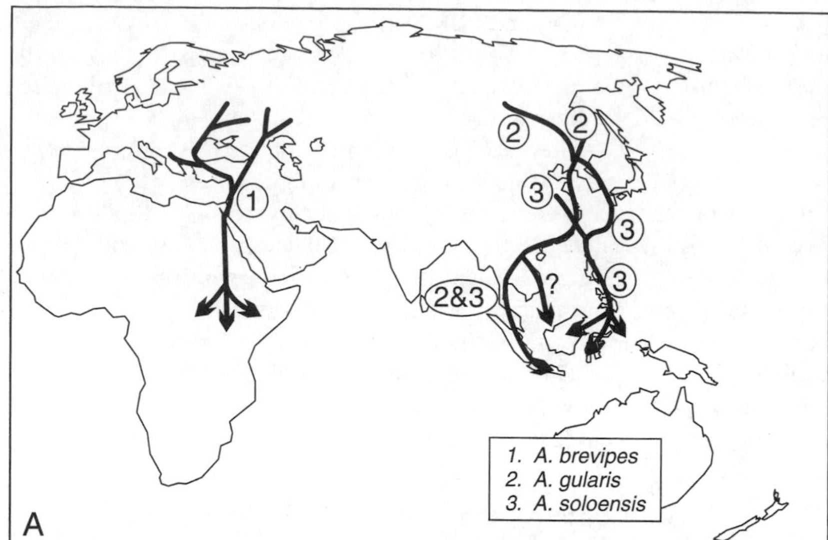

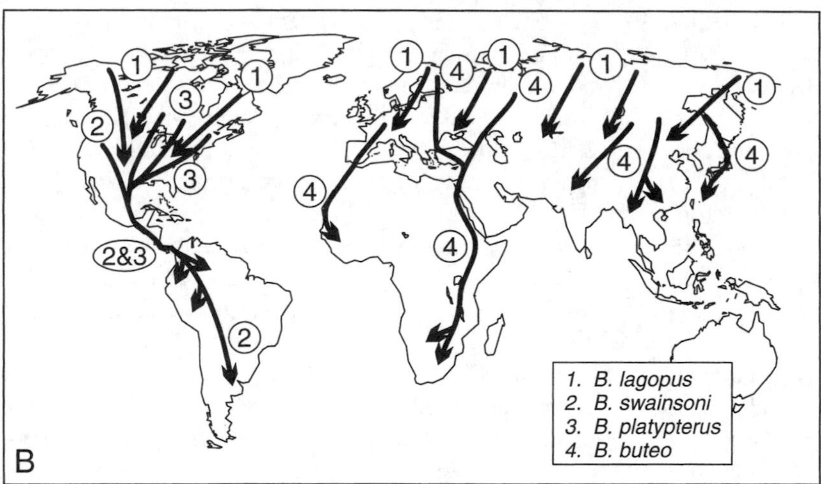

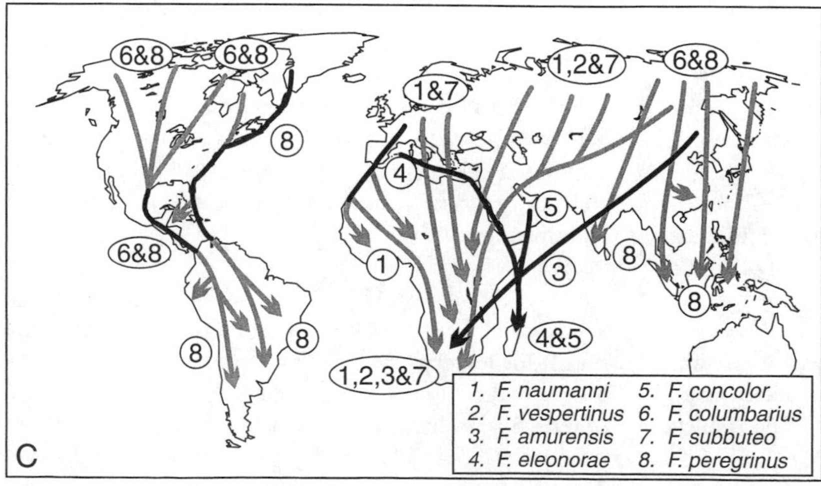

Fig: 13.2. Long-distance migration in accipiters (A), buteos (B), and falcons (C). Numbers refer to principal routes used by long-distance migrants.

Table 13.1 Migration characteristics of *Accipiter, Buteo,* and *Falco*

Number of species (%)	*Accipiter*	*Buteo*	*Falco*
Total in genus[a]	50	28	37
Migratory[b]	20 (40%)	25 (89%)	30 (81%)
Complete migrants[b]	2 (4%)	3 (11%)	6 (16%)
Partial migrants[b]	13 (26%)	14 (50%)	17 (46%)
Irruptive or irregular migrants[b]	5 (10%)	8 (29%)	7 (19%)
Long-distance migrants[c]	3 (6%)	4 (14%)	9 (24%)
Transequatorial migrants[d]	2 (4%)	3 (11%)	7 (19%)
Intermediate-water-crossing species[e]	3 (6%)	1 (4%)	8 (22%)
Long-water-crossing species[e]	1 (2%)	0 (0%)	8 (22%)
Flocking migrants[f]	3 (6%)	4 (14%)	6 (16%)
Old World species	39 (78%)	9 (32%)	29 (78%)
New World species	10 (20%)	17 (61%)	5 (14%)
Cosmopolitan species	1 (2%)	1 (4%)	4 (11%)
Wholly tropical species	25 (50%)	5 (18%)	5 (14%)
Island-restricted species[g]	21 (42%)	4 (14%)	6 (16%)
Restricted-range endemics[h]	11 (22%)	4 (14%)	2 (5%)
Country endemics[i]	9 (18%)	1 (4%)	6 (16%)

[a]Taxonomy based on del Hoyo et al. 1994.

[b]All migratory species, including complete migrants (species in which >90% of all individuals leave the breeding range outside of the breeding season), partial migrants (species in which ≤90% of all individuals leave the breeding range), and irruptive and irregular migrants (species whose movements are less regular than the above) (sensu Zalles and Bildstein 2000).

[c]Species in which more than 20% of all individuals (sometimes entire subspecies) regularly migrate more than 1,500 km.

[d]Long-distance migrants in which more than 20% of all individuals (sometimes entire subspecies) regularly migrate across the equator.

[e]Species that regularly undertake intermediate (25–100 km) or long (>100 km) water crossings during migration (as updated from Kerlinger 1989).

[f]Species that regularly migrate in flocks.

[g]Species found only on islands (including continental islands, oceanic islands, and archipelagos).

[h]Species whose historic and current breeding ranges are larger than 50,000 km^2 (based on Stattersfield et al. 1998).

[i]Species whose historic and current breeding ranges although restricted to a single country are larger than 50,000 km^2.

1989) interrupted by slope-soaring while migrating alone and in small groups of from two to six birds (Appendix to this chapter). Most accipiters appear to feed regularly en route (Shelley and Benz 1985), often on migrating passerines (Belopolski 1971; Rosenfield and Bielefeldt 1993; McCanch 1997; Bildstein and Meyer 2000).

Unlike "typical" *Accipiter* migrants, three Old World species, the Levant Sparrowhawk (*A. brevipes*), Chinese Goshawk (*A. soloensis*), and Japanese Sparrowhawk (*A. gularis*), are highly synchronous, gregarious long-distance migrants that sometimes travel in flocks of more that 1,000 birds (Appendix to this chapter).

Both the Japanese Sparrowhawk and Chinese Goshawk are transequatorial migrants. The Levant Sparrowhawk engages in considerable thermal soaring and nocturnal flapping flight (Stark and Liechti 1993; Shirihai et al. 2000). The Chinese Goshawk and the Japanese Sparrowhawk probably do so as well.

The Japanese Sparrowhawk, which migrates along the East Asian Continental Flyway, undertakes several over-water passages of more than 25 km in southern Malaysia and western and central Indonesia en route (fig. 13.2). The Chinese Goshawk also migrates along this flyway, as well as along the East Asian Oceanic Flyway, where it makes numerous water crossings of 300 km or less (fig. 13.2). Both species presumably fuel the undulating-flight portions of their migrations with a combination of preflight fat deposition and, at least for the Japanese Sparrowhawk, considerable feeding en route.

The only other long-distance accipiter migrant, the Levant Sparrowhawk, uses northern and central portions of the Eurasian–East African Flyway. This species, which occurs primarily in mixed and open habitats, is similar to the Chinese Goshawk in having long, pointed wings and shortish tails. Both species migrate synchronously and frequently in large soaring flocks (Shirihai et al. 2000). The Levant Sparrowhawk presumably manages to complete its 5,000- to 6,000-km journey across considerable inhospitable desert with a combination of solar-powered soaring flight and fat-powered undulating flight.

BUTEO. With three complete migrants (11% of all buteos) and 14 (50%) partial migrants, buteos are one of the most migratory of all raptor genera (table 13.1). Most migratory buteos, including the Red-tailed Hawk (*B. jamaicensis*) and nominate race of the Eurasian Common Buzzard (*B. b. buteo*), are short- to medium-distance, partial migrants that rely heavily upon both slope- and thermal-soaring to complete their migrations. Most travel alone or in small groups of up to ten birds (Appendix to this chapter). Although some feed regularly en route, others appear to do so only opportunistically.

Long-distance migrants in the group include the New World Broad-winged Hawk (*B. platypterus*) and Swainson's Hawk (*B. swainsoni*), the circumboreal Rough-legged Hawk (*B. lagopus*), the northernmost race of the Eurasian (Steppe) Buzzard (*B. b. vulpinus*), and an East Asian, *japonicus,* race of the Eurasian Buzzard (Appendix to this chapter).

Three of these, the Broad-winged Hawk, Swainson's Hawk, and Steppe Buzzard, are transequatorial migrants that depend on soaring flight to complete their journeys (Bildstein 1999). All three rarely undertake all but the shortest overwater crossings (Goodrich et al. 1996; England et al. 1997; Spaar and Bruderer 1997), and all three migrate only along flyways that are exclusively overland (or nearly so) with considerable open habitats (fig. 13.2). Broad-winged and Swainson's Hawks migrate along the overland Trans-american Flyway; the Steppe Buzzard migrates along the Eurasian–East African Flyway (Bildstein and Zalles 2001; Ferguson-Lees and Christie 2001) (fig. 13.2). *Buteo b. japonicus* migrates between northeastern and southeastern Asia along northern portions of the East Asian Continental Flyway, but not along the flyway's southern half, where passage would require water crossings of more than 25 km. *Japonicus* also migrates along the northern, largely overland portion of the East Asian Oceanic Flyway, but only irregularly farther south, where passage would require even longer water crossings.

All long-distance *Buteo* movements involve prolonged passage over desert or steppe, during which the birds soar almost exclusively, and all three transequatorial migrants overwinter in open habitats or in mosaics of open habitat and forest (see Bailey 1989; Ferguson-Lees and Christie 2001).

FALCO. With six complete migrants (16% of all falcons) and 17 (46%) partial migrants, *Falco* also is one of the most migratory of all raptor genera (table 13.1). Many migratory falcons, including the circumboreal Merlin, cosmopolitan Peregrine Falcon, Western Hemisphere American Kestrel (*F. sparverius*), Old World Saker Falcon (*F. cherrug*), and Eurasian Kestrel (*F. tinnunculus*), are short- to medium-distance migrants that travel alone or in small groups and rely on a combination of powered flight and feeding en route to complete their migrations.

Seven species of falcons undertake long-distance transequatorial migrations. Three of these, the Old World Lesser Kestrel, Northern Hobby, and Amur Falcon, are synchronous and moderately to highly gregarious migrants. Two additional species make long-distance movements entirely within the Northern Hemisphere.

Unlike buteos and accipiters, whose long-distance migrations are largely constrained by the presence of continental land masses and archipelagos, six of the seven transequatorial *Falco* migrants (Appendix to this chapter) fly long distances (i.e., >100 km) over open water when traveling between breeding and wintering areas. One species, *amurensis,* undertakes the longest regular overwater passage of any raptor, a crossing of the Indian Ocean of more than 4,000 km between southwestern India and tropical East Africa that includes nocturnal flight. This species, which breeds in central to northeastern Asia and overwinters in East Africa, first travels south to northeastern India and Bangladesh, where it apparently fattens (Ali and Ripley 1978) while staging for overland and then overwater flights across peninsular India and the Indian Ocean, respectively. The latter passage, which occurs in late November–early December, is assisted by northern-winter monsoonal tail winds, which are then in place. The species is an elliptical migrant (sensu Kerlinger 1989), and its return migration in spring occurs largely overland north and west of its outbound passage.

Eleonora's Falcon (*F. eleonorae*), which breeds in the Mediterranean Basin and overwinters in Madagascar, is another exceptional migrant. Rather than flying overland across Africa, most *eleonorae* migrate east, along the Mediterranean, and then south along the Red Sea, before "shortcutting" inland across Somalia to avoid circumnavigating the Horn of Africa. They then continue south along the coast of East Africa. The species presumably feeds en route.

All nine long-distance migratory falcons have been recorded at watchsites along at least one of the world's principal raptor flyways, and two species (*columbarius* and *peregrinus*) have been recorded at watchsites along all five flyways. Even so, falcons appear to be far less constrained by continental geography than accipiters and buteos (fig. 13.2).

Relationships between Migration Geography and Endemism

ACCIPITER. Two accipiters, the Japanese Sparrowhawk and Chinese Goshawk, are long-distance, transequatorial migrants. Both overwinter in Southeast Asia and on islands in the South Pacific Ocean and Andaman Sea. All 11 range-restricted endemic accipiters (Stattersfield et al. 1998) also occur in the region, and 10 of the 11 are restricted to areas east of Wallace's Line. The only range-restricted endemic west of Wallace's Line, the Nicobar Sparrowhawk (*A. butleri*), is generally considered an allospecies of the highly migratory Chinese Goshawk. Many of the remaining endemics differ considerably among themselves in plumage, size, and wing formulae (Wattel 1973), and some are considered allospecies of several superspecies. At least some of these endemics appear to have speciated via mechanisms suggested for many of the birds in the region (see Mayr and Diamond 2001). Of the nine country endemics (Appendix to this chapter), five are from this region, compared with three from Africa and one from Central and South America.

BUTEO. Three buteos, Broad-winged Hawk, Swainson's Hawk, and Steppe Buzzard, are long-distance, transequatorial migrants. The first two overwinter in Central and South America, and the third overwinters in Africa. Three of four range-restricted endemics in the genus, Galapagos Hawk (*B. galapagoensis*), Ridgway's Hawk (*B. ridgwayi*), and Archer's Buzzard (*B. archeri*), also occur in these regions. The fourth range-restricted endemic is the Hawaiian Hawk (*B. solitarius*). Mayr (1943) initially described the Hawaiian Hawk as a geographic representative of *swainsoni* as well, and two molecular phylogenies indicate that *solitarius* forms a clade with both *galapagoensis* and *swainsoni,* as well as with the Central and South American Short-tailed Hawk (*B. brachyurus* [Fleischer and McIntosh 2001; Riesing et al. 2003]). The only country endemic is the Madagascar Buzzard (*B. brachypterus*) (fig. 13.2).

FALCO. Seven *Falco* species, *naumanni, vespertinus, amurensis, eleonorae, concolor, subbuteo,* and *peregrinus,* are long-distance, transequatorial migrants. The first three overwinter exclusively or almost exclusively in Africa, the fourth in Madagascar, the fifth in both Africa and Madagascar, and the sixth in Africa and southern Asia. *Peregrinus* overwinters throughout the Southern Hemisphere. All seven species occur along the world's major flyways. However, many are broad frontal migrants that engage in long-distance water crossings. *Falco* has two range-restricted endemics, *punctatus* and *araea,* and six country endemics, *newtoni, moluccensis, zoniventris, hypoleucos, novaeseelandiae,* and *subniger.* Four of these are Malagasy, one is Moluccan, two are Australian, and one is from New Zealand.

DISCUSSION

Our analyses demonstrate clear and consistent differences in the global geography of long-distance migration in *Accipiter, Buteo,* and *Falco,* most, if not all, of which can be ascribed to intergeneric differences in migration ability. They also reveal substantial linkages between migration patterns and centers of endemism in each of the three genera (fig. 13.2).

In the New World, the hourglass configuration of continental landmasses, together with the north-south orientation of mountain ranges, has created opportunities for long-distance soaring migration and speciation in buteos. In sub-Saharan Africa, the vast open habitats are associated with long-distance falcon migration and speciation. And the largely forested, fringing archipelagos of the South Pacific region are associated with long-distance accipiter migration and speciation in this part of the world.

Taken together, these relationships suggest that long-distance migration ability and continental-scale geography and ecology are important determinants in migration geography and patterns of species distributions in raptors. Below we describe a speciation mechanism we call "*migration dosing*" that we believe contributes to species distribution in raptors, and we discuss this mechanism in terms of global biodiversity and conservation biology.

MIGRATION DOSING AND RAPTOR SPECIATION. One of the unintended consequences of large-scale post-breeding raptor migration is that each year some of the migrants, particularly inexperienced juveniles (Agostini and Logozzo 1995; Agostini et al. 2002; Hake et al. 2003; Thorup et al. 2003), are diverted from their principal migration routes, either by weather or by failed navigational systems. Once displaced, some of the misguided migrants are likely to wind up in areas that are geographically isolated from the species' traditional wintering areas. These vagrants (sensu Newton 2003) face three potential outcomes: death in the new-found area before breeding, successful reorientation and subsequent return to the breeding grounds in spring, or, along with other simultaneously diverted vagrants, successful breeding in the new area. The last possibility results in migration dosing.

As we see it, migration dosing is orchestrated vagrancy that occurs when "doses" of potential colonists in the form of diverted long-distance migrants: (1) simultaneously arrive in areas tangential to or beyond their major migration flyways, (2) subsequently fail to return to their normal destinations the following season, and (3) eventually speciate in isolation. For speciation to occur via this mechanism, dosing must be highly irregular and must occur at well-spaced intervals. Most likely, inexperienced juveniles constitute the majority of "dosed" propagules, as they often travel together, are especially prone to wind drift (Kerlinger 1989; Hake et al. 2003; Thorup et al. 2003), have a higher likelihood of developing new migratory habits (Viverette et al. 1996), and, for some species, are less likely than adults to leave overwintering areas the following spring (Ferguson-Lees and Christie 2001).

Unlike *adaptive radiation,* which involves the "simultaneous divergence of numerous lines from much the same adaptive type into different [and] also diverging adaptive zones" (Simpson 1953), migration dosing refers to changes from relatively *specialized* highly migratory continental forms into other specialized or, in some instances, *generalized* sedentary insular or continental forms. (See Brown and Kodric-Brown 1977, Snow 1978, and Hubbell 2001 for arguments supporting aspects of this type of speciation mechanism, and Grinnell 1922 for the role that accidentals or vagrants can play in avian biogeography.)

Although the phenomenon of migration dosing has yet to be appreciated as a speciation mechanism, conditions favoring it, including long-distance movements, migration in large flocks, vulnerability to wind drift, and high rates of vagrancy among migratory birds (Alerstam 1990; Newton 2003), as well as numerous examples of speciation that are most simply explained by this process, are widespread in the avian literature (Snow 1978; Bildstein 2004).

Long-distance, transequatorial migratory raptors are ideal candidates for migration dosing for several reasons. Most prey on resources that are "nutritionally substitutable," rather than on specific taxa, and this, together with the reduced intraguild predation and interference competition they are likely to encounter on most isolated islands, enhances the likelihood of groups of founders surviving there (Pimm 1991; Schulter 2000; Blackburn and Duncan 2001). Also, many raptors soar on migration, increasing the chance of "wind drift" en route, particularly among first-year birds (Kerlinger 1989), and especially within the tropical-cyclone regions that most transequatorial migrants travel through (NOAA 1988). Finally, most long-distance migrants travel in large flocks, substantially increasing the likelihood of simultaneous mis-transport of potential colonizing propagules (Williamson 1996).

Although migration dosing most likely occurs when raptors are blown off course and onto relatively isolated islands from which return may be difficult, we believe that it also happens on continental land masses when selection favors the elimination of migration behavior among certain populations of migratory species. This would include situations in which extrinsic environmental change or growing population densities within traditional breeding areas select against existing migration strategies (Berthold 1999).

Having established a theoretical framework for migration dosing, we now provide evidence to support its occurrence in *Accipiter, Buteo,* and *Falco.* The most detailed analysis of geographical differentiation in *Accipiter* concludes that the genus originated in Eurasia and spread from there (Wattel 1973). If this is true, then the most likely route of accipiter colonization in the South Pacific would be from northwest to southeast (i.e., from continental Asia outward across Pacific archipelagos). Although some might suggest that "faunal dominance" (sensu Mayr and Diamond 2001), together with the archipelago nature of the region, is responsible for the origin and distribution of the large number of *Accipiter* species in the South Pacific (fig. 13.3), we suggest that migration dosing also has played an important role in speciation events there.

Each autumn, more than 400,000 Chinese Goshawks and about 100,000 Japanese Sparrowhawks (Ferguson-Lees and Christie 2001; Chong 2000) migrate along the East Asian Continental and Oceanic Flyways into peninsular Malaysia, the Philippines, the Indonesian Archipelago, and their geographically associated islands. Both species, and in particular *soloensis,* often travel in multithousand-bird flocks. In most years, both species complete their southbound journeys in autumn with the aid of seasonal monsoonal northwesterly winds, and their northwesterly springtime returns are aided by easterly trade winds (Lam and Williams 1994; Krishnamurti 1996). Each autumn, however, their migrations take them through an active tropical-cyclone region (NOAA 1988), and during the springs of El Niño–Southern Oscillation events, strong westerly winds replace the region's easterly trade winds (Glantz 2001).

We suggest that migration geography and climatic conditions such as these provide the backdrop for infrequent but inevitable episodes of misdirected migrations during which flocks of *gularis* and *soloensis* are blown off course and onto islands in the region's archipelagos from which, because of adverse winds, they are unable to return to their breeding grounds. These dosed propagules then provide potential seed stock for regional endemism.

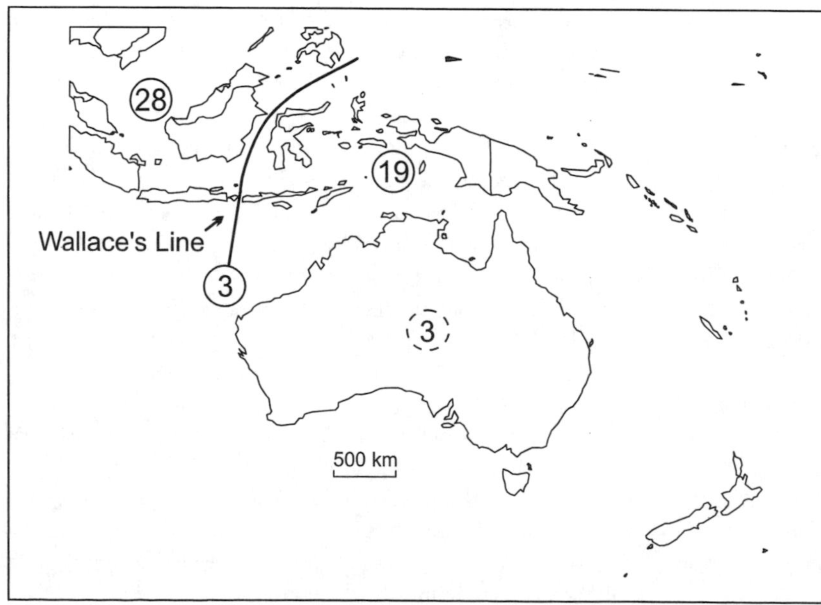

Fig. 13.3. Global distribution of the world's 50 species of accipiters relative to Wallace's Line. Twenty-eight species occur either only in the New World or west of Wallace's Line in the Old World, or both. Nineteen species occur only east of Wallace's Line, either in Wallacea or in Australasia, and three of these occur only in Australia. Three additional species occur on both sides of the line.

That migration dosing has contributed to the high-level *Accipiter* endemism in the region is supported by the coincidental lack of long-distance *Buteo* migration and endemism there, as well as by the relative paucity of long-distance *Accipiter* migration and endemism elsewhere in the world (figs. 13.2, 13.3). We do not claim that all or even most of the region's endemic accipiters resulted directly from migration dosing. Two of the region's endemics, the Nicobar Sparrowhawk and the Sulawesi Dwarf Sparrowhawk (*A. nanus*), however, almost certainly have resulted from this mechanism (see Ferguson-Lees and Christie 2001).

An additional example of migration dosing in *Accipiter* includes three Central and South American species (*chionogaster, ventralis,* and *erythronemius*) that are often considered allospecies of the partially migratory North American Sharp-shinned Hawk (*A. striatus*), a species that regularly migrates into Mexico and northern Central America (Ferguson-Lees and Christie 2001).

Examples of migration dosing in *Buteo* include the range-restricted endemic Galapagos Hawk and, possibly, Hawaiian Hawk, both of which are closely related to the highly migratory Swainson's Hawk (Riesing et al. 2003). *Falco* examples include the Oriental Hobby (*F. severus*) and the Australian Hobby (*F. longipennis*), which often are considered allospecies of the highly migratory Northern Hobby.

At the subspecies level, five nonmigratory Caribbean races of the Broad-winged Hawk, *B. platypterus* (*cubanensis, brunnescens, insulicola, rivierei, antillarum*), and six races of the Eurasian Buzzard (*bannermani* [Cape Verdes], *insularum* [Canaries], *rothschildi* [Azores], *arrigonii* [Corsica-Sardinia], *toyoshimai* [Izu and Bonin Islands], and *oshiroi* [Daito-jima]), occur on islands that are along or at the ends of migration flyways regularly used by these migratory buteos. Similarly, four nonmigratory, insular endemic subspecies of the Eurasian Kestrel (*canariensis* [Madeira and western Canary Islands], *dacotiae* [eastern Canaries], *neglectus,* [northern Cape Verde Island], and *alexandri* [southeast Cape Verde]), as well as four Caribbean races of the American Kestrel (*sparveroides, dominicensis, brevipennis,* and *caribaearum*) occur on islands along or near the ends of major migration corridors for these species.

Although it is possible that these insular subspecies represent examples of incipient migration-dosed speciation, we believe that they more likely reflect circumstances in which extensive and regular migratory intrusions by continental propagules act to prevent speciation (Brown and Kodric-Brown 1977). (See Rowlett 1980 and Hagar 1988 for why the latter may be true for *platypterus.*)

CONCLUSIONS AND FUTURE DIRECTIONS

Our comparisons of the migration abilities and species geography in *Accipiter, Buteo,* and *Falco* demonstrate substantial linkages between the two phenomena. Within these three groups of birds of prey, differences in long-distance migration ability appear to have shaped regional, continental, and global patterns of species richness. In fact, we suggest that migration ability and its potential for colonization events via *migration dosing* explains regional differences in species richness in the three groups at least as well as do regional differences in ecological resources and ecosystem function. This latter point has important implications for conservation biology.

Although conservation biologists continue to focus considerable attention on describing and protecting biological diversity at various geographic scales, the discipline remains remarkably ignorant of the natural forces that shape and maintain such diversity (Hubbell 2001). This ignorance continues to compromise our ability to protect existing levels of biological diversity as well as to reconstruct historic levels. To date, most investigations regarding the creation and maintenance of natural biological diversity have centered on the roles that local and regional resource availability, geography, ecosystem process, and dispersal play in shaping that diversity (see Gaston 2000, Primack 2002). In contrast, little if any attention has been paid to the role that long-distance migration might play in creating and maintaining geographic patterns of biological diversity.

Our findings suggest that migration plays an important role in shaping patterns of species richness in several genera of Falconiformes. It seems reasonable to conclude that the associations we have uncovered in raptors also may occur in other groups of birds, and we advocate investigating this possibility. Specifically, we recommend that researchers examine whether long-distance flocking migrants are more likely than nonflocking migrants and nonmigrants to have closely related insular and endemic continental forms and, if so, whether such forms are more likely to occur along and beyond the principal migration routes of the migrants in question. We also recommend that systematists working on molecular phylogenies consider migration dosing when assessing the relationships they uncover. Finally, we recommend that migration dosing also be considered when researchers attempt to explain geographic patterns of species diversity in other groups of migratory animals, including insects. We think that such investigations will yield valuable insights into the creation and maintenance of species diversity, and should improve our ability to protect that diversity.

APPENDIX: MIGRATION CHARACTERISTICS OF INDIVIDUAL SPECIES OF *ACCIPITER*, *BUTEO*, AND *FALCO*

Species	Type migrant[a]	Long-distance migrant[b]	Trans-equatorial migrant[c]	Water crossing behavior[d]	Flocking behavior[e]	World distribution[f]	Wholly tropical species[g]	Island-restricted species[h]	Endemic species[i]
Accipiter									
poliogaster	P	N	N	N	N	NW	N	N	N
trivirgatus	I			N	N	OW	N	N	N
griseiceps	NM					OW	Y	Y	CE
toussenelii	NM					OW	Y	N	N
tachiro	I			N	N	OW	N	N	N
castanilius	NM					OW	Y	N	N
badius	P	N	N	N	N	OW	N	N	N
butleri	NM					OW	Y	Y	RRS
brevipes	C	Y	N	I	R(>1000)	OW	N	N	N
soloensis	C	Y	Y	L	R(>100)	OW	N	N	N
francesii	NM					OW	N	Y	CE
trinotatus	NM					OW	Y	Y	CE
novaehollandiae	NM					OW	N	N	N
fasciatus	P	N	N	I	N	OW	N	N	N
melanochlamys	NM					OW	Y	Y	CE
albogularis	NM					OW	Y	Y	RRS
haplochrous	NM					OW	Y	Y	RRS
rufitorques	NM					OW	Y	Y	RRS
henicogrammus	NM					OW	Y	Y	RRS
luteoschistaceus	NM					OW	Y	Y	RRS
imitator	NM					OW	Y	Y	RRS
poliocephalus	NM					OW	Y	Y	CE
princeps	NM					OW	Y	Y	RRS
superciliosus	NM					OW	N	N	N
collaris	NM					NW	Y	N	N
erythropus	NM					OW	Y	N	N
minullus	I			N	N	OW	N	N	N
gularis	P	Y	Y	I	R(>100)	OW	N	N	N
virgatus	P	N	N	N	N	OW	N	N	N
nanus	NM					OW	Y	Y	RRS
erythrauchen	NM					OW	Y	Y	RRS
cirrhocephalus	I			N	N	OW	N	N	N
brachyurus	NM					OW	Y	N	RRS
rhodogaster	NM					OW	Y	Y	CE
madagascariensis	NM					OW	N	Y	CE
ovampensis	P	N	N	N	N	OW	N	N	N
nisus	P	N	N	I	O(<10)	OW	N	N	N
rufiventris	I			N	N	OW	N	N	N
striatus	P	N	N	S	O(<10)	NW	N	N	N
chionogaster	NM					NW	Y	N	N
ventralis	NM					NW	Y	N	N
erythronemius	NM					NW	N	N	N
cooperii	P	N	N	S	O(<10)	NW	N	N	N
gundlachi	NM					NW	Y	Y	CE
bicolor	P	N	N	N	N	NW	N	N	N
chilensis	P	N	N	N	N	NW	N	N	N
melanoleucus	P	N	N	N	N	OW	N	N	N
henstii	NM					OW	Y	Y	CE
gentilis	P	N	N	S	N	C	N	N	N
meyerianus	NM					OW	Y	Y	N
Buteo									
nitidus	P	N	N	N	N	NW	N	N	N
magnirostris	I			N	N	NW	N	N	N
lineatus	P	N	N	S	O(<10)	NW	N	N	N
ridgwayi	NM					NW	Y	Y	RRS
platypterus	C	Y	Y	S	R(>1000)	NW	N	N	N
leucorrhous	I			N	N	NW	N	N	N
brachyurus	P	N	N	N	O(<10)	NW	N	N	N
albigula	P	N	N	N	N	NW	N	N	N
swainsoni	C	Y	Y	S	R(>1000)	NW	N	N	N
albicaudatus	P	N	N	N	O(<10)	NW	N	N	N
galapagoensis	NM					NW	Y	Y	RRS

Species	Type migrant[a]	Long-distance migrant[b]	Trans-equatorial migrant[c]	Water crossing behavior[d]	Flocking behavior[e]	World distribution[f]	Wholly tropical species[g]	Island-restricted species[h]	Endemic species[i]
polyosoma	P	N	N	N	N	NW	N	N	N
poecilochrous	I			N	N	NW	N	N	N
albonotatus	P	N	N	N	O(<10)	NW	N	N	N
solitarius	I			N	N	HI	Y	Y	RRS
jamaicensis	P	N	N	S	O(>10)	NW	N	N	N
ventralis	I			N	N	NW	N	N	N
buteo	P	Y	Y	S	R(>1000)	OW	N	N	N
oreophilus	P	N	N	N	N	OW	N	N	N
brachypterus	I			N	N	OW	N	Y	CE
rufinus	P	N	N	N	O(<10)	OW	N	N	N
hemilasius	P	N	N	N	N	OW	N	N	N
regalis	P	N	N	N	O(<10)	NW	N	N	N
lagopus	C	Y	N	I	R(>10)	C	N	N	N
auguralis	P	N	N	N	O(<10)	OW	Y	N	N
augur	I	N	N	N	1(<10)	OW	N	N	N
archeri	NM					OW	Y	N	RRS
rufofuscus	I			N	N	OW	N	N	N
Falco									
naumanni	C	Y	Y	L	R(>1000)	OW	N	N	N
tinnunculus	P	N	N	I	O(<100)	OW	N	N	N
newtoni	NM					OW	N	Y	CE
punctatus	NM					OW	Y	Y	RRS
araea	NM					OW	Y	Y	RRS
moluccensis	NM					OW	Y	Y	CE
cenchroides	P	N	N	I	R(<10)	OW	N	N	N
sparverius	P	N	N	I	O(<10)	NW	N	N	N
rupicoloides	I			N	N	OW	N	N	N
alopex	P	N	N	N	N	OW	Y	N	N
ardosiaceus	P	N	N	N	N	OW	Y	N	N
dickinsoni	NM					OW	N	N	N
zoniventris	NM					OW	N	Y	CE
chicquera	P	N	N	N	N	OW	N	N	N
vespertinus	C	Y	Y	L	R(>1000)	OW	N	N	N
amurensis	C	Y	Y	L	R(>1000)	OW	N	N	N
eleonorae	C	Y	Y	L	R(>10)	OW	N	N	N
concolor	C	Y	Y	L	O(<10)	OW	N	N	N
femoralis	P	N	N	S	N	NW	N	N	N
columbarius	P	Y	N	L	O(<10)	C	N	N	N
rufigularis	I			N	N	NW	N	N	N
deiroleucus	I			N	N	NW	N	N	N
subbuteo	C	Y	Y	I	R(<100)	OW	N	N	N
cuvierii	I			N	O(<10)	OW	N	N	N
severus	P	N	N	S	O(<10)	OW	N	N	N
longipennis	P	N	N	I	N	OW	N	N	N
novaeseelandiae	P	N	N	I	N	OW	N	Y	CE
berigora	P	N	N	I	O(<10)	OW	N	N	N
hypoleucos	I			N	N	OW	N	N	CE
subniger	I			N	N	OW	N	N	CE
biarmicus	P	N	N	N	N	OW	N	N	N
jugger	I	N	N	N	N	OW	N	N	N
cherrug	P	Y	N	I	N	OW	N	N	N
rusticolus	P	N	N	L	N	C	N	N	N
mexicanus	P	N	N	N	N	NW	N	N	N
peregrinus	P	Y	Y	L	N	C	N	N	N
fasciinucha	NM					OW	N	N	N

[a]NM = nonmigrants, C = complete migrants, P = partial migrants, and I = irruptive or irregular migrants (based on Zalles and Bildstein 2000).

[b]Complete or partial migrants, 20% of whose populations (sometimes entire subspecies) regularly migrate more than 1,500 km.

[c]Complete or partial long-distance migrants, at least 20% of whose populations (sometimes entire subspecies) regularly migrate across the equator.

[d]N = none; S = short, <25 km; I = intermediate, 25–100 km; L = long, >100 km (Kerlinger 1989, with updates).

[e]N = never or hardly ever; O = occasional, often with heterospecifics; R = regular, typically with conspecifics; (maximum flock size) (Kerlinger 1989, with updates).

[f]OW = Old World species, NW = New World species (i.e., species found only in the Old World or New World, respectively), C = cosmopolitan species (i.e., species found in both the Old World and New World), HI = Hawaiian Islands.

[g]Species distributed only in the Tropics.

[h]Species found only on islands (including continental islands, oceanic islands, and archipelagos).

[i]N = Not an endemic, CE = Country endemics (i.e., species restricted to a single country), RRS = Restricted-range species (i.e., species whose historic and current breeding ranges are larger than 50,000 km^2) (based on Stattersfield et al. 1998).

ACKNOWLEDGMENTS

We thank our many *Hawk Aloft Worldwide* cooperators for providing previously unpublished data needed to construct our raptor migration flyways. Martin Riesing, Anita Gamauf, and their co-workers shared their unpublished molecular phylogeny of buteos. Hawk Mountain Sanctuary provided the intellectual climate in which to develop our ideas, and Peter Marra and Russ Greenberg organized the forum in which to present them. Manuscript referees Lloyd Kiff and Ian Newton and proceedings editors Marra and Greenberg and copyeditor Natasha Atkins helped improve our presentation considerably. We thank them all. This is Hawk Mountain Sanctuary contribution to conservation science number 100.

LITERATURE CITED

Agostini, N., C. Coleiro, F. Corbi, G. Di Lieto, F. Pinos, and M. Panuccio. 2002. Water-crossing tendency of juvenile Honey Buzzards during migration. Avocetta 26:41–43.

Agostini, N., and D. Logozzo. 1995. Autumn migration of Honey Buzzards in southern Italy. Journal of Raptor Research 29: 275–277.

Alerstam, T. 1990. Bird Migration. Cambridge University Press, Cambridge.

Ali, S., and S. D. Ripley. 1978. Handbook of the Birds of India and Pakistan, Vol. 1 (second ed.). Oxford University Press, Delhi and London.

Bailey, R. G. 1989. Ecoregions of the continents: map and explanatory supplement. Environmental Conservation 16:307–309.

Bednarz, J. C., and P. Kerlinger. 1989. Monitoring hawk populations by counting migrants. Pages 328–342 *in* Proceedings of the Northeast Raptor Management Symposium and Workshop (B. G. Pendleton, ed.). National Wildlife Federation, Washington, D.C.

Belopolski, L. O. 1971. Migration of Sparrowhawk on the Courland Spit. Notatki Ornithologiczne 12:1–12.

Berthold, P. 1999. A comprehensive theory of the evolution, control and adaptation of avian migration. Ostrich 70:1–11.

Berthold, P. 2001. Bird Migration: A General Survey (second ed.). Oxford University Press, Oxford.

Bildstein, K. L. 1999. Racing with the sun: the forced migration of the Broad-winged Hawk. Pages 79–102 *in* Gatherings of Angels: Migrating Birds and their Ecology (K. P. Able, ed.). Cornell University Press, Ithaca.

Bildstein, K. L. 2004. Raptor migration in the Neotropics: patterns, processes, and consequences. Ornitología Neotropical 15(Suppl.):83–99.

Bildstein, K. L., and K. Meyer. 2000. Sharp-shinned Hawk (*Accipiter striatus*). The Birds of North America, no. 482 (A. Poole and F. Gill, eds.). The Birds of North America, Inc., Philadelphia.

Bildstein, K. L., W. Schelsky, J. Zalles, and S. Ellis. 1998. Conservation status of tropical raptors. Journal of Raptor Research 32:3–18.

Bildstein, K. L., and J. I. Zalles. 2001. Raptor migration along the Mesoamerican Land Corridor. Pages 119–141 *in* Hawk-watching in the Americas (K. L. Bildstein and D. Klem Jr., eds.). Hawk Migration Association of North America, North Wales, Penn.

Blackburn, T. M., and R. P. Duncan. 2001. Determinants of establishment success in introduced birds. Nature 414:195–197.

Brown, J. H., and A. Kodric-Brown. 1977. Turnover rates in insular biogeography: effect of immigration on extinction. Ecology 58:445–449.

Brown, J. H., and M. V. Lomolino. 1998. Biogeography (second ed.). Sinauer Associates, Sunderland, Mass.

Brown. L., and D. Amadon. 1968. Eagles, Hawks and Falcons of the World. McGraw-Hill, New York.

Cade, T. J. 1982. The Falcons of the World. Cornell University Press, Ithaca.

Chong, M. 2000. Asian raptor migrations areas and population survey (Asian Raptor M. A. P. S.). Asian Raptors 1:12–15.

Cox, C. B., and P. D. Moore. 2000. Biogeography: An Ecological and Evolutionary Approach (sixth ed.). Blackwell Science, Oxford.

Darlington, P. J. 1957. Zoogeography: The Geographical Distribution of Animals. John Wiley and Sons, New York.

del Hoyo, J. A. Elliot, and J. Sargatal. 1994. Handbook of the Birds of the World, Vol. 2. Lynx Ediciones, Barcelona.

England, A. S., M. J. Bechard, and C. S. Houston. 1997. Swainson's Hawk (*Buteo swainsoni*). The Birds of North America, no. 265 (A. Poole and F. Gill, eds.). The Birds of North America, Inc. Philadelphia.

Ferguson-Lees, J., and D. A. Christie. 2001. Raptors of the World. Houghton-Mifflin, Boston.

Fleischer, R. C., and C. E. McIntosh. 2001. Molecular systematics and biogeography of Hawaiian avifauna. Studies in Avian Biology 22:51–60.

Gaston, K. J. 2000. Global patterns in biodiversity. Nature 405:220–227.

Geyr von Schweppenburg, H. F. 1963. Zur Terminologie und Theorie der Leitlinie. Journal of Ornithology 104:191–204.

Glantz, M. H. 2001. Currents of Change: Impacts of El Niño and La Niña on Climate and Society. Cambridge University Press, Cambridge.

Goodrich, L. J., S. C. Crocoll, and S. E. Senner. 1996. Broad-winged Hawk (*Buteo platypterus*). The Birds of North America, no. 218 (A. Poole and F. Gill, eds.). The Birds of North America, Inc., Philadelphia.

Grinnell, J. 1922. The role of the "accidental." Auk 39:373–380.

Hagar, J. A. 1988. Migration (Broad-winged Hawk). Pages 12–25 *in* Handbook of North American Birds, Vol. 5 (R. S. Palmer, ed.). Yale University Press, New Haven.

Hake, M., N. Kjellén, and T. Alerstam. 2003. Age-dependent migration strategy in honey buzzards *Pernis apivorus* tracked by satellite. Oikos 103:385–396.

Hubbell, S. P. 2001. The Unified Neutral Theory of Biodiversity and Biogeography. Princeton University Press, Princeton.

Kerlinger, P. 1989. Flight Strategies of Migrating Hawks. University of Chicago Press, Chicago.

Krishnamurti, T. N. 1996. Monsoons. Pages 512–515 *in* Encyclopedia of Climate and Weather (S. H. Schneider, ed.). Oxford University Press, Oxford.

Lam, C. Y., and M. Williams. 1994. Weather and bird migration in Hong Kong. Hong Kong Bird Report 1993:139–169.

Mayr, E. 1943. The zoogeographic position of the Hawaiian Islands. Condor 45:45–48.

Mayr, E., and J. Diamond. 2001. The Birds of Northern Melanesia: Speciation, Ecology, and Biogeography. Oxford University Press, Oxford.

McCanch, N. V. 1997. Sparrowhawk *Accipiter nisus* passage through the Calf of Man. Ringing & Migration 18:1–13.

Newton, I. 2003. The speciation and biogeography of birds. Academic Press, New York.

Newton, I., and L. C. Dale. 1996. Bird migration at different latitudes in eastern North America. Auk 113:626–635.

NOAA [National Oceanic and Atmospheric Administration]. 1988. Constructed World-wide Tropical Cyclones 1871–1988. Publication TD-9636. U.S. Department of Commerce, Washington, D.C.

Pimm, S. L. 1991. The Balance of Nature? University of Chicago Press, Chicago.

Primack, R. B. 2002. Essentials of Conservation Biology (third ed.). Sinauer Associates, Sunderland, Mass.

Riesing, J. J., L. Kruckenhauser, A. Gamuf, and E. Haring. 2003. Molecular phylogeny of the genus *Buteo* (Aves: Accipitridae) based on mitochondrial marker sequences. Molecular Phylogenetics and Evolution 27:328–342.

Rosenfield, R. N., and J. Bielefeldt. 1993. Cooper's Hawk (*Accipiter cooperii*). The Birds of North America, no. 75 (A. Poole and F. Gill, eds.). The Birds of North America, Inc., Philadelphia.

Rowlett, R. A. 1980. Migrant Broad-winged Hawks in Tobago. Journal of the Hawk Migration Association of North America 2:54.

Schulter, D. 2000. The Ecology of Adaptive Radiation. Oxford University Press, Oxford.

Shelley, E., and S. Benz. 1985. Observations of aerial hunting, food carrying and crop size of migrant raptors. Pages 299–301 *in* Conservation Studies on Raptors (I. Newton and R. D. Chancellor, eds.). International Council for Bird Preservation, Cambridge, U.K.

Shirihai, H., R. Yosef, D. Alon, G. M. Kirwan, and R. Spaar. 2000. Raptor Migration in Israel and the Middle East. International Birding and Research Center, Eilat, Israel.

Simpson, G. G. 1953. The Major Features of Evolution. Columbia University Press, New York.

Snow, D. W. 1978. Relationships between European and African avifaunas. Bird Study 5:134–148.

Spaar, R., and B. Bruderer. 1997. Optimal flight behavior in soaring migrants: a case study of migrating Steppe Buzzards *Buteo buteo vulpinus*. Behavioral Ecology 8:288–297.

Stark, H., and F. Liechti. 1993. Do Levant Sparrowhawks *Accipiter brevipes* also migrate at night? Ibis 135:233–236.

Stattersfield, A. J., M. J. Crosby, A. J. Long, and D. C. Wrege. 1998. Endemic Bird Areas of the World. BirdLife International, Cambridge, U.K.

Thorup, K., T. Alerstam, M. Hake, and K. Kjellén. 2003. Bird orientation: compensation for wind drift in migrating raptors is age dependent. Proceedings of the Royal Society of London (Supplement) Biology Letter 7:S1–S4.

Viverette, C. B., S. Struve, L. J. Goodrich, and K. L. Bildstein. 1996. Decreases in migrating Sharp-shinned Hawks (*Accipiter striatus*) at traditional raptor-migration watch-sites in eastern North America. Auk 113:32–40.

Wattel, J. 1973. Geographical Distribution in the Genus *Accipiter.* Nuttall Ornithological Club, Cambridge, Mass.

Williamson, M. 1996. Biological Invasions. Chapman and Hall, London.

Zalles, J. I., and K. L. Bildstein. 2000. Raptor Watch: A Global Directory of Raptor Migration Sites. BirdLife International, Cambridge, U.K., and Hawk Mountain Sanctuary, Kempton, Penn.

14

R. TERRY CHESSER

Seasonal Distribution and Ecology of South American Austral Migrant Flycatchers

SOUTH AMERICAN AUSTRAL MIGRANTS are birds that breed in southern parts of the continent during the austral, or southern, summer and migrate north, toward or into tropical South America, for the austral winter. A distinguishing characteristic of austral migration is the numerical predominance of a single family, the Tyrannidae or tyrant-flycatchers, which constitutes more than half of austral migrant passerine species. Detailed examination of the seasonal and geographical distribution of the more than 70 species of austral migrant flycatchers revealed a wide variety of distributional patterns. Examination of overall seasonal distributions indicated that highest species numbers of breeding migrants occur in northern Argentina and that highest species numbers of wintering migrants occur in and around southwestern Amazonia. Although flycatchers wintering in Amazonia undergo a substantial seasonal habitat change, microhabitats occupied in Amazonia tend to resemble those of their breeding grounds. Elevational ranges of many austral migrant flycatchers change seasonally, and routes of migration may differ between spring and fall. Migration tends to occur during September through November (southern spring) and March through May (southern fall). Latitude and temperature are the factors most closely associated with the breeding distribution of migrant flycatchers, and temperature has also been proposed as an important factor in wintering distributions. Austral migration presents unique opportunities for the study of migration, and potential future research directions are outlined.

INTRODUCTION

South American austral migrants are birds that breed in southern parts of the continent during the austral, or southern, summer and migrate north, toward or into tropical South America, for the austral winter. Although seasonal movements of south temperate breeding birds were noted as far back as Azara (1802–1805), details of these migrations have been slow in accumulating. Indeed, the first clues to the winter ranges of many tropical-wintering species were published only in the 1930s (Zimmer 1931–1955, 1938), and the brief discussions of Zimmer (1938) and Sick (1968) were until recently the only general discussions of the austral system.

The past two decades have seen a dramatic increase in interest in South American austral migration, including publication of the first overview of the entire system (Chesser 1994). Continuing the tradition of earlier observers in Argentina and Chile, contributions focused on migrant birds of Brazil (Willis 1988), Paraguay (Hayes et al. 1994), and Bolivia (Chesser 1997) have provided much valuable information on austral migrants, as have more general regional works (e.g., Belton 1984, 1985) and country monographs (e.g., Hilty and Brown 1986). Several detailed studies of the seasonal and geographical distribution of particular migrant species have appeared (e.g., Marini and Cavalcanti 1990; Marantz and Remsen 1991; Chesser and Marín 1994), as well as papers addressing the evolution of austral migration (e.g., Chesser and Levey 1998; Chesser 2000; Joseph 2003a) and studies relating distributions of austral migrants to climatic or other environmental variables (e.g., Joseph 1996, 1997, 2003b; Chesser 1998; Joseph and Stockwell 2000). Austral migrants, together with intratropical migrants, were also accorded a chapter in a recent comprehensive monograph on Neotropical birds (Stotz et al. 1996).

At least 220 species are South American austral migrants (Chesser 1994; Stotz et al. 1996) and more than half of these are passerines (table 14.1). The tyrant-flycatchers (Tyrannidae) are the largest group of migrants, and finches (Emberizidae), ducks (Anatidae), and swallows (Hirundinidae) are also prominently represented. Seasonal distributions of most austral migrants are characterized by partially overlapping breeding and wintering ranges, and distances migrated tend to be short relative to those of migrants breeding in the Northern Hemisphere (Chesser 1994) (table 14.1). The lack of east-west geographical barriers in South America (table 14.1) and the relatively large areas of land at tropical and subtropical latitudes (compared with temperate latitudes) appear to greatly influence these patterns of austral migration (Zimmer 1938; Chesser 1994).

The Tyrannidae (tyrant-flycatchers) constitute a remarkable one-third of austral migrants and more than one-half of austral migrant passerines, a predominance unparalleled in other migration systems (Chesser 1994). Key to this predominance has been their ability to evolve into ecological, behavioral, and morphological positions occupied in other migration systems by a variety of taxonomic groups. For example, many elaeniine austral migrant flycatchers are small, foliage-gleaning insectivores, convergent with Nearctic wood-warblers (Parulinae) and various migratory Old World groups. Similarly, aerial foraging tyrannids, such as those of the genus *Knipolegus*, are convergent with a variety of migratory Old World flycatchers. Ground-foraging migrants of the fluvicoline genera *Muscisaxicola*, *Lessonia*, and others are convergent with thrushes and chats (Turdinae), and the relatively large, generalized predators of the genus *Agriornis* with shrikes (Laniidae). Thus, tyrant-flycatchers represent a diversified sample of migratory birds, and in particular a highly representative sample of South American austral migrants.

In this chapter I discuss the biogeography and ecology of South American austral migration, focusing on the tyrant-flycatchers, and investigate what these reveal about the biogeography and ecology of migration generally. Subjects covered include patterns of seasonal and geographical distribution, elevational movements, migration routes, timing of migration, habitat and resource use patterns, the influence of environmental and climatic factors on migrant distributions, and future research directions.

SEASONAL DISTRIBUTION OF AUSTRAL MIGRANT TYRANNIDS

An extensive literature search indicated 85 species of south-temperate-breeding tyrannids believed to be migratory in at least some part of their range. Detailed specimen-based ex-

Table 14.1 Characteristics of South American passerine austral-tropical migrants compared to those of passerine Nearctic-Neotropical migrants

Characteristic	Passerine Austral-Neotropical migrants	Passerine Nearctic-Neotropical migrants
Number of migratory species	122	211
Predominant migratory group	Tyrannidae	Parulinae
Average distance migrated	9.2° of latitude	22.5° of latitude
Breeding and wintering ranges	Overlapping for most species	Disjunct for most species
Geographical barriers to migration	No major east-west barriers	Gulf of Mexico / Caribbean Sea
Predominant breeding habitats	Open country and scrub	Woodland and forest

Note: Taken from Chesser (1994). The number of Nearctic-Neotropical migrants was revised following Stotz et al. (1996).

amination of the seasonal and geographical distribution of these species (after Zimmer 1938; Lanyon 1978; Marini and Cavalcanti 1990; Marantz and Remsen 1991; Chesser and Marín 1994) revealed that 74 show evidence of seasonal movements at the species or subspecies level (Chesser 1995; Appendix to this chapter). Among these 74 migrants are 24 species of elaeniine flycatchers, including six species of the genus *Elaenia,* two of the genus *Serpophaga,* and three species of *Pseudocolopteryx;* 32 species of fluvicolines, including two species each of the genera *Xolmis, Neoxolmis,* and *Agriornis,* eight species of *Muscisaxicola,* both species of *Lessonia,* and four species of *Knipolegus;* 14 species of tyrannines, including three species of *Myiarchus,* both species of *Empidonomus,* and three species of *Tyrannus;* and four species of tityrines, including two species each of the genera *Pachyramphus* and *Tityra.*

Species for which no evidence of seasonal distributional change was detected included *Phyllomyias burmeisteri, Myiopagis caniceps, Elaenia flavogaster, Serpophaga nigricans, Pseudocolopteryx sclateri, Machetornis rixosa, Sirystes sibilator,* and *Pitangus sulphuratus,* all of which have been considered austral migrants (cf. Chesser 1994; Stotz et al. 1996). Although species-wide distributions of these taxa appear to undergo no seasonal changes, this does not mean that individual birds do not migrate.

As is typical of austral migrants, seasonal movements of tyrant-flycatchers tend to involve partially overlapping breeding and wintering ranges. Fifty-four of the 74 species of migrant tyrannids show this pattern, their breeding and winter distributions ranging from slightly to extensively overlapping. However, breeding and winter ranges of 20 austral migrant tyrannids are either disjunct (12 taxa) or more-or-less parapatric, showing no overlap but little or no disjunction (eight taxa).

Disjunct migrants travel the longest distances of any austral migrant tyrannids; the five longest-distance migrants are disjunct migrants, and nine of the 12 disjunct migrants are in the top 20 in distance migrated (table 14.2). Five of the eight parapatric migrants are in the top 20 in distance migrated, as are six migrants with partially overlapping breeding and wintering ranges. The fact that one of these latter migrants, *Muscisaxicola m. maculirostris,* was not listed as migratory by either Chesser (1994) or Stotz et al. (1996) gives some indication of the incipient state of our knowledge of austral migrant birds relative to those of the Northern Hemisphere.

Patterns of Distribution

Patterns of geographical distribution of individual austral migrant tyrannids are complex and varied. Distributions are to some extent unique for each taxon; nevertheless, distributions of many austral migrant tyrannids follow discernible general patterns, due to large-scale similarities in habitat, elevational range, and other characteristics.

Most efforts to classify patterns of austral migration have focused on a predominantly tropical wintering group and a predominantly temperate wintering group. In regional

Table 14.2 Twenty longest-distance austral migrant flycatchers, with average distance migrated and relationship between breeding and wintering ranges

Species or subspecies	Average distance migrated	Breeding/wintering ranges
Elaenia albiceps chilensis	33°08′	Disjunct
Muscisaxicola albilora	32°45′	Disjunct
Elaenia strepera	32°34′	Disjunct
Empidonomus v. varius	28°18′	Disjunct
Tyrannus s. savana	25°22′	Disjunct
Elaenia parvirostris	24°20′	Parapatric
Myiarchus s. swainsoni / ferocior	23°53′	Parapatric
Muscisaxicola flavinucha	23°01′	Parapatric
Empidonomus a. aurantioatrocristatus	22°08′	Disjunct
Muscisaxicola maclovianus mentalis	21°48′	Overlapping
Muscisaxicola capistratus	21°46′	Disjunct
Elaenia s. spectabilis	19°43′	Parapatric
Knipolegus hudsoni	19°34′	Disjunct
Attila phoenicurus	18°16′	Disjunct
Sublegatus modestus brevirostris	15°23′	Overlapping
Muscisaxicola m. maculirostris	14°50′	Overlapping
Myiodynastes maculatus solitarius	14°18′	Overlapping
Muscisaxicola frontalis	14°00′	Parapatric
Pyrocephalus r. rubinus	13°56′	Overlapping
Agriornis m. micropterus	12°56′	Overlapping

Note: Distance of migration was measured in degrees of latitude from the midlatitude of the breeding range to the midlatitude of the winter range.

works on Uruguay and Paraguay, for example, the distinction was drawn between species present in these countries mainly or exclusively during the breeding season and those present mainly or exclusively during winter, termed summer visitors and winter visitors, respectively (Gore and Gepp 1978), or northern austral migrants and southern austral migrants (Hayes et al. 1994). Using a continental perspective, Joseph (1996, 1997) codified these distinctions and introduced the terms South American Temperate-Tropical Migration, referring to species wintering in the warm humid lowlands of the Neotropics, and South American Cool Temperate Migration, referring to species that winter in cooler or temperate areas of all latitudes, including the tropical high Andes. Joseph (1996, 1997) noted that although species in the temperate-tropical category could be considered a unified group, the cool temperate category likely included a variety of distinct patterns or migration subsystems.

Stotz et al. (1996) took a different approach, classifying austral migrants by the southern extent of their breeding range. Five categories were identified: (1) migrants breeding as far south as Patagonia or the southern Andes; (2) migrants breeding as far south as the *chaco* or *pampas* regions of Argentina; (3) migrants breeding as far south as the Atlantic forests of southeastern Brazil, eastern Paraguay, and northeastern Argentina; (4) migrants breeding in the Andean forests of northwestern Argentina; and (5) the single species *Larus modestus* of the central Pacific coast. Despite the difference in approach, more than 90% of species were considered migrants of category 1 or 2, which correspond roughly to the cool temperate and temperate-tropical groups denoted above.

Biogeographical Regions

Classification of distributions of austral migrant flycatchers using recognized biogeographic regions (e.g., Hueck and Seibert 1972; Cabrera and Willink 1973; Brown and Gibson 1983; Nores 1987) or combinations of regions revealed a wide variety of migration patterns (Chesser 1995). Six major patterns characterized the breeding distributions (including regions of both permanent and summer residence) of five or more migrant tyrannids (table 14.3): (1) the *chaco*, which contains 11 taxa that breed predominantly in this region; (2) the Andean/Fuegian open alpine zone (nine species); (3) a combination of central South American woodland and forest regions, including the *chaco*, Atlantic forest, *yungas* forest, and *cerrado* zones (nine species); (4) the Atlantic forest of southeastern Brazil, northeastern Argentina, and eastern Paraguay (six species); (5) the *monte* of central Argentina (five species), and (6) a combination of Atlantic forest, the *chaco*, northwestern Argentine *yungas* forest, and the *caatinga* or *cerrado* zones of Brazil (five species).

Wintering areas (including regions of permanent and winter residence) shared by five or more migrant flycatchers corresponded to five major patterns (table 14.4): (1) the *chaco*, which contains eight taxa that winter predominantly in this region; (2) the Andean/Fuegian open alpine zone (eight species); (3) a combination of northern and north-central South American lowland woodland and forest regions, including much of Amazonia and the *cerrado* of Brazil (six species); (4) a combination of *cerrado* and subtropical/tropical forest zones from Paraguay or western Bolivia to eastern or northeastern Brazil, and extending into southern or southwestern Amazonia (five species); and (5) the Atlantic forest of southeastern Brazil, northeastern Argentina, and eastern Paraguay (five species).

Although most austral migrant flycatchers breed and winter in different regions or combinations of regions, many others remain in the same biogeographic area throughout the year. Twenty-six migratory tyrannids undergo essentially no seasonal biogeographical transition, their breeding and nonbreeding ranges occurring predominantly within the same biogeographical region or combination of regions. For example, most migrant flycatchers breeding mainly within the *chaco* also winter there, with six of 11 taxa undergoing a northward shift in distribution primarily within the region or around its borders. This same pattern holds for migrants breeding within the Andean/Fuegian alpine zone (eight of nine) and the Atlantic forest zone (five of six), as well as for widespread breeders of northern and central South America (three of four), breeders in a combined Atlantic forest and *cerrado* zone (three of three), and non-alpine Andean breeders (two of three), and nearly so for migrants breeding in central Argentine *monte* (two of five). Most of these taxa, as might be expected, have extensively overlapping breeding and wintering ranges, although migrants that move within the alpine life zone, which extends nearly the entire length of the continent, migrate longer distances and are characterized by completely or extensively nonoverlapping breeding and wintering ranges.

The remaining 48 taxa undergo a seasonal biogeographic transition, migrating between zones or among combinations of zones. Patterns represented by these taxa are notable for their diversity; indeed, 24 migratory tyrannids display unique distribution patterns. These include *Elaenia s. spectabilis*, which breeds primarily in the *chaco* border regions and winters mainly in Amazonia (there is also a separate population in northeastern Brazil); *Elaenia albiceps chilensis*, which breeds in the *Nothofagus* forest and Mediterranean zones of southern Chile and Argentina (also in the southern and western *chaco*) and has two apparently disjunct wintering ranges, one in western Amazonia and the other in eastern Amazonia and the *caatinga* and *cerrado* zones; *Elaenia strepera*, which breeds in Andean forest in Bolivia and Argentina and winters in a restricted area of northern Venezuela (see Marantz and Remsen 1991); *Neoxolmis rufiventris*, which breeds in Patagonia steppe and winters primarily in the *pampas* region of eastern Argentina; *Muscisaxicola macloviana mentalis*, which breeds in the alpine zone of Patagonia and the southern Andes and winters predominantly along the Pacific arid zone of Chile and Peru and in the lowlands of central Argentina; *Tyrannus albogularis*, southern populations of which breed in the *cerrado*

Table 14.3 Six regions or combinations of regions characteristic of five or more breeding austral migrant tyrannids, with lists of species breeding in those areas

1. *Chaco*	
Sublegatus modestus brevirostris	*Pseudocolopteryx dinelliana*
Suiriri s. suiriri	*Knipolegus striaticeps*
Elaenia s. spectabilis	*K. a. aterrimus*
Serpophaga munda	*Myiarchus t. tyrannulus*
S. s. subcristata	*Xenopsaris albinucha*
Inezia inornata	
2. Andean or Fuegian open alpine zone	
Muscisaxicola m. maculirostris	*M. c. cinereus*
M. maclovianus mentalis	*M. flavinucha*
M. capistratus	*M. frontalis*
M. r. rufivertex / pallidiceps	*Lessonia oreas*
M. albilora	
3. Widespread in central South American woodland and forest regions, including the *chaco*, Atlantic forest, *yungas* forest, and *cerrado* zones	
Camptostoma o. obsoletum / bolivianum	*Empidonomus v. varius*
Elaenia parvirostris	*E. a. aurantioatrocristatus*
Satrapa icterophrys	*Tyrannus s. savana*
Casiornis rufus	*Pachyramphus polychopterus spixii*
Myiarchus s. swainsoni / ferocior	
4. Atlantic forest of southeastern Brazil, northeastern Argentina, and eastern Paraguay	
Phyllomyias fasciatus brevirostris	*Knipolegus cyanirostris*
Elaenia mesoleuca	*Muscipipra vetula*
Contopus c. cinereus	*Attila phoenicurus*
5. *Monte* of central Argentina	
Stigmatura budytoides flavocinerea	*Neoxolmis rubetra*
Anairetes f. flavirostris	*Knipolegus hudsoni*
Xolmis coronatus	
6. Atlantic forest (pattern 4 above), but extending westward into the eastern *chaco* and in a corridor across Paraguay or adjacent Bolivia or Brazil, to the eastern Andean slopes, and south from there into the northwestern Argentine *yungas* forest zone; also extending northeastward into northeastern Brazil	
Myiopagis v. viridicata	*Lathrotriccus e. euleri / argentinus*
Euscarthmus m. meloryphus	*Pachyramphus v. validus / audax*
Myiophobus fasciatus flammiceps / auriceps	

Note: Breeding distributions are centered on these regions, although in many cases they extend beyond these regions in one or more directions. Breeding regions include areas of both summer and permanent residence.

zone of Brazil and Bolivia and winter on river islands in Amazonia; and *Xenopsaris albinucha*, which breeds mainly in the *chaco* and winters in various non-Amazonian lowlands of Bolivia and Brazil.

The other 24 migrant tyrannids show biogeographical transition patterns characteristic of two or three taxa. Apart from the obvious latitudinal differences associated with seasonal migratory movements, the feature uniting many of these patterns is a movement in winter into more humid biogeographical zones, primarily Amazonia. Some taxa (*Myiarchus s. swainsoni / ferocior, Empidonomus a. aurantioatrocristatus*) completely exchange their relatively dry breeding regions for humid Amazonian winter grounds, and others (*Elaenia parvirostris, Empidonomus v. varius, Tyrannus s. savana*) vacate a variety of dry and semihumid breeding regions and occupy a winter range composed again of a variety of regions, but predominantly Amazonia. Most taxa, however, exhibit overlapping breeding and winter ranges, and in winter simply vacate a relatively dry region (e.g., the *chaco*) at their southern breeding extremity, and seasonally occupy a humid region at the northern limit of their winter range. Exceptions to this pattern of relatively humid win-

Table 14.4 Five regions or combinations of regions characteristic of five or more wintering austral migrant tyrannids, with lists of species wintering in those areas

1. *Chaco*	
Suiriri s. suiriri	*Agriornis micropterus*
Serpophaga munda	*Agriornis murinus*
S. s. subcristata	*Knipolegus striaticeps*
Pseudocolopteryx dinelliana	*K. a. aterrimus*
2. Andean open alpine zone, extending in some species to open habitat at lower elevation	
Muscisaxicola m. maculirostris	*M. c. cinereus*
M. capistratus	*M. flavinucha*
M. r. rufivertex/pallidiceps	*M. frontalis*
M. albilora	*Lessonia oreas*
3. Widespread in northern or north-central South America	
Elaenia chiriquensis albivertex	*Megarynchus p. pitangua*
Pyrocephalus r. rubinus	*Myiodynastes maculatus solitarius*
Fluvicola pica albiventer	*Tyrannus m. melancholicu*
4. *Cerrado* and subtropical/tropical forest zones from Paraguay or western Bolivia to eastern and northeastern Brazil, extending northwest into southern or southwestern Amazonia	
Euscarthmus m. meloryphus	*Hirundinea ferruginea bellicosa*
Myiophobus fasciatus flammiceps/auriceps	*Pachyramphus v. validus/audax*
Cnemotriccus fuscatus bimaculatus	
5. Atlantic forest zone, principally in southeastern and eastern Brazil	
Phyllomyias fasciatus brevirostris	*Knipolegus cyanirostris*
Elaenia mesoleuca	*Muscipipra vetula*
Contopus c. cinereus	

Note: Wintering distributions are centered on these regions, although in many cases they extend beyond these regions in one or more directions. Wintering regions include areas of both nonbreeding and permanent residence.

tering grounds are taxa found in dry habitats throughout the year, breeding in *monte* and wintering in the *chaco/espinal/pampa* zone (*Xolmis coronatus, Neoxolmis rubetra*), or taxa that breed mainly in *Nothofagus* forest and winter mainly in the Chilean Mediterranean zone (*Colorhamphus parvirostris, Xolmis p. pyrope*), whose winter range is drier on average than their breeding range.

General Trends

Investigation of cumulative patterns of summer residence, based on seasonal occurrence within 200 × 200-km blocks, revealed that migrant flycatcher diversity is greatest in northern Argentina, with high numbers of migrants also breeding in eastern Paraguay and parts of southeastern Brazil and southern Bolivia. Up to 27 summer resident flycatchers are found in single 200 × 200 blocks along the western border of the *chaco* and eastern slopes of the Andes, a region of considerable elevational and habitat turnover. Other regions of relatively high species diversity are transitional areas between the southern *chaco* and the *espinal* (up to 24 species per block), and between the eastern *chaco* and the Atlantic forest region (up to 24 species per block). Gradual declines in species numbers were found in all directions from the *chaco* border regions, including toward the interior of the *chaco*. Areas of exclusively summer residence occur as far north as 3° S, where migratory northeastern Brazilian populations of *Legatus l. leucophaius* are present seasonally.

Investigation of areas of exclusively winter residence revealed that the highest diversity of wintering migrants occurs in and around southwestern Amazonia, where numbers of wintering flycatchers per block range from 14 to 19. Progressively lower numbers of wintering species occur outside this region, although eight to ten species may still be found in single blocks as far south as Buenos Aires Province, Argentina. Moderate numbers of migrants (11–13 species per block) winter in central Amazonia, but as few as three species per block winter in far eastern Amazonia. Areas of exclusively winter residence are located as far south as eastern Río Negro Province, Argentina, where *Muscisaxicola maclovianus mentalis* occurs in winter, and wintering tyrannids regularly occur to the north virtually throughout South America.

Elevational Movements

Thirty-three of the 74 austral migrant tyrannids show evidence of a seasonal difference in elevation of 500 m or more. Twenty-four of these undergo a typical elevational change, with winter ranges at lower elevations on average than breeding ranges. The other nine undertake the opposite type of elevational movement, wintering at higher elevations than those at which they breed. Eight species migrate north along the Andes to winter in life zones similar to those of their breeding grounds, but which occur at higher elevation in the north; and the final migrant, *Polystictus p. pectoralis,* breeds at elevations up to 300 m in the lowlands, primarily of Argentina and Paraguay, and extends its range north in winter to the central Brazilian plateau, where it is found to ca. 800 m.

The elevational breeding and wintering ranges of three migrants—*Muscisaxicola capistratus, M. r. rufivertex,* and *Elaenia strepera*—are completely elevationally disjunct. *Muscisaxicola capistratus* winters at higher elevation than it breeds, whereas the other two species are more typical migrants, wintering at lower elevation. *Muscisaxicola capistratus* breeds from sea level to ca. 1,500 m in southern Chile and Argentina and winters from 3,000 to 4,600 m in the Andes of northern Argentina and Chile, Bolivia, and southern Peru. *Muscisaxicola r. rufivertex* breeds in the Chilean Andes from 2,000 to 3,500 m, but winters north along the Chilean coast from sea level to 800 m (although the conspecific migrant *M. r. pallidiceps* winters as high as 4,400 m). *Elaenia strepera* breeds on eastern Andean slopes in Argentina and Bolivia from 900 to 2,700 m, but winters in northern Venezuela only up to 450 m. Breeding and wintering elevations of one other species, *Muscisaxicola albilora,* overlap only slightly; *M. albilora* breeds from ca. 200 to 2,500 m in Argentina and Chile and winters from 2,375 to ca. 4250 m in Bolivia, Peru, and Ecuador.

Migration Routes

Most South American migrants have overlapping or contiguous breeding and wintering ranges, making it nearly impossible to determine areas of passage migration from specimen records; however, some data on routes, or areas of concentration, are available for selected austral migrant tyrannids. The available data from specimens and observations support two points: (1) that migration routes may differ in spring and fall, and (2) that concentrations of migrants likely occur along the Andes.

That migration routes may differ between spring and fall is supported most clearly by the well-documented passage migrations of *Elaenia albiceps chilensis* (see Marini and Cavalcanti 1990). This migrant breeds mainly in Chile and western Argentina, and its migrations north after breeding are concentrated in two distinct areas: one along the eastern Andes, where it passes through Bolivia en route to wintering grounds as far north as Colombia, and the other along the Atlantic coastal plain, where it passes through Buenos Aires Province, Argentina; Uruguay; and eastern Brazil en route to wintering areas in northeastern Brazil and eastern Amazonia. Although the Andean route is also used during spring migration, when the species returns south, only a single specimen is known from spring for the Atlantic route delineated above (RTC, unpubl. data). Instead, there are multiple specimens during spring migration from interior areas of South America, in south-central Brazil, Paraguay, and north-central Argentina. Further support for differences in spring and fall migration routes is provided by the observations of Davis (1993) in lowland Bolivia, who noted two of four long-distance passage migrants (*Elaenia s. spectabilis* and *Myiarchus swainsoni*) only during spring migration (September to November).

Suggestive of the Andes as a migration route, and perhaps of differences in fall and spring routes, are the many long-distance migrants for which the highest elevational records have occurred in the Bolivian Andes during fall passage migration (Chesser 1997). Of the 11 flycatchers that migrate an average distance greater than 20° of latitude (table 14.2), four belong to the genus *Muscisaxicola* and regularly occur at extremely high elevations. The remaining seven species breed and winter at low or middle elevations, ranging from sea level to ca. 2,500 m. High elevational records for six of these seven species (*Elaenia strepera, E. albiceps chilensis, Myiarchus swainsoni* [records for both *swainsoni* and *ferocior*], *Empidonomus v. varius, E. a. aurantioatrocristatus,* and *Tyrannus s. savana;* excluding only *Elaenia parvirostris*) have occurred from mid to late March, at 2,700–3,760 m, in the Bolivian Andes, a highly significant result (binomial test; $p < 0.0005$). Although it seems unlikely that the high Andes form part of the regular migration routes of these species, T. A. Parker (pers. comm.) and B. M. Whitney (pers. comm.) have noted "fallouts" of migrants, especially tyrant-flycatchers, during fall migration along the Andean foothills in the Department of Santa Cruz, Bolivia, indicating that the eastern slopes of the Andes may be a key fall migration route. That the Bolivian Andes and northern Argentine Cordillera Oriental lie along a direct route between much of Argentina and western Amazonia adds geographical plausibility to this hypothesis.

Timing of Migration

Spring migration of austral migrant flycatchers occurs mainly from September through November, and fall migration from March through May, although variation in timing, even intraspecific variation, may be extensive (Chesser 1997). Although little is known of the migratory physiology of South American flycatchers, heavy subcutaneous fat deposits have been noted in specimens of many taxa (e.g., *Sublegatus modestus brevirostris, Myiopagis v. viridicata, Elaenia spectabilis, E. parvirostris, Empidonomus a. aurantioatrocristatus, Tyrannus m. melancholicus, T. s. savana*) during periods of spring or fall migration.

Departures from the general pattern of timing of migration are evident among several groups of austral migrant flycatcher. Some species, especially those of the fluvicoline genera *Agriornis, Neoxolmis, Muscisaxicola,* and *Lessonia,* tend to be early migrants. The pattern of early fall migration is well established in *Agriornis micropterus, A. murinus, Neoxolmis rufiventris, Muscisaxicola maclovianus, M. capistratus,* and *M. rufivertex;* early arrivals on the wintering grounds have been collected in February or March. This is particularly noticeable in *Lessonia rufa,* which is already on its wintering grounds by January. Many of these species also depart from their wintering grounds early, with *N. rufiventris, M. capistratus,* and *M. rufivertex* not recorded after September. Although spring arrivals of these ground-tyrants are not especially early relative to austral migrants as a group, probably because of their relatively longer-distance migrations, they generally precede other species breeding at similar latitudes and migrating long distances. For example, Humphrey et al. (1970) noted that *Muscisaxicola capistratus* arrives on Isla Grande, Tierra del Fuego, in early September, whereas the earliest record for *Elaenia albiceps chilensis* is 14 October.

Tyrannus s. savana and *Tyrannus albogularis* are likewise early spring and fall migrants, both regularly arriving on their breeding areas as early as August and their wintering grounds as early as February. *Tyrannus s. savana,* for example, gathers in flocks in Argentina at the end of January and begins migration in early February (Wetmore 1926). This species has been collected in areas of passage migration as early as February and August, and large flocks (1,000–1,500 birds) of this migrant have been observed passing through southeastern Peru as early as February (O'Neill 1974).

Pseudocolopteryx acutipennis, in contrast, undertakes extremely late migrations. Southern populations of this species (there are apparent permanent resident populations in the northern Andes) breed in the highlands of Bolivia and south along the Andes and at lower elevation from northwestern Argentina south to Mendoza and Córdoba Provinces, from 2,200 to 3,550 m in Bolivia and Jujuy Province, Argentina, and from 100 to 3,500 m south of Jujuy. Specimens have been collected in this area from 25 November to 4 May. The southern populations winter from the lowlands of Bolivia (at elevations up to 700 m) northwest to southern Peru (at elevations to 1,800 m), where specimens have been collected from 3 June to 5 December. Specimens of apparently transient birds have been taken from Paraguay and eastern Bolivia in December and from late April through mid-June. These data are supported by observations that the species arrives in the Cachi area of Salta Province, Argentina, in mid-December (B. M. Whitney, pers. comm.) and that it is still fledging young in early April in Cochabamba Department, Bolivia (B. M. Whitney, pers. comm.), although there is a report of nesting in northwestern Argentina in November (Fjeldså and Krabbe 1990).

Spring migration of *Elaenia s. spectabilis* also appears to be rather late. Although earlier arrival dates for the breeding grounds have been recorded, most spring migration of this species appears to occur only in late October and November, with a late wintering record of 20 November. Specimens with no subcutaneous fat have been collected from the Bolivian wintering grounds in early to mid October, whereas specimens with heavy or moderate fat have been collected in late October and November.

Timing of spring migration has presumably been selected to maximize reproductive success, and deviations from the general pattern are presumably related to timing of breeding or other activities, such as territory settlement, closely associated with breeding. These activities appear to be tied to peaks in food availability (Lack 1954; Perrins 1970; Martin 1987); thus, one might predict variation in timing of migration of tyrant-flycatchers to be closely associated with abundance of their food resources. Indeed, late migrations in some other austral migrants (e.g., *Sporophila* seedeaters) have been proposed to be related to seasonality of food resources on their breeding grounds (Remsen and Hunn 1979). The relatively late migration and nesting of *P. acutipennis* do not appear to be shared by other Andean and temperate marsh-nesting species such as *Tachuris rubrigastra* and *Phleocryptes melanops* (see Peña 1987; Fjeldså and Krabbe 1990); comparative breeding biology of these species would be an interesting avenue of study.

ECOLOGICAL AND ENVIRONMENTAL CONSIDERATIONS

Habitat and Resource Use Patterns

Analysis of breeding habitats of austral migrant flycatchers indicated that these tyrannids are most diverse in scrubby and woodland habitats and least diverse in primary forest, reflecting to some extent the relative availability of these habitats in temperate South America. Flycatchers differ principally from other passerine austral migrants in their preference for woodlands (e.g., *chaco* woodland, gallery woodland) relative to dry open habitats such as grasslands and shrub-steppe. Tyrannids are the predominant migrants in woodlands, constituting 34 (76%) of the 45 migrant passerine species breeding in these habitats. Although many migrant flycatchers breed in open areas of Patagonia and the southern Andes, tyrannid species are significantly less common in dry open habitats, constituting only 19 (32%) of the 59 migrant passerine species breeding in these areas ($\chi^2 = 19.199$; d.f. = 1; $p < 0.0001$).

Although winter habitats of most austral migrant tyrannids are similar to their breeding habitats, a sizable number of flycatchers move partially or wholly into Amazonian rain forest during winter. As noted above, this is particularly common in southwestern and western Amazonia, where 26 species occur mainly or exclusively in winter (table 14.5). These wintering species tend to breed in scrub or woodland habitats; thus, those individuals wintering in Amazonian rain forest undergo a significant seasonal habitat change. However, the microhabitats occupied by migrant flycatchers in Amazonia tend to be secondary in nature, rather than the more complex parts of primary Amazonian forest (table 14.5). Secondary microhabitats regularly occupied by wintering tyrannids include tropical forest edge, second growth, cane (*zabolo*), seasonally flooded forest (*varzea*), forest clearings, river edge, river islands, and lake margins. A few species, such as *Myiopagis v. viridicata,* are also found in primary forest, including primary forest understory, but these species tend to occur in primary forest canopy, which is similar to secondary habitats in seasonal resource availability patterns (Stiles 1980; Levey and Stiles 1992). Resident Amazonian flycatchers, in contrast, are found as often in primary forest understory as in secondary habitats; the residents at Cocha Cashu, Peru (Terborgh et al. 1990), for instance, occupy significantly different habitats than wintering migrants ($\chi^2 = 0.009$; d.f. = 1).

Nearctic-Neotropical migrants occupy mainly secondary and edge habitats on their tropical, principally Central American, wintering grounds (e.g., Slud 1960; Willis 1966; Karr 1976), although some species occur in primary forest (Terborgh 1980; Rappole and Morton 1985; Lynch 1989). Pearson (1980) and Robinson et al. (1988) likewise suggested that a principal strategy of wintering migrant birds in Amazonia is use of secondary and edge habitats. Their conclusions were derived mainly from Nearctic-Neotropical migrants in Amazonia but included observations of a limited number of austral migrant species. The results presented above support their conclusions for a wide range of austral migrant tyrannids at a variety of Amazonian sites.

The apparent preference of many austral migrant tyrannids for riverine habitats (table 14.5) suggests that this is a particularly seasonally productive environment. Robinson et al. (1988) found austral migrant flycatchers primarily in riverine *Tessaria* and cane, as they did Nearctic-Neotropical migrants. Although these habitats contain sporadically abundant resources, they are flooded for weeks at a time from November to May (Robinson et al. 1988) and thus are often unavailable during the wet season. These habitats do not flood during the dry season, however, and must represent a relatively stable resource while austral migrants are present (primarily May through September).

The traditional explanation for the patterns of resource and habitat use observed among wintering tropical migrants was that they use ephemeral resources and seasonally productive habitats because few resident tropical species use them; thus, they are "extra" resources available to migrants. Stiles (1980) and Levey and Stiles (1992), however, found that species that undergo local or elevational movements in the Tropics also use these variable resources, and they suggested that tropical migrants display these resource use patterns not because they are temperate species finding a way to fit in with the diverse Neotropical community, but because these are precisely the resource use patterns that were characteristic of their original tropical lineages. Thus, Levey and Stiles (1992) proposed that certain habitat and resource use patterns—use of edge, canopy, or secondary (nonbuffered) habitat, and at least partial frugivory—predisposed particular lineages to develop long-distance migration. Chesser and Levey (1998) tested this hypothesis with austral migrant passerines and found significant support for the use of nonbuffered habitat as an evolutionary precursor to migration, although frugivory was not significantly supported.

Unlike many Nearctic-Neotropical migrants, the vast majority of austral migrants, including both tyrannids and non-tyrannids, breed in habitats other than primary forest (Chesser 1994). The relationship between temperate and tropical habitat and resource use patterns is thus more straightforward among austral migrants than among Nearctic-Neotropical migrants, because edge, secondary, and scrub habitats are frequently occupied during the breeding season. Thus, despite the obvious macrohabitat differences between the breeding and wintering grounds of many species, wintering-season microhabitats of austral migrant flycatchers resemble those occupied during the breeding season.

Resemblance between breeding and wintering microhabitat is characteristic of most Palearctic-African migrants (Moreau 1952) as well. Species that inhabit wooded habitats during the breeding season also do so during winter; nightingales (*Luscinia* spp.), for instance, reside in deep thickets in Africa, just as they do in Europe. Furthermore, many Palearctic-African migrants may change macrohabitats between breeding and nonbreeding seasons, while maintaining their preferred microhabitat, much as austral migrants wintering in Amazonia. For example, during winter some wheatears (e.g., *Oenanthe isabellina, O. oenanthe*) occupy more humid country than their summer steppe habitat, but they prefer bare or recently burned areas within these more humid winter macrohabitats (Moreau 1952, 1972). Many Palearctic migrants thus tolerate a variety of vegetational and climatic conditions, so long as their primary microhabitat requirements are satisfied.

Influence of Environmental and Climatic Factors

Habitat complexity and climatic variables such as temperature have been proposed as significant influences on the distribution patterns of migratory birds (e.g., MacArthur 1959; Willson 1976; Herrera 1978; Newton and Dale 1996a, 1996b). Chesser (1998) investigated patterns of distribution of breeding austral migrant tyrannids relative to 15 ecological, biogeographical, and climatic variables, including habitat complexity and various measures of temperature. Proportion of migrant species was most closely associated with latitude and two temperature variables, mean temperature of the coldest month and relative annual range of temperature. Percentage of migrants was positively related

Table 14.5 Breeding habitat and wintering microhabitat of migrant tyrannids present in western Amazonia exclusively or mainly during winter

Species	Predominant breeding habitat	Microhabitat in Amazonia
Camptostoma obsoletum bolivianum	Second-growth scrub, secondary woodland	River edge, *zabolo*
Sublegatus modestus brevirostris	Arid lowland scrub, gallery woodland	"Willow bars," clearings, edge, bamboo
Myiopagis v. viridicata	Gallery woodland, secondary woodland, forest borders	Edge, canopy, or understory of hillside or lowland forest, or upland forest adjacent to *varzea, zabolo,* sometimes with mixed-species flocks
Elaenia s. spectabilis	Gallery woodland, secondary woodland, forest borders, shrubby clearings	Edge of tropical forest, river-edge second growth, river islands, *zabolo*
Elaenia albiceps chilensis	*Nothofagus* forest, woodland, scrub	Second growth, shrubby areas
Elaenia parvirostris	Gallery woodland, secondary woodland, forest borders, shrubby clearings	Young river islands, *zabolo,* clearings, forest edge, shrubby areas
Inezia inornata	Scrub, gallery woodland	Canopy of lowland or hill forest, *zabolo,* roadside edges, often with mixed-species flocks
Pseudocolopteryx acutipennis	Freshwater marshes, sedges, riparian thickets	"Willow bars," marshes
Euscarthmus m. meloryphus	Arid and second-growth scrub, shrubby thickets, *cerrado*	*Zabolo,* roadside scrub, bamboo in *varzea*
Myiophobus fasciatus auriceps	Arid and second-growth scrub, shrubby clearings, lighter woodland, riparian thickets, forest borders	Low in second growth of tropical forest, or river-edge forest
Lathrotriccus e. euleri / argentinus	Humid forest, forest borders, secondary woodland	River-edge forest, bamboo, primary forest understory
Cnemotriccus fuscatus bimaculatus	Woodland and forest borders, gallery woodland, secondary woodland	Second growth, *varzea,* river-edge forest, bamboo, upland forest adjacent to *varzea,* streamside vine tangles
Pyrocephalus r. rubinus	Pasture, riparian thickets, second-growth scrub, borders of light woodland	River edge, *varzea,* forest edge and clearings in tropical hill or lowland forest, river islands, second growth
Fluvicola pica albiventer	Freshwater marshes, riparian areas	Lake margins, marshes
Satrapa icterophrys	Pasture, open shrubby areas, second-growth scrub, *monte,* gallery woodland and woodland edge	River margins
Attila phoenicurus	Humid forest, secondary woodland	Forest canopy
Casiornis rufus	Deciduous and gallery woodland, *cerrado, chaco* woodland and scrub	*Varzea,* vine tangles, edge, or understory of tropical lowland or hill forest, river edge, occasionally with mixed-species flocks
Myiarchus s. swainsoni / ferocior	Gallery woodland, secondary woodland, forest borders	River edge, *varzea, zabolo,* river islands, clearings, canopy of low forest
Myiarchus t. tyrannulus	Gallery woodland, secondary woodland, arid scrub, *cerrado, caatinga*	Forest edge, clearings, *varzea,* roadsides, canopy or subcanopy of lowland or hill forest
Myiodynastes maculatus solitarius	Gallery and secondary woodland, forest borders	Forest canopy, advanced second growth, forest edge
Empidonomus v. varius	Forest borders, secondary and gallery woodland	Forest clearings, forest edge, river islands
Empidonomus a. aurantioatrocristatus	Gallery woodland, lighter woodland and scrub	Forest edge, forest clearings, *varzea,* river edge, roadsides
Tyrannus albogularis	Gallery woodland, shrubby areas, woodland edge	Canopy of river islands, scrub, clearings
Tyrannus s. savana	Grasslands and pastures with scattered trees, second-growth scrub	River edge, river islands, lake margins
Pachyramphus polychopterus spixii	Gallery and secondary woodland, forest borders	Second growth, river edge, forest edge or subcanopy of tropical hill or upland forest, often with mixed-species flocks
Pachyramphus v. validus / audax	Woodland and forest borders, secondary woodland	Vine tangles in upland forest

Note: Data from literature (e.g., O'Neill 1974; Pearson 1980; Hilty and Brown 1986; Narosky and Yzurieta 1987; Rosenberg 1990; Parker et al. 1994; Ridgely and Tudor 1994; Stotz et al. 1996) and from habitat descriptions on specimen tags, primarily from birds at the LSU Museum of Natural Science and the Field Museum of Natural History.

to latitude and annual range of temperature, and negatively related to mean temperature of the coldest month. Thus, higher proportions of breeding migrant species were associated with higher latitudes, broader annual range of temperature, and lower mean temperature of the coldest month.

The association of latitude and temperature with the breeding distribution of South American austral migrant tyrannids was consistent with the results of Willson (1976) and Newton and Dale (1996b) for migrants breeding in North America and those of Herrera (1978) and Newton and Dale (1996a) for European migrants, suggesting that temperature and latitude, presumably a surrogate for one or more environmental variables, are significant factors in the worldwide breeding distribution of migratory birds.

It has also been proposed that temperature plays a significant role in the distribution of wintering austral migrants (Joseph 1996, 2003b). Joseph found that wintering distributions of the "temperate breeding, tropical wintering" group of austral migrants (Joseph 1996, 1997) tended to occur in areas with July daily mean temperatures (DMT) greater than ca. 20°C, whereas those of the "temperate breeding, temperate wintering" group occur in areas with July DMTs below 20°C. Furthermore, for selected austral migrants, January and July DMTs of the temperate-tropical group were similar and generally greater than 20°C, whereas those of the temperate-temperate group were relatively dissimilar (Joseph 2003b).

Although winter distributional patterns were not analyzed in terms of the relative influence of climatic, biogeographical, and ecological variables, precipitation may play a larger role during winter than during the breeding season. The highest numbers of wintering austral migrant flycatchers occur in and around northern Bolivia, coincident with the southwestern limit of Amazonian rain forest. As suggested by Blondel (1969) for European migrants and by Gauthreaux (1982) for Nearctic-Neotropical migrants, migrants in general seem to go only as far as necessary to reach suitable survival areas during winter, and the humid environment of southwestern Amazonia may provide some minimum level of resources necessary for large numbers of species to winter regularly.

The data on transitions between biogeographical regions, which indicate that many transitions occur between relatively dry breeding and relatively humid wintering grounds, combined with the strong east-west trans-Amazonian gradient in numbers of wintering migrant flycatchers, are consistent with the idea that precipitation may be an important factor in winter distributions.

FUTURE DIRECTIONS

South American austral migration provides exceptionally fertile ground for expanding our knowledge of bird migration, with possible studies ranging from basic investigations of distribution, behavior, ecology, and physiology, to the testing of hypotheses generated from studies of other migration systems (e.g., Chesser 1998; Chesser and Levey 1998). Austral migration has been the subject of relatively little study and much fundamental information is lacking. For instance, it is not known whether austral migrants are primarily nocturnal migrants, as is the case for many Nearctic and Palearctic breeding birds, or whether many or most are diurnal migrants, as might be expected in a system dominated by tyrant-flycatchers and relatively short-distance movements. Likewise, the influence of weather on the timing, routes, and other aspects of austral migration is an untapped area, although anecdotal evidence suggests that southern cold fronts (*surs* or *surazos*) may be an influential factor (pers. obs.; T. A. Parker and B. M. Whitney, pers. comm.).

Year-round fieldwork on austral migrants, especially in tropical and subtropical areas (e.g., Davis 1993), is a particularly pressing need. Fieldwork will prove useful in further elucidating seasonal distributional patterns, migration routes, timing of migration, and aspects of the ecology and behavior of individual migrant species in areas of breeding, wintering, or transience. Relating timing of migration to resource use and life history strategies, as mentioned above, would be a rewarding avenue of study, as would the comparative ecology of suites of migrants on breeding or wintering grounds.

Austral migration is particularly well suited to questions relating to intraspecific differences in migratory behavior, because many austral migrant species occur throughout the year in parts of their range. In localities in which resident individuals are supplemented seasonally by migrants, comparative wintering or breeding biology of migrant versus resident individuals would be a topic of great interest.

Conservation is another potential area of study. Although few austral migrants are of current conservation concern, several species that breed and winter in grasslands of central South America are critically endangered. These include the tyrannid species *Alectrurus risora,* which has disappeared from much of its former breeding range, and several little-known species of the seedeater genus *Sporophila* (Collar et al. 1992).

Geographical distribution of migrants, although explored in more detail than many other aspects of austral migration, remains an open avenue of research. Winter ranges of some species (e.g., *Attila phoenicurus*) are still poorly known, and additional fieldwork and museum study will undoubtedly modify what we know about the identity and distribution of austral migrants. Distributional data may also be applied to additional areas of research, for biogeographical syntheses or for studies of environmental variables and migration. Application of molecular and other types of data promises to further document seasonal movements of migrants, to clarify links between breeding and wintering populations, and to provide insight into the evolution of austral migration.

APPENDIX: SPECIES LIST OF SOUTH AMERICAN AUSTRAL MIGRANT TYRANNIDS

Scientific name	English name	Scientific name	English name
Phyllomyias fasciatus brevirostris	Planalto Tyrannulet		
Camptostoma o. obsoletum / bolivianum	Southern Beardless-Tyrannulet	*Muscisaxicola capistratus*	Cinnamon-bellied Ground-Tyrant
Phaeomyias murina ignobilis	Mouse-colored Tyrannulet	*Muscisaxicola r. rufivertex / pallidiceps*	Rufous-naped Ground-Tyrant
Sublegatus modestus brevirostris	Southern Scrub-Flycatcher	*Muscisaxicola albilora*	White-browed Ground-Tyrant
Suiriri s. suiriri	Suiriri Flycatcher	*Muscisaxicola c. cinereus*	Cinereous Ground-Tyrant
Myiopagis v. viridicata	Greenish Elaenia	*Muscisaxicola flavinucha*	Ochre-naped Ground-Tyrant
Elaenia s. spectabilis	Large Elaenia	*Muscisaxicola frontalis*	Black-fronted Ground-Tyrant
Elaenia albiceps chilensis	White-crested Elaenia	*Lessonia oreas*	Andean Negrito
Elaenia parvirostris	Small-billed Elaenia	*Lessonia rufa*	Austral Negrito
Elaenia mesoleuca	Olivaceous Elaenia	*Knipolegus striaticeps*	Cinereous Tyrant
Elaenia strepera	Slaty Elaenia	*Knipolegus hudsoni*	Hudson's Black-Tyrant
Elaenia chiriquensis albivertex	Lesser Elaenia	*Knipolegus cyanirostris*	Blue-billed Black-Tyrant
Serpophaga munda	White-bellied Tyrannulet	*Knipolegus a. aterrimus*	White-winged Black-Tyrant
Serpophaga s. subcristata	White-crested Tyrannulet	*Hymenops p. perspicillatus / andinus*	Spectacled Tyrant
Inezia inornata	Plain Tyrannulet	*Fluvicola pica albiventer*	Pied Water-Tyrant
Stigmatura budytoides flavocinerea	Greater Wagtail-Tyrant	*Alectrurus risora*	Strange-tailed Tyrant
Anairetes f. flavirostris	Yellow-billed Tit-Tyrant	*Satrapa icterophrys*	Yellow-browed Tyrant
Anairetes p. parulus / patagonicus	Tufted Tit-Tyrant	*Hirundinea ferruginea bellicosa*	Cliff Flycatcher
Tachuris r. rubrigastra	Many-colored Rush-Tyrant	*Muscipipra vetula*	Shear-tailed Gray Tyrant
Polystictus p. pectoralis	Bearded Tachuri	*Attila phoenicurus*	Rufous-tailed Attila
Pseudocolopteryx dinelliana	Dinelli's Doradito	*Casiornis rufus*	Rufous Casiornis
Pseudocolopteryx acutipennis	Subtropical Doradito	*Myiarchus tuberculifer atriceps*	Dusky-capped Flycatcher
Pseudocolopteryx flaviventris	Warbling Doradito	*Myiarchus s. swainsoni / ferocior*	Swainson's Flycatcher
Euscarthmus m. meloryphus	Tawny-crowned Pygmy-Tyrant	*Myiarchus t. tyrannulus*	Brown-crested Flycatcher
Myiophobus fasciatus flammiceps / auriceps	Bran-colored Flycatcher	*Megarynchus p. pitangua*	Boat-billed Flycatcher
		Myiodynastes maculatus solitarius	Streaked Flycatcher
Contopus c. cinereus	Tropical Pewee	*Legatus l. leucophaius*	Piratic Flycatcher
Lathrotriccus e. euleri / argentinus	Euler's Flycatcher	*Empidonomus v. varius*	Variegated Flycatcher
Cnemotriccus fuscatus bimaculatus	Fuscous Flycatcher	*Empidonomus a. aurantioatrocristatus*	Crowned Slaty-Flycatcher
Pyrocephalus r. rubinus	Vermilion Flycatcher	*Tyrannus albogularis*	White-throated Kingbird
Colorhamphus parvirostris	Patagonian Tyrant	*Tyrannus m. melancholicus*	Tropical Kingbird
Xolmis p. pyrope	Fire-eyed Diucon	*Tyrannus s. savana*	Fork-tailed Flycatcher
Xolmis coronatus	Black-crowned Monjita	*Xenopsaris albinucha*	Xenopsaris
Neoxolmis rubetra	Rusty-backed Monjita	*Pachyramphus polychopterus spixii*	White-winged Becard
Neoxolmis rufiventris	Chocolate-vented Tyrant	*Pachyramphus v. validus / audax*	Crested Becard
Agriornis m. micropterus	Gray-bellied Shrike-Tyrant	*Tityra cayana braziliensis*	Black-tailed Tityra
Agriornis murinus	Lesser Shrike-Tyrant	*Tityra i. inquisitor*	Black-crowned Tityra
Muscisaxicola m. maculirostris	Spot-billed Ground-Tyrant		
Muscisaxicola maclovianus mentalis	Dark-faced Ground-Tyrant		

ACKNOWLEDGMENTS

I thank Van Remsen, Pete Marra, Tom Schulenberg, and Richard Schodde for helpful comments on the manuscript. I am grateful to the following curators and museum staff for granting access to or providing data on specimens under their care and for numerous other kindnesses: Mark Robbins, David Agro, and Christopher Thompson (Academy of Natural Sciences of Philadelphia); George Barrowclough, Mary LeCroy, François Vuilleumier, and John Bates (American Museum of Natural History, New York); Robert Prys-Jones and Peter Colston (British Museum [Natural History], Tring); Kenneth Parkes and Robin Panza (Carnegie Museum, Pittsburgh); Gene Hess (Delaware Museum of Natural History, Wilmington); Scott Lanyon, Tom Schulenberg, Melvin Traylor and David Willard (Field Museum of Natural History, Chicago); Estela Alabarce and Claudio Laredo (Fundación Miguel Lillo, Tucumán, Argentina); Ricardo Ojeda (Instituto de Investigaciones de las Zonas Aridas, Mendoza, Argentina); Kimball Garrett (Los Angeles County Museum); Van Remsen (Louisiana State University Museum of Natural Science, Baton Rouge); Jorge Navas (Museo Argentina de Ciencias Naturales "Bernardino Rivadavia," Buenos Aires); Nelly Bó and Aníbal Camperi (Museo de La Plata, Argentina); Douglas Stotz (Museu de Zoologia da Universidade de São Paulo, Brazil); Dante Teixeira and Jorge Nacinovic (Museu Nacional de Rio de Janeiro, Brazil); David Oren (Museu Paraense Emílio Goeldi, Belém, Brazil); Raymond Paynter (Museum of Comparative Zoology, Harvard University, Cambridge); Gary Graves (National Museum of Natural History, Washington, D.C.); Julio Contreras (Programa de Biología Básica y Aplicada Subtropical,

Corrientes, Argentina); Janet Hinshaw and Robert Storer (University of Michigan Museum of Zoology, Ann Arbor); and Fred Sibley (Yale Peabody Museum, New Haven). I also thank John O'Neill, Van Remsen, and Jacques Viellard for kindly providing information on specimens housed in the Museo de Historia Natural de la Universidad Nacional Mayor de San Marcos (Lima, Peru), the Colección Boliviana de Fauna (La Paz), and the Museu de Biologia Mello Leitão (Espírito Santo, Brazil), respectively; and José Maria Cardoso da Silva and Jon Fjeldså for answering queries concerning specimens in the Swedish Museum of Natural History, Stockholm. This research could not have been completed without the gratefully acknowledged financial support of the American Museum of Natural History Frank M. Chapman Fund, the American Ornithologists' Union, Charles Fugler, the Louisiana State University Museum of Natural Science, the Georgia Ornithological Society, J. William Eley, and Sigma Xi.

LITERATURE CITED

Azara, F. de. 1802–1805. Apuntamientos para la Historia Natural de los Paxaros del Paraguay y Río de la Plata, Vols. 1–3. Imprenta de la Viuda de Ibarra, Madrid.

Belton, W. 1984. Birds of Rio Grande do Sul, Brazil: Pt. 1. Rheidae through Furnariidae. Bulletin of the American Museum of Natural History 178:369–636.

Belton, W. 1985. Birds of Rio Grande do Sul, Brazil: Pt. 2. Formicariidae through Corvidae. Bulletin of the American Museum of Natural History 180:1–241.

Blondel, J. 1969. Les migrations transcontinentales d'oiseaux vues sous l'angle écologique. Bulletin de la Société Zoologique de France 94:577–598.

Brown, J. H., and A. C. Gibson. 1983. Biogeography. C. V. Mosby, St. Louis.

Cabrera, A. L., and A. Willink. 1973. Biogeográfia de America Latina. Organización de los Estados Americanos, Washington, D.C.

Chesser, R. T. 1994. Migration in South America: an overview of the austral system. Bird Conservation International 4: 91–107.

Chesser, R. T. 1995. Biogeographic, ecological, and evolutionary aspects of South American austral migration, with special reference to the family Tyrannidae. Ph.D. dissertation, Louisiana State University, Baton Rouge.

Chesser, R. T. 1997. Patterns of seasonal and geographical distribution of austral migrant flycatchers (Tyrannidae) in Bolivia. Ornithological Monographs 48:171–204.

Chesser, R. T. 1998. Further perspectives on the breeding distribution of migratory birds: South American austral migrant flycatchers. Journal of Animal Ecology 67:69–77.

Chesser, R. T. 2000. Evolution in the high Andes: the phylogenetics of *Muscisaxicola* ground-tyrants. Molecular Phylogenetics and Evolution 15:369–380.

Chesser, R. T., and D. J. Levey. 1998. Austral migrants and the evolution of migration in New World birds: diet, habitat, and migration revisited. American Naturalist 152:311–319.

Chesser, R. T., and M. Marín A. 1994. Seasonal distribution and natural history of the Patagonian Tyrant (*Colorhamphus parvirostris*). Wilson Bulletin 106:649–667.

Collar, N. J., L. P. Gonzaga, N. Krabbe, A. Madroño Neito, L. G. Naranjo, T. A. Parker III, and D. C. Wege. 1992. Threatened Birds of the Americas: The ICBP/IUCN Red Data Book: Pt. 2. International Council for Bird Preservation, Cambridge, U.K.

Davis, S. E. 1993. Seasonal status, relative abundance, and behavior of birds of Concepción, Departamento Santa Cruz, Bolivia. Fieldiana, Zoology (N. S.) 71:1–33.

Fjeldså, J., and N. Krabbe. 1990. Birds of the High Andes. Zoological Museum, University of Copenhagen, and Apollo Books, Svendborg, Denmark.

Gauthreaux, S. A. 1982. The ecology and evolution of avian migration systems. Pages 93–168 *in* Avian Biology, Vol. 6 (D. S. Farner, J. R. King, and K. C. Parkes, eds.). Academic Press, New York.

Gore, M. E. J., and A. R. M. Gepp. 1978. Las Aves del Uruguay. Mosca Hnos., Montevideo.

Hayes, F. E., P. A. Scharf, and R. S. Ridgely. 1994. Austral bird migrants in Paraguay. Condor 96:83–97.

Herrera, C. 1978. On the breeding distribution pattern of European migrant birds: MacArthur's theme reexamined. Auk 95:496–509.

Hilty, S. L., and W. Brown. 1986. A Guide to the Birds of Colombia. Princeton University Press, Princeton.

Hueck, K., and P. Seibert. 1972. Vegetationskarte von Südamerika. Gustav Fischer Verlag, Stuttgart.

Humphrey, P. S., D. Bridge, P. W. Reynolds, and R. T. Peterson. 1970. Birds of Isla Grande (Tierra del Fuego). Preliminary Smithsonian Manual, Smithsonian Institution, Washington, D.C.

Joseph, L. 1996. Preliminary climatic overview of migration patterns in South American austral migrant passerines. Ecotropica 2:185–193.

Joseph, L. 1997. Towards a broader view of Neotropical migrants: consequences of a re-examination of austral migration. Ornitología Neotropical 8:31–36.

Joseph, L. 2003a. Independent evolution of migration on the South American landscape in a long-distance temperate-tropical migratory bird, Swainson's flycatcher (*Myiarchus swainsoni*). Journal of Biogeography 30:925–937.

Joseph, L. 2003b. Predicting distributions of South American migrant birds in fragmented environments: a possible approach based on climate. Ecological Studies 162:263–283.

Joseph, L., and D. Stockwell. 2000. Temperature-based models of the migration of Swainson's Flycatcher (*Myiarchus swainsoni*) across South America: a new use for museum specimens of migratory birds. Proceedings of the Academy of Natural Sciences of Philadelphia 150:293–300.

Karr, J. R. 1976. On the relative abundance of migrants from the north temperate zone in tropical habitats. Wilson Bulletin 88:433–458.

Lack, D. 1954. The Natural Regulation of Animal Numbers. Clarendon Press, Oxford.

Lanyon, W. E. 1978. Revision of the *Myiarchus* flycatchers of South America. Bulletin of the American Museum of Natural History 161:427–628.

Levey, D. J., and G. F. Stiles. 1992. Evolutionary precursors of long-distance migration: resource availability and movement patterns in Neotropical landbirds. American Naturalist 140:447–476.

Lynch, J. F. 1989. Distribution of overwintering Nearctic migrants in the Yucatan Peninsula: Pt. 1. General patterns of occurrence. Condor 91:515–544.

MacArthur, R. H. 1959. On the breeding distribution patterns of North American migrant birds. Auk 76:318–325.

Marantz, C. A., and J. V. Remsen. 1991. Seasonal distribution of the Slaty Elaenia, a little-known austral migrant of South America. Journal of Field Ornithology 62:162–172.

Marini, M. A., and R. B. Cavalcanti. 1990. Migrações de *Elaenia albiceps chilensis* e *Elaenia chiriquensis albivertex* (Aves: Tyrannidae). Boletim do Museu Paraense Emílio Goeldi, série Zoologica 6:59–67.

Martin, T. E. 1987. Food as a limit on breeding birds: a life-history perspective. Annual Review of Ecology and Systematics 18:453–487.

Moreau, R. E. 1952. The place of Africa in the Palaearctic migration system. Journal of Animal Ecology 21:250–271.

Moreau, R. E. 1972. The Palaearctic-African Bird Migration Systems. Academic Press, London.

Narosky, T., and D. Yzurieta. 1987. Guía para la Identificación de las Aves de Argentina y Uruguay. Asociación Ornitológica del Plata, Buenos Aires.

Newton, I., and L. Dale. 1996a. Relationship between migration and latitude among European birds. Journal of Animal Ecology 65:137–146.

Newton, I., and L. Dale. 1996b. Bird migration at different latitudes in eastern North America. Auk 113:626–635.

Nores, M. 1987. Zonas ornitogeográficas de Argentina. Pages 295–303 *in* Guía para la Identificación de las Aves de Argentina y Uruguay (T. Narosky and D. Yzurieta). Asociación Ornitológica del Plata, Buenos Aires.

O'Neill, J. P. 1974. The birds of Balta, a Peruvian dry forest locality, with an analysis of their origins and ecological relationships. Ph.D. dissertation, Louisiana State University, Baton Rouge.

Parker, T. A., III, P. K. Donahue, and T. S. Schulenberg. 1994. Birds of the Tambopata Reserve Zone (Explorer's Inn Reserve). Pages 106–124 *in* The Tambopata-Candamo Reserved Zone of Southeastern Perú: A Biological Assessment (R. B. Foster et al., eds.). RAP Working Papers 6. Conservation International, Washington, D.C.

Pearson, D. L. 1980. Bird migration in Amazonian Peru, Ecuador, and Bolivia. Pages 273–283 *in* Migrant Birds in the Neotropics: Ecology, Behavior, Distribution, and Conservation (A. Keast and E. S. Morton, eds.). Smithsonian Institution Press, Washington, D.C.

Peña, M. de la. 1987. Nidos y Huevos de Aves Argentinas. Published by the author, Santa Fe, Argentina.

Perrins, C. M. 1970. The timing of birds' breeding seasons. Ibis 112:242–255.

Rappole, J. H., and E. S. Morton. 1985. Effects of habitat alteration on a tropical forest community. Pages 1013–1019 *in* Neotropical Ornithology (P. A. Buckley, M. S. Foster, E. S. Morton, R. S. Ridgely, and F. G. Buckley, eds.), AOU Monograph 36. American Ornithologists' Union, Washington, D.C.

Remsen, J. V., and E. S. Hunn. 1979. First records of *Sporophila caerulescens* from Colombia: a probable long distance migrant from southern South America. Bulletin of the British Ornithological Club 99:24–26.

Ridgely, R. S., and G. Tudor. 1994. Birds of South America, Vol. 2, The Suboscine Passerines. University of Texas Press, Austin.

Robinson, S. K., J. Terborgh, and J. W. Fitzpatrick. 1988. Habitat selection and relative abundance of migrants in southeastern Peru. Pages 2298–2307 *in* Proceedings of the 19th International Ornithological Congress (H. Ouellet, ed.). University of Ottawa Press, Ottawa.

Rosenberg, G. H. 1990. Habitat specialization and foraging behavior by birds of Amazonian river islands in northeastern Peru. Condor 92:427–443.

Sick, H. 1968. Vogelwanderungen im kontinentalen Südamerika. Vogelwarte 24:217–243.

Slud, P. 1960. The birds of Finca La Selva, a tropical wet forest locality. Bulletin of the American Museum of Natural History 121:1–148.

Stiles, F. G. 1980. Evolutionary implications of habitat relations between permanent and winter resident landbirds in Costa Rica. Pages 421–435 *in* Migrant Birds in the Neotropics: Ecology, Behavior, Distribution, and Conservation (A. Keast and E. S. Morton, eds.). Smithsonian Institution Press, Washington, D.C.

Stotz, D. F., J. W. Fitzpatrick, T. A. Parker III, and D. Moskovits. 1996. Neotropical Birds: Status and Conservation. University of Chicago Press, Chicago.

Terborgh, J. W. 1980. The conservation status of Neotropical migrants: present and future. Pages 21–30 *in* Migrant Birds in the Neotropics: Ecology, Behavior, Distribution, and Conservation (A. Keast and E. S. Morton, eds.). Smithsonian Institution Press, Washington, D.C.

Terborgh, J., S. K. Robinson, T. A. Parker III, C. A. Munn, and N. Pierpont. 1990. Structure and organization of an Amazonian forest bird community. Ecological Monographs 60:213–238.

Wetmore, A. 1926. Observations on the birds of Argentina, Paraguay, Uruguay, and Chile. Bulletin of the U.S. National Museum 133:1–448.

Willis, E. O. 1966. The role of migrant birds at swarms of army ants. Living Bird 5:187–231.

Willis, E. O. 1988. Land-bird migration in Sao Paulo, southeastern Brazil. Pages 756–764 *in* Proceedings of the 19th International Ornithological Congress Ottawa (H. Ouellet, ed.). University of Ottawa Press, Ottawa.

Willson, M. F. 1976. The breeding distribution of North American migrant birds: a critique of MacArthur (1959). Wilson Bulletin 88:582–587.

Zimmer, J. T. 1931–1955. Studies of Peruvian birds, 1–66. American Museum Novitates (between numbers 500 and 1723 inclusive).

Zimmer, J. T. 1938. Notes on migrations of South American birds. Auk 55:405–410.

SIDNEY A. GAUTHREAUX JR., JENNY E. MICHI, AND CARROLL G. BELSER

The Temporal and Spatial Structure of the Atmosphere and Its Influence on Bird Migration Strategies

GLOBAL CLIMATE CHANGES HAVE ALTERED the distribution and productivity of habitat types in the Temperate Zone and in doing so have dictated the amount of available breeding habitat for birds. Despite the resulting changes in the northward extent of species distributions, it is the operation of the annual climatic cycle that selects for the "southward escape from adversity" segment and the "northward return to breed" segment of the annual bird migration circuit. We examine the relationship between patterns of atmospheric flow associated with seasonal climatic changes and bird migration strategies during spring and fall. We use vector maps of seasonal wind that show patterns of atmospheric flow for different altitudes and simulated trajectory plots that use archived atmospheric data to plot the movement of air parcels over the Earth's surface during a 24-h period. Wind patterns in spring are highly favorable for the movements of birds from tropical wintering quarters to breeding areas in North America. Migrants departing from these areas can follow a "go-with-the-prevailing-flow" strategy. In contrast, prevailing wind patterns in fall are unfavorable for the movements of birds from northern breeding areas to wintering areas to the south, and many migrants may use a "sit-and-wait-for-favorable-winds" strategy. In both seasons migratory birds selectively fly with tailwinds because: (a) it is more fuel efficient, (b) it increases the velocity of movement and consequently decreases the time en route, and (c) it results in birds arriving at their destinations in good physical condition.

INTRODUCTION

The structure of the atmosphere is of paramount importance for all organisms that fly. The spatial and temporal dynamics of atmospheric structure (long-term climatic changes, annual climatic cycle, and short-term synoptic weather changes) can greatly influence flight behavior and atmospheric transport of organisms and ultimately shape foraging, dispersal, and migration strategies (Gauthreaux 1980a; Dingle 1996). This is so because atmospheric flow affects the cost of flying. A tail wind reduces the cost of transport (minimizes energy consumption) and increases the speed across the ground (minimizes time to reach a destination) (Pennycuick 1978; Hedenström and Alerstam 1995). Head winds and crosswinds have opposite effects, and as velocity of crosswinds increases relative to the air speed of an organism, the probability of displacement from a "preferred" trajectory increases (Cochran and Kjos 1985; Richardson 1991; Alerstam and Hedenström 1998; Green and Alerstam 2002). When birds are crossing ecological barriers where landing is not an option and in-flight correction for drift is too costly, displacement or wind drift is likely (Alerstam and Pettersson 1976; Green 2001). Displacement may extend the period of travel, delay the time of arrival on the breeding grounds, affect the energetic state and condition at arrival, and ultimately influence reproductive success and survival (Sandberg 1996). Birds have evolved many strategies to minimize the cost and maximize the benefits of traveling through the atmosphere (Alerstam 1979).

Given the advantages of migrating with following winds, one would predict that migratory pathways have evolved to maximize the probability of migrating with following winds and to minimize the probability of encountering adverse winds. From the standpoint of a migratory bird the best atmospheric conditions would provide following winds for both segments of its migration circuit. In contrast the worst possible conditions would involve *constant* adverse winds during its spring and fall migration. The relationship between migratory routes and atmospheric circulation has been of interest for some time (Landsberg 1948; Evans 1966; Able 1972; Richardson 1976; Gauthreaux 1980b). An initial study of the relationships among seasonal patterns of wind circulation, topography, and patterns of bird migration in North and Central America by Gauthreaux (1980b) confirmed that the seasonal migrations of birds appear to follow a general elliptical pattern in keeping with the general zonal circulation of the atmosphere (easterly winds below 30° N and westerly winds above 30° N), supporting the earlier suggestions of Bellrose and Graber (1963). However, Gauthreaux's findings were based on limited climatic data (mean surface barometric pressure patterns and surface winds).

Recent technological advancements in data acquisition and data processing have greatly increased our knowledge of the statistical properties of the atmosphere over several decades, during the annual climatic cycle, and during synoptic weather events. These advances allow us to explore how flight strategies of migrating birds may have been selected for in relation to the statistical properties of the atmosphere in the past and how atmospheric dynamics continue to influence migration strategies. Here, we examine the spatial and temporal properties of the atmosphere below 3,000 m between 10 and 70° N latitude and between 55 and 140° W longitude during the spring and fall migration periods. We examine the null hypothesis that prevailing winds have not been an important factor in the evolution of migratory pathways of birds that breed in North America above 30° N and spend most of their nonbreeding time south of that latitude.

METHODS

We used both monthly long-term (30-year) climatic maps of wind patterns and atmospheric trajectory models to examine the structure of the atmosphere. The maps of wind patterns are generated by the U.S. National Center for Environmental Prediction and U.S. National Center for Atmospheric Research (NCEP/NCAR) 40-year reanalysis project (Kalnay et al. 1996; Kistler et al. 2001) through the Climate Data Assimilation System I (CDAS-1). Each wind vector measurement is a vector having three components (X, Y, and P), one along each of the dimensions of longitude (zonal, E-W, or U-wind), latitude (meridional, N-S, or V-wind), and pressure (17 different atmospheric pressure layers). In this study we concentrated on patterns of wind vectors at the following pressure layers measured in hectopascal (hPa) or millibar (mb) levels: 1,000 hPa (100 m above mean sea level [MSL]), 925 hPa (800 m above MSL), 850 hPa (1,500 m above MSL), and 700 hPa (3,000 m above MSL). Climatological wind vectors for each pressure level indicate wind direction averaged over the days of a selected month for the period 1961–1990 drawn over a 2.5 × 2.5 degree latitude/longitude grid with speed proportional to the length of the vector (m sec^{-1}).

To examine the role of atmospheric structure on transport we generated trajectory models using the Hybrid Single-Particle Lagrangian Integrated Trajectory (HYSPLIT) model (Draxler and Hess 1997). The HYSPLIT model was developed to facilitate research on dispersal of airborne pollutants. The model applies meteorological data in grid format to conformal map projections by using the following components of the wind field: vertical diffusivity, wind shear, and horizontal deformation (Draxler and Hess 1997).

The HYSPLIT models we used for analysis were run directly on the Realtime Environmental Application and Display System (READY) website (HYSPLIT4 Model 1997). As a registered user we accessed this model within the trajectory model products of HYSPLIT, which simulates passive particle transport around the model field on the basis of mean wind fields and turbulence. The Global Data Assimilation System (GDAS) Model archives (also known as FNL for "final") were selected to generate trajectories with a horizontal resolution of approximately 191 km. We selected a start time of 00:00 UTC, starting heights of 500, 1,000, and 2,500 m above ground level, and a run time of 24 h, and the models were run starting at a chosen latitude and longitude. We calculated the distance and azimuth between the starting and end points of each model run for each starting height. Data were collected for 5 years (1997–2001) from 1 March to 31 May for three locations: Lake Charles, Louisana (30.11° N, 93.21° W), Mérida, Mexico (20.97° N, 89.65° W), and the Caribbean Coast of Colombia (9° N, 76° W). We performed circular statistical analyses with Oriana software (Version 1.0, Kovach Computing Services).

RESULTS

Spring Migration

The patterns of prevailing atmospheric flow change at different altitudes so that trajectories of birds moving with the wind change depending on the altitude of flight. The winds at the 1,000, 925, and 850 hPa levels for April–May for the period 1961–1990 are shown as vector plots in fig. 15.1. In April and May over the Caribbean Sea the winds at the 1,000 hPa level (approximately 50 m above MSL) blow from east to west (fig. 15.1A,B). Birds flying at this altitude over the northern coast of South America and the Antilles would be displaced westward toward Central America. North of the Texas and Louisiana coasts winds from April to May at 1,000 hPa increase in favorability for movements toward the north and northwest. At the 925 hPa level (approximately 850 m above MSL), the winds over the eastern Caribbean shift from blowing from the east in April and blow more from the east-southeast in May (fig. 15.1C,D). From 15 to 25° N winds blow more from the southeast and birds departing from the Antilles and from the Yucatán Peninsula in Mexico would be biased toward the northwest. Once over the waters of the Gulf of Mexico, atmospheric flow would carry most birds to the coasts of Texas and Louisiana. At latitudes greater than 30° N, the 925 hPa winds shift and blow more from the south and southwest, and atmospheric transport at this altitude would bias movements toward the north and northeast over much of the central and eastern United States. At the 850 hPa level (approximately 1,525 m above MSL), birds leaving the northern coast of South America in April would be drifted westward by strong winds from the east, but in May birds flying at this altitude could follow strong winds blowing from the southeast and reach Cuba or the Yucatán Peninsula, and if they move beyond these destinations, the 850-hPa winds would carry them to the northern coast of the Gulf of Mexico and into most of the central and eastern United States (fig. 15.1E,F). The wind patterns for the 700 hPa level (approximately 3,120 m above MSL) are not shown, but the winds are mostly zonal, with winds from the east and southeast south of the Gulf of Mexico and winds from the west north of the Gulf. Birds flying over the northern coast of South America at this altitude would be deflected to Central America, as would birds departing from the Antilles.

The results of the 24-h-trajectory simulations for the spring season for three different geographical locations for three different altitudes are in table 15.1. Birds departing the northern coast of South America and following the wind current at different altitudes would follow different trajectories. In March, trajectories starting at 500 m would move from 353° (toward 173°) after 24 h, whereas those starting at 2,500 m would move from 88° (toward 268°). The pattern is similar for April, and in May the trajectory starting at 500 m moves from 320° (toward 140°) and the trajectory starting at 2,500 m is from 107° (toward 287°). The change in the trajectories beginning at 2,500 m for April and May are in keeping with the changes in the flow patterns over northern South America noted earlier (fig. 15.1).

The 24-h trajectories beginning near Mérida, Yucatán Peninsula, Mexico, also show the trend of movement from the east toward the west for trajectories beginning at an altitude of 500 m and movement from the southeast toward the northwest for trajectories beginning at an altitude of 2,500 m (table 15.1). The shift to a more southeast-northwest trajectory with altitude is also reflected in the May mean wind vector data (fig. 15.1B,D,F). The lengths of the mean vectors indicate low variance in the wind trajectories.

The orientation of the 24-h trajectories for Lake Charles, Louisiana, changes with altitude (table 15.1, fig. 15.2A). Those beginning at 500 m are oriented approximately from south to north for each of the three spring months. The trajectories beginning at 1,000 m are oriented from 197–214° to 17–34°, whereas the trajectories starting at an altitude of 2,500 m are directed from 244–253° to 64–73°. The wind shift to the northeast at higher altitudes is also seen in the wind vector maps at higher latitudes (fig. 15.1A–F). The mean of the resultant vector also reflects the variability of the winds with altitude at Lake Charles (table 15.1). In April at 500 m above ground level (AGL), the mean vector length is 0.14, but the length increases at 1,000 m (0.36) and at 2,500 m (0.72). This is because the cold fronts that penetrate this latitude in spring affect the lower altitudes more than the higher altitudes. In May the mean vectors for 500 and 1,000 m AGL suggest that winds are much less variable.

At the midlatitudes (roughly 35 to 75° N) the atmosphere shows cyclical behavior as synoptic weather systems move across North America from west to east directed by the jet stream. The progression of high-pressure and low-pressure cells greatly influences the pattern of winds aloft. From March through May the frequency and strength of cold fronts reaching the southern United States decreases as the flow of tropical maritime air increasingly penetrates farther northward and eastward (fig. 15.1). During the summer, frontal systems rarely penetrate the southern United States. A 24-h-trajectory model showing favorable flow is plotted in fig. 15.2A, and one showing the effects of a cold front is plotted in fig. 15.2B.

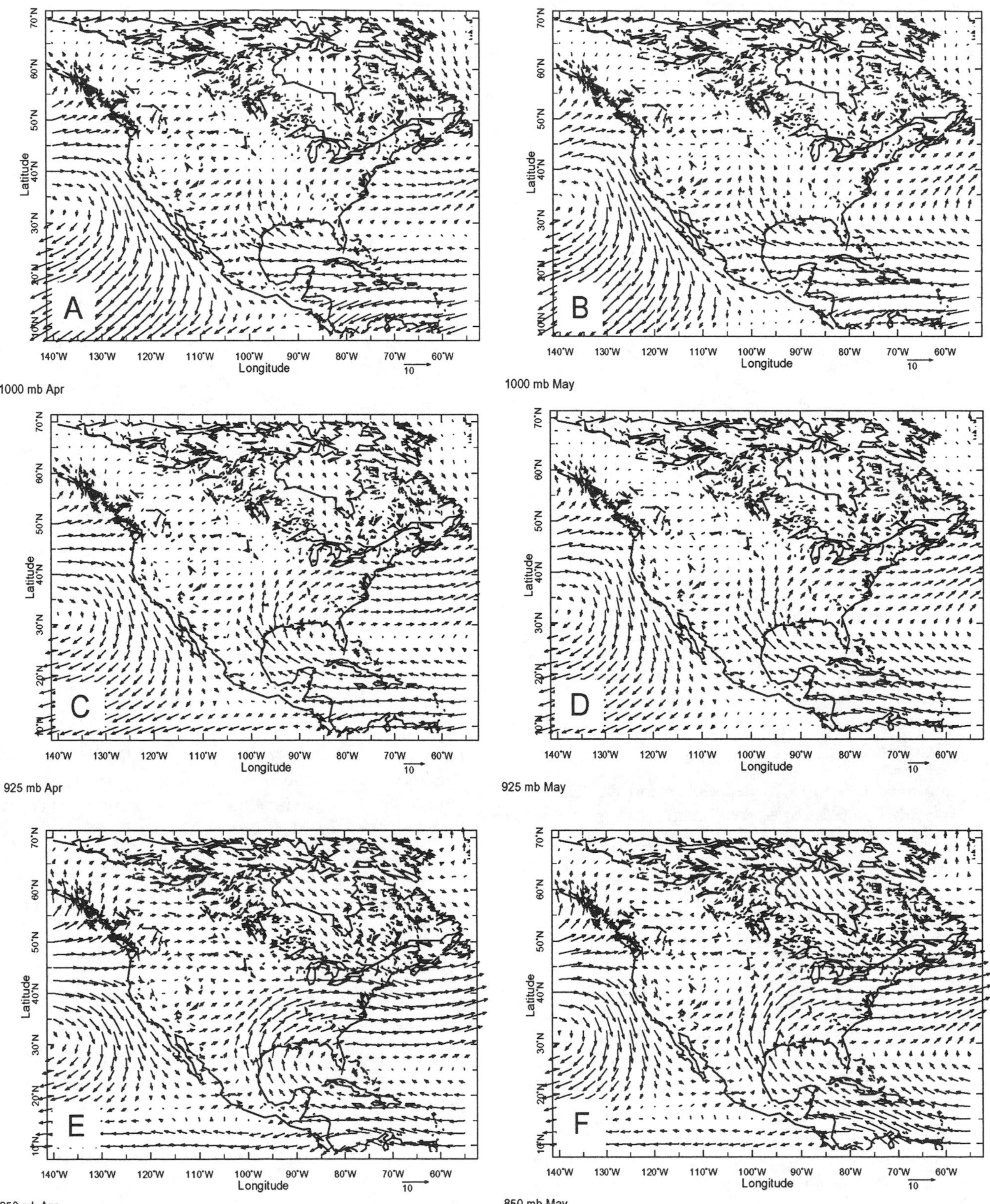

Fig. 15.1. Monthly mean vector plots of winds during April (A, C, E) and May (B, D, F) at the 1,000, 925, and 850 hPa levels for the years 1961–1990. The speed of the resulting wind is equal to the length of the vector.

Table 15.1 Atmospheric trajectories for three locations and three altitudes during spring migration for the period 1997 through 2001

Location	Month	Altitude (m)	*n*	Mean vector[a]	Mean vector length	Circular standard deviation[a]	Standard error[a]
South America	March	500	155	173.68	0.85	32.94	2.63
		1,000	155	202.15	0.81	37.44	2.98
		2,500	155	268.17	0.72	46.32	3.7
	April	500	150	168.6	0.78	40.17	3.24
		1,000	150	201.05	0.74	44.71	3.62
		2,500	150	265.61	0.69	49.15	4.02
	May	500	155	139.85	0.6	58.02	4.85
		1,000	155	173.66	0.37	80.52	8.52
		2,500	155	287.23	0.72	46.57	3.72
Yucatán	March	500	155	265.26	0.66	52.12	4.23
		1,000	155	286.52	0.6	57.51	4.79
		2,500	155	332.19	0.3	89.31	10.72
	April	500	150	268.12	0.77	41.76	3.37
		1,000	150	286.32	0.73	45.75	3.71
		2,500	150	314.76	0.4	77.32	7.87
	May	500	155	268.37	0.83	64.52	2.75
		1,000	155	282.65	0.82	36.17	2.88
		2,500	155	291.67	0.62	55.62	4.59
Lake Charles	March	500	155	11.42	0.25	95.55	12.86
		1,000	155	34.66	0.31	88.09	10.35
		2,500	155	73.23	0.73	45.56	3.63
	April	500	150	0.86	0.14	114.53	24.28
		1,000	150	24.07	0.36	82.11	8.93
		2,500	150	67.75	0.72	46.71	3.8
	May	500	155	12.14	0.69	49.71	4
		1,000	155	17.18	0.63	55.04	4.53
		2,500	155	63.75	0.56	61.58	5.27

[a] Units are in degrees and mean vector is an azimuth direction.

Fall Migration

During the fall months of September and October over most of the United States, the climatic pattern of winds aloft is not favorable for the southward migration of birds. The September pattern of wind vectors for 1,000 hPa (fig. 15.3A) is quite similar to that for April (fig. 15.1A), but the pattern changes a bit in October (fig. 15.3B) as winds along the mid-Atlantic Coast and over the extreme southeastern United States and Gulf of Mexico become favorable for the transport of birds from northeast to southwest. Light winds over the western United States and stronger winds just offshore over the Pacific Ocean blow from the northwest as they do in spring (fig. 15.1). Birds attempting to move south over the Great Plains on average encounter head winds, whereas birds moving over the northeastern United States and eastern Canada encounter winds from the southwest-west. In September the winds blowing from west-southwest are positioned mostly north of the United States (fig. 15.3C,E), but in October the westerly winds are displaced a bit southward and occur over much of the northern states (fig. 15.3D,F). During the fall, mean winds for September and October at all levels over the southeast are light and variable (fig. 15.3), and winds over the Gulf are from the east-southeast during September (fig. 15.3A,C,E), and from the northeast-east during October (fig. 15.3B,D,F). Cold fronts begin to penetrate the northern Gulf Coast in September but they are infrequent. When they do occur, wind trajectories are temporarily favorable for birds attempting to cross the Gulf of Mexico (fig. 15.2C), but the average atmospheric flow in fall on the northern Gulf Coast is unfavorable and most atmospheric trajectories resemble those plotted in fig. 15.2D.

DISCUSSION

Does the geographical pattern of atmospheric circulation serve as a template for migration patterns between breeding and wintering areas? To answer this question we must first consider the characteristics of atmospheric circulation and then consider the ultimate factors that select for movements that increase reproductive success (dispersal) and movements that increase survival (migration).

Characteristics of Atmospheric Circulation

Atmospheric circulation is characterized by its mean behavior and its variations (Lorenz 1967; Panel on Climate

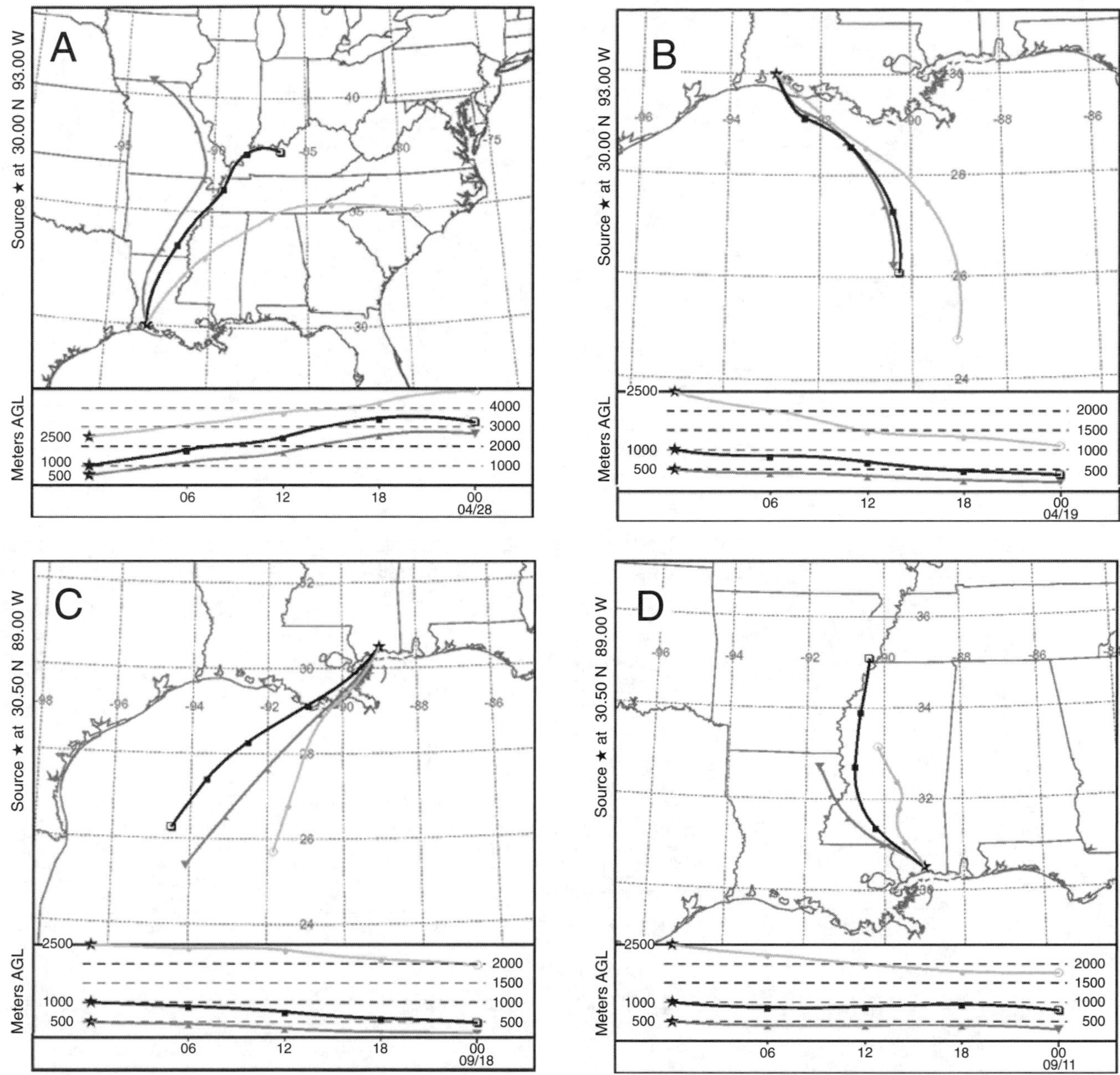

Fig. 15.2. Simulated 24-h trajectories of air parcels generated by the HYSPLIT model for three different altitudes at: (A) Lake Charles, Louisiana, on 27 April 1998; (B) Lake Charles, Louisiana, on 18 April 1999; (C) Biloxi, Mississippi, on 17 September 2000, and (D) Biloxi, Mississippi, on 10 September 2000.

Variability on Decade-to-Century Time Scales 1998:62–63). The mean state and the variability of the atmosphere depend on the annual climatic cycle (seasons), but the basic properties of the atmosphere are present in all seasons. Permanent mean-circulation features have a three-dimensional asymmetry: in the vertical dimension compressibility of air and gravity produce a rapid drop in pressure with altitude. On the N-S axis the Earth's rotation and variations in solar radiation generate a meridional circulation that produces large-scale belts of different weather and climate. On the E-W axis topography and land-sea contrast produces undulations in the atmosphere's strongest flow—the zonal wind.

Zonal winds are dynamically unstable and produce baroclinic and stationary eddies that induce movement of heat and moisture away from the equator and flow of cold dry air toward the equator. Variation in symmetric zonal flow and fluctuation in strength and position of stationary eddies dictate climate patterns. Stationary eddies set up different climate regimes on east and west sides of continents. Transient baroclinic eddies in the midlatitudes (30–60° of latitude) are steered by stationary eddies and induce uplifting of air masses. Baroclinic eddies are extremely important in the maintenance of the average zonal circulation and the stationary waves.

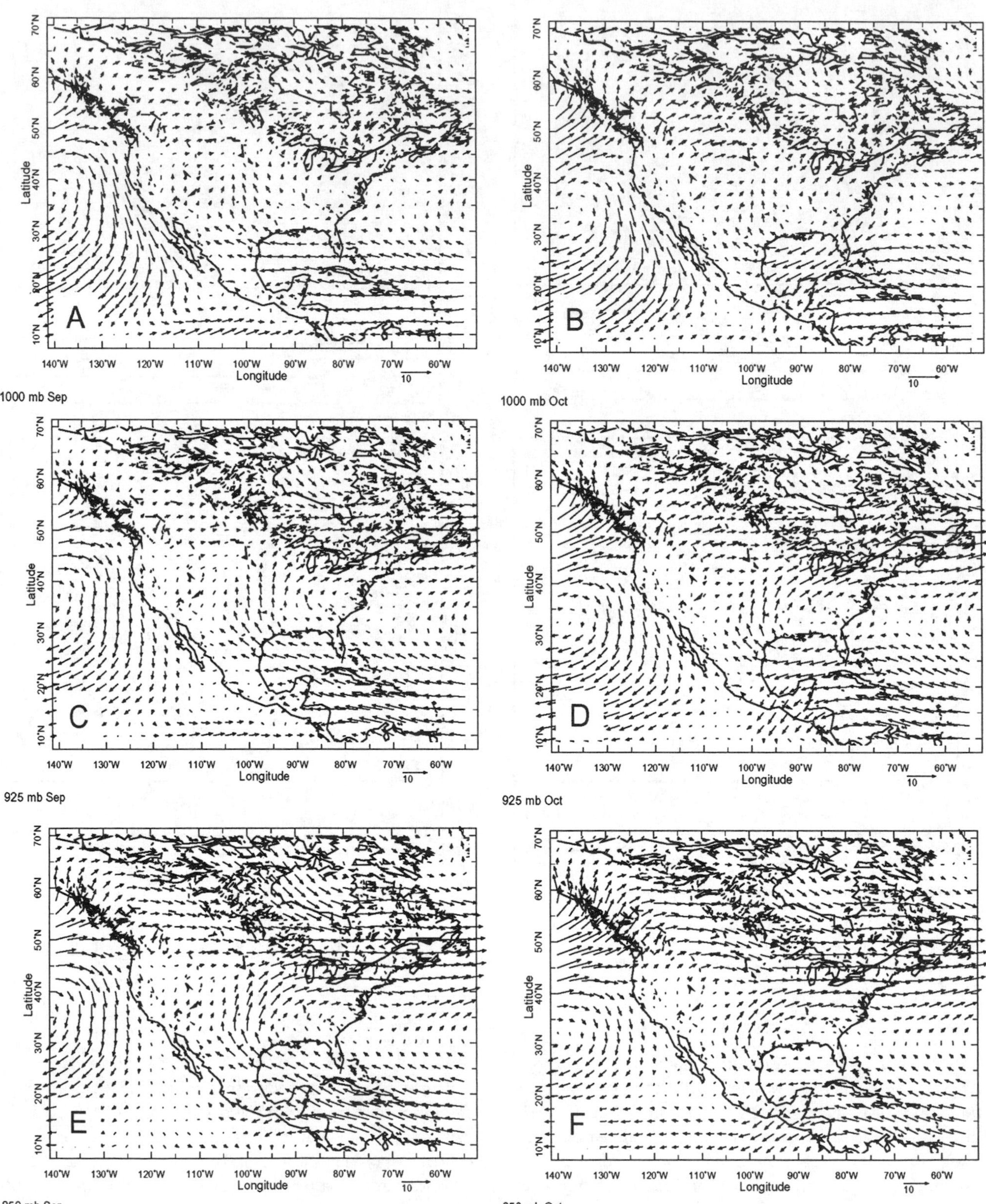

Fig. 15.3. Monthly mean vector plots of winds at the 1,000, 925, and 850 hPa levels for September (A, C, E) and October (B, D, F) of 1961–1990. The speed of the resulting wind is equal to the length of the vector.

The atmosphere is constantly changing in time and space, and in order to discuss the influence of these changes on organisms, one must do so with reference to appropriate temporal and spatial scales. Relatively long-term climate cycles (e.g., glacial and interglacial periods) are responsible for patterns of biotic retreat and colonization—biogeographical migrations (Gauthreaux 1980a). Retreats to refugia characterize the period leading up to glacial maximum. During the last glacial maximum (18,000 ± 1,000 years B.P. on the radiocarbon time scale) the Wisconsin Ice Front extended south of the Great Lakes to central Ohio and Illinois (Frenzel 1973). Growing degree-days (5°C base) were fewer, the mean temperature of the coldest month was 4°–8° lower, and upper level winds increased in speed and were displaced southward of their current position (Whitlock et al. 2001).

The mean annual temperatures characteristic of the Canadian border region at present extended southward to Tennessee and North Carolina, and mixed forests were restricted to the mid-Atlantic and Southeast (Frenzel et al. 1992). In the early Holocene (14,000–6,000 years B.P.) conditions improved as temperatures climbed to 1°C colder in July than present and 1°–2°C colder in January than present, and upper-level winds during this period returned to near their present configuration (Whitlock et al. 2001). With ameliorating conditions recolonization (range expansions) began as the biota moved from refugia northward.

Dispersal, Range Expansion, and Atmospheric Flow

Much has been written about the advantages of breeding in the highly seasonal Temperate Zone of the Western Hemisphere (e.g., increased summer resource levels [Ricklefs 1980]; lower predation [Klopfer et al. 1974; Skutch 1985; Martin 1992]; and longer day length [Hussell 1985]). Gauthreaux (1982) summarized many of the early theories that suggest that tropical birds evolved migration to exploit Temperate Zone resources during the summer season of the annual climatic cycle. Since then several authors have considered the origins of the tropical-temperate migration system (e.g., Levey and Stiles 1992; Rappole and Tipton 1992; Rappole 1995; Safriel 1995; Bell 2000) as the ranges of species of tropical ancestry expanded northward to exploit increased resources and maximize their reproductive output.

Shifts in the geographical distributions of species in response to long-term climatic fluctuations in the Palearctic and Africa have been examined and summarized by Lövei (1989). He noted that between 18,000–7,000 years B.P. as glaciers declined, vegetation belts moved northward and Palearctic migrants expanded about tenfold. In contrast to the Neotropical-Nearctic expansions from refugia, the African-Palearctic expansions from equatorial Africa and the Mediterranean encountered more east–west-oriented geographical barriers (desert, water, and mountains). Also in marked contrast to the Neotropics and the Orient, many Afrotropical species undertake polarized migrations in sub-Saharan Africa, with part of the population moving north of, and part south of, the Tropics (Hockey 2000). This is thought to be so because of the spatial symmetry and large extent of savannas both north and south of the African Tropics, coupled with a lack of north-south dispersal barriers on either side of the Tropics. According to Hockey (2000) these conditions are not replicated in either the Neotropics or the Orient. By using a molecular phylogeographic approach to study the evolution of migration, Joseph et al. (1999) and Joseph (Chap. 2, this volume) add to a growing body of literature suggesting that shifts of breeding distributions are commonly, though not necessarily exclusively, involved in the evolution of migration.

Does wind influence dispersal movements of birds as it influences the dispersal patterns of insects and plants during range expansion? We suspect that wind patterns do bias dispersal, but this hypothesis has not been tested. To adequately do so would require two data sets, one containing information on dispersal directions and range expansion and the other containing information on the direction and speed of prevailing winds. Although dispersal flights tend to be random in direction with respect to the population as a whole, selection should favor individuals with directional tendencies that benefit from a tailwind. These individuals will have an advantage in reaching unexploited resources first, establishing a territory, and breeding. By dispersing with tailwinds it is also possible to cross more easily ecological barriers (desert or large bodies of water) with long-distance flights. Birds moving against the wind or flying in directions that cross the wind do not have such advantages.

During range expansions following the retreat of the glaciers in North America, young (natal dispersers) and prebreeding adults (breeding dispersers) moving northward from refugia in the Tropics, Subtropics, and South Temperate areas might have benefited from using favorable atmospheric flow to reach more temperate areas—some north of water barriers. This would be particularly true if the dispersal and range expansion movements took the birds over the Caribbean Sea or Gulf of Mexico. There is some evidence that birds disperse at night (Beebe 1947; McClure 1974; Bulyuk et al. 2000; Mukhin 2004), and it is likely that dispersing birds would more likely move out over a water barrier during darkness than during daylight. By moving with the parcels of tropical atmosphere, these individuals would reach locations where the maritime tropical air mass increases temperature and rainfall during the spring and summer months and greatly increases biological productivity. Successful breeding would enhance site-attachment (Ketterson and Nolan 1990) to the new location. This is important because individuals forced to escape the adversity of the Temperate Zone winter would be able to return to familiar areas where they bred successfully.

Survival Movements and Atmospheric Flow

As breeding ranges expand northward (via dispersal), an increasing proportion of individuals must leave the breeding grounds during the Temperate Zone winter (partially migratory), and at high latitudes all of the individuals in the population will have to leave (completely migratory) or perish (Gauthreaux 1985). Not surprisingly the percentage of migrant species increases with increasing latitude and severity of winter temperature in North America (Willson 1976; Newton and Dale 1996a), in Europe (Herrera 1978; Newton and Dale 1996b), and in South America (Chesser 1998).

The patterns of atmospheric flow over much of North America in the fall are not favorable for the southward survival movements of birds. Prevailing wind patterns are not favorable for southward flights in the central United States in fall, because the winds are usually from the south (head winds), whereas in the northeastern United States and eastern Canada the winds are from the west / southwest-west. In western Canada winds at the 925 and 850 hPa levels flow from west to east in fall (fig. 15.3C–F). There is evidence that continental populations of Swainson's Thrush (*Catharus ustulatus*) relatively recently have expanded breeding ranges toward the northwest into Canada and Alaska. In the fall the thrushes do not take the shortest route from Alaska and northwestern Canada toward their wintering grounds, but migrate by first moving eastward and then southward to reach wintering grounds in Panama and South America (Ruegg and Smith 2002). Although these authors suggest that the birds are following their range expansion, the route also takes advantage of prevailing winds. Migratory birds can find favorable winds along the Atlantic Coast and in extreme southeastern United States in the fall at the 1,000 and 925 hPa levels (fig. 15.3A–D), but clearly many birds in the central United States cannot just follow the mean flow of the wind and arrive at destinations to the south.

How, then, can birds move southward in the fall and take advantage of following winds? If they require following winds they must wait until a synoptic weather system (baroclinic eddy) passes and winds become temporarily favorable for southward flight. From the perspective of the Temperate Zone, peak numbers of most types of migrants move with following winds relative to their preferred direction (Alerstam 1979; Richardson 1990, 1991; Liechti and Bruderer 1998), and fall migrants have adopted a "sit-and-wait" strategy for favorable winds that will carry them to a suitable nonbreeding area (Åkesson and Hedenström 2000). The sit-and-wait strategy not only allows for building energy stores at stopover areas (Moore 2000), but also allows birds to avoid adverse wind and wait for favorable winds blowing in a preferred direction (Weber et al. 1998).

In central Europe opposing winds prevail, and head winds and tailwinds occur in similar frequencies on the western and eastern edges of the Mediterranean. These winds are highly variable in altitude and time (Liechti and Bruderer 1998). Under these circumstances if a bird waits for favorable winds to migrate it speeds up its flight by 30% on average compared with an individual disregarding wind conditions, and if it selects the best flight altitude it gains an additional 40% in flight speed. Liechti and Bruderer (1998) conclude that if a bird takes wind conditions into account carefully, it can almost double its flight speed and save about half of the energy required for its migration through central Europe and the Mediterranean. They add that for birds attempting a single long-distance flight, selecting favorable winds should be more important than the time required to accumulate departure levels of fat.

How did the linkage between wind direction and migration evolve in the fall? As fall progresses, the frequency and strength of cold fronts increase and ultimately the strong winds from the northwest, north, and northeast bring colder temperatures and with them a decline in the suitability of the habitat. With respect to atmospheric flow, the winds associated with decline in habitat quality also bias the directions birds fly in their escape from the adversity of seasonal climate change. By moving on winds from the northwest, north, and northeast birds ultimately show greater survival. For insectivorous species and possibly frugivorous species the first hard freeze of the fall greatly diminishes the quantity and quality of food resources, and movement to an area south of the freeze line would be highly adaptive. The relationship between synoptic atmospheric flow in the fall and the deterioration of suitable habitat has undoubtedly played an important role in the evolution of the linkage between northwest and north winds and the timing of fall movements from areas where breeding or site attachment occurred. Recent studies by Pulido et al. (2001) suggested that evolutionary changes in the timing of autumn migration may take place over a very short time period and is likely unconstrained by the lack of additive genetic variation.

It is not surprising that migrating birds are highly influenced by winds aloft and prefer to fly at altitudes with favorable winds (Bruderer 1971; Bruderer and Steidinger 1972; Cochran and Kjos 1985; Gauthreaux 1991; Bruderer et al. 1995). Although most bird migration is accomplished at altitudes within 1,000 m above the ground in the Northern Hemisphere (Bellrose and Graber 1963; Able 1970; Bruderer 1971), migration at 2,000–6,000 m above sea level has been recorded over Israel (Bruderer and Liechti 1995), the western Atlantic and Puerto Rico (Richardson 1976), and the Caribbean Sea (Williams and Williams 1978). In southern Israel some migrating wading birds fly downwind within a low-level jet stream at heights of 5,000 to almost 9,000 m above sea level at speeds up to 50 m sec^{-1} (Liechti and Schaller 1999).

In the Western Hemisphere in spring the prevailing structure of atmospheric flow is generally favorable for the return movement of birds from nonbreeding areas to familiar areas on their breeding grounds in the central and eastern United States (fig. 15.1). South of 30° N few baroclinic eddies form and flow is dictated by a large stationary eddy located off the southeastern United States. As spring progresses, synoptic systems rarely penetrate latitudes south of 30° N, so there is relatively little need to "sit and wait" for favorable winds as is the case in fall. Consequently migrants wintering in eastern Mexico and the Antilles could follow a "go-with-the-mean-flow" strategy and reach the southern United States. However, above 30° N latitude synoptic systems continue to move west to east, and under these circumstances, as migrants move to distant northern locations, they are more likely to adopt a "sit-and-wait" strategy and wait for tailwinds that will assist them in reaching their final goal. In the western Mediterranean, wind direction is a determining factor for stopover decisions and resumption of flight for Reed Warblers (*Acrocephalus scirpaceus*) at an intermediate stage of their spring migration where topography governs wind patterns (Barriocanal et al. 2002).

Global climate changes have altered the distribution and productivity of habitat types in North America and have dictated the amount of available breeding habitat for birds in the Temperate Zone (Woodward 1987; Woodward and Williams 1987; Huntley and Webb 1988; Thompson et al. 1999; Dynesius and Jansson 2000). Despite the resulting changes in the northward extent of species distributions, the fact remains that it is the operation of the *annual climatic cycle* that ultimately selects for migratory behavior in birds. The patterns of atmospheric flow in conjunction with seasonal climatic changes have undoubtedly influenced the evolution of strategies during the "southward escape from adversity" leg and the "northward return to a familiar area to breed" leg of the annual migration circuit. In both instances selection has favored minimizing the cost of transport by using winds. The process is ongoing and has certainly operated for thousands of years as the global circulation patterns have remained fairly constant from glacial times to the present (Williams and Webb 1996; Preusser et al. 2002). Global warming will certainly affect bird migration patterns, and according to Berthold (1991), residents will benefit from a decrease in winter mortality, partial migrants will shift to sedentariness, and as a result of increased competition, long-distance migrants with the slowest adaptation rate will show further decline.

FUTURE DIRECTIONS

Have prevailing wind patterns influenced dispersal and range expansions of birds? We really do not know how wind influences the dispersal of birds, so we cannot say much about how wind patterns have influenced range expansions. Our knowledge of dispersal movements (time of day, flight altitude, directionality) is minuscule (e.g., Baker 1993; Bulyuk et al. 2000; Mukhin 2004) and much more research is needed before we can accurately determine whether atmospheric flow biases dispersal movements in expanding populations.

Some of the dispersal patterns and range expansions of Neotropical migratory birds following the retreat of the glaciers in North America involved crossing water barriers (refugia on islands), whereas others were restricted to land (refugia in Central America and the southern United States). Expansion over land likely would be influenced less by wind patterns (unless dispersing birds are biased by wind), but expansion over a sizable water barrier would more likely be downwind in the prevailing wind direction, particularly if the wind pattern shows low variability in direction. We suggest that tests examining the role of wind direction in determining dispersal directions and range expansions should compare range expansion over land versus range expansion over water. We also need to explore the relationship between dispersal behavior and migratory behavior in migratory birds and examine the factors that select for the orientation (heading) of these movements.

Have prevailing wind patterns influenced the evolution of migratory pathways? For Neotropical migratory birds the answer to this question cannot be a simple "yes" or "no" because during the spring the prevailing atmospheric flow is favorable for return migration to the breeding grounds for many migrants breeding in central and eastern North America, but in the fall this is not the case. Consequently, we doubt that all migration pathways have evolved in response to the prevailing wind patterns on the globe. We find that where prevailing winds are adverse for migration there are episodes of favorable wind, and not surprisingly most migration occurs during these periods. Clearly more information is required on how birds assess atmospheric structure and how decisions to migrate are made on the ground with reference to wind conditions (Schaub et al. 2004). We also need to examine how changes in wind patterns influence changes in flight strategies and how the latter changes affect condition, reproductive success, and ultimately survival.

Studies that relate genetic structure and history of population expansions to patterns of migration (e.g., Ruegg and Smith 2002) hold great promise for understanding the role of atmospheric circulation. Studies using genetic markers clearly show that the Dunlin (*Calidris alpina*) has a parallel migration system, with western Palearctic breeding populations wintering mainly in the western part of the wintering range and populations breeding farther east wintering farther east. The results also show that the distance between breeding and wintering area increases eastward in this region (Wennerberg 2001). As more genetic and stable isotope studies on migratory connectivity between breeding and wintering areas are conducted (Webster et al. 2002; see Smith et al., Chap. 18, and Hobson, Chap. 19, both in this volume), it will be easier to test null hypotheses regarding the role of prevailing wind patterns in avian range expansions (dispersal) and the evolution of migration pathways.

ACKNOWLEDGMENTS

Pete Marra's and Russ Greenberg's recommendations were particularly beneficial in revising the manuscript as were the comments and suggestions from three anonymous reviewers. We also appreciate the access to the web-based HYSPLIT model that is restricted to registered users.

LITERATURE CITED

Able, K. P. 1970. A radar study of the altitude of nocturnal passerine migration. Bird-Banding 41:282–290.

Able, K. P. 1972. Fall migration in coastal Louisiana and the evolution of migration patterns in the Gulf region. Wilson Bulletin 84:231–242.

Åkesson, S., and A. Hedenström. 2000. Wind selectivity of migratory flight departures in birds. Behavioral Ecology and Sociobiology 47:140–144.

Alerstam, T. 1979. Wind as selective agent in bird migration. Ornis Scandinavica 10:76–93.

Alerstam, T., and A. Hedenström. 1998. The development of bird migration theory. Journal of Avian Biology 29:343–369.

Alerstam, T., and S.-G. Pettersson. 1976. Do birds use waves for orientation when migrating across the sea? Nature 259:205–207.

Baker, R. R. 1993. The function of post-fledging exploration: a pilot study of three species of passerines ringed in Britain. Ornis Scandinavica 24:71–79.

Barriocanal, C., D. Montserrat, and D. Robson. 2002. Influences of wind flow on stopover decisions: the case of the reed warbler *Acrocephalus scirpaceus* in the Western Mediterranean. International Journal of Biometeorology 46:192–196.

Beebe, W. 1947. Avian migration at Rancho Grande in north-central Venezuela. Zoologica 32:153–168.

Bell, C. R. 2000. Process in the evolution of bird migration and pattern in avian ecogeography. Journal of Avian Biology 31:258–265.

Bellrose, F. C., and R. C. Graber. 1963. A radar study of the flight directions of nocturnal migrants. Pages 362–389 *in* Proceedings of the 13th International Ornithological Congress.

Berthold, P. 1991. Patterns of avian migration in light of current global "greenhouse" effects: a Central European perspective. Pages 780–786 *in* Proceedings of the 22nd International Ornithological Congress.

Bruderer, B. 1971. Radarbeobachtungen über den Frühlingszug im Schweizerischen Mittelland. Ornithologischer Beobachter 68:89–158.

Bruderer, B., and F. Liechti. 1995. Variation in density and height distribution of nocturnal migration in the south of Israel. Israel Journal of Zoology 41:477–487.

Bruderer, B., and P. Steidinger. 1972. Methods of quantitative and qualitative analysis of bird migration with tracking radar. Pages 151–167 *in* Animal Orientation and Navigation: A Symposium (S. R. Galler, K. Schmidt-Koenig, G. J. Jacobs, and R. E. Belleville, eds.). National Aeronautics and Space Administration, Washington, D.C.

Bruderer, B., L. Underhill, and F. Liechti. 1995. Altitude choice of night migrants in a desert area predicted by meteorological factors. Ibis 137:44–55.

Bulyuk, V. N., A. Mukhin, V. A. Fedorov, A. Tsvey, and D. Kishkinev. 2000. Juvenile dispersal in reed warblers *Acrocephalus scirpaceus* at night. Avian Ecology and Behavior 5:45–61.

Chesser, R. T. 1998. Further perspectives on the breeding distribution of migratory birds: South American austral migrant flycatchers. Journal of Animal Ecology 67:69–77.

Cochran, W. C., and C. G. Kjos. 1985. Wind drift and migration of thrushes: a telemetry study. Illinois Natural History Survey Bulletin 33:297–330.

Dingle, H. 1996. Migration: The Biology of Life on the Move. Oxford University Press, New York.

Draxler, R. R., and G. D. Hess. 1997. Description of the HYSPLIT-4 modeling system. NOAA Technical Memorandum ERL ARL-224, December.

Dynesius M., and R. Jansson. 2000. Evolutionary consequences of changes in species' geographical distributions driven by Milankovitch climate oscillations. Proceedings of the National Academy of Science (USA) 97: 9115-9120.

Evans, P. R. 1966. Migration and orientation of passerine night migrants. Journal of Zoology 150:319–369.

Frenzel, B. 1973. Climatic Fluctuations of the Ice Age. [Translated by A. E. M. Nairn.] Case Western Reserve University Press, Cleveland.

Frenzel, B, M. Pécsi, and A. A. Velichko. 1992. Atlas of Paleoclimates and Paleoenvironments of the Northern Hemisphere, Late Pleistocene–Holocene. Geographical Research Institute, Hungarian Academy of Sciences, Budapest, Gustav Fischer Verlag, Budapest and Stuttgart.

Gauthreaux, S. A., Jr. 1980a. The influences of long-term and short-term climatic changes on the dispersal and migration of organisms. Pages 103–174 *in* Animal Migration, Orientation, and Navigation (S. A. Gauthreaux Jr., ed.). Academic Press, New York.

Gauthreaux, S. A., Jr. 1980b. The influences of global climatological factors on the evolution of bird migratory pathways. Pages 517–525 *in* Proceedings of the 17th International Ornithological Congress.

Gauthreaux, S. A., Jr. 1982. The ecology and evolution of avian migration systems. Pages 93–168 *in* Avian Biology, Vol. 6 (D. S. Farmer, J. R. King, and K. C. Parkes, eds.). Academic Press, New York.

Gauthreaux, S. A., Jr. 1985. The temporal and spatial scales of migration in relation to environmental changes in time and space. Contributions in Marine Science 27 (Supplement): 503–515.

Gauthreaux, S. A., Jr. 1991. The flight behavior of migrating birds in changing wind fields: radar and visual analyses. American Zoologists 31:187–204.

Green, M. 2001. Is wind drift in migrating barnacle and brant geese, *Branta leucopsis* and *Branta bernicla,* adaptive or non-adaptive? Behavioral Ecology and Sociobiology 50:45–54.

Green, M., and T. Alerstam. 2002. The problem of estimating wind drift in migrating birds. Journal of Theoretical Biology 218:485–496.

Hedenström, A., and T. Alerstam. 1995. Optimal flight speed of birds. Philosophical Transactions of the Royal Society of London, Series B, Biological Sciences 348:471–487.

Herrera, C. M. 1978. On the breeding distribution pattern of European migrant birds: MacArthur's theme reexamined. Auk 95:496–509.

Hockey, P. A. R. 2000. Patterns and correlates of bird migrations in sub-Saharan Africa. Emu 100:401–417.

Huntley, B., and T. Webb, III. eds. 1988. Vegetation History: Handbook of Vegetation Science, Vol. 7. Kluwer Academic Publishers, Dordrecht.

Hussell, D. J. T. 1985. Clutch size, daylength and seasonality of resources: comments on Ashmole's hypothesis. Auk 102:632–634.

HYSPLIT 4 (Hybrid Single-Particle Lagrangian Integrated Trajectory) Model. 1997. Web address: www.arl.noaa.gov/ready/hysplit4.html, NOAA Air Resources Laboratory, Silver Spring, Md.

Joseph, L., E. P. Lessa, and L. Christidis. 1999. Phylogeny and biogeography in the evolution of migration: shorebirds of the *Charadrius* complex. Journal of Biogeography 26:329–342.

Kalnay, E., M. Kanamitsu, R. Kistler, et al. 1996. The NCEP/NCAR 40-year reanalysis project. Bulletin of the American Meteorological Society 76:437–471.

Ketterson, E. D., and V. Nolan Jr. 1990. Site attachment and site fidelity in migratory birds: experimental evidence from the field and analogies from neurobiology. Pages 118–129 *in* Bird Migration (E. Gwinner, ed.). Springer-Verlag, Berlin.

Kistler, R., E. Kalnay, W. Collins, et al. 2001. The NCEP-NCAR 50-year reanalysis: monthly means CD-ROM and documentation. Bulletin of the American Meteorological Society 82:247–267.

Klopfer, P. H., D. I. Rubenstein, R. S. Ridgely, and R. J. Barnett. 1974. Migration and species diversity in the tropics. Proceedings of the National Academy of Sciences USA 71:339–340.

Landsberg, H. 1948. Bird migration and pressure patterns. Science 108:708–709.

Levey, D. J., and F. G. Stiles. 1992. Evolutionary precursors of long-distance migration: resource availability and movement patterns in Neotropical landbirds. American Naturalist 140:447–476.

Liechti, F., and B. Bruderer. 1998. The relevance of wind for optimal migration theory. Journal of Avian Biology 29:561–568.

Liechti, F., and E. Schaller. 1999. The use of low-level jets by migrating birds. Naturwissenschaften 86:549–551.

Lorenz, E. N. 1967. Nature and theory of the general circulation of the atmosphere. World Meteorological Organization Bulletin 16:74–78.

Lövei, G. 1989. Passerine migration between the Palaearctic and Africa. Current Ornithology 6:143–174.

Martin, T. E. 1992. Interaction of nest predation and food limitation in reproductive strategies. Current Ornithology 9:163–197.

McClure, H. E. 1974. Migration and survival of the birds of Asia. United States Army Medical Component, South-East Asia Treaty Organization (SEATO) Medical Project, Bangkok.

Moore, F. R. (ed.). 2000. Stopover ecology of Nearctic-Neotropical landbird migrants: habitat relations and conservation implications. Studies in Avian Biology no. 20. Cooper Ornithological Society, Camarillo, Calif.

Mukhin, A. 2004. Night movements of young Reed Warblers (*Acrocephalus scirpaceus*) in summer: is it postfledging dispersal? Auk 121:203–209.

Newton, I., and L. Dale. 1996a. Bird migration at different latitudes in eastern North America. Auk 113:626–635.

Newton, I., and L. Dale. 1996b. Relationship between migration and latitude among European birds. Journal of Animal Ecology 65:137–146.

Panel on Climate Variability on Decade-to-Century Time Scales 1998. Decade-to-century-scale climate variability and change: a science strategy. Board on Atmospheric Sciences and Climate, Commission on Geosciences, Environment and Resources; National Research Council. National Academy Press, Washington, D.C.

Pennycuick, C. J. 1978. Fifteen testable predictions about bird flight. Oikos 30:165–176.

Preusser, F., D. Radies, and A. Matter. 2002. A 160,000-year record of dune development and atmospheric circulation in Arabia. Science 296:2018–2020.

Pulido, F., P. Berthold, G. Mohr, and U. Querner. 2001. Heritability of the timing of autumn migration in a natural bird population. Proceedings of the Royal Society of London, Series B, Biological Sciences 268:953–959.

Rappole, J. H. 1995. The ecology of migrant birds: a Neotropical perspective. Smithsonian Institution Press, Washington, D.C.

Rappole, J. H., and A. R. Tipton. 1992. The evolution of avian migration in the Neo-tropics. Ornithologia Neotropical 3:45–55.

Richardson, W. J. 1976. Autumn migration over Puerto Rico and the western Atlantic: a radar study. Ibis 118:309–332.

Richardson, W. J. 1990. Timing of bird migration in relation to weather: updated review. Pages 78–101 *in* Bird Migration (E. Gwinner, ed.). Springer-Verlag, Berlin.

Richardson, W. J. 1991. Wind and orientation of migrating birds: a review. Experientia 46:416–425.

Ricklefs, R. E. 1980. Geographic variation in clutch size among passerine birds: Ashmole's hypothesis. Auk 101:498–506.

Ruegg, K. C., and T. B. Smith. 2002. Not as the crow flies: a historical explanation for circuitous migration in Swainson's thrush (*Catharus ustulatus*). Proceedings of the Royal Society London, Series B, Biological Sciences 269:1375–1381.

Safriel, U. N. 1995. The evolution of Palaearctic migration: the case of southern ancestry. Israel Journal of Zoology 41:417–431.

Sandberg, R. 1996. Fat reserves of migrating passerines at arrival on the breeding grounds in Swedish Lapland. Ibis 138:514–524.

Schaub, M, F. Liechti, and L. Jenni. 2004. Departure of migrating European robins, *Erithacus rubecula,* from a stopover site in relation to wind and rain. Animal Behaviour 67:229–237.

Skutch, A. S. 1985. Clutch size, nesting success, and predation on nests of Neotropical birds, reviewed. Ornithological Monographs 36:575–603.

Thompson, R. S., K. H. Anderson, and P. J. Bartlein. 1999. Atlas of Relations Between Climatic Parameters and Distributions of Important Trees and Shrubs in North America. U.S. Geological Survey Professional Paper 1650 A&B, Online Version 1.0 (http://pubs.usgs.gov/pp/p1650-a/).

Weber, T. P., T. Alerstam, and A. Hedenström. 1998. Stopover decisions under wind influence. Journal of Avian Biology 29:552–560.

Webster, M. S., P. P. Marra, S. M. Haig, S. Bensch, and R. T. Holmes. 2002. Links between worlds: unraveling migratory connectivity. Trends in Ecology & Evolution 17:76–83.

Wennerberg, L. 2001. Breeding origin and migration pattern of dunlin (*Calidris alpina*) revealed by mitochondrial DNA analysis. Molecular Ecology 10:1111–1120.

Whitlock, C., P. Bartlein, V. Markgraf, and A. C. Ashworth. 2001. The mid-latitudes of North and South America during the last glacial maximum and early Holocene: similar paleoclimatic sequences despite differing large-scale controls. Pages 391–416 *in* Interhemispheric Climate Linkages: Present and Past Interhemispheric Climate Linkages in the Americas and Their Societal Effects (V. Markgraf, ed.). Academic Press, New York.

Williams, T. C., and T. Webb III. 1996. Neotropical bird migration during the ice ages: orientation and ecology. Auk 113:105–118.

Williams, T. C., and J. M. Williams. 1978. An oceanic mass migration of land birds. Scientific American 239:138–145.

Willson, M. F. 1976. The breeding distribution of North American migrant birds: a critique of MacArthur (1959). Wilson Bulletin 88:582–587.

Woodward, F. I. 1987. Climate and plant distribution: Cambridge, Cambridge University Press.

Woodward, F. I., and B.G. Williams. 1987. Climate and plant distribution at global and local scales. Vegetation 69:189–197.

Part 4 / CONNECTIVITY

ROBERT C. FLEISCHER

Overview

MIGRATORY BIRDS ARE CLEARLY CAPABLE of moving very long geographical distances from their breeding to their nonbreeding grounds. However, does their ability to travel long distances necessarily mean that migratory birds have high levels of natal and breeding dispersal and thus little or no population structure compared to sedentary species? If population structure is found for a species, can it be effectively used to identify the breeding origins of individuals caught in migration or on nonbreeding grounds? What types of markers can be of use in establishing connectivity—the patterns of connection between populations in different seasons? Finally, can we measure the effect of conditions at nonbreeding sites on aspects of life history, genetics, and behavior at breeding sites, or vice versa? In this set of chapters, authors use a variety of methods to assess population structure of migratory birds and proceed to show the relative value of intrinsic markers of this structure, such as genetic variants, parasites, and isotope signals, for assessing connectivity. They also discuss the importance of documenting population and individual connectivity throughout the annual cycle, *both* for understanding the evolution of migration and other adaptive characteristics and for realistically addressing conservation management of migratory species.

The first step in such studies is to assess *population structure.* Because of their ability to move greater distances, we expect that migratory populations will be less structured than sedentary ones, but this has not always proven to be the case. Banding, other auxiliary markers, and radio transmitters have been successfully used to measure the movements of individuals directly, and from these to infer dispersal. But these studies do not often reveal whether dispersal is effective (i.e., whether gene flow takes place) and they often suffer from sample size problems. Thus, other methodologies (discussed below) may be required to document population structure in migratory birds adequately.

The follow-up is then to use markers of population structure and assignment to *identify the origins of individuals,* that is, to geographically link breeding and nonbreeding ranges and migratory pathways. Again, banding can be useful, but rarely does such an approach yield return or recapture rates that can provide statistical rigor. Likewise, with small songbirds, technology has not yet advanced to the point where satellite transmitters can be used, and conventional radio transmitters do not have sufficient range. Many important intrinsic markers of population structure have been developed and used in only the past few years, including genetic variants, parasite types, and stable isotope and trace element signatures. Each marker can vary in strength and resolution; thus, as often pointed out in the chapters that follow, this goal may also be best reached by a combination of methods.

Genetic markers have become remarkably powerful over the past few years for determining population structure, mating system, and evolutionary relationships. The primary genetic markers include allozymes (protein electrophoresis), mitochondrial DNA sequences, and DNA fragment analyses such as microsatellites and amplified fragment length polymorphism (AFLP). As important as the genetic marker systems are the methods of analyzing variation and inferring structure, including recently developed coalescent and individual-based statistical methods. Mitochondrial DNA is maternally inherited, nonrecombining, and haploid; it is especially well suited for coalescent methods and can be useful for assessing female-biased dispersal. Microsatellites are highly variable, diploid markers, often averaging ten or more alleles per locus and heterozygosities above 90%. Tree-building and classical population genetic analyses (e.g., F_{ST}) are useful methods for uncovering structure, but the true power of microsatellites comes with population assignment and kinship analyses. These methods can often assign individual nonbreeding birds and dispersers to particular source populations and even families. Last, AFLP or RAPD analyses offer the advantage of being able to screen literally thousands of independent markers, which can increase the probability of finding ones that differentiate populations. The markers can also be used in assignment tests, with some assumptions. These markers are usually dominant (presence of a fragment can indicate either a homozygote or heterozygote) without further development and can be of unknown function, origin, or inheritance (they could, e.g., be derived from parasite rather than bird DNA).

Parasites themselves are another potential marker system of population identity and structure. That is, it may be more likely for a bird species to contract particular parasite species or strains in one geographical region than another, and infection by those parasites could serve as an indicator that a bird had been in that region. As noted in the chapter by Ricklefs et al., whether this is feasible depends in part on the distribution and specificity of the parasite and the rapidity with which it spreads and evolves. Clearly, we know little about these systems at present, but some aspects of parasites, such as their tendency to have rapidly evolving genes, do suggest that they hold some promise as markers for resolution of connectivity.

Isotopes and trace elements are perhaps the most promising source of markers for source population identification of migrant birds (see the chapter by Hobson). In this case, the bird picks up the marker from the environment and carries it, for a variable period of time, on its subsequent movements. The most commonly used isotopes are the light elements N, O, S, C, and H (as in what you do at a deli), all taken from food and incorporated into tissues. The methods require calibration; in particular, one needs to know how isotope ratios vary spatially, elevationally, and temporally in the landscape. Another disadvantage is that, unlike genetic markers, an isotopic marker can vary among tissues and can disappear over time, so calibrations such as tissue-specific isotopic fraction and "turnover" rate need to be made. In some cases, studies are more apt to uncover habitat associations rather than explicit geographic ranges, but studies of this resolution can still reveal important information about impacts of wintering habitat on breeding birds. Some markers, such as deuterium in feathers, exhibit a strong, repeatable latitudinal gradient, but other markers are required to provide longitudinal resolution. Trace element marker concentrations also can vary geographically, and the ability to routinely measure them from bird tissues offers another important method to identify sources of migratory birds.

We are just beginning to understand the impacts of migration on a wide range of characteristics of avian life history and evolution. This is a particularly exciting time in that new technologies such as satellite transmitters, powerful DNA markers, and isotope ratios are allowing us to answer questions about migration in far greater detail than we were ever able to with more standard ornithological methods. The chapters that follow survey results to date in their respective fields and ponder further developments that may explain the causes and consequences of amazing transglobal travel by tiny, feathered creatures.

16

MICHAEL S. WEBSTER
AND PETER P. MARRA

The Importance of Understanding Migratory Connectivity and Seasonal Interactions

RECENT TECHNOLOGICAL ADVANCES potentially allow researchers to follow migratory birds throughout the annual cycle, thus opening two exciting avenues for research on the ecology and evolution of migratory birds. First, researchers can examine seasonal interactions—how events or conditions in one season affect reproduction and/or behavior in another. At the population level, events on the wintering grounds may affect population dynamics on the breeding grounds, and these may be particularly important to understand in times of rapid habitat alteration and climate change. At the individual level, events on wintering grounds can affect individual breeding strategies, such as overall reproductive effort, parental care, and possibly extra-pair copulatory behavior. Second, patterns of connectivity (weak vs. strong) are likely to affect the degree to which birds are locally adapted to their wintering grounds and the rate of winter range expansion. Patterns of migratory connectivity seem less likely to affect adaptation to the breeding grounds, and it is not clear whether patterns of migratory connectivity promote or hinder the speciation process. These latter processes are more likely to be affected by natal dispersal (gene flow among breeding populations) than by migratory patterns per se. This is not to say, however, that migration is irrelevant to these processes, as natal dispersal may have a complex relationship with migration patterns, depending on how migratory behavior is transmitted from one generation to the next. On the whole, it is clear that a better understanding of migratory connectivity is needed before we can understand the forces that have shaped and continue to shape the lives of migratory birds.

INTRODUCTION

Many migratory birds spend approximately 3 months of their annual cycle on the breeding grounds, 6 months on their overwintering grounds, and 3 or more months traveling between these locations. Large-scale movements such as these—through a diverse array of habitat types and environmental conditions—create a series of complex ecological interactions that have important consequences for the biology of migratory birds. Simultaneously, these large-scale movements complicate our ability to fully understand these organisms, as it is difficult to follow them across the entire annual cycle. Despite the complexities raised by the movement of migratory individuals between summer and winter locations, recent technological advances now make it possible to detect connections between broad regions of the summer and winter ranges (e.g., Haig et al. 1997; Hobson and Wassenaar 1997; Chamberlain et al. 1997, 2000; Wennerberg 2001; Rubenstein et al. 2002; Ruegg and Smith 2002), and even greater precision appears to be within reach (Webster et al. 2002). These advances open the door for new kinds of studies and a deeper understanding of migratory birds. In this chapter we discuss two important avenues for future research on migratory birds, both of which depend on an ability to track (either directly or indirectly) individuals or populations across the annual cycle.

First, "seasonal interactions" can occur if events and conditions in one region/season affect populations and individuals in another, but these have proven difficult to study in migratory birds because of the geographical separation between summer and winter regions. For example, weather conditions and/or territory quality on the wintering grounds may affect later reproductive success on the breeding grounds (Marra et al. 1998; Sillett et al. 2000). These seasonal interactions may influence individual behavior, reproductive success, and population dynamics. Migratory birds may also face opposing conditions and selective forces on the wintering and breeding grounds, and this may affect both short-term population processes and long-term evolutionary responses.

Second, the patterns of connections among specific summer and winter populations may have important consequences for the ecology and evolution of migratory birds. For any migratory organism, breeding locations (or populations) are connected to wintering locations (populations) via the movement of individuals, and we term these connections "migratory connectivity" (Webster et al. 2002). Many different patterns of connectivity are possible (see Salomonsen 1955). For example, connectivity is strong if most individuals breeding in one area also spend the winter together at a particular location. At the opposite end of the spectrum, migratory connectivity is diffuse if a breeding population in one area is composed of individuals from a large number of separate wintering locations, and vice versa. In this latter case a breeding population is only weakly connected to any particular wintering population. Determining the strength of migratory connectivity is critical to a thorough understanding of the ecology and evolution of such organisms as well as to the development of sound conservation strategies.

In this chapter we illustrate the consequences of both seasonal interactions and migratory connectivity for population dynamics and conservation. In particular, we emphasize the consequences for the behavior of individuals and the evolution of migratory populations.

SEASONAL INTERACTIONS, POPULATION DYNAMICS, AND INDIVIDUAL BEHAVIOR

Evidence is mounting that seasonal interactions can affect the population dynamics of migratory organisms. For example, research on waterfowl indicates that good conditions on the wintering grounds can result in better recruitment into breeding populations (Heitmeyer and Fredrickson 1981; Kaminski and Gluesing 1987; Raveling and Heitmeyer 1989). Similarly, Sillett et al. (2000) found that for the migratory Black-throated Blue Warbler (*Dendroica caerulescens*) recruitment into both wintering and breeding populations was positively correlated with fecundity the preceding summer and was also affected by the El Niño–La Niña climate cycle. Thus, winter events can affect the dynamics of breeding populations (and vice versa) when conditions on the wintering grounds affect the number of individuals that return to the summer grounds to breed.

A timely nonavian example of a seasonal interaction at the population level is that of the monarch butterfly (*Danaus plexippus*). This species spends its summers in the northern United States and southern Canada but overwinters at a small number of locations in Mexico and along the California coastline. In January 2002, unusually cold night-time temperatures at a key wintering site in Mexico resulted in massive mortality, with an estimated 200–300 million butterflies dying in just 2 days (see [www.wwf.org.mx/news_new_monarch.php]). This massive die-off on the wintering grounds will likely result in reduced monarch populations in the eastern United States and Canada. These populations may experience higher levels of reproductive success over the next several years because of reduced density dependence. This is a good example of a seasonal effect of wintering-ground mortality on breeding-ground recruitment. Interestingly, migratory connectivity also plays a role in determining the consequences of this die-off: western populations will be relatively unaffected by the mass mortality event because populations west of the Rocky Mountains overwinter in California rather than Mexico (Hobson et al. 1999). Thus, because of the pattern of migratory connectivity in this species, winter events in Mexico affect some monarch summer populations but not others; these differences may in turn be reflected in milkweed, which is the monarch's principal food resource.

Events or conditions in one season also can have consequences for survival and reproduction of individuals in another season. Consider a male warbler that breeds in the temperate zone and overwinters in the Tropics. Research on the breeding grounds indicates that for many such species, male reproductive success depends on arrival date and condition at the time of arrival, as these could affect the quality of the territory and mate that a male obtains (e.g., Lozano et al. 1996; Hasselquist 1998). In turn, arrival date and condition likely are affected by the quality of the winter territory that a male occupies and the overall conditions, such as weather, at the wintering site. Such effects of winter weather and territory quality on reproductive success during the summer are an example of a seasonal interaction.

To date, only one study provides empirical support for the idea that seasonal interactions may affect reproductive success of individual migratory birds. Using stable-carbon isotopes, Marra et al. (1998) showed that American Redstarts (*Setophaga ruticilla*) overwintering in high-quality habitats arrived on breeding grounds earlier and in better physical condition than individuals overwintering in low-quality habitats. Because arrival date and condition often are positively correlated with reproductive success on the breeding grounds (Lozano et al. 1996; Verboven and Visser 1998; Kokko 1999), this study suggests that winter territory quality has an important effect on summer reproduction. Although seasonal interactions make intuitive sense and are likely pervasive in nature, identifying their magnitude and attributing variation of a biological event to circumstances in a previous season is difficult.

Seasonal interactions may also have subtle effects on male and female behavior. For example, extra-pair copulations (EPC) are common in many populations of migratory birds (e.g., Yezerinac et al. 1995; Perreault et al. 1997; Stutchbury et al. 1997; Webster et al. 2001), and seasonal interactions may play an important role in driving EPC rates by influencing the timing of female migration and the physical condition of females upon arrival at the breeding grounds. We propose three mechanisms by which this might happen. First, seasonal interactions likely influence the timing of arrival, which influences breeding synchrony. If breeding synchrony affects EPC rates (Stutchbury and Neudorf 1998; Chuang et al. 1999), then events on the wintering grounds and/or during migration that affect female arrival date may in turn affect EPC rates. Second, wintering and/or migratory conditions may influence extra-pair behavior of individual females by affecting the costs and benefits of EPC to those females. Specifically, females with poor winter territories may arrive on the breeding grounds late (Marra et al. 1998) and find most high-quality males already paired. These females will be forced to pair socially with low-quality males (Møller 1992) and, assuming that females engage in extra-pair matings to obtain genetic benefits (Jennions and Petrie 2000; Tregenza and Wedell 2000), they will be more likely to engage in EPC than early-arriving females (who had high-quality winter territories). Thus, conditions on the wintering ground, by influencing the timing of female arrival on the breeding grounds, can affect the benefits of EPC to individual females. Finally, females from poor winter territories may arrive on the breeding grounds in poorer condition and obtain poorer breeding territories than females from high-quality winter territories. If females in poor condition have greater difficulty in evading the mate-guarding efforts of their mates (Gowaty 1996), EPC behavior would be affected.

Thus, there are several ways that winter ground events and their influence on the timing and condition of female arrival *could* influence reproductive behavior of individual females and males. Similarly, wintering conditions and/or time needed to migrate may affect the time and energy that an individual can devote to parental care (Myers 1981). We emphasize the speculative nature of these ideas because there is little evidence to support or negate them, in large part because it has been difficult to follow females across the annual cycle. More thorough and innovative tests of these hypotheses will be possible when researchers can follow individuals or populations across seasons.

NATURAL SELECTION ACROSS THE ANNUAL CYCLE

Migration complicates selection not only because the act of migration itself selects for particular morphologies, behaviors, and physiologies, but also because migratory organisms are subject to selective pressures in different geographic locations and habitats throughout the annual cycle. For example, consider the selective pressures acting on body size in a migratory bird. Many North American migrants are sexually dimorphic in body size, with males being slightly larger than females (Dunning 1984). Most explanations regarding the evolution of sexual size dimorphism (SSD) focus on sexual selection during the breeding season (Darwin 1874; Webster 1992; Andersson 1994; Dunn et al. 2001), such as mate choice and male-male competition (Darwin 1874; Campbell 1972). However, natural selection acts year-round on individuals, and SSD is likely the result of multiple selection pressures (i.e., natural and sexual) acting differentially on males and females. For example, Darwin (1874) also proposed that sex-specific ecological adaptations could contribute to SSD (Slatkin 1984; Shine 1989; Webster 1997). Empirical evidence to support the ecological adaptation hypothesis in birds is weak and focuses primarily on sex-specific differences in bill morphology (Selander 1966, 1972; Temeles et al. 2000). In the case of a long-distance migratory bird, events and behaviors occurring during the temperate breeding, tropical winter, and migratory seasons may differ substantially from each other and, as such, may impose very different selective pressures on the sexes.

One example involves the American Redstart, which breeds in terrestrial and riparian woodlands throughout North America and winters in a variety of habitat types in the Caribbean, Central America, and northern South America (Sherry and Holmes 1997). In this species males are territorial during both summer and winter, whereas females are territorial only during winter (Sherry and Holmes 1997). If body size plays some role in territorial defense (Marra 2000), then larger male size and SSD may result because of positive selection for size in males in both summer and winter. Selection for small body size in females also may occur during the breeding season owing to bioenergetic constraints imposed by incubation and parental care (Downhower 1976). However, a consideration of winter ecology complicates this simple interpretation. Specifically, females with territories in high-quality winter habitat (i.e., mangrove) are significantly larger than females in the poor-quality scrub habitat, whereas male body size does not vary across winter habitat types (Marra 2000). Thus, selection during the winter should favor large female size. Clearly, sexual selection during the breeding season is an incomplete explanation for SSD in this and other migratory species, and a more thorough examination of selective pressures, acting throughout the annual cycle, is needed here.

At the opposite end of the spectrum, and somewhat counterintuitively, migratory birds may actually encounter selective pressures that are more constant over the annual cycle than do nonmigratory taxa. One consequence of migratory behavior is that it allows for relatively constant ecological conditions across the annual cycle. For example, during winter, migratory organisms may be able to select and inhabit microhabitats and foraging niches that are relatively similar to their habitats and niches during the breeding season, whereas nonmigratory birds often confront dramatic seasonal differences in food and habitat. Indeed, most temperate residents switch from an insectivorous diet during the breeding season to an omnivorous or entirely granivorous diet in winter, whereas many migratory species do not. Such diet switches require morphologies suited for diverse purposes. The key point here is that a full picture of the selective forces that shape the morphology of migratory species cannot be gained without careful study of such species throughout their annual cycle. Such studies require some knowledge of migratory connectivity, because it is important to know where breeding birds spend their winters and vice versa.

MIGRATORY CONNECTIVITY AND LOCAL ADAPTATION

The above sections examined the consequences of migration and cross-seasonal effects. We now consider migratory connectivity and its effects on local adaptation (i.e., the extent to which migratory organisms can become adapted to local conditions on their breeding and wintering grounds). The relevant issues here can be seen in a simple single-locus population genetic model in which a single breeding population is connected to two different winter sites (fig. 16.1A). In this simple model we assume that this is an annual species with no overlap in generations, we ignore other possible breeding populations, and we consider a single genetic locus with two alleles (*A* and *a*). This locus affects a trait that is under selection on the wintering grounds but not on the breeding grounds. The frequency of the *A* allele in wintering areas X and Y before spring migration to the breeding grounds are p_x and p_y, respectively. The proportion of individuals in the focal breeding population that migrated from wintering area Y is given by m (the remainder migrated from wintering area X). Mating on the breeding grounds is random and, as a consequence, the frequency of the *A* allele in winter population X after fall migration (p_x') is a weighted mean of the frequencies in winter populations X and Y (top equation in fig. 16.1A). At the end of winter just before spring migration, the new frequency of allele *A* in wintering area X (p_x'') depends on the fitnesses of genotypes *AA*, *Aa*, and *aa* (W_{11}, W_{12}, and W_{22}, respectively) in that area (bottom equation).

This simple model has three key features. First, selection acts on this locus, favoring one allele over the other, but only during the winter (for simplicity), such that allele frequencies are affected by selection on the wintering grounds but not the breeding grounds. Second, gene exchange (reproduction) occurs only on the breeding grounds—an assumption that seems reasonable for most migratory birds. Finally, natal dispersal is limited, such that offspring tend to breed in the same general region where their parents bred. As a consequence of these last two assumptions, genetic recombination on the breeding grounds will lead to gene flow from one wintering population to another, but there will be little gene flow from one breeding population to another.

In this situation the equations describing change in allele frequency are exactly the same as those describing the joint activity of selection and gene flow in sedentary populations (Hartl and Clark 1997), with the breeding grounds serving as a conduit for gene flow from one winter population to another (fig. 16.1A). Consider a situation in which one allele (*a*) is favored in wintering area X but the alternative allele (*A*) is favored in wintering area Y (fig. 16.1B). Results from this simple model demonstrate that the ability of a population to adapt to a wintering site (i.e., frequency of the favored allele) depends strongly on the level of migratory connectivity, because this affects gene flow between wintering populations. When connectivity is strong (i.e., gene flow between wintering populations is low), local adaptation occurs quickly and is nearly complete (fig. 16.1B). Conversely, when connectivity is weak (gene flow is high), local adaptation to the wintering grounds is severely restricted.

Although greatly simplified, this model illustrates that local adaptation to the wintering grounds is severely hampered by weak migratory connectivity, because under these conditions gene flow between wintering populations is high. Strong migratory connectivity, on the other hand, limits gene flow between wintering populations and allows for local adaptation to occur. This latter situation would occur because individuals overwintering in one area would also breed together. Note that the conclusions from this simple model depend critically on random mating on the breeding grounds. If individuals mate assortatively with respect to migratory behavior, then local adaptation to the wintering

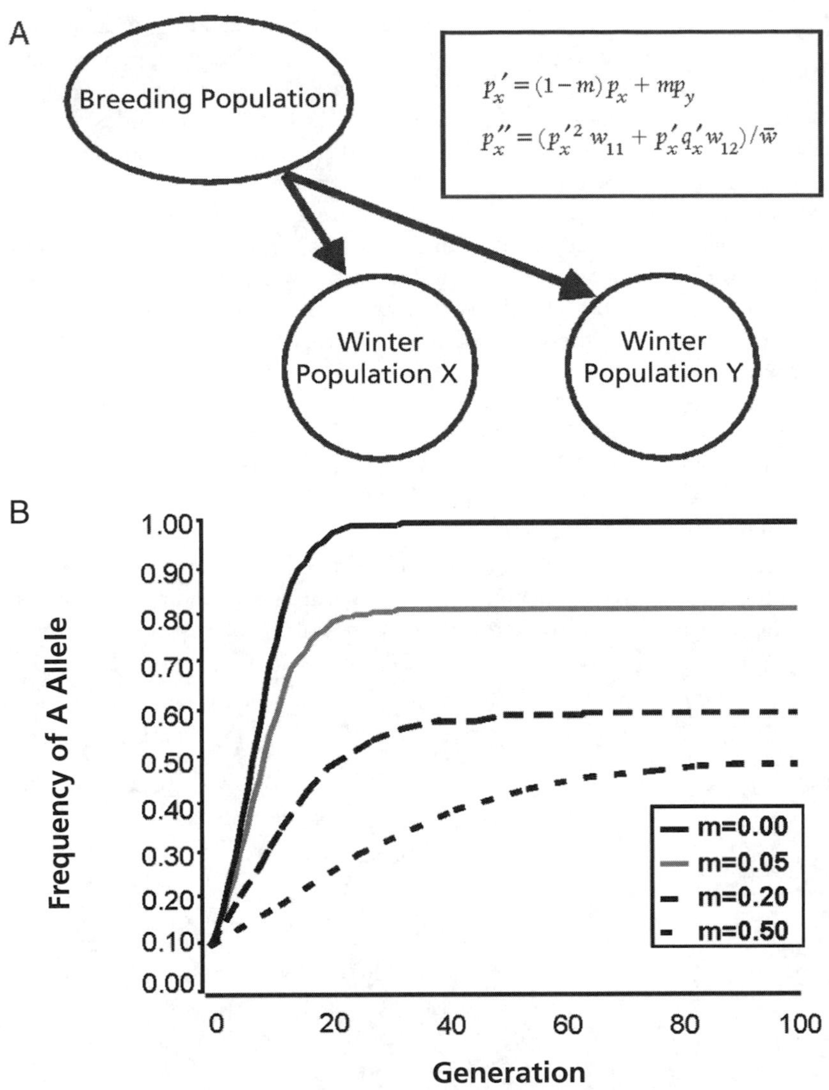

Fig. 16.1. A simple model of migration and local adaptation. Panel (A) shows that individuals from a particular breeding population can potentially migrate to two different wintering locations (X and Y). Equations in the box show the frequency of the *A* allele in winter population X after fall migration (p_x') and at the end of winter just before spring migration (p_x''). Panel (B) shows the results of simulations in which natural selection favors allele *A* in wintering location X but allele *a* in wintering location Y. For these simulations there was no dominance, and fitnesses for the *AA*, *Aa*, and *aa* genotypes were (respectively) 0.9, 0.7, and 0.5 at wintering location X and 0.5, 0.7, and 0.9 at wintering location Y. As can be seen, the population can become well adapted to local conditions (i.e., high frequency of the *A* allele in wintering location X) when migratory connectivity is strong (low *m*, indicating low gene flow), but not when migratory connectivity is weak (high *m* and hence high gene flow).

grounds is enhanced. This could occur, for example, if there is pairing on the wintering grounds or during migration, or if individuals migrating from different locations arrive on the breeding grounds and form pairs at different times.

The strength of migratory connectivity and local winter adaptation might also affect expansion of the wintering range. Theoretical models suggest that range expansion can be impeded by gene flow from the center of the range to the periphery, although this also may depend on specific patterns of selection and gene flow (Holt and Gomulkiewicz 1997; Kirkpatrick and Barton 1997). Under conditions of weak connectivity, gene flow from the center of the wintering range to the periphery would be substantial, and hence expansion of the winter range would be hindered by movement of "maladapted" alleles from the center of the range to the periphery (see Kirkpatrick and Barton 1997). Conversely, under conditions of strong migratory connectivity, local adaptation to conditions at the periphery of the wintering range would be possible, and so expansion of the wintering range would be less constrained and range expansion would depend only on the rate at which migratory individuals can colonize new habitats.

Interestingly, these same arguments do not apply in reverse: connectivity will not affect local adaptation to the breeding grounds or expansion of the summer range because gene exchange does not occur on the wintering grounds (at least in migratory birds; this may not hold for some other migratory organisms, such as insects). The wintering grounds therefore do not act as a conduit for gene flow from one summer population to another. Instead, adaptation to the breeding grounds is affected by the degree to which offspring migrate to the same regions where they were born—that is, natal dispersal and the heritability of migratory behavior (Berthold et al. 1992; Berthold 1996; Pulido et al. 2001). Thus, local adaptation to the breeding grounds depends on the interplay between migration and natal dispersal and not on migratory connectivity per se. Similarly, expansion of the breeding range is not affected by connectivity, but it might depend on the malleability of migratory behavior (Böhning-Gaese et al. 1998; Bensch 1999).

MIGRATORY CONNECTIVITY AND SPECIATION

Let us move from connectivity and local adaptation to a brief consideration of connectivity and speciation. Some, possibly many, migratory bird species show connectivity patterns like that shown in fig. 16.2A, with a strong migratory divide across the breeding range but weak connectivity within each subrange (e.g., Bensch et al. 1999; Rubenstein et al. 2002). Patterns such as this are suggestive of increasing genetic divergence between the breeding populations on either side of the divide and also between the wintering populations that are connected to these diverging breeding populations. This increased divergence may lead ultimately to allopatric speciation, and situations such as this are likely to be fertile ground for the study of speciation in progress. However, speciation can occur even without a migratory divide (i.e., in situations of weak connectivity [fig. 16.2B]) if offspring tend to migrate to the same general breeding region as their parents. The key to speciation in migratory species is likely the interaction between migration and natal dispersal (i.e., gene flow between breeding populations). Note that sympatric speciation could occur in migratory species if individuals mate assortatively according to migratory behavior. Again, this could occur if pairing occurs before individuals arrive on the breeding ground, or if individuals from different areas arrive at the same breeding location but at different times. As the degree of migratory connectivity is not known for most species of migratory birds, it is currently difficult to evaluate these hypothetical considerations.

MIGRATORY CONNECTIVITY, POPULATION DYNAMICS, AND CONSERVATION

Determining the magnitude of migratory connectivity is fundamental to advancing our understanding of many aspects of the conservation biology of migratory animals.

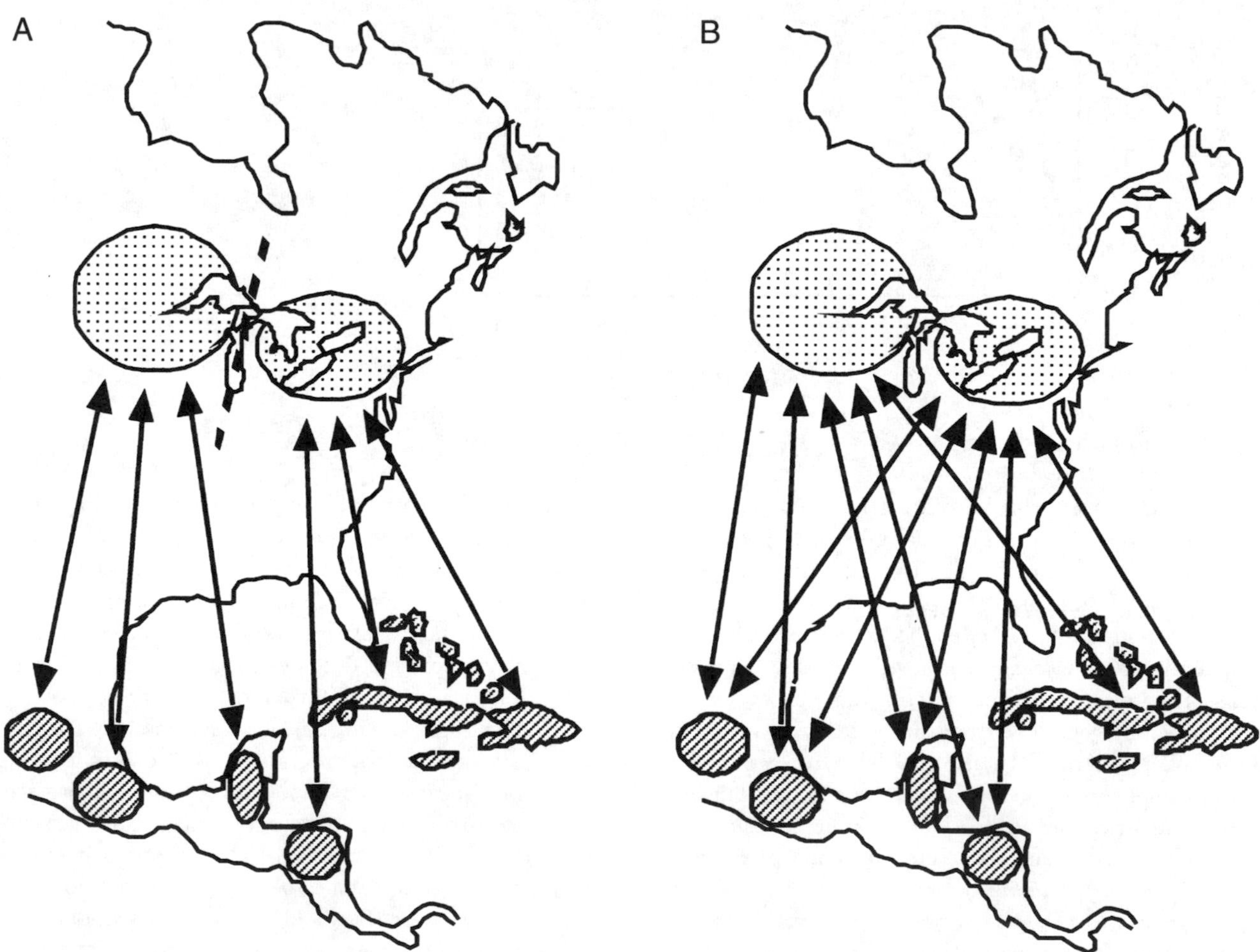

Fig. 16.2. Patterns of migratory connectivity for a hypothetical Nearctic-Neotropical migrant that breeds in eastern North America. Breeding populations are stippled, overwintering populations are cross-hatched, and arrows show migratory connections. Migratory connectivity for this species might be moderate (A), with a "migratory divide" separating western breeding populations (which overwinter in Mexico) from eastern breeding populations (which overwinter in eastern Central America and islands in the Caribbean). Alternatively, migratory connectivity may be weak (B), with both western and eastern breeding populations overwintering throughout the wintering range of the species. In this latter case birds from separate (eastern and western) breeding populations mix on the wintering grounds.

Habitat change induced by human activities, at both local and global scales, has become the primary threat for most organisms on our planet (Pimm et al. 1995; Palumbi 2001). Although both nonmigratory and migratory organisms face habitat alteration, the response of migratory organisms to anthropogenic habitat change is complicated by the geographical scale of their annual cycle. Simply put, migratory organisms must contend with human-induced habitat change on their breeding grounds, their wintering grounds, and the areas used while migrating between the two. Patterns of migratory connectivity are likely to affect the ability of migratory birds to respond to habitat alterations.

Aside from going extinct, migratory organisms may respond to anthropogenic habitat change in two general ways, both of which may be affected by migratory connectivity. First, they may accommodate the change through phenotypic plasticity or genetic adaptation. The latter can occur if habitat change is gradual rather than abrupt, as might be expected for climate change (although see Penuelas and Filella 2001; Walther et al. 2002). As an example, some migratory bird populations appear to be responding to global climate change by breeding earlier in the spring (Crick et al. 1997; McCleery and Perrins 1998; Dunn and Winkler 1999), a likely example of behavioral plasticity. As discussed above, understanding local adaptation of migratory organisms is complicated by their complex annual cycles, and it is unclear how easily migratory birds will be able to adapt to changing habitat pressures. Although adaptive responses to climate change can be rapid (Pulido et al. 2001; Warren et al. 2001), in many cases genetic adaptation will be too slow to track climate change (e.g., Etterson and Shaw 2001). This will be particularly true for migratory organisms if gene flow from peripheral populations is strong—as in the case of weak connectivity and adaptation to the wintering grounds.

Second, organisms can respond to climate change by shifting their ranges to track shifts in suitable habitat (Parmesan 1996). For migratory organisms, such changes in species range will require changes in the migratory program itself (i.e., the genetic program underlying the direction, distance, and timing of migration). Some empirical evidence suggests that such changes in the migratory program can be rapid (e.g., Berthold et al. 1992), but other evidence suggests that the migratory program may constrain range shifts or expansions (Böhning-Gaese et al. 1998; Bensch 1999) as well as other aspects of breeding biology (Both and Visser 2001).

In an important contribution, Dolman and Sutherland (1994; see also Sutherland 1996) explored interactions between habitat loss, population regulation, and the evolution of migration behavior in response to gradual loss of a wintering site. Their model considered a simple migratory species (fig. 16.3[top]) in which two separate breeding populations (A and B) migrate to two geographically separate wintering sites (X and Y). Migratory behavior is genetically determined at a single locus—individuals in breeding population X with the dominant allele (genotypes *AA* and *Aa*) migrate to winter location X, whereas individuals homozygous for the alternative allele (genotype *aa*) migrate to location Y. Initially the populations are at equilibrium and the frequency of the alternative allele in population A (p_0) is low (i.e., most individuals in population A migrate to area X). The model considers the effect of the gradual loss (i.e., complete loss over course of 100 years) of wintering location X on the size of breeding population A. This model is applicable to situations in which habitat becomes unsuitable for an organism, whether through direct habitat modification by humans or indirect habitat change via anthropogenic climate change.

A key result from Dolman and Sutherland's model is that when connectivity is strong, such that almost all individuals from breeding population A go to wintering site X, then breeding populations can be severely affected by loss of winter habitat (i.e., reduced to extremely low numbers) and likely lost (fig. 16.3[bottom]). In contrast, when connectivity is weak, such that some individuals in the breeding pop-

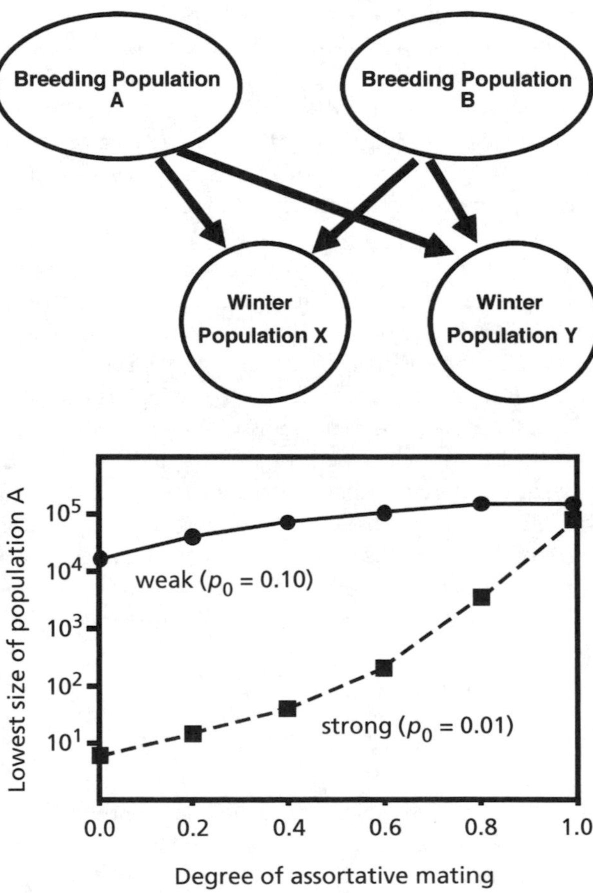

Fig. 16. 3. Modeling the flexibility of migratory connectivity. The top panel shows a simple migratory species in which two separate breeding populations (A and B) migrate to two geographically separate wintering sites (X and Y). The lower panel shows the lowest size that population A reaches as a function of the degree of assortative mating (0.0 = random mating, 1.0 = completely assortative mating as a function of migratory type) and initial frequency of the alternative allele (p_0). Lines show results of simulations under conditions of strong connectivity (initial frequency of alternative migration allele $p_0 = 0.01$) and relatively weak connectivity ($p_0 = 0.10$). Bottom panel redrawn from Dolman and Sutherland (1994).

ulation migrate to winter site Y, then the effect of loss of a wintering site on the focal breeding population is greatly reduced, the reason being that birds in the breeding population quickly evolve to migrate to alternative wintering sites. The critical point here is that when connectivity is weak, substantial genetic variation for migratory behavior exists within a breeding population, thus allowing for rapid evolutionary responses of the migratory program. Note that a little bit of connectivity goes a long way: Dolman and Sutherland (1994) found that the focal breeding population is little affected by loss of a wintering site when the initial frequency of the "alternative" migration allele was only $p_0 = 0.10$ (fig. 16.3[bottom]), even though under this situation only 1% of the initial population migrates to area Y (assuming random mating). Thus, some populations may be able to evolve new patterns of migratory connectivity in response to habitat loss and climate change, except in cases where connectivity is very strong (i.e., situations in which virtually no individuals migrate to alternative sites). This conclusion, of course, relies on the assumption that some of the wintering locations used by a breeding population are less severely affected by human activities. Also important to consider is the effect of assortative mating on the example shown in fig. 16.3[bottom]. When individuals mate assortatively according to migratory behavior, new patterns of migratory connectivity will evolve more rapidly and the breeding population is unlikely to be reduced to very low numbers.

Finally, patterns of migratory connectivity also have implications for conservation and management plans (see Esler 2000 for discussion). For example, when connectivity is strong it may be possible to manage or maintain particular breeding or winter populations by protecting critical breeding sites. An excellent example of where such an approach has worked is management of the Kirtland's Warbler by protecting the remaining coniferous breeding habitat. Interestingly, little is known of the winter ecology of this species. In contrast, in cases where migratory connectivity is weak, protection efforts will be more challenging and must include larger geographic areas for the protection of a species.

FUTURE DIRECTIONS

In this chapter we have discussed the importance of understanding migratory connectivity and seasonal interactions. We have argued that such seasonal interactions are likely to be common and to have important consequences at the level of both the population and the individual. At the population level, events on the wintering grounds may affect population dynamics on the breeding grounds (e.g., Sillett et al. 2000), and these may be particularly important to understand in times of rapid habitat alteration and climate change (Saether et al. 2000). At the individual level, events on the wintering ground can potentially affect individual breeding strategies, such as overall reproductive effort (Marra et al. 1998), parental care (Myers 1981), and possibly extra-pair copulatory behavior. Other areas of research, such as an examination of selective pressures and habitat choice throughout the annual cycle, similarly require a better understanding of migratory connectivity.

We have also argued that patterns of connectivity (weak vs. strong) are likely to affect the degree to which birds are locally adapted to their wintering grounds and the rate of winter range expansion. In contrast, patterns of migratory connectivity seem unlikely to affect adaptation to the breeding grounds, and it is not clear that patterns of migratory connectivity promote or hinder speciation. These latter processes are more likely to be affected by natal dispersal (gene flow between breeding populations) than by migratory patterns per se. This is not to say, however, that migration is irrelevant to these processes, as natal dispersal may have a complex relationship with migration patterns depending on how migratory behavior is transmitted from one generation to the next. On the whole, it is clear that a better understanding of migratory connectivity is needed before we understand the forces that have shaped and continue to shape the lives of migratory birds.

Our review has revealed several important avenues for future research. First, recent studies have suggested that events in one season can affect populations in another (e.g., Marra et al. 1998; Sillett et al. 2000; Gill et al. 2001), but in most cases we can only guess at the mechanisms underlying these seasonal interactions. Research in this area has been hampered by the logistic difficulties of tracking individuals across seasons. Thanks to recent technological advances (Webster et al. 2002), we will soon be able to follow individuals, or at least populations, throughout the annual cycle.

Second, the potential effects of winter conditions and events on breeding-season behavior have received little empirical or theoretical attention. Most of the work on extra-pair copulations, for example, has focused on potential genetic benefits of EPC to females, but far less attention has been paid to extrinsic factors that might affect a female's propensity to seek EPC (Griffith et al. 2002). Winter conditions, such as the quality of a female's winter territory, may play a significant role here by affecting the timing of arrival on the breeding grounds and/or the female's condition during breeding. Similarly, winter conditions may also affect individual parental care and other breeding-season behaviors, and thereby provide a link to population dynamics as described above. These possibilities remain unexplored. We must also gain a better understanding of how the breeding season influences events during fall migration and the subsequent winter period.

Finally, seasonal interactions are likely to influence the evolution of migratory organisms and vice versa. Theoretical genetic models suggest that migratory connectivity should affect local adaptation and range expansion, but empirical data from migratory birds are lacking. Böhning-Gaese et al. (1998) and Bensch (1999) have examined the possibility that migratory behavior itself affects range size, but these analyses are preliminary and more detailed examination is necessary. In particular, the interplay between migration and local adaptation requires examination in the same way that the relationship between gene flow and local adaptation has been explored for nonmigratory populations (Dias and Blondel 1996; Smith et al. 1997; Blondel et al. 1999). In summary, understanding the population connectivity of migratory animals, and ultimately how the periods of the annual cycle interact, will undoubtedly provide unexpected advances in our knowledge of the basic ecology and evolution of these organisms.

ACKNOWLEDGMENTS

We thank the National Science Foundation for support, and R. Greenberg, R. T. Holmes, R. Norris, R. Sandberg, and J. Wunderle for insightful comments that greatly improved the clarity and usefulness of this chapter.

LITERATURE CITED

Andersson, M. 1994. Sexual Selection. Princeton University Press, Princeton.

Bensch, S. 1999. Is the range size of migratory birds constrained by their migratory program? Journal of Biogeography 26:1225–1235.

Bensch, S., T. Andersson, and S. Åkesson. 1999. Morphological and molecular variation across a migratory divide in willow warblers, *Phylloscopus trochilus*. Evolution 53:1925–1935.

Berthold, P. 1996. Control of Bird Migration. Chapman and Hall, London.

Berthold, P., A. J. Helbig, G. Mohr, and U. Querner. 1992. Rapid microevolution of migratory behaviour in a wild bird species. Nature 360:668–669.

Blondel, J., P. C. Dias, P. Perret, M. Maistre, and M. M. Lambrechts. 1999. Selection-based biodiversity at a small spatial scale in a low dispersing insular bird. Science 285:1399–1402.

Böhning-Gaese, K., L. I. González-Guzmán, and J. H. Brown. 1998. Constraints on dispersal and the evolution of the avifauna of the Northern Hemisphere. Evolutionary Ecology 12:767–783.

Both, C., and M. E. Visser. 2001. Adjustment to climate change is constrained by arrival date in a long-distance migrant bird. Nature 411:296–298.

Campbell, B. (ed.). 1972. Sexual Selection and the Descent of Man 1871–1971. Aldine Publishing, Chicago.

Chamberlain, C. P., S. Bensch, X. Feng, S. Åkesson, and T. Andersson. 2000. Stable isotopes examined across a migratory divide in Scandinavian willow warblers (*Phylloscopus trochilus trochilus* and *Phylloscopus trochilus acredula*) reflect their African winter quarters. Proceedings of the Royal Society of London, Series B, Biological Sciences 267:43–48.

Chamberlain, C. P., J. D. Blum, R. T. Holmes, X. Feng, T. W. Sherry, and G. R. Graves. 1997. The use of isotope tracers for identifying populations of migratory birds. Oecologia 109:132–141.

Chuang, H. C., M. S. Webster, and R. T. Holmes. 1999. Extra-pair paternity and local synchrony in the Black-throated Blue Warbler. Auk 116:726–736.

Crick, H. Q. P., C. Dudley, D. E. Glue, and D. L. Thomson. 1997. UK birds are laying eggs earlier. Nature 388:526.

Darwin, C. 1871. The Descent of Man, and Selection in Relation to Sex. Reprinted by Princeton University Press, Princeton (1981).

Dias, P. C., and J. Blondel. 1996. Local specialization and maladaptation in the Mediterranean blue tit (*Parus caeruleus*). Oecologia 107:79–86.

Dolman, P. M., and W. J. Sutherland. 1994. The response of bird populations to habitat loss. Ibis 137 (Supplement):S38–S46.

Double, M., and A. Cockburn. 2000. Pre-dawn infidelity: females control extra-pair mating in superb fairy-wrens. Proceedings of the Royal Society of London, Series B, Biological Sciences 267:465–470.

Downhower, J. F. 1976. Darwin's finches and the evolution of sexual dimorphism in body size. Nature 263:558–563.

Dunn, P. O., L. A. Whittingham, and T. E. Pitcher. 2001. Mating systems, sperm competition, and the evolution of sexual dimorphism in birds. Evolution 55:161–175.

Dunn, P. O., and D. W. Winkler. 1999. Climate change has affected the breeding date of tree swallows throughout North America. Proceedings of the Royal Society of London, Series B, Biological Sciences 266:2487–2490.

Dunning, J. B. 1984. Body Weights of 686 Species of North American Birds. Western Bird Banding Association, Eldon Publishing, Cave Creek, Ariz.

Esler, D. 2000. Applying metapopulation theory to conservation of migratory birds. Conservation Biology 14:366–372.

Etterson, J. R., and R. Shaw. 2001. Constraint to adaptive evolution in response to global warming. Science 294:151–154.

Gill, J. A., K. Norris, P. M. Potts, T. G. Gunnarsson, P. W. Atkinson, and W. J. Sutherland. 2001. The buffer effect and large-scale population regulation in migratory birds. Nature 412:436–438.

Gowaty, P. A. 1996. Battles of the sexes and origins of monogamy. Pages 21–52 *in* Partnerships in Birds: The Study of Monogamy (J. M. Black, ed.). Oxford University Press, Oxford.

Griffith, S. C., I. P. F. Owens, and K. A. Thuman. 2002. Extra pair paternity in birds: a review of interspecific variation and adaptive function. Molecular Ecology 11:2195–2211.

Haig, S. M., C. L. Gratto-Trevor, T. D. Mullins, and M. A. Colwell. 1997. Population identification of western hemisphere shorebirds throughout the annual cycle. Molecular Ecology 6:413–427.

Hartl, D. L., and A. G. Clark. 1997. Principles of Population Genetics (third ed.). Sinauer, Sunderland, Mass.

Hasselquist, D. 1998. Polygyny in Great Reed Warblers: a longterm study of factors contributing to male fitness. Ecology 79:2376–2390.

Heitmeyer, M. E., and L. H. Fredrickson. 1981. Do wetland conditions in the Mississippi Delta hardwoods influence mallard recruitment? Transactions of the North American Wildlife and Natural Resources Conference 46:44–57.

Hobson, K. A., and L. I. Wassenaar. 1997. Linking breeding and wintering grounds of neotropical migrant songbirds using stable hydrogen isotopic analysis of feathers. Oecologia 109:142–148.

Hobson, K. A., L. I. Wassenaar, and O. R. Taylor. 1999. Stable isotopes (delta D and delta C-13) are geographic indicators of natal origins of monarch butterflies in eastern North America. Oecologia 120:397–404.

Holt, R. D., and R. Gomulkiewicz. 1997. How does immigration influence local adaptation? A reexamination of a familiar paradigm. American Naturalist 149:563–572.

Jennions, M. D., and M. Petrie. 2000. Why do females mate multiply? A review of the genetic benefits. Biological Reviews of the Cambridge Philosophical Society 75:21–64.

Kaminski, R. M., and E. A. Gluesing. 1987. Density- and habitat-related recruitment in mallard s. Journal of Wildlife Management 51(1):141–148.

Kirkpatrick, M., and N. H. Barton. 1997. Evolution of a species' range. American Naturalist 150:1–23.

Kokko, H. 1999. Competition for early arrival in migratory birds. Journal of Animal Ecology 68:940–950.

Lozano, G. A., S. Perreault, and R. E. Lemon. 1996. Age, arrival date and reproductive success of male American Redstarts (*Setophaga ruticilla*). Journal of Avian Biology 27:164–170.

Marra, P. P. 2000. The role of behavioral dominance in structuring patterns of habitat occupancy in a migrant bird during the non-breeding period. Behavioral Ecology 11:299–308.

Marra, P. P., K. A. Hobson, and R. T. Holmes. 1998. Linking winter and summer events in a migratory bird by using stable-carbon isotopes. Science 282:1884–1886.

Marra, P. P., and R. T. Holmes. 2001. Consequences of dominance-mediated habitat segregation in a migrant passerine bird during the non-breeding season. Auk 118:92–104.

McCleery, R. H., and C. M. Perrins. 1998. Temperature and egg-laying trends. Nature 391:30–31.

Møller, A. P. 1992. Frequency of female copulations with multiple males and sexual selection. American Naturalist 139:1089–1101.

Myers, J. P. 1981. Cross-seasonal interactions in the evolution of sandpiper social systems. Behavioral Ecology and Sociobiology 8:195–202.

Palumbi, S. R. 2001. Humans as the world's greatest evolutionary force. Science 293:1786–1790.

Parmesan, C. 1996. Climate and species' range. Nature 382: 765–766.

Penuelas, J., and I. Filella. 2001 Responses to a changing world. Science 294:793–795.

Perreault, S., R. E. Lemon, and U. Kuhnlein. 1997. Patterns and correlates of extra-pair paternity in American redstarts (*Setophaga ruticilla*). Behavioral Ecology 8:612–621.

Pimm, S. L., G. J. Russell, J. L. Gittleman, and T. M. Brooks. 1995. The future of biodiversity. Science 269:347–350.

Pulido, F., P. Berthold, G. Mohr, and U. Querner. 2001. Heritability of the timing of autumn migration in a natural bird population. Proceedings of the Royal Society of London, Series B, Biological Sciences 268:885–993.

Raveling, D. G., and M. E. Heitmeyer. 1989. Relationships of population size and recruitment of pintails to habitat condition and harvest. Journal of Wildlife Management 53(4):1088–1103.

Rubenstein, D. R., C. P. Chamberlain, R. T. Holmes, M. P. Ayres, J. R. Waldbauer, G. R. Graves, and N. C. Tuross. 2002. Linking breeding and wintering ranges of a migratory songbird using stable isotopes. Science 295:1062–1065.

Ruegg, K. C., and T. B. Smith. 2002. Not as the crow flies: a historical explanation for circuitous migration in the Swainson's Thrush (*Catharus ustulatus*). Proceedings of the Royal Society of London, Series B, Biological Sciences 269: 1375–1381.

Sæther, B.-E., J. Tufto, S. Engen, K. Jerstad, O. W. Røstad, and J. E. Skåtan. 2000. Population dynamical consequences of climate change for a small temperate songbird. Science 287:854–856.

Salomonsen, F. 1955. The evolutionary significance of bird migration. Biologiske Meddelelser 22:1–62.

Selander, R. K. 1966. Sexual dimorphism and differential niche utilization in birds. Condor 68:113–151.

Selander, R. K. 1972. Sexual selection and dimorphism in birds. Pages 180–230 *in* Sexual Selection and the Descent of Man 1871–1971 (B. Campbell, ed.). Aldine Publishing, Chicago.

Sherry, T. W., and R. T. Holmes. 1997. American Redstart (*Setophaga ruticilla*). The Birds of North America, no. 277 (A. Poole and F. Gill, eds.). The Birds of North America, Inc., Philadelphia.

Shine, R. 1989. Ecological causes for the evolution of sexual dimorphism: a review of the evidence. Quarterly Review of Biology 64:419–461.

Sillett, T. S., R. T. Holmes, and T. W. Sherry. 2000. Impacts of a global climate cycle on population dynamics of a migratory songbird. Science 288:2040–2042.

Slatkin, M. 1984. Ecological causes of sexual dimorphism. Evolution 38:622–630.

Smith, T. B., R. K. Wayne, D. J. Girman, and M. W. Bruford. 1997. A role for ecotones in generating rainforest biodiversity. Science 276:1855–1857.

Stutchbury, B. J. M., and D. L. Neudorf. 1998. Female control, breeding synchrony, and the evolution of extra-pair mating systems. Pages 103–123 *in* Avian Reproductive Tactics: Female and Male Perspectives (P. G. Parker and N. T. Burley, eds.). Ornithological Monographs 49. American Ornithologists' Union, Washington, D.C.

Stutchbury, B. J. M., W. H. Piper, D. L. Neudorf, S. A. Tarof, J. M. Rhymer, G. Fuller, and R. C. Fleischer. 1997. Correlates of extra-pair fertilization success in hooded warblers. Behavioral Ecology and Sociobiology 40:119–126.

Sutherland, W. J. 1996. Predicting the consequences of habitat loss for migratory populations. Proceedings of the Royal Society of London, Series B, Biological Sciences 263:1325–1327.

Temeles, E. J., I. L. Pan, J. L. Brennan, and J. N. Horwitt. 2000. Evidence for ecological causation of sexual dimorphism in a hummingbird. Science 289:441–443.

Tregenza, T., and N. Wedell. 2000. Genetic compatibility, mate choice and patterns of parentage: invited review. Molecular Ecology 9:1013–1027.

Verboven, N., and M. E. Visser. 1998. Seasonal variation in local recruitment of Great Tits: the importance of being early. Oikos 81:511–524.

Walther, G.-R., E. Post, P. Convey, A. Menzel, C. Parmesan, T. J. C. Beebee, J.-C. Fromentin, O. Hoegh-Guldberg, and B. Bairlein. 2002. Ecological responses to recent climate change. Nature 416:389–395.

Warren, M. S., J. K. Hill, J. A. Thomas, J. Asher, R. Fox, B. Huntley, D. B. Roy, M. G. Telfer, S. Jeffcoate, P. Harding, G. Jeffcoate, S. G. Willis, J. N. Greatorex-Davies, D. Moss, and C. D. Thomas. 2001. Rapid responses of British butterflies to opposing forces of climate and habitat change. Nature 414:65–69.

Webster, M. S. 1992. Sexual dimorphism, mating system and body size in New World blackbirds (Icterinae). Evolution 46:1621–1641.

Webster, M. S. 1997. Extreme sexual size dimorphism, sexual selection and the foraging ecology of Montezuma Oropendolas (Aves: Icteridae). Auk 114:570–580.

Webster, M. S., H. C. Chuang-Dobbs, and R. T. Holmes. 2001. Microsatellite identification of extra-pair sires in a socially monogamous warbler. Behavioral Ecology 12: 439–446.

Webster, M. S., P. P. Marra, S. M. Haig, S. Bensch, and R. T. Holmes. 2002. Links between worlds: unraveling migratory connectivity. Trends in Ecology and Evolution 17:76–83.

Wennerberg, L. 2001. Breeding origin and migration pattern of dunlin (*Calidris alpina*) revealed by mitochondrial DNA analysis. Molecular Ecology 10:1111–1120.

Yezerinac, S. M., P. J. Weatherhead, and P. T. Boag. 1995. Extra-pair paternity and the opportunity for sexual selection in a socially monogamous bird (*Dendroica petechia*). Behavioral Ecology and Sociobiology 37:179–188.

17

ROBERT E. RICKLEFS,
SYLVIA M. FALLON, STEVEN C. LATTA,
BETHANY L. SWANSON,
AND ELDREDGE BERMINGHAM

Migrants and Their Parasites

A Bridge between Two Worlds

MIGRATING BIRDS CARRY A VARIETY of external and internal parasites and can transmit pathogens, such as West Nile virus, between host populations in the Tropics and temperate regions. The rapid spread of diseases such as West Nile virus and mycoplasmal conjunctivitis (*Mycoplasma gallisepticum*) in eastern North America and the devastating effects of avian malaria (*Plasmodium relictum*) on the native Hawaiian avifauna illustrate the potential for emerging diseases carried by migrants to become major management and conservation problems. Recent molecular studies using PCR and DNA sequencing to identify infections and characterize lineages of avian malaria (*Plasmodium* and *Haemoproteus*) show that migrants transport these parasites between tropical and temperate regions. Some lineages of *Plasmodium* have been recovered from both tropical and temperate resident host populations, as well as migrants, indicating that migrants have been responsible for dispersal of the parasites between unrelated hosts in geographically distant locations. Analysis of parasite and host phylogenies shows that most parasite lineages have strong host phylogenetic conservatism, but switching among distantly related hosts is not infrequent. Moreover, because host switching appears to be haphazard, it is unlikely that cases of emerging diseases can be predicted. Nonetheless, we must acknowledge the ability of birds to spread diseases widely and undertake further empirical and experimental studies of disease distribution and transmission in wild bird populations.

INTRODUCTION

Each year billions of birds leave their wintering areas in tropical and subtropical regions of the Earth and migrate poleward to summer breeding grounds. Each migrant carries a variety of symbiotic organisms, many of them potentially pathogenic to individuals of other species. These global movements should cause concern about the potential of migrants to spread novel diseases worldwide (McClure 1974; Service 1991; Daniels 1995; Lundstrom 1999), even reaching isolated regions where native species may have reduced immunological defenses against pathogens (Van Riper et al. 1986; Massey et al. 1996). In addition to host movement, physical conditions in the external environment and the availability of suitable vectors and alternative hosts should also influence the distribution of parasitic organisms (e.g., Super and Van Riper 1995; Randolph and Rogers 2000).

The biogeography of parasites, especially microparasites, such as viruses, bacteria, and protozoa, is not well understood. Parasite populations typically are sparsely sampled compared with their hosts, and the taxonomy and systematic relationships of many groups of parasites are poorly known. With the advent of PCR and DNA sequencing, however, we can now characterize parasite lineages and describe their distributions unambiguously with respect to geography and hosts. From the distribution of naturally occurring parasites, we can begin to understand parasite dispersal and host switching as historical occurrences, and we can perhaps gain useful insights concerning diseases in host populations.

Our laboratories have recently begun to characterize avian malaria parasites in eastern North America and the Caribbean Basin using sequences of the mitochondrial cytochrome *b* gene (Bensch et al. 2000; Perkins and Schall 2002; Ricklefs and Fallon 2002; Fallon et al. 2003). Although our data are preliminary at this point, our results provide a glimpse of the complex patterns of geographic and host distributions of parasite lineages. Drawing on our own work and on the literature, we address the following questions: (1) What is the impact of disease organisms on host populations? Of particular interest is the effect of disease on the performance and survival of birds during migration. (2) What is the role of host dispersal in the movement of parasite organisms? Specifically, do migrants form a parasite bridge between resident species on the wintering and breeding grounds? It is not sufficient that migrants carry disease organisms; the parasites require suitable hosts, vectors, and physical conditions to become established in new areas. (3) Recognizing the importance of connecting wintering and breeding populations, we also ask whether parasites can serve as markers for the geographic origins of migrant individuals.

THE IMPACT OF DISEASES ON HOST POPULATIONS

We expect the consequences of pathogens for conservation, wildlife management, agriculture, and public health to vary with time following a population's first exposure. Host and parasite populations tend to coevolve toward a point at which disease organisms maintain themselves at low levels controlled by reduced parasite virulence and increased parasite resistance (Ewald 1994; Hill 2001). Although endemic parasites can have profound consequences for host populations, as shown, for example, in the case of *Trichostrongylus tenuis* infecting Red Grouse (*Lagopus lagopus*) in Scotland (Dobson and Hudson 1995; Hudson and Dobson 1997; Hudson et al. 1998), newly introduced, or "emerging," diseases often have even more dramatic effects upon initial exposure.

The impact of the malaria parasite *Plasmodium relictum* on the Hawaiian avifauna has been documented by Van Riper et al. (1986) and in more recent experimental studies by Atkinson and his colleagues (e.g., Atkinson et al. 1995, 2000, 2001a, 2001b). *Plasmodium relictum* was most likely introduced to Hawaii in the early 1900s; a competent vector, the mosquito *Culex quinquefasciatus,* has been present in Hawaii since 1826. Although many introduced birds in Hawaii acquire *Plasmodium* infections (Shehata et al. 2001), the disease rarely develops to the clinical levels observed in many species of the native Hawaiian avifauna (summarized by Jarvi et al. 2001). Even native birds vary in their resistance to malaria. Hawaiian thrushes (the Omao [*Myadestes obscurus*]) survive experimental infection whereas various species of Hawaiian honeycreepers (Drepanididae) exhibit 50–100% mortality. The Hawaii Amakihi (*Hemignathus virens*) shows evidence of recently evolved resistance. Individuals from low elevation, where mosquito vectors spread the disease, now resist infection better than individuals from high elevations that lack the mosquito and the disease (Van Riper et al. 1986; Atkinson et al. 2000). Although isolated island faunas might be particularly vulnerable to new disease organisms, emerging diseases also have spread rapidly through continental wildlife populations, as in the case of rinderpest in African ungulates (Plowright 1982) and West Nile virus in North American birds (Marfin and Gubler 2001).

Because migration imposes a tremendous stress, birds may be especially vulnerable to the disease effects of parasites during this period of the annual cycle. Apapane (*Himatione sanguinea*) infected with *Plasmodium relictum* showed reduced locomotory activity and feeding and probably would not have been physiologically capable of long-distance migration (Yorinks and Atkinson 2000). Even when an infection is not intense enough to interfere directly with oxygen transport in the blood, malaria parasites elicit a strong immune response (Apanius et al. 2000; Atkinson et al. 2001a, 2001b), which may impair physiological function (Fair et al. 1999), including flight performance. Avian hematozoa may be a particular problem for young birds controlling their first infections, which are likely to be contracted on the breeding grounds either in the nest or shortly after fledging. Many passerine nestlings and fledglings brought to the St. Louis Wild Bird Rehabilitation Center are infected with hematozoans (Ricklefs et al., unpubl.). Although this may be a biased sample, it does highlight the potential vulnerability of young birds to disease prior to their first long-distance migration.

A particularly dramatic effect of parasitic organisms on survival of migration can be inferred from the failure of individual Palm Warblers (*Dendroica palmarum*) infected with *Knemidokoptes* mites on their wintering grounds in Hispaniola to return the following winter (Latta 2003). In this study, individuals with mite infestations, which cause abnormal hypertrophy of scales on the legs, lost weight as the winter progressed, although comparisons with uninfested individuals revealed no effect on site fidelity within the wintering area. In subsequent years of the study, however, none of the mite-infested individuals returned to their wintering grounds, compared with return rates of 40–70% among noninfested birds. Although the direct cause and timing of mortality are unknown, the physiological costs of mite infestation probably resulted in mortality during migration. A path analysis, which evaluates hypotheses concerning causal relationships, showed that mortality was directly related to mite infestation rather than to the secondary effect of parasitism on condition. Clearly, further information on the stress of migration as a selective agent on disease resistance is needed (see Sheldon and Verhulst 1996; Råberg et al. 1998).

EMERGING DISEASES AND THE SPREAD OF INFECTIONS

Recently, we have witnessed in North America the rapid spread of two novel diseases over large areas of the continent. This most likely resulted from movements by infected hosts and shows the power of dispersal and, by extension, migration in the spread of infectious agents. Mycoplasmal conjunctivitis, or House Finch eye disease, first appeared in populations of House Finches (*Carpodacus mexicanus*) in the vicinity of Washington, D.C., in January 1994. The responsible organism, the bacterium *Mycoplasma gallisepticum,* is the agent of a widespread endemic disease (MG) of poultry over most of the United States. The emergence of the disease in House Finches may have resulted from their contact with contaminated food and refuse after seeking food in poultry barns. Within 2 years, the disease had spread as far as Georgia, Illinois, and Quebec, and by 1997 it was broadly distributed throughout the entire range of the House Finch in eastern North America (Dhondt et al. 1998; Roberts et al. 2001). The particular form of MG contracted by House Finches appears not to have infected other birds, except for several American Goldfinches (*Carduelis tristis*) that contracted the disease in 1995 and 1996 in the southeastern United States (J. Cook [www.members.aol.com/FinchMG/IntroBac.htm]).

West Nile virus (WNV) has received more attention because it infects a broad spectrum of birds and mammals, including humans. The virus first appeared in North America in the vicinity of New York City in 1999. By 2001, it had spread west as far as St. Louis and south to Florida and the Gulf Coast. WNV has been isolated from more than 100 species of birds and causes substantial mortality of corvids (Anderson et al. 1999; Ebel et al. 2001; Marfin and Gubler 2001). Human cases of encephalitis caused by WNV have appeared as far south as the Cayman Islands (August 2001 [www.carec.org/data/alerts/011017.htm]), possibly having been carried by migrant birds. In Europe, epidemics of WNV are associated with high populations of *Culex* mosquitoes, the primary vector species, after heavy rains and flooding, but rarely persist more than a couple of years (Hubalek 2000), suggesting that high vector densities are required for persistence of a WNV outbreak, as in the similar case of an outbreak of Japanese encephalitis virus in northern Australia (Hanna et al. 1996).

DO MIGRANTS CARRY DISEASE ORGANISMS?

Observations of parasites in migrating birds strongly suggest that migratory birds carry disease organisms (McClure and Ratanaworabhan 1973). Migrants commonly carry ixodid ticks, many of which are infected by *Ehrlichia* pathogens and the spirochete bacterium *Borellia,* which causes Lyme disease (Smith et al. 1996; Fubito et al. 2000; Alekseev et al. 2001; Scott et al. 2001). Hemoproteids, including avian malaria, are frequently found in the peripheral blood of migrants (Rintamaki et al. 1998; Bensch et al. 2000). Outbreaks of West Nile virus and related viruses in the Middle East and southern and eastern Europe are thought to have been brought from parts of tropical Africa (where the pathogen probably is endemic) by migrants returning to their northern breeding grounds (Hubalek 2000; Miller et al. 2000; Zeller and Murgue 2001; Weissenböck et al. 2002).

Regardless of the potential of migrants to carry disease organisms between regions, the establishment of new diseases as the result of migration depends on several factors. First, the probability of a disease spreading by migrants should increase in relation to the number of arriving individuals with high parasitemias (Anderson and May 1991). High parasitemias may be more likely among migrants if the stress of migration suppresses immune function (Ots and Hõrak 1996; Råberg et al. 1998; Webster et al. 2002) and results in higher intensities of infection, although data on this point are lacking. Second, suitable vectors must be present to transmit the disease and maintain the infection in the population (Shroyer 1986; Mitchell 1991). Third, resident species must be susceptible to infection. The prospects for the spread of a disease vary greatly depending on the complexity of the life cycle, the duration and requirements of free-living stages, and the degree of specialization of vectors and the disease organism itself. How well particular parasite-host systems fulfill these criteria is not well understood. Viral diseases spread by physical inoculation by mosquitoes or other biting insects might encounter few barriers to their spread other than the immune defenses of potential hosts. To the extent that this is true, pathogens such as West Nile virus are of great concern, particularly for isolated areas with few endemic pathogens to maintain strong immune defenses in potential hosts.

MALARIA IN NORTH AMERICAN AND THE CARIBBEAN BASIN

Malaria is caused in birds by apicomplexan parasites belonging primarily to the genera *Plasmodium, Haemoproteus,* and *Leucocytozoon* (Atkinson and Van Riper 1991). *Plasmodium* is transmitted by various species of *Culex* mosquitoes, whereas *Haemoproteus* is transmitted by biting midges (Ceratopogonidae) or louse flies (Hippoboscidae) (Atkinson and Van Riper 1991). The related *Leucocytozoon* infects leucocytes and erythrocytes in the peripheral blood and is transmitted by blackflies of the genus *Simulium* and by *Culicoides* midges (Fallis et al. 1974).

Malaria infections in birds traditionally have been identified on thin blood smears (Godfrey et al. 1987; Fedynich et al. 1995). Typically ca. 10,000 red blood cells are scanned and infections are reported as the prevalence of a parasite in a population, that is, the proportion of individuals infected of those examined. The intensity of infection in a single individual—its parasitemia—is typically reported as the number of infected red blood cells per 10,000, or other base number, examined.

Several compilations of hematozoan prevalence are summarized in fig. 17.1 for: (1) both migrant and resident birds sampled in North America, (2) Neotropical migrants in the Tropics, and (3) endemic Neotropical species. Because blackflies are uncommon in tropical regions, *Leucocytozoon* is present only at low levels, both among endemic or resident species and among Neotropical migrants. The prevalence of *Leucocytozoon* is much higher in North America, particularly at more northern latitudes within this region (Greiner et al. 1975). *Haemoproteus* is also an abundant parasite in North America, but the prevalence among North American migrants on their wintering grounds is low and comparable to that of tropical residents. *Plasmodium* is less frequent than *Haemoproteus* and its overall prevalence does not vary between regions.

Variation in prevalence can be caused by variation in the rate of infection or in the control of infection by host individuals. Analysis of blood smears results in false negatives (parasites present but not detected) when the intensity of infection drops below the detection limit due to sampling of about 10^{-4} cells. Recently, a number of PCR-based tests have been developed to detect infections (Feldman et al. 1995; Li et al. 1995; Bensch et al. 2000; Jarvi et al. 2002; Ricklefs and Fallon 2002), but any one of these is not fully reliable across host species (Richard et al. 2002). We now screen blood samples with several PCR primers based on protein-coding and mitochondrial RNA-coding sequences, which can detect infections having parasitemias as low as 10^{-5} infected cells across a wide variety of species (Fallon et al. 2003). We compared infection rates based on blood smears (ca. 10,000 cells) and several primer pairs in a winter sample of 100 individuals from the Guanica Forest in Puerto Rico, including both migrants and residents. Blood smears indicated an overall malaria prevalence of 28%, whereas the corresponding value based on PCR screening was 42%. Thus, parasite prevalence may typically be higher than reported from smears (Richard et al. 2002). However, if the disease is not present in the peripheral circulation, or is present at very low levels, PCR screening may also fail to detect its presence (Jarvi et al. 2002). If many infections are maintained at low levels, this also implies that a component of the variation in prevalence among samples might reflect immune control of infections rather than presence or absence of infection (Applegate 1970; Richie 1988). Thus, migrants might carry potentially pathogenic parasites cryptically.

Malaria parasites are distinguished and named according to morphological characters visible on blood smears, in-

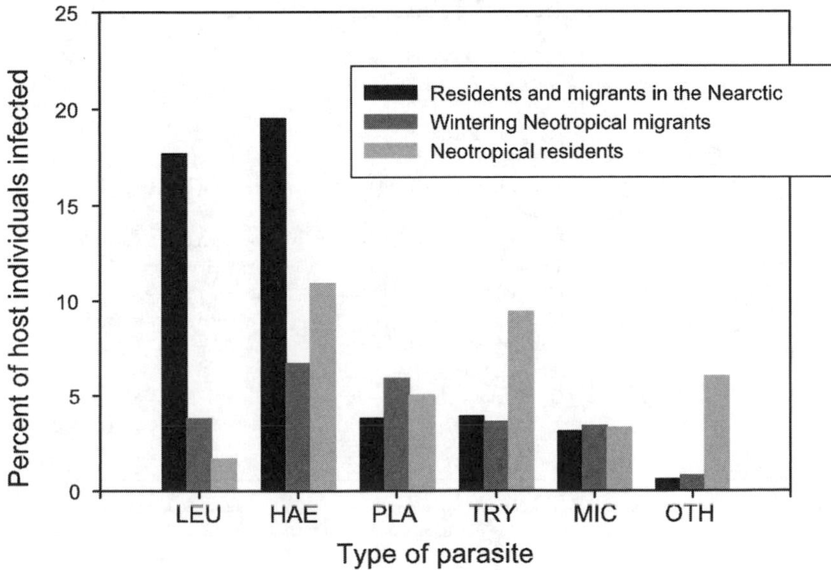

Fig. 17.1. Prevalence of several types of blood parasites, including the malaria parasites *Plasmodium* (PLA) and *Haemoproteus* (HAE), observed in blood smears of birds in the Nearctic (residents and migrants considered together) and in the Neotropics (residents and migrants separated). LEU = *Leucocytozoon;* TRY = trypanosomes; MIC = microfilariae; OTH = other. Based on summaries in Greiner et al. (1975) and White et al. (1978).

cluding the size, shape, and number of pigment granules in mature gametocyte (infective) forms of the parasites in red blood cells (Bennett et al. 1993, 1994; Peirce and Bennett 1996). Because malaria parasites are believed to be specialized (Atkinson and Van Riper 1991), host taxonomic group is also used as a diagnostic character. Analyses of phylogenetic relationships based on the mitochondrial cytochrome *b* gene are beginning to reveal substantial diversity of parasite lineages and suggest that current taxonomy does not adequately reflect either the diversity of, or relationships among, malaria parasites (Escalante et al. 1998; Bensch et al. 2000; Perkins 2000; Ricklefs and Fallon 2002; Waldenström et al. 2002). Phylogenetic analyses show that avian and mammalian malaria parasites are distinct clades and that avian *Haemoproteus* is a distinct clade nested within paraphyletic lineages of avian *Plasmodium* (Bensch et al. 2000; Perkins and Schall 2002; Ricklefs and Fallon 2002).

The most straightforward evidence for host switching in avian malaria is the presence of a single lineage in more than one host. In the analysis of Ricklefs and Fallon (2002), which was based on haphazard sampling of hosts, there were ten such cases. Two of these involved hosts in the same genus (*Sialia* and *Piranga*), seven more involved more distantly related hosts in the same family, and the tenth involved hosts in the same superfamily (in the sense of Sibley and Ahlquist 1990). Analysis of host distributions in closely related parasite lineages, including a tree-based analysis of cospeciation and host switching (Page 1995; Ronquist 1997), reinforced the general impression that although host switching is conservative taxonomically, malaria parasites occasionally cross fairly large taxonomic distances to infect unrelated hosts (see also Bensch et al. 2000; Waldenström et al. 2002; Ricklefs et al. 2004). With the limited information currently available, these instances appear to be infrequent and unpredictable. However, like the broad host distribution of one or a limited number of strains of *Plasmodium relictum* in the Hawaiian avifauna, they emphasize the possibility of parasites switching hosts and potentially causing disease epidemics.

MIGRANTS AS BRIDGES BETWEEN NORTH AMERICA AND THE CARIBBEAN

The most straightforward evidence that migrants have transmitted parasites between birds in tropical and temperate regions would be the presence of a genetically identifiable lineage of parasite in both resident and migrant populations (Ricklefs and Fallon 2002; Waldenström et al. 2002). We are sequencing parasite cytochrome *b* from large numbers of individuals sampled in North America and the West Indies and have found several cases of shared parasite lineages (tables 17.1, 17.2). Table 17.1 includes nine malaria parasite lineages from Puerto Rico and the Lesser Antilles identified in three species of fringillid (Bananaquit [*Coereba flaveola*]; Lesser Antillean Bullfinch [*Loxigilla noctis*]; Black-

Table 17.1 The distribution of five lineages of *Haemoproteus* and four lineages of *Plasmodium* commonly found in four species of Lesser Antillean birds, among Neotropical migrants and Nearctic residents in Missouri, Alabama, and Michigan

		Tropical residents					
Parasite	A	NANA	LABU	BFGR	BWVI	Neotropical migrants	North Temperate residents
Haemoproteus	A	51	99	10	2	Red-eyed Vireo	
	B		1		7	Red-eyed Vireo	
	C	52	2	1			
	D	1	21	2	2	Northern Parula	
	E				10	Red-eyed Vireo (2)	
Plasmodium	A	24	19	28		Yellow-breasted Chat (3)	Northern Cardinal
						Black-and-White Warbler	
	B	5	2	3		Indigo Bunting (3)	Northern Cardinal (4)
						Yellow-breasted Chat	Carolina Wren
						Hooded Warbler	Eastern Towhee
						Blackpoll Warbler	Northern Mockingbird
						Rose-breasted Grosbeak	
						Scarlet Tanager	
						Gray Catbird	
	C	1			3	Red-eyed Vireo	
	D	7					

Note: Numbers in the columns for tropical residents are the numbers of individuals carrying each parasite lineage. Numbers of Nearctic birds found with each of the parasite lineages are indicated in parentheses if more than one. From Fallon et al. (2003) and Ricklefs et al. (unpubl.). NANA = Bananaquit; LABU = Lesser Antillean Bullfinch; BFGR = Black-faced Grassquit; BWVI = Black-whiskered Vireo.

Table 17.2 Additional examples of closely related parasite lineages present in small numbers of West Indian residents and also in either Nearctic residents or Neotropical migrants sampled in North America

		Tropical residents	Neotropical migrants	Temperate residents
Haemoproteus	F	Scaly-breasted Thrasher Pearly-eyed Thrasher (Montserrat)		Gray Catbird (Missouri)
	G	Bananaquit (Trinidad)	Prairie Warbler (Missouri)	
	H	Puerto Rican Bullfinch	Scarlet Tanager Summer Tanager Kentucky Warbler Red-eyed Vireo (Missouri)	
Plasmodium	E	Brown Trembler (Guadeloupe) Bananaquit (Jamaica)	Worm-eating Warbler Magnolia Warbler Yellow-throated Warbler Hooded Warbler Black-and-White Warbler (Missouri) Common Yellowthroat (Michigan)	Tufted Titmouse (Missouri)

faced Grassquit [*Tiaris bicolor*]; and the Black-whiskered Vireo [*Vireo altiloquus*]). Seven of the nine lineages have also been found in Neotropical migrants and two have been recovered from resident songbirds in southern Missouri and elsewhere in eastern North America. Although Lesser Antillean parasite lineages are present in migrant birds in Missouri, they are relatively uncommon. This may be due to nonoverlap of the particular migrant and tropical resident populations examined, as relatively few Neotropical migrants reach the Lesser Antilles, and migrants breeding in Missouri tend to winter in the Greater Antilles and Mesoamerica. It should also be noted that we have sampled relatively few North Temperate residents and so the sparse appearance of Lesser Antillean parasite lineages in these species is not informative. Some sharing of parasite lineages between winter migrant and resident host species has similarly been observed in Nigeria, indicating parasite transmission between migrants and residents on the wintering grounds (Waldenström et al. 2002). Table 17.2 shows additional examples of parasite connections from smaller samples among a variety of other hosts.

Clearly, migrants can form an effective bridge for parasites between tropical and temperate resident birds. These connections tend to involve hosts from the same general taxonomic groups (e.g., fringillids or mimids); however, two of the temperate residents sharing parasite lineages (Carolina Wren [*Thryothorus ludovicianus*] and Tufted Titmouse [*Baeolophus bicolor*]) are distantly related to the migrant or tropical resident co-hosts. We have not yet analyzed a large number of samples from North America to determine the extent to which parasite lineages occurring there have also been recovered from resident West Indian birds. Nor have we yet analyzed the large number of migrant individuals that we have sampled in the Greater Antilles during the northern winter. When these studies are completed, we shall be able to determine the degree to which the parasites of Neotropical migrants are unique to them as opposed to being characteristic of the regions in which they spend the winter or summer months.

As previously mentioned, the ability of a parasite to switch to a new host depends on the presence of infective host individuals, suitable vectors, and susceptible new hosts. Whether a potential host is susceptible to a new lineage of parasite depends on antigen-immune function interactions between the parasite and the host. Parasites such as West Nile virus (Rappole et al. 2000; Marfin and Gubler 2001) and some strains of malaria parasites (this study) can infect a broad range of host species. Others appear to be more specialized (table 17.1). The outcome of the parasite-host interaction may depend on genetic factors that change rapidly in host and parasite populations, perhaps much faster than the appearance of new genetic variation in many of the gene sequences used in phylogenetic analysis (Fallon et al. 2003). This labile evolutionary interaction between parasite and host populations is indicated by host species-times-island interactions in the prevalence of malaria parasites in the Lesser Antilles (fig. 17.2). The proportion of infected individuals of four species varied independently across islands, suggesting that each island population of host responded independently to the parasite lineages present on each island (Apanius et al. 2000; Fallon et al. 2003).

As can be seen in the case of the Lesser Antillean Bullfinch, increasing intensity of infections parallels the proportion of individuals with infections detected in blood smears. Thus, it is not clear whether prevalence determined on blood smears represents the proportion of individuals infected or the ability of the immune system to control infections. Regardless, prevalence is indicative of the balance between parasite virulence and host resistance. The host-species-times-island interaction illustrates the dynamic nature of the parasite-host interaction and the importance of genetic factors in both populations to the susceptibility of individuals to infection by a particular lineage of parasite.

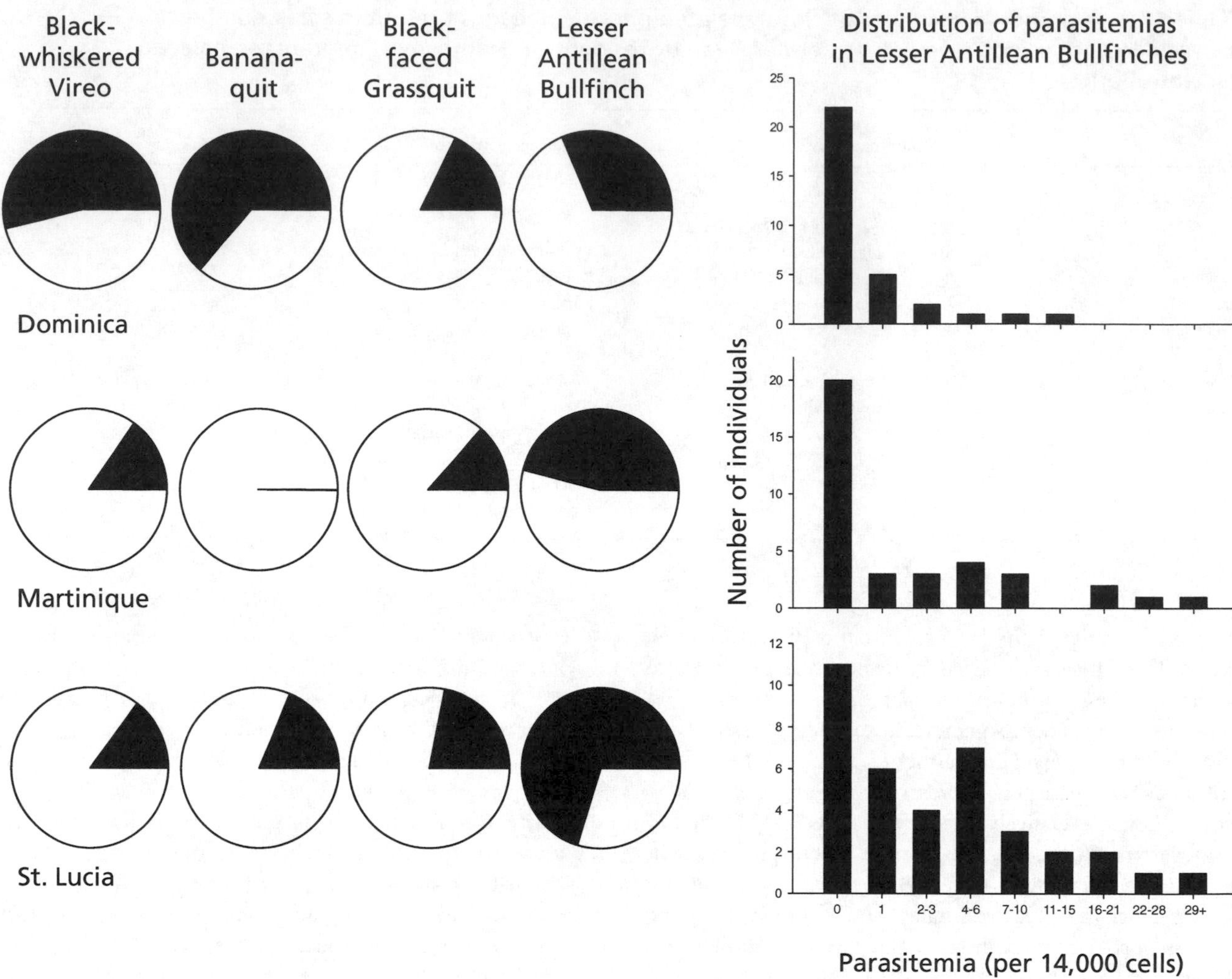

Fig. 17.2. Pie diagrams show the prevalence of malaria parasites revealed in blood smears in four species of birds on three islands in the Lesser Antilles. The host-species-times-island interaction is highly significant, indicating evolutionary independence of the host-parasite interactions (Apanius et al. 2000). The bar diagrams at right show the distribution of parasitemias for Lesser Antillean bullfinches on each of the three islands, illustrating the positive correlation between parasitemia and prevalence.

This is shown quite clearly in table 17.1 in the heterogeneous distribution of malaria parasite lineages among the four host species in the Lesser Antilles. We cannot predict whether a particular parasite lineage will infect a particular species. It is also unclear whether the single cases of *Haemoproteus* lineage C in a Black-faced Grassquit or of *Haemoproteus* D in a Bananaquit represent populations of these parasite lineages established in these hosts, spillover infections originating in the common host, or the rare manifestation of generally well-controlled infections.

PARASITE LINEAGES AS INDICATORS FOR LOCATION

To understand the causes of population trends in migrant birds, it is essential to identify the wintering and breeding areas of each population. So few birds banded in North America are recovered on the wintering grounds that the geographical connections between summering and wintering migrants have proved elusive. Attempts to localize populations of migrants by genetic markers have been successful in some species (Haig et al. 1997; Wennerberg 2001), but mostly have been disappointing because of the general homogenization of host genotypes throughout their North American ranges (Merila et al. 1997; Mila et al. 2000; Webster et al. 2002). Recently, new applications of technologies, especially in stable isotope chemistry, have allowed some discrimination of the areas of origin of migrant birds (Chamberlain et al. 1997; Marra et al. 1998; Chamberlain et al. 2000; Hobson et al. 2001).

Several authors have suggested that parasite lineages may be sufficiently locality- specific to indicate the origins of migrant birds. For example, Rintamaki et al. (1998) found that Willow Warblers (*Phylloscopus trochilus*) passing through an autumn stopover site in southern Finland carried different blood parasites (*Leucocytozoon* vs. *Haemoproteus*) at different times during the migration period. Using morphological evidence as well, the authors concluded that birds carrying the different parasites became infected in different areas and likely originated in different breeding populations (see also Thul et al. 1980 and Pung et al. 1997).

Parasites may be useful for distinguishing distinct populations occupying different types of habitats, but several factors argue against their general application for localizing individuals in broadly distributed populations. Such populations tend to lack genetic spatial structure, and the movement of individuals that prevents such structure from developing might be sufficient to homogenize parasite lineages, as suggested by surveys of Sehgal et al. (2001) on trypanosomes in central African birds. Rapid spread of diseases such as House Finch conjunctivitis and West Nile virus, at rates of hundreds of kilometers per year, provides little hope for local differentiation and endemism of parasite lineages. We have found only limited evidence for localization of parasite lineages on islands in the Lesser Antilles, where movement of individuals between islands is certainly less than across similar distances in North America (Fallon et al. 2003, unpubl.). Malaria parasite lineages tend to be localized to continents (Ricklefs and Fallon 2002), but this scale is not useful for localizing the origins of migratory individuals within populations.

The genetic markers used in most phylogenetic studies of parasites evolve slowly compared with the rate of dispersal of parasites through the host geographic range. Rapidly evolving genetic markers, which produce new mutations faster than they can disperse through the population, might provide worthwhile information on geographic origin. Such markers might possess transient value as spatially sensitive markers. For example, Anderson et al. (2001) surveyed a portion of the genome of West Nile virus that encodes envelope and membrane proteins and appears to be liable to rapid accumulation of nucleotide substitutions. They identified one mutation localized within 50 miles of Stamford, Connecticut. Surface proteins of malaria parasites, which mutate rapidly to avoid suppression by the host immune system, may be good candidates for geographically explicit markers.

DISCUSSION

Birds do not leave their parasites behind when they migrate from their wintering to their breeding grounds and back. Although there is some evidence that many infected individuals may not be physiologically capable of long-distance migration (Yorinks and Atkinson 2000; Latta 2003), some infected individuals do complete migration. In fact, migrant birds are flying sources of potential infection and we might wonder why they do not cause more disease epidemics than they appear to. Intensity of infection, suitable vectors, and susceptible potential hosts all appear to be important in determining whether the diseases of migrants will become established in resident populations in either the breeding or wintering areas. Parasites such as West Nile virus appear to infect a broad spectrum of hosts, although not all express strong symptoms and parasites may be asymptomatic in many species.

In our study of malaria parasites of birds in the West Indies and North America, we are finding that parasites common in resident populations in the Lesser Antilles occasionally are recovered from migrant and resident populations in Missouri and elsewhere in eastern North America. Because of their low prevalence, it is difficult to say whether these parasites are established in these alternative host populations or merely represent spillover cases from normal hosts. However, the connections emphasize the potential for migrants to carry potentially epidemic disease organisms from one region to another. Analyses of parasites in resident populations in the Greater Antilles and Mesoamerica may be more informative, as there is more overlap of particular populations of migrating birds.

Because the vulnerability of a resident population to a novel parasite must depend on many genetic and other factors, it is pointless to try to predict the emergence of new diseases. Where knowledge of the capacity of a pathogen to infect a particular population is critical, the most reasonable course may be experimental infection in safe laboratories to determine vulnerability of potential host populations (Atkinson et al. 2000). It is possible that migrants, such as the Bobolink (*Dolichonyx oryzivorous*) will carry West Nile virus to the Galápagos Islands. Because this eventuality presents a potential management problem, a logical course would be to test various species of the endemic Galápagos avifauna in continental laboratories to determine their susceptibility, and thus the potential magnitude of the threat, before disaster strikes.

Phylogenetic analyses of parasite and host populations can provide estimates of the prevalence of host switching over the history of a parasite clade (Page and Hafner 1996; Ronquist 1997; Atkinson et al. 2000). Analyses of cospeciation between malaria parasites and their avian hosts suggest that parasites switch between unrelated hosts frequently enough to obscure deep historical relationships between lineages of hosts and their parasites (Ricklefs and Fallon 2002; Waldenström et al. 2002). This appears to be particularly true of *Plasmodium* compared with *Haemoproteus* (see tables 17.1 and 17.2). We have nonetheless found several lineages of *Plasmodium* in West Indian residents, Neotropical migrants, and temperate residents, indicating that migrants might play an important role in carrying disease organisms between distant populations, or even in maintaining diseases in resident populations in North America.

The region of origin of parasite populations shared by residents and migrants can be determined in principle when only tropical or temperate residents are infected. Several of the lineages identified in our work appear primarily in West Indian resident birds and occasionally in Neotropical migrants, especially the Red-eyed Vireo (*Vireo olivaceous*), which has been well sampled in our Missouri study area (tables 17.1 and 17.2, *Haemoproteus* lineages A, B, D, E). We assume that these lineages are primarily endemic to the West

Indies, and that they are occasionally picked up by Neotropical migrants but are infrequently transmitted to other migrants or residents in northern breeding areas. In contrast, several lineages (*Plasmodium* B and E) occur in clades in which most closely related parasites were recovered from temperate regions, in both residents and migrants. Thus, it is likely that these parasite lineages are temperate in origin and have been carried to the West Indies by migrants and transmitted to local resident populations.

Although predicting emerging disease is unlikely to become a precise science, several types of studies would help us to understand the process of host switching generally and may provide insights into effective management practices. If intense infections are required for the transmission of some parasites, then studies of the effects of environmental contaminants on immune system function might identify potential hazards. Several studies have highlighted the importance of dense populations of vectors in the transmission of vector-borne diseases, such as West Nile fever or malaria (Hanna et al. 1996; Hubalek 2000). Studies of vector transmission of disease have provided valuable information for the control of malaria in human populations (e.g., Shroyer 1986; Gurtler et al. 1997; Lundstrom 1999) and presumably would yield insights for the control of particular diseases of wildlife populations.

Finally, the potential of genetic markers of parasites to localize host populations of long-distance migrants should be explored further. Suitable parasite markers should undergo nucleotide substitution rapidly, compared with their spread geographically through a host population. Although we would not expect the spatial distribution of such markers to remain unchanged for long periods, they might provide transient indicators of the origins of migrant populations.

Studies on the special relationship of parasites to migrants are just beginning. Although monitoring of emerging diseases is an important component of the migrant research program, it is also important to undertake basic studies of disease transmission, vector populations, and susceptibility of populations to understand the general conditions that promote host switching and the emergence of novel pathogens.

ACKNOWLEDGMENTS

Our research on avian malaria is generously supported by the U.S. National Science Foundation. We thank Pete Marra, Russ Greenberg, and several reviewers for helpful comments on the manuscript.

LITERATURE CITED

Alekseev, A. N., H. V. Dubinina, A. V. Semenov, and C. V. Bolshakov. 2001. Evidence of ehrlichiosis agents found in ticks (Acari: Ixodidae) collected from migratory birds. Journal of Medical Entomology 38:471–474.

Anderson, J. F., T. G. Andreadis, C. R. Vossbrinck, S. Tirrell, E. M. Wakem, R. A. French, A. E. Garmendia, and H. J. Van Kruiningen. 1999. Isolation of West Nile virus from mosquitoes, crows, and a Cooper's hawk in Connecticut. Science 286:2331–2333.

Anderson, J. F., C. R. Vossbrinck, T. G. Andreadis, A. Iton, W. H. Beckwith, and D. R. Mayo. 2001. A phylogenetic approach to following West Nile virus in Connecticut. Proceedings of the National Academy of Sciences USA 98:12885–12889.

Anderson, R. M., and R. M. May. 1991. Infectious Diseases of Humans: Dynamics and Control. Oxford University Press, Oxford.

Apanius, V., N. Yorinks, E. Bermingham, and R. E. Ricklefs. 2000. Island and taxon effects in parasitism and resistance of Lesser Antillean birds. Ecology 81:1959–1969.

Applegate, J. E. 1970. Population changes in latent avian malaria infections associated with season and corticosterone treatment. Journal of Parasitology 56:439–443.

Atkinson, C. T., R. J. Dusek, and J. K. Lease. 2001a. Serological responses and immunity to superinfection with avian malaria in experimentally-infected Hawaii Amakihi. Journal of Wildlife Diseases 37:20–27.

Atkinson, C. T., R. J. Dusek, K. L. Woods, and W. M. Iko. 2000. Pathogenicity of avian malaria in experimentally-infected Hawaii Amakihi. Journal of Wildlife Diseases 36:197–204.

Atkinson, C. T., J. K. Lease, B. M. Drake, and N. P. Shema. 2001b. Pathogenicity, serological responses, and diagnosis of experimental and natural malarial infections in native Hawaiian thrushes. Condor 103:209–218.

Atkinson, C. T., and C. Van Riper III. 1991. Pathogenicity and epizootiology of avian haematozoa: *Plasmodium, Leucocytozoon,* and *Haemoproteus.* Pages 19–48 *in* Bird-Parasite Interactions: Ecology, Evolution, and Behavior (J. E. Loye and M. Zuk, eds.). Oxford University Press, New York.

Atkinson, C. T., K. L. Woods, R. J. Dusek, L. S. Sileo, and W. M. Iko. 1995. Wildlife disease and conservation in Hawaii: pathogenicity of avian malaria (*Plasmodium relictum*) in experimentally infected iiwi (*Vestiaria coccinea*). Parasitology 111 (Supplement):S59–S69.

Bennett, G. F., M. A. Bishop, and M. A. Peirce. 1993. Checklist of the avian species of *Plasmodium* Marchiafava and Celli, 1885 (Apicomplexa) and their distribution by avian family and Wallacean life zones. Systematic Parasitology 26:171–179.

Bennett, G. F., M. A. Peirce, and R. A. Earlé. 1994. An annotated checklist of the valid avian species of *Haemoproteus, Leucocytozoon* (Apicomplexa, Haemosporida) and *Hepatozoon* (Apicomplexa, Haemogregarinidae). Systematic Parasitology 29:61–73.

Bensch, S., M. Stjernman, D. Hasselquist, O. Ostman, B. Hansson, H. Westerdahl, and R. T. Pinheiro. 2000. Host specificity in avian blood parasites: a study of *Plasmodium* and *Haemoproteus* mitochondrial DNA amplified from birds. Proceedings of the Royal Society of London, Series B, Biological Sciences 267:1583–1589.

Chamberlain, C. P., S. Bensch, X. Feng, S. Akesson, and T. Andersson. 2000. Stable isotopes examined across a migratory divide in Scandinavian willow warblers (*Phylloscopus trochilus trochilus* and *Phylloscopus trochilus acredula*) reflect their African winter quarters. Proceedings of the Royal Society of London, Series B, Biological Sciences 267:43–48.

Chamberlain, C. P., J. D. Blum, R. T. Holmes, X. H. Feng, T. W. Sherry, and G. R. Graves. 1997. The use of isotope tracers for identifying populations of migratory birds. Oecologia 109: 132–141.

Daniels, P. W. 1995. Australian-Indonesian collaboration in veterinary arbovirology: a review. Veterinary Microbiology 46:151–174.

Dhondt, A. A., D. L. Tessaglia, and R. L. Slothower. 1998. Epidemic mycoplasmal conjunctivitis in house finches from eastern North America. Journal of Wildlife Diseases 34:265–280.

Dobson, A., and P. Hudson. 1995. The interaction between the parasites and predators of red grouse *Lagopus lagopus scoticus.* Ibis 137:S87–S96.

Ebel, G. D., A. P. Dupuis, K. Ngo, D. Nicholas, E. Kauffman, S. A. Jones, D. Young, J. Maffei, P. Y. Shi, K. Bernard, and L. D. Kramer. 2001. Partial genetic characterization of West Nile Virus strains, New York State 2000. Emerging Infectious Diseases 7:650–653.

Escalante, A. A., D. E. Freeland, W. E. Collins, and A. A. Lal. 1998. The evolution of primate malaria parasites based on the gene encoding cytochrome b from the linear mitochondrial genome. Proceedings of the National Academy of Sciences USA 95:8124–8129.

Ewald, P. W. 1994. Evolution of Infectious Disease. Oxford University Press, Oxford.

Fair, J. M., E. S. Hansen, and R. E. Ricklefs. 1999. Growth, developmental stability and immune response in juvenile Japanese quails (*Coturnix coturnix japonica*). Proceedings of the Royal Society of London, Series B, Biological Sciences 266:1735–1742.

Fallis, A. M., S. S. Desser, and R. A. Khan. 1974. On species of leucocytozoon. Advances in Parasitology 12:1–67.

Fallon, S. M., E. Bermingham, and R. E. Ricklefs. 2003. Island and taxon effects in parasitism revisited: avian malaria in the Lesser Antilles. Evolution 57: 606–615.

Fallon, S. M., R. E. Ricklefs, B. L. Swanson, and E. Bermingham. 2003. Detecting avian malaria: an improved polymerase chain reaction diagnostic. Journal of Parasitology 89:1044–1047.

Fedynich, A. M., D. B. Pence, and R. D. Godfrey. 1995. Hematozoa in thin blood smears. Journal of Wildlife Diseases 31:436–438.

Feldman, R. A., L. A. Freed, and R. L. Cann. 1995. A PCR test for avian malaria in Hawaiian birds. Molecular Ecology 4:663–673.

Fubito, I., T. Nobuhiro, M. Toshiyuki, and F. Takako. 2000. Prevalence of Lyme disease *Borrelia* spp. in ticks from migratory birds on the Japanese mainland. Applied & Environmental Microbiology 66:982–986.

Godfrey, R. D. J., A. M. Fedynich, and D. B. Pence. 1987. Quantification of hematozoa in blood smears. Journal of Wildlife Diseases 23:558–565.

Greiner, E. C., G. F. Bennett, E. M. White, and R. F. Coombs. 1975. Distribution of the avian hematozoa of North America. Canadian Journal of Zoology 53:1762–1787.

Gurtler, R. E., J. E. Cohen, M. C. Cecere, and R. Chuit. 1997. Shifting host choices of the vector of chagas-disease, *Triatoma infestans,* in relation to the availability of hosts in houses in north-west Argentina. Journal of Applied Ecology 34:699–715.

Haig, S. M., C. L. Grattotrevor, T. D. Mullins, and M. A. Colwell. 1997. Population identification of Western Hemisphere shorebirds throughout the annual cycle. Molecular Ecology 6:413–427.

Hanna, J., S. Ritchie, D. Phillips, J. Shield, M. C. Bailey, J. S. MacKenzie, M. Poidinger, B. J. McCall, and P. J. Mills. 1996. An outbreak of Japanese encephalitis in the Torres Strait, Australia, 1995. Medical Journal of Australia 165:256–260.

Hill, A. V. S. 2001. The genomics and genetics of human infectious disease susceptibility. Annual Review of Genomics & Human Genetics 2:373–400.

Hobson, K. A., K. P. McFarland, L. I. Wassenaar, C. C. Rimmer, and J. E. Goetz. 2001. Linking breeding and wintering grounds of Bicknell's thrushes using stable isotope analyses of feathers. Auk 118:16–23.

Hubalek, Z. 2000. European experience with the West Nile virus ecology and epidemiology: could it be relevant for the New World? Viral Immunology 13:415–426.

Hudson, P. J., and A. P. Dobson. 1997. Transmission dynamics and host-parasite interactions of *Trichostrongylus tenuis* in Red Grouse (*Lagopus lagopus scoticus*). Journal of Parasitology 83:194–202.

Hudson, P. J., A. P. Dobson, and D. Newborn. 1998. Prevention of population cycles by parasite removal. Science 282: 2256–2258.

Jarvi, S. I., C. T. Atkinson, and R. C. Fleischer. 2001. Immunogenetics and resistance to avian malaria in Hawaiian honeycreepers (Drepanidinae). Studies in Avian Biology 2:254–263.

Jarvi, S. I., J. J. Schultz, and C. T. Atkinson. 2002. PCR diagnostics underestimate the prevalence of avian malaria (*Plasmodium relictum*) in experimentally-infected passerines. Journal of Parasitology 88:153–188.

Latta, S. C. 2003. The effects of scaley-leg mite infestations on body condition and site fidelity of migratory warblers. Auk 120: 730–743.

Li, J., R. A. Wirtz, G. A. McConkey, J. Sattabongkot, A. P. Waters, M. J. Rogers, and T. F. McCutchan. 1995. *Plasmodium:* genus-conserved primers for species identification and quantitation. Experimental Parasitology 81:182–190.

Lundstrom, J. O. 1999. Mosquito-borne viruses in western Europe: a review. Journal of Vector Ecology 24:1–39.

Marfin, A. A., and D. J. Gubler. 2001. West Nile encephalitis: an emerging disease in the United States. Clinical Infectious Diseases 33:1713–1719.

Marra, P. P., K. A. Hobson, and R. T. Holmes. 1998. Linking winter and summer events in a migratory bird by using stable-carbon isotopes. Science 282:1884–1886.

Massey, J. G., T. K. Graczyk, and M. R. Cranfield. 1996. Characteristics of naturally acquired *Plasmodium relictum capistranoae* infections in naïve Hawaiian crows (*Corvus hawaiiensis*) in Hawaii. Journal of Parasitology 82:182185.

McClure, H. E. 1974. Migration and Survival of the Birds of Asia. U.S. Army Component, SEATO Medical Project, Bangkok.

McClure, H. E., and N. Ratanaworabhan. 1973. Some Ectoparasites of the Birds of Asia. U.S. Army Medical Component, SEATO Medical Project, Bangkok.

Merila, J., M. Bjorklund, and A. J. Baker. 1997. Historical demography and present day population structure of the greenfinch, *Carduelis chloris:* an analysis of mtDNA control-region sequences. Evolution 51:946–956.

Mila, B., D. J. Girman, M. Kimura, and T. B. Smith. 2000. Genetic evidence for the effect of a postglacial population expansion on the phylogeography of a North American songbird. Proceedings of the Royal Society of London, Series B, Biological Sciences 267:1033–1040.

Miller, B. R., R. S. Nasci, M. S. Godsey, H. M. Savage, J. J. Lutwama, R. S. Lanciotti, and C. J. Peters. 2000. First field evidence for natural vertical transmission of West Nile virus in *Culex univittatus* complex mosquitoes from Rift Valley Province, Kenya. American Journal of Tropical Medicine & Hygiene 62:240–246.

Mitchell, C. J. 1991. Vector competence of North and South American strains of *Aedes albopictus* for certain arboviruses: a review. Journal of the American Mosquito Control Association 7:446–451.

Ots, I., and P. Hõrak. 1996. Great tits *Parus major* trade health for reproduction. Proceedings of the Royal Society of London, Series B, Biological Sciences 263:1443–1447.

Page, R. D. M. 1995. Parallel phylogenies: reconstructing the history of host-parasite assemblages. Cladistics 10:155–173.

Page, R. D. M., and M. S. Hafner. 1996. Molecular phylogenies and host-parasite cospeciation: gophers and lice as a model system. Pages 255–270 *in* New Uses for New Phylogenies (P. H. Harvey, A. J. L. Brown, J. Maynard Smith, and S. Nee, eds.). Oxford University Press, Oxford.

Peirce, M. A., and G. F. Bennett. 1996. A revised key to the avian subgenera of *Plasmodium* Marchiafava and Celli, 1885 (Apicomplexa). Systematic Parasitology 33:31–32.

Perkins, S. L. 2000. Species concepts and malaria parasites: detecting a cryptic species of *Plasmodium*. Proceedings of the Royal Society of London, Series B, Biological Sciences 267: 2345–2350.

Perkins, S. L., and J. J. Schall. 2002. A molecular phylogeny of malarial parasites recovered from cytochrome b gene sequences. Journal of Parasitology 88:972–978.

Plowright, W. 1982. The effects of rinderpest and rinderpest control on wildlife in Africa. Symposium of the Zoological Society of London 50:1–28.

Pung, O. J., N. E. Maxwell, E. C. Greiner, J. R. Robinette, and J. E. Thul. 1997. *Haemoproteus greineri* in wood ducks from the Atlantic flyway. Journal of Wildlife Diseases 33: 355–358.

Råberg, L., M. Grahn, D. Hasselquist, and E. Svensson. 1998. On the adaptive significance of stress-induced immunosuppression. Proceedings of the Royal Society of London, Series B, Biological Sciences 265:1637–1641.

Randolph, S. E., and D. J. Rogers. 2000. Fragile transmission cycles of tick-borne encephalitis virus may be disrupted by predicted climate change. Proceedings of the Royal Society of London, Series B, Biological Sciences 267:1741–1744.

Rappole, J. H., S. R. Derrickson, and Z. Hubalek. 2000. Migratory birds and spread of West Nile virus in the Western Hemisphere. Emerging Infectious Diseases 6:319–328.

Richard, F. A., R. N. M. Sehgal, H. I. Jones, and T. B. Smith. 2002. A comparative analysis of PCR-based detection methods for avian malaria. Journal of Parasitology 88:819–822.

Richie, T. L. 1988. Interactions between malaria parasites infecting the same vertebrate host. Parasitology 96:607–639.

Ricklefs, R. E., and S. M. Fallon. 2002. Diversification and host switching in avian malaria parasites. Proceedings of the Royal Society of London, Series B, Biological Sciences 269:885–892.

Ricklefs, R. E. S. M. Fallon, and E. Bermingham. 2004. Evolutionary relationships, cospeciation, and host switching in avian malaria parasites. Systematic Biology 53:111–119.

Rintamaki, P. T., W. Ojanen, H. Pakkala, and M. Tynjala. 1998. Blood parasites of migrating willow warblers (*Phylloscopus trochilus*) at a stopover site. Canadian Journal of Zoology 76:984–988.

Roberts, S. R., P. M. Nolan, L. H. Lauerman, L. Q. Li, and G. E. Hill. 2001. Characterization of the mycoplasmal conjunctivitis epizootic in a house finch population in the southeastern USA. Journal of Wildlife Diseases 37:82–88.

Ronquist, F. 1997. Phylogenetic approaches in coevolution and biogeography. Zoologica Scripta 26:313–322.

Scott, J. D., K. Fernando, S. N. Banerjee, L. A. Durden, S. K. Byrne, M. Banerjee, R. B. Mann, and M. G. Morshed. 2001. Birds disperse ixodid (Acari: Ixodidae) and *Borrelia burgdorferi*-infected ticks in Canada. Journal of Medical Entomology 38:493–500.

Sehgal, R. N. M., H. I. Jones, and T. B. Smith. 2001. Host specificity and incidence of *Trypanosoma* in some African rainforest birds: a molecular approach. Molecular Ecology 10:2319–2327.

Service, M. W. 1991. Agricultural development and arthropod-borne diseases: a review. Revista de Saude Publica 25:165–178.

Shehata, C., L. Freed, and R. L. Cann. 2001. Changes in native and introduced bird populations on O'ahu: infectious diseases and species replacement. Studies in Avian Biology 22:264–273.

Sheldon, B. C., and S. Verhulst. 1996. Ecological immunology: costly parasite defences and trade-offs in evolutionary ecology. Trends in Ecology & Evolution 11:317–321.

Shroyer, D. A. 1986. *Aedes albopictus* and arboviruses: a concise review of the literature. Journal of the American Mosquito Control Association 2:424–428.

Sibley, C. G., and J. E. Ahlquist. 1990. Phylogeny and Classification of the Birds of the World. Yale University Press, New Haven.

Smith, R. P., P. W. Rand, E. H. Lacombe, S. R. Morris, D. W. Holmes, and D. A. Caporale. 1996. Role of bird migration in the long-distance dispersal of *Ixodes dammini,* the vector of Lyme disease. Journal of Infectious Diseases 174:221–224.

Super, P. E., and C. Van Riper III. 1995. A comparison of avian hematozoan epizootiology in two California coastal scrub communities. Journal of Wildlife Diseases 31:447–461.

Thul, J. E., D. J. Forrester, and E. C. Greiner. 1980. Hematozoa of wood ducks (*Aix sponsa*) in the Atlantic flyway. Journal of Wildlife Diseases 16:383–390.

Van Riper, C., III, S. G. Van Riper, M. L. Goff, and M. Laird. 1986. The epizootiology and ecological significance of malaria in Hawaiian land birds. Ecological Monographs 56:327–344.

Waldenström, J., S. Bensch, S. Kiboi, D. Hasselquist, and U. Ottosson. 2002. Cross-species infection of blood parasites between resident and migratory songbirds in Africa. Molecular Ecology 11:1545–1554.

Webster, M. S., P. P. Marra, S. M. Haig, S. Bensch, and R. T. Holmes. 2002. Links between worlds: unraveling migratory connectivity. Trends in Ecology & Evolution 17:76–83.

Weissenböck, H., J. Kolodziejek, A. Url, H. Lussy, B. Rebel-Bauder, and N. Nowotny. 2002. Emergence of Usutu virus, and African mosquito-borne flavivirus of the Japanese encephalitis virus group, Central Europe. Emerging Infectious Diseases 8:652–656.

Wennerberg, L. 2001. Breeding origin and migration pattern of dunlin (*Calidris alpina*) revealed by mitochondrial DNA analysis. Molecular Ecology 10:1111–1120.

White, E. M., E. C. Greiner, G. F. Bennett, and C. M. Herman. 1978. Distribution of the hematozoa of Neotropical birds. Revista de Biología Tropical 26 (Supplement)1:43–102.

Yorinks, N., and C. T. Atkinson. 2000. Effects of malaria on activity budgets of experimentally infected juvenile Apapane (*Himatione sanguinea*). Auk 117:731–738.

Zeller, H. G., and B. Murgue. 2001. The role of migrating birds in the West Nile virus epidemiology [French]. Médecine et Maladies Infectieuses 31:168S–174S.

18

THOMAS B. SMITH, SONYA M. CLEGG,
MARI KIMURA, KRISTEN C. RUEGG,
BORJA MILÁ, AND IRBY J. LOVETTE

Molecular Genetic Approaches to Linking Breeding and Overwintering Areas in Five Neotropical Migrant Passerines

DEMOGRAPHIC STUDIES OF NEOTROPICAL MIGRANT songbirds have been limited by the difficulty of following them through a complete annual cycle. As population regulation conceivably may occur in either the breeding or wintering areas, or on migration routes, determining levels of connectivity between breeding and wintering areas is fundamental to understanding the dynamics of Neotropical migrant populations. Recent studies have explored the potential for genetic markers to determine the breeding origin of migrating and overwintering birds. The utility of this method is dependent upon the level of geographic structure in a particular species. The finer the scale of geographic structure resolved by a particular genetic marker, the more useful it is in resolving breeding origins. We assessed the utility of mitochondrial DNA (mtDNA) markers in determining breeding origins of five long-distance Neotropical migrants: the Yellow-breasted Chat (*Icteria virens*), Nashville Warbler (*Vermivora ruficapilla*), Common Yellowthroat (*Geothlypis trichas*), Wilson's Warbler (*Wilsonia pusilla*), and Swainson's Thrush (*Catharus ustulatus*) and used and contrasted both mtDNA and microsatellite analyses in Wilson's Warbler. We assessed the extent of mtDNA phylogeographic structure and used these data to assign individuals captured on wintering sites in Mexico and Central and South America to their respective breeding areas. Genetic structure on the breeding grounds was found on a broad continent-wide scale for all five of these species, thus enabling the assignment of overwintering individuals to either eastern or western breeding lineages. Patterns of genetic divergence were not always in concordance with morphological subspecies definitions. The degree of admixture of genetic lineages on overwintering grounds varied for each species, with high geographic segregation for the Yellow-breasted Chat and Swainson's Thrush and more geographic mixing of lineages for the Common Yellowthroat, Nashville Warbler, and Wilson's Warbler. The suggested distribution of morphological subspecies on wintering grounds was not always supported by the genetic analysis. The methods used here allowed associations of breeding and wintering grounds at a broad scale. The ability to link populations at a finer geographic scale may be possible when molecular genetic techniques are combined with other sources of information on geographic origin.

INTRODUCTION

Characterizing levels of population connectivity between breeding and overwintering areas has proven to be a challenge for the majority of migratory songbirds. The lack of specific information on levels of connectivity has limited integrated studies of life history and population regulation in migrant species (Webster et al. 2002). The few studies that have been able to examine demographic processes on both breeding and wintering areas have revealed previously unappreciated interactions and relationships (e.g., Marra et al. 1998; Sillett et al. 2000; Gill et al. 2001). Documenting levels of connectivity has the potential to aid in understanding the relationship between migratory behavior and gene flow (e.g., Arguedas and Parker 2000) and its theorized importance in speciation (Winker 2000). In a conservation context, if links between breeding and wintering areas could be resolved to a fine enough geographic scale, demographic trends and land use changes could be related, thereby informing management decisions. Finally, because migratory birds may act as intermediate hosts in some human diseases, studies of connectivity may aid in understanding the epidemiology of diseases (Rappole et al. 2000; Alekseev et al. 2001).

The identification of markers that reveal the origin of an individual or link populations at different stages of the annual cycle is an essential first step in studies of population connectivity (Wenink and Baker 1996; Haig et al. 1997; Webster et al. 2002). Although traditional banding studies are useful for some avian taxa, particularly shorebirds and waterfowl, they have typically been of limited utility for migrant songbirds, for which return rates are often extremely low (Berthold 1993; Webster et al. 2002). For example, of the more than 140,000 individuals of Wilson's Warbler (*Wilsonia pusilla*) banded in the United States and Canada to date, only three have been recovered on their wintering areas in Latin America, yielding a dismal return rate of 0.002% (Bird Banding Laboratory, Laurel, Maryland). Similarly, although radio and satellite telemetry are valuable for determining movements of large-bodied migrants capable of carrying heavy transmitters (Ristow et al. 2000), most passerines are too small to carry the necessary transmitter and battery to be tracked efficiently over large distances. An alternative to marking and tracking individuals is to use population-specific genetic markers. A major advantage of this approach is that it relies on the association among individuals in a population, and therefore a particular individual does not have to be recaptured or followed.

A population-based molecular approach is a potentially powerful tool for assessing levels of connectivity between breeding and overwintering sites (Webster et al. 2002). For example, molecular genetic markers have been used successfully to examine connectivity in shorebirds (Wenink and Baker 1996; Haig et al. 1997) and more recently in some small passerines (Buerkle 1999; Milot et al. 2000; Kimura et al. 2002; Ruegg and Smith 2002; Clegg et al. 2003; Lovette et al. 2004). To apply molecular methodology effectively to the question of connectivity, genetic variation in populations needs to be geographically structured and the chosen molecular marker must be sensitive enough to detect existing structure. Molecular markers vary widely in their capacity to detect variation at a given phylogenetic level (i.e., species, subspecies, or populations); thus, considerable care must be taken when selecting markers for studying connectivity.

The main classes of genetic marker that have been used to study connectivity in birds include allozymes, random amplified polymorphic DNA (RAPD), mitochondrial DNA (mtDNA) and, more recently, microsatellites (Haig et al. 1997; Webster et al. 2002; Clegg et al. 2003). These markers evolve at different rates (in general, lowest for allozymes and highest for microsatellites), and therefore have the potential to provide different levels of geographically structured genetic variation (Avise 1994). Studies to date using the various classes of markers indicate that levels of genetic variation are generally low to negligible in Neotropical migrant species, especially those with geographic distributions in formerly glaciated areas (Ball and Avise 1992; Seutin et al. 1995; Buerkle 1999; Arguedas and Parker 2000; Gibbs et al. 2000; Milá et al. 2000; Winker 2000; Lovette and Bermingham 2001; Kimura et al. 2002). Several studies have contrasted migratory and sedentary species, documenting higher magnitudes of phylogeographic variation in the latter (Gill et al. 1993; Klein 1994; Zink 1997; Lovette et al. 1998). This difference has been attributed to higher gene flow in migrants (e.g., Arguedas and Parker 2000) and the genetic effects of rapid demographic expansions following recent glaciation events (e.g., Milá et al. 2000). The application of molecular markers to studies of connectivity in migrant songbirds might be affected by these factors, and further comparative studies are needed to make generalizations regarding the amount of structure expected in this group. Additionally, it is important to explore the potential of new molecular techniques, such as using amplified fragment length polymorphisms (AFLP) to find informative single-nucleotide polymorphisms (SNP) (Bensch et al. 2002), and of combining molecular data with other sources of information such as data from chemical isotopes (Hobson, Chap. 19, this volume), banding data (Ruegg and Smith 2002), remote sensing (Szép and Møller, Chap. 29, this volume), and disease information (Rintamaki et al. 2000; Ricklefs et al., Chap. 17, this volume), and how combined approaches could augment studies of connectivity.

Here, we summarize results on five species of Neotropical migrants. Our specific objectives are to: (1) describe the population structure on breeding grounds by using genetic markers, particularly mtDNA, and in one species also microsatellites; (2) assess the scale at which these markers can be successfully used to study connectivity between breeding and overwintering areas; (3) discuss how these data can contribute to life history and demographic studies; and (4) discuss and illustrate how molecular genetic data might be integrated with other sources of data to better assess connectivity.

METHODS

Study Species

Two main criteria were used in choosing the species we examined. First, to maximize the scope of geographic coverage and the potential for detecting structure if it existed, we chose North American songbird species with subspecific variation and broad continental breeding ranges. Secondly, we selected species for which sample sizes were sufficient and distributed widely across both breeding and wintering areas.

The Yellow-breasted Chat (*Icteria virens*) is found throughout eastern deciduous forests and western riparian habitats. The two recognized subspecies differ subtly in size, plumage, and song characteristics (Eckerle and Thompson 2001). The eastern subspecies, *I. v. virens,* breeds from the eastern Great Plains to the eastern United States and is thought to winter from eastern Mexico to Central America (Eckerle and Thompson 2001). The western subspecies, *I. v. auricollis,* has a more fragmented distribution, breeding from the western Great Plains westward and is thought to winter from western Mexico to central Guatemala (Eckerle and Thompson 2001).

The Nashville Warbler (*Vermivora ruficapilla*) has a disjunct breeding range in North America. The two recognized subspecies differ in morphology and plumage; the western race is brighter in plumage and has a longer tail (Williams 1996). The eastern subspecies, *V. r. ruficapilla,* breeds from the northern hardwood and boreal forest of the eastern United States, central Quebec, and westward to parts of central Manitoba. The suggested overwintering range is in eastern Mexico and Guatemala (Williams 1996). The western subspecies, *V. r. ridgwayi,* has a patchy distribution from southern British Columbia south to parts of the western United States. The suggested overwintering range of the western subspecies is California and western Mexico (Williams 1996).

The Common Yellowthroat (*Geothlypis trichas*) is the most widespread wood-warbler in North America. This species shows a complex pattern of subspecific variation, with some authors recognizing as many as 13 subspecies and considerable clinal variation (Lowery and Monroe 1968). The breeding range of this species spans most of continental North America. Both wintering and breeding populations occur in the southern United States and parts of central Mexico, and strictly wintering populations are found in Baja California, parts of western Mexico, eastern Mexico to Panama, and portions of the West Indies and Bermuda (Guzy and Ritchison 1999).

Wilson's Warbler (*Wilsonia pusilla*) is a common wood-warbler associated with wet habitats. The three recognized subspecies differ subtly in coloration and size (Lowery and Monroe 1968; Pyle et al. 1997). *W. p. pusilla* breeds from the boreal forests of eastern Canada west to British Columbia; *W. p. pileolata* breeds from Alaska to parts of Montana and Idaho; and *W. p. chryseola* breeds along the Pacific Coast to south-central California. The three subspecies have a wintering range extending from eastern and western Mexico (including southern Baja) and from parts of southern Louisiana and Texas to Panama (Ammon and Gilbert 1999).

Swainson's Thrush (*Catharus ustulatus*) breeds in interior forest, secondary growth, and riparian thickets. Although a number of subspecies have been described, two main groups are recognized on the basis of plumage coloration. The olive-backed group, *C. u. alame* and *C. u. swainsoni,* is found in continental regions of Canada to western Alaska, and the russet-backed group, *C. u. ustulatus* and *C. u. oedicus,* is spread along the Pacific Coast (Bond 1963). It is suggested that the olive-backed group winters primarily in South America, whereas the russet-backed group winters primarily in southern Mexico and Central America (Bond 1963; Ramos and Warner 1980).

Sampling and Molecular Approaches

Blood and feather samples were collected from adult birds mist-netted at breeding sites in Canada and the United States and at overwintering sites in Mexico, Central America, and South America. Sampling locations and sample sizes for each species are shown in table 18.1. Blood samples were obtained by sub-brachial venipuncture, and feather samples by plucking the outermost two rectrices. See Kimura et al. (2002), Ruegg and Smith (2002), and Lovette et al. (2004) for methods of DNA extraction, sequencing, and restriction enzyme digests, and Clegg et al. (2003) for methods of genotyping individuals using microsatellites in Wilson's Warblers.

For each species, we first used samples from across the breeding range to reconstruct a phylogeny based on mtDNA sequence (see above papers). We then identified restriction enzymes that were diagnostic of statistically well-supported, geographically defined lineages (table 18.2). Enzyme assays were used to screen samples from individuals captured on overwintering areas to assign them to geographically defined breeding areas.

Table 18.1 Sampling localities in breeding and overwintering areas, locality codes (as shown in figs. 18.2 and 18.3), and numbers of genetic samples obtained for the five species in the study

Breeding localities	Figure code	YBCH	NAWA	COYE	WIWA	SWTH
Juneau, Alaska	AK1	—	—	—	—	5
Denali National Park, Alaska	AK2	—	—	—	15	—
Tetlin National Wildlife Refuge, Fort Yukon, Alaska	AK3	—	—	—	—	15
Yukon Flats National Wildlife Refuge Alaska	AK4	—	—	—	—	20
Kotzebue, Alaska	AK5	—	—	—	—	7
Pitt Lake, British Columbia, CAN	BC1	—	—	2	—	—
Queen Charlotte Island, British Columbia, CAN	BC2	—	—	—	—	10
Squamish, British Columbia, CAN	BC3	—	—	—	—	11
Pemberton, British Columbia, CAN	BC4	—	—	—	—	19
Revelstoke, British Columbia, CAN	BC5	—	—	—	—	11
Quesnel, British Columbia, CAN	BC6	—	—	—	—	11
Hinton, Alberta, CAN	AB	—	—	—	14	—
Mt. Baker National Forest, Washington	WA	—	—	3	12	—
Paisley, Oregon	OR1	1	—	—	—	—
Williams, Oregon	OR2	—	4	—	—	—
Umatilla National Forest, Oregon	OR3	—	—	—	5	24
Siuslaw National Forest, Oregon	OR4	—	—	—	16	20
Boise, Idaho	ID	—	3	—	—	—
Shasta, California	CA1	4	—	—	—	—
Bolinas, California	CA2	—	—	1	—	15
Los Banos, California	CA3	—	—	3	—	—
Foresthill, California	CA4	—	2	—	—	—
Tahoe National Forest, California	CA5	—	—	—	15	—
Kings Canyon National Park, California	CA6	—	—	—	12	—
Pillar Point, California	CA7	—	—	—	17	—
Big Sur, California	CA8	—	—	—	8	—
Ruby Lake, Nevada	NV1	4	—	—	—	—
Lake Mead, Nevada	NV2	—	—	3	—	—
Holter Dam, Montana	MT1	1	—	4	—	—
Denton, Montana	MT2	1	—	—	—	—
Flathead National Forest, Montana	MT3	—	—	—	—	20
Atlantic City, Wyoming	WY	3	—	1	—	—
Manila, Utah	UT1	2	—	—	—	—
Grantsville, Utah	UT2	9	—	—	—	—
Mt. Timpanogos, Utah	UT3	—	—	—	—	11
Grand Mesa, Colorado	CO	—	—	—	14	—
Junction City, Kansas	KS	1	—	4	—	—
Fort Leonard Wood, Missouri	MO	4	—	—	—	—
Owensburg, Indiana	IN	3	—	4	—	—
Fort Knox, Kentucky	KY	3	—	—	—	—
Cuyahoga, Ohio	OH	—	—	3	—	—
Seney National Wildlife Refuge, Michigan	MI1	—	—	1	—	—
Dearborn, Michigan	MI2	—	4	—	—	—
Bristol, Tennessee	TN	4	—	—	—	—
Charleston, South Carolina	SC	2	—	—	—	—
Finland, Minnesota	MN	—	2	—	—	—
Kakabeka, Ontario, CAN	ON1	—	4	—	—	—
Hilliardton, Ontario, CAN	ON2	—	—	2	4	—
Thunder Bay, Ontario, CAN	ON3	—	—	—	—	9
Kakabeka Falls, Ontario, CAN	ON4	—	—	—	—	10
Camp Myrica, Quebec, CAN	PQ1	—	3	5	16	—
Charlevoix, Quebec City, Quebec, CAN	PQ2	—	—	—	—	2
Fredericton, New Brunswick, CAN	NB	—	1	4	4	—
Rochester, New York	NY	—	4	—	—	—
South Britain, Connecticut	CT	—	—	2	—	—
Truro, Massachusetts	MA	—	—	4	—	—
Fort Polk, Louisiana	LA	1	—	—	—	—

continued

Table 18.1 (continued)

Overwintering localities	Figure code	YBCH	NAWA	COYE	WIWA	SWTH
Los Cabos, Baja California Sur, MEX	BCS	7	—	11	7	—
Chupaderos, Sinaloa, MEX	SIN	2	12	—	—	—
Sierra de Manatlán Biosphere Reserve, Autlan, Jalisco, MEX	JAL	1	30	—	22	—
Nevado de Colima, Colima, MEX	COL	—	2	—	—	—
La Maria, Colima, MEX	COL	—	—	—	25	—
Huautla, Morelos, MEX	MOR	1	7	—	—	—
Zacualtipán, Hidalgo, MEX	HGO	—	5	—	—	—
El Cielo Biosphere Reserve, Tamaulipas, MEX	TAM	—	—	—	13	—
Coatepec, Veracruz, MEX	VER1	1	—	—	19	—
Catemaco, Veracruz, MEX	VER2	5	—	—	7	—
Chila, Oaxaca, MEX	OAX1	5	6	—	—	—
Animas de Trujano, Oaxaca, MEX	OAX2	2	23	9	15	33
Hidalgo, Oaxaca, MEX	OAX3	—	6	—	—	—
El Triunfo Reserve, Chiapas, MEX	CHS	—	—	—	—	20
El Ocote Reserve, Chiapas, MEX	CHS	7	—	—	6	18
San Ignacio, BZ	BZ	13	—	9	—	—
Cockscomb Basin, BZ	BZ	—	—	—	1	—
El Boqueron Volcano, ES	ES	2	2	—	—	2
San Salvador, ES	ES	—	—	—	15	—
Tegucigalpa, HON	HON	—	—	—	25	4
Esteli, NIC	NIC	—	—	—	9	—
Santa Elena, CR	CR1	—	—	—	10	4
San Vito, CR	CR2	—	—	—	12	5
Cerro Jefe, PN	PN	—	—	—	—	3
Mindo, EC	ECU	—	—	—	—	18
Nuevo Peru, PE	PE	—	—	—	—	6
Santa Cruz, BOL	BOL	—	—	—	—	2
Totals		*89*	*120*	*75*	*338*	*335*

Note: Country abbreviations: CAN = Canada; MX = Mexico; ES = El Salvador, BZ = Belize; HON = Honduras; NIC = Nicaragua; CR = Costa Rica; PN = Panama; EC = Ecuador; PE = Peru; BOL = Bolivia. Species abbreviated: YBCH = Yellow-breasted Chat; NAWA = Nashville Warbler; COYE = Common Yellowthroat; WIWA = Wilson's Warbler; SWTH = Swainson's Thrush.

RESULTS AND DISCUSSION

Patterns of Variation on the Breeding Grounds

Several common patterns in population genetic structure are evident among all five species (fig. 18.1A–E). The most obvious similarity is that each species is divided into two main haplotype groups associated to varying degrees with eastern and western sampling sites (fig. 18.2A–E). The level of divergence between these groups was between 0.5 and 2%, consistent with a late-Pleistocene divergence (e.g., Avise and Walker 1998; Kimura et al. 2002; Ruegg and Smith 2002; Lovette et al. 2004). Another similarity among the five species was the relative lack of structure within eastern and western haplotype groups. Low levels of variation could be due to current or historical gene flow or past demographic events (e.g., Milá et al. 2000). The high level of homogeneity across broad geographic areas was most evident in the eastern lineage of all sufficiently sampled species, suggesting that eastern and western lineages may have had different demographic histories. For example, the broad distribution of very similar haplotypes within the east could suggest that these lineages may have experienced a more rapid demographic expansion following a Pleistocene glaciation event than occurred in western regions. In general, there was a slightly higher level of phylogenetic structure within western groups (fig. 18.1), possibly stemming from a less

Table 18.2 Summary of the mtDNA regions and restriction enzymes used to discriminate among eastern and western forms of each species

Species	mtDNA region	Restriction enzyme
Yellow-breasted Chat	ATPase	East: *Rsa*I West: *Dpn*II
Common Yellowthroat	ATPase	East: *Tsp45* West: *Bst*NI
Nashville Warbler	ATPase	East: *Hinc*II West: *Stu*I
Swainson's Thrush	Control region I	West: *Sfc*I
Wilson's Warbler	Control region I & cytochrome b	East: *Nsi*I West: *Hinc*II

Note: The eastern lineage of Swainson's Thrush was defined by the absence of the *Sfc*I site.

A. Yellow-breasted Chat

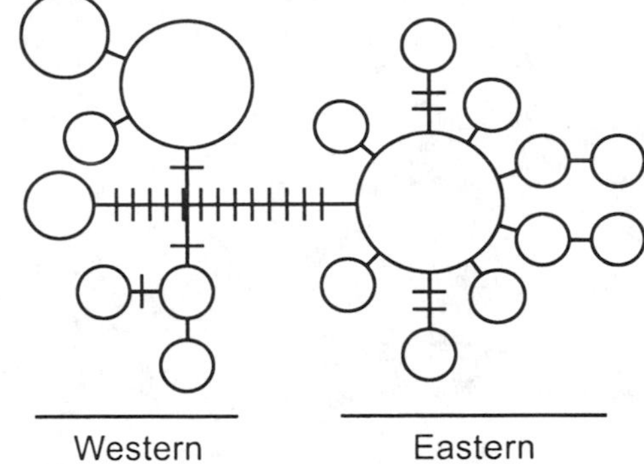

B. Nashville Warbler

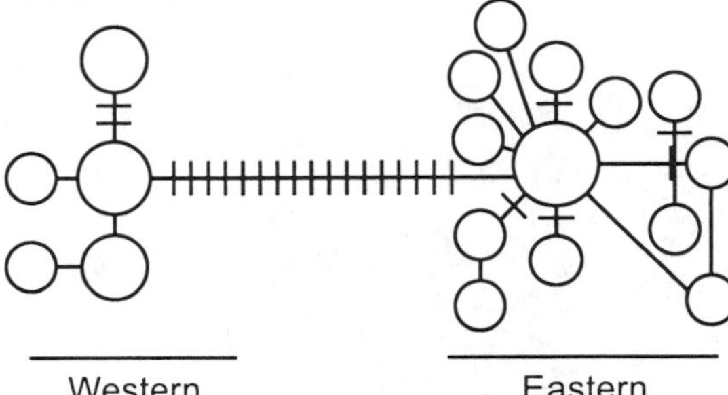

D. Wilson's Warbler

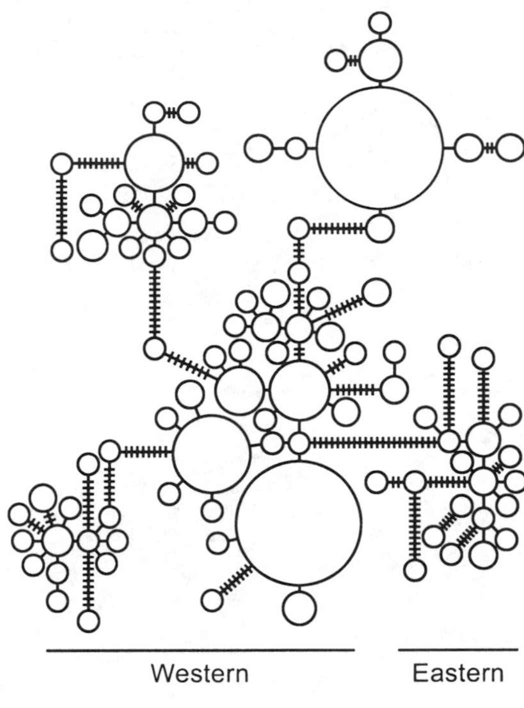

C. Common Yellowthroat

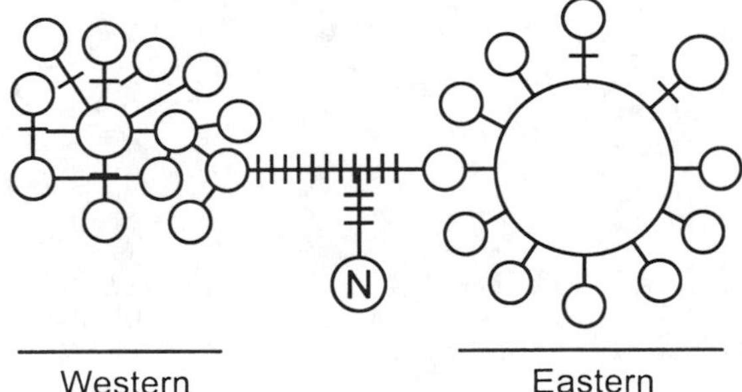

E. Swainson's Thrush

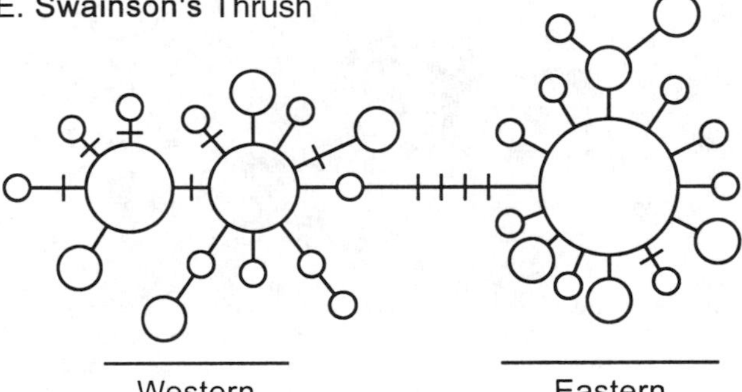

Fig. 18.1. Minimum-spanning network with each unique haplotype indicated by a circle and area proportional to the number of individuals sampled. Hatch marks along branches indicate inferred haplotype differences. Eastern and western geographic lineages are indicated below each network. (A) Yellow-breasted Chat: mtDNA ATPase sequences were obtained from 34 individuals, including 11 eastern and 7 western individuals. A total of 18 unique haplotypes with 17 nucleotide substitutions (1.8% sequence divergence) between eastern and western populations (Lovette et al. 2004). (B) Nashville Warbler: sequences obtained from 27 individuals, including 18 eastern and 9 western individuals. Eastern and western haplotypes differed by 16 to 22 substitutions, 1.7–2.3 % sequence divergence (Lovette et al. 2004). (C) Common Yellowthroat: sequences from 47 individuals with a maximum of 19 nucleotide substitutions (2%) (see Lovette et al. 2004). Divergent Nevada haplotype indicated by "N." (D) Wilson's Warbler: mtDNA control region sequences from 200 individuals. A total of 94 unique haplotypes were identified, and eastern and western haplotypes differed by a minimum of 22 substitutions (see Kimura et al. 2002). (E) Swainson's Thrush: mtDNA control region sequences from 183 individuals with a net sequence divergence between lineages of 0.69% (Ruegg and Smith 2002).

severe effect of glaciation in the west and/or the maintenance of a higher level of population subdivision over long periods.

Although these species share the general patterns described above, some clear species-specific differences in phylogeographic patterns are evident. The geographic location of the east-west split in each species differed (fig. 18.2). In the Yellow-breasted Chat, Nashville Warbler, and Wilson's Warbler, the two lineages corresponded well to sampling locations in eastern and western North America (although this is not conclusive for Wilson's Warbler because of sampling gaps) (fig. 18.2A,B,D). However, in the

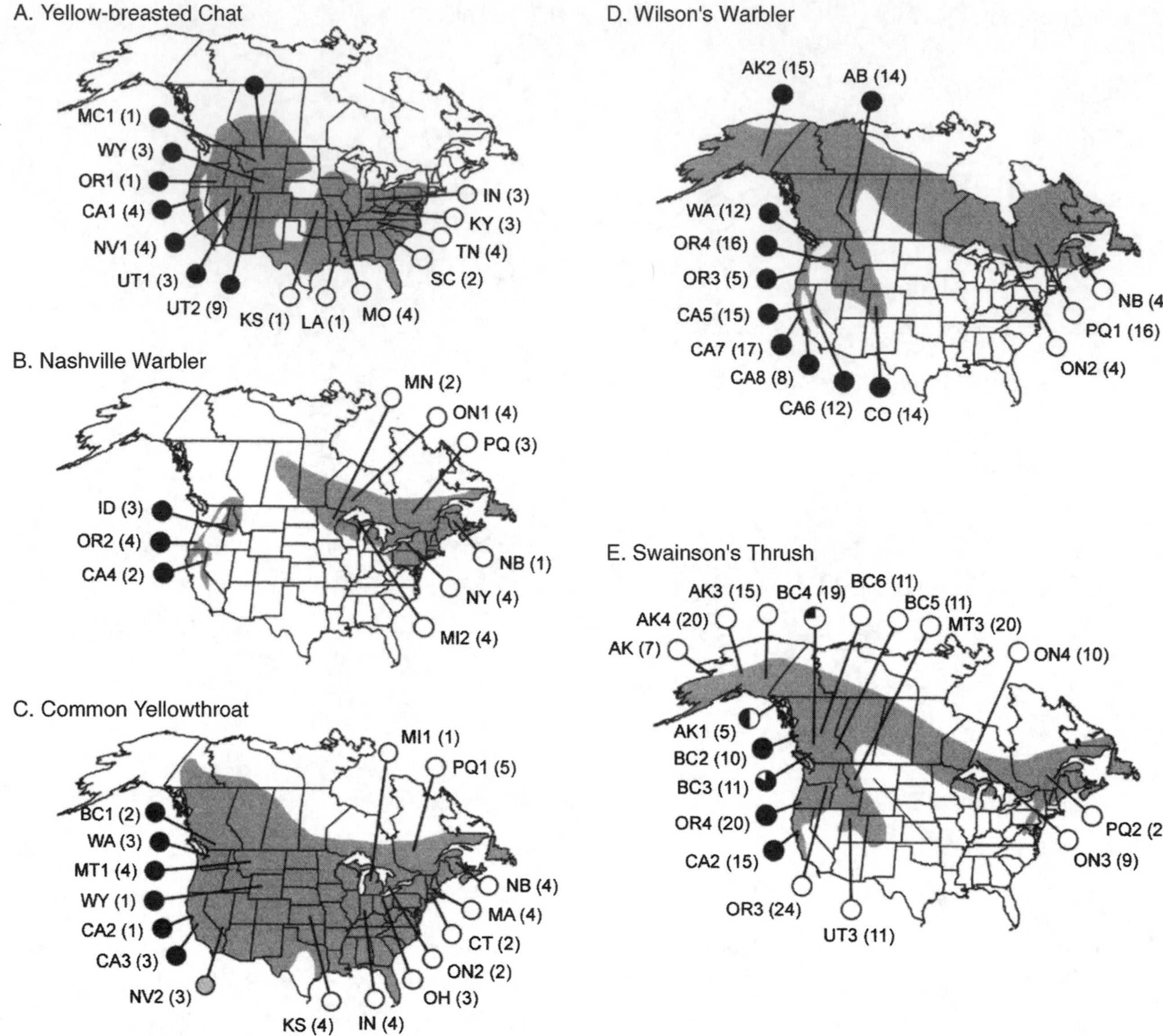

Fig. 18.2. Distribution of eastern and western haplotypes in Yellow-breasted Chat (A), Nashville Warbler (B), Common Yellowthroat (C), Wilson's Warbler (D), and Swainson's Thrush (E), superimposed on their breeding distribution (shaded area). Western and eastern haplotypes are shown in black and white, respectively. Gray circle for Nevada sample of Common Yellowthroat corresponds to haplotype N in fig. 18.1. Numbers in parentheses indicate sample sizes. See table 18.1 for location abbreviations.

Common Yellowthroat, the eastern lineage extends west to central Montana (Lovette et al. 2004) (fig. 18.2C) and in Swainson's Thrush, the eastern lineage extends to central British Columbia and western Alaska (Ruegg and Smith 2002) (fig. 18.2E). Therefore, post-glacial climate and vegetation changes may have affected current lineage ranges differently in each species.

In addition, some species showed hints of greater phylogenetic structure that are important to note. In the Common Yellowthroat, a divergent haplotype from Nevada was separated from the eastern group by seven to nine substitutions and from the western by 12–16 substitutions (fig. 18.1C). This population begs further investigation and may represent a distinct migratory population or possibly a nonmigratory population that extends southward beyond where we sampled (Lovette et al. 2004). In Wilson's Warbler more structure was detected among western populations than for the other species. An analysis of molecular variance (AMOVA) revealed both significant within-population and between-population variation (Kimura et al. 2002). It is possible, however, that similar complexities could be revealed in the other species if sampling were conducted with an intensity similar to that for these western Wilson's Warbler populations. Further examination of variation in Wilson's Warbler using five microsatellite loci also showed a genetic difference between the one eastern population and all other western populations (pairwise F_{ST} values shown in table 18.3). However, despite the extra statistical power af-

Table 18.3 Pairwise F_{ST} between sampled Wilson's Warbler populations based on analysis of five microsatellite loci

	AK	BC	AB	WA	OR	CA	CO	PQ
BC	0.026	—	—	—	—	—	—	—
AB	0.020	0.002	—	—	—	—	—	—
WA	0.005	0.014	0.002	—	—	—	—	—
OR	0.018	0.035*	0.019	0.006	—	—	—	—
CA	0.012	0.013	−0.010	−0.007	0.004	—	—	—
CO	0.030*	0.026	0.017	0.007	0.007	0.017	—	—
PQ	0.134*	0.156*	0.145*	0.125*	0.130*	0.129*	0.138*	-—

Note: Asterisk denotes values significantly different from zero following table-wide corrections for multiple comparisons. Abbreviations: AK = Alaska; BC = British Columbia; AB = Alberta; WA = Washington; OR = Oregon; CA = California; CO = Colorado; PQ = Quebec.

forded by multiple loci, further structure within the western portion of the species range was not detected. Population structure in terms of F_{ST} values was minimal, no isolation by distance relationship was detected, and model-based clustering methods failed to identify genetically similar groups within the western samples (Clegg et al. 2003).

Distribution of Genetic Lineages at Overwintering Sites

The distribution of eastern and western lineages on the wintering grounds differed among species (fig. 18.3A–E). These differences ranged from complete segregation to some geographic mixing of eastern and western groups at locations on the wintering grounds. In the Yellow-breasted Chat there was no evidence of mixing of eastern and western groups at wintering locations, although samples for any given site were small (fig. 18.3A). Overwintering western groups were distributed from southern Baja California to Oaxaca, Mexico. Eastern groups were found from Veracruz south through Chiapas, and at sites in Belize and El Salvador. Samples for the Common Yellowthroat were restricted to only three sites but nevertheless are informative (fig. 18.3C). Only western individuals were found in southern Baja, a mixed population was found in Oaxaca, and only eastern individuals were found in Belize. In contrast, haplotype distributions for Nashville Warblers revealed only two out of nine sites with western birds (a site in Sinaloa with nine individuals and a site in Oaxaca with one individual), whereas eastern individuals were distributed throughout the wintering range (fig. 18.3B). This discrepancy could be explained by a difference in the population sizes of the two subspecies; the western subspecies has a more restricted breeding distribution (Williams 1996) and may be less common. In Swainson's Thrush there was a nearly complete segregation of eastern and western groups on the wintering grounds (fig. 18.3E). Eastern groups were found primarily from Panama to northern South America, whereas western groups were found in southern Mexico and Central America. In the Wilson's Warbler, limited mixing of breeding lineages at overwintering sites was evident, mostly in Veracruz and Chiapas. Western haplotypes predominated throughout the wintering range (Figure 18.3D).

Morphological Subspecies and Genetic Variation

Information on morphological subspecies has been useful in the context of establishing connectivity in cases where differences between groups are clear (Webster et al. 2002). However, in many species, morphological differentiation is lacking, or subspecies differ in a very gradual and subtle manner, making it difficult to unequivocally identify a wintering-ground individual as belonging to one particular morphological subspecies or another. In the latter case, genetic information could be useful to verify identification of wintering ground individuals if the subspecies exhibit consistent genetic differences.

In the species examined in detail here, there are varying levels of concordance between the distribution of morphologically recognized subspecies and the distribution of genetic groups on breeding and wintering grounds. In Yellow-breasted Chats, our molecular results were concordant with subspecific variation and the distribution of the subspecies on both the breeding and wintering grounds (Lowery and Monroe 1968; Eckerle and Thompson 2001; Lovette et al. 2004). In the Nashville Warbler, the allopatric eastern and western subspecies (*V. r. ruficapilla* and *V. r. ridgwayi*, respectively) were genetically divergent, but these groups were not always found in their predicted overwintering sites (Williams 1996). Eastern-breeding individuals appear to utilize a broader overwintering range than previously thought, being found throughout eastern and western Mexico. We found very few western individuals in samples from western Mexico where this subspecies is thought to winter according to subspecies descriptions, despite having relatively large sample sizes.

It is possible that the lack of genetically western individuals found on the wintering grounds is due to biases in sampling if those birds occupy particular habitats that were not

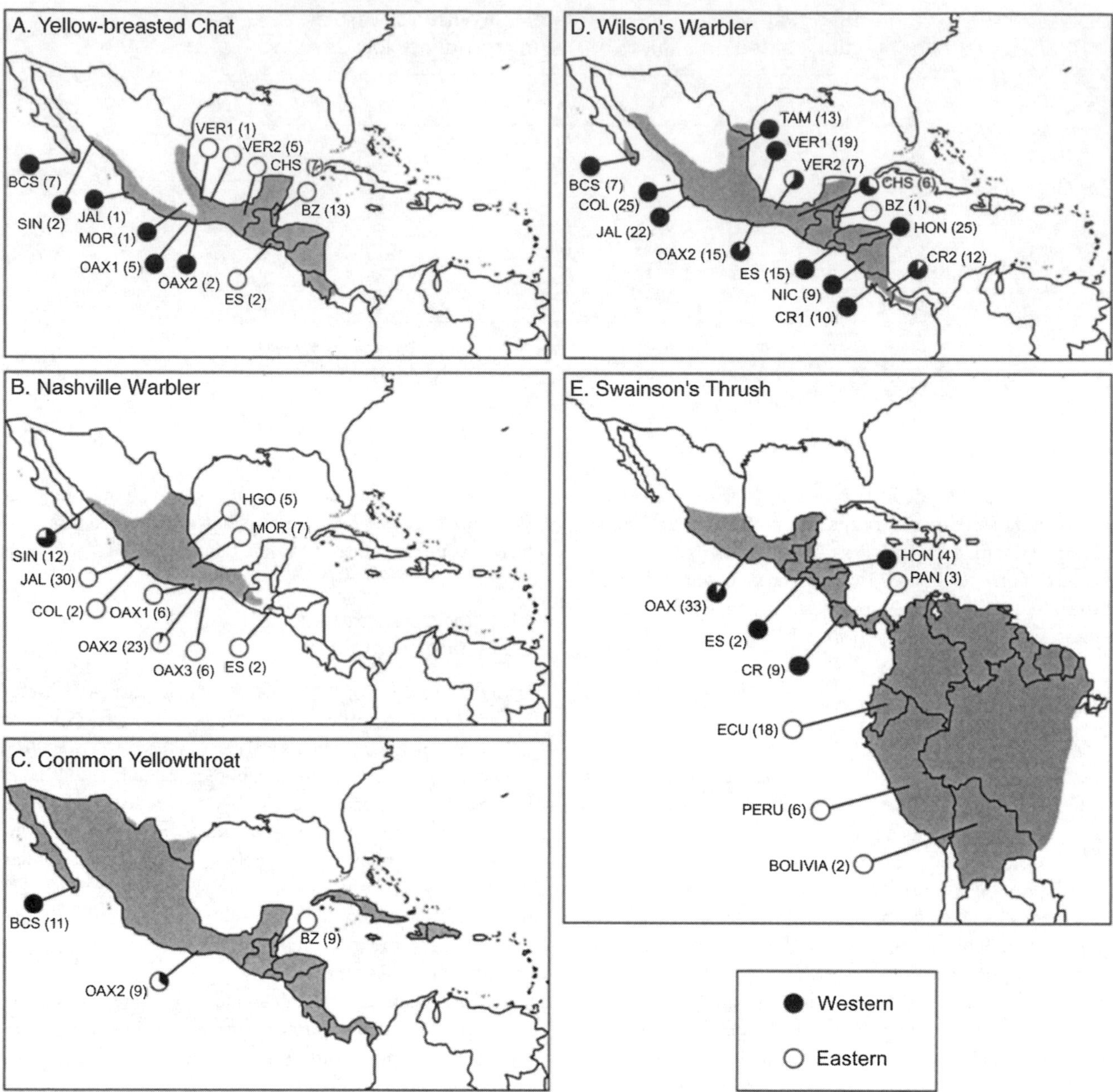

Fig. 18.3. Distribution of eastern and western haplotypes in Yellow-breasted Chat (A), Nashville Warbler (B), Common Yellowthroat (C), Wilson's Warbler (D), and Swainson's Thrush (E), superimposed on their overwintering distribution (shaded area). Western and eastern haplotypes are shown in black and white, respectively. Numbers in parentheses indicate sample sizes. See table 18.1 for location abbreviations. (Modified from Kimura et al. 2002, Ruegg and Smith 2002, and Lovette et al. 2004.)

sampled. Despite this, it seems that the morphological differences between the Nashville Warbler subspecies are too subtle to allow objective identification of the forms on the wintering grounds. In the case of the Common Yellowthroat, sampling was insufficient to comment on the concordance between morphological subspecies and genetic groups. In the Wilson's Warbler, there was partial concordance between subspecies designations and genetic groups. The eastern subspecies *W. p. pusilla* formed a well-supported lineage, but mtDNA molecular markers could not distinguish the two western subspecies (*W. p. chryseola* and *W. p. pileolata*) (Kimura et al. 2002). We are unable to comment on the concordance between morphological subspecies and genetic groups on the wintering grounds in the Wilson's Warbler, as subspecies on the wintering grounds were not previously described (Ammon and Gilbert 1999). In the Swainson's Thrush, there was genetic divergence between morphological subspecies groups (olive-backed group and russet-backed group), and these groups wintered in the locations predicted on the basis of morphology (Ramos and Warner 1980).

Overall, the use of morphological traits to identify breeding-ground origins of wintering-ground birds is species specific. For several of the species in our study, differences among subspecies do not provide reliable markers of breeding-ground origin. There are several potential sources of error when using subspecies to link wintering and breeding populations. For example, morphological differences in measurable traits such as wing and tail length and plumage characteristics can be differentially shaped by natural selection and can swamp the effects of gene flow (Rice and Hostert 1993; Orr and Smith 1998). Thus morphological differences among subspecies may result from differential selection, even in the presence of gene flow. Here, populations would not be demographically independent despite morphological differentiation.

In contrast, consider two populations occurring in the same habitat but separated by a high mountain range. Here, one might find substantial genetic divergence in populations across the divide, with each demographically independent of the other, but an identical pattern of morphological variation owing to the similarity of selection pressures in like habitats on each side of the mountain. Under these circumstances, although it would potentially be possible to classify wintering individuals by breeding population using genetic techniques, it would not be possible using morphologic characters important in fitness. In fact, in this instance using morphology alone to link wintering individuals with breeding populations would result in combining wintering individuals from demographically independent populations from either side of the mountain. Another source of potential error could arise if particularly plastic traits are used for establishing connectivity. Some avian traits are known to be seasonally variable. For example, mandible length has been shown to change seasonally because of wear (Gosler 1986). Thus, using subspecies designations alone to study connectivity may be misleading.

Life History, Demography, and Microevolutionary Processes

The use of molecular markers offers some exciting possibilities for the study of life history, demography, and evolutionary processes. By using a simple restriction enzyme assay, we can now distinguish between eastern and western breeding lineages in several passerine species. The ability to easily distinguish between eastern- and western-origin individuals offers the possibility of studying the relationship between regional breeding origin and life history and ecology. For example, are arrival and departure times on wintering grounds similar or different for eastern- and western-breeding individuals? Do eastern and western individuals winter in different locations? In wintering locations where eastern and western individuals co-occur, are there fine-scale differences in habitat use and behavior? The use of molecular tools makes it possible to study these and other factors associated with migration—factors that may have significant fitness consequences.

Molecular techniques can also provide interesting insights into the demographic history of a species. For example, several recent studies have demonstrated how low levels of genetic differentiation in North American passerines may be due to a rapid demographic expansion following a late Pleistocene glacial event (Milá et al 2000; Ruegg and Smith 2002). In the Swainson's Thrush, the molecular data suggested that both eastern and western lineages had undergone recent demographic expansions (Ruegg and Smith 2002). Furthermore, when molecular data were combined with data on banding returns, they suggested that individuals from east of the Coast Mountains in Alaska and British Columbia may be retracing their post-glacial expansion routes during their yearly migration.

The cases where there is a lack of concordance between subspecific variation and molecular data provide an interesting avenue for investigation in itself. Because of the effects of differential selection, morphological differences among subspecies or populations may not match patterns of neutral genetic variation (Orr and Smith 1998; Smith et al. 2001). This has been observed in a number of passerine species, such as Swamp Sparrows (*Melospiza georgiana*) (Greenberg et al. 1998) and Pied Flycatchers (*Ficedula hypoleuca*) (Haavie et al. 2000).

Utility of Molecular Markers and Future Directions

Results from the five passerine species examined suggest that mitochondrial DNA variation can resolve connectivity between breeding and wintering areas only at large geographic scales. However, the very real possibility remains that the lack of genetic structure found at finer geographic scales is due to limitations of markers. For the Wilson's Warbler, the application of microsatellite loci did not improve the level of geographic resolution above that found by using mtDNA, despite having five independent markers and the opportunity to conduct frequency-based analyses (Clegg et al. 2003). Nevertheless, the use of other, more variable molecular markers or a higher number of markers may ultimately increase resolution and the ability to link populations at a finer scale. In particular, a pioneering approach by Bench et al. (2002) used the amplified fragment length polymorphism (AFLP) method (which produces dominant markers that do not lend themselves to standard population genetic analyses because homozygous individuals cannot be distinguished from heterozygous individuals) to find informative single nucleotide polymorphisms (SNP), which are codominant and thus allow populations to be analyzed with standard population genetic methods. Using this approach, Bensch et al. (2002) were successful in distinguishing migratory populations of Willow Warblers (*Phylloscopus trochilus*) that could not be differentiated by mtDNA and microsatellite markers.

Although the choice of genetic marker or suite of markers should be considered, which works best also depends on life history characteristics. For example, mtDNA markers have been quite successful in establishing connectivity in some waders (Haig et al. 1997; Wennerberg 2001), suggesting that strong philopatry along with subdivided breeding distributions likely make some species more amenable to genetic methods than others.

Innovative new analytical approaches may also help in defining genetic structure. For example, multivariate molecular analyses developed and applied successfully to studying plant population structure show promise (Gram and Sork 2001), but to date have not been used to examine structure in vertebrates. Moreover, population assignment tests that use frequency data on multiple markers might also hold considerable promise.

Using molecular genetic markers to study population connectivity will ultimately be most successful when combined with other types of data such as banding returns, morphologically based subspecific variation, stable isotope markers, radio and satellite telemetry, and variation in disease strains. Ruegg and Smith (2002) combined molecular genetic markers, band returns, and descriptions of subspecific variation for the Swainson's Thrush to show that eastern and western groups had different migration routes and overwintering locations. The use of stable isotope ratios to link breeding and wintering ranges of some migratory songbirds at broad geographic levels has already shown promise (Hobson and Wassenaar 1997; Marra et al. 1998; Rubenstein et al. 2002; Webster et al. 2002; Hobson, Chap. 19, this volume). Stable isotope markers and molecular genetic markers may provide complementary sources of information. Molecular genetic markers are useful in distinguishing east-west patterns of variation, and some isotope ratios are useful in distinguishing north-south variation (Hobson and Wassenaar 1997). Thus, combining the two methods offers the possibility of better geographical delimitation of breeding populations. The combination of genetic and isotopic methods has been used to shed light on migratory patterns in Wilson's Warblers. First, individuals sampled from the wintering grounds could be identified by their genetic haplotypes as being of eastern or western breeding origin. Second, information from stable hydrogen isotopes could be employed to show that among the individuals from western breeding ranges, individuals from the northern part of the western range occupied the most southerly overwintering habitats, whereas individuals from more southerly breeding areas in the west occupied the northern parts of the overwintering range (Clegg et al. 2003). This demonstration of leapfrog migration using both molecular and isotopic information is an example of the power of using multiple approaches.

Radio and satellite telemetry has been a useful technique for tracking larger birds (Ristow et al. 2000). The development of smaller, lighter transmitters in the future may enable this method to be used for small-bodied passerines. Finally, using variation in disease strains to distinguish patterns of connectivity may hold promise (Ricklefs et al., Chap. 17, this volume). For example, PCR-based assays of avian blood parasites allow not only for easy and inexpensive ways to detect the presence of disease, but also for haplotypes of a given pathogen to be differentiated (Sehgal et al. 2001). Combining multiple techniques will likely be the way forward to maximize the resolution of connectivity between breeding and wintering migrant passerines and to link demographic processes across different stages of the annual cycle.

ACKNOWLEDGMENTS

We thank D. DeSante and the Institute for Bird Populations, the Monitoring Avian Productivity and Survivorship Program (MAPS) banders, the Point Reyes Bird Observatory, and the many independent banders across the hemisphere that donated genetic samples. For sharing their study sites or facilitating fieldwork, we thank Y. Aubry, P. Bichier, R. Carlson, A. Cruz, O. Cruz, R. Dickson, O. Figueroa, T. Gavin, M. Grosselet, K. Holl, L. Imbeau, O. Komar, J. Montejo, B. Murphy, A. Oliveras, E. Rodríguez, E. Ruelas, E. Santana, W. Schaldach, C. Spytz, P. Thorn, and the Belize Audubon Society. We would like to thank P. Marra, R. Greenberg, and two anonymous reviewers for comments that greatly improved the manuscript. This work was supported by grants from San Francisco State University, National Institutes of Health Office of Research on Minority Health (5P20RR11805), The Turner Foundation, the Environmental Protection Agency (R827109-01-0), and the National Science Foundation (DEB-9726425 and IRCEB-9977072) to T. Smith.

LITERATURE CITED

Alekseev, A. N., H. V. Dubinina, A. V. Semenov, and C. V. Bolshakov. 2001. Evidence of ehrlichiosis agents found in ticks (Acari: Ixodidae) collected from migratory birds. Journal of Medical Entomology 38:471–474.

Ammon, E. M., and W. M. Gilbert. 1999. Wilson's Warbler (*Wilsonia pusilla*). Pages 1–28 *in* The Birds of North America, no. 478 (A. Poole and F. Gill, eds.). The Birds of North America, Inc., Philadelphia.

Arguedas, N., and P. G. Parker. 2000. Seasonal migration and genetic population structure in house wrens. Condor 102:517–528.

Avise, J. C. 1994. Molecular Markers, Natural History and Evolution. Chapman and Hall, New York.

Avise, J. C., and D. Walker. 1998. Pleistocene phylogeographic effects on avian populations and the speciation process. Proceedings of the Royal Society of London, Series B, Biological Sciences 265:457–463.

Ball, R. M., and J. C. Avise. 1992. Mitochondrial DNA phylogeographic differentiation among avian populations and the evolutionary significance of subspecies. Auk 109:626–636.

Bensch S., S. Åkesson, and D. E. Irwin. 2002. The use of AFLP to find an informative SNP: genetic differences across a migratory divide in willow warblers. Molecular Ecology 11: 2359–2366.

Berthold, P. 1993. Bird Migration: A General Survey. Oxford University Press, New York.

Bond, G. M. 1963. Geographic variation in the thrush *Hylocichla ustulata*. Proceedings of the U.S. National Museum 114: 373–387.

Buerkle, C. A. 1999. The historical pattern of gene flow among migratory and non-migratory populations of prairie warblers (Aves: Parulinae). Evolution 53:1915–1924.

Clegg, S. M., J. F. Kelly, M. Kimura, and T. B. Smith. 2003. Combining genetic markers and stable isotopes to reveal leapfrog migration in a Neotropical migrant, Wilson's warbler (*Wilsonia pusilla*). Molecular Ecology 12: 819–830.

Eckerle, K. P., and C. F. Thompson. 2001. Yellow-breasted chat (*Icteria virens*). Pages 1–28 *in* The Birds of North America, no. 572 (S. Poole and F. Gill, eds.). The Birds of North America, Inc., Philadelphia.

Gibbs, H. L., R. J. G. Dawson, and K. A. Hobson. 2000. Limited differentiation in microsatellite DNA variation among northern populations of the yellow warbler: evidence for male-biased gene flow? Molecular Ecology 9:2137–2147.

Gill, F. B., A. M. Mostrom, and A. L. Mack. 1993. Speciation in North American chickadees: I. Patterns of mtDNA genetic divergence. Evolution 47:195–212.

Gill, J. A., K. Norris, P. M. Potts, T. G. Gunnarsson, P. W. Atkinson, and W. J. Sutherland. 2001. The buffer effect and large-scale population regulation in migratory birds. Nature 412: 436–438.

Gosler, A. G. 1986. Pattern and process in the bill morphology of the Great Tit *Parus major.* Ibis 129:451–476.

Gram, W. K., and V. L. Sork. 2001. Association between environmental and genetic heterogeneity in forest tree populations. Ecology 82:2012–2021.

Greenberg, R., P. J. Cordero, S. Droege, and R. C. Fleischer. 1998. Morphological adaptation with no mitochondrial DNA differentiation in the coastal plain Swamp Sparrow. Auk 115: 706–712.

Guzy, M. J., and G. Ritchison. 1999. Common Yellowthroat (*Geothlypis trichas*). Pages 1–24 *in* The Birds of North America, no. 48 (A. Poole and F. Gill, eds.). The Birds of North America, Inc., Philadelphia.

Haavie, J., G.-P. Sætre, and T. Moum. 2000. Discrepancies in population differentiation at microsatellites, mitochondrial DNA and plumage colour in the pied flycatcher: inferring evolutionary processes. Molecular Ecology 9:1137–1148.

Haig, S. M., C. L. Gratto-Trevor, T. D. Mullins, and M. A. Colwell. 1997. Population identification of western hemisphere shorebirds throughout the annual cycle. Molecular Ecology 6:413–427.

Hobson, K. A., and L. I. Wassenaar. 1997. Linking breeding and wintering grounds of neotropical migrant songbirds using stable hydrogen isotopic analysis of feathers. Oecologia 109: 142–148.

Kimura, M., S. M. Clegg, I. J. Lovette, K. R. Holder, D. J. Girman, B. Milá, P. Wade, and T. B. Smith. 2002. Phylogeographic approaches to assessing demographic connectivity between breeding and overwintering regions in a Nearctic-Neotropical warbler (*Wilsonia pusilla*). Molecular Ecology 11:1605–1616.

Klein, N. 1994. Intraspecific molecular phylogeny in the yellow warbler (*Dendroica petechia*), and implications for avian biogeography in the West Indies. Evolution 48:1914–1932.

Lovette, I. J., and E. Bermingham. 2001. Mitochondrial perspective on the phylogenetic relationships of the Parula Wood-warblers. Auk 118:211–215.

Lovette, I. J., E. Bermingham, G. Seutin, and R. E. Ricklefs. 1998. Evolutionary differentiation in three endemic West Indian warblers. Auk 115:890–903.

Lovette, I. J., S. M. Clegg, and T. B. Smith. 2004. Limited utility of mtDNA markers for determining connectivity among breeding and overwintering locations in three Neotropical migrant birds. Conservation Biology 18:156–166.

Lowery, G. H., and B. L. Monroe. 1968. Family Parulidae. Pages 5–93 *in* Checklist of Birds of the World (R. A. Paynter, ed.). American Ornithologists' Union, Cambridge, Mass.

Marra, P. P., K. A. Hobson, and R. T. Holmes. 1998. Linking winter and summer events in a migratory bird by using stable-carbon isotopes. Science 282:1884–1886.

Milá, B., D. J. Girman, M. Kimura, and T. B. Smith. 2000. Genetic evidence for the effect of a postglacial population expansion on the phylogeography of a North American songbird. Proceedings of the Royal Society of London, Series B, Biological Sciences 267:1033–1040.

Milot, E., H. L. Gibbs, and K. A. Hobson. 2000. Phylogeography and genetic structure of northern populations of the yellow warbler (*Dendroica petechia*). Molecular Ecology 9:667–681.

Orr, M. R., and T. B. Smith. 1998. Ecology and speciation. Trends in Ecology and Evolution 13:502–506.

Pyle, P. 1997. Identification Guide to North American Birds. Slate Creek Press, Bolinas, Calif.

Ramos, M. A., and D. W. Warner. 1980. Analysis of North American subspecies of migrant birds wintering in Los Tuxlas, southern Veracruz, Mexico. Pages 173–180 *in* Migrant Birds in the Neotropics: Ecology, Behavior, Distribution, and Conservation (E. A. Keast and E. S. Morton, eds.). Smithsonian Institution Press, Washington, D.C.

Rappole, J. H., S. R. Derrickson, and Z. Hubalek. 2000. Migratory birds and spread of West Nile virus in the Western Hemisphere. Emerging Infectious Diseases 6:319–328.

Rice, R. R., and E. E. Hostert. 1993. Laboratory experiments on speciation: what have we learned in 40 years? Evolution 47:1637–1653.

Rintamaki, P. T., M. Ojanen, H. Pakkala, M. Tynjala, and A. Lundberg. 2000. Blood parasites of juvenile Willow Tits *Parus montanus* during autumn migration in northern Finland. Ornis Fennica 77:83–87.

Ristow, D., P. Berthold, D. Hashmi, and U. Querner. 2000. Satellite tracking of Cory's Shearwater migration. Condor 102: 696–699.

Rogers, A. R., and H. Harpending. 1992. Population growth makes waves in the distribution of pairwise genetic differences. Molecular Biology and Evolution 9:552–569.

Rubenstein, D. R., C. P. Chamberlain, R. T. Holmes, M. P. Ayres, J. R. Waldbauer, G. R. Graves, and N. C. Tuross. 2002. Linking breeding and wintering ranges of a migratory songbird using stable isotopes. Science 295:1062–1065.

Ruegg, K. C., and T. B. Smith. 2002. Not as the crow flies: an historical explanation for circuitous migration in Swainson's thrush (*Catharus ustulatus*). Proceedings of the Royal Society of London, Series B, Biological Sciences 269: 1375–1381.

Schneider, S., and L. Excoffier. 1999. Estimation of past demographic parameters from the distribution of pairwise differences when the mutation rates vary among sites: application to human mitochondrial DNA. Genetics 152:1079–1089.

Sehgal, R. N. M., H. I. Jones, and T. B. Smith. 2001. Host specificity and incidence of Trypanosoma in some African rainforest birds: a molecular approach. Molecular Ecology 10: 2319–2327.

Seutin, G., L. M. Ratcliffe, and P. T. Boag. 1995. Mitochondrial DNA homogeneity in the phenotypically diverse redpoll finch complex (Aves: Carduelinae: *Carduelis flammea-hornemanni*). Evolution 49:962–973.

Sillett, T. S., R. T. Holmes, and T. W. Sherry. 2000. Impacts of a global climate cycle on population dynamics of a migratory songbird. Science 288:2040–2042.

Smith, T. B., C. J. Schneider, and K. Holder. 2001. Refugial isolation versus ecological gradients: testing alternative mechanisms of evolutionary divergence in four rainforest vertebrates. Genetica (Dordrecht) 112/113:383–398.

Webster, M. S., P. P. Marra, S. M. Haig, S. Bensch, and R. T. Holmes. 2002. Links between worlds: unraveling migratory connectivity. Trends in Ecology and Evolution 17:76–83.

Wenink, P. W., and A. J. Baker. 1996. Mitochondrial DNA lineages in composite flocks of migratory and wintering dunlins (*Calidris alpina*). Auk 113:744–756.

Wiens, J. A. 1989. The Ecology of Bird Communities. Cambridge University Press, New York.

Williams, J. M. 1996. Nashville Warbler (*Vermivora ruficapilla*). Pages 1–20 *in* The Birds of North America, no. 205 (A. Poole and F. Gill, eds.). The Birds of North America, Inc., Philadelphia.

Winker, K. 2000. Migration and speciation. Nature 404:36.

Zink, R. 1997. Phylogeographic studies of North American birds. Pages 301–324 *in* Avian Molecular Evolution and Systematics (D. P. Mindell, ed.). Academic Press, San Diego.

Zink, R. M., A. E. Kessen, T. V. Line, and R. C. Blackwell-Rago. 2001. Comparative phylogeography of some aridland bird species. Condor 103:1–10.

19

KEITH A. HOBSON

Flying Fingerprints

Making Connections with Stable Isotopes and Trace Elements

STABLE ISOTOPE AND TRACE ELEMENT analysis of tissues of migratory birds provides an alternative or complementary means of making connections among breeding, wintering, and stopover sites. This field has seen remarkable growth in the last decade, especially in the use of deuterium isotope analyses to determine approximate latitude of origin of birds in North America. Technological advances have also resulted in a resurgence of trace element approaches. This chapter provides a brief review of the strengths and weaknesses of these technologies and provides suggestions for future research and applications. Additionally, new information is presented on blood ^{13}C and ^{15}N signatures in whole blood of five species of Neotropical migrant passerines and one short-distance migrant arriving in spring in southern Manitoba together with a model of ^{13}C turnover in blood protein for a hypothetical migrant undergoing fasting and refueling.

INTRODUCTION

The ability to track migratory birds between breeding, stopover, and wintering locations forms the basis of our understanding of migration and the development of paradigms involving life history theory and physiology as well as conservation and management. Our success in making such connections varies tremendously among species and regions. In North America, banding operations have provided reasonable information on movements of game birds but poor information on migrant songbirds (Hobson 2002). In Europe more information derived from banding efforts is available for linkages associated with European-African songbird migration but large gaps in knowledge remain for many species (Bairlein 1998). Leg bands and other extrinsic markers have great utility for monitoring population trends through constant-effort mist-netting operations (Dunn and Hussell 1995) or examining spatial/temporal or energetic aspects of migration (Bairlein 2001) but are clearly inadequate to provide basic connectivity for most populations. Radio or satellite markers similarly have current size and weight limitations and are prohibitively expensive as a viable widespread technique for use on large numbers of individuals. Such limitations have resulted in tremendous interest in the use of intrinsic markers such as population-level genetic markers and stable isotope and trace element profiles in bird feathers and other tissues (Webster et al. 2001; Rubenstein and Hobson 2004). The fundamental advantage of using intrinsic markers is that information on origins can be obtained without the bias and cost associated with mark-recapture (Wassenaar and Hobson 2001). In this sense, every capture becomes a recapture. The basis of all tracking techniques using intrinsic markers is that an individual organism acquires a molecular, isotopic or elemental profile in a tissue at one location that differs from that expected at another location. If we know how these expected profiles differ spatially and further choose a tissue with a sufficient "memory" to reflect the origin of interest, then we can infer connectivity between sites.

In this chapter, I review the use of stable isotope and trace element profiles to establish origins of migrant birds. This is a rapidly developing field and any review becomes quickly outdated, usually before publication. Following on from previous reviews (Hobson 1999a, 2002; Rubenstein and Hobson 2004) this presentation thus focuses more on assumptions inherent in the use of these techniques and examines future broad research directions. The interested reader is directed to texts such as Rundel et al. (1988) and Lajtha and Michener (1994) for general background theory on stable isotope applications to ecological studies and to Hoeffs (1997) for background on geochemistry.

STABLE ISOTOPES

Stable isotope abundance of any element is typically expressed as a ratio of the rarer, heavy form to that of the more common, lighter form. Stable isotopes of light elements of most interest to ecological applications are those of carbon ($^{13}C/^{12}C$), nitrogen ($^{15}N/^{14}N$), sulfur ($^{34}S/^{32}S$), hydrogen ($^{2}H/^{1}H$), and oxygen ($^{18}O/^{16}O$). Those of heavier elements such as strontium (^{87}Sr) and lead (^{210}Pb) are also particularly useful but require more involved analytical procedures. Stable isotope ratios of the light elements are measured with isotope-ratio mass spectrometry (IRMS) and are usually expressed in abundance relative to international standards in delta (δ) notation and reported as parts-per-thousand deviation from those standards. This is an extremely well-established field in analytical chemistry and highly accurate measurements are routinely achieved in most laboratories. Fortunately, various biogeochemical processes in nature result in materials that differ in their stable isotope abundance and these differences can be exploited to infer origins of organisms that come into equilibrium with local food webs. Such a phenomenon is based on the now-familiar "you are what you eat, plus a few parts per thousand" adage that can be expressed as follows:

$$\delta X_t = \delta X_d + \delta\Delta_{dt} \tag{1}$$

where X is the isotope of interest, t the tissue of interest, d the diet, and $\Delta\delta_{dt}$ the isotopic fractionation factor between diet and tissue. This simple relationship brings up two very important principles of applying stable isotope measurements to food webs in general and to migratory tracking in particular. First, the diet-tissue isotopic fractionation factor can be tissue specific and these specifications may need to be established experimentally (Hobson and Clark 1992b). Second, for metabolically active tissues, this relationship is not static, but is based on equilibrium time constants related to elemental turnover rates in the tissue (Hobson and Clark 1992a). The choice of tissue is of fundamental importance, then, when deciphering isotopic information. Three key components must be considered when inferring origins of migrant birds: (1) the isotopic signature of the source and how this varies spatially and/or temporally, (2) the isotopic fractionation associated with the tissue being used to reflect that source, and (3) the isotopic turnover rate of that tissue.

USEFUL ISOTOPIC PATTERNS IN NATURE

Fortunately, several isotopic patterns known in nature can be exploited to infer origins of migratory birds and other organisms. These vary according to individual stable isotope and how they behave in various biogeochemical reactions. For our purposes, these patterns can be grouped into dietary signals that are related to local biome or climatic conditions and those related to larger-scale isotopic patterns based on underlying geology or continental patterns in precipitation.

Undoubtedly the most studied and well-known stable isotopic pattern in nature is that of stable carbon isotope signatures associated with photosynthetic pathways. Plants with a C-3 photosynthetic pathway have tissues that are more depleted in their $\delta^{13}C$ values than those with a C-4 or CAM pathway. It is fairly straightforward to trace isotopically the relative dependence of a consumer on C-3 versus C-4 or C-3 versus CAM-based food webs where such distinctions occur in sources of available carbon to migrants. Such an approach has been used to infer the origins of migrating Snow Geese (*Chen caerulescens*) in North America because several populations exploit agricultural sources of corn, a C-4 plant in an otherwise agricultural or natural

C-3 biome (Alisauskas et al. 1998; Féret et al. 2003). Similarly, Wassenaar and Hobson (2000) demonstrated how different populations of Red-winged Blackbirds (*Agelaius phoeniceus*) depended differentially on agricultural corn versus C-3 plants. In principle, if it is possible to establish regional or continental patterns of corn production in places where birds could access these crops as food, then migratory patterns or origins of birds using these sources can be inferred. However, in an attempt to distinguish agricultural production from the relative occurrence of natural C-3 versus C-4 plants, Hobson and Wassenaar (2001) demonstrated that wintering Loggerhead Shrikes (*Lanius ludovicianus*) in the southern United States and northern Mexico originated from areas with food webs ranging from pure C-3 to pure C-4. Because we do not have useful spatial resolution of the distribution of C-3 versus C-4 biomes throughout much of the range for most species, such information will be quite limited in inferring origins of birds such as shrikes.

CAM plants are relatively rare in North America but are well represented in dry areas by cactus. Tracing CAM-based carbon input to a migratory animal in North America was first accomplished by Flemming et al. (1993), who examined *Leptonycteris curasoae,* a nectar-feeding bat whose spring movement north from Mexico was regulated by the flowering phenology of CAM plants. Since then, Wolf and Martinez del Rio (2000) and Wolf et al. (2002) have examined the dependence of White-winged (*Zenaida asiatica*) and Mourning Doves (*Zenaida macroura*) on Saguaro Cactus (*Carnegiea gigantea*) and are currently using this as a marker for populations of doves originating in the American Southwest.

In addition to having lower $\delta^{13}C$ values as a result of the photosynthetic pathway per se, C-3 plants show remarkable variation in $\delta^{13}C$ signature based on mechanisms associated with water-use efficiency (reviewed by Lajtha and Marshall 1994). The net result is that C-3 plants generally become more enriched in ^{13}C under more xeric conditions than under cooler or more mesic conditions (Farquhar et al. 1989). Marra et al. (1998) determined that American Redstarts (*Setophaga ruticilla*) wintering in moister, more favorable, habitat in Jamaica and Honduras had stable-carbon isotope values in their tissues that were more depleted in ^{13}C than those of birds wintering in secondary, more xeric habitat. They were able to test the hypothesis that birds first to arrive on the breeding grounds had come from the more favorable habitats on the wintering grounds. That study was the first to provide direct isotopic linkage between avian habitat use on the wintering grounds and the consequences of that use on subsequent arrival phenology on the breeding ground. Further isotopic research on redstarts is now underway to determine what ultimate consequences on reproductive success relate to habitat use on the wintering grounds habitat use (Norris et al. 2004; P. Marra, pers. comm.).

In their isotopic investigation of population structure in Willow Warblers (*Phylloscopus trochilus*), Chamberlain et al. (2000) used a similar approach to the redstart work of Marra et al. (1998). The subspecies *P. t. trochilus* winters in West Africa, whereas *P. t. acredula* winters in central sub-Saharan Africa. These birds molt flight feathers on their African wintering grounds and the analysis of $\delta^{13}C$ and $\delta^{15}N$ values in these feathers obtained on the European breeding grounds demonstrated a distinct enrichment in both isotopes crossing their geographical zone of segregation. This isotopic shift corresponds to the drier wintering biome of *acredula* compared with that of *trochilus.* At the broader level, these sorts of studies demonstrate the value of the stable isotope technique for inferring winter habitat use rather than as a means of tracking migrants or delineating population segregation per se. However, if such isotopic patterns could be demonstrated for large geographic areas of the wintering or breeding grounds, this information could be used to delineate populations and differential migration routes. Clearly, more isotopic investigations are required in various biomes of major wintering areas of Africa, Central America, and northern South America before we can assess with confidence wintering biomes and the possible geographic origins of wintering populations that migrate to breed in Europe and North America, respectively (Pain et al. 2004; Møller and Hobson 2004).

Within terrestrial systems, land-use practices can influence stable isotope abundance in food webs. Most notably, agricultural practices can alter $\delta^{15}N$ values in both upland and wetland systems. Soil nitrogen can be isotopically variable within and among sites but two processes can result in agricultural soils being more enriched in ^{15}N than temperate forest soils. These are the presence of animal-based fertilizers and the greater volatilization of isotopically lighter nitrogenous compounds such as ammonia from agricultural soils as a result of tillage and their lower acidity (Nadelhoffer and Fry 1994). Hobson (1999b) demonstrated that feathers from insectivorous boreal forest passerines were more depleted in ^{15}N than feathers from similar species occupying riparian habitats in agricultural areas of the Canadian prairies, and Hebert and Wassenaar (2001) found that ducks produced in agricultural wetlands had feathers more enriched in ^{15}N than those from boreal sites. However, it is currently not clear how scale-dependent these biome or area markers might be since Alexander et al. (1996) found high variability in $\delta^{15}N$ values of tissues of shorebirds breeding in an agricultural matrix.

Stable-carbon isotope abundance in food webs is more enriched in marine versus terrestrial and freshwater biomes, a phenomenon also seen with stable N, H, O, and S isotopes. For birds that move seasonally between marine and freshwater habitats, stable isotope methods may be used to delineate populations. For example, several species of sea ducks, including Common Goldeneye (*Bucephala clangula*), scoters (*Melanitta* spp.) and Long-tailed Duck (*Clangula hyemalis*), winter in coastal marine areas in eastern North America and also on the Great Lakes. By examining tissues of birds on their breeding grounds that correspond to relevant winter-ground habitat (e.g., body feathers grown prior to spring migration), it should be possible to trace winter-ground origins to these two categories (B. Braune et al., unpubl.).

Within marine systems, only a few geographical isotopic

markers appear to be useful for tracking migratory birds. The subtropical convergence of the Southern Ocean is an area of marked changes in food web $\delta^{13}C$ and $\delta^{15}N$ values, and this delineation has been found in the baleen plates of Southern Right Whales (*Eubalaena australis*) that migrate annually across this isotopic anomaly (Best and Schell 1996). Seabirds that move seasonally north and south of this convergence may similarly acquire useful isotopic markers in their tissues grown in these regions (Cherel et al. 2000). In the Bering Sea, Schell et al. (1998) have shown that zooplankton associated with this area of intense upwelling have $\delta^{13}C$ and $\delta^{15}N$ values that are higher than adjacent regions and generally higher than those in the eastern Arctic. Such a phenomenon may well be useful in determining origins of sea ducks such as eiders (*Somateria* sp.) that winter on both the northeast and northwest coasts of North America and breed across the Arctic (Mehl et al., in press). Other areas of upwelling include coastal Peru and California, and Minami and Ogi (1997) speculated that high $\delta^{15}N$ values in muscle of immature Short-tailed Shearwaters (*Puffinus tenuirostris*) resulted from feeding in those areas, thereby differentiating those birds from others that move along the western Pacific.

DEUTERIUM

Without question, the single isotope that has shown the greatest potential for helping to elucidate origins of migratory birds in North America is deuterium. Its usefulness is based on the fact that stable-hydrogen isotope ratios in precipitation show a continent-wide pattern with a general gradient from enriched values in the southeast to more depleted values in the northwest (fig. 19.1A). This pattern was known for 30 years before the concept of applying this phenomenon to migratory bird research was realized (e.g., Sheppard et al. 1969). After the first avian applications by Chamberlain et al. (1997) and Hobson and Wassenaar (1997), several studies confirmed the strong association between growing-season average δD values in precipitation and those in the feathers of birds grown at those locations (Kelly et al. 2001; Meehan et al. 2001; Wassenaar and Hobson 2001; Rubenstein et al. 2002). The growing-season deuterium precipitation map was recently constructed for Europe (Hobson 2002) (fig. 19.1B), and similar applications to migratory bird research on that continent will undoubtedly follow (Hobson et al. 2004).

The strong latitudinal pattern apparent in the deuterium growing-season contour map in North America, however, was recognized as a disadvantage, because without a longitudinal marker, much ambiguity might occur in assigning individuals to molt origin on the basis of feather δD measurements alone. This shortcoming has led to a good deal of interest in investigating additional isotopic markers to provide longitudinal segregation. Recent work on Black-throated Blue Warblers (*Dendroica caerulescens*) has shown that ^{87}Sr and ^{13}C may be useful here (Chamberlain et al. 1997; Rubenstein et al. 2002), and the combined use of δD and $\delta^{13}C$ measurements has been instrumental in providing natal information for Monarch Butterflies (*Danaus plexippus*) in eastern North America (Hobson et al. 1999b). Other approaches have been to combine east-west genetic markers with stable isotope techniques to better clarify origins (reviewed by Webster et al. 2001; Rubenstein and Hobson 2004). Ultimately, the utility of additional isotopic, elemental, and genetic markers will need to be demonstrated on a case-by-case basis.

Further complications or opportunities with the deuterium approach to tracking migrants include the well-known effect of altitude on δD measurements (reviewed by Poage and Chamberlain 2001). Deuterium abundance in precipitation tends to decrease with altitude. If this is not taken into account, birds from higher elevations may be assigned erroneously to higher latitudes in North America. On the other hand, at more local scales, such isotopic patterns may assist in examining questions of dispersal and altitudinal migration (Hobson et al. 2003). The recent isotopic investigation of Black-throated blue Warblers by Graves et al. (2002) demonstrated that because of the isotopic consequences of plant responses to local temperature and growing conditions (Körner et al. 1991) elevation may also be a factor influencing $\delta^{13}C$ values of feathers. However, $\delta^{13}C$ and $\delta^{15}N$ values of insectivorous birds can vary with diet in addition to other factors related to altitude, so it is not clear yet how useful such stable isotopes will be for inferring altitudinal position. Overall, the application of stable isotope tracking techniques to populations in areas where elevation is a factor may need to take these influences into consideration.

A common question arising from the application of growing-season average precipitation contour maps for deuterium based on the International Atomic Energy Agency (IAEA) data base is the robustness of the kriged (spatially interpolated) relationships. In any given year, how much variation in these patterns might be expected? This is not an easy question to answer because the geographical and temporal coverage in sampling sites for this data base is so variable. The patterns depicted by Hobson and Wassenaar (1997) and Hobson (2002) are based on about a 35-year IAEA record. However, several considerations increase our confidence in these relationships, at least qualitatively. The first is that short-term variation in precipitation signals will, to some extent, be smoothed out by the longer-term averaging of the growing season itself. Thus, in many areas, each feather measurement will, in effect, represent the average of many rainfall events and so will tend to smooth short-term fluctuations. This will not necessarily be the case in areas or times of lower precipitation or in areas that are subject to single or synoptic rainfall events (Hobson et al. 2004; Bowen et al., in press). Nor will it apply to areas where groundwater or reservoirs form a significant source of hydrogen for local food webs. For a group of European sites where long-term data are available, several showed extremely small interyear variation in average growing-season

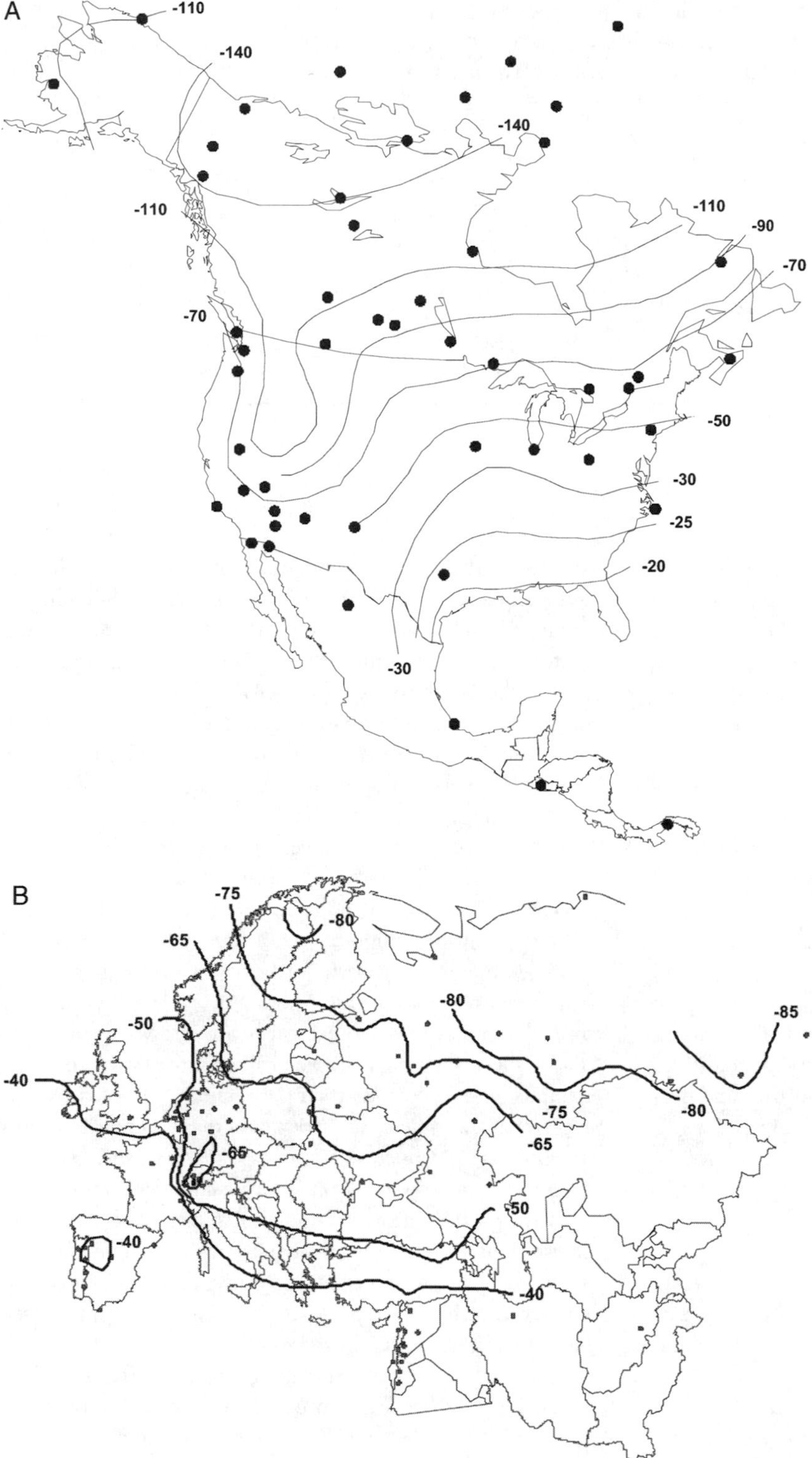

Fig. 19.1. Patterns of growing-season average deuterium values for precipitation based on the IAEA data base (1969–2000) for (A) North America (from Hobson and Wassenaar 1997) and (B) Europe (from Hobson 2002).

δD in precipitation, of the order of measurement error, whereas others, notably coastal sites, showed variation at least three to four times as high (table 19.1). Despite numerous potential sources of error, it is remarkable how well long-term average values of δD in precipitation are correlated with δD values of feathers grown in any given year, a relationship now demonstrated independently by several research groups (see references cited above). How well this relationship holds in future given climate change scenarios is, of course, unknown and will be an important area of research. An alternative to using the long-term average contour maps is the direct measurement of isotopic patterns of interest for a particular year of interest (e.g., Hobson et al. 1999b).

Table 19.1 Variation (mean and standard deviation) in growing-season average deuterium values in precipitation for a selection of European sites having continuous long-term IAEA data

Location	Country	Number of years	Mean δD (‰)	SD (‰)
IsFjord	Norway	7	−60	12.5
Ny Alesund	Norway	7	−80	10.5
Naimakka	Sweden	5	−97	9.7
Reyjavik	Iceland	13	−57	5.8
Valentia Obs	Ireland	25	−36	3.7
Lista	Norway	10	−47	3.3
De Kooy	Netherlands	7	−47	3.0
Wallingford	England	15	−50	2.9

FRACTIONATION

As indicated by equation (1), the determination of average isotopic values for diet using measurements of avian tissues requires that we know isotopic fractionation factors for tissues and diets of interest. Although this may not be absolutely necessary in investigations of origins of migratory birds, the better refinement of such values will be of use in a number of other areas. Raising birds in captivity on known diets provides the experimental control necessary to establish these values, and the work of Hobson and Clark (1992a), Mizutani et al. (1991, 1992), and Hobson and Bairlein (2003) has provided some good baseline information for $\delta^{13}C$ and $\delta^{15}N$ measurements. Hobson et al. (1999a) and Wassenaar and Hobson (2001) have similarly provided estimates of deuterium isotopic fractionation between diet and feathers. However, it would be particularly useful to repeat these sorts of studies for insectivorous passerines and to include additional keratinous tissues like toenail. In addition, we have little idea how different diets such as fruit and nectar may influence isotopic fractionation compared with, say, insect diets (Pearson et al. 2003).

CHOOSING THE RIGHT TISSUE

When considering which tissue to use for isotopic measurement we are faced with metabolically active and metabolically inactive tissues, and both have advantages and disadvantages. Metabolically active tissues provide a "moving window" of past origins and the width of that window depends on the elemental turnover rate associated with that tissue. For fast metabolic rate tissues like liver or blood plasma, the window is short, on the order of a week, because the elemental half-life of these tissues is of the order of 3 days (Hobson and Clark 1992b). Muscle and whole blood have slower turnover rates and dietary information can be derived for a period up to about 6 weeks for larger birds. Bone collagen has a very slow turnover rate and so can provide dietary information averaged over years. The problem currently facing researchers who wish to use metabolically active tissues to infer origins of migratory birds is that precise metabolic turnover rates for wild, migrating birds are essentially unknown. Such a lack of information may undermine studies like that of Marra et al. (1998), who examined muscle tissue of newly arrived American Redstarts on the breeding grounds in New Hampshire to infer habitat use on the wintering grounds. However, further field testing has provided indirect support that the muscle stable isotope values measured by Marra et al. (1998) indeed provided relative information on wintering ground habitat use prior to spring migration (Norris et al. 2004).

Whole blood has a turnover rate similar to that of muscle (Hobson and Clark 1993). Blood samples from five species of Neotropical migrant and one species of short-distance migrant songbirds were obtained as they moved through the Delta Marsh Bird Observatory in southern Manitoba, Canada, during May 2001. The results of $\delta^{13}C$ and $\delta^{15}N$ analyses of these birds are shown in fig. 19.2. Those species having isotopic profiles more typical of xeric habitats were Yellow Warbler (*Dendroica petechia*) and Least Flycatcher (*Empidonax minimus*), whereas Ovenbird (*Seiurus aurocapillus*), Swainson's Thrush (*Catharus ustulatus*), and White-throated Sparrow (*Zonotrichia albicollis*) were associated more with moister, cooler habitats. The American Redstart occupied an intermediate position. These results are in general agreement with our understanding of the habitats used by these species on the wintering grounds (see papers in Hagan and Johnston 1992).

Isotopic turnover in a metabolically active tissue such as whole blood or muscle can be modeled for a hypothetical migrant bird that leaves one isotopic regime and then enters another while moving toward its breeding grounds. Such a species undergoes a series of migratory flights punctuated by stopover in habitats where they refuel with food that differs isotopically from their wintering diet. Previous diet-switch experiments with captive birds have provided baseline turnover rates of elements in avian tissues (Hobson and Clark 1992a; Bearhop et al. 2002; Hobson and Bairlein 2003; Evans-Ogden et al. 2004), but these are not particularly useful when considering models of isotopic change in tissues of a migrating bird that punctuates periods of fasting during

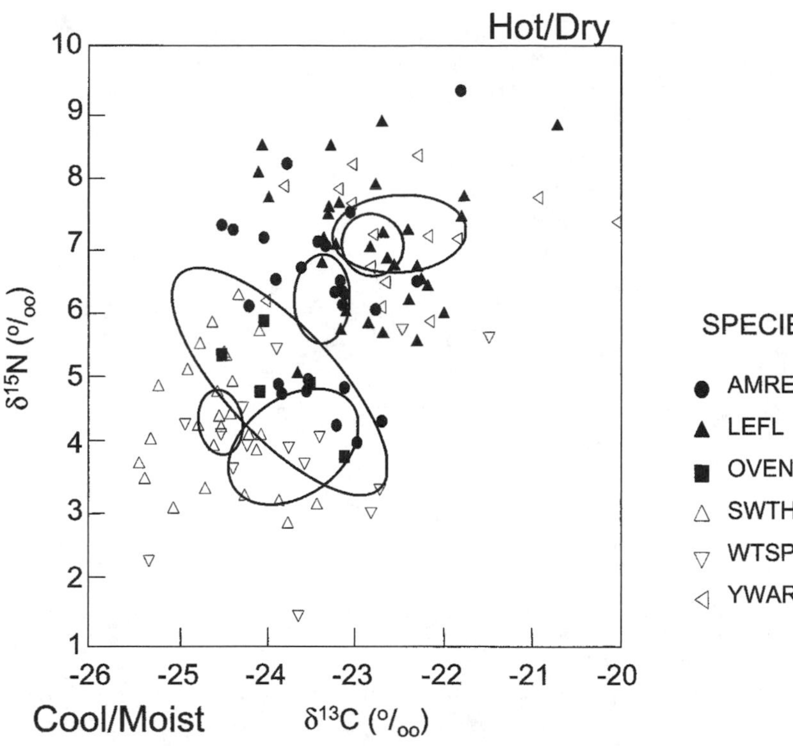

Fig. 19.2. Pattern of stable-carbon and nitrogen isotope values for whole blood of spring migrants moving through the Delta Marsh Bird Observatory, Manitoba, Canada, in 2000.

migratory flight with refueling at stopover sites. Much of the energy required for flight is provided by lipid stores and these will isotopically represent food webs at stopover sites. However, the protein requirements for flight are derived not from discrete stores per se but from muscle and organs (e.g., Lindström et al. 2000). Assuming protein stores are replenished at stopover sites, this will result in a stepwise change in the isotopic composition of muscle or blood protein. This process can be modeled for a hypothetical case of a bird moving from a wintering-ground food web that is 5‰ more enriched in ^{13}C than the stopover environments, given protein contributions of varying proportions (fig. 19.3). Protein contributions to migratory flight is on the order of 5% (Jenni and Jenni-Eiermann 1998) but may occasionally be higher. Figure 19.3 shows what we might typically expect from protein metabolism alone. Such a modeling approach could be modified to meet different migratory scenarios and isotopic changes as well as overall metabolic rate effects (Hobson and Bairlein 2003). The situation with ^{15}N may be more complicated but the effects of migratory fasting on ^{15}N protein values are currently unknown (Hobson et al.1993).

Metabolically inactive tissues like the keratin of feathers and toenails present information on origins typical of the period of growth (assuming no endogenous reserves are used in their formation). For cases involving birds whose molt schedules are well known, the isotopic measurement of a single feather can be a powerful tool to determine migratory connectivity. The disadvantage to using feathers is that if they are lost they can be replaced at locations other than those where they are assumed to have grown. In addition, for several species, we still do not understand molt schedules well enough and it is possible (but difficult to corroborate) that failed breeders might leave the breeding grounds early and molt en route or may be the first to molt on breeding grounds. The good news is that stable isotope methods can be used to determine molt patterns as well as breeding origins. Wassenaar and Hobson (2001) confirmed that adult Swainson's Thrush molted flight feathers south of their actual breeding grounds. Toenails of birds arriving in the spring may give good isotopic information on environments occupied on the wintering grounds because they grow relatively slowly (of the order of 0.04 mm per day; Bearhop et al. 2003) and so will represent diet on the order of the previous weeks to months.

TRACE ELEMENTS

Patterns of trace elements in feathers are ultimately derived from diet which in turn are influenced strongly by surficial geology, and so are expected to provide spatial information. This was a comparatively early approach to using endogenous signatures in avian migration tracking (early reviews by Means and Peterle 1982; Kelsall 1984). The method has great intuitive appeal because it is possible to measure relative abundance of numerous elements in feathers and so the chances of acquiring a unique signature for an individual or population is increased. Recent developments in analytical techniques allow the routine measurement of concentrations in feathers of numerous elements, including As, Cd, Mg, Mn, Mo, Se, Sr, Co, Fe, Zn, Li, P, Ti, V, Ag, Cr, Ba, Hg, Pb, S, Ni, and Cu. However, despite the potential of this

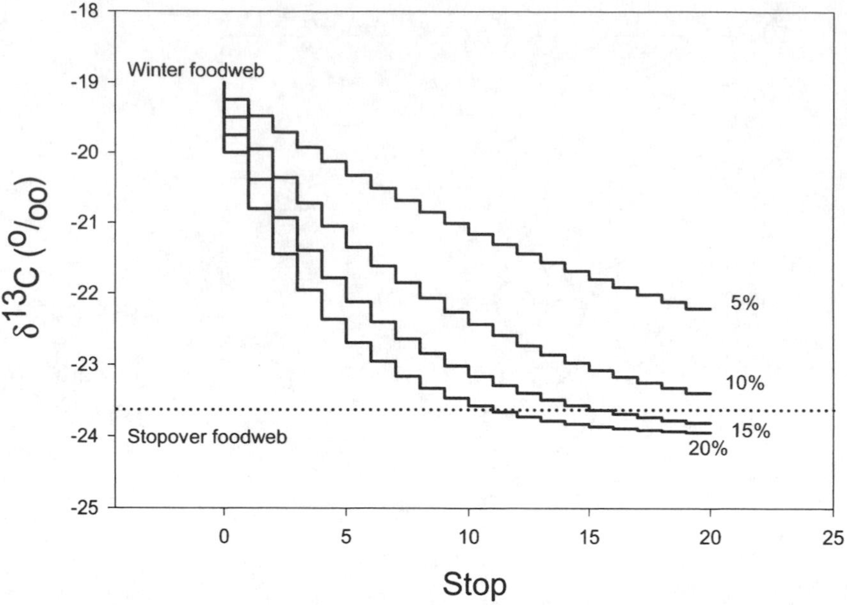

Fig. 19.3. Theoretical pattern of isotopic change in metabolically active tissue protein of a migrant bird moving from a wintering site where the food web is 5‰ more enriched than the stopover sites. Depicted here are curves corresponding to different proportions of protein contribution to fat metabolism during flight. This model is based on the assumption that the bird replenishes protein lost at each stopover and that tissue isotopic change does not result from fasting between stops.

technique, the field was largely abandoned a decade ago owing to several concerns over its reliability. Some of these criticisms have since been addressed through improvement in sample-preparation and measurement techniques that made elemental measurements much more reliable, but the stigma remains.

The first attempts to use trace element analysis to infer geographical origin were in waterfowl (e.g., Devine and Peterle 1968; Hanson and Jones 1976; Kelsall and Calaprice 1972; Kelsall et al. 1975; Kelsall and Burton 1979). These studies met with variable success but were followed by the good paper of Parish et al. (1983), who clearly distinguished three natal populations of Peregrine Falcons (*Falco peregrinus*) by measuring as few as five trace elements in feathers (see also Barlow and Bortolotti 1988). However, several studies presented evidence of considerable intrapopulation variation in feather elemental profiles related to age (Hanson and Jones 1976) and sex (Hanson and Jones 1974; Kelsall and Burton 1979; Bortolotti and Barlow 1988). The causes of such differences are poorly understood but are likely related to hormonal and metabolic mechanisms influencing secretion of trace elements into feathers. As pointed out by Bortolotti et al. (1990), such variation has been problematic because it usually makes discrimination among populations difficult or may create results that are artifacts of sampling biases.

In addition to doubts raised over intrapopulational variation in elemental profiles, a more fundamental issue that has not been addressed adequately is how such profiles change among disparate populations. For example, Bortolotti et al. (1989) found that Spruce Grouse (*Falcipennis canadensis*) from similar forest types hundreds of kilometers apart had similar feather elemental compositions, whereas those from adjacent populations occupying different forest types were quite different. Similarly, Szép et al. (2003) determined that feathers from populations of Sand Martins (*Riparia riparia*) grown at locations across Europe varied with age within colonies, and they also showed that similarity and differences in elemental profiles were not related to distance separating colonies. This situation obviously reflects the ways in which the relative abundance of trace elements can change with soil substrate. The utility of using trace element profiles to make connections between breeding, wintering, and stopover sites will therefore depend on the case in question and on how spatially discrete the populations of interest may be.

For highly colonial species, it may well be possible to characterize the different colonies according to trace element composition in feathers. If we are fortunate, such colonies may have useful elemental fingerprints. For other, more dispersed species it may simply be impossible to describe the trace element profile patterns across the range well enough to reach unambiguous conclusions about origins. This is not to suggest that this field of research will not prove to be fruitful. Rather, in contrast to the use of continental deuterium precipitation maps, it will simply be difficult to make a priori predictions about expected trace element profiles, especially at regional scales, without detailed geological information. However, trace element profiles in bird feathers may well be useful for less traditional applications. For example, Szép et al. (2003) suggest that because trace element analysis is sensitive to micro-geographical differences among individuals, this approach might be better suited to elucidating migration or wintering behavior at the level of the individual or small group. Bortolotti et al. (1990) even suggested that if the effects of age and sex on trace element profiles were well known, then population demographic information might be gleaned from elemental patterns within study populations.

Measurement techniques for establishing trace element profiles in tissues have advanced tremendously over the last several decades. Some approaches such as Inductively Coupled Plasma (ICP) techniques require the dissolution of the sample to a liquid form prior to spectral analysis, whereas others such as the neutron activation technique require that the sample be irradiated but not destroyed. Both approaches have advantages and disadvantages. The recent development of ICP-MS technology, which interfaces a mass spectrometer with an ICP machine to provide isotopic measurements of a suite of elements certainly holds great promise for migration tracking studies; by increasing the number of elements and species of isotopes that can be examined, it presumably allows for much greater resolution and for tracing isotope signatures hitherto impossible by more conventional MS techniques.

FUTURE DIRECTIONS

The use of endogenous trace element and stable isotope markers to track migrant birds better is an extremely active area of current research and future prospects are encouraging. Undoubtedly, progress will continue to be made on the analytical front that will allow analyses of more elements and isotopic species at more affordable cost. Sample size requirements are also decreasing and sensitivity is increasing so that several analyses within feathers or toenails, such as correspondences to different periods of growth, are already feasible. Because technology will likely not be the limiting factor, it behooves us to reflect more on which areas of ecological research are most needed.

The documentation of reliable isotopic and trace element patterns in nature that can be used to infer origins of birds is still an important area of research. For deuterium in precipitation, ongoing sampling will result in refinement of the relationships shown in fig. 19.1 (e.g., [www.nrel.colostate.edu/projects/usnip]; [ecophys.biology.utah.edu/LabFolks/glowen/pages/isomaps]). For other elements, one possibility might be to establish a sampling regime involving the collection of feathers or other tissues from birds produced at known locations at a continental scale and to subject these tissues to intensive analysis in the hopes of finding useful patterns. However, it would be difficult to fund such an approach without well-defined a priori expectations, although such "fishing expeditions" might certainly be feasible at smaller geographic scales. One direction that shows promise is the more formal collaboration between ecologists and experts in the geological and earth sciences. Patterns of surficial geology or the behavior of stable isotopes of elements in relation to landscape-scale processes are areas of research directly related to investigations of migration markers. For some elements, however, surficial geology alone may be a poor predictor of local food web values, because several metals may enter food webs through atmospheric deposition (see review by Pacyna and Pacyna 2001).

Further research on how to use trace element or stable isotope assays for assigning origin to individuals will require the development of more appropriate statistical methods. For investigations of the composition of wintering populations this also involves the application of expected probability of occurrence based on breeding range and how that relates to the chemical pattern of interest. A useful starting point here would be to use the North American pattern of deuterium in rainfall during the growing season to model expected patterns on wintering grounds. In addition to delineating population structure, this direction of research could use expected proportions of birds from areas with different δD values to establish zones of relative breeding productivity (Royle and Rubenstein, in press).

It is increasingly obvious that museum specimens and other archived materials are proving to be immensely valuable in isotopic and trace element studies, and their continued existence and support must be encouraged (Smith et al. 2003). Fortunately, required sample sizes are becoming impressively small but curators will necessarily have to deal with multiple requests for the same material. Researchers interested in chemical techniques applied to migration tracking should always be aware of the potential use of materials such as salvaged carcasses and feathers from a variety of sources. This topic also underlines the need for the academic and government research communities to agree on the few large-scale projects requiring museum material that need to be accomplished. There may well be no point to having detailed isotopic or elemental baseline maps generated at the continental scale more than once.

Putting out the cry for more controlled laboratory experiments is as relevant as ever (Gannes et al. 1997). Some of the next generation of these studies will be conducted on birds in wind tunnels, where migration can be simulated (e.g., Lindström et al. 1999). In this way, more precise turnover rates of stable isotopes in metabolically active tissues will be determined and the relative influences of fueling and fasting established. In addition, more refined studies dealing with the differential effects of foods with varying macromolecular content on isotopic fractionation factors will be especially useful as we learn more about the ways in which migrants can dramatically change diets (e.g., from insect to fruits) during migration.

ACKNOWLEDGMENTS

Thanks to Peter Marra and Russ Greenberg for their kind invitation to present this material at the Birds of Two Worlds Symposium and to the numerous attendees who provided stimulating discussions. Tibor Szép, Anders Møller, and G. Bortolotti provided important published and unpublished manuscripts on trace element analyses of feathers. D. Mazerolle and S. van Wilgenburg provided discussion and technical assistance. C. Page Chamberlain, Peter Marra, and an anonymous reviewer provided valuable comments on an earlier manuscript draft.

LITERATURE CITED

Alexander, S. A., K. A. Hobson, C. L. Gratto-Trevor, and A. W. Diamond. 1996. Conventional and isotopic determinations of shorebird diets at an inland stopover: the importance of invertebrates and *Potamogeton pectinatus* tubers. Canadian Journal of Zoology 74:1057–1068.

Alisauskas, R. T., E. E. Klaas, K. A. Hobson, and C. D. Ankney. 1998. Stable-carbon isotopes support use of adventitious color to discern winter origins of lesser snow geese. Journal of Field Ornithology 69:262–268.

Bairlein, F. 1998. The European-African songbird migration network: new challenges for large-scale study of bird migration. Biological Conservation of Fauna 102:12–27.

Bairlein, F. 2001. Results of bird ringing in the study of migration routes. Ardea 89:7–19.

Barlow, J. C., and G. R. Bortolotti. 1988. Adaptive divergence in morphology and behavior in some New World island birds: with special reference to *Vireo altiloquus*. Pages 1535–1549 *in* Proceedings of the 19th International Ornithological Congress.

Bearhop, S., R. W. Furness, G. H. Hilton, S. C. Votier, and S. Waldron. 2003. A forensic approach to understanding diet and habitat use from stable isotope analysis of (avian) claw material. Functional Ecology 17:270–275

Bearhop, S., S. Waldron, S. C. Votier, and R. W. Furness. 2002. Factors influencing assimilation and fractionation of nitrogen and carbon stable isotopes in avian blood and feathers. Physiological and Biochemical Zoology 75:451–458.

Best, P. B., and D. M. Schell. 1996. Stable isotopes in southern right whale (*Eubalaena australis*) baleen as indicators of seasonal movements, feeding and growth. Marine Biology 124:483–494.

Bortolotti, G. R., and J. C. Barlow. 1988. Some sources of variation in the elemental composition of Bald Eagle feathers. Canadian Journal of Zoology 63:2707–2718.

Bortolotti, G. R., K. J. Szuba, B. J. Naylor, and J. F. Bendell. 1989. Mineral profiles of spruce grouse feathers show habitat affinities. Journal of Wildlife Management 48:853–866.

Bortolotti, G. R., K. J. Szuba, B. J. Naylor, and J. F. Bendell. 1990. Intrapopulation variation in mineral profiles of feathers of Spruce Grouse. Canadian Journal of Zoology 68:585–590.

Bowen, G. J., L. I. Wassenaar, and K. A. Hobson. In Press. Application of stable hydrogen and oxygen isotopes to wildlife forensic investigations at global scales. Oecologia.

Chamberlain, C. P., S. Bensch, X. Feng, S. Åkesson, and T. Andersson. 2000. Stable isotopes examined across a migratory divide in Scandinavian willow warblers (*Phylloscopus trochilus trochilus* and *Phylloscopus trochilus acredula*) reflect African winter quarters. Proceedings of the Royal Society of London, Series B, Biological Sciences 267:43–48.

Chamberlain, C. P., J. D. Blum, R. T. Holmes, X. Feng, T. W. Sherry, and G. R. Graves. 1997. The use of isotope tracers for identifying populations of migratory birds. Oecologia 109:132–141.

Cherel Y., K. A. Hobson, and H. Weimerskirch. 2000. Using stable-isotope analysis of feathers to distinguish moulting and breeding origins of seabirds. Oecologia 122:155–162.

Devine, T., and T. J. Peterle. 1968. Possible differentiation of natal areas of North American waterfowl by neutron activation analysis. Journal of Wildlife Management 32:274–279.

Dunn, E. H., and D. J. T. Hussell. 1995. Using migration counts to monitor landbird populations: review and evaluation of current status. Current Ornithology 12:43–88.

Evans-Ogden, L., K. A. Hobson, and D. B. Lank. 2004. Blood isotopic ($\delta^{13}C$ and $\delta^{15}N$) turnover and diet-tissue fractionation factors in captive Dunlin: implications for dietary analysis of wild birds. Auk 121:170–177.

Farquhar, G. D., K. T. Hubick, A. G. Condon, and R. A. Richards. 1989. Carbon isotope fractionation and plant water-use efficiency. Pages 21–40 *in* Stable Isotopes in Ecological Research (P. W. Rundel, J. R. Ehleringer, and K. A. Nagy, eds.). Springer-Verlag, New York.

Féret, M., G. Gauthier, A. Béchet, J.-F. Giroux, and K. A. Hobson. 2003. Impact of a spring hunt on nutrient storage by Greater Snow Geese in southern Québec. Journal of Wildlife Management 67:796–807.

Flemming, T. H., R. A. Nunez, and L. S. L. Sternberg. 1993. Seasonal changes in the diets of migrant and non-migrant nectarivorous bats as revealed by carbon stable isotope analysis. Oecologia 94:72–75.

Gannes, L. Z., D. O'Brien, and C. Martinez del Rio. 1997. Stable isotopes in animal ecology: assumptions, caveats, and a call for more laboratory experiments. Ecology 78: 1271–1276.

Graves, G. R., C. S. Romanek, and A. R. Navarro. 2002. Stable isotope signature of philopatry and dispersal in a migratory songbird. Proceedings of the National Academy of Sciences USA 99:8096–8100.

Hagan, J. M., III, and D.W. Johnston (eds.). 1992. Ecology and Conservation of Neotropical Migrant Landbirds. Smithsonian Institution Press, Washington, D.C.

Hanson, H. C., and R. L. Jones. 1974. An inferred sex differential in copper metabolism in Ross' Geese (*Anser rossi*): biogeochemical and physiological considerations. Arctic 27:111–120.

Hanson, H. C., and R. L. Jones. 1976. The biogeochemistry of Blue, Snow, and Ross' geese. Illinois Natural History Survey Special Publication no. 1.

Hebert, C. E., and L. I. Wassenaar. 2001. Stable nitrogen isotopes in waterfowl feathers reflect agricultural land use in western Canada. Environmental Science and Technology 35:3482–3487.

Hobson, K. A. 1999a. Tracing origins and migration of wildlife using stable isotopes: a review. Oecologia 120:314–326.

Hobson, K. A. 1999b. Stable-carbon and nitrogen isotope ratios of songbird feathers grown in two terrestrial biomes: implications for evaluating trophic relationships and breeding origins. Condor 101:799–805.

Hobson, K. A. 2002. Making migratory connections with stable isotopes. Pages 379–391 *in* Avian Migration (P. Berthold and P. Gwinner, eds.). Springer-Verlag, Berlin.

Hobson, K. A., R. T. Alisauskas, and R. G. Clark. 1993. Stable-nitrogen isotope enrichment in avian tissues due to fasting and nutritional stress: implications for isotopic analyses of diet. Condor 95:388–394.

Hobson, K. A., L. Atwell, and L. I. Wassenaar. 1999a. Influence of drinking water and diet on the stable-hydrogen isotope ratios of animal tissues. Proceedings of the National Academy of Sciences USA 96:8003–8006.

Hobson, K. A., and F. Bairlein. 2003. Isotopic discrimination and turnover in captive Garden Warblers (*Sylvia borin*): implications for delineating dietary and migratory associations in wild passerines. Canadian Journal of Zoology 81:1630–1635.

Hobson, K. A., G. Bowen, L. Wassenaar, Y. Ferrand, and H. Lormee. 2004. Using stable hydrogen isotope measurements of feathers to infer geographical origins of migrating European birds. Oecologia 141:477–488.

Hobson, K. A., and R. W. Clark. 1992b. Assessing avian diets using stable isotopes: Pt. 1. Turnover of carbon-13 in tissues. Condor 94:181–188.

Hobson, K. A., and R. W. Clark. 1992a. Assessing avian diets using stable isotopes: Pt. 2. Factors influencing diet-tissue fractionation. Condor 94:189–197.

Hobson, K. A., and R. G. Clark. 1993. Turnover of ^{13}C in cellular and plasma fractions of blood: implications for nondestructive sampling in avian dietary studies. Auk 110: 638–641.

Hobson, K. A., K. P. McFarland, L. I. Wassenaar, C. C. Rimmer, and J. E. Goetz. 2001. Linking breeding and wintering grounds of Bicknell's Thrushes using stable isotope analyses of feathers. Auk 118:16–23.

Hobson, K. A., and L. Wassenaar. 1997. Linking breeding and wintering grounds of Neotropical migrant songbirds using stable hydrogen isotopic analysis of feathers. Oecologia 109:142–148.

Hobson, K. A., and L. I. Wassenaar. 2001. A stable isotope approach to delineating population structure in migratory wildlife in North America: an example using the Loggerhead Shrike. Ecological Applications 11:1545–1553.

Hobson, K. A., L. I. Wassenaar, B. Milá, I. Lovette, C. Dingle, and T. B. Smith. 2003. Stable isotopes as indicators of altitudinal distributions and movements in an Ecuadorean hummingbird community. Oecologia 136:302–308.

Hobson, K. A., L. I. Wassenaar, and O. R. Taylor. 1999a. Stable isotopes (δD and $\delta^{13}C$) are geographic indicators of natal origins of monarch butterflies in eastern North America. Oecologia 120:397–404.

Hoeffs, J. 1997. Stable Isotope Geochemistry. Springer-Verlag, New York.

Jenni, L., and S. Jenni-Eiermann. 1998. Fuel supply and metabolic constraints in migrating birds. Journal of Avian Biology 29:521–528.

Kelly, J. F., V. Atudorei, Z. D. Sharp, and D. M. Finch. 2001. Insights into Wilson's Warbler migration from analyses of hydrogen stable-isotope ratios. Oecologia 130:216–221.

Kelsall, J. P. 1984. The use of chemical profiles from feathers to determine the origins of birds. Pages 501–515 *in* Proceedings of the 5th Pan-African Ornithological Congress, Lilongwe, Malawi 1980 (J. Ledger, ed.). South African Ornithological Society, Johannesburg.

Kelsall, J. P., and R. Burton. 1979. Some problems in identification of origins of lesser snow geese by chemical profiles. Canadian Journal of Zoology 57:2292–2302.

Kelsall, J. P., and J. R. Calaprice. 1972. Chemical content of waterfowl plumage as a potential diagnostic tool. Journal of Wildlife Management 36:1088–1097.

Kelsall, J. P., W. J. Pannekoek, and R. Burton. 1975. Chemical variability in plumage of wild lesser snow geese. Canadian Journal of Zoology 53:1369–1375.

Körner, C., G. D. Farquhar, and S. C. Wong. 1991. Carbon isotope discrimination by plants follows latitudinal and altitudinal trends. Oecologia 88:30–40.

Lajtha, K., and J. D. Marshall. 1994. Sources of variation in the stable isotopic composition of plants. Pages 1–21 *in* Stable Isotopes in Ecology and Environmental Sciences (K. Lajtha and R. H. Michener, eds.). Blackwell Scientific, Oxford.

Lajtha, K., and R. H. Michener (eds.). 1994. Stable Isotopes in Ecology and Environmental Sciences. Blackwell Scientific, Oxford.

Lindström, A., M. Klaassen, and A. Kvist. 1999. Variation in energy intake and basal metabolic rate of a bird migrating in a wind tunnel. Functional Ecology 13:352–359.

Lindström, A., A. Kvist, T. Piersma, A. Dekinga, and M. W. Dietz. 2000. Avian pectoral muscle size rapidly tracks body mass changes during flight, fasting and refueling. Journal of Experimental Biology 203:913–919.

Marra, P. P., K. A. Hobson, and R. T. Holmes. 1998. Linking winter and summer events in a migratory bird using stable carbon isotopes. Science 282:1884–1886.

Means , J. W., and T. J. Peterle. 1982. X-ray microanalysis of feathers for obtaining population data. Pages 465–473 *in* Transactions of 14th International Congress of Game Biologists.

Meehan, T. D., C. A. Lott, Z. D. Sharp, R. B. Smith, R. N. Rosenfield, A. C. Stewart, and R. K. Murphy. 2001. Using hydrogen isotope geochemistry to estimate the natal latitudes of immature Cooper's Hawks migrating through the Florida Keys. Condor 103:11–20.

Mehl, K. R., R. T. Alisauskas, K. A. Hobson, and F. R. Merkel. In Press. Linking breeding and wintering grounds of king eiders: making use of polar isotopic gradients. Journal of Wildlife Management.

Minami, H., and H. Ogi. 1997. Determination of migratory dynamics of the sooty shearwater in the Pacific using stable carbon and nitrogen isotope analysis. Marine Ecology Progress Series 158:249–256.

Mizutani, H., M. Fukuda, and Y. Kabaya. 1992. ^{13}C and ^{15}N enrichment factors of feathers of 11 species of adult birds. Ecology 73:1391–1395.

Mizutani, H., Y. Kabaya, and E. Wada. 1991. Nitrogen and carbon isotope compositions relates linearly in cormorant tissues and its diet. Isotopenpraxis 27:166–168.

Møller, A. P., and K. A. Hobson. 2004. Heterogeneity in stable isotope profiles predicts coexistence of two populations of barn swallows *Hirundo rustica* differing in morphology and reproductive performance. Proceedings of the Royal Society (London) 271:1355–1362.

Nadelhoffer, K. J., and B. Fry. 1994. Nitrogen isotope studies in forest ecosystems. Pages 22–44 *in* Stable Isotopes in Ecology and Environmental Science (K. Lajtha and R. H. Michener, eds.). Blackwell Scientific, Oxford.

Norris, D. R., P. P. Marra, T. K. Kyser, T. W. Sherry, and L. M. Ratcliff. 2004. Tropical winter habitat limits reproductive success on the breeding grounds in a migratory bird. Proceedings of the National Academy of Sciences 271: 59–64.

Pacyna, J. M., and E. G. Pacyna. 2001. An assessment of global and regional emissions of trace metals to the atmosphere from anthropogenic sources worldwide. Environmental Reviews 9:269–298.

Pain, D. J., R. E. Green, B. Giebing, A. Kozulin, A. Poluda, U. Ottosson, M. Flade, and G. M. Hilton. 2004. Using stable isotopes to investigate migratory connectivity of the globally threatened aquatic warbler *Acrocephalus paludicola*. Oecologia 138:168–174.

Parrish, J. R., D. T. Rogers Jr., and F. P. Ward. 1983. Identification of natal locales of Peregrine Falcons (*Falco peregrinus*) by trace element analysis of feathers. Auk 100:560–567.

Pearson, D. F., D. J. Levey, C. H. Greenberg, and C. Martinez del Rio. 2003. Effects of elemental composition on the incorporation of dietary nitrogen and carbon isotopic signatures in an omnivorous songbird. Oecologia 135:516–523.

Poage, M. A., and C. P. Chamberlain. 2001. Empirical relationships between elevation and the stable isotope composition of precipitation and surface waters: considerations for studies of paleoelevation change. American Journal of Science 301:1–15.

Royle, J. A., and D. R. Rubenstein. In Press. The role of species abundance in determining the breeding origins of migratory birds using stable isotopes. Ecological Applications.

Rubenstein, D. R., C. P. Chamberlain, R. T. Holmes, M. P. Ayres, J. R. Waldbauer, G. R. Graves, and N. C. Tuross. 2002. Linking breeding and wintering ranges of a migratory songbird using stable isotopes. Science 295:1062–1065.

Rubenstein, D. R., and K. A. Hobson. 2004. From birds to butterflies: animal movement patterns and stable isotopes. Trends in Ecology and Evolution 19:256–263.

Rundel, P. W., J. R. Ehleringer, and K. A. Nagy (eds.). 1988. Stable Isotopes in Ecological Research. Springer-Verlag, New York.

Schell, D. M., B. A. Barnett, and K. Vinette. 1998. Carbon and nitrogen isotope ratios in zooplankton of the Bering, Chukchi and Beaufort Seas. Marine Ecology Progress Series 162:11–23.

Sheppard, S. M. F., R. L. Nielsen, and H. P. Taylor. 1969. Oxygen and hydrogen isotope ratios of clay minerals from porphyry copper deposits. Economic Geology 64:755–777.

Smith, T. B., P. Marra, M. S. Webster, I. Lovette, L. Gibbs, R. T. Holmes, K. A. Hobson, and S. Rohwer. 2003. A call for feather sampling. Auk 67:796–807.

Szép, T., A. P. Møller, J. Vallner, B. Kovacs, and D. Norman. 2003. Use of trace elements in feathers of Sand Martins *Riparia riparia* to identify molting areas: effects of molting origin and age on the composition of elements. Journal of Avian Biology 34:307–320.

Szép, T., A. P. Møller, J. Vallner, B. Kovacs, and D. Norman. 2003. Use of trace elements in feathers of sand martins *Riparia riparia* to identify molting areas: effects of molting origin and age on the composition of elements. Journal of Avian Biology 34:307–320.

Wassenaar, L.I., and K.A. Hobson. 2000. Stable-carbon and hydrogen isotope ratios reveal breeding origins of Red-winged Blackbirds. Ecological Applications 10:911–916.

Wassenaar, L. I., and K. A. Hobson. 2001. A stable-isotope approach to delineate geographical catchment areas of avian migration monitoring stations in North America. Environmental Science and Technology 35:1845–1850.

Webster, M. S., P. P. Marra, S. M. Haig, S. Bensch, and R. T. Holmes. 2001. Links between worlds: unraveling migratory connectivity. Trends in Ecology and Evolution 17:76–83.

Wolf, B. O., and C. Martinez del Rio. 2000. Use of saguaro fruit by white-winged doves: isotopic evidence of a tight ecological association. Oecologia 124:536–543.

Wolf, B. O., C. Martinez del Rio, and J. Babson. 2002. Stable isotopes reveal that saguaro fruit provides different resources to two desert dove species. Ecology 83:1286–1293.

Part 5 / MIGRATION ITSELF

ÅKE LINDSTRÖM

Overview

ALTHOUGH THIS VOLUME is about migratory birds, little so far has focused on migration itself. Up to this point in the narrative, authors have emphasized the evolution of migration and how being migratory influences all aspects of the biology of species. However, it is impossible to develop a complete understanding of the evolution of migratory birds without a clear understanding of the demands of their travels. More importantly, students of migration need to integrate how these demands influence performance at other points of the annual cycle. It is with an eye toward integration into the entire annual cycle that the following chapters examine how birds prepare for and make their tremendous journeys.

How to understand the significance of migration? Some people may have rather romantic images of migration, such as large flocks of geese heading north with the first mild winds in spring or thrushes calling in starry October nights, fleeing the first frost. Others consider it the most stressful and hazardous period of the birds' lives, when, within the bird populations, the wheat is separated from the chaff. Others again may think of migration as something the birds "just do" in order to get from winter to breeding quarters, where more important activities await. Whatever one's view, it is clear that migration is a multifaceted life history trait involving many specific behavioral and physiological adaptations. For some long-distance migrants who spend as much as 5 to 6 months on migration every year, it is clearly a dominant part of their lives.

Migration is foraging. About 90% of the time and 66% of the energy spent on migration is spent at stopover sites, where foraging for fuel deposition is the main activity. Because foraging rates determine fueling rates, they greatly influence migration speed. Rapid migration is likely to be essential for most birds, so successful foraging at stopover sites is crucial.

Migration is flight. During relatively short bursts of high-intensity work, the preparations during the last days or weeks at the stopover sites are converted into distance, the ultimate goal of migration. Efficient flight depends on finding the best day (or night) for migration, when winds and conditions for orientation and navigation are optimal, and flying at the optimal flight speed and altitude. Some birds can take advantage of flying in flocks, though possibly at the cost of having to trade-off their own readiness for that of their flock mates.

Migration is physiological changes. Through natural selection, the bodies of migrants have evolved to become highly efficient migratory "machines," adapted to the repeated transitions from flight to foraging and back to flight again. The birds continually rebuild their muscles and organs from "eating machines" to " flying machines," in addition to building up and burning off fat stores. Physiological flexibility, from the level of the cell to whole sets of organs, may well be a most important attribute of a successful migrant.

Migration is decision making. Migrants must make numerous important decisions every day—where to land for fueling, what to feed on, how much fuel to put on, and how long to stay. Migrants may have evolved migration strategies to minimize travel time, energy expended, and risk of predation during migration. However, understanding the role that these factors play is complicated by the fact that they are measured in different currencies (e.g., units of time, energy, as well as mortality rates), and, moreover, the importance of these currencies to overall fitness varies with species and season. Perhaps by hypothesizing models for decision making that require birds to optimize across these currencies, a difficult task for even the most mathematically able scientist, we are expecting too much sophistication from the birds. Maybe they use instead rather crude rules of thumb for their decision making. Understanding migration, therefore, is to a large extent understanding how decisions are made and what the ultimate consequences are.

The following chapters address many of these important issues of bird migration. Moore, Smith, and Sandberg give a broad overview of the various problems to be solved during migration—how to find an appropriate habitat for efficient fueling, to avoid predation and competition, and to find their precise way to their goal, all while coping with the yet poorly understood threat from parasites and pathogens. Further, Moore et al. present examples of how migration performance may be linked to reproductive performance, a crucial link for our understanding of migration itself. In two species of tropical migrants, they found that both early arrival to the breeding grounds and relatively large fuel loads at arrival were correlated with important breeding parameters, such as early start of breeding and more and larger eggs. Clearly, when migration is carried out rapidly and efficiently, the most sought-after reward—higher fitness through enhanced reproductive success—may well await at the breeding quarters.

Many stopover studies have been carried out at a single study site using a unique study setup, often leaving doubt as to the generality of the results. Few studies have collected comparable data at many sites along the migration route of a given species. But Piersma et al. show the potential strength of this approach, taking advantage of a worldwide web of researchers studying a worldwide web of migration routes of Red Knots (*Calidris canutus*). Collating data on fueling rates of Red Knots at important staging sites from all corners of the world, they found that fueling rates differed markedly among sites, being higher at high latitudes than near the equator. The most likely explanation is that the food basis for fueling is much poorer near the equator, negating the positive effect tropical temperatures should have on energy budgets and hence fueling rate. Does this mean that the Tropics are a real hurdle for long-distance migrating shorebirds? Piersma et al. also found that faster fueling rates resulted in higher fuel loads at departure, suggesting that knots aim to maximize speed of migration. But, surprisingly, departure fuel loads did not reflect the length of the ensuing flight, indicating that factors other than pure distance influence the fueling decisions of migrating knots.

For technical and practical reasons it is still a dream for most bird migration researchers to follow single individuals throughout their journey, getting repeated and detailed data on stopover and flight performance in parallel. Cochran and Wikelski may have come closest to fulfilling this dream. Building on results from Cochran's pioneering work on radio-tagged thrushes followed by car and airplane through parts of their migration, extended with cage experiments and recent detailed energetic measurements, they produce a fascinating insight into the migratory details of free-flying birds. For example, thrushes often migrate during only parts of the night, flights are surprisingly cheap, birds may make considerable deviations toward areas with rain, and some flights seem highly suboptimal (e.g., birds apparently do not always take unfavorable wind speed and direction into account). A combination of high-quality empirical data and careful experiments, in which causal relationships can be established, seems a most promising path for future research.

Migration involves large, rapid, and repeated physiological changes. Holberton and Dufty review the hormonal and metabolic basis for behavioral and physiological adaptations during migration. Although some facts are known, there are also many loose ends. For example, which physiological mechanism stimulates a surge in food intake for migratory fueling? Is it possible that the neuropeptide Y mediates this gluttony and that during migration the birds' brains become extra sensitive to Y? How do the birds know when they have stored enough fuel for the next flight? A relatively recent discovery is the protein hormone leptin, which has the potential to be the "fuel tank" signal, giving a direct measure of how much fat has been stored. Clearly many exciting discoveries are yet to be made concerning the biochemical processing that enables birds to travel the world.

20

FRANK R. MOORE, ROBERT J. SMITH,
AND ROLAND SANDBERG

Stopover Ecology of Intercontinental Migrants

En Route Problems and Consequences for Reproductive Performance

HOW WELL MIGRATORY BIRDS SATISFY energetic requirements and meet exigencies that arise during passage determines the success of their migration, and a successful migration is measured in terms of survival and reproductive performance. We first present evidence of the reproductive consequences of time of arrival and energetic condition upon arrival for American Redstarts (*Setophaga ruticilla*) and Pied Flycatchers (*Ficedula hypoleuca*), both intercontinental songbird migrants. Birds that arrive earlier commence breeding activity sooner, and those individuals that arrive on the breeding grounds with surplus fat stores experience enhanced reproductive performance. We then consider how problems encountered during the migratory period, including acquisition of food, predator avoidance, and competition, might affect a songbird migrant's time of arrival on the breeding grounds and its energetic condition upon arrival.

INTRODUCTION

Some have argued that long-distance, intercontinental migrants experience the best of two worlds by virtue of their migratory strategy: increased reproductive success by breeding in food-rich, competitor-poor temperate areas and increased survival by spending temperate winter in the Tropics (Greenberg 1980). Be that as it may, traveling long distances across areas that vary physiographically comes with considerable risks, and the mortality associated with migration, though difficult to estimate, may be substantial, especially among hatching-year birds (e.g., Lack 1946; Greenberg 1980; Ketterson and Nolan 1983; Sillett and Holmes 2002). Visualize a Red-eyed Vireo (*Vireo olivaceus*) gleaning small caterpillars from the edge of hackberry leaves in the middle of the long, narrow chenier near Johnson's Bayou, Louisiana, following a flight across the Gulf of Mexico in mid-April. Now consider the many problems with which it must cope. Besides the energetic cost of transport (Berthold 1975; Blem 1980), the bird must adjust to unfamiliar surroundings (Aborn and Moore 1997; Petit 2000), satisfy nutritional demands under time constraints (Loria and Moore 1990), compete with other migrants and resident birds for limited resources (Moore and Yong 1991; Carpenter et al. 1993a, 1993b), avoid predation and balance conflicting demands between predator avoidance and food acquisition (Lindström 1989; Moore 1994; Cimprich and Moore 1999), cope with unfavorable weather (Richardson 1978), and determine an appropriate direction for the next migratory flight (e.g., Moore 1987, 1990).

A successful migration depends on solving these and other problems (Moore et al. 1995), and the solutions are measured in units of time and condition (see Alerstam and Lindström 1990). For example, if our hypothetical migrant stays longer than usual at a stopover site, a penalty may be attached to late arrival at the next stopover site if resource levels have been depressed by earlier migrants (see Schneider and Harrington 1981; Moore and Yong 1991). If it does not make up lost time, arrival on the wintering or breeding grounds is necessarily delayed. Migrants that arrive late on the breeding grounds, for example, may jeopardize opportunities to secure a territory or a mate. Moreover, if a migrant departs a stopover site with lower than usual fat stores, it will have a smaller "margin of safety" to buffer the effect of adverse weather on the availability of food supplies at the next stopover. If a bird expects to "catch up" with the overall time schedule of migration and maintain a "margin of safety" vis-à-vis anticipated energetic demands, it must refuel faster than average during the next stopover—and a domino effect may ensue (Piersma 1990).

The extent to which migrants solve en route problems is measured in units of time and condition upon arrival at the destination as well as during passage. Not surprisingly, Darwin (1871) stressed the consequences of time and condition in relation to reproductive performance: "Let us take any species, a bird for instance, and divide the females inhabiting a district into two equal bodies: the one consisting of the more vigorous and better-nourished individuals, and the other of the less vigorous and healthy. The former, there can be little doubt, would be ready to breed in the spring before the others. . . . There can also be no doubt that the most vigorous, healthy, and best-nourished females would on average succeed in rearing the largest number of offspring." Seasonal declines in clutch size, not to mention reduction in the probability of recruitment, are well documented in passerines (Perrins 1970; Nilsson 1994; Rowe et al. 1994; van Noordwijk et al. 1995). Just a few days delay in onset of breeding can have important fitness consequences (Nilsson 1994; van Noordwijk et al. 1995). Early nesting individuals typically lay more and larger eggs and produce heavier nestlings and fledglings than delayed nesters (see Carey 1996).

It is also well established that the pre-breeding nutritional condition of parents affects reproductive success (Drent and Daan 1980; Price et al. 1988; Rowe et al. 1994). Although it is unlikely that a small songbird migrant could accumulate energy stores sufficient to produce a clutch of eggs (Perrins 1970), experiments in which food has been supplemented prior to egg laying provide compelling evidence that parental condition is a determinant of clutch size and laying date in passerines (reviews in Davies and Lundberg 1985; Arcese and Smith 1988; Daan et al. 1988). Hence, the availability of resources in the form of endogenous fat stores acquired prior to arrival on the breeding grounds should improve parental condition and influence reproductive success among landbird migrants (Sandberg and Moore 1996).

Despite the intuitive appeal of the above argument, empirical support is limited (Smith and Moore 2003). King and his colleagues (1963) speculated that substantial lipid accumulation for spring migration in the White-crowned Sparrow (*Zonotrichia leucophrys gambelii*) represented an adaptation for a quicker migration and for confronting inclement weather on the breeding grounds. The arrival of female Pied Flycatchers (*Ficedula hypoleuca*) on breeding grounds in northern Finland with fat stores estimated at 14% body mass prompted Ojanen (1984) to suggest that "[A]rriving with plenty of reserves thus assists with energy requirements for competition over nest-holes and for a rapid onset of breeding activities." Møller (1994) reported that male Barn Swallows (*Hirundo rustica*) with larger energy stores were better able to capture food or sustain periods of food shortage during times of adverse weather in early spring. He further suggested that females may discriminate against males in poor condition during mate choice.

Our intention here is twofold: First, we report on the reproductive performance of female American Redstarts (*Setophaga ruticilla*) and female Pied Flycatchers (*Ficedula hypoleuca*), both intercontinental songbird migrants that breed at high latitudes, in relation to their endogenous fat stores upon arrival. Although we do not know where or when these fat stores were deposited, it is reasonable to assume that events during passage influence condition upon arrival. We test two predictions that follow from arguments advanced by Sandberg and Moore (1996): (1) Females with more fat on arrival will initiate clutches sooner, as evidenced by decreasing the interval between arrival date and clutch initiation; and (2) females arriving with endogenous fat stores will experience enhanced reproductive success. Second, we consider how events during passage might influence a songbird migrant's time of arrival on the breeding grounds and the migrant's energetic condition upon arrival.

METHODS

We collected data on American Redstarts arriving to breed in Michigan's eastern Upper Peninsula (46°2′ N, 84°35′ W) and Pied Flycatchers arriving at their breeding grounds in

Ammarnäs, Sweden (65°58′ N, 16°17′ E). Mist-nets (12×2.6 m, 30-mm mesh) were used to trap redstarts from 1997 through 1999 and flycatchers from 1997 through 2001. The study site in Michigan was characterized by lowland coniferous forest dominated by Northern White Cedar (*Thuja occidentalis*) (see Smith et al. 1998), and the Swedish site was subalpine birch forest dominated by European White Birch (*Betula pendula*).

After capture each bird was fitted with a numbered aluminum band and a unique color-band combination. Standard measurements taken from birds included body mass and wing chord. We visibly quantified arrival fat loads in both species. Visible fat classification avoids problems associated with weight-related variation caused by differences in gut, water, and protein content and is independent of absolute body size (Rogers 1991). Further, this commonly used method provides a repeatable method of indexing fat stores as long as comparisons are made within species (Krementz and Pendleton 1990). To estimate arrival fat load in redstarts we used the six-point ordinal scale of Helms and Drury (1960) and for flycatchers we used the seven-point ordinal scale (Pettersson and Hasselquist 1985; see also Sandberg 1996). All fat scoring (and other measures) was performed by trained individuals who were checked periodically to ensure validity and repeatability of measures.

Birds remaining at each study site were relocated during the breeding season to determine date of laying the first egg, clutch size, and in the case of redstarts, egg mass. For redstarts, we computed mean egg mass for each clutch, and these averages were used in subsequent analyses. We used correlational analyses to explore relations between arrival timing and arrival fat and reproductive performance. Some of our redstart data did not meet the assumptions of normality so we used nonparametric correlations. Tests are one-tailed when we made a priori predictions and are so indicated.

An important assumption of this study is that fat load at first capture is representative of actual arrival fat. We believe this to be valid for several reasons. First, our use of passive and active netting was designed to maximize the probability of capturing birds on the day of arrival. Further, field observations, especially of singing males, support this assumption. Because we were recording foraging behavior (as part of other studies) and looking for newly arrived, unbanded individuals, the study site and surrounding area was under intense daily observation. Newly arrived individuals were typically confirmed as captured (observed with bands) within a few hours of the original observation.

RESULTS

Arrival date is related to when a female initiated a clutch in both American Redstarts (Pearson's $r = 0.74$, one-tailed $p < 0.001$) and Pied Flycatchers (Spearman's $\rho = 0.64$, one-tailed $p < 0.001$) (fig. 20.1). There was no relationship between female arrival date and clutch size in redstarts (Pearson's $r = 0.29$, one-tailed $p = 0.079$, $n = 25$) nor was there a relationship between arrival fat and clutch initiation date (Spearman's $\rho = 0.08$, one-tailed $p = 0.172$, $n = 26$). Female redstarts arriving with more fat did not appear to breed sooner than females in poor condition. In Pied Flycatchers, however, the negative relationship between arrival fat and clutch initiation date was significant (Spearman's $\rho = -0.37$, one-tailed $p < 0.001$, $n = 49$).

Clutch initiation date was not associated with average per clutch egg mass in American Redstarts (Spearman's $\rho = 0.01$, $p = 0.979$, $n = 26$). There was a significant correlation between female arrival fat and average per clutch egg mass in American Redstarts (Spearman's $\rho = 0.71$, one-tailed $p = 0.005$, $n = 12$ [fig. 20.2A]). After controlling for clutch initiation date (clutch size was related to clutch initiation date: Spearman's $\rho = 0.38$, one-tailed $p = 0.035$, $n = 24$), we found that female redstarts fatter on arrival laid more eggs (Kendall's partial $T = 0.32$, $n = 24$, one-tailed $p = 0.026$). Finally, there was a significant correlation between female arrival fat and the number of hatched eggs in flycatchers (Spearman's $\rho = 0.46$, $n = 47$, one-tailed $p < 0.001$ [fig. 20.2B]).

DISCUSSION

Seasonal declines in clutch size, along with an associated reduction in the probability of recruitment, are well documented in passerines (Perrins 1970; Nilsson 1994; Rowe et al. 1994; van Noordwijk et al. 1995). Individual delays in the onset of breeding can have important fitness consequences, which may become evident after just a few days' delay in onset of breeding (Nilsson 1994; van Noordwijk et al. 1995). Early-nesting individuals typically lay more and larger eggs and have heavier nestlings and fledglings than delayed nesters (see Carey 1996). Our results indicate that arrival schedules influence breeding schedules among American Redstarts and Pied Flycatchers. Birds that arrive earlier commence breeding activity sooner. Birds that depart wintering grounds late and/or experience en route delays in their migration schedules may arrive later to breed and may suffer reproductive consequences.

Our results also reveal that American Redstarts and Pied Flycatchers that arrive on the breeding grounds with surplus fat stores experience enhanced reproductive performance. Egg production is energetically expensive, costing up to 41% of a female's daily basal metabolic rate (Carey 1996). The significant relationships between arrival fat and reproductive performance in both species provide constructive replication and strengthen the inference that migrants experience reproductive advantages by arriving on the breeding grounds with surplus fat (Sandberg and Moore 1996). Egg number, egg mass, and nestling mass have all been shown to enhance the number of young recruited into the breeding population (Martin 1987; Magrath 1991, 1992; Carey 1996; Smith and Bruun 1998).

Although it is unlikely that female American Redstarts and Pied Flycatchers utilize the added energy from endogenous stores as a direct aid in forming eggs or increasing egg size (see Perrins 1970; Klaassen et al. 2001), we cannot rule

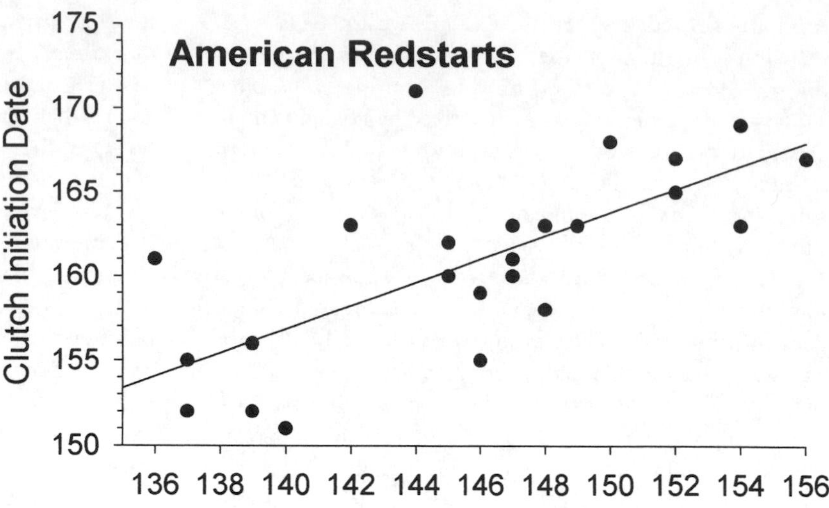

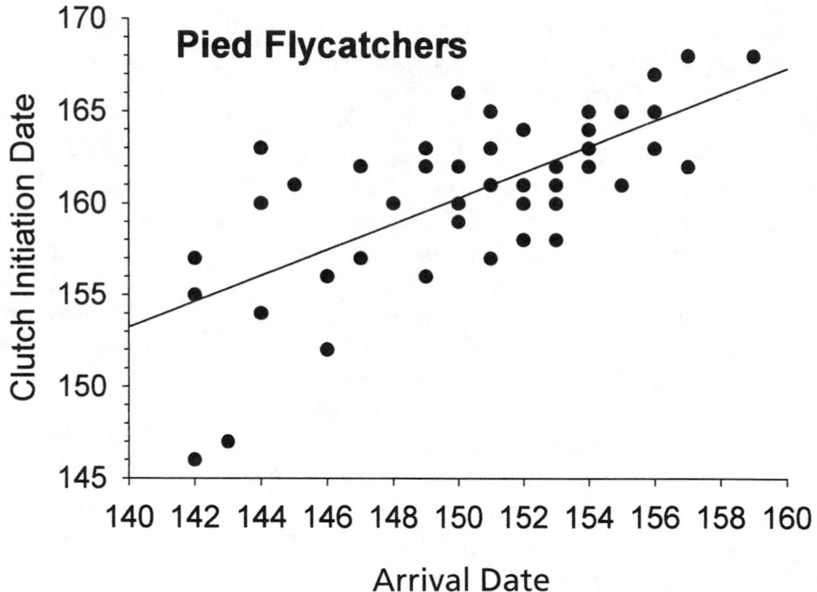

Fig. 20.1. For both American Redstarts and Pied Flycatchers, female arrival date at the breeding grounds was significantly associated with when she initiated her clutch. Dates are Julian (1 May = Julian Day 121).

out the possibility. It is more likely, however, that arrival fat may allow females (a) to devote more time to territory or mate assessment or both during the compressed arrival period or (b) to acquire specific nutrients in preparation for breeding. Arrival fat might also be a consequence of individual quality, reflecting numerous factors such as an individual's ability to offset the energetic costs of migration or the ability to compete for and maintain a quality wintering territory.

Perhaps one of the most obvious benefits of arriving with extra fat stores is insurance against predictably variable environmental conditions encountered upon early arrival (e.g., Widmer and Biebach 2001; Smith and Moore 2005). Landbird migrants often outpace phenological development of vegetation and terrestrial invertebrates as they move north during spring migration (Slagsvold 1976; Ewert and Hamas 1995). Consequently, birds may arrive at high-latitude breeding grounds when food abundance is low. Early arrival may increase the potential for exposure to poor weather conditions such as late-season snowstorms, low temperatures, or extended periods of rain. Food limitation and/or poor environmental conditions may lead to substantial mortality (e.g., Dence 1946; Zumeta and Holmes 1978; Brown and Brown 2000), reverse migration (Alerstam 1978; Terrill and Ohmart 1984; Åkesson et al. 1996), or shifts in foraging behavior as birds attempt to overcome food limitation and offset increased thermoregulatory requirements (Zumeta and Holmes 1978; Martin and Karr 1990). If migrants encounter unfavorable circumstances during this transition period, fat stores accumulated during passage would serve to overcome unpredictable foraging situations (e.g., Møller 1994), sustaining an individual until the environment becomes more suitable.

Evidence presented here suggests that the reproductive performance of American Redstarts and Pied Flycatchers is influenced by when females arrive on their breeding

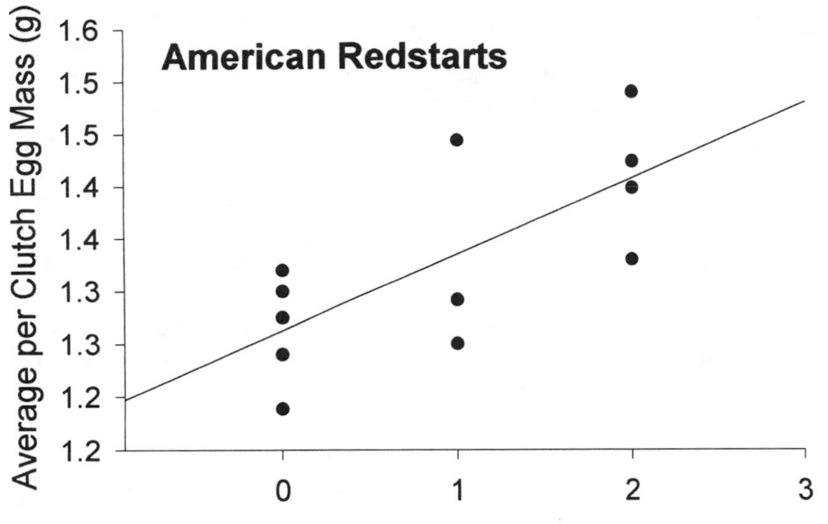

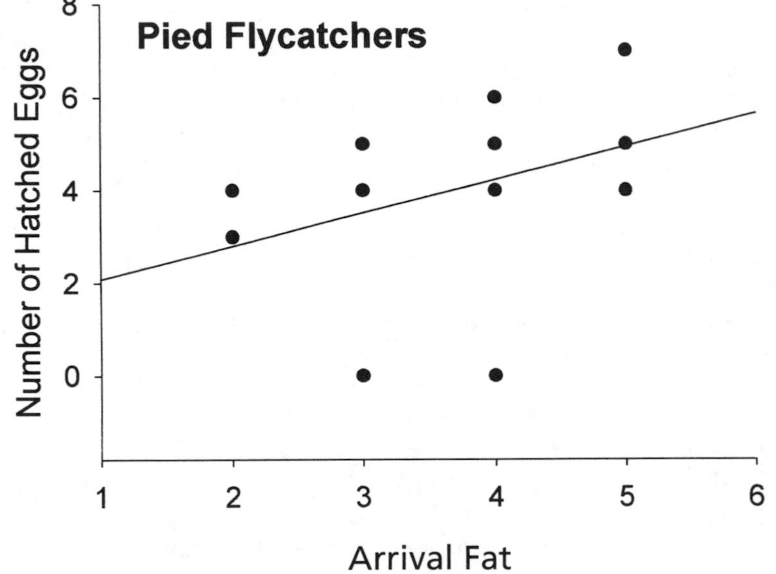

Fig. 20.2. The amount of fat with which a female arrived at the breeding grounds was significantly associated with reproductive performance in American Redstarts and Pied Flycatchers.

grounds and by their energetic state upon arrival. Time of arrival on the breeding grounds and reproductive performance are known to be linked to habitat quality on the wintering grounds (Marra et al. 1998; Norris et al 2004). We should expect that the consequences of winter habitat quality will be evident when migrants stop over en route to their breeding grounds (i.e., linkage between winter ground events and stopover biology). Moreover, we should expect events during passage not only to influence the migrant's condition and schedule of passage established upon departure but also to be responsible for differences in condition and schedule among migrants that departed at the same time and in the same condition. Although the causal linkage between events during migration and other phases of the annual cycle remains to be determined, we need to consider how problems encountered during the migratory period might affect a songbird migrant's time of arrival on the breeding grounds and her energetic condition upon arrival.

HABITAT SELECTION. One of the first "stopover" decisions a migrant must make revolves around where to settle at the end of a migratory flight, and that decision has consequences in relation to time and condition. Although we might expect migrants to settle in habitats on the basis of relative suitability, and several lines of evidence suggest that migratory species exhibit selective use of locally available habitats during stopover (Moore et al. 1995; Moore and Aborn 2000; Petit 2000), that outcome is not assured. Favorable en route habitat, where migrants can safely and rapidly meet nutritional needs, is probably limited in an absolute sense (Hutto 1985), or effectively so because migrants have limited time to search for the "best" stopover site.

Although the effect of habitat use on fitness (i.e., survival and reproductive success) is difficult to estimate during migration, more immediate consequences of habitat use can be measured in relation to how well migrants satisfy energy demand and meet other exigencies that arise in a timely fashion. For example, Bibby and Green (1983) found that Sedge Warblers (*Acrocephalus schoenobaenus*) and Reed Warblers (*A. scirpaceus*) gained mass during fall passage at a French marshland near Le Migron, but failed to do so at a second stopover site near Passay, France. Similarly, Sedge Warblers gained 0.40 and 0.55 g/day in 1973 and 1975, respectively, but only 0.05 g/day in 1974 when aphid abundance was very low in the French marshland (Bibby et al. 1976). When stopover biology of migrants was studied simultaneously in spring at two sites along the northern coast of the Gulf of Mexico, the probability of replenishing fat stores was higher and the length of stay shorter at the site where food abundance was estimated to be greater (Kuenzi et al. 1991).

FORAGING BEHAVIOR. Arguably the most important constraint during migration is to acquire sufficient nutrients to meet energetic requirements. Lipids are the primary source of energy for migratory flight. When fat stores have been estimated following long, nonstop flights, substantial individual variation characterizes samples (e.g., Bairlein 1985; Biebach et al. 1986; Moore and Kerlinger 1987). The consequences of arriving in a fat-depleted condition are several, and none are favorable. First, lean migrants have a smaller "margin of safety" to buffer the effect of conditions that reduce the availability of food during stopover. Second, efforts to satisfy energy demand may expose fat-depleted migrants to increased risk of predation. Third, if lean birds remain longer at a stopover site than birds in better energetic condition and do not make up the lost time, they will arrive later than normal at their destination.

Generally, lean migrants stay longer than birds that have not mobilized fat stores (e.g., Bairlein 1985; Pettersson and Hasselquist 1985; Biebach et al. 1986; Moore and Kerlinger 1987; Kuenzi et al. 1991). Even among lean birds the probability of staying is dependent on habitat suitability (e.g., Rappole and Warner 1976; Graber and Graber 1983; Kuenzi et al. 1991) and subject to time constraints (e.g., Alerstam and Lindström 1990; Winker et al. 1992). Birds that rapidly restore fat loads to levels appropriate for the next leg of their journey minimize the time spent en route (see Schaub and Jenni 2001), and field studies show that free-ranging migrants are capable of restoring fat loads at rates approaching 10% of their lean body mass per day (e.g., Dolnik and Blyumental 1967; Bairlein 1985; Biebach et al. 1986; Moore and Kerlinger 1987). Moreover, evidence reveals that migrants adjust foraging behavior in response to energetic demand during stopover (Loria and Moore 1990). As a consequence, lean birds are more likely to gain fat during stopover than birds carrying unmobilized stores (Yong and Moore 2005). If energetically stressed migrants increase their rate of food acquisition, a favorable energy budget is achieved more quickly, length of stay decreases, and the speed of migration increases.

COMPETITION. Migrants with similar food requirements and heightened energy demand are often locally concentrated in unfamiliar areas during stopover, which increases the likelihood of competition. Competition for available food resources could reduce the rate at which migrants meet nutritional requirements, lead to nutritional deficits, and delay passage. If competition does occur during stopover (see Rappole and Warner 1976; Kodric-Brown and Brown 1978; Bibby and Green 1980; Carpenter et al. 1983; Mehlum 1983; Viega 1986), migrants should experience a decreased rate at which energy stores are replenished with an increase in the number of migrants present, either because the availability of food is depressed or because migrants interfere with each other's intake rates (Schneider and Harrington 1981; Hansson and Pettersson 1989; Moore and Yong 1991; Carpenter et al. 1993a, 1993b). When migrants stop over along the coast of Louisiana each spring, they feed on geometrid larvae that defoliate the hackberry trees that dominant chenier habitat (Moore 1999). When fine-mesh crop netting was used to create exclosures that kept the birds away from the foliage (and their food) but permitted caterpillars to move freely about the tree, numbers of caterpillars within exclosures were significantly higher relative to control areas without crop netting (Moore and Yong 1991). Moreover, the rate at which several migrant species replenished fat stores slowed considerably when the density of birds was high in the chenier (Moore and Yong 1991).

Social asymmetries often mediate competitive interactions, and subordinate status could be a handicap to the deposition of energy stores (Terrill 1987; see also Woodrey 2000), not to mention the schedule of migration. Younger birds are often less proficient foragers (Burger 1988; Heise and Moore 2003) and usually socially subordinate to adults (Gauthreaux 1978), which may explain why immature American Redstarts deposit less fat than adults when captured during stopover along the northern coast of the Gulf of Mexico in fall (Woodrey and Moore 1997). If young, inexperienced migrants are at a competitive disadvantage, they may leave stopover areas in response to increased density and delay passage. Observations of reverse movements inland away from coastal areas have been interpreted as an age-dependent response to competition (Alerstam 1978; Lindström and Alerstam 1986; Sandberg et al. 1988).

Intersexual competition during migration may also lead to a slower rate of migration for the subordinate sex due to lower rates of mass gain at stopover sites. Empirical evidence for sex-based asymmetries during passage is mixed. Male Wilson's Warblers (*Wilsonia pusilla*) carry more fat, gain mass faster, and are captured less often than females during spring stopover in riparian habitat near Albuquerque, New Mexico (Yong et al. 1998). In contrast, Izhaki and Maitav (1998) found no difference in the mass or fat load of male and female Blackcaps (*Sylvia atricapilla*) stopping over at two sites in Israel.

If intersexual competition is important in shaping sex-specific migration schedules, we would expect to observe differences in the performance of individuals in same-sex and opposite-sex environments. Behavioral study of captive Pied Flycatchers reveals that both males and females are disposed to defend limited resources during migration (Moore et al. 2003). Furthermore, females are more likely to challenge dominants and successfully access resources in a same-sex environment than in an opposite-sex environment. From a female's perspective, it would be beneficial to avoid migrating concurrently with males, moving instead among a predominately same-sex cohort.

PREDATION. Predation constitutes a significant hazard to migrating birds (e.g., Lindström 1989; Moore et al. 1990). Moreover, the need to avoid predators must be balanced against the need to acquire food to meet the energetic demands of migration. Although many animals face this difficulty, the problem is particularly complex for birds during migratory stopover. This complexity arises from the conditions imposed on birds by migration: (1) Predation risk is variable and unpredictable during migration, (2) migrants often carry relatively large fat stores, (3) migrants experience elevated foraging demands, (4) there is pressure to travel quickly, and (5) migrants lack information concerning predation risks and foraging opportunities. This combination of factors creates a complex and shifting environment within which migrants may trade off food acquisition to reduce predation risk at the expense of time and condition (see Cimprich and Moore 1999).

PATHOGENS AND PARASITES. Given the energetic demands of passage, migration is no time to be fighting an infection, yet intercontinental migratory birds are exposed to a diverse array of pathogens by virtue of their travels. For example, protozoan blood parasites are among the most common infectious agents in passerine birds (Greiner et al. 1975). When migrant passerines were sampled during spring migration at sites along the northern coast of the Gulf of Mexico (M. C. Garvin et al., unpubl. data), over 20% of the birds were infected with one or more species of four genera of blood parasites (*Haemoproteus, Plasmodium, Leucocytozoon,* and *Trypanosoma*), but the intensity of infection varied widely within and among species. Risk of parasite infections is usually density-dependent (e.g., Begon et al. 1986), and migrants sometimes find themselves concentrated in high numbers during stopover. Whether migrants suffer adverse consequences by virtue of blood parasite infections remains to be determined.

Field study of migrants that stop over along the northern coast of the Gulf of Mexico following trans-Gulf migration does suggest that immune response is condition-dependent (J. Owen and Moore, unpubl. data). Migrants that stop over in an energetically stressed condition or at stopover sites of poor quality may be more susceptible to infection and disease. Although it has been suggested that migrants may allocate more resources to immune function than resident species because they are exposed to a diverse parasite fauna (Møller and Erritzoe 1998), immunocompromised individuals may nonetheless experience difficulty satisfying nutritional requirements and deposit less fat than healthier individuals or do so at a slower rate. A migrant fighting an infection may rest longer than usual at a stopover site to conserve energy needed for an immune response (cf. Svensson et al. 1998), yet a migrant that stays longer and does not make up time arrives later at its destination.

SLEEP. Most bird species are active during the day and sleep at night, except during migration, when the migratory flights of most songbird species occur at night. Generally, a migratory bird begins a night's flight shortly after sunset and flies for about half the night. By migrating at night a Red-eyed Vireo experiences decreased predatory pressure and improved atmospheric conditions for flight, and is able to allocate more time to feeding during the day (Kerlinger and Moore 1989). Nevertheless, nocturnal migration presents a time budget problem. If birds normally sleep at night, then a nocturnal migrant experiences loss of sleep unless she compensates in some way. If loss of sleep has negative consequences, the migrating bird should show compensatory adjustments to sleep loss. Compensatory sleep during the day is not without costs for a migratory bird: time available for other activities, such as foraging, is reduced, which could have consequences in units of time and condition.

ORIENTATION. Although free-flying migrants are seldom observed to be disoriented, flight in seemingly inappropriate directions is not an uncommon observation (e.g., Moore 1987). For small songbirds, displacement by wind is a real possibility (Richardson 1978). If migrants do not compensate for displacement or correct for an error in their orientation, the time required to regain their migratory pathway may translate into an untimely arrival at their destination, not to mention more time spent coping with en route problems (see Moore 1990).

It is also important to recognize that orientation decisions are made in an ecological context by migrants that vary in their nutritional condition (Alerstam 1978; Sandberg and Moore 1996). Energetic condition is known to influence the expression of migratory activity (e.g., Biebach 1985; Yong and Moore 1993) and the orientation of that activity (Able 1977; Sandberg et al. 1988). When the migratory activity of Red-eyed Vireos about to cross the Gulf of Mexico was examined during fall passage, those migrants with sufficient fat stores to fly across the Gulf of Mexico and beyond oriented in a seasonally appropriate southerly direction, whereas lean birds oriented inland or parallel to the coastline (Sandberg and Moore 1996), a response that would delay progress toward their destination.

FUTURE DIRECTIONS

A successful migration depends on solving problems that arise during the vulnerable third of the annual cycle spent in passage. Although most of the problems encountered en route are not different in kind from those that occur at other times and places during the annual cycle, their perplexity is exaggerated by virtue of the context, and we are only beginning to understand how migrants resolve these problems (see Alerstam and Hedenstrom 1998; Moore 2000).

It is often assumed that migrant populations are highly adaptable vis-à-vis the exigencies of migration, yet the extent to which migratory behavior and physiology are plastic warrants more attention. For example, when the foraging ecology of landbird migrants has been examined during migration (e.g., Parnell 1969; Graber and Graber 1983; Hutto 1985; Loria and Moore 1990; Martin and Karr 1990) considerable variation is revealed, which may come as little surprise given the different vegetation structures, wide variations in resource quality and quantity, and changes in competitive pressures and community composition experienced at that time. Such variability may represent adaptive plasticity that permits the migrant to occupy successfully a diverse array of habitat types as well as respond to novel circumstances. On the other hand, a migrant's ability to respond to changing circumstances is probably a matter of degree and subject to endogenous constraints (e.g., Kullberg et al. 2003). In any case, evidence of phenotypic plasticity and/or specialization has clear implications for a migrant's ability to adjust to habitat loss and degradation throughout the annual cycle (see last point below).

Migration is an energetically demanding, high-risk event that is thought to take its toll in the form of increased mortality, especially among birds of the year. In fact, Sillett and Holmes (2002) argue that mortality rates for songbird migrants may be 15 times higher during migration than during stationary periods. What are the sources of mortality during migration? Is mortality density-dependent? Population limitation and regulation may occur not only on breeding and wintering grounds; resource limitation and competition at stopover sites during migration can also restrict population sizes and distributions of migratory birds. Moreover, efforts to model migrant population dynamics have failed to include explicitly density-dependent and density-independent migration events.

Connectivity between phases of the annual cycle is a central theme of this volume. Although we present evidence here that time of arrival and energetic condition upon arrival on the breeding grounds have reproductive consequences for two intercontinental songbird migrants, and we point out that problems encountered en route affect timing of passage and energetic condition, the causal linkage between events during migration and other phases of the annual cycle remains to be determined. For example, it would be informative to "intercept" migrants at stopover sites that carry isotopic signatures identifying occupancy of winter habitats that vary in quality (e.g., Marra et al. 1998). The consequences of winter-habitat quality should be evident when migrants stop over en route to their breeding grounds, and possibly of more interest: How might stopover-habitat quality influence condition and schedule of passage established upon departure? To what extent are events during stopover responsible for variation in condition and schedule among migrants that departed at the same time and in the same condition from high-quality wintering habitat? Stable isotope signatures as well as other markers are also shedding light on migratory routes and patterns of differential migration—essential information in determining the nature and degree of connectivity among populations across the annual cycle.

Available long-term data sets reveal population declines among many intercontinental landbird migrants over the past quarter-century (Askins et al. 1990; Baillie and Peach 1992; Sauer et al. 1996). Our ability to discern the causes of population declines is complicated by the very life history characteristic that permits them to exploit seasonal environments, namely migration. Although the causes of population decline remain to be resolved, attention has focused largely on events associated with the breeding and wintering phases of the annual cycle (Rappole and McDonald 1994; Faaborg 2002). The ecological diversity of migratory species, coupled with the often variable weather patterns that steer migratory movements, makes an assessment of habitat requirements and the development of conservation strategies aimed at the en route period particularly difficult. Indeed, the complexity of this issue, and the fact that the abundance of migrants found at individual stopover sites can vary dramatically from year to year, makes it tempting to devalue the migratory period when developing conservation programs for these birds. However, conservation measures focused on temperate breeding grounds and/or tropical wintering areas will be compromised unless habitat requirements during migration are also met. Protecting migratory landbirds during passage depends on identifying and characterizing important landbird stopover sites, no easy task given the spatial and temporal variability that characterizes landbird migration.

ACKNOWLEDGMENTS

We thank members of the University of Southern Mississippi Migratory Bird Group for their constructive thoughts on the manuscript. The research was supported by the National Science Foundation (DEB-0073190 and LTREB 0078189), the Swedish Foundation for International Cooperation in Research and Higher Education, the Animal Behavior Society, Eastern Bird Banding Association, Sigma Xi, the Frank M. Chapman Memorial Student Research Fund, the Kalamazoo Audubon Society, and the U.S. Fish and Wildlife Service. Hiawatha National Forest personnel granted permission to perform research with American Redstarts on Forest Service property, and the Lund University provide facilities for fieldwork in Ammarnas, Sweden.

LITERATURE CITED

Able, K. 1977. The orientation of passerine nocturnal migrants following offshore drift. Auk 94:320–330.

Aborn, D. A., and F. R. Moore. 1997. Pattern of movement by summer tanagers (*Piranga rubra*) during migratory stopover: a telemetry study. Behaviour 134:1077–1100.

Akesson, S., L. Karlsson, G. Walinder, and T. Alerstam. 1996. Bimodal orientation and the occurrence of temporary reverse bird migration during autumn in south Scandinavia. Behavioral Ecology and Sociobiology 38:293–302.

Alerstam, T. 1978. Reoriented bird migration in coastal areas: dispersal to suitable resting grounds? Oikos 30:405–408.

Alerstam, T., and A. Hedenstrom (eds.). 1998. Optimal migration. Journal of Avian Biology 29:337–636.

Alerstam, T., and A. Lindström. 1990. Optimal bird migration: the relative importance of time, energy and safety. Pages 331–352 *in* Bird Migration: Physiology and Ecophysiology (E. Gwinner, ed.). Springer-Verlag, New York.

Arcese, P., and J. N. M. Smith. 1988. Effects of population density and supplemental food on reproduction in Song Sparrows. Journal of Animal Ecology 57:119–136.

Askins, R. A., J. F. Lynch, and R. Greenberg. 1990. Population declines in migratory birds in eastern North America. Current Ornithology 7:1–57.

Baillie, S. R., and W. J. Peach. 1992. Population limitation in Palearctic-African migrant passerines. Ibis 134:120–132.

Bairlein, F. 1985. Body weights and fat deposition of Palearctic passerine migrants in the central Sahara. Oecologia 66:141–146.

Begon, M., J. L. Harper, and C. R. Townsend. 1986. Ecology: Individuals, Populations, and Communities. Sinauer Associates, Sunderland, Mass.

Berthold, P. 1975. Migration: control and metabolic physiology. Pages 77–128 *in* Avian Biology, Vol. 5 (D. S. Farner and J. R. King, eds.). Academic Press, New York.

Bibby, C. J., and R. Green. 1980. Foraging behaviour of migrant pied flycatchers, *Ficedula hypoleuca,* on temporary territories. Journal of Animal Ecology 49:507–521.

Bibby, C. J, and R. E. Green. 1983. Autumn migration strategies of Reed and Sedge Warblers. Ornis Scandinavica 12:1–12.

Bibby, C. J., R. E. Green, G. R. M. Pepler, and P. A. Pepler. 1976. Sedge Warbler migration and reed aphids. British Birds 69:384–398.

Biebach, H. 1985. Sahara stopover in migratory flycatchers: fat and food affect the time program. Experientia 41: 695–697.

Biebach, H., W. Friedrich, and G. Heine. 1986. Interaction of bodymass, fat, foraging and stopover period in trans-Sahara migrating passerine birds. Oecologia 69:370–379.

Blem, C. R. 1980. The energetics of migration. Pages 175–224 *in* Animal Migration, Orientation, and Navigation (S. A. J. Gauthreaux, ed.). Academic Press, New York.

Brown, C. R., and M. B. Brown. 2000. Weather-mediated natural selection on arrival time in cliff swallows (*Petrohelidon pyrrhonota*). Behavioral Ecology and Sociobiology 47:339–345.

Burger. J. 1988. Effects of age on foraging in birds. Pages 1127–1140 *in* Proceedings of the 19th International Ornithological Congress (H. Ouellet, ed.). University of Ottawa Press, Ottawa.

Carey, C. 1996. Female reproductive energetics. Pages 324–374 *in* Avian Energetics and Nutritional Ecology (C. Carey, ed.). Chapman and Hall, New York.

Carpenter, F. L., M. Hixon, R. Russell, D. Paton, and E. Temeles. 1993a. Interference asymmetries among age-classes of rufous hummingbirds during migratory stopover. Behavioral Ecology and Sociobiology 33:297–304.

Carpenter, F. L., M. Hixon, E. Temeles, R. Russell, and D. Paton. 1993b. Exploitative compensation by subordinate age-classes of migrant rufous hummingbirds. Behavioral Ecology and Sociobiology 33:305–312.

Carpenter, F. L., D. Paton, and M. Hixon. 1983. Weight gain and adjustment of feeding territory size in migrant hummingbirds. Proceedings of the National Academy of Sciences USA 80:7259–7263.

Cimprich, D. A., and F. R. Moore. 1999. Energetic constraints and predation pressure during stopover. *In* Proceedings of the 22nd International Ornithological Congress, Durban (N. J. Adams and R. H. Slotow, eds.). Ostrich 69:834–846.

Daan, S., C. Dijkstra, R. Drent, and T. Meijer. 1988. Food supply and the annual timing of avian reproduction. Pages 342–352 *in* Proceedings of the 19th International Ornithological Congress (H. Ouellet, ed.). University of Ottawa Press, Ottawa.

Darwin, C. R. 1871. The Descent of Man, and Selection in Relation to Sex. Appleton, New York.

Davies, N. B., and A. Lundberg. 1985. The influence of food on time budgets and timing of breeding in the Dunnock *Prunella modularis.* Ibis 127:100–110.

Dence, W. A. 1946. Tree Swallow mortality from exposure during unseasonable weather. Auk 63:440.

Dolnik, V. R., and T. I. Blyumental. 1967. Autumnal premigratory and migratory periods in the Chaffinch (*Fringilla coelebs coelebs*). Condor 69:435–468.

Drent, R. H., and S. Daan. 1980. The prudent parent: energetic adjustments in avian breeding. Ardea 68:225–252.

Ewert, D. N., and M. J. Hamas. 1995. Ecology of migratory landbirds during migration in the Midwest. Pages 200–208 *in* Management of Midwestern Landscapes for the Conservation of Neotropical Migratory Birds (F. R. Thompson, ed.). General Technical Report NC-187. North Central Forest Experiment Station, Forest Service, U.S. Department of Agriculture, Detroit.

Faaborg, J. 2002. Saving Migrant Birds. University of Texas Press, Austin.

Gauthreaux, S. A. 1978. The ecological significance of behavioral dominance. Perspectives in Ethology 3:17–54.

Graber, J. W., and R. R. Graber. 1983. Feeding rates of warblers in spring. Condor 85:139–150.

Greenberg, R. 1980. Demographic aspects of long-distance migration. Pages 493–516 *in* Migrants in the Neotropics: Ecology, Behavior, Distribution, and Conservation (A. Keast and E. S. Morton, eds.). Smithsonian Institution Press, Washington, D.C.

Greiner, E. C, G. Bennett, E. White, and R. Coombs. 1975. Distribution of the avian haematozoa of North America. Canadian Journal Zoology 53:1762–1787.

Hansson, M., and J. Pettersson. 1989. Competition and fat deposition in goldcrests (*Regulus regulus*) at a migration stopover site. Vogelwarte 35:21–31.

Heise, C. D., and F. R. Moore. 2003. Age-related differences in foraging efficiency, molt, and fat deposition of Gray Catbirds prior to autumn migration. Condor 105:496–504.

Helms, C. W. 1968. Food, fat and feathers. American Zoologist 8:151–167.

Helms, C. W., and W. H. J. Drury. 1960. Winter and migratory weight and fat field studies on some North American buntings. Bird-Banding 31:1–40.

Hutto, R. L. 1985. Habitat selection by nonbreeding, migratory land birds. Pages 455–476 *in* Habitat Selection in Birds (M. L. Cody, ed.). Academic Press, San Diego.

Izhaki, I., and A. Maitav. 1998. Blackcaps *Sylvia atricapilla* stopping over at the desert edge: physiological state and flight-range estimates. Ibis 140:223–233.

Kerlinger, K., and F. R. Moore. 1989. Atmospheric structure and avian migration. Current Ornithology 6:109–142.

Ketterson, E., and V. J. Nolan. 1983. The evolution of differential bird migration. Pages 357–402 *in* Current Ornithology, Vol. 3 (R. F. Johnston, ed.). Plenum Press, New York.

King, J. R. 1972. Adaptive periodic storage by birds. Pages 200–217 *in* Proceedings of the 15th International Ornithological Congress, Leiden (K. H. Voous, ed.). E. J. Brill, Leiden.

King, J. R., S. Barker, and D. Farner. 1963. A comparison of energy reserves during autumnal and vernal migratory periods in the White-crowned Sparrow, *Zonotrichia leucophrys gambelii*. Ecology 44:513–521.

Klaassen M., A. Lindström, H. Meltofte, and T. Piersma. 2001. Arctic waders are not capital breeders. Nature 413:794.

Kodric-Brown, A., and J. Brown. 1978. Influence of economics, interspecific competition, and sexual dimorphism on territoriality of migrant rufous hummingbirds. Ecology 59:285–296.

Krementz, D. G., and G. W. Pendleton. 1990. Fat scoring: sources of variability. Condor 92:500–507.

Kuenzi, A. J., F. R. Moore, and T. R. Simons. 1991. Stopover of Neotropical landbird migrants on East Ship Island following trans-Gulf migration. Condor 93:869–883.

Kullberg, C., J. Lind, T. Fransson, S. Jakobsson, and A. Vallin. 2003. Magnetic cues and time of season affect fuel deposition in migratory thrush nightingales (*Luscinia luscinia*). Proceedings of the Royal Society of London, Series B, Biological Sciences 270:373–378.

Lack D. 1946. Do juvenile birds survive less well than adults? British Birds 39:258–264.

Lindström, Å. 1989. Finch flock size and risk of hawk predation at a migratory stopover site. Auk 106:225–232.

Lindström, Å. 1990. The role of predation risk in stopover habitat selection in migrating bramblings, *Fringilla montifringilla*. Behavioral Ecology 1:102–106.

Lindström, Å, and T. Alerstam. 1986. The adaptive significance of reoriented migration of chaffinches *Fringilla coelebs,* and bramblings, *F. montifringilla,* during autumn in southern Sweden. Behavioral Ecology and Sociobiology 19:417–424.

Loria, D., and F. R. Moore. 1990. Energy demands of migration on red-eyed vireos, *Vireo olivaceus.* Behavioral Ecology 1:24–35.

Magrath, R. D. 1991. Nestling weight and juvenile survival in the Blackbird, *Turdus merula*. Journal of Animal Ecology 60:335–351.

Magrath, R. D. 1992. The effect of egg mass on the growth and survival of blackbirds: a field experiment. Journal of Zoology 227:639–653.

Marra, P. P., K. A. Hobson, and R. T. Holmes. 1998. Linking winter and summer events in a migratory bird by using stable-carbon isotopes. Science 282:1884–1886.

Martin, T. E. 1987. Food as a limit on breeding birds: a life history perspective. Annual Review of Ecology and Systematics 18:453–487.

Martin, T. E., and J. R. Karr. 1990. Behavioral plasticity of foraging maneuvers of migratory warblers: multiple selection periods for niches? Studies in Avian Biology 13:353–359.

Mehlum, F. 1983. Weight changes in migrating robins (*Erithacus rubecula*) during stop–over at the island of Store Faerder, Outer Oslofjord, Norway. Fauna Norvegica, Series C, Cinclus 6:57–61.

Møller, A. P. 1994. Phenotype-dependent arrival time and its consequences in a migratory bird. Behavioral Ecology and Sociobiology 35:115–122.

Møller, A. P., and J. Erritzoe. 1998. Host immune defence and migration in birds. Evolutionary Ecology 12:945–953.

Moore, F. R. 1987. Sunset and the orientation behaviour of migrating birds. Biological Reviews 62:65–86.

Moore, F. R. 1990. Evidence for redetermination of migratory direction following wind displacement. Auk 107:425–428.

Moore, F. R. 1994. Resumption of feeding under risk of predation: effect of migratory condition. Animal Behaviour 48:975–977.

Moore, F. R. 1999. Cheniers of Louisiana and the stopover ecology of migrant landbirds. Pages 51–62 *in* A Gathering of Angels: Ecology and Conservation of Migrating Birds (K. P. Able, ed.). Cornell University Press, Ithaca.

Moore, F. R. (ed.). 2000. Stopover ecology of Nearctic-Neotropical Landbird Migrants: Habitat Relations and Conservation Implications. Studies in Avian Biology no. 20.

Moore, F. R., and D. Aborn. 2000. Mechanisms of en route habitat selection: how do migrants make habitat decisions during stopover? Studies in Avian Biology 20:34–42.

Moore, F. R., S. A. J. Gauthreaux, P. Kerlinger, and T. R. Simons. 1995. Habitat requirements during migration: important link in conservation. Pages 121–144 *in* Ecology and Management of Neotropical Migratory Birds (T. E. Martin and D. M. Finch, eds.). Oxford University Press, New York and Oxford.

Moore, F. R., and P. Kerlinger. 1987. Stopover and fat deposition by North American wood-warblers (Parulinae) following spring migration over the Gulf of Mexico. Oecologia 74:47–54.

Moore, F. R., P. Kerlinger, and T. E. Simons. 1990. Stopover on a Gulf coast barrier island by spring trans-Gulf migrants. Wilson Bulletin 102:487–500.

Moore, F. R., S. Mabey, and M. Woodrey. 2003. Priority access to food in migratory birds: age, sex and motivational asymmetries. Pages 281–292 *in* Avian Migration (P. Berthold, E. Gwinner, and E. Sonnenschein, eds.). Springer-Verlag, Berlin, Heidelberg, and New York.

Moore, F. R., and W. Yong. 1991. Evidence of food-based competition during migratory stopover. Behavioral Ecology and Sociobiology 28:85–90.

Morse, D. H. 1971. The insectivorous bird as an adaptive strategy. Annual Review of Ecology and Systematics 2:177–200.

Nilsson, J.-Å. 1994. Energetic bottle-necks during breeding and the reproductive cost of being too early. Journal of Animal Ecology 63:200–208.

Norris, D. R., P. P. Marra, K. K. Kyser, T. W. Sherry, and L. M. Ratcliffe. 2004. Tropical winter habitat controls reproductive success on the temperate breeding grounds in a migratory bird. Proceedings of the Royal Society of London, Series B, Biological Sciences 271:59–64.

Ojanen, M. 1984. The relation between spring migration and the onset of breeding in the Pied Flycatcher *Ficedula hypoleuca* in northern Finland. Annals Zoologica Fennici 21:205–208.

Parnell, H. F. 1969. Habitat relations of the Parulidae during spring migration. Auk 86:505–521.

Perrins, C. M. 1970. The timing of birds' breeding seasons. Ibis 112:242–255.

Petit, D. R. 2000. Habitat use by landbirds along Nearctic-neotropical migration routes: implications for conservation of stopover habitats. Studies in Avian Biology 20:15–33.

Pettersson, J., and D. Hasselquist. 1985. Fat deposition and migration capacity of Robins *Erithacus rebecula* and Goldcrest *Regulus regulus* at Ottenby, Sweden. Ringing and Migration 6:66–75.

Piersma, T. 1990. Pre-migratory "fattening" usually involves more than the deposition of fat alone. Ringing and Migration 11:113–115.

Price, T. D., M. Kirkpatrick, and S. J. Arnold. 1988. Directional selection and the evolution of breeding date in birds. Science 240:798–799.

Rappole, J., and M. V. McDonald. 1994. Cause and effect in population declines of migratory birds. Auk 111:652–660.

Rappole, J., and R. Warner. 1976. Relationships between behavior, physiology and weather in avian transients at a migration stopover site. Oecologia 26:193–212.

Richardson, W. J. 1978. Timing and amount of bird migration in relation to weather: a review. Oikos 30:224–272.

Rogers, C. M. 1991. An evaluation of the method of estimating body fat in birds by quantifying visible subcutaneous fat. Journal of Field Ornithology 62:349–356.

Rowe, L., D. Ludwig, and D. Schluter. 1994. Time, condition, and the seasonal decline of avian clutch size. American Naturalist 143:698–722.

Sandberg, R. 1996. Fat reserves of migrating passerines at arrival on the breeding grounds in Swedish Lapland. Ibis 138:514–524.

Sandberg, R., and F. R. Moore. 1996. Fat stores and arrival on the breeding grounds: reproductive consequences for passerine migrants. Oikos 77:577–581.

Sandberg, R., J. Pettersson, and T. Alerstam. 1988. Why do migrating robins, *Erithacus rubecula,* captured at two nearby stop-over sites orient differently? Animal Behaviour 36:865–876.

Sauer, J. R., G. W. Pendleton, and B. G. Peterjohn. 1996. Evaluating causes of population change in North American insectivorous songbirds. Conservation Biology 10:465–478.

Schaub, M., and L. Jenni. 2001. Stopover durations of three warbler species along their autumn migration route. Oecologia 128:217–227.

Schneider, D. C., and B. A. Harrington. 1981. Timing of shorebird migration in relation to prey depletion. Auk 98:801–811.

Sillett, T. S., and T. S. Holmes. 2002. Variation in survivorship of a migratory songbird throughout its annual cycle. Journal of Animal Ecology 71:296–308.

Slagsvold, T. 1976. Arrival of birds from spring migration in relation to vegetational development. Norwegian Journal of Zoology 24:161–173.

Smith, H. G., and M. Bruun. 1998. The effect of egg size and habitat on starling nestling growth and survival. Oecologia 115:59–63.

Smith R. J., M. J. Hamas, M. J. Dallman, and D. N. Ewert. 1998. Spatial variation in foraging of the Black-throated Green Warbler along the shoreline of northern Lake Huron. Condor 100:474–484.

Smith R. J., and F. R. Moore. 2003. Arrival fat and reproductive performance in a long-distance passerine migrant. Oecologia 134:325–331.

Smith, R. J. and F. R. Moore. 2005. Fat stores of American redstarts *Setophaga ruticilla* arriving at northerly breeding grounds. Journal of Avian Biology 36:1–10.

Svensson, E., L. Raberg, C. Koch, and D. Hasselquist. 1998. Energetic stress, immunosuppression and the costs of antibody response. Functional Ecology 12:912–919.

Terrill, S. B. 1987. Social dominance and migratory restlessness in the dark-eyed junco (*Junco hyemalis*). Behavioral Ecology and Sociobiology 21:1–11.

Terrill, S. B., and R. D. Ohmart. 1984. Facultative extension of fall migration by Yellow-rumped Warblers (*Dendroica coronata*). Auk 101:427–438.

van Noordwijk, A. J., R. H. McCleery, and C. M. Perrins. 1995. Selection for the timing of great tit breeding in relation to caterpillar growth and temperature. Journal of Animal Ecology 64:451–458.

Viega, J. P. 1986. Settlement and fat accumulation by migrant Pied Flycatchers in Spain. Ringing and Migration 7:85–98.

Webster, M. S., P. P. Marra, S. M. Haig, S. Bensch, and R. T. Holmes. 2002. Links between worlds: unraveling migratory connectivity. Trends in Ecology and Evolution 17:76–83.

Widmer, M., and H. Biebach. 2001. Changes in body condition from spring migration to reproduction in the Garden Warbler *Sylvia borin:* a comparison of a lowland and a mountain population. Ardea 89:57–68.

Winker, K., D. Warner, and A. R. Weisbrod. 1992. Daily mass gains among woodland migrants at an inland stopover site. Auk 109:853–862.

Woodrey, M. 2000. Age-dependent aspects of stopover biology of passerine migrants. Studies in Avian Biology 20:43–52.

Woodrey, M. S., and F. R. Moore. 1997. Age-related differences in the stopover of fall landbird migrants on the coast of Alabama. Auk 114:695–707.

Yong, W., D. M. Finch, F. R. Moore, and J. F. Kelly. 1998. Stopover ecology and habitat use of migratory Wilson's Warblers. Auk 115:829–842.

Yong, W, and F. R. Moore. 1993. Relation between migratory activity and energetic condition. Condor 95:934–943.

Yong, W. and F. R. Moore. 2005. Long-distance bird migrants adjust their foraging behavior in relation to energy stores. Acta Zoologica Sinica. In press.

Zumeta, D. C., and R. T. Holmes. 1978. Habitat shift and roadside mortality of Scarlet Tanagers during a cold wet New England spring. Wilson Bulletin 90:575–586.

21

THEUNIS PIERSMA, DANNY I. ROGERS,
PATRICIA M. GONZÁLEZ, LEO ZWARTS,
LAWRENCE J. NILES, INÊS DE LIMA
SERRANO DO NASCIMENTO,
CLIVE D. T. MINTON,
AND ALLAN J. BAKER

Fuel Storage Rates before Northward Flights in Red Knots Worldwide

Facing the Severest Ecological Constraint in Tropical Intertidal Environments?

RED KNOTS (*CALIDRIS CANUTUS*) ARE birds of many worlds. These sandpipers breed only on High Arctic tundra but move south from their disjunct, circumpolar breeding areas to nonbreeding sites on the coasts of all continents (apart from Antarctica), between latitudes 58° N and 53° S. Because of their specialized sensory capabilities, Red Knots generally eat hard-shelled prey found on intertidal, mostly soft, substrates. As a consequence, ecologically suitable coastal sites are few and far between, so they must routinely undertake flights of many thousands of kilometers. We present a review of fuel storage rates at 14 staging sites during northward migration, based on published and unpublished data collected in South and North America, Europe and Africa, and Australia and New Zealand. Fuel storage rates are interpreted in relation to latitude and associated ecological factors (climate, food abundance, and prey quality). In contrast to prediction (based on the low costs of living and thus the freedom to allocate nutrients to fuel storage), Red Knots at tropical intertidal sites have lower fueling rates than birds at more southern or northern latitudes. The highest fueling rates occur at the most northerly sites, before the flight to the tundra for breeding. These northern fueling areas offer smaller benthic biodiversity but greater harvestable biomass than circa-tropical sites, at least in spring. Such high biomasses may enable Red Knots to fuel up quickly not only for onward flight, but for survival on an initially food-free tundra as well. As predicted for time-minimizing migrants, site-average body mass at departure is a positive function of fueling rate. Surprisingly, however, departure body mass appears uncorrelated with the estimated length of the ensuing flight. The association between fueling rates of the highly food- and habitat-specialized Red Knots and latitude provides a first empirical assessment of the intimate interplay between migratory schedules and ecological factors on a world scale.

INTRODUCTION

Long-distance migrant birds capitalize on our seasonal world (Alerstam 1990). From a Northern Hemisphere perspective they seem to optimize the use of northern habitats that become available in summer (for breeding) and Southern Hemisphere habitats that offer favorable environmental conditions during the northern winter. Typical landbirds such as passerines tend to use relatively similar habitats during the breeding and the nonbreeding seasons (Rappole 1995); the sandpipers that are the focus of this chapter, however, certainly use habitats in summer that contrast with those they use the rest of the year (Piersma et al. 1996). For breeding they use dry marshlands and tundras, areas that offer protected nesting sites and plentiful food resources for the chicks. During the rest of the year sandpipers use the relatively scarce wetlands of the world, often with a tidal regime, where habitat suitable for nesting is scarce and arthropod food for chicks would be difficult to find.

Because potential stopover sites are as scarce as good wintering sites, most sandpiper species must cover the many thousands of kilometers between the southern nonbreeding and the northern breeding grounds in a few long-distance flights (Piersma 1987). Thus, the birds need to store large fuel loads to cover the distances between suitable sites, which in turn requires them to obtain extra energy above the usual requirement for maintaining stable body mass. Fueling rates are arguably the key variable determining the options open to migrant birds (e.g., Alerstam and Lindström 1990; Ens et al. 1994; Alerstam and Hedenström 1998). Daily food intake, and hence fueling rates, may be affected by limited foraging time and/or food availability (Zwarts et al. 1990a; Kvist and Lindström 2000), restricted heat production due to external heat loads in the Tropics (Verboven and Piersma 1995; Battley et al. 2003), organismal design constraints related to the size and capacity of the organ systems (Piersma 2002; Battley and Piersma 2004), predation (van Gils et al. 2004), and quality of the prey (van Gils et al. 2003a).

Here we capitalize on the great volume of work done on the worldwide migration system of one of the largest sandpipers, the Red Knot (*Calidris canutus* [Scolopacidae]) (fig. 21.1), to see whether a comparison between fueling rates at a wide range of coastal sites reveals constraints on daily food intake and fueling rate. For example, in the absence of foraging constraints we would predict fueling rates to be a negative function of other cost factors such as thermoregulation. Thus, fueling rates should be lower where temperatures are lower: the far south and north. Given the limits of current data on the birds, the analysis is restricted to the northward migration. Also, at this point comparative data on harvestable food abundance are still too sparse to bring them fully into the analysis (but see Piersma et al. 1993a).

With this assessment we aim to provide a solid empirical contribution to the development of "stopover ecology," the study of the selective forces acting on migrant birds, especially at stopover sites (Lindström 1995). To the best of our knowledge, this is the first presentation and interpretation of fueling rates of a single species at locations worldwide.

STUDY SYSTEM: THE WWW OF RED KNOTS

The worldwide web of migration routes spun by Red Knots finds its origin in the circumpolar breeding grounds of this High Arctic breeding species (Piersma et al. 1991, 1996; Harrington 1996; Piersma and Baker 2000). Red Knots breed only on the northernmost lands of the world, often within sight of the Arctic Ocean, where vegetation is usually sparse and environmental conditions even in summer are quite extreme. On the tundra they eat mostly spiders and arthropods, which they obtain by surface pecking (Tulp et al. 1998). Outside the breeding season, Red Knots occur only on coastal sites that offer large expanses of intertidal (preferably soft) substrates, where they feed on hard-shelled prey such as bivalves, gastropods, and sometimes small crustaceans obtained by high-frequency probing (Piersma et al. 1993a, 1993b). Prey is ingested whole and crushed in a strong muscular stomach (Zwarts and Blomert 1992; Piersma et al. 1993c, 1999a). The nonbreeding habitat and diet of Red Knots may relate to their specific sensory capacities (a highly specialized bill-tip organ to find hard objects in soft sediments [Piersma et al. 1995, 1998]) and, perhaps, the peculiarities of their immune system (Piersma 1997, 2003). We interpret the remarkable range in wintering latitudes as reflecting the scarcity, worldwide, of suitable coastal intertidal habitats, that is, habitats rich in high-quality and shallow buried mollusc prey. To optimize intertidal patch use, Red Knots are skilled users of self-collected information on prey abundance (van Gils et al. 2003b)

Although Red Knots have a circumpolar breeding distribution, their range is not continuous. Circling the Arctic Ocean, six separate breeding areas host different populations, all of which are now formally recognized as subspecies based on body size and plumage differences (Piersma and Davidson 1992; Tomkovich 2001). From their respective breeding grounds, the different subspecies spread out south to winter in distinct nonbreeding regions. There appears to be little overlap in occurrence between any combination of subspecies except for the temporary overlap of some *canutus* and *islandica* subspecies knots in the Wadden Sea, an extensive area of intertidal flats shared by The Netherlands, Germany, and Denmark (Piersma et al. 1995; Nebel et al. 2000), and of *roselaari* and *rufa* in Delaware Bay in the eastern United States. Of particular relevance for the comparisons between sites and subspecies made in this chapter is the finding that the extant Red Knots shared a common an-

Fig. 21.1. Three plates to illustrate the annual cycle and the habitats used by Red Knots. The top photograph shows a flock of nonbreeding Red Knots in gray-colored basic plumage foraging on an intertidal flat in the Dutch Wadden Sea. The middle photograph shows a flock in which many birds have already molted into the rusty-red alternate plumage alighting on a high-tide roost on the on the Banc d'Arguin, Mauritania in April. The bottom photograph shows a flock in mid June, just after arrival on the snow-covered tundra breeding grounds of Taymyr Peninsula, Russia. All photographs by Jan van de Kam.

cestor as recently as within the last 20,000 years or so (Baker et al. 1994; Baker and Marshall 1997). As a result of this recent expansion from a severely bottlenecked stock, the Red Knot subspecies show little genetic divergence across their worldwide range (A. J. Baker and D. M. Buehler, pers. comm.).

Just as the geographic occurrence of Red Knots is centered on the Arctic Ocean, the timing of everything that happens during their annual cycle is centered on the brief breeding season in the High Arctic. Red Knots arrive on the tundra as soon as, or even before, the snow begins to melt (Meltofte 1985; Morrison and Davidson 1990; Whitfield and Tomkovich 1996; Tulp et al. 1998). In northeast Canada and northern Greenland the first Red Knots may arrive on the tundra in the last days of May, but on Taymyr Peninsula in central Siberia, the tundra does not become free from snow until mid-June. The arrival of Red Knots is consequently later (see fig. 21.1). After 1 to 2 weeks of pair formation and territory establishment and 3 weeks of incubation (shared by the partners), the females return to southern coastal nonbreeding areas as soon as the eggs hatch (or the clutch is lost). They are followed almost 4 weeks later by the successful males (who take care of the chicks) and the first fledglings (Whitfield and Brade 1991). By then it is late July and early August, a period with declining arthropod abundances (Schekkerman et al. 2003) and an increasing likelihood of snowfall. Red Knots usually, if not always, migrate in flocks of 10 to 100 birds, usually leaving for long-distance flight in the late afternoon and early evening (Dick et al.

1987; Piersma et al. 1990b). Departures from wintering and staging sites, especially in spring, tend to be highly synchronized at the population level (e.g., Dick et al. 1987).

A complete review of the latitudinal locations of the sites used in sequence by the six different subspecies of Red Knots during northward migration is presented in fig. 21.2. There is great variability in the distance traveled by the different subspecies as wintering areas range from the north temperate zone (northwest Europe, subspecies *islandica*) to the temperate far south (Tierra del Fuego, *rufa*). Two subspecies (*canutus* and *piersmai*) winter in rather tropical climates, with some *canutus* also going as far south as southern Africa. In addition to the location of the wintering, staging, and breeding areas, fig. 21.2 presents the distances covered by Red Knots in single long-distance flights after departing from the respective areas. Flight distances are always more than 1,000 km, in several cases being more than 6,000 km. To be capable of making such flights, Red Knots need to store considerable quantities of fuel consisting mostly, but not exclusively, of fat (Lindström and Piersma 1993; Piersma 1998; Piersma et al. 1999b; Battley et al. 2000). All else being equal, we would predict fuel stores at departure to be a function of the length of the flights, and fueling rate a function of fuel stores (e.g., Alerstam and Lindström 1990). Data on five of the six subspecies are used here to examine the rates of fuel storage during northward migration at 14 different sites spanning the whole range of latitudes.

MATERIAL AND METHODS

We assembled data on body masses of adult Red Knots captured during the period of northward migration (February–May) from as many sites as possible. We did not correct for body mass losses after capture, and, in contrast to passerines, the skin of most shorebirds is too thick to enable the use of visual subcutaneous fat scores (but see Wiersma and Piersma 1995 for the use of indices based on the shape of the abdomen). Only data sets collected over much of the fueling or stopover period would yield estimates of daily rate of mass gain (rate of fueling). As explained in Appendix to this chapter, data sets were screened to remove data points for possible subadult individuals, including individuals that failed to molt into an alternate plumage. Given the large overlap in body size between males and females (Baker et al. 1999), any differences between them have to remain unexamined at this point. However, data on Red Knots staging in Iceland suggest no sex differences in either timing of stopover or fueling rates (Piersma et al. 1999b). Females do store some skeletal calcium, whereas males do not (Piersma et al. 1996a).

Fueling rates (in g/day) were based on: (1) regression of individual mass values on the number of days before the average date of northward departure date (for relatively small data sets, usually collected in a single season); (2) regression of average body mass values per catch on the number of days before the average date of northward departure date, with catches often collected over a series of years; (3) average rates of body mass gain over time for individuals recaptured in different years (in cases where catch averages were not very informative because of relatively low among-individual synchrony); and (4) a simple comparison between average body mass values early and late in the (re)fueling period. The data used for eight of the 14 sites are presented in fig. 21.3; most of the other data have been published before (see the Appendix to this chapter). In most cases there was a clear break-point in the data to indicate the time-point at which fueling started (especially clear for Delaware Bay, Rio Grande do Sul, and the Wash).

The lack of standardization in the data collection and analyses among sites makes it impossible to present confidence limits with the values. In all cases we tried hard to assess the robustness of the values obtained by testing alternative methods of analysis (e.g., regression of individual data points on time vs. calculating average change in body mass for individuals recaptured either in different years or in the same year). We also note that although there are small body-size differences among the subspecies (with the two subspecies wintering in Australasia being somewhat smaller than the others [Tomkovich 1992, 2001]), we have not tried to account for these structural size differences. Without proper analyses of the relationships of body size to mass and to fuel stores, such an exercise would only add further noise to the data. Given the eightfold range in estimated fueling rates (table 21.1), we believe that the estimates are robust enough to make latitudinal comparisons.

RESULTS

The average site-specific fueling rates of Red Knots varied from 0.6 g/day (Golfo San Matias, Argentina) to 4.6 g/day (Delaware Bay, USA) (table 21.1, fig. 21.3), both estimates being based on the same *rufa* population. Clearly, the refueling rates achieved at the final stopover sites before the flight to the breeding grounds (ranging from 2.7 to 4.6 g/day) were higher than the fueling rates before the first leg of the northward migration from the wintering grounds (0.6–2.8 g/day). Two data points for early stopover sites showed quite a contrast, with *rufa* knots at Golfo San Matias, Argentina, not gaining more than 0.6 g/day over a month of refueling, and *rufa* knots in Rio Grande do Sul, Brazil, achieving a refueling rate of 3.1 g/day in the second half of April. The low rate of body mass gain in Argentina was associated with an intense body molt from basic to alternate plumage, combined with birds undertaking a rather short northward flight to northern Argentina and/or southern Brazil (fig. 21.2) (P. M. González et al., unpubl. data).

Fueling rates at the low latitudes (around the equator) were lower than at the high latitudes (table 21.1). Perhaps surprisingly, this was not a function of the type of site (initial vs. staging): the fueling rate was correlated with latitude for both the wintering sites as a group as for all sites (fig. 21.4A; the variance explained by absolute latitude was 71%

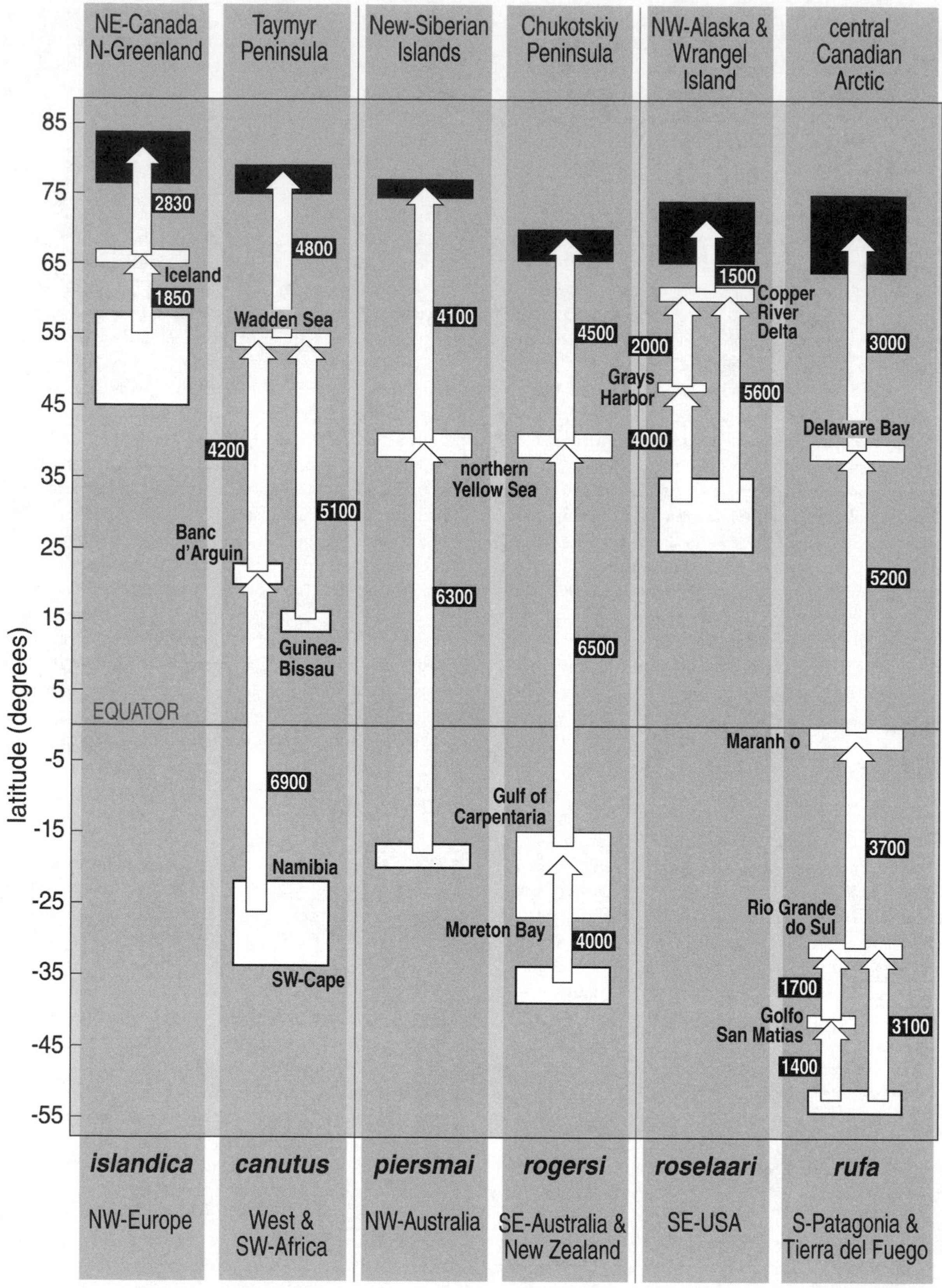

Fig. 21.2. Schematic resume of the long-distance flights for the six subspecies of Red Knots during northward migration. Nonbreeding areas (thick-lined blocks), staging areas (thin-lined blocks), and breeding areas (shaded blocks) are given in relation to latitude; flight lengths are denoted by arrows and are in kilometers. The subspecies are grouped according to three major flyways, from left to right: the East Atlantic Flyway (*islandica, canutus*), the East Asian–Australasian Flyway (*piersmai, rogersi*), and the West-Atlantic (American) Flyway (*roselaari, rufa*). Note that flight distances are great circle distances, except for the flights from Iceland to northern Ellesmere Island in northeast Canada and the flight from the Wadden Sea to Taymyr Peninsula, both of which are known to more closely follow the rhumbline.

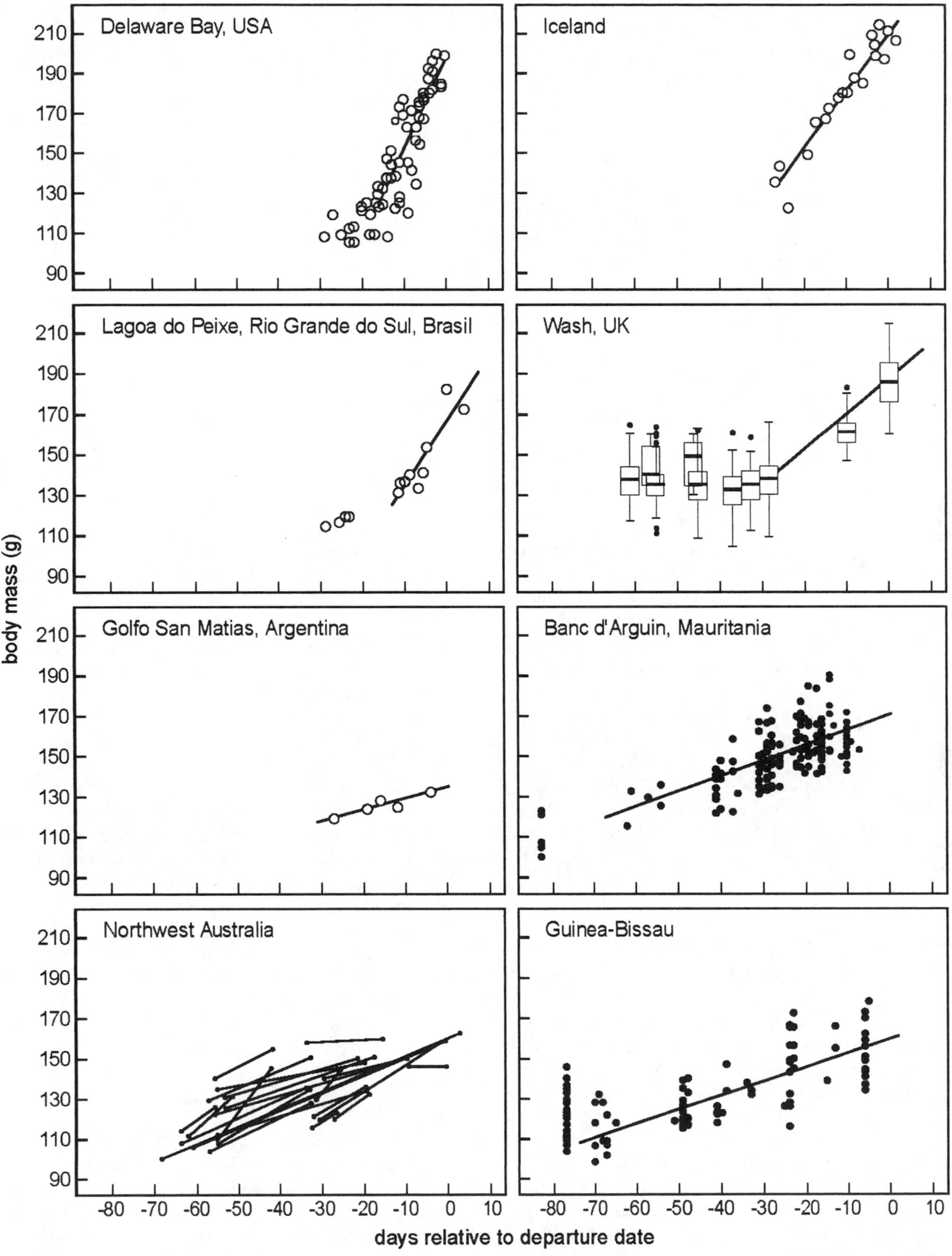

Fig. 21.3. Body mass increases of Red Knots at eight nonbreeding and staging areas before northward migration. Small dots represent individual data points; open circles represent the averages of catches; points joined by lines indicate individual mass trajectories based on catches in different years. These data are summarized in table 21.1, along with the information for the remaining six sites (in five cases based on published information).

Table 21.1 Summary table of the fueling rates, departure mass, and dates of departure in Red Knots of five subspecies studied at 14 different sites

Subspecies	Site	Type of site	Latitude	Rate of refueling (g/day)	Body mass at departure (g)	Day of departure (since 1 January)	Length of fueling period (days)
islandica	Wadden Sea, Germany	Wintering	53	2.8	191	128	20
islandica	Wash, U.K.	Wintering	53	1.7	187	123	30
islandica	Balsfjord, Norway	Final stopover	69	2.7	190	146	21
islandica	Iceland	Final stopover	65	2.9	211	149	28
canutus	South Africa	Wintering	−33	1.5	191	108	40
canutus	Guinea-Bissau	Wintering	11	0.9	157	122	70
canutus	Mauritania	Wintering	21	0.7	168	122	60
canutus	Wadden Sea, Germany	Final stopover	54	3.0	210	152	28
piersmai	NW-Australia	Wintering	−18	0.9	160	117	57
rogersi	New Zealand	Wintering	−37	1.2	185	78	55
rogersi	Victoria, Australia	Wintering	−38	1.6	180	83	40
rufa	Golfo S. Matias, Argentina	Early stopover	−41	0.6	140	91	37
rufa	Rio Grande do Sul, Brazil	Early stopover	−31	3.1	180	121	16
rufa	Delaware Bay, USA	Final stopover	39	4.6	197	151	14

Note: Details on the data sources and analyses are presented in the Appendix to this chapter.

for wintering sites only, $F_{1,6} = 14.6, p = 0.01$, and 27% for all sites, $F_{1,12} = 4.4, p < 0.06$). The estimated body mass at departure on subsequent northward flights was also a function of latitude (fig. 21.4B; variance explained by absolute latitude was 42%, $F_{1,12} = 8.8, p = 0.01$), but, surprisingly, was not correlated with the estimated length of the ensuing flight (from fig. 21.2; $F_{1,12} = 0.03, p = 0.86$). Interestingly, the correlation between fueling mass and fueling rate was relatively strong (fig. 21.4C; variance explained was 55%, $F_{1,12} = 14.5, p = 0.003$). Finally, both fueling rate and departure mass were strongly correlated with the estimated length of the fueling period (variances explained were 69%, $F_{1,12} = 26.9, p < 0.001$ and 33%, $F_{1,12} = 5.9, p = 0.03$, respectively).

DISCUSSION

Is Comparing Gross Fueling Rates among Sites Justifiable?

This review is based on a comparison of whole body mass values. Because the contributions of wet protein and fat to the body mass, gains may vary among species (Lindström and Piersma 1993; Piersma 1998), and even within species depending on the time of year (Piersma and Jukema 2002), the fueling rates may not be directly comparable in terms of energy equivalents. For Red Knots staging in Iceland we know that fat contributed 78% to the total mass increase during the stopover (Piersma et al. 1999b) and a similar estimate is now available for birds staging in Delaware Bay (T. Piersma et al., unpubl. data). However, for all the other staging sites this information is lacking.

Variability in fuel composition among sites could be compounded by temporal variability in the composition of fuel stored and variability in storage rates in the course of a stopover. Studies of *islandica* knots from the Icelandic site (Piersma et al. 1999b) reveal that fueling rates are relatively low during the first week of stopover and consist mainly of wet protein (reflecting the buildup of "digestive" organs such as the gut and liver). This is followed by a phase of rapid mass gain (much of which consists of fat), followed by a final phase of a slower mass gain during which wet protein is lost (reflecting the balance between the breakdown of the digestive systems and the buildup of "exercise" organs such as pectoral muscles and heart) and only fat is gained. As the energy equivalent of fat is eight times higher than that of wet protein (Jenni and Jenni-Eiermann 1998), differences in fuel composition among sites and times should ideally be taken into account in discussions of the ecological factors promoting differences in fueling rates. For now, we assume that the composition of the fuel stored during northward migration is similar among sites and, thus, that the energy intake necessary to deposit a gram of fuel is a constant proportion of the estimated fueling rate across all sites.

Evidence for the Use of Time-Minimization Strategies

As outlined by Alerstam and Lindström (1990), migrants using a time-minimization strategy are predicted to have fuel loads at departure (departure masses) that are a positive function of fueling rates. Such a relationship should be absent in energy-minimizing migrants. In a comparison between species-average values for shorebirds and passerines, Alerstam and Lindström (1990) found positive correlations between fuel loads at departure and fueling rates in both groups of birds, which they interpreted as evidence for the

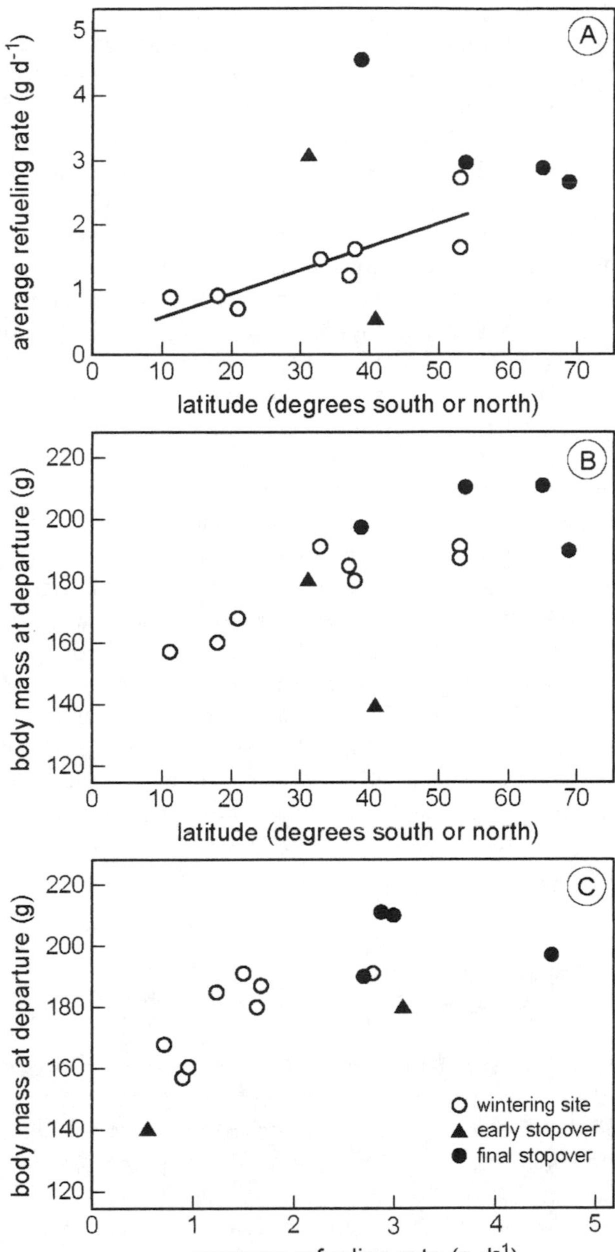

Fig. 21.4. Relationships between (A) population average fueling rates and absolute latitude, (B) average departure mass and absolute latitude, and (C) average departure mass and refueling rate for Red Knots (based on data in table 21.1). Note that open circles indicate wintering sites from which northward migration is started, triangles indicate early stopover sites, and filled circles indicate final stopover sites before the flight to the breeding grounds. The regression line in the plot of fueling rate versus absolute latitude is for the wintering sites only.

quite general use of time-minimization strategies. In a follow-up study, Lindström and Alerstam (1992) were able to obtain qualitative agreement between the predictions based on the time-minimization strategy and the individual fueling rates and departure masses in two passerine species. Our data on estimated population-average take-off masses and fueling rates for Red Knots belonging to different subspecies at different sites occupy the middle stratum of comparison (between individual and species-specific values), but confirm once again the predicted positive correlation between fuel load at departure and fueling rate expected for time-selected migrants.

Optimal migration theory (Alerstam and Hedenström 1998) would also predict that the fuel load at departure, all else being equal, would be a function of the length of the ensuing migration flight. As this relationship appears absent, we must conclude that "all else" is not equal, and that the degree of wind assistance (Piersma and van de Sant 1992; Green 2003), the type of flight (transoceanic or not, Piersma 1998), the extent of organ mass reductions before departure (Piersma 1998), the degree to which an increasing fuel load increases the cost of flight (Kvist et al. 2001), as well as the expected ecological conditions upon arrival vary among flights and perhaps among populations.

Latitudinal Trends

The thermal environment for Red Knots would appear to be rather more favorable at tropical latitudes than at (north) temperate latitudes. In fact, there is a strong negative correlation between estimated average air temperatures and absolute latitude for the 14 fueling events reported here, with air temperatures going down by 0.4°C for every degree of latitude ($r = 0.93$, $F_{1,12} = 78.7$, $p < 0.001$). At tropical latitudes Red Knots experience average air temperatures of 25°–30°C, whereas in Iceland and northern Norway they have do deal with averages of 5°–10°C. This implies a twofold difference in the maintenance costs between high- and low-latitude fueling sites (Wiersma and Piersma 1994; Piersma 2002). Everything else being equal (time-activity budgets, cost of food acquisition, handling and processing, harvestable food abundance, energy density of stored tissue, efficiency of night- and daytime feeding), one would expect Red Knots at circa-tropical sites to have higher, rather than lower, fueling rates than Red Knots at high-latitude sites (Lindström 1991).

The beauty of ecological studies of Red Knots is that these birds have highly specific habitat requirements, which leads to ecological similarities between most wintering and staging sites. The diet consists mainly of molluscs, and birds follow a tidal rhythm with two 6-h bouts of foraging during low tide. The exception (to both rules) is in Delaware Bay, where knots eat the eggs of horseshoe crabs (*Limulus polyphemus* [Tsipoura and Burger 1999, pers. obs.]), during the 12 h of daytime (pers. obs.). At Lagoa do Peixe, Rio Grande do Sul, Brazil, Red Knots make use of ocean beaches as well as semitidal lagoon; here birds could be less constrained by time and tide.

With diet and time budgets being so comparable between high- and low-latitude sites, it makes sense to argue that the surprising trend between latitude and fueling rates must be due to external factors. Although at midday Red Knots in the Tropics may incur some heat stress (Battley et al. 2003), it is unlikely that external heat loads over the entire day could put a constraint on foraging activity (which itself leads to heat production [Verboven and Piersma 1995; Bruinzeel and Piersma 1998]). The remaining ecological factor that could explain the observed pattern in fueling rates is that there may be a constraint on the rate of food intake that becomes more severe at the lower latitudes.

Red Knots fueling in captivity have been shown to achieve fueling rates far above what have been documented in the field, suggesting that food intake in the wild may be constrained at all sites. When *canutus* knots that are migrating through the Baltic (and are therefore highly motivated to eat) are put in cages where they can eat as much high-quality soft food (mealworms *Tenebrio* sp.) as they want, they routinely achieve fueling rates of 10 g/day and higher (Kvist and Lindström 2003). In view of the significant positive correlation between the total benthic biomass at intertidal flats and absolute latitude (for 14 sites where Red Knots occur, $r = 0.56$, $F_{1,12} = 5.3$, $p = 0.04$; most data are from Piersma et al. 1993a [table 21.1]) and the finding that the few available direct estimates of intake rate indeed indicate that food, from the point of view of the Red Knots, is most abundant at southern and northern high-latitude sites (G. B. Escudero, unpubl. data), the severity of a constraint on food intake may be stronger in the Tropics than elsewhere. However, establishing the extent to which higher intake rates at high-latitude staging sites can more than compensate for the increase in energy expenditure (due to a less favorable thermal environment) requires detailed energy budget reconstructions for Red Knots at a good range of sites (see Alerstam et al. 1992; Piersma et al. 1994; González et al. 1996). In addition, variations in prey quality (e.g., the amount of digestible flesh per unit ingested shell mass) determines the extent to which the birds face digestive rather than gross-intake bottlenecks (van Gils et al. 2003a).

FUTURE DIRECTIONS: DISENTANGLING ECOLOGICAL CAUSES AND PHYSIOLOGICAL EFFECTS

Red Knots must arrive on their High Arctic breeding grounds at the right time and with the right remaining energy and nutrient stores. These stores appear to be used for survival (note that late springs have led to major mortality [Boyd 1992]) rather than the actual production of eggs (Klaassen et al. 2001), although the calcium for eggshell *is* strategically stored (Piersma et al. 1996b). In fitness terms, everything else in the annual cycle may be subservient to arrival timing (Drent et al. 2003). Because the staging times are shortest and the fueling rates are highest at the last stopover sites before the High Arctic, there appears to be rather little "slack" time at late stages in the migration. Does this indicate that the time-energy budgets are also tight earlier in the year (restricting the departure times from the south)? Or does it reflect the presence of late-opening "time windows of opportunity" when food is abundant (Myers 1986) and/or dangerous predators (especially for fully fueled birds) are relatively scarce (see Ydenberg et al. 2002)? Or do the patterns reflect simple fitness optimization rules based on assumptions on the relative "fueling qualities" of the different sites (Weber et al. 1998)?

To determine whether changing food abundance at staging sites affects fueling performance and even the subsequent use of stopover sites, we need large-scale experiments. Actually, the time seems ripe to take scientific advantage of unfortunate human-induced alterations to the staging sites of Red Knots, the usual story of overfishing and the concomitant loss of marine resources (e.g., Piersma and Baker 2000; Piersma et al. 2001), by monitoring both food abundance and bird responses. In addition, we may be able to determine which fueling rates are preferred by individual Red Knots at particular sites and times of the year by experimentally measuring rates of food intake and mass gain under captive conditions with an unlimited high-quality food supply (as in the study of Kvist and Lindström 2003). Such a combination of experimental field and laboratory studies in the context of the descriptive framework set up here may for the first time enable us to disentangle the ecological and physiological factors that determine fueling rates of shore- and other birds, assess the patterns in a theoretical context (Alerstam and Hedenström 1998), and determine how ecology and physiology combine to affect the fitness of long-distance migrants. For example, how and to what extent is physiological flexibility used to buffer ecological variance?

APPENDIX: DETAILS OF DATA SOURCES AND ANALYSES

WADDEN SEA, GERMANY. Data for both the subspecies *islandica* and *canutus* were collected from 1978 to 1987 (usually based on large cannon-net catches) and summarized by Prokosch (1988).

WASH, UK. Based on a regression on date of 525 data points for adult Red Knots captured within 30 days of the departure date, having excluded individuals with little evidence of molt into an alternate plumage (that were always lightweight). Data were made available from the period March–May 1969–1991 by the Wash Wader Ringing Group (WWRG and J. Clark et al., unpubl. data; Y. Verkuil, pers. comm.).

BALSFJORD, NORWAY. Data obtained on the basis of a series of cannon-net catches in May 1985 that were summarized by Davidson and Evans (1986).

ICELAND. All date-specific average body mass values obtained between 2 and 30 May (in 1970, 1971, and 1972 and the mid-1980s) and published in Gudmundsson et al. (1991:fig. 4) and Piersma et al. (1999b:table 1) were used as a basis for a linear regression. See also Wilson and Morrison (1992).

SOUTH AFRICA. Data were obtained between 1970 and 1977 (mostly based on mist-net catches) and were summarized for monthly or biweekly intervals by Summers and Waltner (1978:fig. 4).

GUINEA-BISSAU. Based on the original data collected by WIWO-volunteers and associates in 1993 (see Wolff 1998). We regressed body mass values for adults on time of year (excluding individuals not molting into an alternate plumage) and used observations of departing flocks to calculate average date of departure.

BANC D'ARGUIN, MAURITANIA. Based on the original data collected by WIWO-volunteers and associates in 1985, 1986, and 1988 (see Zwarts et al. 1990b). We regressed body mass values for adults on time of year (excluding late birds with very little alternate plumage and aberrantly low body masses) and used observations of departing flocks (Piersma et al. 1990a) to obtain an estimate of the average date of departure.

NW AUSTRALIA. Based on 21 body mass values of adult birds captured with cannon-nets in Roebuck Bay near Broome in different years at least 6 days apart (data collected between 1985 and 2001). Data were collected by the Australasian Wader Study Group.

NEW ZEALAND. Based on a comparison of average masses from cannon-net data collected between 1987 and 1993 summarized by Battley (1999).

VICTORIA, AUSTRALIA. Based on a comparison of average masses in (different season) cannon-net catches on 21 February and 22 March made near Melbourne (data collected between 1987 and 1999). Data were collected by the Victorian Wader Study Group.

GOLFO SAN MATIAS, ARGENTINA. Linear regression of individual data points based on a series of five cannon-net catches of a total of 890 adult birds near the town of San Antonio Oeste, Rio Negro, in March 1998.

LAGOA DO PEIXE, RIO GRANDE DO SUL, BRASIL. Based on body mass values of Red Knots mist-netted by teams of CEMAVE near Lagoa do Peixe in April 1997, 1999, 2001, only using the data from the latter half of the month when the population shows a steep (synchronous) rate of refueling.

DELAWARE BAY, USA. Based on data collected during 63 cannon-net catches by an international consortium from 1997 to 2001 in New Jersey as well as Delaware (Niles et al., unpubl. data). A linear regression on date of average body mass values per catch-day between 14 and 30 May was used to estimate the overall rate of refueling.

ACKNOWLEDGMENTS

Red Knots in all parts of the world have attracted many marvelous, dedicated field workers working in both amateur and professional capacity. Even if this chapter shows that we need further research efforts, it is a tribute to their diligence. We sincerely thank all helpers in the field and the funding agencies that over many years made the work possible. The first author thanks Pete Marra, Russ Greenberg, and the Smithsonian Institution for enabling this contribution, and the Netherlands Organization for Scientific Research (NWO) for funding shorebird and benthic research over an extended period of time with a PIONIER grant. Similarly, work in the Western Hemisphere on *rufa* knots has been funded on a continuing basis by the Royal Ontario Museum Foundation and the Natural Sciences and Engineering Research Council of Canada (NSERC) to AJB. We thank Åke Lindström, Phil Battley, the editors, and two anonymous referees for comments on drafts, Dick Visser for drawing the final figures, and Jan van de Kam for use of his photographs.

LITERATURE CITED

Alerstam, T. 1990. Bird Migration. Cambridge University Press, Cambridge.

Alerstam, T., G. A. Gudmundsson, and K. Johanneson. 1992. Resources for long distance migration: intertidal exploitation of *Littorina* and *Mytilus* by Knots *Calidris canutus* in Iceland. Oikos 65:179–189.

Alerstam, T., and A. Hedenström. 1998. The development of bird migration theory. Journal of Avian Biology 29:343–369.

Alerstam, T., and Å. Lindström. 1990. Optimal bird migration: the relative importance of time, energy and safety. Pages 331–351 *in* Bird Migration: Physiology and Ecophysiology (E. Gwinner, ed.). Springer-Verlag, Berlin.

Baker, A. J., and H. D. Marshall. 1997. Mitochondrial control region sequences as tools for understanding evolution. Pages 51–82 *in* Avian Molecular Evolution and Systematics (D. P. Mindell, ed.). Academic Press, New York.

Baker, A. J., T. Piersma, and A. D. Greenslade. 1999. Molecular versus phenotypic sexing in Red Knots. Condor 101:887–893.

Baker, A. J., T. Piersma, and L. Rosenmeier. 1994. Unraveling the intraspecific phylogeography of Knots *Calidris canutus:* a progress report on the search for genetic markers. Journal für Ornithologie 135:599–608.

Battley, P. F. 1999. Seasonal mass changes of Lesser Knots (*Calidris canutus*) in New Zealand. Notornis 46:143–153.

Battley, P. F., and T. Piersma. 2004. Adaptive interplay between feeding ecology and features of the digestive tract in birds. *In* Physiological and ecological adaptations to feeding in vertebrates (J. M. Starck and T. Wang, eds.). Science Press, Enfield, Conn.

Battley, P. F., T. Piersma, M. W. Dietz, S. Tang, A. Dekinga, and K. Hulsman. 2000. Empirical evidence for differential organ reductions during trans-oceanic bird flight. Proceedings of the Royal Society of London, Series B, Biological Sciences 267:191–196.

Battley, P. F., D. I. Rogers, T. Piersma, and A. Koolhaas. 2003. Behavioural evidence for heat load problems in shorebirds in tropical Australia fuelling for a northward flight of 5,500 km. Emu 103:97–103.

Boyd, H. 1992. Arctic summer conditions and British Knot numbers: an exploratory analysis. Wader Study Group Bulletin 64 (Supplement):144–152.

Bruinzeel, L.W., and T. Piersma. 1998. Cost reduction in the cold: heat generated by terrestrial locomotion partly substitutes for thermoregulation costs in Knot *Calidris canutus*. Ibis 140:323–328.

Davidson, N. C., and P. R. Evans (eds.). 1986. The ecology of migrant Knots in North Norway during May 1985. Report SRG86/1, Department of Zoology, University of Durham.

Dick, W. J. A., T. Piersma, and P. Prokosch. 1987. Spring migration of the Siberian Knots *Calidris canutus canutus:* results of a co-operative Wader Study Group project. Ornis Scandinavica 18:5–16.

Drent, R., C. Both, M. Green, J. Madsen, and T. Piersma. 2003. Pay-offs and penalties of competing migratory schedules. Oikos 101:274–292.

Ens, B. J., T. Piersma, and R. Drent. 1994. The dependence of waders and waterfowl migrating along the East Atlantic Flyway on their coastal food supplies: what is the most profitable research programme? Ophelia 6 (Supplement):127–151.

Gonzalez, P. M., T. Piersma, and Y. Verkuil. 1996. Food, feeding and refuelling of Red Knots during northward migration at San Antonio Oeste, Rio Negro, Argentina. Journal of Field Ornithology 67:575–591.

Green, M. 2003. Flight Strategies in Migrating Birds: When and How to Fly. Ph.D. dissertation, Lund University, Lund.

Gudmundsson, G. A., Å. Lindström, and T. Alerstam. 1991. Optimal fat loads and long-distance flights by migrating Knots *Calidris canutus,* Sanderlings *C. alba* and Turnstones *Arenaria interpres.* Ibis 133:140–152.

Harrington, B. A. 1996. The Flight of the Red Knot. Norton, New York.

Jenni, L., and S. Jenni-Eiermann. 1998. Fuel supply and metabolic constraints in migrating birds. Journal of Avian Biology 29:521–528.

Klaassen, M., Å. Lindström, H. Meltofte, and T. Piersma. 2001. Arctic waders are not capital breeders. Nature 413:794.

Kvist, A., and Å. Lindström. 2000. Maximum daily energy intake: it takes time to lift the metabolic ceiling. Physiological and Biochemical Zoology 73:30–36.

Kvist, A., and Å. Lindström. 2003. Gluttony in migratory waders: unprecedented energy assimilation rates in vertebrates. Oikos 101:397–402.

Kvist, A., Å. Lindström, M. Green, T. Piersma, and G. H. Visser. 2001. Carrying large fuel loads during sustained bird flight is cheaper than expected. Nature 413:730–732.

Lindström, Å. 1991. Maximum fat deposition rates in migrating birds. Ornis Scandinavica 22:12–19.

Lindström, Å. 1995. Stopover ecology of migrating birds: some unsolved questions. Israel Journal of Zoology 41:407–416.

Lindström, Å., and T. Alerstam. 1992. Optimal fat loads in migrating birds: a test of the time-minimization hypothesis. American Naturalist 140:477–491.

Lindström, Å., and T. Piersma. 1993. Mass changes in migrating birds: the evidence for fat and protein storage re-examined. Ibis 135:70–78.

Meltofte, H. 1985. Populations and breeding schedules of waders, Charadrii, in high arctic Greenland. Meddelelser om Grønland, Bioscience 16:1–43.

Morrison, R. I. G., and N. C. Davidson. 1990. Migration, body condition and behaviour of shorebirds during spring migration at Alert, Ellesmere Island, N.W.T. Pages 544–567 *in* Canada's Missing Dimension (C. R. Harington, ed.). Science and History on the Canadian Arctic Islands. Canadian Museum of Nature, Ottawa.

Myers, J. P. 1986. Sex and gluttony on Delaware Bay. Natural History 95:68–77.

Nebel, S., T. Piersma, J. van Gils, A. Dekinga, and B. Spaans. 2000. Length of stopover, fuel storage and a sex-bias in the occurrence of Red Knots *Calidris c. canutus* and *C. c. islandica* in the Wadden Sea during southward migration. Ardea 88:165–176.

Piersma, T. 1987. Hop, skip or jump? Constraints on migration of arctic waders by feeding, fattening, and flight speed. Limosa 60:185–194.

Piersma, T. 1997. Do global patterns of habitat use and migration strategies co-evolve with relative investments in immunocompetence due to spatial variation in parasite pressure? Oikos 80:623–631.

Piersma, T. 1998. Phenotypic flexibility during migration: optimization of organ size contingent on the risks and rewards of fueling and flight? Journal of Avian Biology 29:511–520.

Piersma, T. 2002. Energetic bottlenecks and other design constraints in avian annual cycles. Integrative and Comparative Biology 42:51–67.

Piersma, T. 2003. 'Coastal' versus 'inland' shorebird species: interlinked fundamental dichotomies between their life- and demographic histories? Wader Study Group Bulletin 100:5–9.

Piersma, T., and A. J. Baker. 2000. Life history characteristics and the conservation of migratory shorebirds. Pages 105–124 *in* Behaviour and Conservation (L. M. Gosling and W. J. Sutherland, eds.). Cambridge University Press, Cambridge.

Piersma, T., and N. C. Davidson. 1992. The migrations and annual cycles of five subspecies of Knots in perspective. Wader Study Group Bulletin 64 (Supplement):187–197.

Piersma, T., M. W. Dietz, A. Dekinga, S. Nebel, J. van Gils, P. F. Battley, and B. Spaans. 1999a. Reversible size-changes in stomachs of shorebirds: when, to what extent, and why? Acta Ornithologica 34:175–181.

Piersma, T., R. Drent, and P. Wiersma. 1991. Temperate versus tropical wintering in the world's northernmost breeder, the Knot: metabolic scope and resource levels restrict subspecific options. Pages 761–772 *in* Proceedings of the 20th International Ornithological Congress, Christchurch, Vol. 2.

Piersma, T., P. de Goeij, and I. Tulp. 1993a. An evaluation of intertidal feeding habitats from a shorebird perspective: towards comparisons between temperate and tropical mudflats. Netherlands Journal of Sea Research 31:503512.

Piersma, T., G. A. Gudmundsson, N. C. Davidson, and R. I. G. Morrison. 1996a. Do Arctic-breeding Red Knots (*Calidris canutus*) accumulate skeletal calcium before egg laying? Canadian Journal of Zoology 74:2257–2261.

Piersma, T., G. A. Gudmundsson, and K. Lilliendahl. 1999b. Rapid changes in the size of different functional organ and muscle groups during refueling in a long-distance migrating shorebird. Physiological and Biochemical Zoology 72:405–415.

Piersma, T., R. Hoekstra, A. Dekinga, A. Koolhaas, P. Wolf, P. F. Battley, and P. Wiersma. 1993b. Scale and intensity of intertidal habitat use by Knots *Calidris canutus* in the western Wadden Sea in relation to food, friends and foes. Netherlands Journal of Sea Research 31:331–357.

Piersma, T., and J. Jukema. 2002. Contrast in adaptive mass gains: Eurasian Golden Plovers store fat before midwinter and protein before prebreeding flight. Proceedings of the Royal Society of London, Series B, Biological Sciences 269:1101–1105.

Piersma, T., M. Klaassen, J. H. Bruggemann, A.-M. Blomert, A. Gueye, Y. Ntiamoa-Baidu, and N. E. van Brederode. 1990a. Seasonal timing of the spring departure of waders from the Banc d'Arguin, Mauritania. Ardea 78:123–134.

Piersma, T., A. Koolhaas, and A. Dekinga. 1993c. Interactions between stomach structure and diet choice in shorebirds. Auk 110:552–564.

Piersma, T., A. Koolhaas, A. Dekinga, J. J. Beukema, R. Dekker, and K. Essink. 2001. Long-term indirect effects of mechanical cockle-dredging on intertidal bivalve stocks in the Wadden Sea. Journal of Applied Ecology 38:976–990.

Piersma, T., R. van Aelst, K. Kurk, H. Berkhoudt, and L. R. M. Maas. 1998. A new pressure sensory mechanism for prey detection in birds: the use of principles of seabed dynamics? Proceedings of the Royal Society of London, Series B, Biological Sciences 265:1377–1383.

Piersma, T., and S. van de Sant. 1992. Pattern and predictability of potential wind assistance for waders and geese migrating from West Africa and the Wadden Sea to Siberia. Ornis Svecica 2:55–66.

Piersma, T., J. van Gils, P. de Goeij, and J. van der Meer. 1995. Holling's functional response model as a tool to link the food-finding mechanism of a probing shorebird with its spatial distribution. Journal of Animal Ecology 64:493–504.

Piersma, T., J. van Gils, and P. Wiersma. 1996b. Family Scolopacidae (sandpipers, snipes and phalaropes). Pages 444–533 *in* Handbook of the Birds of the World, Vol. 3, Hoatzin to Auks (J. del Hoyo, A. Elliott, and J. Sargatal, eds.). Lynx Ediciones, Barcelona.

Piersma, T., Y. Verkuil, and I. Tulp. 1994. Resources for long-distance migration of Knots *Calidris canutus islandica* and *C. c. canutus:* how broad is the temporal exploitation window of benthic prey in the western and eastern Wadden Sea? Oikos 71:393–407.

Piersma, T., L. Zwarts, and J. H. Bruggemann. 1990b. Behavioural aspects of the departure of waders before long-distance flights: flocking, vocalizations, flight paths and diurnal timing. Ardea 78:157–184.

Prokosch, P. 1988. Das Schleswig-Holsteinische Wattenmeer als Frühjahrs-Aufenthaltsgebiet arktischer Watvogel-Populationen am Beispiel von Kiebitzregenpfeifer (*Pluvialis squatarola,* L. 1758), Knutt (*Calidris canutus,* L. 1758) und Pfuhlschnepfe (*Limosa lapponica,* L. 1758). Corax 12:273–442.

Rappole, J. H. 1995. The Ecology of Migrant Birds: A Neotropical Perspective. Smithsonian Institution Press, Washington, D.C.

Schekkerman, H., I. Tulp, T. Piersma, and G. H. Visser. 2003. Mechanisms promoting higher growth rate in arctic than in temperate shorebirds. Oecologia 123:332–342.

Summers, R. W., and M. Waltner. 1978. Seasonal variations in the mass of waders in southern Africa, with special reference to migration. Ostrich 50:21–37.

Tomkovich, P. S. 1992. An analysis of the geographic variability in Knots *Calidris canutus* based on museum skins. Wader Study Group Bulletin 64 (Supplement):17–23.

Tomkovich, P. S. 2001. A new subspecies of Red Knot *Calidris canutus* from the New Siberian islands. Bulletin British Ornithologists' Club 121:257–263.

Tsipoura, N., and J. Burger. 1999. Shorebird diet during spring migration stopover on Delaware Bay. Condor 101:635–644.

Tulp, I., H. Schekkerman, T. Piersma, J. Jukema, P. de Goeij, and J. van de Kam. 1998. Breeding waders at Cape Sterlegova, northern Taimyr, in 1994. WIWO-Report 61:1–87.

van Gils, J. A., P. Edelaar, G. Escudero, and T. Piersma. 2004. Carrying capacity models should not use fixed prey density thresholds: a plea for using more tools of behavioural ecology. Oikos 104:197–204.

van Gils, J. A., T. Piersma, A. Dekinga, and M. W. Dietz. 2003a. Cost-benefit analysis of mollusc-eating in a shorebird: Pt. 2. Optimizing gizzard size in the face of seasonal demands. Journal of Experimental Biology 206:3369–3380.

van Gils, J. A., I. W. Schenk, O. Bos, and T. Piersma. 2003b. Incompletely informed shorebirds (Red Knot, *Calidris canutus*) that face a digestive constraint maximize net energy gain when exploiting patches. American Naturalist 161:777–793.

Verboven, N., and T. Piersma. 1995. Is the evaporative water loss of Knot *Calidris canutus* higher in tropical than in temperate climates? Ibis 137:308–316.

Weber, T. P., B. J. Ens, and A. I. Houston. 1998. Optimal avian migration: a dynamic model of fuel stores and site use. Evolutionary Ecology 12:377–402.

Whitfield, D. P., and J. J. Brade. 1991. The breeding behaviour of the Knot *Calidris canutus.* Ibis 133:246–255.

Whitfield, D. P., and P. S. Tomkovich. 1996. Mating system and timing of breeding in Holarctic waders. Biological Journal of the Linnean Society 57:277–290.

Wiersma, P., and T. Piersma. 1994. Effects of microhabitat, flocking, climate and migratory goal on energy expenditure in the annual cycle of Red Knots. Condor 96:257–279.

Wiersma, P., and T. Piersma. 1995. Scoring abdominal profiles to characterize migratory cohorts of shorebirds: an example with Red Knots. Journal of Field Ornithology 66:88–98.

Wilson, J. R., and R. I. G. Morrison. 1992. Staging studies of Knots *Calidris canutus islandica* in Iceland in the early 1970s: body mass patterns. Wader Study Group Bulletin 64 (Supplement):17–23.

Wolff, W. J. (ed.). 1998. The end of the East-Atlantic Flyway, Waders in Guinea-Bissau: Report on a co-operative research project in the estuaries of Guinea-Bissau October 1992–May 1992. WIWO-Report 39:1—94.

Ydenberg, R. C., R. W. Butler, D. B. Lank, C. G. Guglielmo, M. Lemon, and N. Wolf. 2002. Trade-offs, condition dependence and stopover site selection by migrating sandpipers. Journal of Avian Biology 33:47–55.

Zwarts, L., and A.-M. Blomert. 1992.Why Knots *Calidris canutus* take medium-sized *Macoma balthica* when six prey species are available. Marine Ecology Progress Series 83:113–128.

Zwarts, L., A.-M. Blomert, and R. Hupkes. 1990a. Increase of feeding time in waders preparing their spring migration from the Banc d'Arguin, Mauritania. Ardea 78:237–256.

Zwarts, L., B. J. Ens, M. Kersten, and T. Piersma. 1990b. Moult, mass and flight range of waders ready to take off for long-distance migrations. Ardea 78:339–364.

22

WILLIAM W. COCHRAN
AND MARTIN WIKELSKI

Individual Migratory Tactics of New World *Catharus* Thrushes

Current Knowledge and Future Tracking Options from Space

RADIOTELEMETRY ENABLES US TO FOLLOW individual songbirds during migration while simultaneously conducting physiological measurements. In the future, radiotelemetry from space will provide the technology to study connectivity between breeding and wintering grounds in songbirds. We have studied *Catharus* thrushes migrating through the midwestern United States for about 35 years. Their migratory behavior may be sufficiently explained by six simple rules of thumb: (1) Stop over and forage until fat levels are above a score of 1 (body mass ≥ ~32 g); in stopover habitat, establish small (~100 m radius) foraging area. (2) Migrate at night when maximum daily air temperature is at least 21°C and wind at takeoff is less than 10 km/h, regardless of its direction. (3) Calibrate magnetic compass every clear afternoon/at dusk against sunset direction or overhead polarized sky light. (4) Keep constant magnetic heading during entire migration. Birds only deviate by preferentially flying toward lightning/thunderstorms, even if this sometimes means reversing migratory direction. (5) Fly until body mass is back to level at previous morning or until daybreak, whichever comes first, then select wooded stopover habitat with water (stream) to land. Flight durations range from 45 min to 8 h, and northward advances range from –10 to 850 km/night, depending on wind speed and direction. (6) Switch from northward migration to east or west search flights for breeding habitat once specific latitude is reached. Overland migration apparently poses no excessive energetic demand: Doubly labeled water and heart-rate measurements show that migratory flight costs are 15.5 kJ/h (= 4.3 watts); thus the entire northward migration of a Swainson's Thrush (*C. ustulatus*) costs about 4,500 kJ or 0.9

kJ/km (flight costs ~1,280 kJ). Wing beat frequency varies between 570 and 780 beats/min throughout flights, with wing beat pauses being more frequent toward the end of flights. The relationship between wing beat and respiration is largely constant at 4/1. Assuming these decision rules and physiological constraints, birds sometimes behave suboptimally and, for example, are displaced backward in strong head winds. Future studies may use traditional radiotelemetry optimized for signal longevity while tracking birds from space. Although location accuracy would be low (±20 km), benefits of following individual songbirds between continents to link breeding and wintering grounds outweigh such shortcomings.

INTRODUCTION

Much of what we know about songbird migration in the wild is derived from investigations collecting single samples or isolated data points of birds along their migratory path (Alerstam 1991a; Alerstam and Hedenstöm 1998; Berthold 1998, but see Marra et al. 1998). For example, we have some confidence in understanding stopover decisions (Moore and Simons 1992), migratory refueling (Bairlein and Gwinner 1994; Biebach 1998), or general flight behavior from radar observations (Bruderer et al. 1999; Bruderer and Boldt 2001). And what we know about flight physiology comes from isolated captures either en route (Jenni and Jenni-Eiermann 1998) or immediately after landing (Piersma and Lindström 1997). Orientation mechanisms (Alerstam 1991; Able 1996; Wiltschko and Wiltschko 1999) and physiological capabilities of songbirds are also largely known from laboratory or wind tunnel studies (Klaassen et al. 1997; Lindström et al. 1999), but not from the wild.

In the wild, migratory directions and routes of songbirds are primarily known from band recoveries and are therefore available for only a handful of species (Berthold 1993; see also Payne and Payne 1990). Thus we understand the general pattern of songbird migration on a population level, but we know little about individual behavior and decision rules during migration along the entire path of an individual migrant. However, we need to connect what individual songbirds do between breeding and wintering grounds (Marra et al. 1998) to understand where migratory bottlenecks exist (Sillett and Holmes 2002), be they physiological, morphological, ecological, or genetic (Pulido et al. 2001). In particular, devising measures for international songbird conservation (Greenberg et al. 2000) ultimately requires precise quantification of the mortality risks for songbirds along their entire migration route (Rubenstein et al. 2002; Webster et al. 2002).

One of the authors (WWC) invented methods to study migratory behavior in individual birds more than 40 years ago (Lord et al. 1962) and has used and refined this technology until the present. Here we use data on New World thrushes (*Catharus* spp.) to illustrate what can be learned from following individual birds during their migration and to highlight what more could be learned if the present technology were applied to study connectivity between breeding and wintering grounds in songbirds via the use of radiotelemetry monitoring devices in orbit, such as on the International Space Station (ISS). Technology already exists to study such parameters in large (> ~500 g) birds and has revealed spectacular insights (Weimerskirch et al. 1993; Butler et al. 2000). Here we propose a system to track the approximate location of small songbirds on a global scale from space.

To achieve a comprehensive picture of individual migratory decisions of *Catharus* thrushes during their overland migration in the midwestern United States, we summarize data sets gathered over the last 35 years, partly published but largely unpublished, as well as present new data and experiments. Our intention in putting this data set together was to present an overview of our current knowledge of the spring migration of one group of species, as well as to offer an outlook for future research directions, but not necessarily to show all the original data.

METHODS

Studies on *Catharus* Thrushes in the Midwestern United States

Catharus thrushes (mostly Swainson's [*C. ustulatus*] and Hermit [*C. guttatus*]) were caught in mist-nets at the University of Illinois South Farm and adjacent woodlots in Urbana (40°04.82′ N, 88°12.7′ W), central Illinois during the spring migratory period (early April to early June [Graber et al. 1971]). This woodlot measures about 200 × 400 m and has several footpaths across it. It consists of planted deciduous and some evergreen trees with extensive underbrush. The age of the trees in the wood lot is now about 50 years. Upon capture, birds were weighed and measured (wing, tarsus) and the fat score determined (Wingfield et al. 1996). Most birds were fitted with radio transmitters directly, released, and followed (Cochran 1972). Some birds were transported to the laboratory in cloth bags and held briefly, or overnight in 40 × 40 × 50 cm cages and fed with crickets, mealworms, blueberries, and water ad libitum. Birds regularly took food during experimental capture periods and were generally heavier at release than at capture.

Heart Rate Transmitters

We used heart rate transmitters by Sparrow Systems (Dewey, Illinois; Model SP2000-HR). The transmitter weighed 0.7–1 g, measured 0.2 × 0.7 × 0.3 cm, and used a wire antenna that was 9 cm long and 0.2 mm in diameter. It emitted a continuous signal amplitude modulated (AM) by a ca. 1,800-Hz subcarrier oscillator that was frequency modulated (FM) by heart muscle potentials. The transmitters we used were electrically similar to the design of Morrow and Taylor (1976), but were much smaller and gave better life and range owing to the use of modern components. We used the regular radio-tracking transmitter attachment developed by Raim (1978) but modified it slightly for heart rate transmitters. In short, the transmitter is affixed with threads and/or a long bead of rubber cement to a rectangle of thin cloth (1.8 × 1 cm) in such a way that the transmitter adds little rigidity to the cloth across its short dimension. This flexible package is then glued to the back of the bird near its center of gravity between the wings with eyelash adhesive. To prepare the area for the cloth we cut the back body plumage over an area the size of the cloth, leaving feather stumps of about 1.5 mm to enhance adherence of the cloth and adhesive. To attach heart rate transmitters, birds were anesthetized with a mixture of Isoflurane and air, in order to keep their distress at a minimum and not cause pain. After preparing the feathers for attachment as described above, we made a 1-mm incision in the skin centered on the area the cloth was to be glued. The 0.1 mm diameter enameled soft copper electrode wires were threaded through the cloth at a place that would be at or near the incision after attachment. We inserted the conducting end of a wire electrode into the hollow end of a 27-gauge injection needle, which was filed blunt and sterilized before use. The needle was carefully worked under the skin until its tip was at the desired electrode position, where finger pressure held the wire in place while the needle was withdrawn. The other electrode was inserted in the same manner. After the electrode wires were in place, the incision was covered with surgical glue. Then the small amount of slack electrode wire was folded under the cloth, the eyelash adhesive applied, and the cloth pressed against the back covering the incision. The transmitter was then held in place for about 5 min to allow time for the adhesive to set. We used about a 3-cm separation between electrode wire ends. The anterior electrode was situated close to the neck of the bird and the posterior electrode between the rump and the center of gravity. Although all our electrode placements, intentionally posterior-anterior aligned, provided good heart rate (HR) signals, the wing beat muscle signals varied. We suspect that inadvertent lateral displacement from the posterior-anterior line provided stronger muscle signals. Thus, if wing beat and respiration are desired, an intentional lateral offset may be useful. When birds naturally lost a transmitter, which happened in two cases after several days of attachment, the electrode wires slid out from under the skin; no irritation was observed when these birds were recaptured (1 day later).

Upon release, two of the experimental birds were video recorded to determine wing beat frequency visually while recording heart rate simultaneously via the transmitter signal. We analyzed wing beat frequency using frame-by-frame analysis of flight videos. During such short flights, wing beat frequency (~600–780 beats/min) was considerably slower than heart rate (~840 beats/min) in all cases and showed conspicuous pauses, which were not apparent in the heart rate recording. Thus, we are confident that our recording was indeed heart rate and was not influenced by flight muscle signals (the latter may have happened in a study by Eliassen [1963]).

Data Collection and Tracking

We used Falcon V (as designed by WWC) or AR 3000 receivers (AOR Ltd., Tokyo, Japan) in AM mode to recover the ~1,800-Hz tone; the FM-modulated tone was recorded with a Sony TD7 (Sony Inc., Tokyo, Japan) digital tape recorder. The recorded tone was converted to computer 60-sec wave files (8-bit, 48,000 samples/sec) via a PC sound card and Cool Edit 2000 software (Syntrillium Software Corp., Phoenix). Cool Edit also provided band-pass filtering to improve the signal–to-noise ratio, which was useful for weak signals. Heart beat, wing beat, and respiration frequencies show clearly in the power spectrum as separate peaks and were all recorded. Only wing beat frequency and respiration were harmonically related, confirming a separate rhythm for the heart rate. We used a whip antenna on the receiver for laboratory measurements and a six-element Yagi antenna on a tracking vehicle to gather data during migration, as described earlier (Cochran 1972; see also below for determination of heading of the bird during migration).

Metabolic Measurements

We determined the birds' energy expenditure by measuring oxygen consumption and carbon dioxide production in an open-flow, push-through respirometry system (Withers 1977) while simultaneously recording heart rate. Six birds were measured in the laboratory and their heart rates were increased by cooling the ambient air around them. All animal handling methods were approved by the University of Illinois or the Princeton University Animal Care Committee and adhere to the standards set for the use of wild birds in research. We also conducted measurements on the daily energy expenditure of birds in the wild using the doubly labeled water (DLW) method (Masman and Klaassen 1987; Nagy et al. 1999). For this we injected 0.1 ml of deuterium- and oxygen^{18}-enriched water intraperitoneally in birds captured during the morning hours and released the birds. At dusk, we waited with two trucks equipped with mobile radio-receiving equipment for the birds' takeoff. If the birds took off for a migratory flight, we followed their migration route until they landed in a new stopover habitat (six flights). During the next morning, all DLW-injected birds, regardless of whether they had migrated or not, were relocated and recaptured via mist-netting (100% successful for all birds whose stopover site was located; six of 13 birds that migrated were recaptured as well as 26 birds that did not migrate [Wikelski et al. 2003]). Our flight data from DLW-injected birds were not different from previously published data on unmanipulated birds (Kjos and Cochran 1970; Cochran and Kjos 1985; Cochran 1972, 1987). However, for obvious reasons we do not have control data for birds that were not fitted with transmitters.

Orientation Experiments

Eight Swainson's Thrushes, netted in the early evening during spring migrations in 1978, 1979, and 1984, were held in cages with a full view of the sky from about sunset until the end of nautical twilight when they were released. Only one bird was followed per night. While in the cages birds were subjected to a magnetic field with the horizontal component shifted 70° clockwise (toward the east), but of the same magnitude as the natural magnetic field; the vertical component was not altered. Because of the wide spread of natural headings for this species (fig. 1B), a flight on the night of release and a subsequent control flight begun under natural conditions were required; the difference or lack of difference in the headings between the experimental and control flight allowed testing three mutually exclusive propositions: (1) If the magnetic field experienced by the thrush during evening twilight had no effect on migratory heading, the heading flown on the second flight would be the same as that of the release flight. (2) If the magnetic field was used to calibrate the celestial background, the heading of the second control flight would be 70° counterclockwise compared with the experimental flight. (3) However, if the bird's magnetic compass was calibrated by solar cues either directly or by twilight polarized skylight (as suggested by Cochran 1987), the second control flight would be shifted 70° clockwise compared to the experimental flight immediately after treatment in the Helmholtz coils. The necessary condition that Swainson's Thrushes normally fly the same heading from night to night is documented (Cochran and Kjos 1985; Cochran 1987) (fig. 2B). In this experiment the heading of the bird was important; the path, wind-drifted or not, was not important because *Catharus* thrushes exhibit little or no compensation for drift by lateral winds (Cochran and Kjos 1985).

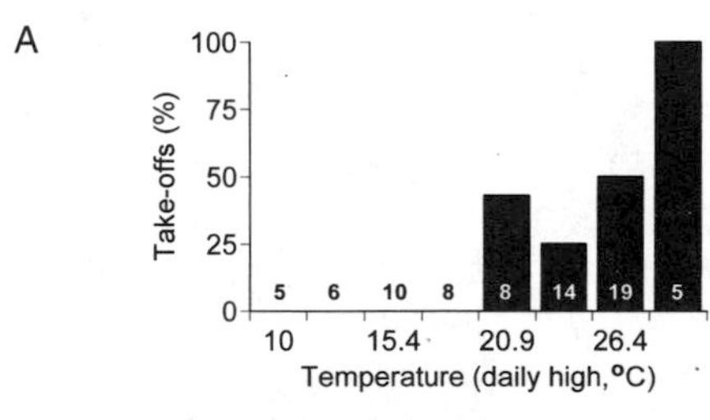

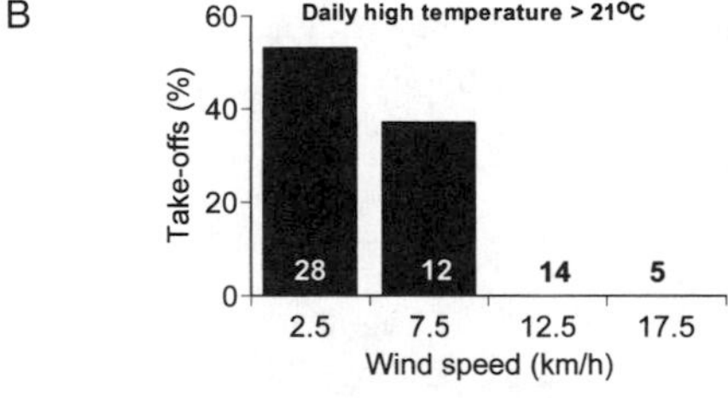

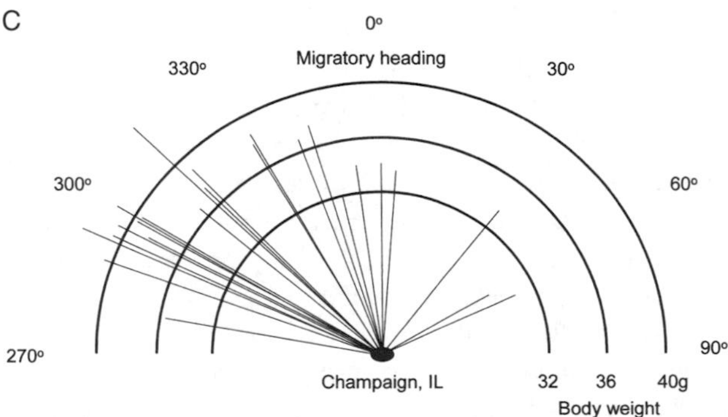

Fig. 22.1. Migration behavior of Swainson's Thrushes in central Illinois. Birds take off on migratory flights when (A) the maximum daily air temperatures exceed ca. 20°C and when (B) the wind speed on warm days is less than 10 km/h. (C) The heading of migratory birds is generally northerly, but heavier birds have more northwesterly, lighter birds more northerly or northeasterly, headings. Data from 1965/1966 and 1999/2000 were combined for all figures.

RESULTS AND SYNOPSIS

Stopover and Migration Behavior

Swainson's and Hermit Thrushes captured during stopover at the University of Illinois South Farm woodlot weighed between 27 and 42 g. Subcutaneous fat was not visible in birds that weighed less than 30 g. Subsequent recaptures of either banded birds or those that were fitted with radio transmitters showed that none of the birds that weighed less than 30 g ever left the woodlot and migrated (all light-weight birds were recaptured or resighted).

We carefully followed 18 birds that were fitted with radio transmitters during their daily foraging trips through the woodlot. These birds foraged within areas of about 100 m diameter. Within their foraging areas, birds leisurely advanced while picking insects or berries from the ground or low branches. Birds were usually active for several minutes, then sat for similar amounts of time (see below). When individually known birds had migrated and settled again in their new stopover habitat, they again established more or less circular foraging areas similar in size to the ones they had had in the prior stopover habitat (the Southfarm woodlot). Such stereotypic behaviors enabled us to recapture individuals after a nocturnal migration flight in a new stopover habitat. Mist-nets were set up at a place where the bird had passed during one of the previous circles and we then followed the bird during one of the subsequent rounds to herd it into the row of mist-nets.

With very few exceptions, only birds that had at least a fat score of 1 during the morning capture hours (0600–1000) migrated at night, weather conditions permitting (see below). The body mass of migrating birds (32.0 ± 1.0 g before, 30.12 ± 0.9 g after migration) significantly decreased by 6% (–1.9 g), whereas that of nonmigrating birds (31.3 ± 0.4 g at capture, 30.4 ± 0.4 g at recapture) did not change significantly (–0.9 g). Fat scores did not change in migrating (1.3 ± 0.3 before, 1.3 ± 0.2 after migration) or nonmigrating birds

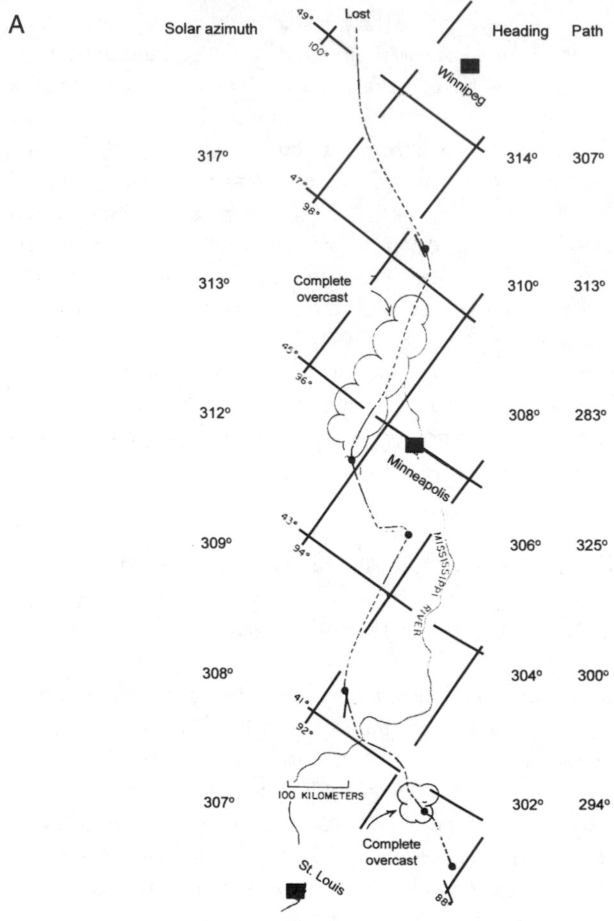

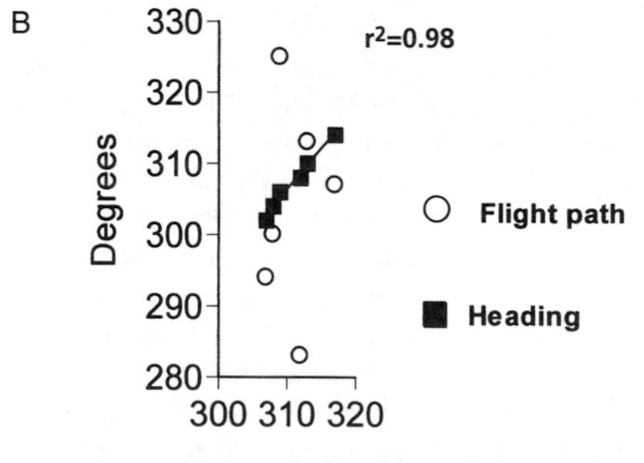

Fig. 22.2. Migration behavior of a single Swainson's Thrush between 13 and 20 May 1973, as followed for 1,512 km (Cochran 1987). The flight path is indicated by lines (solid line: heading information available; dotted line: no heading information). The six stopover sites are depicted by circles along the path. The bird kept a constant heading throughout its flight even when flying under completely overcast skies (nights 2 and 5) and landed when it hit cold fronts, such as during 13/14 May (first night) and 15/16 May (third night). The heading of the bird (second right column) always showed a constant relationship toward the solar azimuth (left column), whereas the realized flight path (right column) showed no relationship to the solar azimuth. The relationship between solar azimuth, heading, and flight path is further depicted in (B), showing an almost complete congruence of the bird's heading and the solar azimuth (heading was always 3°–4° less than the solar azimuth). Note that solar azimuth changes with latitude, as did the bird's heading.

(1.27 ± 0.23 capture, 1.1 ± 0.16 recapture; data from Wikelski et al. 2003). Migrating *Catharus* thrushes ($n = 12$) flew at average ground speeds of 54 ± 5 km/h (range: 26–91 km/h), indicating the influence of tailwinds when compared with their less variable air speeds in still air (~35 km/h [Cochran and Kjos 1985]).

Environmental Conditions Conducive for Takeoff

The takeoff decisions of 47 Hermit and Swainson's thrushes fitted with radio transmitters were observed over the period of 30 years and linked to weather conditions (Cochran and Wikelski, in prep.). We found that two environmental factors were associated with more than 95% of all takeoff decisions: birds took off when (1) the maximum daily shaded air temperature was above 21°C and (2) the surface wind speed at the time of takeoff (usually around 2000–2400) was lower than 10 km/h (fig. 22.1A,B). Interestingly, neither the surface wind direction during the takeoff period nor the surface wind direction or speed during the day influenced the decision to take off (Cochran and Wikelski, in prep.). Thus, birds did not fly in higher winds, even if those were coming straight from the south. However, the winds at flight altitude (400–1,500 m above ground) are usually much stronger and often very different in direction from the surface winds during takeoff. Furthermore, wind direction and speed aloft often change dramatically as a bird changes altitude or as it encounters different weather along its path (Cochran and Kjos 1985). Thus, surface wind speed influences the decision to take off, and weather can strongly influence flight path (see below).

Orientation and Other Behaviors during Migration Flight

The heading of a bird, determined by the relative signal strength of the radio transmitter at various angles from the migrating birds, generally appeared to remain the same during an entire nocturnal migratory flight (Cochran and Kjos 1985). The heading of Swainson's Thrushes migrating through central Illinois depended upon their body weight, with heavier birds showing a more westerly heading compared with lighter birds (fig. 22.1C). Such differences in heading may indicate that heavier birds are on their way toward more distant areas in western Canada and Alaska, whereas lighter birds are heading toward nearer breeding grounds north of the Great Lakes, and even lighter birds head toward breeding grounds in northeastern United States and Canada.

We found a remarkable consistency in migratory headings within flights and among individuals. For example, a bird that was followed for seven nocturnal flights during its northward migration did not change its heading relative to sunset direction (Cochran 1987). As a consequence of a constant heading, therefore, speed and direction of winds aloft strongly influenced flight paths. A graphical example is depicted in fig. 22.2A (redrawn from Cochran 1987). The

Swainson's Thrush started in Champaign, Illinois, on 13 May and was lost west of Winnipeg, Canada, on 20 May. The solar azimuth during seven migratory flights increased from 307° to 317°, similar to the bird's heading, which increased from 302° to 314° (fig. 22.2B). The flight path, on the other hand, varied between 283° and 325°, depending on weather, and showed no systematic change during the migration.

The only instances when birds changed and even reversed migratory heading was to fly toward thunderstorms indicated by lightning flashes visible from cloud reflections at distances of up to 100 km. Upon reaching the area with rain, birds landed within about half an hour. Heading changes by more than 90° were sometimes required to reach an area of rain. One Swainson's Thrush completely reversed its migration route in fall and instead of flying south flew about 95 km north, thereby leaving a dry area (Champaign, Illinois) and landing in heavy rain south of Kankakee, Illinois. Lightning flashes to the north were visible from the takeoff site, which never received rain because of an easterly storm path. We suggest that birds alter migratory direction to get to a site that will be moist on landing. We consider it unlikely that songbirds try to get into an area of updrafts for greater lift, a tactic that could be important for soaring birds.

Although many flights continued until dawn, many also ended with landing after a few hours of flight. Some of these predawn landings remain unexplained on any environmental basis. However, most early landings occurred when birds encountered a cold front whereupon they landed within minutes (fig. 22.2B). Even weak fronts (temperature drop <2°C and wind shear negligible) elicited the landing response. The absolute temperature during migration apparently did not matter: Birds often migrated all night with gradually decreasing air temperatures in the 4°–10°C range but landed when encountering a sudden drop as small as from 12.7° to 11°C.

Orientation Mechanisms for Nocturnal Flights

Of the eight Swainson's Thrushes held in and released from cages with a 70° clockwise shifted magnetic field, five made migratory flights but only two of these were tracked on subsequent flights (1 and 4 days later). The headings on the first (experimental) fight of these birds were 63° and 78° counterclockwise (more westerly) from the headings later taken on flights initiated under natural conditions. These results are consistent with migratory orientation by a solar-calibrated magnetic compass (Cochran 1987) during the major distance-covering portion of migration. Four additional Swainson's Thrushes were subjected to the same treatment in 2003 and also shifted their control flight counterclockwise, confirming previous results (Cochran et al. 2004).

Physiological Measurements during Natural Migration

The 24-h energy expenditure of migrating birds based on DLW measurements is best described by a linear regression with increasing energetic costs according to flight duration (fig. 22.3; migration energy expenditure = 70 ± 9 kJ + 12.5 ± 2.1 (hour); $r^2 = 0.89$, $p = 0.004$ [after Wikelski et al. 2003]). The daily energy expenditure of birds that did not migrate was on average 88 ± 5 kJ/day. The daily energy expenditure of birds that migrated for only relatively short periods of time (less than about 3 h) was not significantly elevated over that of birds that did not migrate (fig. 22.3).

Advanced radio transmitter and signal processing techniques will allow us to better understand the flight behaviors and physiological constraints during natural migration. Although we do not want to give a formal analysis of physiological parameters here, we do want to outline the type of additional information that can be gained by using such techniques. Using an EKG radio transmitter, we recorded heart rate, respiration, and wing beat frequency from several Swainson's Thrushes during migration (an example is shown in fig. 22.4A). The bird had an average nocturnal heart rate of about 612 beats/min before migration, with occasional increases up to 720 beats/min. Such heart rate peaks could resemble metabolic arousals as suggested during laboratory respirometry measurements by Buttemer et al. (1991). When the bird took off, the heart rate immediately increased to 840 beats/min, gradually slowing as the bird ascended to cruising altitude. During the landing phase, when the bird was apparently searching for suitable foraging habitat, heart rate increased again dramatically (to 840 beats/min). After landing, average heart rate slowly but steadily decreased, presumably also reflecting the increase in ambient temperature in the early morning hours after sunrise. The bird started to forage at about 0600 and had several approximately 10-min-long bouts of intensive foraging activity, interspersed with resting phases (as confirmed by focal observation). During migratory flight, wing

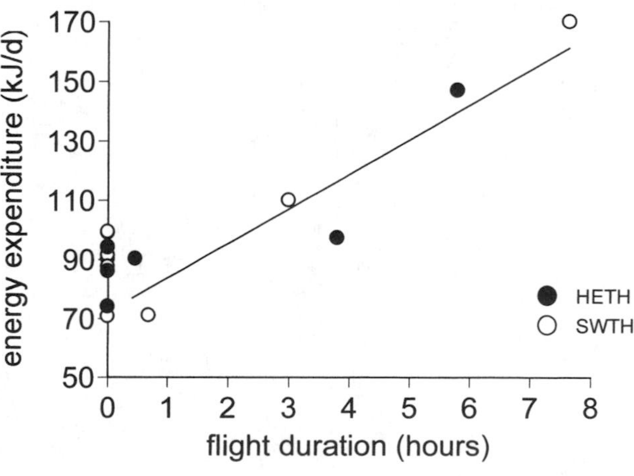

Fig. 22.3. Daily energy expenditure of *Catharus* thrushes during the migratory period, as measured with the doubly labeled water method (DLW). Data are standardized for a 24-h period. Birds that did not migrate during the measurement period are indicated as symbols at a flight duration of zero hours. The linear regression line represents migrating birds only. HETH = Hermit Thrushes, SWTH = Swainson's Thrushes (redrawn after Wikelski et al. 2003).

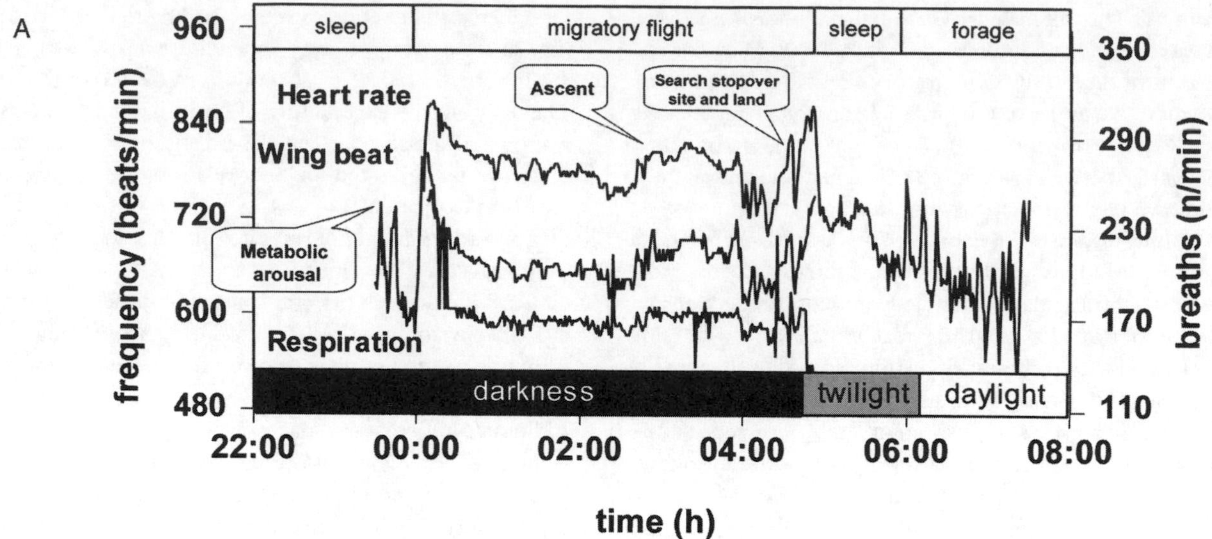

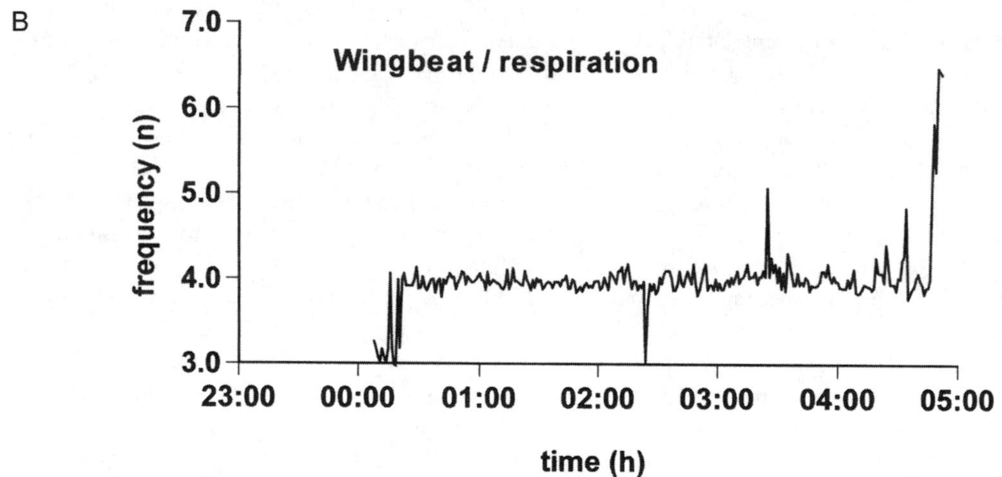

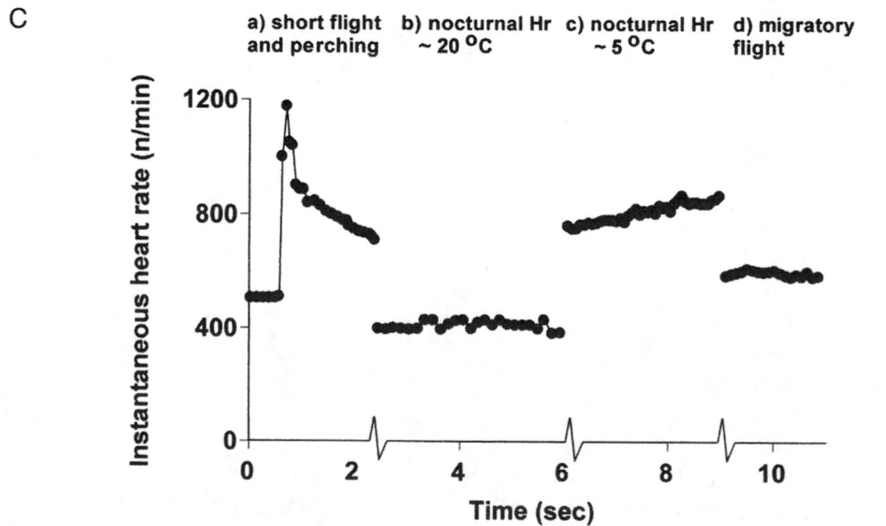

Fig. 22.4. Detailed physiological or behavioral measurements during migratory flights as recorded from radio transmitters attached to the birds. (A) Heart rate (top line), wing beat frequency (middle line), and respiration rate (lower line) of a Swainson's Thrush during a migratory flight that lasted from midnight to 0500. Callouts indicate interpretations of the bird's migratory behavior or observations. The shaded bar below the diagram shows light levels; the bar above shows the bird's behavioral states. (B) Diagram showing the relationship between wing beat frequency and respiration during the entire flight. (C) Short traces of heart rate recordings from another Swainson's Thrush during different activities.

beat frequency and heart rate tracked each other in detail, wing beats running about 120 beats/min lower than heart rate. At the same time, respiration rates tracked wing beat frequency (with isolated exceptions) during cruising flight at a ratio of 1:4. During takeoff and landing phases, ratios of, respectively, 1:3 and >1:4 were observed (fig. 22.4B).

Figure 22.4C depicts heart rate traces for another bird during four different activities: short between-perch flights during foraging bouts (up to 1,200 beats/min), resting at night at ambient temperatures of 20°C and 5°C (400 vs. 800 beats/min), and during the cruising phase of the migratory flight (~600 beats/min). These differences in heart rate are likely to be reflected in the energy expenditure of the bird (Butler et al. 2000). Our calibration of heart rate against oxygen consumption in a respirometer revealed a very strong correlation (linear regression: oxygen consumption [ml/h] $= -394 + 0.55$ heart rate $+ 9.9$ body mass, $n = 38$, $r^2 = 0.91$, $p < 0.001$; Wikelski and Cochran, in prep.). However, it is not yet clear if the same formula describes the relationship between oxygen and heart rate during natural flight as it does during cold exposure in the respirometer (Butler et al. 2000). Nevertheless, it is promising that heart rate can serve as a close proximate indicator for instantaneous energy expenditure in the wild. At the same time, our DLW data from natural flights confirmed the heart rate and respirometry calculations (see above).

Fine-scale patterns in the three parameters, wing beat frequency, heart rate, and respiration, could yield important information about natural flight in songbirds. For example, fig. 22.5A depicts the wing beat frequency of a Swainson's Thrush during its 11-h nocturnal migratory flight in fall. The apparent wing beat changes with an approximate period length of about 30 min are so far unexplained. It is conceivable that sleep deficit during the migration period could account for the repeated pattern of high and decelerating wing beat frequencies ("napping"). Our recent development of an EEG-transmitter (Cochran and Wikelski, in prep.) raises the possibility of asking questions such as the occurrence of unihemispheric sleep during migratory flight in the wild (Rattenborg et al. 2000). An alternative explanation for the rhythmic pattern is that the bird went through cycles of ascent and descent either in pursuit of the best air layers or because the descent allows for a more relaxed flight, recuperating flight muscles, for example. Other explanations such as changes in wind speed are possible and cannot be ruled out at present. All these hypotheses are entirely testable on free-flying birds using currently available technology. Figure 22.5B documents another flight of a *Catharus* thrush and shows the variability in wing beat frequency as well as the wing beat pause percentage. Apparently, *Catharus* thrushes do not show a consistent pattern of wing beat frequencies throughout a migratory flight, as all birds we observed ($n = 11$) paused frequently, sometimes as much as 50% of the time (contrary to assertions in Diehl and Larkin 1998). It is conceivable that birds ride weather fronts and at times do not have to produce a strong uplift by continuously beating their wings at a high frequency. Such connections between low wing beat frequencies and low energy expenditure (i.e., low heart rates) can also be detected in the traces in fig. 22.4A. Heart rate traces of a bird that flew for about 30 min through a major thunderstorm and lightning are not yet fully analyzed. Unfortunately, at present we do not have good heart rate traces for birds that enter thunderstorms.

Combined data on wing beat frequency and wind direction allow us to hypothesize what environmental cues birds are able to perceive during nocturnal flight. We found that birds flying in head winds tend to increase their wing beat frequency relative to other times during their flight, whereas birds flying with supporting tailwinds tend to decrease their wing beat frequency relative to flight periods with neither head wind nor tailwind (fig. 22.5C). However, this effect was not very strong: instead of beating their wings 36,000 times/h during no-wind cruising speed, birds would increase their wing beat by 1,000 beats/h, to about 37,000 beats/h in head winds. The decrease in wing beat frequency may be accompanied by a concurrent decrease in mean duration of wing beat pause (150 msec to 100 msec) and an overall decrease in the number of wing beat pauses (by about 10% [Cochran and Wikelski, in prep.]). Taken together, these data indicate that the thrushes had some knowledge of either their ground speed, by visual ground reference, or the wind vector relative to their heading, by sensing acceleration from small-scale directional air turbulence (for discussions of air-ground speed issues see, e.g., Liechti et al. 2000)

Testing the Decision Rules of Migrating *Catharus* Thrushes

The data from individual thrushes followed by Cochran (1987) are representative of the population of *Catharus* thrushes migrating through central Illinois. Figure 22.6 shows the approximate one-night paths of 24 *Catharus* thrushes whose headings were measured as northerly (±20°) by radio tracking (lines radiating mostly northward from Champaign, Illinois, center of the map). Those birds that had tailwinds generally migrated for longer distances per night than birds that had head winds or strong side winds, which pushed them off in either westerly or easterly directions. Several of these birds, heading northward, were blown toward 270° (due west) or 90° (due east) by strong east-northeasterly or west-northwesterly winds, respectively. These adverse winds aloft were not apparent on the ground at the takeoff site, where surface winds had been variable in direction and low in strength (<10 km/h).

On the basis of the individual fight paths (fig. 22.2A) and migration decisions of *Catharus* thrushes (fig. 22.1A), we suggest that six simple rules of thumb could sufficiently explain most of the migratory behavior of *Catharus* thrushes: Migrate (1) when sufficiently fat, (2) when the daily maximum air temperature is above 21°C, (3) when surface wind speed at takeoff is less than 10 km/h, regardless of direction; then (4) use constant heading based on sun-calibrated mag-

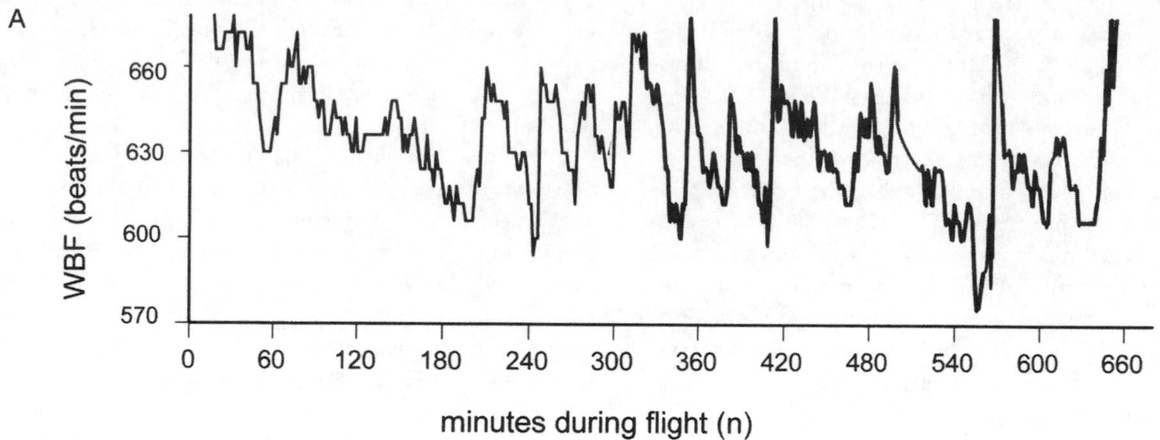

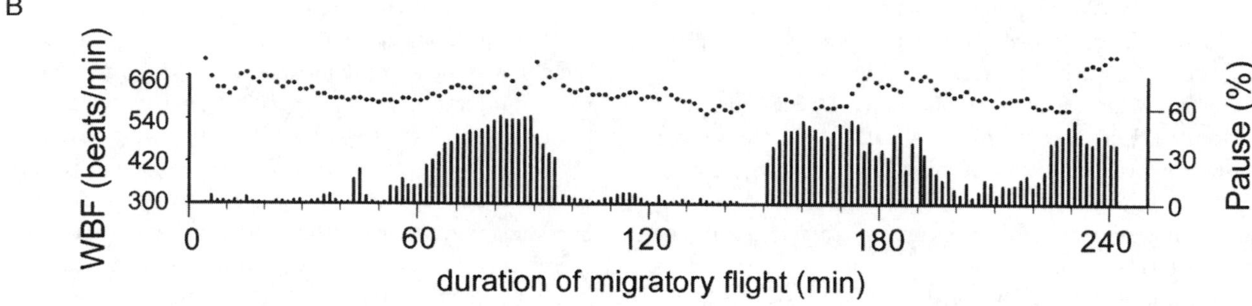

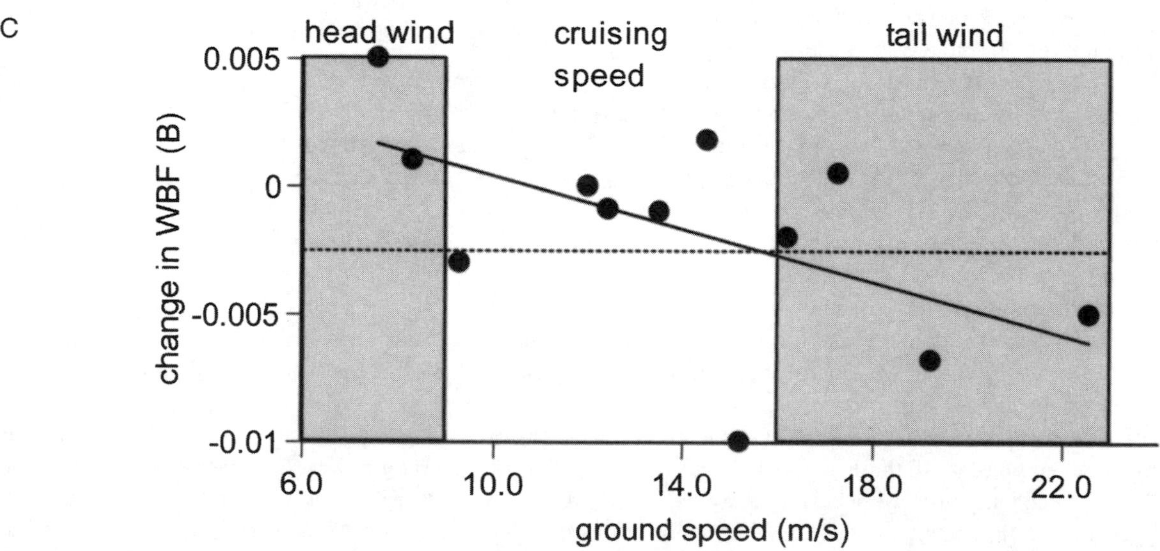

Fig. 22.5. (A) Example of traces of wing beat frequencies from migrating Swainson's Thrushes during an 11-h fall nocturnal flight. Note that wing beat frequencies are accurate to a single beat but averaged for each minute. Variation thus reflects true differences in the birds' behavior. Typical wing beat patterns include high values at the beginning and end of flights. (B) Wing beat frequencies and the percent wing beat pauses during one entire fall nocturnal migratory flight of a Swainson's Thrush. Note that both wing beat frequency and wing beat pauses vary a lot during the flight, potentially allowing important insights into the natural flight behavior of the bird. (C) Summary graph of changes in wing beat frequencies for 12 *Catharus* thrushes during natural migratory flight. The change in wing beat frequency was determined as the change in the regression coefficient (*B*) of a linear regression of wing beat frequency (Hz) against time of flight (min). Thus, a change of $B = 0.005$ accounts for a change of about 1,100 beats/h.

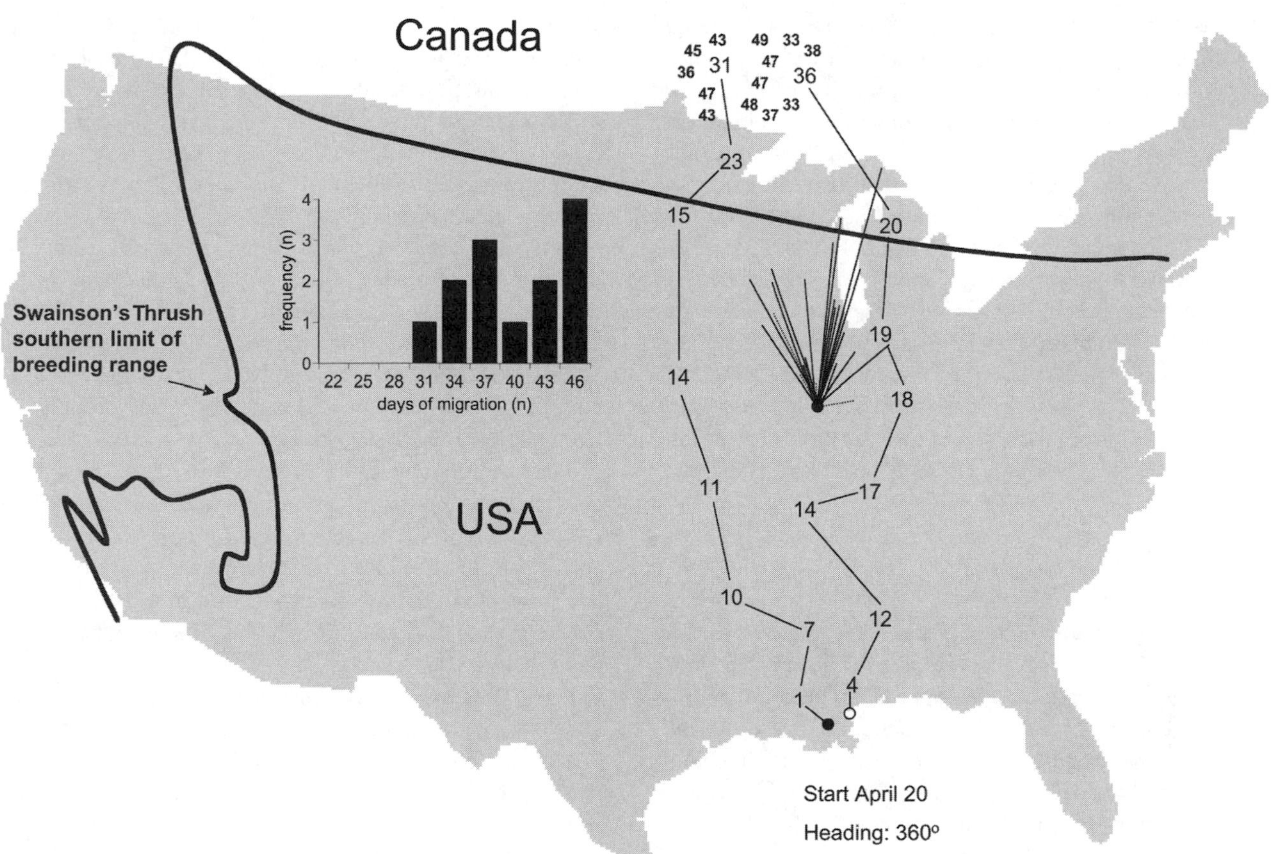

Fig. 22.6. Map of the continental United States including actually observed nocturnal migration flights starting from Champaign-Urbana, Illinois, as well as imaginary flights for two birds (out of a larger sample of 15 migrants) if they were to follow the six rules of thumb for overland migration flights. For observed migratory flights (lines radiating northward), solid lines indicate that birds were located and/or recaptured in their new stopover habitat, and broken lines indicate that birds could not be located on the ground but had landed within 50 km. For imaginary birds, we assumed they appeared on the Gulf Coast in migratory condition on 20 April. Numbers along the migratory pathway indicate days since 20 April, and lines depict approximate migration routes under prevailing weather conditions during spring 1965. We assume that both birds started east-west search flights for breeding habitat as soon as they reached their breeding latitudes in Canada (on days 31 and 36 of the migration, respectively). Additional numbers indicate the number of days en route and arrival locations for an additional 13 birds that started in the same area. Inset frequency histogram shows the distribution of total overland migration days. The approximate breeding range of Swainson's Thrushes (as an example for *Catharus* Thrushes) is depicted above the solid line.

netic compass, (5) land when body reserves approach levels of previous morning or when you hit a cold front, and (6) adjust flying efforts (wing beat frequency) to wind direction, but do not change heading (i.e., no compensation for wind drift).

Below we provide an example how the migratory path of *Catharus* thrushes throughout the continental United States could potentially be predicted solely on the basis of these rules (fig. 22.6). We are aware, however, that for a rigid test of the birds' decision rules we would need a detailed knowledge of ground conditions and bird condition over the years, data that are not available at present (but see Future Directions below). To predict birds' migratory path in relation to climate we collected weather information for spring (April, May, June) of the years 1965 and 1966. We then predicted flight paths for 15 north-heading *Catharus* thrushes starting in Louisiana in 1965. To calculate the birds' flight paths we used the above rules of thumb in relation to daily climate conditions. We assumed that all birds landed within an 80-km stretch of coastline at the Louisiana shores in mid-April. We only use two birds here to illustrate the discrete flight paths and not to confound the figure, but simulations could identify isoclines of most likely flight paths (see below). On 20 April both birds were supposed to be in sufficiently good body condition to start northward migration, weather permitting. Provided both birds would follow only the same generalized decision rules outlined above, the westerly bird migrated on 21 April (day 1), but the easterly bird did not migrate because of locally cold and windy weather. However, on 24 April (day 4) the easterly bird encountered warm weather and low wind speeds and took off. Once in the air, the wind was modestly strong from the north and became stronger throughout the night, permitting only a short flight distance during the first northward migratory flight. As indicated in fig. 22.6, both birds needed eight nocturnal migratory flights to cross the United States and ended up in their breeding range in Canada on 21 May and 27 May, respectively. Although their paths ranged from

285° to 78°, their locations after 8 days of flight deviated only 4° and 10°, respectively, from where they would have migrated had their paths been exactly north on every flight. Birds that start their ca. 2,200-km-long migration in an 80-km-long area in Louisiana would end it in a restricted but overall somewhat larger area in central Canada. In our simulations, the average migration duration for northerly migrants after landfall in North America is 40 ± 2 (SE) days.

Overall we suggest that a few simple rules of thumb are apparently sufficient to explain what appears as a complex and diverse migratory pattern. We are encouraged in our interpretation by the fact that the predicted migration routes drawn in fig. 22.6 closely resemble those actually observed in the midwestern United States (compare length and direction of migratory paths in imaginary and real nocturnal flights).

DISCUSSION

Our data, observations, and experiments indicate that a small set of simple migratory decision rules may be sufficient to explain much of what appears to be complex migratory behavior of New World *Catharus* thrushes. This finding is encouraging because, particularly in songbirds that presumably do not migrate in groups, hatch-year birds already need to have a genetic template for their long migration toward the tropical wintering grounds (Pulido et al. 2001). However, simple decision rules, if unmodified by environmental input, are inherently error-prone. Thus, it is expected that at least young birds display errors in their behavioral decisions, for example, flying in unfavorable winds or progressing northward too far during unseasonably favorable weather and then facing an inadequate food supply. We believe that the migration behavior of songbirds contains such "errors," which are inherent in a fixed program and that these errors remove some individuals from the breeding population temporarily or permanently.

Nevertheless, a simple set of decision rules, potentially genetically inherited, will on average and in spite of high variety, lead to the successful completion of two journeys of thousands of kilometers every year by enough individuals to sustain a viable population. So far we have little information about how the postulated decision rules may be modified, for example, by experience, or temporarily altered, for example, by a physical condition of a bird. In particular it will prove interesting to determine how migrating birds deal with the apparent sleep deficit due to nocturnal (migratory) and diurnal (foraging) activities (Rattenborg et al. 2000). If birds get all the sleep they need during short naps between foraging bouts, this would be an exciting and innovative way to satisfy sleep needs.

Our data suggest that birds react at least slightly to changes in wind direction, thus increasing their flight efforts during head winds and decreasing it during tailwinds. The mechanism of how birds sense wind speeds while in the air still needs to be determined. Our data further show that on average *Catharus* thrushes achieve ground speeds (54 ± 5 km/h) greater than air speeds (~35 km/h) that have been reported for songbirds in general during migratory flights (Bruderer and Boldt 2001). Thus, simple decision rules of the sort we suggest apparently provide a much better than random choice of nights to migrate. These rules, however, sometimes result in flight with winds aloft that cause significant lateral drift (fig. 22.2) (Cochran and Kjos 1985). Nevertheless, the lateral drift averages out over a number of flights so that progress is generally in the direction of the heading (figs. 22.2, 22.6). Such averaging may create a bias in the overall direction of migration. We suggest that the birds' intrinsically preferred headings are selected to account for prevailing winds over the entire migration route from South America to Canada.

The development of radio transmitter technology that allows for the in-flight measurement of detailed behavioral and physiological parameters might be one avenue to advance our understanding of migratory tactics in songbirds. We were astonished to learn from our doubly labeled water measurements that migratory flights of less than about 3 h did not show up in the daily (24-h) energy budget of *Catharus* thrushes when compared with birds stopping over and experiencing cold nights (Wikelski et al. 2003). Our heart rate recordings of wild birds allow us to speculate that nocturnal thermoregulation costs during cold spring stopover nights account for most of the high-energy budget during no-flight nights. Thus stopover thrushes that did not fly on cold nights had elevated thermoregulation costs that were as high as if birds had made an approximately 3-h migration flight. Our measurements also enable us to calculate the total energetic costs for the spring migration of *Catharus* thrushes. For instance, birds leaving Panama during early April and arriving in Canada in late May would travel about 4,800 km. At an average nocturnal flight distance of 265 km (±192 km SD) and an average of 4.6 h (±2.6 h SD) in flight, the birds would need to make 18 nocturnal flights. During the flight days, a bird expends on average 130 KJ; thus for 18 days, this would amount to 2,340 KJ. The flight costs alone would be 71 KJ per night for 18 nights, thus 1,278 KJ. The stopover period would cover a total of 24 days at an average of 88 KJ, thus totaling 2,112 KJ. The total amount of energy necessary during a migration from Panama to Canada would therefore be about 4,450 KJ, whereas the total cost for the active flight period would amount to about 1,278 KJ. This calculation, however, does not take account of weather conditions, which could change the energy balance.

We do not yet understand why birds do not fly when the spring weather is too cold. The data just discussed suggest that birds could be better off flying on cold nights, because that is energy they would have to spend in thermoregulation if they stayed on the ground. It is, however, possible that in cold weather their in-flight heat loss would be too great. Alternatively, the decision rule may take into account the probability that cold weather at the present stopover site predicts even colder, and thus potentially detrimental (low food and/or low temperatures), weather farther north and thus dictate abstaining from northward flights for a while. Our present energy calculations do not acknowledge that wind patterns might be different in different areas of the migration route (see Gauthreaux et al., Chap. 15, this volume).

From our wing beat measurements we can also calculate the overall number of wing beats during migratory flight from Panama to Canada. As a *Catharus* thrush makes 18 nocturnal flights of about 4.6 h duration and beats its wings at approximately 650 beats/min, it would beat its wings about 3.2 million times during its entire cross-continental migratory flight.

Achieving "Connectivity" between Breeding and Wintering Grounds

Advances in radiotelemetry will also enable us to follow individual birds directly from breeding to wintering grounds and back. Questions of what small birds do, where they go, and particularly where, why, and how many die when they leave either breeding or wintering areas have been major unknowns obstructing progress in avian conservation ecology (but see Rubenstein et al. 2002; Sillett and Holmes 2002; Webster et al. 2002). Similarly, we know little about the seemingly amazing orientation capabilities of intercontinental migrants because we largely lack the tools to study such mechanisms in small birds in the wild.

To address some of these questions we propose to install a radio receiver on the ISS that would allow us to listen to radio signals from transmitters that are attached to songbirds (fig. 22.7A,B). Even the 1-g radio transmitters we use for terrestrial tracking could be detected at ISS altitudes (fig. 22.7C).

Possibilities for Global Tracking of Small (30- to 200-g) Birds

Satellite tracking of animals with the ARGOS satellites at an accuracy of ten to hundreds of meters has become common in recent years (Weimerskirch et al. 1993; Bevan et al. 1995). Over the past decade improvements in technology and design have resulted in transmitter weight reductions from more than 100 g to about 20 g. The basic limitation of the ARGOS system is that it requires animal transmitters that radiate about 100 milliwatts (+20 dBm). A new series of satellites may allow use of 50 milliwatt (+17 dBm) transmitters weighing perhaps 14 g (Howey 2003). These could be used on birds weighing about 500 g if 3% of body weight is tolerable according to U.S. Fish and Wildlife Service regulations. Further transmitter weight reduction is unlikely because of battery weight and system power requirements (Paul Howey, Microwave Telemetry, pers. comm. 2003).

Alternatively, bird transmitters for conventional terrestrial tracking of songbirds weigh about 0.4–1 g, have radiated power of about 0.1 milliwatt (–10 dBm), and have a working life of about 1 month. Life can be extended to several months by slowing the pulse rate or much longer by increasing weight to 2 g. ARGOS transmitters require several hundred times as much power.

The Feasibility of Tracking from the ISS: A Simple Example

Feasibility of detection of –10 dBm transmitters from satellite altitudes can be demonstrated by calculations based on more than 40 years of radio-tracking experience. At 220 MHz, path loss over a distance of 400 km (height of the ISS) is 132 dB (fig. 22.7C) (Crombie 1989); that is, a transmitter radiating –10 dBm (isotropically) will excite an isotropic receiving antenna 400 km away with a signal of –142 dBm. A receiver antenna with 6 dBi linear polarization gain (a 9-dBi-gain helix) will improve the received signal to –136 dBm (Kraus 1988). A –136 dBm pulsed signal (18-ms pulse width and several seconds or shorter pulse interval) is easily audible in current wildlife tracking receivers. These numbers are valid for transmitters directly under the satellite.

A 9-dBi-gain helix antenna has a half-power (–3 dB) beam width of about 25°, and therefore a gain of 3 dBi around the circumference of a circle 185 km in diameter (tan 25 × 400 km) centered under the satellite. Adjusting for both the 3-dB off-axis loss due to the antenna pattern and the increased distance (400/cos 25 = 442 km), the received signal level will be –139.8 dBi for –10-dBm transmitters on this circumference; this is still an easily detected level for wildlife receivers. These receivers typically have a noise figure of 3 dB and bandwidth common to SSB reception (ca. 2.5 KHz). The detection is marginal at –143 dBm, but still possible even at –145 dBm (albeit with errors).

The detection footprint for terrestrial transmitters from the ISS is thus a traveling circle 370 km in diameter centered under the satellite. This circle will, at its maximum diameter, pass over an animal transmitter in about 45 sec or in enough time for about 22 pulses for a transmitter pulse interval of 2 sec. The pass-over time will be less for off-center passes. For instance, if the satellite's closest approach is 130 km, the time to detect a transmitter signal is reduced to 31 sec or just enough time for about 15 pulses. At 160 km off-center, the time allowed to detect a transmitter signal drops to about 22 sec (11 pulses). Taking 11 pulses as a minimum for reliable detection, we thus have the satellite sweeping a 320-km-wide path. Such a path is covered at least every 5 days. Thus, in the above example, a receiver installed at the ISS will detect a songbird fitted with a transmitter approximately 1.5 times a week.

The number of pulses that are detected by the ISS receiver is thus a crude measure of how far the animal is off to the side. Subsequent passes could be used to resolve the to-the-left versus to-the-right ambiguity. The time of the middle-of-the-set pulses is a measure of closest approach. These crude measures will allow closer estimation of position, perhaps to a radius of 40 km.

An even higher accuracy can be attained if more than one antenna can be installed at the ISS. Higher-directivity receiver antennas would improve performance by providing stronger signals. An additional advantage of a smaller footprint would also be that less man-made noise is received. The smaller footprint also reduces the width of the path

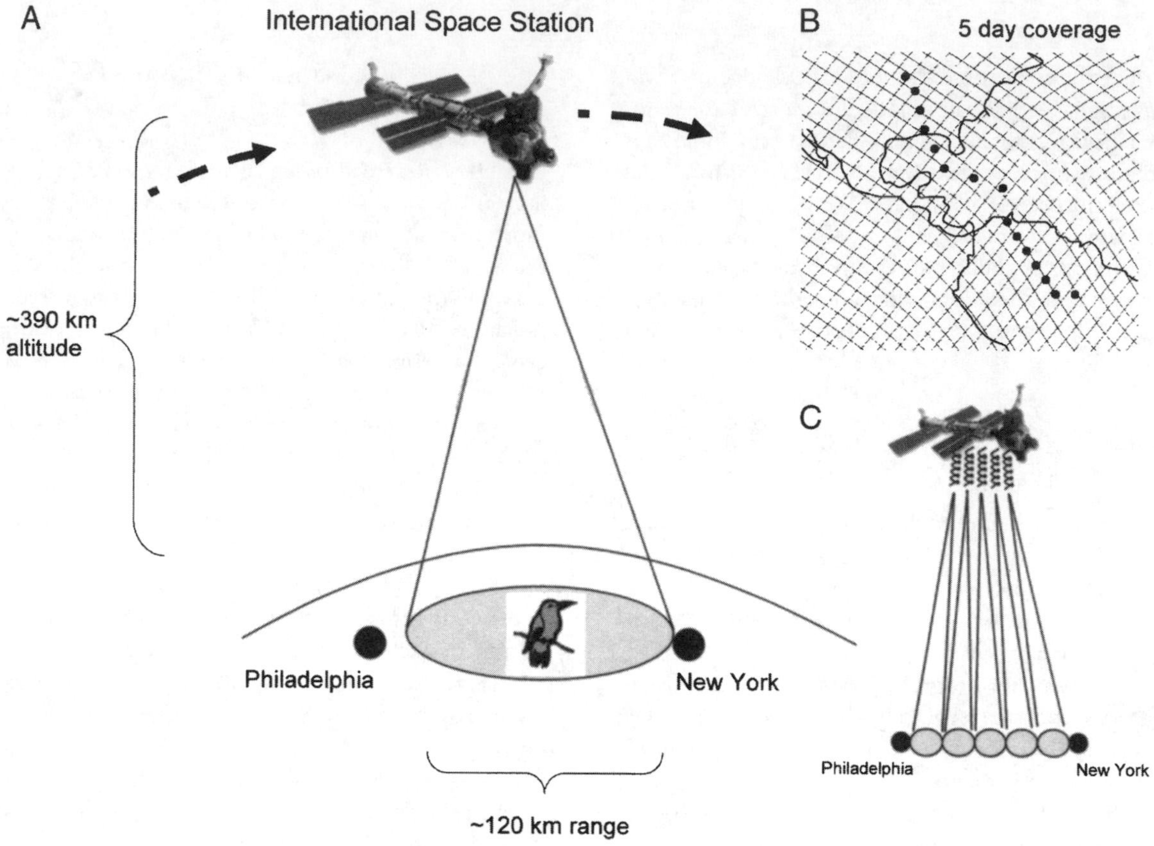

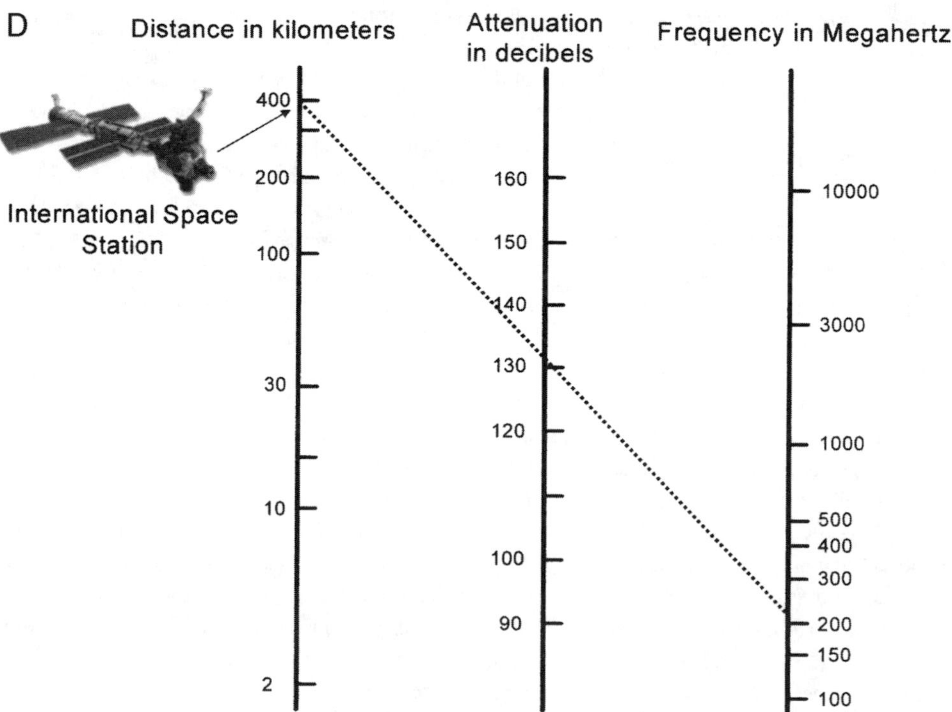

Fig. 22.7. The approximate migration routes of small birds can be tracked by using small, currently available radio transmitters and a receiver at the ISS. (A) The ISS orbits the Earth at an altitude of ca. 390 km and its projected track covers the globe in a tightly woven network. Radio transmitters for space tracking would have a fixed frequency but differ in their pulse coding, allowing concurrent tracking of several hundred animals per continent (B). Radio tracking birds from the ISS would not allow pinpoint accuracy (range ca. ±120 km with one antenna, ca. ±40 km with multiple helix antenna, (C). However, such low accuracy would still be highly useful in tracking migrants from temperate zones to the tropics (points in B). (D) The physical basis for radiotelemetry from space: the electromagnetic wave propagation diagram (Crombie 1989) documents the signal attenuation over distance.

swept and the number of pulses per pass. However, the small size of a footprint could easily be increased if five separate receivers, each with a higher-directivity antenna, were installed. Each antenna would provide near ideal, improved signals, more accurate locations, and less noise without reducing total footprint size. Once birds are located from space within a circle of 40 km they could, if desired, be searched for by ground-tracking vehicles or small planes, which is routinely done in terrestrial radio tracking.

Which Transmitter Frequency Should Be Used?

The frequency should be chosen to reduce interference from Earth transmitters and gadgetry as much as possible. Observations while tracking animals from airplanes over the midwestern United States show that frequencies in the 150–170-MHz band are heavily used, as are all the television frequencies (170–213 and 480 and up) and the mobile services in the 450–470-MHz band. The frequencies from 300 to 390 MHz are rapidly becoming polluted with illegal garage door superegenerative receivers and "wireless" household light and coffee-maker controllers.

Although the proposed space tracking would provide relatively poor location accuracy by terrestrial standards, we suggest that the advantages of studying the connectivity between breeding and wintering grounds in songbirds would outweigh such shortcomings. In addition, space tracking of songbirds would allow us to investigate long-distance dispersal phenomena that are currently largely beyond scientific reach (but see Winkler, Chap. 30, this volume).

FUTURE DIRECTIONS

Our long-term data combined with modern ecophysiological methods suggest that a detailed mechanistic understanding of individual migratory behavior of wild songbirds can be achieved. Many important questions could be addressed in the future: (1) How representative are *Catharus* thrushes for New World migrating songbirds? Are decision rules likely to apply generally across taxa and in other migration systems? How and when might they differ? The development of automated radiotelemetry systems (ARTS) (Cochran et al. 1965; Wikelski and Kays 2003) allows for the continuous monitoring of tens of individuals per stopover site. Many additional migrant species could be radio-tagged and followed during stopover and migration. (2) How representative are our data for migratory behavior of *Catharus* thrushes elsewhere and at other times during their migration? Again, ARTS systems established at several areas along the migration routes of thrushes would provide information about the within-species variability in migration tactics. In particular, more information is needed about how migration tactics change when large geographical barriers such as deserts or water need to be overflown. A comparison of data from overland migration flights in the New World with data on the crossing of ecological barriers in the Old World could be very informative. Migration tactics in the New World conceivably could differ from those in the Old World in many ways, in particular because of the large ecological barriers (Sahara, Alps, Mediterranean) in Europe and Africa. It is therefore possible that optimization criteria for overland migrants (our study) with regard to fat deposition, digestion, and speed of migration are different from birds that repeatedly have to cross ecological barriers. (3) How do birds adjust their in-flight behavior to atmospheric conditions? Modern atmospheric simulation models could be combined with in-flight heart rate, wing beat, and respiration measurements, with simultaneous determinations of the birds' flight altitude. We hypothesize that changes in physiological parameters should be closely tied to local atmospheric conditions. We also expect birds to adjust their altitude to get into different ambient temperatures. (4) What are the behavioral and physiological mechanisms that allow migrating songbirds to find their way from breeding to wintering grounds? The establishment of a receiver setup at the ISS will allow us to conduct orientation experiments on a grand scale in the wild. Birds could be relocated during their migration and the outcome of such manipulations could be followed on intercontinental scales. Along the same lines, one could conduct clock-shift experiments on circannual clocks and determine how important such seasonal cues are in the wild. (5) Where do conservation-relevant migrant songbird species go when they disappear from the North Temperate map? For many species such as the European Aquatic Warbler (*Acrocephalus paludicola*) or even common birds like the House Martin (*Delichon urbica*), it is very important to identify major wintering sites. In general, achieving connectivity between breeding and wintering grounds in songbirds would solve many long-standing mysteries of bird migration that cannot be addressed with current methodologies.

ACKNOWLEDGMENTS

We thank George Swenson, Herbert Biebach, Ebo Gwinner, Ela Hau, Marty Martin, Thomas Rödl, Alex Scheuerlein, Scott Robinson, Ron Larkin, Robb Diehl, and Elisa Tarlow for helpful discussions. WWC also sincerely thanks Richard R. Graber (deceased), who in 1965 initiated radiotelemetric studies of songbirds. This study was supported by the National Geographic Society, and in part by the University of Illinois, the Illinois Natural History Survey, Princeton University, and the National Science Foundation International Programs. We thank Arlo Raim, Karin Nelson, and many students for help during fieldwork. We are particularly grateful to Henk Visser for isotopic analyses.

LITERATURE CITED

Able, K. P. 1996. Large-scale navigation. Journal of Experimental Biology 199:1–2.

Alerstam, T. 1991a. Bird flight and optimal migration. Trends in Ecology and Evolution 6:210–215.

Alerstam, H. 1991b. Ecological causes and consequences of bird orientation. Pages 202–225 *in* Orientation in Birds (P. Berthold, ed.). Birkhäuser Verlag, Basel.

Alerstam, T., and A. Hedenstrom. 1998. Optimal migration. Journal of Avian Biology 29:339–340.

Bairlein, F., and E. Gwinner. 1994. Nutritional mechanisms and temporal control of migratory energy accumulation in birds. Annual Revue of Nutrition 14:187–215.

Baker, R. R. 1984. Bird Navigation. Holmes and Meier, New York.

Berthold, P. 1993. Bird Migration: A General Survey. Oxford University Press, Oxford.

Berthold. P. 1998. Spatiotemporal aspects of avian long-distance migration. Pages 103–118 *in* Spatial Representation in Animals (S. Healy, ed.). Oxford University Press, New York.

Berthold, P., and A. J. Helbig. 1992. The genetics of bird migration-stimulus, timing, and direction. Ibis 134:35–40.

Bevan, R. M., P. J. Butler, A. J. Woakes, and P. A. Prince. 1995. The energy expenditure of free-ranging black-browed albatrosses. Philosophical Transactions of the Royal Society of London, Series B, Biological Sciences 350(1332):119–131.

Biebach, H. 1998. Phenotypic organ flexibility in Garden Warblers *Sylvia borin* during long-distance migration. Journal of Avian Biology. 29:529–535.

Bruderer, B., and A. Boldt. 2001. Flight characteristics of birds: Pt. 1. Radar measurements of speeds. Ibis 143:178–204.

Bruderer, B., D. Peter, and T. Steuri. 1999. Behaviour of migrating birds exposed to X-band radar and a bright light beam. Journal of Experimental Biology 202:1015–1022.

Butler, P. J., A. J. Woakes, R. M. Bevan, and R. Stephenson. 2000. Heart rate and rate of oxygen consumption during flight of the barnacle goose, *Branta leucopsis*. Comparative Biochemistry and Physiology A 126:379–385.

Buttemer, W. A., L. B. Astheimer, and J. C. Wingfield. 1991. The effect of corticosterone on standard metabolic rates of small passerine birds. Journal of Comparative Physiology B 161:427–431.

Cochran, W. W. 1972. Long-distance tracking of birds. Pages 39–59 *in* Animal Orientation and Navigation (K. Schmidt-Koenig, G. J. Jacobs, S. R. Galler, and R. E. Belleville, eds.). U.S. Government Printing Office, Washington, D.C.

Cochran, W. W. 1987. Orientation and other migratory behaviors of a Swainson's thrush followed for 1500 km. Animal Behavior 35:927–929.

Cochran, W. W., R. R. Graber, and G. G. Montgomery. 1967. Migratory flights of *Hylocichla* thrushes. Living Bird 6:213–225.

Cochran, W. W., and C. J. Kjos. 1985. Wind drift and migration of thrushes: a telemetry study. Illinois Natural History Survey Bulletin 33:297–330.

Cochran, W. W., H. Mouritsen, and M. Wikelski. 2004. Migrating songbirds recalibrate their magnetic compass daily from twightlight cues. Science 304:405–408.

Cochran, W. W., D. W. Warner, J. R. Tester, and V. B. Kuechle. 1965. Automatic radio-tracking system for monitoring animal movements. Bioscience 15:98–103.

Crombie, D. D. 1989. Electromagnetic-wave propagation. Chapter 33 *in* Reference Handbook for Engineers: Radio, Electronics, Computer, and Communication (seventh ed.) (E. C. Jordan, ed.). Sams & Company, Indianapolis.

Diehl, R. H., and R. P. Larkin. 1998. Wing beat frequency of *Catharus* thrushes during nocturnal migration, measured via radiotelemetry. Auk 115:591–601.

Eliassen, E. 1963. Preliminary results from new methods of investigating the physiology of birds during flight. Ibis 105:234–237.

Graber, R. R., J. W. Graber, and E. L. Kirk. 1971. Illinois birds: Turdidae. Biological Notes no. 75. Illinois Natural History Survey:1–45.

Greenberg, R., P. Bichier, A. C. Angon, C. MacVean, R. Perez, and E. Cano. 2000. The impact of avian insectivory on arthropods and leaf damage in some Guatemalan coffee plantations. Ecology 81:1750–1755.

Howey, P. 2003. www.cls.fr/html/argos/documents/newsletter/nslan52/bird_tracking_en.html.

Jenni, L., and S. Jenni-Eiermann. 1998. Fuel supply and metabolic constraints in migrating birds. Journal of Avian Biology 29:521–528.

Kjos, C. G., and W. W. Cochran. 1970. Activity of migrant thrushes determined by radio-telemetry. Wilson Bulletin 82:225–226.

Klaassen, M., A. Lindstrom, and R. Zijlstra. 1997. Composition of fuel stores and digestive limitations to fuel deposition rate in the long-distance migratory Thrush Nightingale, *Luscinia luscinia*. Physiological Zoology 70:125–133.

Kraus, J. D. 1988. Antennas. McGraw-Hill, New York.

Liechti, F., M. Klaassen, and B. Bruderer. 2000. Predicting migratory flight altitudes by physiological migration models. Auk 117:205–214.

Lindström, A., M. Klaassen, and A. Kvist. 1999. Variation in energy intake and basal metabolic rate of a bird migrating in a wind tunnel. Functional Ecology 13:352–359.

Lord, R. D., F. C. Bellrose, and W. W. Cochran. 1962. Radio telemetry of the respiration of a flying duck. Science 137:39–40.

Marra, P. P., K. A. Hobson, and R. T. Holmes. 1998. Linking winter and summer events in a migratory bird by using stable-carbon isotopes. Science 282:1884–1886.

Masman, D., and M. Klaassen. 1987. Energy expenditure during free flight in trained and free-living Eurasian Kestrels (*Falco tinnunculus*). Auk 104:603–616.

Moore, F. R., and T. R. Simons. 1992. Habitat suitability and stopover ecology of Neotropical landbird migrants. Pages 345–355 *in* Ecology and Conservation of Neotropical Migrant Landbirds (J. M. Hagan III and D. W. Johnston, eds.). Smithsonian Institution Press, Washington, D.C.

Morrow, J., and D. H. Taylor. 1976. A telemetry system for recording heart rates of unrestrained game birds. Journal of Wildlife Management 40:359–360.

Nagy, K. A., I. A. Girard, and T. K. Brown. 1999. Energetics of free-ranging mammals, reptiles, and birds. Annual Revue in Nutrition 19:247–277.

Payne, R. B., and L. L. Payne. 1990. Survival estimates of Indigo Buntings-comparison of banding recoveries and local observations. Condor 92:938–946.

Pennycuick, C. J., M. Klaassen, A. Kvist, and A. Lindström, 1996. Wing beat frequency and the body drag anomaly-wind-tunnel observations on a Thrush Nightingale (*Luscinia luscinia*) and a Teal (*Anas crecca*). Journal of Experimental Biology 1999:2757–2765.

Piersma, T., and A. Lindström. 1997. Rapid reversible changes in organ size as a component of adaptive behaviour. Trends in Ecology and Evolution 12:134–138.

Pulido, F., P. Berthold, G. Mohr, and U. Querner. 2001. Heritability of the timing of autumn migration in a natural bird population. Proceedings of the Royal Society of London, Series B, Biological Sciences 268:953–959.

Raim, A. 1978. A radio transmitter attachment for small passerine birds. Bird Banding 49:326–332.

Rattenborg, N. C., C. J. Amlaner, and S. L. Lima. 2000. Behavioral, neurophysiological and evolutionary perspectives on unihemispheric sleep. Neuroscience and Biobehavioral Reviews 24:817–842.

Rubenstein, D. R., C. P. Chamberlain, R. T. Holmes, M. P. Ayres, J. R. Waldbauer, G. R. Graves, and N. C. Tuross. 2002. Linking breeding and wintering ranges of a migratory songbird using stable isotopes. Science 295:1062–1065.

Sillett, T. S., and R. T. Holmes. 2002. Variation in survivorship of a migratory songbird throughout its annual cycle. Journal of Animal Ecology 71:296–308.

Webster, M. S., P. P. Marra, S. M. Haig, S. Bensch, and R. T. Holmes. 2002. Links between worlds: unraveling migratory connectivity. Trends in Ecology and Evolution 17:76–83.

Weimerskirch, H., M. Salamolard, F. Sarrazin, and P. Jouventin. 1993. Foraging strategy of wandering albatrosses through the breeding season: a study using satellite telemetry. Auk 110:325–342.

Wikelski, M., and R. Kays. 2003. www.princeton.edu/~tracking.

Wikelski, M., E. M. Tarlow, A. Raim, R. H. Diehl, R. P. Larkin, and G. H. Visser. 2003. Costs of migration in free-flying songbirds. Nature 423:704.

Wiltschko, R., and W. Wiltschko. 1999. The orientation system of birds: Pt. 4. Evolution. Journal für Ornithologie 140:393–417.

Wingfield, J. C., T. P. Hahn, M. Wada, L. B. Astheimer, and S. Schoech. 1996. Interrelationship of day length and temperature on the control of gonadal development, body mass, and fat score in white-crowned sparrows, *Zonotrichia leucophrys gambelii*. General and Comparative Endocrinology 101:242–255.

Withers, P. C. 1977. Measurement of VO2, VCO2, and evaporative water loss with a flow-through mask. Journal of Applied Physiology 42:120–123.

23

REBECCA L. HOLBERTON
AND ALFRED M. DUFTY JR.

Hormones and Variation in Life History Strategies of Migratory and Nonmigratory Birds

LIFE ON OUR PLANET IS intimately tied to its changing seasons. Migratory birds exemplify this with their seasonal comings and goings. Past research programs tended to focus on a single season of the avian annual cycle, but within the last decade we have come to understand that the stages of the annual cycle are deeply interconnected. To understand how an individual fares within a particular stage, we must look at that individual's past and how it prepares for the future. Ecologists, behaviorists, and physiologists now work together to uncover the capabilities and constraints of migratory birds. In this chapter, we look into some of the ecophysiological aspects of the transitional lifestyle of migratory birds, with an emphasis on songbirds because that group has received most of the attention in endocrine studies. It is not our goal to develop a comprehensive survey of avian life history strategies. However, when possible, we describe various hormonal mechanisms involved with activities in different stages of the annual cycle and compare migrants with nonmigrants, and tropical with temperate species. We first present an overview of the stationary periods on the breeding and wintering grounds, and for brevity's sake we have limited this chapter to the well-known monogamous system with biparental care. We then discuss behavioral and physiological aspects of energy demand during the periods of transition when all birds prepare for or recover from breeding, and as migrants move between the wintering and breeding quarters. Finally, we suggest future horizons for endocrine studies on birds residing in, and moving between, two worlds.

THE STATIONARY PERIOD OF BREEDING

Migrants live in multiple worlds and face unique behavioral and physiological challenges by virtue of their transitory occupancy of more than one habitat. For example, upon arriving at their spring or autumn destinations, individuals may compete for breeding or wintering territories. In addition, vernal migrants face the additional need to establish pair bonds. We know much more about the hormonal basis of territorial behavior on the breeding grounds than during other times of year, and most of what is known about territorial behavior, in general, comes from studies on males. Therefore, in the following discussion of the stationary breeding and wintering period, we focus on patterns of hormone secretion in males.

The timing of seasonal breeding is inextricably linked to predictable changes in photoperiod, with gonadal development stimulated by exposure to long day lengths (for review, see Dawson et al. 2001). Although a long-day photoperiod activates the hypothalamic-pituitary-gonadal (HPG) axis, it also leads to the development of photorefractoriness, whereby birds become unresponsive to long days and the gonads regress (for review, see Dawson et al. 2001). In many species, a minimum number of relatively short days is needed before birds become photosensitive again, allowing the HPG axis to respond once more to increasing day length (Dawson and Goldsmith 1997).

The relationship between day length and the cycle of photosensitivity and photorefractoriness is critical to understanding reproductive time constraints across a latitudinal gradient. For example, the length of the stimulatory photoperiod (i.e., the proportion of light within a 24-h cycle) affects how soon birds become photorefractory (Dawson and Goldsmith 1983). For high-latitude Arctic breeders, therefore, very long summer days result in a relatively rapid development of photorefractoriness (i.e., short breeding periods), whereas mid- and low-latitude birds experience shorter maximal day length and become photorefractory more slowly (i.e., long breeding periods).

Although day length is the major reproductive cue for temperate breeders, additional cues, such as food availability, humidity, rainfall, temperature, and the presence of a mate, also affect the rate of gonadal maturation and the onset of breeding (for review, see Wingfield et al. 2000). However, high-latitude breeders may be less sensitive than mid- and low-latitude breeders to these supplementary cues. For example, Gambel's White-crowned Sparrow (*Zonotrichia leucophrys gambelii*) breeds at high latitudes and is much less sensitive to temperature effects on gonadal development than the lower-latitude-breeding Puget Sound White-crowned Sparrow (*Z. l. pugetensis* [Wingfield et al. 1996, 1997]). The reduced sensitivity of high-latitude breeders to some environmental cues may enable these populations to pursue reproduction when facing limited opportunities to do so (Holberton and Wingfield 2003).

In most avian species, males vie for breeding territories and access to mates, and the male endocrine profiles reflect these activities. Testosterone (T), directly or indirectly (i.e., through its conversion to estradiol), facilitates both courtship and territorial behavior, and plasma T levels are elevated during these periods of social flux (Wingfield et al. 1990). In monogamous species with bi-parental care, T levels decline after territories are established and pairs have formed. This allows males, who would otherwise spend considerable time patrolling and being vigilant against intruders, to contribute sufficiently to parental care (for review, see Wingfield et al. 1987). The reduction in T may also help avoid the hormone's immunosuppressive effects and to minimize the risk of injury and mortality due to continued T-mediated aggression (Dufty 1989; Hillgarth and Wingfield 1997).

The general pattern of T production over the breeding season is similar in resident and migrant populations across a wide range of latitudes (Wingfield and Hahn 1994), suggesting conservation of the basic gonadal hormone response mechanisms. The greatest latitudinal differences lie in the timing and amplitude of endocrine events. High-latitude breeders tend to have shorter periods of elevated T, as short as a single day (Hunt et al. 1995), compared with those at lower latitudes. At high latitudes, the sharp, short-lived peak of T accompanies the rapid, nearly simultaneous appearance and settlement of males that compete for limited resources, a process that is less intense at more southern latitudes (Hunt et al. 1995). Little is known about within-species latitudinal differences in the relationship between T and territoriality during the breeding season. Two subspecies of white-crowned sparrows (*Z. l. gambelii* and *Z. l. pugetensis*) show different T responses to simulated territorial intrusions during incubation. The Arctic-breeding *Z. l. gambelii* does not increase its T secretion when faced with a simulated conspecific male, whereas *Z. l. pugetensis*, which breeds at midlatitudes and has a longer breeding season, does (Wingfield and Hahn 1994; Meddle et al. 2002). Interestingly, when testicular responsiveness to gonadotropin-releasing hormone injections was compared among these two *Zonotrichia* species and a tropical congener (*Z. capensis*, the Rufous-collared Sparrow), it was *Z. capensis* that showed the strongest response (Moore et al. 2002), but the evolutionary significance of these differences remains unclear. Although experimentally elevated T levels may enable paired males to attract additional females or to gain extra-pair fertilizations, they also inhibit male parental care, and net reproductive success in T-treated males is similar to that of control males (Wingfield 1984; Raouf et al. 1997).

To date, seasonal endocrine patterns in the Tropics have been examined in only a few species. Wikelski et al. (1999a) noted that although some tropical species are territorial year-round, T levels associated with territorial behavior are lower in tropical species than in birds breeding at higher latitudes. Interestingly, in contrast to temperate breeders, the tropical Spotted Antbird (*Hylophylax n. naevioides*) can elevate plasma T levels without undergoing significant gonadal recrudescence (Wikelski et al. 2000). How antbirds produce high levels of T with regressed gonads is unknown (but see below). Nonetheless, testosterone remains relatively low throughout the year in these birds, except during periods of social instability when T levels increase regardless of gonadal state (Wikelski et al. 1999b).

The contrast between the T patterns of tropical breeders such as antbirds and those of temperate breeders provides an excellent example of differences in the endocrine basis of territoriality in birds breeding in two very different worlds. Stutchbury and Morton (2001) suggest that these differences in T secretion may be linked to differences in the likelihood of extra-pair copulations (EPCs). That is, given the lower degree of breeding synchrony in the Tropics compared with higher latitudes, fewer neighboring males are physiologically capable of engaging in EPCs in tropical species, reducing the selective pressure on tropical males to maintain T-related vigilance. Additional information on the occurrence of EPCs in tropical species is needed to test this intriguing idea.

Although much is known about the various cues that influence the onset of breeding in the temperate regions, extremely little is known about the extent to which tropical residents use seasonal cues to regulate the timing of their breeding activities. Gwinner (1996) suggested that temperate breeding migrants overwintering in the Tropics rely on endogenous, circannual rhythms rather than on changes in photoperiod to initiate the endocrine pathways leading to reproduction (but see Dawson et al. 2001). Sharp (1996) noted that continued exposure to long days, not short days, terminates photorefractoriness in some species, a mechanism that may be used by migratory and tropical species alike to prepare for seasonal breeding. Tropical breeders that experience a relatively constant photoperiod can grow their gonads in response to annual changes in photoperiod of only a few minutes (Hau et al. 1998) or to seasonal changes in light intensity (Gwinner and Scheuerlein 1998). Thus photoperiod, directly or indirectly, remains an important cue for at least some tropical breeders. Temperate breeders may respond more robustly to day length during the immediate prebreeding period because, for them, seasonal changes in photoperiod reliably portend changes in resource availability in the period leading up to breeding (Wingfield et al. 1992). For tropical species, resources may be less tightly linked to photoperiod, and birds may monitor changes in food availability directly (Hau et al. 2000b) or cues that influence it such as rainfall (Hau 2001). Clearly, selection pressures affecting the timing of reproduction vary significantly along a north-south gradient. Although the Tropics may appear to offer less dramatic changes in environmental cues throughout an annual cycle, tropical breeders have developed the ability to respond more readily to less intense seasonal cues than birds breeding at higher latitudes (e.g., Wright and Cornejo 1990; Leigh et al. 1996). In spite of their respective abilities to adjust their timing for reproduction, compared with tropical breeders, high-latitude breeders may face greater demands on time and energy because they must recover from the rigors of migration rapidly before energy can be invested in reproduction. Unfortunately, there has been little work systematically comparing breeding constraints between temperate and tropical species or within species across a broad geographical range (see, however, Weathers 1977; Breuner et al. 2003).

Energy regulation poses one of the greatest challenges to breeding birds, regardless of where they breed. One of the hormones that has received much attention recently for its role in regulating energy reserves during breeding is corticosterone, a glucocorticosteroid produced by adrenocortical tissue that plays a key role in regulating lipid synthesis and protein use in birds (Holmes and Phillips 1976). A quick increase in plasma corticosterone levels in response to a variety of potential stressors, such as storms, can facilitate changes in foraging behavior (increased food searching activity and food intake) and help an individual regulate its energy reserves in response to challenges to its energy demand (Wingfield 1994). However, although this hormone response to environmental stressors may be immediately beneficial to an individual, it may have its costs. If such a response were to be needed repeatedly or over a longer period of time, the energetic costs of survival may come at the expense of breeding. Birds unable to maintain an energy balance during breeding may be forced to reduce parental care or abandon territories when the risk to the individual outweighs immediate reproductive benefits (Wingfield et al. 1983; see references in Wingfield and Ramenofsky 1999).

The trade-off between an immediate survival benefit and the cost in immediate breeding success may not be the same for everyone, however, and the adrenocortical response during breeding could be modulated in such a way that reducing it might improve an individual's chances of producing offspring (Wingfield et al. 1994a, 1994b). Several recent studies on high-altitude or high-latitude breeders have shown that the strength of the adrenocortical response can track changes in parental effort or investment, with the greatest reduction occurring when reproductive effort may be most critical, or in the sex with the greatest reproductive effort (Wingfield et al. 1995; Holberton and Wingfield 2003). It may be that temperate or tropical breeders, in the absence of severe time constraints, can retain a robust adrenocortical response throughout the breeding season because they have greater opportunities for renesting after reproductive failure (Silverin et al. 1997). Given the great diversity of mating systems across and within bird species, and the broad geographic regions over which birds breed, this area of inquiry holds great potential for understanding the costs and benefits of different breeding strategies and how variation in endocrine mechanisms can be reflected by diverse environmental conditions under which birds breed (Breuner et al. 2003).

THE STATIONARY PERIOD OF WINTER

Winter offers challenges to migrants and residents alike. While migrants arriving on their tropical wintering grounds must learn how to locate food resources and avoid predators that may differ from those on the breeding grounds, migrants and residents both must contend with the increased competition that might ensue. Such interactions can result in different habitat use among members of the two groups (see Matthysen 1993; Pérez-Tris and Tellería 2002). Both migrants and residents can establish and defend territories during the nonbreeding period. However, although aggressive behaviors may be overtly similar in nonbreeding and breeding contexts, the scant information to date suggests that how hormones regulate territoriality varies across seasons and among species. For example, some year-round residents of temperate regions exhibit aggression mediated by changes in T production, whereas others do not (Logan and Wingfield 1990; Wingfield and Hahn 1994). Territorial male Song Sparrows (*Melospiza melodia morphna*) show aggression during the nonbreeding season that is independent of plasma T levels, and estradiol converted from T appears to be the actual cue that regulates winter territorial behavior (Soma et al. 1999, 2000). Thus, testosterone can be used locally in the brain as a precursor (for estradiol production) needed for aggressive behavior without activating other T-sensitive mechanisms (Soma et al. 2000 and references therein). The source of this testosterone may be dehydroepiandrosterone, an androgen precursor found in the plasma of Song Sparrows during the nonbreeding season (Soma et al. 2002). It remains to be determined if a similar mechanism enables tropical antbirds with regressed gonads to express aggressive behaviors in the absence of elevated T (see also Hau et al. 2000a). Furthermore, many long-distance migrants establish winter territories in the Tropics (Rappole and Warner 1980; Stutchbury 1994). Little is known about the hormonal basis of this behavior, although Canoine and Gwinner (2002) noted that in a Palearctic migrant, the endocrine mechanisms differ between breeding and winter territoriality.

In winter, the adrenocortical response can also be modulated in relation to the probability of encountering adverse conditions. For example, Dark-eyed Juncos (*Junco hyemalis hyemalis*) wintering at higher latitudes where weather and food are less predictable show a more rapid adrenocortical response than juncos wintering in more benign environments farther south, a response that may enable more-northern wintering populations to respond more quickly to the rapidly changing conditions likely to be encountered there (Rogers et al. 1993; Holberton and Able 2000). Within a given latitude, the adrenocortical response can also be modulated to reflect differences in winter habitat quality. Marra and Holberton (1998) found that patterns of corticosterone secretion in American Redstarts (*Setophaga ruticilla*), a species with sexual habitat segregation on the wintering grounds, track changes in habitat quality throughout the winter. Birds that overwinter in the more favorable and preferred habitat are able to maintain body mass and low baseline corticosterone levels, as well as a robust adrenocortical response. In contrast, redstarts forced to overwinter in poorer-quality habitats lose mass, have elevated baseline corticosterone levels, and exhibit a reduced adrenocortical response, perhaps as a way to protect energy reserves. It remains to be seen if elevated baseline corticosterone levels are acting to facilitate the higher foraging effort that might be needed by birds wintering in suboptimal areas. It would be important to know, however, if the chronically elevated, but necessary, levels of corticosterone experienced by some individuals during winter occur at the expense of adequate immune function and overall health. It is clear that, in addition to losing mass and possibly working harder to meet energy needs, individuals wintering in poorer habitats may ultimately experience lower reproductive success if they are unable to prepare as readily for spring migration (i.e., gain fat reserves) and depart for the breeding grounds at the same time as those who wintered in better habitats (Marra et al. 1998; Norris et al. 2003).

THE PERIOD OF MOLT

Most adult passerines undergo a complete prebasic molt prior to autumnal migration, at stopover sites en route, or upon reaching the winter quarters (see references in Hall and Fransson 2000). The obvious benefit to migrating birds of an aerodynamically sound plumage renders molt a very important process, one that may even affect the number of reproductive attempts (Evans Ogden and Stutchbury 1996). Hormones are involved in the timing of molt, although their precise roles are under debate. For example, thyroid hormones play an important function. Several studies (including those of tropical and subtropical species) have demonstrated that thyroidectomy can eliminate prebasic molt if performed at the appropriate time in the photoperiodic cycle, and molt can be restored if thyroidectomy is combined with thyroxine replacement therapy (Dawson et al. 2001; Dawson and Thapliyal 2001; Kuenzel 2003). Questions remain as to whether thyroxine is permissive (i.e., must be present but has no direct role) or plays a central role in regulating the process (Dawson et al. 2001). A second hormone, prolactin, peaks near the end of the breeding season in conjunction with gonadal regression (Sharp and Sreekumar 2002). It also appears to be important in initiating molt, for if its release from the anterior pituitary is blocked, then prebasic molt is inhibited (Dawson and Sharp 1998).

Molt may be energetically demanding (Murphy and King 1992) and most migrants must wait until the termination of breeding before beginning to molt. Reproductive hormones, which are elevated during breeding, delay the onset of molt until breeding is nearly completed (Dawson 1994). Interestingly, late-onset molt, which may occur in renesting birds, can be accelerated and completed in some species before departure, thereby keeping them on track for autumnal migration (Hall and Fransson 2000). However, an attempt to stay on schedule by accelerating the molt process may incur a fitness cost, as rapid feather replacement rate can occur at the expense of feather quality (Dawson et al. 2000). In spring, some species exhibit an extensive prealternate molt into the breeding plumage. In contrast to prebasic molt, pre-alternate molt requires gonadal hormones for its completion (Hannon and Wingfield 1990; Peters et al. 2000). The timing of prebasic molt in tropical species is much more variable than in temperate breeders, and in many—but not all—species it may overlap with breeding activities (Gwinner and Dittami 1990). This suggests that the selective pressures associated with molt in tropical residents are somewhat relaxed, or at least different, from those experienced by temperate breeders.

THE TRANSITION PERIODS OF SPRING AND FALL MIGRATION

Unlike their resident temperate or tropical neighbors, twice each year migratory birds experience a time in which their social and physical environment becomes less familiar and much less predictable. For many species, the period of migration poses a significant risk to annual survivorship compared with other stages of the annual cycle (Butler 2000; Sillett and Holmes 2002). The loss of critical energy reserves en route can influence how long an individual needs to replenish them at stopover sites and, ultimately, a migrant's timely progress along the migration route (Moore and Simons 1992). While events during the premigratory and migratory periods may delay a migrant's arrival at its destination, spring migrants face the additional demands of allocating energy to reproduction, with males experiencing significant gonadal activity and the development of T-dependent accessory tissues such as the cloacal protuberance. Male passerines can also produce sperm and may even copulate before reaching the breeding grounds (Moore and McDonald 1993; Briskie 1996). Although females of most species tend to arrive after their male counterparts, they, too, face a transect of increasing environmental stochasticity as they move northward. Female arrival date and energetic condition greatly influence the timing of egg production and, thus, the breeding success for both sexes (Winkler and Allen 1996; Smith and Moore 2003).

A critical distinction between migrants and nonmigrants is that the former experience dramatic changes in feeding patterns, metabolic activity, body composition, and behavior during periods of the annual cycle associated with migration. Migratory condition is characterized by a dramatic shift in food preference, increased food intake rate (hyperphagia), fat deposition, and, in some species, muscle hypertrophy (Marsh 1984; Gaunt et al. 1990; McWilliams and Karasov 2001; Long and Stouffer 2003). In addition, captive birds in migratory condition often show increased locomotor activity, expressed at night by nocturnal migrants and during the day by diurnal ones. This migratory unrest, called *Zugunruhe,* has been used extensively as a tool for captive studies on migratory birds (e.g., Berthold 1996; Gwinner 1996). However, despite its widespread use, virtually nothing is known about the basic biology of *Zugunruhe:* Where within the central nervous system is it generated? What is its neuroendocrine basis, and how do different stimuli, such as photoperiod, food availability, and change in body mass, influence its expression? It is surprising that we know so little about a tool that has played a central role for over a century in pivotal studies on migratory behavior, genetics, and physiology.

With an increase in day length in spring, birds at similar latitudes, whether they are migratory or sedentary, wintering in the Temperate Zone or the Tropics, experience the same external cues. However, only the migrants undergo the changes in neural and metabolic pathways regulating the development of migratory condition. For many migrants, hyperphagia, lipogenesis, and the impetus for extended flight are stimulated not only at initial departure, but repeatedly throughout the journey as they stop to replenish energy reserves. In terms of energy, migration is a series of stages characterized by energy intake and storage, energy use, and ultimately, rest / recovery, which is usually repeated more than once until a bird reaches its destination (e.g., Loria and Moore 1990; Kuenzi et al. 1991).

An important prerequisite for energy storage is energy intake—the type and amount of food—and the rate at which it is ingested. Migrants must increase their food intake not only to meet the additional demands of migratory flight but also to have reserves should they encounter periods of poor environmental conditions. Carbohydrates, lipids, and proteins provide the energy for cellular respiration. Plasma glucose is the most readily accessible energy source, and the pancreatic hormones, insulin and glucagon, work together to regulate glucose availability. Insulin is essential for storing glucose in liver and muscle as glycogen, and glucagon converts glycogen back into glucose to be released back into the blood when plasma glucose levels decline. In mammals, decreased plasma glucose levels stimulate food intake and increased levels suppress feeding. However, unlike mammals, birds do not use plasma glucose levels to regulate feeding activity (Boswell et al. 1995a, 1997). Instead, they maintain fairly constant plasma glucose levels even when undergoing short (overnight) or extended (days) periods of fasting (Le Maho et al. 1981). By avoiding dramatic fluctuations of available energy in the blood, birds are able to support higher mass-specific metabolic rates than the other homeotherms and experience rapid growth and energetically expensive activities such as flight.

If birds do not use glucose as a feeding cue, then what stimulates refueling behavior? Feeding signals include those made in the brain and those that reach the brain from the digestive system (for review, see Boswell 2001). Studies on migratory birds held on short-day photoperiod showed that experimental administration of insulin or glucagon had the common effect of inhibiting feeding and raising plasma concentrations of free fatty acids (FFAs) (Boswell et al. 1995a, 1997). Because FFAs contain more energy per gram than glucose and fluctuate with lipid utilization, birds may use levels of plasma FFA and other lipid metabolites as more reliable indicators of their energy reserves. However, there have been so few studies on migratory birds that there is still much to learn about the signals they use in regulating their feeding behavior and how signaling systems may vary with different migration strategies.

Chemical signals from the digestive system and brain that regulate vertebrate feeding behavior comprise an "alphabet soup" of protein/peptide hormones, and only a few have been investigated in birds. The biosynthesis of neuropeptide Y (NPY) in the feeding centers of the hypothalamus of migratory birds increases during fasting periods to stimulate feeding behavior (Richardson et al. 1995; Boswell et al. 2002). Not only does NPY stimulate increased food intake and, possibly, genes encoding lipogenic enzymes (Woods et al. 1998), but the brain's sensitivity to NPY increases as migrants come into migratory condition (Richardson et al. 1995). An altered sensitivity to NPY and other neurochemical feeding signals may represent one of the fundamental changes in the brain that occur with the development of migratory condition. Systematic studies of these changes are sorely needed in migrants and nonmigrants alike to better understand the seasonal divergence between the two distinct life history strategies.

Migrants end a feeding bout in response to information from the gut signaling its "fullness." The peptide cholecystokinin (CCK), released as food passes through the gut, reduces feeding behavior (Richardson et al. 1993; Boswell and Li 1998). In nonmigratory birds, CCK acting in the gut influences the signaling activity of the neural pathways connecting the digestive system to the appetite-regulating centers in the brain. How the activity of this signaling mechanism is influenced by changes in appetite and digestion during migration remains to be explored.

The distances that migrants travel and the behavioral and physiological strategies they use to reach their destinations are diverse. Variation in time and energy constraints is reflected in the different choices that migrants make in habitat preference, their differences in feeding behavior, and the metabolic processes needed by them to accumulate and use energy reserves (Jenni and Jenni-Eiermann 1992; Schaub and Jenni 2000). For any migrant, the ability to fly nonstop for hours or days requires significant changes in the structures that power flight and the metabolic activities that support it. Assimilation efficiency is enhanced during migration by significant changes in the digestive system (for review, see McWilliams and Karasov 2001; Bairlein 2002). Through changes in digestive activity and intermediary metabolism, migrants are able to increase body mass more quickly and to become more efficient at glucose utilization during the migratory period (Totzke et al. 1998). Because liver and muscle glycogen stores would be used up during the first hour of flight (Farner et al. 1961), migrants rely almost exclusively on lipids as an energy source and often shift their diets to foods high in fatty acids to support greater fat deposition rates (Jenni-Eiermann and Jenni 1996; Bairlein 2002). Enzyme activity associated with lipid synthesis and storage is higher during migration (Guglielmo et al. 1998; Williams et al. 1999; Egeler et al. 2000), and flight muscles increase their ability to oxidize free fatty acids as they become available from the gut or from lipid stores (Marsh 1981). The skeletal muscles of the legs and pectoralis can become an important source of energy for migrants when fat reserves drop below a threshold (Jenni et al. 2000; Lindström et al. 2000), and migrants become more efficient at sparing this source of protein during the migratory period (Jenni-Eiermann and Jenni 1996).

Chemical signals from the gut and blood can provide information about energetic state on a relatively short time scale throughout the day. But, how does a migrant know when it has stored enough energy to fuel a bout of migratory flight? This information is critical for migrants, particularly for those preparing to cross large areas of land or water where refueling is impossible. Signals from energy-storing tissues about their status could act as cues for many components of migratory behavior: the cessation of feeding, the feeling of "readiness" or restlessness, or even an increased sensitivity to sensory cues needed for orientation. One such candidate, the protein hormone leptin, may hold promise for studies of migration physiology. It is widely accepted that, in mammals, leptin secreted primarily from adipose tissue provides a reliable signal to the brain of an individual's fat content (Zhang et al. 1994). Leptin receptors have been found in the mammalian brain, particularly in the areas regulating feeding behavior. Also in mammals, leptin and insulin together act as powerful signals for influencing food intake, with increasing blood concentrations stimulating the expression of catabolic neuropeptides within the brain, and, conversely, declining concentrations stimulating the biosynthesis of anabolic neuropeptides (Woods et al. 1998; Ellacott et al. 2002). However, there is much work to be done in this area within the field of avian biology. The molecular structure of leptin has yet to be characterized in non-mammalian vertebrates, with the exception of sequences from the chicken that have created some controversy (Friedman-Einat et al. 1999; Dunn et al. 2001). Several recent studies have reported a decrease in food intake in domestic and wild birds treated with either mammalian leptin or leptin derived from chickens (Denbow et al. 2000; Løhmus et al. 2003a). But although these results suggest that migrants could use leptin as a "fuel tank" signal, more work is needed to determine if leptin does, indeed, provide valuable information to birds, especially to migrants who must make critical decisions about when and where to go based on their energy reserves.

Another protein hormone, prolactin, is secreted primarily from the anterior pituitary. In birds, prolactin plays an important role in parental care, osmoregulation, and feeding behavior. Several studies have found that low plasma prolactin levels observed under short-day, winterlike conditions rise in response to increasing day length in spring, when migrants begin to eat more and to put on fat reserves (Boswell et al. 1995b; Dawson et al. 2001; Holberton et al., unpubl. data). Prolactin stimulates food intake in nonmigratory and migratory species during both the nonmigratory and migratory periods (Buntin and Tesch 1985; Boswell et al. 1995b). However, although prolactin has been implicated in stimulating migratory fattening, evidence suggests that it plays only an indirect role through its stimulation of food intake, production of growth factors (Mick and Nicoll 1985) and lipid storage enzymes (Garrison and Scow 1975), as well as hypertrophy of the gut during migration (see references in Boswell et al. 1995b).

Corticosterone has been linked to the development of migratory condition since the 1960s (Meier et al. 1965). More recent studies have shown that plasma baseline levels of corticosterone can be elevated above normally low baseline levels during the migratory period, particularly if the birds are in the process of putting on fat (Holberton et al. 1996, 1999; Holberton 1999; but see Romero et al. 1997). Laboratory studies on several species of short- and long-distance migrant songbirds have shown that corticosterone plays an important role in the fattening process and a minor, if any, part in facilitating migratory hyperphagia. When the seasonal elevation in baseline corticosterone was chemically blocked in Dark-eyed Juncos and Yellow-rumped Warblers (*Dendroica coronata*) during the migratory period, the birds were unable to fatten in spite of the fact that they showed the same increase in food intake normally seen when birds come into migratory condition (Holberton, unpubl. data). These results suggest that intermediate levels of corticosterone act as an important regulator of one or more of the metabolic processes involved with the accumulation of migratory fat reserves (e.g., lipid synthesis, transport, and storage). However, corticosterone's link with increased food intake, as observed in a nonmigratory context (Wingfield and Silverin 1986; Gray et al. 1990), either becomes less direct or is lost altogether during migration.

Unfortunately, most of the experimental work manipulating endogenous corticosterone levels in migratory birds is limited to songbirds. Researchers are beginning to uncover additional roles for corticosterone during migration. For example, several recent studies have shown that for birds nearing the end of a stopover bout, corticosterone may serve as a departure (Piersma et al. 2000) and orientation cue (Lõhmus et al. 2003b). Although the concentration- and context-dependent aspects of corticosterone as an energy regulator in birds make it a challenging hormone for behavioral and physiological studies, carefully designed studies both in the field and in the laboratory hold promise for uncovering the hormone's links with the great diversity of migration strategies. The wide array of migration strategies within and across species provides a rich palette from which a clearer picture can be constructed to illustrate the complex role this hormone plays in regulating migratory behavior.

Energy reserves fluctuate during migration, and, as seen during the breeding and wintering periods, the strength of the adrenocortical stress response often varies with energetic condition (Jenni et al. 2000; Long and Holberton 2004; but see Romero et al. 1997). Several studies on songbirds have found that during migration the adrenocortical response can be reduced in such a way that might help protect skeletal muscle from catabolic activity for as long as possible (Holberton et al. 1996, 1999; Holberton 1999; Jenni et al. 2000). When fat reserves are depleted, migrants may secrete high, emergency levels of corticosterone to break down skeletal muscle protein for energy (Schwabl et al. 1991; Holberton et al. 1996, 1999; Jenni et al. 2000). Although high levels of corticosterone may help an individual survive through extreme conditions, survival may be at the expense of energy that could otherwise be invested in preparing for the breeding season and in the immune system (Wingfield et al. 1994a). During spring migration, migrants in poorer body condition on stopover often showed higher levels of corticosterone and had delayed gonadal development compared with their cohorts carrying more energy reserves (Holberton et al. 1999). Poor body condition and delayed gonadal development, as well as the effects that chronically elevated corticosterone might have on immunity and general health, may ultimately compound the effects of late arrival on the breeding grounds. In contrast, essentially nothing is known about body condition and arrival time on the wintering grounds and their impact on overwintering success. It is clear that we must have a better understanding of the interrelationships among different stages of the annual cycle and how conditions encountered by birds during migration influence their subsequent survival and ultimate reproductive success (Loria and Moore 1990; Holmes et al. 1992; Smith and Moore 2003).

FUTURE DIRECTIONS

Migrants and residents alike have developed behavioral and physiological capabilities to survive in the multiple worlds they experience in space and time. In spite of hundreds of years of human interest in the movements of birds, we are just beginning to uncover some of the changes that occur during the transition period between seasons. However, we know much more about where and when migrants travel than we do about the underlying mechanisms that help the birds reach their destinations. Similarly, we know little about how residents cope with the impact that the arrival of migrants has on the social and physical environment of residents. To date, the majority of migration studies have focused on Neotropical-Nearctic and Paleotropical-Palearctic species, leaving the ecophysiology of austral migrants essentially terra incognita. Surprisingly, although there may be relatively few altitudinal migrants in the western Northern Hemisphere, altitudinal movements may be a common life history strategy for a great number of species on other continents. But, essentially nothing is known about the cues used to time seasonal events in birds that move along an altitudinal gradient, nor do we understand what factors have shaped the development of this strategy. Where does one begin in the effort to understand the basis of variation in migration strategies? Comparative studies can provide lots of insight into the degree to which species or populations share characteristics of a life history strategy. Ideally, within-species comparisons avoid possible confounding effects of phylogeny. A partial migration system, whereby some individuals within a population migrate and others do not, would provide rich opportunities to examine physiological differences between migrants and nonmigrants within the same species. In North America, several such candidates exist. For example, the eastern population of House Finches (*Carpodacus mexicanus*) developed migratory behavior in just a few generations after the ancestral nonmigrant population was first introduced to the eastern United States (Able and Belthoff 1998). This population has been shown to be a partial migrant (Belthoff and Gauthreaux 1991), but how the two eastern phenotypes differ in response to similar environmental conditions is, as yet, unknown.

The selection of closely related co-occurring species with different life history strategies can provide an excellent setting for looking at the ecophysiology of migratory behavior. For example, RLH has been studying the ecological, behavioral, and physiological basis of two such species of wood-warblers in North America. The Blackpoll Warbler (*Dendroica striata*) exhibits the most dramatic autumn migration of any Neotropical passerine. This small bird (11–15 g lean body mass) can double its body mass with lipid reserves in preparation for a 4- to 5-day nonstop transoceanic flight over the North Atlantic to winter in South America (Nisbet 1970; Williams and Williams 1990; Nisbet et al. 1995). It has been estimated that Blackpolls have the greatest fuel efficiency during migration compared with other warblers (Hussell and Lambert 1980). In contrast, the Yellow-rumped Warbler (*D. coronata*), whose breeding range overlaps greatly with that of the Blackpoll Warbler, migrates initially over a shorter distance. Yellow-rumped Warblers can then make facultative movements throughout the fall and winter in response to poor weather and decreased food supply if needed (Terrill and Ohmart 1984). In autumn, both species stop over in the same habitats but show dramatically different physiological preparations for their respective journeys. Recent work suggests that the Blackpoll's ability to put on such extensive fat reserves (see fig. 23.1) is rooted in a dramatic change in lipid synthesis and storage not experienced by Yellow-rumped Warblers. Figure 23.2 illustrates that when the two species were sampled on the breeding grounds prior to migration (sampling time 1), they showed similar triglyceride levels (A), fat score (B), and body mass (C) (Blackpoll Warbler, $n = 46$; Yellow-rumped Warbler, $n = 19$; ages and sexes pooled for both species, Mann-Whitney *U*-test, $p > 0.05$ for species comparisons). However, when sampled during the migratory period (fig. 23.2, sampling times 2 and 3, in captivity), Blackpolls ($n = 9$) had significantly greater body mass, fat score, and plasma triglyceride levels than Yellow-rumped Warblers ($n = 17$) (Mann-Whitney *U*-test, $p < 0.05$, for species comparisons, ages and sexes pooled). Surprisingly, although both species showed an increase in daily food intake while coming into migratory condition in captivity, Blackpolls did not eat more than Yellow-rumped Warblers (data not shown). Higher plasma triglyceride levels could represent greater digestive ability to absorb and assimilate fats and greater lipogenic activity by the liver, and work is underway to uncover how these two closely related and co-occurring species are able to support their divergent migration strategies.

Another area that deserves additional investigation is the relationship among hormones, migration, and the immune system. Hormones associated with migration, such as glucocorticoids, are known to affect immunoresponsiveness (Sapolsky 1992), and immunocompetence has recently been linked to the evolution of a variety of life history traits (Tella et al. 2002). However, little work has been done that explores hormones and immunological responses of birds during migration or to examine the fitness ramifications of those responses as compared with resident species.

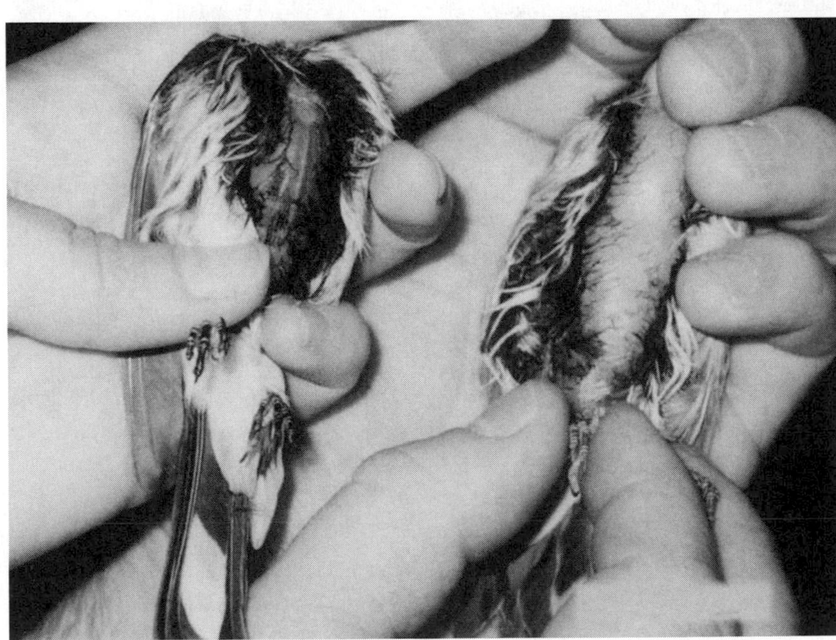

Fig. 23.1. This photo illustrates the extensive migratory fat reserves put on by the long distance migrant, the Blackpoll Warbler (right), shown here at 32.0 g, compared with the short-distance migrant, the Yellow-rumped Warbler (left), shown here at 13.0 g, during autumn migration. © R. Holberton.

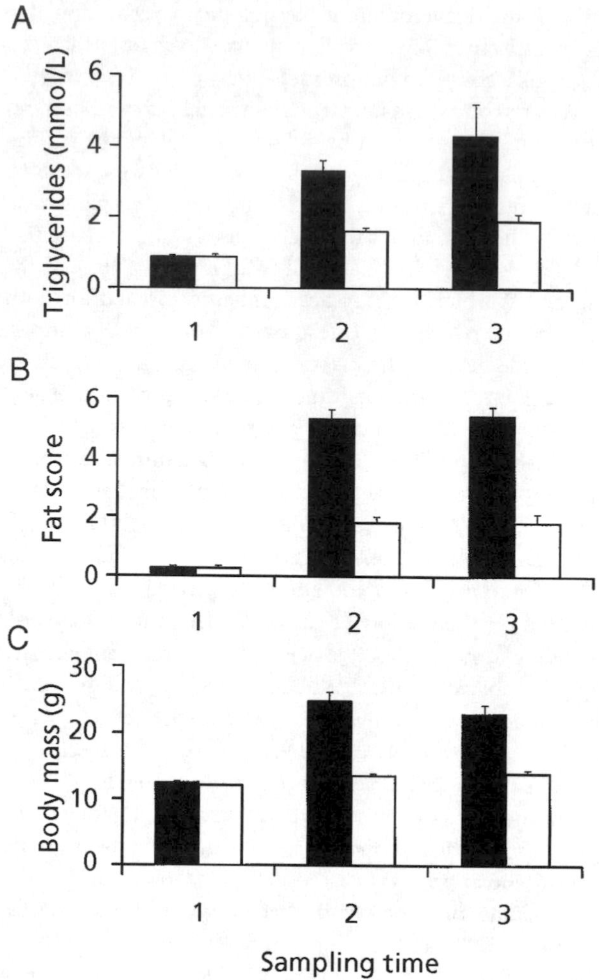

Fig. 23.2. Plasma triglycerides levels (A), fat score (B), and body mass (C) in Blackpoll (black bars) and Yellow-rumped Warblers (white bars) before departure from the breeding grounds (sampling time 1, Churchill, Manitoba, 1 July–20 August 1999, 2000) and in captivity during the autumn migration period (sampling times 2 and 3, 1 October–30 November 1999). Before migration, both species had similar triglyceride levels (A), fat score (B), and body mass (C) (sampling time 1: Blackpoll Warbler, $n = 46$; Yellow-rumped Warbler, $n = 19$; ages and sexes pooled for both species, Mann-Whitney *U*-test, $p > 0.05$ for all comparisons). When sampled during migration (sampling times 2 and 3, in captivity), Blackpolls ($n = 9$) had significantly greater body mass, fat reserves, and plasma triglyceride levels than Yellow-rumped Warblers ($n = 17$) (Mann-Whitney *U*-test, $p < 0.05$, for species comparisons, ages and sexes pooled).

Ecophysiological studies can be done at any level of organization, from the whole animal down to the regulation of the genes. Within- and across-season patterns of hormones circulating in the bloodstream are still not known for a great many of these endocrine signals. Hormones regulate cellular processes from the initiation of zygote development, through growth and development, and throughout the life of the individual (Dufty et al. 2002), but how these processes relate to differences in life history strategies of birds is largely uncharted territory. The basic cellular or molecular mechanisms through which environmental cues affect changes in neuroendocrine and endocrine pathways regulating feeding and social behavior, body composition, flight performance, digestive responses, biological clocks, and sensory mechanisms related to orientation information wait to be uncovered. We are just beginning to discover some of the neuroendocrine pathways of migratory feeding and metabolic aspects of energy use, but only a few species have been investigated to date. Taking inquiry across the various levels of the whole organism and, ultimately, down to the molecular level will enhance our understanding of the evolutionary bases for differences between migratory and nonmigratory birds.

Our approach to behavioral endocrine studies has been to integrate field and laboratory studies. Although field studies initially provide an opportunity to uncover naturally occurring patterns between variables of interest, laboratory studies allow methods for measuring or manipulating hormones of interest to be developed, variables of interest to be "calibrated" in the absence of confounding ones, and the testing of hypotheses derived from patterns observed in the field. Ideally, these would be followed up by more field studies in which experimental methods can be applied or further observations made. One need not be a "master of all trades"; collaborations with researchers whose strengths in other areas balance one's own can be richly rewarding, and current trends within some funding agencies favor those projects that integrate different levels of inquiry.

Finally, birds serve as bioindicators of the health of our environment and more work is needed to understand how they serve as vectors of disease, how disease susceptibility and transmission may change during the annual cycle, and how these changes relate to endocrine activity. Clearly, understanding the constraints on the adaptive capabilities of birds could allow us to predict how migrants and residents alike will cope with a world of dramatic global change.

ACKNOWLEDGMENTS

We thank Jennifer Long, Timothy Boswell, and an anonymous reviewer for helpful comments on an earlier version of this manuscript. RLH has been supported by the National Science Foundation, the Northern Research Fund (Churchill Northern Studies Centre), and the Humboldt Institute of Field Research at Eagle Hill.

LITERATURE CITED

Able, K. P., and J. R. Belthoff. 1998. Rapid "evolution" of migratory behaviour in the introduced house finch of eastern North America. Proceedings of the Royal Society of London, Series B, Biological Sciences 265:2063–2071.

Bairlein, F. 2002. How to get fat: nutritional mechanisms of seasonal fat accumulation in migratory songbirds. Naturwissenschaften 89:1–10.

Belthoff, J. R., and S. A. Gauthreaux Jr. 1991. Partial migration and differential winter distribution of house finches in the eastern United States. Condor 93:374–382.

Berthold, P. 1996. Control of Bird Migration. Chapman and Hall, London.

Boswell, T. 2001. Regulation of feeding by neuropeptide Y. Pages 349–360 *in* Avian Endocrinology (A. Dawson and C. M. Chaturvedi, eds.). Narosa Publishing House, New Delhi.

Boswell, T., T. L. Lehman, and M. Ramenofsky. 1997. Effects of plasma glucose manipulations on food intake in white-crowned sparrows. Comparative Biochemistry and Physiology, Series A, Comparative Physiology 118:721–726.

Boswell, T., and Q. Li. 1998. Cholecystokinin induces *fos* expression in the brain of Japanese quail. Hormones and Behavior 34:56–66.

Boswell, T., Q. Li, and S. Takeuchi. 2002. Neurons expressing neuropeptide mRNA in the infundibular hypothalamus of Japanese quail are activated by fasting and co-express agouti-related protein mRNA. Molecular Brain Research 100:31–42.

Boswell, T., R. D. Richardson, R. J. Seeley, M. Ramenofsky, J. C. Wingfield, M. I. Friedman, and S. C. Woods. 1995a. Regulation of food intake by metabolic fuels in white-crowned sparrows. American Journal of Physiology 269:R1462–R1468.

Boswell, T., P. J. Sharp, M. R. Hall, and A. R. Goldsmith. 1995b. Migratory fat deposition in European quail: a role for prolactin? Journal of Endocrinology 146:71–79.

Breuner, C. W., M. Orchinik, T. P. Hahn, S. L. Meddle, I. P. Moore, N. T. Owen-Ashley, T. S. Sperry, and J. C. Wingfield. 2003. Differential mechanisms for regulation of the stress response across latitudinal gradients. American Journal of Physiology (Regulatory, Integrative and Comparative Physiology) 285:R594–R600.

Briskie, J. V. 1996. Lack of sperm storage by female migrants and the significance of copulations *en route.* Condor 98:414–417.

Buntin, J. D., and D. Tesch. 1985. Effects of intracranial prolactin administration on maintenance of incubation readiness, ingestive behavior, and gonadal condition in ring doves. Hormones and Behavior 19:188–203.

Butler, R. W. 2000. Stormy seas for some North American songbirds: are declines related to severe storms during migration? Auk 117:518–522.

Canoine, V., and E. Gwinner. 2002. Seasonal differences in the hormonal control of territorial aggression in free-living European stonechats. Hormones and Behavior 41:1–8.

Dawson, A. 1994. The effects of daylength and testosterone on the initiation and progress of moult in starlings, *Sturnus vulgaris.* Ibis 136:335–340.

Dawson, A., and A. R. Goldsmith. 1983. Plasma prolactin and gonadotrophins during gonadal development and the onset of photorefractoriness in male and female starlings (*Sturnus vulgaris*) on artificial photoperiods. Journal of Endocrinology 97:253–260.

Dawson, A., and A. R. Goldsmith. 1997. Changes in gonadotrophin-releasing hormone (GnRH-I) in the pre-optic area and median eminence of starlings (*Sturnus vulgaris*) during the recovery of photosensitivity and during photostimulation. Journal of Reproduction and Fertility 111:1–6.

Dawson, A., S. A. Hinsley, P. N. Ferns, R. H. C. Bonser, and L. Eccleston. 2000. Rate of moult affects feather quality: a mechanism linking current reproductive effort to future survival. Proceedings of the Royal Society of London, Series B, Biological Sciences 267:2093–2098.

Dawson, A., V. M. King, G. E. Bentley, and G. F. Ball. 2001. Photoperiodic control of seasonality in birds. Journal of Biological Rhythms 16:366–381.

Dawson, A., and P. J. Sharp. 1998. The role of prolactin in the development of reproductive photorefractoriness and postnuptial molt in the European starling (*Sturnus vulgaris*). Endocrinology 139:485–490.

Dawson, A., and J. P. Thapliyal. 2001. The thyroid and photoperiodism. Pages 141–154 *in* Avian Endocrinology (A. Dawson and C. M. Chaturvedi, eds.). Narosa Publishing House, New Delhi.

Denbow, D. M., S. Meade, A. Robertson, J. P. McMurtry, M. Richards, and C. Ashwell. 2000. Leptin-induced decrease in food intake in chickens. Physiology and Behavior 69:359–362.

Dufty, A. M., Jr. 1989. Testosterone and survival: a cost of aggressiveness? Hormones and Behavior 23:185–193.

Dufty, A. M., Jr., J. Clobert, and A. P. Møller. 2002. Hormones, developmental plasticity and adaptation. Trends in Ecology and Evolution 17:190–196.

Dunn, I. C., G. Girishvarma, R. T. Talbot, D. Waddington, T. Boswell, and P. J. Sharp. 2001. Evidence for low homology between the chicken and mammalian leptin genes. Pages 327–336 *in* Avian Endocrinology (A. Dawson and C. M. Chaturvedi, eds.). Narosa Publishing House, New Delhi.

Egeler, O., T. D. Williams, and C. G. Guglielmo. 2000. Modulation of lipogenic enzymes, fatty acid synthase and delta-9-desaturase, in relation to migration in the western sandpiper (*Calidris mauri*). Journal of Comparative Physiology, Series B, Biochemical, Systematic, and Environmental Physiology 170:169–174.

Ellacott, K. L., C. B. Lawrence, N. J. Rothwell, and S. M. Luckman. 2002. PRL-Releasing peptide interacts with leptin to reduce food intake and body weight. Endocrinology 143: 368–374.

Evans Ogden, L. J., and B. J. M. Stutchbury. 1996. Constraints on double brooding in a Neotropical migrant, the Hooded Warbler. Condor 98:736–744.

Farner, D. S., A. Oksche, F. I. Kamenoto, J. R. King, and H. E. Cheyney. 1961. A comparison of the effect of long daily photoperiods on the pattern of energy storage in migratory and non-migratory finches. Comparative Biochemistry and Physiology 2:125–142.

Friedman-Einat, M., T. Boswell, G. Horev, G. Girishvarma, I. C. Dunn, R. T. Talbot, and P. J. Sharp. 1999. The chicken leptin gene: has it been cloned? General and Comparative Endocrinology 115:354–363.

Garrison, M. M., and R. O. Scow. 1975. Effect of prolactin on lipoprotein lipase in crop sac and adipose tissue of pigeons. American Journal of Physiology 228:1542–1544.

Gaunt, A. S., R. S. Hikida, J. R. Jehl Jr., and L. Fenbert. 1990. Rapid atrophy and hypertrophy of an avian flight muscle. Auk 107:649–659.

Gray, J., D. Yarian, and M. Ramenofsky. 1990. Corticosterone, foraging behavior, and metabolism in dark-eyed juncos, *Junco hyemalis.* General and Comparative Endocrinology 79:375–384.

Guglielmo, C. G., N. H. Haunerland, and T. D. Williams. 1998. Fatty acid binding protein, a major protein in the flight muscle of migrating Western Sandpipers. Comparative Biochemistry and Physiology, Series B, Metabolic and Transport Functions 119:549–555.

Gwinner, E. 1996. Circadian and circannual programmes in avian migration. Journal of Experimental Biology 199:39–48.

Gwinner, E., and J. Dittami. 1990. Endogenous reproductive rhythms in a tropical bird. Science 249:906–908.

Gwinner, E., and A. Scheuerlein. 1998. Seasonal changes in daylight intensity as a potential *Zeitgeber* of circannual rhythms in equatorial Stonechats. Journal für Ornithologie 139:407–412.

Hall, K. S. S., and T. Fransson. 2000. Lesser Whitethroats under time-constraint moult more rapidly and grow shorter wing feathers. Journal of Avian Biology 31:583–587.

Hannon, S. J., and J. C. Wingfield. 1990. Endocrine correlates of territoriality, breeding stage, and body molt in free-living willow ptarmigan of both sexes. Canadian Journal of Zoology 68:2130–2134.

Hau, M. 2001. Timing of breeding in variable environments: tropical birds as model systems. Hormones and Behavior 40:281–290.

Hau, M., M. Wikelski, K. K. Soma, and J. C. Wingfield. 2000a. Testosterone and year-round territorial aggression in a tropical bird. General and Comparative Endocrinology 117:20–33.

Hau, M., M. Wikelski, and J.C. Wingfield. 1998. A Neotropical forest bird can measure the slight changes in tropical photoperiod. Proceedings of the Royal Society of London, Series B, Biological Sciences 265:8995.

Hau, M., M. Wikelski, and J. C. Wingfield. 2000b. Visual and nutritional food cues fine-tune timing of reproduction in a Neotropical rainforest bird. Journal of Experimental Zoology 286:494–504.

Hillgarth, N., and J. C. Wingfield. 1997. Parasite-mediated sexual selection: endocrine aspects. Pages 78–104 *in* Host-Parasite Evolution (D. H. Clayton and J. Moore, eds.). Oxford University Press, Oxford.

Holberton, R. L. 1999. Changes in patterns of corticosterone secretion concurrent with migratory fattening in a Neotropical migratory bird. General and Comparative Endocrinology 116:49–58.

Holberton, R. L., and K. P. Able. 2000. Differential migration and an endocrine response to stress in wintering dark-eyed juncos (*Junco hyemalis*). Proceedings of the Royal Society of London, Series B, Biological Sciences 267:1889–1896.

Holberton, R. L., P. P. Marra, and F. R. Moore. 1999. Endocrine aspects of physiological condition, weather and habitat quality in landbird migrants during the non-breeding period. Pages 847–866 *in* Proceedings of the 22nd International Ornithological Congress (N. Adams and R. Slotow, eds.). BirdLife South Africa, Johannesburg.

Holberton, R. L., J. D. Parrish, and J. C. Wingfield. 1996. Modulation of the adrenocortical stress response in Neotropical migrants during autumn migration. Auk 113:558–564.

Holberton, R. L., and J. C. Wingfield. 2003. Modulating the corticosterone stress response: a mechanism for balancing individual risk and reproductive success in Arctic breeding sparrows? Auk 120:1140–1150.

Holmes, R. T., T. W. Sherry, P. P. Marra, and K. E. Petit. 1992. Multiple-brooding, nesting success, and annual productivity of a Neotropical migrant, the Black-throated Blue Warbler, *Dendroica caerulescens,* in an unfragmented temperate forest. Auk 109:321–333.

Holmes, W. N., and J. G. Phillips. 1976. The adrenal cortex of birds. Pages 293–420 *in* General and Comparative Endocrinology of the Adrenal Cortex (I. Chester-Jones and I. Henderson, eds.). Academic Press, New York.

Hunt, K., J. C. Wingfield, L. B. Astheimer, W. A. Buttemer, and T. P. Hahn. 1995. Temporal patterns of territorial behavior and circulating testosterone in the Lapland longspur and other Arctic passerines. American Zoologist 35:274–284.

Hussell, D. J. T., and A. B. Lambert. 1980. New estimates of weight loss in birds during nocturnal migration. Auk 97:547–558.

Jenni, L., and S. Jenni-Eiermann. 1992. Metabolic patterns of feeding, overnight fasted and flying night migrants during autumn migration. Ornis Scandinavica 23:251–259.

Jenni, L., S. Jenni-Eiermann, F. Spina, and H. Schwabl. 2000. Regulation of protein breakdown and adrenocortical response to stress in birds during migratory flight. American Journal of Physiology (Regulatory, Integrative and Comparative Physiology) 278:R1182–R1189.

Jenni-Eiermann, S., and L. Jenni. 1996. Metabolic differences between the postbreeding, moulting and migratory periods in feeding and fasting passerine birds. Functional Ecology 10:62–72.

Kuenzel, W. J. 2003. Neurobiology of molt in avian species. Poultry Science 82:981–991.

Kuenzi, A. J., F. R. Moore, and T. R. Simons. 1991. Stopover of Neotropical landbird migrants on East Ship Island following trans-Gulf migration. Condor 93:869–883.

Leigh, E. C., A. S. Rand, and D. M. Windsor (eds.). 1996. The Ecology of a Tropical Forest: Seasonal Rhythms and Long-term Changes. Smithsonian Institution Press, Washington, D.C.

Le Maho, Y., H. Vu Van Kha, H. Koubi, G. Dewasmes, J. Girard, P. Ferre, and M. Cagnard. 1981. Body composition, energy expenditure, and plasma metabolites in long-term fasting geese. American Journal of Physiology 241:E342–E354.

Lindström, Å, A. Kvist, T. Piersma, A. Dekinga, and M. W. Dietz. 2000. Avian pectoral muscle size rapidly tracks body mass changes during flight, fasting, and fuelling. Journal of Experimental Biology 203:913–919.

Logan, C. A., and J. C. Wingfield. 1990. Autumnal territorial aggression is independent of plasma testosterone in mockingbirds. Hormones and Behavior 24:568–581.

Lõhmus, M., L.F. Sundström, M. El Halawani, and B. Silverin. 2003a. Leptin depresses food intake in great tits (*Parus major*). General and Comparative Endocrinology 131:57–61.

Lõhmus, M., R. Sandberg, R. L. Holberton, and F. R. Moore. 2003b. Corticosterone levels in relation to migratory readiness in red-eyed vireos (*Vireo olivaceus*). Behavioral Ecology and Sociobiology 54:233–239.

Long, J. A., and R. L. Holberton. 2004. Corticosterone secretion, energetic condition, and a test of the Migration Modulation Hypothesis in the Hermit Thrush (*Catharus guttatus*): a short-distance migrant. Auk 121:1094–1102.

Long, J. A., and P.C. Stouffer. 2003. Diet and the preparation for spring migration in captive Hermit Thrushes. Auk 120: 323–330.

Loria, L., and F. R. Moore. 1990. Energy demands of migration on red-eyed vireos, *Vireo olivaceus.* Behavioral Ecology 1:24–35.

Marra, P. P., K. A. Hobson, and R. T. Holmes. 1998. Linking winter and summer events in a migratory bird using stable-carbon isotopes. Science 282:1884–1886.

Marra, P. P., and R. L. Holberton. 1998. Corticosterone levels as indicators of habitat quality: effects of habitat segregation in a migratory bird during the non-breeding season. Oecologia 116:284–292.

Marsh, R. L. 1981. Catabolic enzyme activities in relation to premigratory fattening and muscle hypertrophy in the Gray Catbird, *Dumetella carolinensis.* Journal of Comparative Physiology 141:417–423.

Marsh, R. L. 1984. Adaptations of the Gray Catbird, *Dumetella carolinensis,* to long distance migration: flight muscle hypertrophy associated with elevated body mass. Physiological Zoology 57:105–117.

Matthysen, E. 1993. Nonbreeding social organization in migratory and resident birds. Pages 93–141 *in* Current Ornithology, Vol. 11 (D. M. Power, ed.). Plenum Press, New York.

McWilliams, S. R., and W. H. Karasov. 2001. Phenotypic flexibility in digestive system structure and function in migratory birds and its ecological significance. Comparative Biochemistry and Physiology, Series A, Comparative Physiology 128:579–593.

Meddle, S. L., L. M. Romero, L. B. Astheimer, W. A. Buttemer, I. T. Moore, and J. C. Wingfield. 2002. Steroid hormone interrelationships with territorial aggression in an Arctic-breeding songbird, Gambel's white-crowned sparrow, *Zonotrichia leucophrys gambelii*. Hormones and Behavior 42:212–221.

Meier, A. H., D. S. Farner, and J. R. King. 1965. A possible endocrine basis for migratory behavior in the white-crowned sparrow, *Zonotrichia leucophrys gambelii*. Animal Behaviour 13:453–465.

Mick, C. C. W., and C. S. Nicoll. 1985. Prolactin directly stimulates the liver *in vivo* to secrete a factor (synlactin) which acts synergistically with the hormone. Endocrinology 116: 2049–2053.

Moore, F. R., and M. V. McDonald. 1993. On the possibility that intercontinental landbird migrants copulate *en route*. Auk 110:157–160.

Moore, F. R., and T. R. Simons. 1992. Habitat suitability and stopover ecology of Neotropical landbird migrants. Pages 345–355 *in* Ecology and Conservation of Neotropical Migrant Landbirds (J. M. Hagan III and D. W. Johnston, eds.). Smithsonian Institution Press, Washington, D.C.

Moore, I. T., N. Perfito, H. Wada, T. S. Sperry, and J. C. Wingfield. 2002. Latitudinal variation in plasma testosterone levels in birds of the genus *Zonotrichia*. General and Comparative Endocrinology 129:13–19.

Murphy, M. E., and J. R. King. 1992. Energy and nutrient use during moult by white-crowned sparrows, *Zonotrichia leucophrys gambelii*. Ornis Scandinavica 23:304–313.

Nisbet, I. C. T. 1970. Autumn migration of the Blackpoll warbler: evidence for long flight provided by regional survey. Bird-Banding 41:207–240.

Nisbet, I. C. T., D. B. McNair, W. Post, and T. C. Williams. 1995. Transoceanic migration of the Blackpoll warbler: summary of scientific evidence and response to criticisms by Murray. Journal of Field Ornithology 66:612–622.

Norris, D. R., P. P. Marra, K. K. Kyser, T. Sherry, and L. M. Ratcliffe. 2003. Tropical winter habitat limits reproductive success on the temperate breeding grounds in a migratory bird. Proceedings of the Royal Society of London, Series B, Biological Sciences 271:59–64.

Pérez-Tris, J., and J. L. Tellería. 2002. Migratory and sedentary blackcaps in sympatric non-breeding grounds: implications for the evolution of avian migration. Journal of Animal Ecology 71:211–224.

Peters, A., L. B. Astheimer, C. R. J. Boland, and A. Cockburn. 2000. Testosterone is involved in acquisition and maintenance of sexually selected male plumage in Superb Fairywrens, *Malurus cyaneus*. Behavioral Ecology and Sociobiology 47:438–445.

Piersma, T., J. Reneerkens, and M. Ramenofsky. 2000. Baseline corticosterone peaks in shorebirds with maximal energy stores for migration: a general preparatory mechanism for rapid behavioural and metabolic transitions? General and Comparative Endocrinology 120:118–126.

Raouf, S. A., P. G. Parker, E. D. Ketterson, V. Nolan Jr., and C. Ziegenfus. 1997. Testosterone affects reproductive success by influencing extra-pair fertilizations in male dark-eyed juncos (Aves: *Junco hyemalis*). Proceedings of the Royal Society of London, Series B, Biological Sciences 264:1599–1603.

Rappole, J. H., and D. W. Warner. 1980. Ecological aspects of migrant bird behavior in Veracruz, Mexico. Pages 353–393 *in* Migrant Birds in the Neotropics: Ecology, Behavior, Distribution, and Conservation (A. Keast and E. S. Morton, eds.). Smithsonian Institution Press, Washington, D.C.

Richardson, R. D., T. Boswell, B. D. Raffety, R. J. Seeley, J. C. Wingfield, and S. C. Woods. 1995. NPY increases food intake in white-crowned sparrows: effect in short and long photoperiods. American Journal of Physiology 268: R1418–R1422.

Richardson, R. D., T. Boswell, S. C. Weatherford, J. C. Wingfield, and S. C. Woods. 1993. Cholecystokinin octapeptide decreases food intake in white-crowned sparrows. American Journal of Physiology 264:R852–R856.

Rogers, C. M., M. Ramenofsky, E. D. Ketterson, V. Nolan Jr., and J. C. Wingfield. 1993. Plasma corticosterone, adrenal mass, winter weather, and season in nonbreeding populations of dark-eyed juncos, *Junco hyemalis hyemalis*. Auk 110:279–285.

Romero, L. M., M. Ramenofsky, and J. C. Wingfield. 1997. Season and migration alters the corticosterone response to capture and handling in an Arctic migrant, the White-crowned Sparrow, *Zonotrichia leucophrys gambelii*. Comparative Biochemistry and Physiology, Series C, Pharmacology, Toxicology, and Endocrinology 116:171–177.

Sapolsky, R. M. 1992. Neuroendocrinology of the stress response. Pages 287–324 *in* Behavioral Endocrinology (J. B. Becker, S. M. Breedlove, and D. Crews, eds.). MIT Press, Cambridge.

Schaub, M., and L. Jenni. 2000. Fuel deposition of three passerine bird species along the migration route. Oecologia 122:306–317.

Schwabl, H., F. Bairlein, and E. Gwinner. 1991. Basal and stress-induced corticosterone levels of Garden Warbler, *Sylvia borin*, during migration. Journal of Comparative Physiology, Series B, Biochemical, Systematic, and Environmental Physiology 161:576–580.

Sharp, P. J. 1996. Strategies in avian breeding cycles. Animal Reproduction Science 42:505–513.

Sharp, P. J., and K. P. Sreekumar. 2002. Photoperiodic control of prolactin secretion. Pages 245–255 *in* Avian Endocrinology (A. Dawson and C. M. Chaturvedi, eds.). Narosa Publishing House, New Delhi.

Sillett, T. S., and R. T. Holmes. 2002. Variation in survivorship of a migratory songbird throughout its annual cycle. Journal of Animal Ecology 71:296–308.

Silverin, B., B. Arvidsson, and J. Wingfield. 1997. The adrenocortical responses to stress in breeding willow warblers, *Phylloscopus trochilus*, in Sweden: effects of latitude and gender. Functional Ecology 11:376–384.

Smith, R. J., and F. R. Moore. 2003. Arrival fat and reproductive performance in a long-distance passerine migrant. Oecologia 134:325–331.

Soma, K. K., K. Sullivan, and J. C. Wingfield. 1999. Combined aromatase inhibitor and antiandrogen treatment decreases territorial aggression in a wild songbird during the nonbreeding season. General and Comparative Endocrinology 115:442–453.

Soma, K. K., A. D. Tramontin, and J. C. Wingfield. 2000. Oestrogen regulates male aggression in the non-breeding season. Proceedings of the Royal Society of London, Series B, Biological Sciences 267:1089–1096.

Soma, K. K., A. M. Wissman, E. A. Brenowitz, and J. C. Wingfield. 2002. Dehydroepiandrosterone (DHEA) increases territorial song and the size of an associated brain region in a male songbird. Hormones and Behavior 41:203–212.

Stutchbury, B. J. 1994. Competition for winter territories in a Neotropical migrant: the role of age, sex and color. Auk 111:63–69.

Stutchbury, B. J. M., and E. S. Morton. 2001. Behavioral Ecology of Tropical Birds. Academic Press, New York.

Tella, J. L., A. Scheuerlein, and R. E. Ricklefs. 2002. Is cell-mediated immunity related to the evolution of life-history strategies in birds? Proceedings of the Royal Society of London, Series B, Biological Sciences 269:1059–1066.

Terrill, S. B., and R. D. Ohmart. 1984. Facultative extension of fall migration by Yellow-rumped Warblers (*Dendroica coronata*). Auk 101:427–438.

Totzke, U., A. Hübinger, and F. Bairlein. 1998. Glucose utilization rate and pancreatic hormone response to oral glucose loads are influenced by the migratory condition and fasting in the garden warbler (*Sylvia borin*). Journal of Endocrinology 158:191–196.

Weathers, W.W. 1977. Climatic adaptation in avian standard metabolic rate. Oecologia 42:8189.

Wikelski, M., M. Hau, W. D. Robinson, and J. C. Wingfield. 1999a. Seasonal endocrinology of tropical passerines: a comparative approach. Pages 1224–1241 *in* Proceedings of the 22nd International Ornithological Congress, Durban (N. J. Adams and R. H. Slotow, eds.). BirdLife South Africa, Johannesburg.

Wikelski, M., M. Hau, and J. C. Wingfield. 1999b. Social instability increases plasma testosterone in a year-round territorial Neotropical bird. Proceedings of the Royal Society of London, Series B, Biological Sciences 266:551–556.

Wikelski, M., M. Hau, and J. C. Wingfield. 2000. Seasonality of reproduction in a Neotropical rain forest bird. Ecology 81:2458–2472.

Williams, T. C., and J. M. Williams. 1990. The orientation of transoceanic migrants. Pages 7–21 *in* Bird Migration: Physiology and Ecophysiology (E. Gwinner, ed.). Springer-Verlag, Berlin.

Williams, T. D., C. G. Guglielmo, O. Egeler, and C. J. Martyniuk. 1999. Plasma lipid metabolites provide information on mass change over several days in captive western sandpipers. Auk 116:994–1000.

Wingfield, J. C. 1984. Androgens and mating systems: testosterone-induced polygyny in normally monogamous birds. Auk 101:665–671.

Wingfield, J. C. 1994. Modulation of the adrenocortical response to stress in birds. Pages 520–528 *in* Perspectives in Comparative Endocrinology (K. G. Davey, R. E. Peter, and S. S. Tobe, eds.). National Research Council of Canada, Ottawa.

Wingfield, J. C., G. F. Ball, A. M. Dufty Jr., R. E. Hegner, and M. Ramenofsky. 1987. Testosterone and aggression in birds. American Scientist 75:602–608.

Wingfield, J. C., P. Deviche, S. Sharbaugh, L. B. Astheimer, R. Holberton, R. Suydam, and K. Hunt. 1994a. Seasonal changes of the adrenocortical responses to stress in redpolls, *Acanthis flammea*, in Alaska. Journal of Experimental Zoology 270:372–380.

Wingfield, J. C., and T. P. Hahn. 1994. Testosterone and territorial behavior in sedentary and migratory sparrows. Animal Behaviour 47:77–89.

Wingfield, J. C., T. P. Hahn, R. Levin, and P. Honey. 1992. Environmental predictability and control of gonadal cycles in birds. *In* Biology of the Chordate Testis (H. Grier and R. Cochran, eds.). Journal of Experimental Zoology 261:214–231.

Wingfield, J. C., T. P. Hahn, M. Wada, L. B. Astheimer, and S. Schoech. 1996. Interrelationship of day length and temperature on the control of gonadal development, body mass and fat depots in white-crowned sparrows, *Zonotrichia leucophrys gambelii*. General and Comparative Endocrinology 101:242–255.

Wingfield, J. C., T. P. Hahn, M. Wada, and S. Schoech. 1997. Effects of day length and temperature on gonadal development, body mass and fat depots in white-crowned sparrows, *Zonotrichia leucophrys pugetensis*. General and Comparative Endocrinology 107:44–62.

Wingfield, J. C., R. E. Hegner, A. M. Dufty Jr., and G. F. Ball. 1990. The "challenge hypothesis": theoretical implications for patterns of testosterone secretion, mating systems, and breeding strategies. American Naturalist 136:829–846.

Wingfield, J. C., J. D. Jacobs, A. D. Tramontin, N. Perfito, S. Meddle, D. L. Maney, and K. Soma. 2000. Toward an ecological basis of hormone-behavior interactions in reproduction of birds. Pages 85–128 *in* Reproduction in Context: Social and Environmental Influences on Reproductive Physiology and Behavior (K. Wallen and J. E. Schneider, eds.). MIT Press, Cambridge.

Wingfield, J. C., M. C. Moore, and D. S. Farner. 1983. Endocrine response to inclement weather in naturally breeding populations of White-crowned Sparrows, *Zonotrichia leucophrys pugetensis*. Auk 100:56–62.

Wingfield, J. C., K. M. O'Reilly, and L. B. Astheimer. 1995. Modulation of the adrenocortical responses to acute stress in Arctic birds: a possible ecological basis. American Zoologist 35:285–294.

Wingfield, J. C., and M. Ramenofsky. 1999. Hormones and the behavioral ecology of stress. Pages 1–51 *in* Stress Physiology in Animals (P. H. M. Balm, ed.). Sheffield Academic Press, Sheffield, U.K.

Wingfield, J. C., and B. Silverin. 1986. Effects of corticosterone on territorial behavior of free living male song sparrows, *Melospiza melodia*. Hormones and Behavior 20:405–417.

Wingfield, J. C., R. Suydam, and K. Hunt. 1994b. Adrenocortical responses to stress in snow buntings and Lapland longspurs at Barrow, Alaska. Comparative Biochemistry and Physiology 108:299–306.

Winkler, D. W., and P. E. Allen. 1996. The seasonal decline in tree swallow clutch size: physiological constraint or strategic adjustment? Ecology 77:922–932.

Woods, S. C., R. J. Seeley, D. Porte Jr., and M. W. Schwartz. 1998. Signals that regulate food intake and energy homeostasis. Science 280:1378–1383.

Wright, S. J., and F. H. Cornejo. 1990. Seasonal drought and leaf fall in a tropical forest. Ecology 71:1165–1175.

Zhang, Y., R. Proenca, M. Maffei, M. Barone, L. Leopold, and J. M. Friedman. 1994. Positional cloning of the mouse obese gene and its human homologue. Nature 372:425–432.

Part 6 / BEHAVIORAL ECOLOGY

MICHAEL S. WEBSTER

Overview

THE BEHAVIOR OF MIGRATORY BIRDS has been the subject of much research, in part because these birds have served as model organisms for developing and testing much of the theory in behavioral and evolutionary ecology. Given this high level of attention, it is almost embarrassing to admit that most of this research has focused on a very limited portion of the annual cycle. Most migratory birds spend about 4 months on the breeding grounds each year, the remaining time being spent on the wintering grounds and in transit, and yet virtually all behavioral and ecological studies of these birds have been conducted during the breeding season. The behavior of migratory birds during the nonbreeding season and the effect of migration and wintering conditions on breeding behavior have been largely overlooked. We therefore remain relatively ignorant about many aspects of migratory bird behavior: What do these birds do during the nonbreeding season? Why do they do it? And how do behaviors in the breeding and nonbreeding seasons interact?

The chapters in this section attempt to rectify this situation by presenting behavioral data gathered on the wintering grounds and by analyzing the scant previously published data that are available. In so doing, these chapters provide new insights into the evolution and ecology of migratory birds, suggest some surprising differences between Old and New World migrants, and unveil a host of new questions that will keep ornithologists busy for years to come.

Price and Gross examine the foraging behavior and ecology of migratory birds on the Indian subcontinent. They focus their attention on a key question: How do foraging adaptations in one season interact with and shape foraging behavior in another? Using *Phylloscopus* warblers as their model, they argue that diversification was driven by adaptations to conditions during the breeding season (e.g., adaptations to foraging at particular altitudes in the Himalayas) and that these adaptations affect the foraging ecology of birds on their wintering grounds. For example, Price and Gross find that species breeding at high altitude tend to hop more while foraging, and this appears to affect the height at which these birds forage during the winter. These patterns appear to have evolved in parallel multiple times, opening the question of how summer and winter foraging adaptations interact in other avian taxa.

As highlighted in the chapter by Greenberg and Salewski, the social behavior of migrants in the winter is poorly known, but these authors nevertheless review the evidence that is available to determine whether intraspecific and/or interspecific variation is related to key ecological variables. They find that most migratory birds appear to be solitary during the winter, but that gregariousness is associated with diet and foraging style. These and other patterns suggest that flocking may be allowed when intraspecific competition and/or interference is reduced. Interestingly, much of the variation in winter sociality occurs at high taxonomic levels—among subfamilies, for example—suggesting that winter sociality is associated with a few key ecological variables that diversified early in the evolutionary history of these groups.

Stutchbury, Morton, and Pitcher also examine sociality, in both the winter and the summer, and focus in particular on interactions and conflicts between the sexes. With respect to the summer, these authors find (surprisingly) that rates of extra-pair paternity tend to be higher in Nearctic-Neotropical migrants than in Palearctic-African migrants. The reasons underlying this apparent difference are not clear; although migratory behavior may promote extra-pair mating in several ways (see the chapter by Webster and Marra), there is no obvious reason why it should do so more in the New World. With respect to the nonbreeding season, intersexual interactions vary substantially across species, ranging from complete habitat segregation to joint defense of a territory by male/female pairs. Stutchbury et al. find that females are relatively large in species where they compete directly with males, and they conclude that winter territoriality facilitates the evolution of large female size. However, it is not clear whether competition with males selects for large female size, whether joint territoriality favors sexual size differences (to facilitate niche partitioning), or whether large female size evolves in some other context (e.g., as a breeding adaptation) and then allows females to compete more directly with males.

The selective consequences of winter territoriality are also addressed in the chapter by Froehlich, Rohwer, and Stutchbury, but the focus here is on plumage coloration. The authors find, perhaps surprisingly, that bright winter coloration is not strongly associated with winter territoriality. Rather, bright winter plumage appears to result from a constraint imposed by the costs of a spring molt. Thus, bright plumage appears to be a good example of how adaptation to one season can lead to maladaptation in the other. This amplifies the point made by Price and Gross, and indeed by many other authors in this volume: migratory birds must live in two different worlds, and adaptation to one may complement or constrain adaptation to the other.

The patterns revealed in these chapters are tantalizing, but they should be considered preliminary, as they often are based on small sample sizes. As virtually all of these authors point out, we need far more data on social behavior during the winter before robust patterns can emerge or firm conclusions can be drawn. Some of the patterns suggested in these chapters may become clearer, or possibly even disappear, as more data come in. Take, for example, the difference in mating patterns between Palearctic-African and Nearctic-Neotropical migrants, as discussed by Stutchbury et al. This pattern is based on a small sample size (14 or fewer species from each geographic area), and additional data may either sharpen or negate the pattern. Indeed, the testis size data reported by Stutchbury et al. (see their fig. 24.2A) suggest that there may not be such a pronounced difference between the two regions in sperm competition among migratory species (if anything, Nearctic residents appear to have unusually low levels of sperm competition). Similarly, our understanding of winter foraging and social behavior is based on studies of a few model organisms. Do the patterns seen in these species also occur in other migratory taxa? Only further studies will tell.

As highlighted by the chapters in this section, the behavioral ecology of migratory birds is far more complex than can be captured by a focus on the breeding season alone. Ornithologists are just now beginning to expand their scope to include behavior during the winter, and this broader perspective is unveiling a host of important questions. What ecological and social factors shape behavior in winter? How do winter behavior and migration itself shape behavioral traits? How do traits shaped by selection in one season shape or constrain behavior in the other? The chapters in this section are a good first step toward answering these questions, but firm answers will come only after we gather more data on migratory birds outside of the breeding season.

BRIDGET J. M. STUTCHBURY,
EUGENE S. MORTON,
AND TREVOR E. PITCHER

Sex Roles in Migrants

Extra-Pair Mating Systems and Winter Social Systems

WE EXAMINE EXTRA-PAIR MATING SYSTEMS of migrant passerines and male-female competition for resources during the non-breeding season. High levels of extra-pair paternity (EPP) are typical for Nearctic-Neotropical migrants. Extra-pair young (EPY) account for an average of 32% of nestlings in a population and 46% of broods ($n = 13$ species). Palearctic-African migrants also typically have an extra-pair mating system, though the levels of EPP are much lower (11.6% EP young and 21.5% EP broods; $n = 14$ species) than for Nearctic-Neotropical migrants. Species that are residents in, or migrate largely within, North America have significantly lower levels of extra-pair paternity (16.6% of nestlings, 29.1% of broods) than Nearctic-Neotropical migrants. In the Palearctic-African migration system, however, both residents and migrants have similar levels of extra-pair paternity. These results are supported by both species level and phylogenetically controlled analyses of relative testis mass, an indicator of the intensity of sperm competition. The low EPP observed in Palearctic breeders cannot be explained by low breeding synchrony in these species. High genetic variation can favor the evolution of extra-pair mating systems, and we found that Nearctic-Neotropical species were significantly more genetically polymorphic than Nearctic residents. We expect that Palearctic migrants and residents have relatively low levels of genetic polymorphism, but insufficient data existed to test this prediction. Winter social system varies greatly among migrants, and we discuss how male-female competition for resources may vary with the type of winter social system. Nearctic-Neotropical migrants that are territorial on their wintering grounds have less sexual dimorphism in wing length than non-territorial species, suggesting that there may be selection for relatively larger size in females who compete directly with males for winter territories. We also explore what behavioral tactics females could use to overcome a size disadvantage when competing directly with males.

INTRODUCTION

Sex roles refer to the relative participation of males and females in activities like territoriality and parental care. The early view of breeding sex roles in migratory passerines used to be one dominated by monogamy and a cooperative division of labor (e.g., Lack 1968). Males defend territories to protect nest sites and food resources for breeding, while females build nests, lay eggs, and incubate the eggs. Sex roles are often similar when it comes to feeding the young and sometimes nest defense as well. Male-female competition and conflict were thought to be low. Little was known about the winter ecology of migrants or how males and females compete for similar resources. Our views on sex roles during the breeding season began to change with Trivers's (1972) revelation that socially monogamously mated males should pursue a mixed mating strategy. In recent years genetic studies have revealed high levels of extra-pair paternity in monogamous birds. Similarly, detailed behavioral studies of wintering migrants in the past decade have shown that the nature of intersexual competition is highly variable among species, and within a species there can be intense selection on male and female competitive ability, which can result in sex-specific tactics (Marra 2000). Our goals in this chapter are to (1) compare the extra-pair mating systems of Nearctic-Neotropical versus Palearctic-African migrants and (2) consider how the winter social system affects male-female competition.

At the time of the last Migratory Bird Symposium held in 1990; the first DNA fingerprinting studies on birds were just being completed and published (Gibbs et al. 1990; Gyllensten et al. 1990; Morton et al. 1990; Westneat 1990). Although early allozyme studies surprised us with high levels of cuckoldry (e.g., Westneat 1987), these studies revealed just the tip of the iceberg. The popularity of paternity studies is enormous, and by now papers have been published on extra-pair paternity (EPP) for over 130 species of birds (reviewed in Møller and Cuervo 2000 and Griffith et al. 2002). Many of these studies have focused on species that breed in the Nearctic or Palearctic regions, including migrants to the Tropics. Extra-pair mating systems are important from an evolutionary perspective, because they create strong sexual selection in socially monogamous species and can drive the divergence of sex roles (Andersson 1994). Here we show that Nearctic-Neotropical migrants have higher levels of EPP than species that winter within the Nearctic and, unexpectedly, that Nearctic-Neotropical migrants have higher levels of EPP than Palearctic-African migrants.

Second, we consider the sex roles of migrants in the nonbreeding season. The research effort on this aspect of a migrant's life cycle is still small compared with that on breeding behavior, but many important advances have been made in the past decade (Rappole 1995). In many species of migrants, females must go head-to-head with males in competition for territories during the nonbreeding season (e.g., Stutchbury 1994; Marra 2000). We examine how male-female competition varies with the type of winter social system and predict that females are relatively large in species that are strongly territorial as a result of selection on females to compete directly with males for resources. We also consider behavioral strategies females could use to circumvent or reduce direct competition with larger males.

EXTRA-PAIR MATING SYSTEMS

We obtained data on extra-pair paternity in socially monogamous species from the recent review by Griffith et al. (2002) and supplemented this with recently published studies and several unpublished studies by our colleagues (Appendix to this chapter). We excluded from our analysis species with high levels of social polygyny, such as the Red-winged Blackbird (*Agelaius phoeniceus* [Westneat 1993]), Corn Bunting (*Miliaria calandra* [Hartley et al. 1993]), and Dusky Warbler (*Phylloscopus fuscatus* [Fortsmeier 2002]).

We categorized each species as a resident and/or migrant largely within the temperate region versus a migrant mainly to tropical regions (Appendix to this chapter). Although some species are easily defined as Nearctic-Neotropical migrants because the entire population overwinters in the Tropics, other species are more problematic because some portion of the population overwinters in the southern United States rather than strictly in the Neotropics. Our definition of Nearctic-Neotropical migrant does not include species with a large part of their winter range in the Nearctic, such as the Eastern Phoebe (*Sayornis phoebe*), Tree Swallow (*Tachycineta bicolor*), House Wren (*Troglodytes aedon*), and Savannah Sparrow (*Passerculus sandwichensis*). We note that Rappole (1995) considers these species to be Nearctic-Neotropical migrants, which illustrates the difficulty of clearly dichotomizing the group. Our main conclusions do not change when these four species are categorized as migrants.

Among Nearctic-Neotropical migrants that have a socially monogamous mating system, most species (11 of 13) studied to date have over 20% of extra-pair young (EPY) within the population (fig. 24.1). These migrant species average a remarkable 31.6% (SD = 13.6, n = 13) EPY, with 46.0% (SD = 16.8) of broods containing at least one EPY. This means that in a typical Nearctic-Neotropical migrant species, almost half the females in a population are laying eggs fertilized by an extra-pair male. This group of 13 species spans the taxonomic diversity of Nearctic passerines, and includes flycatchers, swallows, vireos, and various fringillids (warblers, buntings, and orioles).

We can ask whether it is the Nearctic breeding region that favors high EPP, or whether some other ecological or be-

A) Extra-pair Young

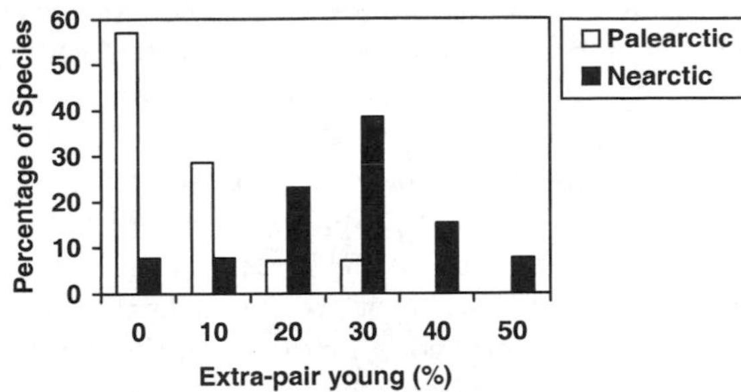

B) Broods with Extra-pair Young

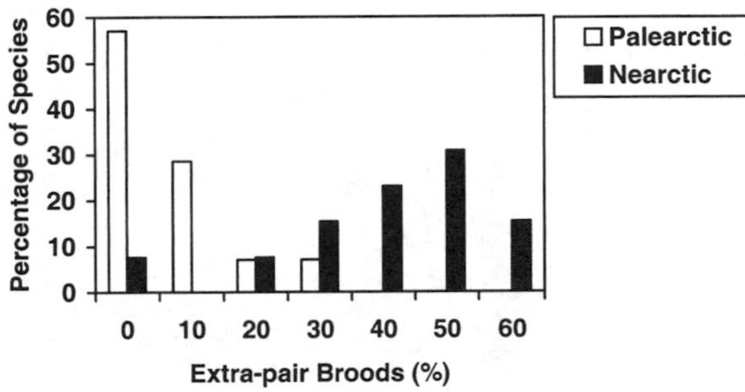

Fig. 24.1. Frequency of distribution of percentage extra-pair young (A) and percentage of broods with extra-pair young (B) found in socially monogamous passerine species for Nearctic-Neotropical migrants ($n = 13$) and Palearctic-African migrants ($n = 14$).

havioral factor is driving the high levels of EPP in the Nearctic-Neotropical migrants. We compared the extra-pair paternity of Nearctic-Neotropical migrants versus Nearctic species that are residents or migrate largely within North America (Appendix to this chapter). The species that overwinter largely within the Nearctic average only 16.6% (SD = 13.4, $n = 13$) extra-pair young and 29.2% (SD = 19.2) extra-pair broods. This differs significantly from migrants for both extra-pair young (Mann-Whitney *U*-test, $Z = 2.74$, $p = 0.006$) and EP broods ($Z = 2.45$, $p = 0.014$). This suggests that the Nearctic-Neotropical migrants as a group are under stronger selection for extra-pair mating behavior, despite having a breeding range and a habitat similar to those of residents and North American migrants.

Nearctic-Neotropical migrants also have much higher EPP than Palearctic-African migrants (fig. 24.1). Extra-pair paternity is typical for Palearctic passerines but occurs at a lower frequency compared with Nearctic-Neotropical migrants. Socially monogamous Palearctic-African migrants average only 11.6% (SD = 9.7, $n = 14$) extra-pair young and 21.5% (SD = 15.4) extra-pair broods. Compare this with the much higher averages of 31.6% EPY and 46.0% broods for Nearctic-Neotropical migrants! The differences are highly significant for both EPY (Mann-Whitney *U*-test, $Z = 3.35$, $p = 0.001$) and extra-pair broods ($Z = 3.21$, $p = 0.001$). Only two of 14 Palearctic-African migrant species feature consistently high levels of extra-pair paternity above 20% EPY: Barn Swallow (*Hirundo rustica* [Møller and Tegelström 1997; Saino et al. 1997]) and Bluethroat (*Luscinia svecica* [Krokene et al. 1996]). Several individual populations of Palearctic-African migrants have high levels of EPP, for example, Pied Flycatcher (*Ficedula hypoleuca* [Gelter and Tegelström 1992]), and Willow Warbler (*Phylloscopus trochilus* [Bjornstad and Lifjeld 1997]), but the overall weighted average for the species is much lower (Appendix to this chapter) because other populations have lower EPP. We discuss this is in more detail in the section below on breeding synchrony.

Extra-pair paternity was similar for Palearctic species that migrate to the Tropics (11.6% EPY, 21.5% broods) and those that overwinter largely within the Palearctic (14.2% EPY, SD = 14.8, $n = 15$ and 31.3.4 % broods, SD = 26.4). These values were not significantly different ($Z = 0.35$, $p = 0.73$; and $Z = 0.58$, $p = 0.56$, respectively).

Testis Mass and Sperm Competition

To expand our comparison to include species for which paternity data are not available, we used relative testis mass during the breeding season as an index of sperm competition in each species (Møller and Briskie 1995; Pitcher and Stutchbury 1998; Dunn et al. 2001). Testis mass, relative to body size, is strongly and positively correlated with levels of extra-pair paternity, reflecting the high levels of sperm competition that occur when males are competing to inseminate females. All testis size data were taken from specimens known to be in breeding condition and during the appropriate dates for the main breeding season. At lower latitudes, where breeding seasons are long and breeding less synchronous, it is possible that specimens from a given date could include both males who are actively nesting and males who have not yet begun, or have finished, breeding, resulting in a low average testis size for species at low latitudes. However, for six Nearctic-breeding species, there was no relationship between the coefficient of variation in testis mass and latitude (Pitcher and Stutchbury 1998), indicating that breeding asynchrony does not result in greater variance in samples of testis mass.

Relative testis mass was estimated from the residuals of the regression of testis mass on total body mass (Dunn et al. 2001). Both testis and total mass were log10-transformed prior to analysis. Testis mass for each socially monogamous species was obtained from published compilations (Møller 1991; Møller and Briskie 1995; Stutchbury and Morton 1995; Pitcher and Stutchbury 1998; Dunn et al. 2001) or data from tags of museum specimens, which consisted of testis length and width measurements. Testis mass was estimated from these measurements by using Møller's (1991) formula: testis mass (g) $= 2 \times 1.087 \text{g} \cdot \text{cm}^{-3} 1.33\pi[a(\text{cm})]^2 b(\text{cm})$, where *a* and *b* are the width and length of each testis (see also Møller and Briskie 1995). Average testis mass was calculated as the mean testis value from at least ten breeding males, but typically 20 or more breeding males were used (e.g., Pitcher and Stutchbury 1998; Dunn et al. 2001).

Residual testis mass of Nearctic-Neotropical migrants was higher than residents and species that migrate within North America (fig. 24.2A; unpaired *t*-test, $t = 3.31$, d.f. = 178, $p = 0.0007$). For Palearctic species, residual testis mass was similar for migrants and residents, and the difference was not statistically significant ($t = 0.332$, d.f. = 34, $p = 0.74$). As we saw with the paternity data, Nearctic-Neotropical migrants as a group appear to have much higher levels of sperm competition than residents and Nearctic migrants, but a similar difference is not seen in Palearctic species.

Comparisons across species may potentially be confounded by common phylogenetic ancestry (Harvey and Pagel 1991; Bennett and Owens 2002). However, because differences between results using raw species data and phylogenetic methods may provide some biological insight, we have analyzed our data by using comparative methods that control for phylogeny (independent contrasts) and the raw species values (see Price 1997; Martins 2000; Dunn et al. 2001). To produce data that were phylogenetically independent, we used Comparative Analysis of Independent Contrasts (CAIC) by Purvis and Rambaut (1995) to calculate standardized linear contrasts (Felsenstein 1985; Harvey and Pagel 1991). Contrasts were standardized assuming that lengths of branches in the phylogeny were equal in length, which represents a punctuated model of evolution (see Dunn et al. 2001 for details). Our phylogeny was based on the molecular phylogeny of Sibley and Ahlquist (1990), which provides a branching pattern to the level of family, subfamily, or tribe, depending on the clade. When we had more than two species below the lowest level in Sibley and Ahlquist's phylogeny, we used recent phylogenetic analyses to complete the phylogeny to the species level (see Dunn et al. 2001 for details).

Using the linear contrasts analysis, we found the same patterns in testis mass even after controlling for phylogeny (fig. 24.2B). Nearctic-Neotropical migrants had larger testes

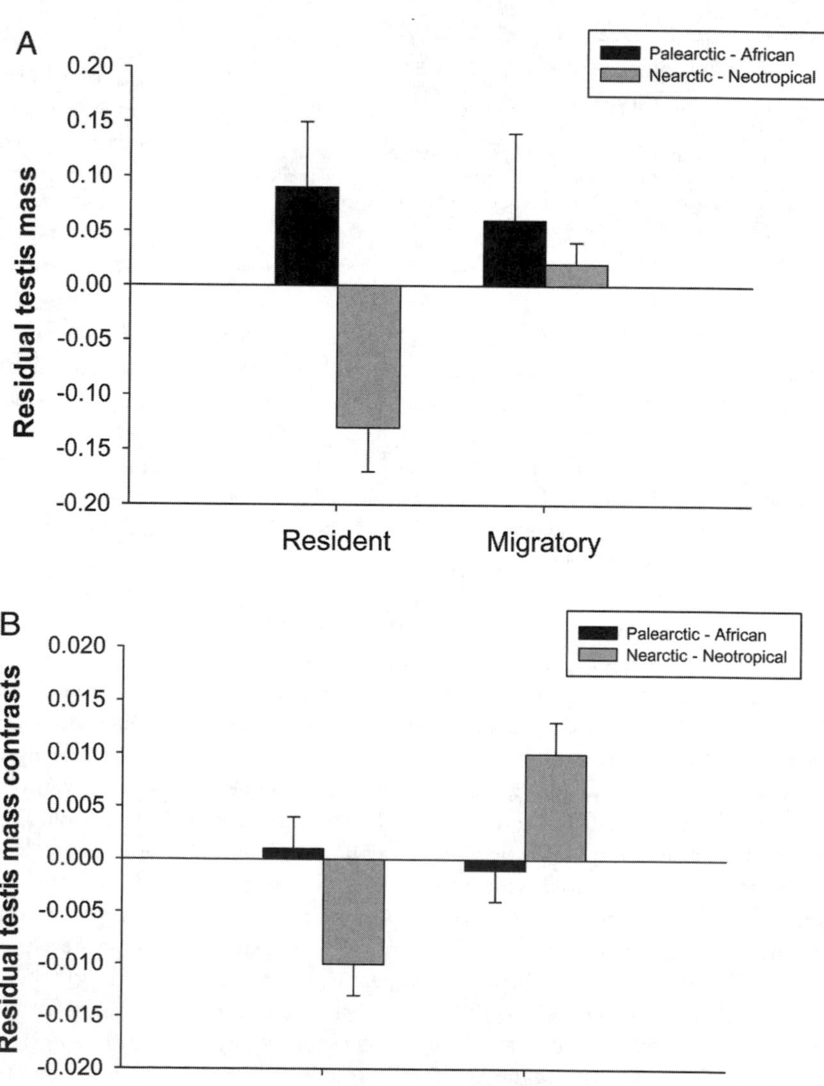

Fig. 24.2. (A) Residual testis mass for socially monogamous passerines that are Nearctic-Neotropical migrants ($n = 133$) versus residents and species that migrate within the Nearctic ($n = 47$) and Palearctic-African migrants ($n = 17$) versus migrants and species that migrate within the Palearctic ($n = 19$). A positive value means the species has larger testes than expected given its body mass as determined by the regression of testis mass on body mass. (B) Independent contrasts in residual testis mass calculated using comparative methods that control for phylogeny.

than residents ($t = 2.9$, $n = 83$ contrasts, $p = 0.006$), but Palearctic-African migrants did not have larger testes than residents ($t = 0.63$, $n = 31$ contrasts, $p = 0.53$).

The difference in EPP between Nearctic-Neotropical migrants and their counterparts in the Palearctic is so striking that it begs an explanation. Why should males in the Palearctic be less able to pursue extra-pair copulations (EPCs), and why should females be less willing to pursue and/or accept EPCs? Several ecological and behavioral factors may explain interspecific variation in extra-pair mating systems (reviewed in Griffith et al. 2002), including breeding synchrony (Stutchbury and Morton 1995), sex roles in mate guarding and parental care (Møller and Ninni 1998; Møller and Cuervo 2000), and genetic variability (Petrie et al. 1998). Breeding density is not a likely explanation because it is not a good predictor of extra-pair paternity (Westneat and Sherman 1997). Here, we consider whether breeding synchrony can explain the EPP difference between migrants in the Nearctic and Palearctic.

Breeding Synchrony and Extra-Pair Paternity

Extra-pair paternity in passerines is positively correlated with breeding synchrony (Stutchbury and Morton 1995; Stutchbury 1998). Synchrony may give females more opportunities to compare displaying males, and in synchronously breeding populations more fertile females are available to males at a given time. Both aspects lower costs relative to benefits and thus favor the evolution of extra-pair mating systems. It is possible that Palearctic breeders have low EPP because they have relatively low breeding synchrony. The breeding synchrony index (Kempenaers 1993) reflects the average percentage of females in the population that are fertile on the same day. Breeding synchrony is often highest for populations breeding at high altitude or latitude, because the breeding season is so short (Stutchbury and Morton 1995; Bjornstad and Lifjeld 1997).

The Palearctic-African migrants in our sample have a synchrony index of 49.5% (SD = 23.6, $n = 6$), which is not lower than for Nearctic-Neotropical migrants (SI = 32 %, SD = 9.6, $n = 9$). The correlation between extra-pair paternity and breeding synchrony is strong for the Nearctic passerines (Spearman rank correlation, $r = 0.61$, $n = 18$, $p = 0.008$; fig. 24.3), but not for Palearctic passerines ($r = 0.483$, $n = 12$, $p = 0.112$). We had never before realized that in the overall correlation between EPP and synchrony (Stutchbury 1998), most of the points with high synchrony but relatively low EPP were Palearctic species. Several of the synchrony indices in the small sample of Palearctic species are from highly synchronous populations breeding at high altitude and/or latitude. These populations of Willow Warbler (Bjornstad and Lifjeld 1997), Bluethroat (Krokene et al. 1996), and House Martin (Whittingham and Lifjeld 1995) have high extra-pair paternity (19–32% EPY). With the exception of the Reed Bunting (*Emberiza schoeniclus* [Dixon et al. 1994]), for most Palearctic passerines EPP is low relative to Nearctic passerines even for species with a synchrony index of 30–60% (fig. 24.3). There must be some other ecological or behavioral factor influencing EPP that differs between Nearctic and Palearctic breeders, an interesting puzzle for future research.

A) Nearctic Breeders

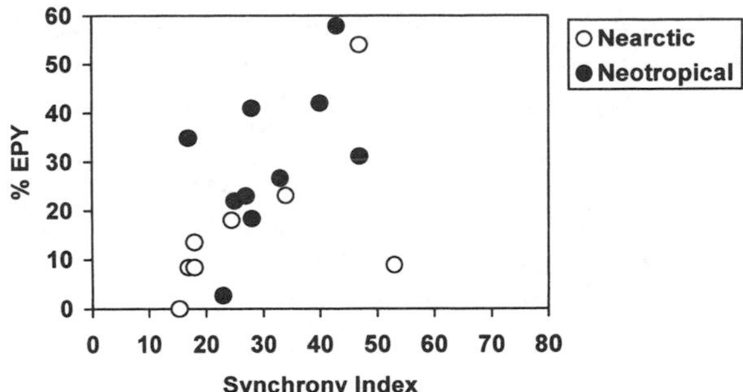

B) Palearctic Breeders

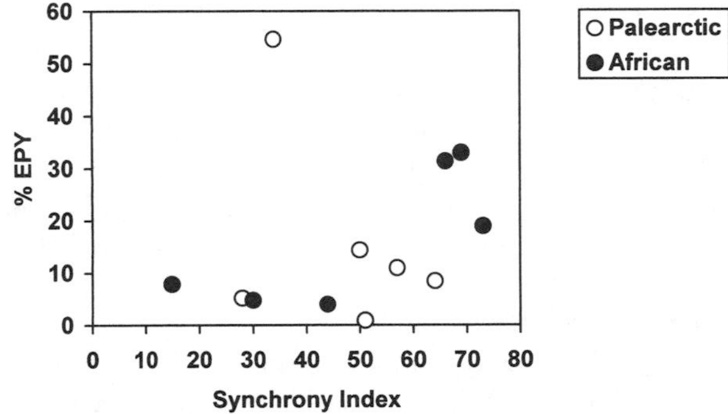

Fig. 24.3. Relationship between percentage extra-pair young in socially monogamous species and their breeding synchrony index for species that breed in the Nearctic (A) and Palearctic (B). For each group, we also indicate the long-distance migrants who overwinter in the Tropics versus the residents and species that migrate within the temperate zone.

SEX ROLES AND COMPETITION ON THE WINTERING GROUNDS

Although sex roles are highly divergent during the breeding season, they converge in the nonbreeding season because sexual selection is greatly reduced and females compete directly with males when food resources are limited. Females may nevertheless have different tactics for winter competition than males, because sexual dimorphism in body size puts females at a disadvantage in competing with males. Sexual dimorphism results largely from sexual selection during the breeding season, and it is driven by the social mating system and to a lesser extent the extra-pair mating system (Dunn et al. 2001). How do females compete directly with larger males for winter resources?

We were frustrated at the lack of detailed information on winter social behavior for many migrants. Many reviews simply list species as being "territorial" or "non-territorial," which is an important first step in understanding winter ecology. But winter social systems are more diverse than that, and understanding female versus male roles in the winter social system requires detailed behavioral studies (e.g., Rappole and Warner 1980; Rappole 1988; Morton 1990; Stacier 1992; Wunderle 1992; Marra 2000). The different winter social systems can be arranged in a gradient of increasing intersexual competition, from flocking species to those in which males and females defend similar territories (fig. 24.4).

In the Nearctic-Neotropical system, species that are obligately territorial (sensu Morton 1980) are those in which most individuals are seen defending a territory for virtually the entire time they are on their winter site. In some species males and females occur on adjacent territories, for example, Kentucky Warbler (*Oporornis formosus* [Mabey and Morton 1992]) and Black-and-White Warbler (*Mniotilta varia* [Lopez Ornat and Greenberg 1990]) with little or no habitat segregation. Females in this social system are expected to face the highest levels of intersexual competition (fig. 24.4). In these species, where sex roles are virtually identical, we expect to find strong selection on females for large body size to enhance competitive ability.

We collected data on dimorphism in size (wing length) from both museum specimens in North American and Europe and the literature (Ridgway 1901–1946; Cramp and Simmons 1977–1983; Cramp 1985–1992; Cramp and Perrins 1993–1994). We did not analyze body mass because all data come from the breeding season and body mass dimorphism may differ between the seasons. This is less likely to be true for wing length dimorphism because flight feathers molt only once per year in most species. Wing length was log10-transformed prior to analysis. Dimorphism in wing length was analyzed by using the residuals from the regression of male wing length on female wing length, which avoids the statistical pitfalls of ratios (Zar 1999). Thus a species with a large positive residual is one in which males are larger than expected, whereas a species with a large negative residual is one in which females are larger than expected.

We obtained information on winter social systems from the literature (Cramp 1977–1994; Rappole and Warner 1980; Lopez Ornat and Greenberg 1990; Morton 1992; Rappole 1995; see also Froehlich et al., Chap. 25, this volume). Despite the broad range of winter social systems that exist for migrants (fig. 24.4), for most species we only have enough information to classify species as territorial versus non-territorial. The category "territorial" includes species that are "obligately" territorial, with most individuals defending territories all winter, and those that are "facultatively" territorial, where territory defense is shorter term. It does not include species that defend resources (e.g., flowers) for only hours or days or species with occasional reports of conspecific aggression.

For Nearctic-Neotropical migrants, species that are territorial during the nonbreeding season were less dimorphic in wing length than species that are non-territorial (unpaired t-test, d.f. $= 43$, $t = 2.06$, $p = 0.04$ [fig. 24.5]). In other words, territorial species tended to fall below the regression line of male versus female size (e.g., females larger than expected), whereas flocking species tended to fall above the regression line (e.g., males larger than expected). We expected to find the same to be true for Palearctic-African migrants, but this was not the case. Territorial species and non-territorial species had almost identical residual wing length ($t = 0.97$, d.f. $= 51$, $p = 0.37$). We found the same results when controlling for phylogeny. Wing dimorphism of winter territorial versus non-territorial species differed for Nearctic-

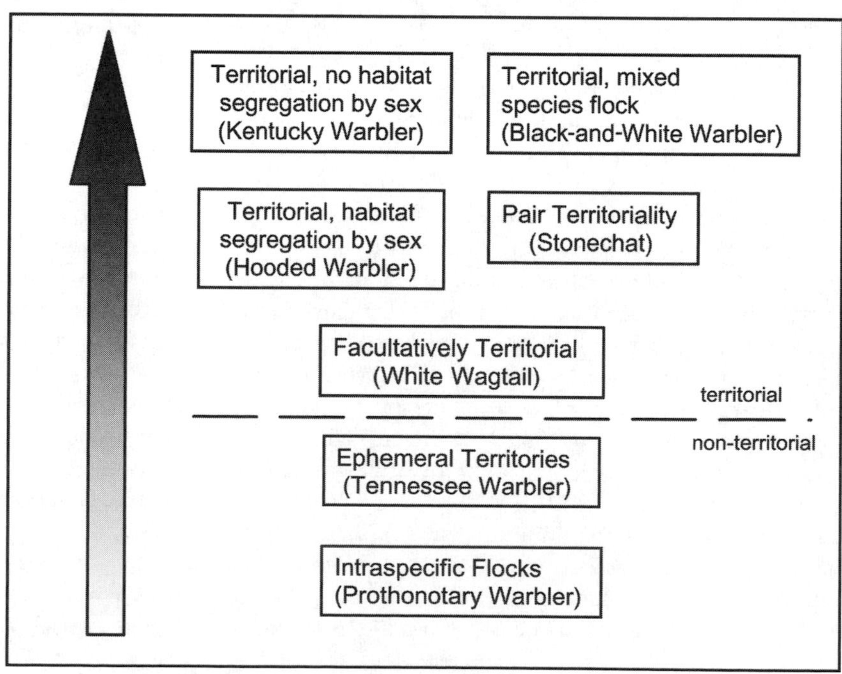

Fig. 24.4. Diversity of winter social systems in migrants to the Tropics, arranged to show a hypothesized increase in male-female competition.

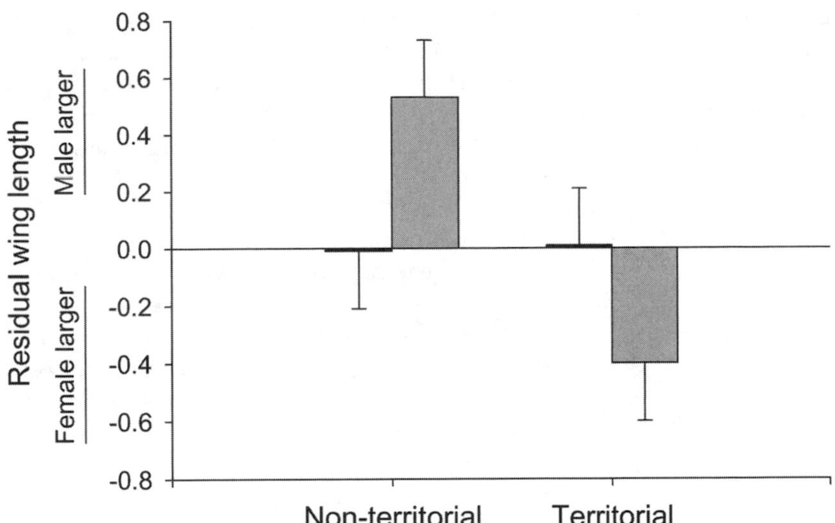

Fig. 24.5. Residual wing length of Nearctic-Neotropical migrant species (gray bars) that are non-territorial ($n = 18$) versus territorial ($n = 27$) in the Tropics, and Palearctic-African migrants species (black bars) that are non-territorial ($n = 26$) versus territorial ($n = 27$). Bars show mean and standard deviation. Positive values indicate that males are larger than expected, negative values indicate that females are larger than expected.

Neotropical migrants ($t = 3.1$, $n = 21$ contrasts, $p = 0.0005$ [fig. 24.5]) but not for Palearctic-African migrants ($t = 0.98$, $n = 15$, $p = 0.19$).

These results suggest that there may be strong selection on female size in species with winter territoriality. Even given this larger size, females may have a suite of behavioral strategies to compete with larger males. In Nearctic-Neotropical migrants, habitat segregation by sex is common and females are usually the sex that occupies early-successional habitat (Lynch et al. 1985; Lopez Ornat and Greenberg 1990; Morton 1990; Marra 2000; Murphy et al. 2001). This appears to be less common (or less studied?) in the Palearctic-African migrants (Herremans 1997), perhaps because territoriality itself is a less common winter social system, owing to the relative scarcity of forested habitat (Rappole 1995). The consistent occurrence of habitat segregation by sex raises the question of whether females use different habitats because they are forced out of high-quality habitat by larger males, or whether females prefer a different habitat and can forage and survive well in early successional habitat (Lynch et al. 1985).

Female occupancy of lower-quality habitat can have a high cost in terms of poor body condition during the non-breeding season and a later departure for the breeding grounds (Marra and Holmes 2001). Even if females are actively excluded from high-quality habitat by dominant males, as is the case for American Redstarts (*Setophaga ruticilla* [Marra 2000; Marra and Holmes 2001]), females presumably face a choice of "floating" in high-quality habitat versus settling in poor-quality habitat. In redstarts, females opt to occupy poor-quality habitat because a floater strategy is not successful (Marra 2000). Females rarely can claim vacant territories in optimal habitat, so clearly they are making the best of a bad situation.

If females prefer different habitat than males to begin with, then the intensity of male-female competition may not be as severe as in the case of American Redstarts. In Hooded Warblers (*Wilsonia citrina*), there is evidence that habitat segregation occurs by female preference. Females and males differ in their innate habitat preference and use verticality of the vegetation as their cue for habitat choice (Morton 1990). Males prefer vertical elements and females prefer oblique elements, a difference that results in habitat segregation with male-biased sex ratios in habitats with vertical elements (e.g., mature forest). Females do not occupy territories after male removals in forest habitat (Morton et al. 1987). In mixed habitat, female floaters are just as likely as male floaters to take over vacant territories, and females are able to defend these territories despite the presence of territorial and floater males (Stutchbury 1994). Thus it appears that Hooded Warbler females are not at a severe competitive disadvantage, at least compared with female American Redstarts.

Another tactic females may use for gaining access to winter resources is to join forces with a male on his territory. Pair territoriality is reported for some Palearctic-African species (e.g., Stonechat [Gwinner et al. 1994]) in which some, but not all, females live on the winter territory of a male. Males may allow cohabitation by females because it reduces the individual costs of territory defense and may increase vigilance in detecting predators (Gwinner et al. 1994). Many Palearctic-African migrants occupy open habitat in the winter, rather than forest (reviewed in Rappole 1995). This territorial system has not (so far?) been reported for a Neotropical-Nearctic migrant, perhaps because greater use of forest habitat makes teamwork less efficient.

Many species are territorial, but not obligately so. In other words, some individuals defend territories all winter, whereas others do not. These alternative strategies can co-occur within a population, and the frequency of territorial behavior can vary between populations. In the Red-backed Shrike (*Lanius collurio*) males are more likely to be territorial than females, and females who are territorial typically have lower-quality territories (Herremans 1997). Thus the costs and benefits of territory defense differ between the sexes. Some Nearctic-Neotropical migrants defend ephemeral food resources, like flowers (e.g., Tennessee Warbler [*Vermivora peregrina*] [Morton 1992]), for short periods of time. It is not known whether males are more likely than females to defend ephemeral territories, and what the costs might be to females of being excluded from food resources that are not as limiting as insects.

FUTURE DIRECTIONS

It is impressive that paternity studies have been done for dozens of migrant species in the past decade. As Nearctic-Neotropical biologists, we had taken for granted the high levels of EPP in the species we are most familiar with. We are not surprised to find 41% EPY in a newly studied species, like the Acadian Flycatcher (*Empidonax virescens* [Woolfenden et al., 2005]). But this would surprise someone studying a Palearctic-African migrant, a group that so far has consistently lower levels of EPP (fig. 24.1).

Palearctic species have relatively low levels of EPP but we can rule out breeding synchrony as an explanation. Mate-guarding intensity is one behavioral trait that can affect EPP. Breeding synchrony can have opposing effects on opportunities for seeking EPCs, depending on how intensely males guard their fertile mates (Birkhead and Biggins 1997; Stutchbury and Neudorf 1998). Intense mate guarding may result in low EPP in a synchronously breeding population because males do not seek EPCs until their own mate is incubating, and by that time few females in the population are fertile (Birkhead and Biggins 1987). In contrast, if males pursue extra-pair matings even when their own mate is fertile, then breeding synchrony results in a huge EPC opportunity for males (Stutchbury and Morton 1995). The combination of high synchrony and intense mate guarding can therefore restrict extra-pair opportunities for both males and females.

For several Palearctic passerines, intense mate guarding by males may limit extra-pair matings (Leisler and Wink 2000; Davies et al. 2003), especially during the early part of the breeding season when breeding is most synchronous (Currie et al. 1998). Mate guarding has been reported for many of the Palearctic-African migrants in our sample (e.g., Krokene et al. 1996; Møller and Tegelström 1997), but it would be premature to conclude that this is responsible for lower EPP levels. In the Bluethroat, females can apparently escape male paternity guards and EPP is high (Johnsen et al. 1998). EPC opportunities are limited by male mate guarding only if females do not seek out male extra-pair partners and if males forego EPC attempts during their own mate's fertile period (Currie et al. 1998). For most species we do not yet know the details of how males, and particularly females, obtain extra-pair fertilizations.

High levels of male parental care, particularly male incubation, may also limit a male's ability to pursue EPCs (Ketterson and Nolan 1994; Schwagmeyer et al. 1999; Møller and Cuervo 2000; but see Pitcher and Stutchbury 2000). We used data from Møller and Cuervo (2000) and more recent empirical studies to assess whether the male's role in parental care for Palearctic species in our data set differed consistently from the Nearctic species. Male incubation occurs in one out of 15 Nearctic-Neotropical migrants (Blue-headed Vireo) and four out of 13 Palearctic-African migrants (House Martin, Bank Swallow, Marsh Warbler, and Reed Warbler). But recall that the highly synchronously breeding population of House Martins (*Delichon urbica*) has high EPP (19% EPY), despite male incubation (Whittingham and Lifjeld 1995). Average male feeding effort (% trips by male) to nestlings was similar for Nearctic-Neotropical migrants (44.8, SD = 14.4, n = 10) and Palearctic-African migrants (51.1, SD = 5.7, n = 12). There do not appear to be any conspicuous differences in sex roles in parental care between the Nearctic and Palearctic species in our sample.

Finally, species of birds with high levels of EPP tend to have higher genetic variability (Petrie et al. 1998). Females may be more likely to pursue EPCs if males differ greatly in genetic quality, because a relatively homogeneous genetic pool of males would reduce the "good genes" benefits females stand to gain from EPCs. Thus we would expect to find that Neotropical-Nearctic migrants have a high level of genetic variation compared with both Nearctic residents and Palearctic species. Estimates of polymorphism were collected from Petrie et al. (1998) and from personal communications with M. Petrie. First, we compared Nearctic-Neotropical migrants with Nearctic residents. The proportion of polymorphic loci is a quantification of genetic variation that measures the number of loci found to be polymorphic in any one sample (a locus is typically considered to be polymorphic when the frequency of the most common allele is not greater than 0.95). The range of polymorphic loci for the socially monogamous passerine species we examined varied from 0 to 53.3% (mean [SE] = 20.7% ± 1.3). We arcsine square-rooted the proportional polymorphism data in order to satisfy all underlying statistical assumptions. Nearctic-Neotropical species were significantly more polymorphic than resident species (21.8% ± 1.4 [n = 57] versus 16.3% ± 3.1 [n = 14]; $t = -2.06$, $p = 0.04$). Insufficient data existed to examine the Palearctic-African system, though we would expect these species to have relatively low levels of polymorphism, which may contribute to the low levels of EPP in this group.

Comparisons with other migration systems would be extremely valuable in terms of understanding what feature of the Nearctic-Neotropical migration system selects for such high levels of EPP. The Palearctic-Asian system is more similar to the Nearctic-Neotropical system than is the Palearctic-African system (Rappole 1995). Intratropical migration systems within the Neotropics are not well studied, but we have predicted (Stutchbury and Morton 2001) that intratropical migrants will feature high EPP.

Studies on migrant winter social systems are still too few and far between. We have written elsewhere (Stutchbury and Morton 2001) about the Temperate Zone bias in behavioral ecology and how, as a consequence, relatively few behavioral ecology studies have been performed with run-of-the-mill tropical species. Although it is true that some spectacularly interesting work has been done with migrant winter ecology in recent years, we are still a long way from being able to answer some fundamental and important questions about the evolution of sex roles in winter social systems. For territorial species, selection for larger female size may reduce the costs of competing with males and/or increase a female's ability to gain valuable winter resources. We know little about the fitness consequences to females of pursuing alternative territorial tactics like habitat segregation (whether by choice or male dominance), floating, and pair territoriality. These details are important not just for understanding the evolution of sex roles, but also for understanding how habitat loss is expected to affect population dynamics (e.g., Marra and Holmes 2001).

APPENDIX: THE RATE OF EXTRA-PAIR PATERNITY FOR SOCIALLY MONOGAMOUS PASSERINES THAT BREED IN THE NEARCTIC (N) OR PALEARCTIC (P) REGIONS

Family	Species	Common name	Breeding location	Wintering location	% EPP offspring	% EPP broods	SI (%)	References
Tyrannidae	*Empidonax virescens**	Acadian Flycatcher	N	NT	41	58	28	Woolfenden et al. 2005
	*Empidonax minimus**	Least Flycatcher	N	NT	39	62		Tarof et al. 2005
	Sayornis phoebe	Eastern Phoebe	N	N	11.8	20.0		Conrad et al. 1998
	Tyrannus tyrannus	Eastern Kingbird	N	NT	42	60	40	Rowe et al. 2001
Laniidae	*Lanius minor**	Lesser Gray Shrike	P	NT	0	0		Valera et al. 2003
Vireonidae	*Vireo olivaceus*	Red-eyed Vireo	N	NT	57.9	57.1	43	Morton et al. 1998
	Vireo solitarius	Blue-headed Vireo	N	NT	2.7	6.3	23	Morton et al. 1998
Corvidae	*Corvus monedula*	Jackdaw	P	P	0.9	2.9	51	Henderson et al. 2000
Muscicapidae	*Luscinia svecica*	Bluethroat	P	A	31.4	50.7	66	Krokene et al. 1996
	Ficedula albicollis	Collared Flycatcher	P	A	15.5	32.9		Sheldon and Ellegren 1999
	Ficedula hypoleuca	Pied Flycatcher	P	A	8.4	19.2		Ratti et al. 1995; Lifjeld et al. 1991
		Pied Flycatcher			4.0	15	44	Lifjeld et al. 1991
	Oenanthe oenanthe	Northern Wheatear	P	A	7.4	22.2		Currie et al. 1998
	Sialia mexicana	Western Bluebird	N	N	18.8	45.0		Dickinson and Arke 1998
	Sialia sialis	Eastern Bluebird	N	N	8.4	23.8	18	Meek et al. 1994
Sturnidae	*Sturnus vulgaris*	European Starling	N,P	P	9.1	30.6		Pinxten et al. 1993
Cethidae	*Troglodytes aedon*	House Wren	N	N	8.4	26.7		Soukop and Thompson 1997
	Thryothorus ludovicianus	Carolina Wren	N	N	0.0	0.0	15.4	Haggerty et al. 2001
Paridae	*Parus ater*	Coal Tit	P	P	25.3	75.0		Schmoll et al. 2003
	Parus caeruleus	Blue Tit	P	P	10.9	40.2		Kempenaers et al. 1995
		Blue Tit			11	31	57	Kempenaers et al. 1995
	Parus cristatus	Crested Tit	P	P	11.0	30.0		Lens et al. 1997
	Parus major	Great Tit	P	P	7.3	31.0		Gullberg et al. 1992; Verboven and Mateman 1997; Blakey 1994
		Great Tit			8.5	40.0	64	Stronbach et al. 1998
	Parus montanus	Willow Tit	P	P	0.9	4.2		Orell et al. 1997
	Panurus biarmicus	Bearded Tit	P	P	14.4	29.5	50	Hoi and Hoi-Leitner 1997
	Parus atricapillus	Black-capped Chickadee	N	N	8.9	29.3	53	Otter et al. 1998
Hirundinidae	*Tachycineta bicolor*	Tree Swallow	N	N	54.0	77.0	47	Lifjeld et al. 1991; Dunn et al. 1994
	Hirundo rustica	Barn Swallow	N,P	A	28.2	—		Møller and Tegelström 1997
	Progne subis	Purple Martin	N	NT	18.3	—	28	Wagner et al. 1996
	Riparia riparia	Sand Martin	N,P	A	14.0	36.0		Alvez and Bryant 1998
	Delichon urbica	House Martin	P	A	18.6	33.3		Riley et al. 1995; Whittingham and Lifjeld 1995
		House Martin			19	35	73	Whittingham and Lifjeld 1995
Sylviidae	*Acrocephalus arundinaceus*	Great Reed Warbler	P	A	4.8	—		Hasselquist et al. 1996
		Great Reed Warbler			3.1	5.4	30	Hasselquist et al. 1996

continued

Family	Species	Common name	Breeding location	Wintering location	% EPP offspring	% EPP broods	SI (%)	References
Sylviidae	*Acrocephalus palustris*	Marsh Warbler	P	A	3.1	9.1		Leisler and Wink 2000
	Acrocephalus schoenobaenus	Sedge Warbler	P	A	7.9	—	15	Langefors et al. 1998
	*Acrocephalus scirpaceus**	Reed Warbler	P	A	6.0	15.0		Davies et al. 2003
	Phylloscopus sibilatrix	Wood Warbler	P	A	0.0	0.0		Gyllensten et al. 1990
	Phylloscopus trochilus	Willow Warbler	P	A	17.2	—		Gyllensten et al. 1990; Bjornstad and Lifjeld 1997; Fridolfsson et al. 1997
		Willow Warbler			33	45	69	Bjornstad and Lifjeld 1997
Passeridae	*Passer domesticus*	House Sparrow	N,P	P	10.1	20.4		Wetton and Parkin 1991; Wetton et al. 1995
	Anthus spinoletta	Water Pipit	P	P	5.2	12.4	28	Reyer et al. 1997
Fringillidae	*Fringilla coelebs*	Chaffinch	P	P	17.0	23.1		Sheldon and Burke 1994
	Carpodacus mexicanus	House Finch	N	N	8.4	14.3	17	Hill et al. 1994
	Emberiza citrinella	Yellowhammer	P	P	37.4	68.8		Sundberg and Dixon 1996
	Emberiza schoeniclus	Reed Bunting	P	P	54.6	86.2	34	Dixon et al. 1994
	Passerculus sandwichensis	Savannah Sparrow	N	N	23.1	40.0	34	Freeman-Gallant 1997
	Zonotrichia albicollis	White-throated Sparrow	N	N	18	31.3	24.5	Tuttle 2003
	Junco hyemalis	Dark-eyed Junco	N	N		28.3		Raouf et al. 1997
	Dendroica petechia	Yellow Warbler	N	NT	31.1	59.8	47	Yezerinac et al. 1995
	Dendroica caerulescens	Black-throated Blue Warbler	N	NT	23.0	34.2	27	Webster et al. 2001
	Wilsonia citrina	Hooded Warbler	N	NT	26.7	35.3	33	Stutchbury et al. 1997
	Setophaga ruticilla	American Redstart	N	NT	39.8	59.4		Perreault et al. 1997
	Cardinalis cardinalis	Northern Cardinal	N	N	13.5	15.8	18	Ritchison et al. 1994
	Passerina cyanea	Indigo Bunting	N	NT	34.9	48.0	17	Westneat 1990
	Icterus galbula bullockii	Bullock's Oriole	N	NT	32.0	46.0		Richardson and Burke 2001
	Carduelis tristis	American Goldfinch	N	N	14.3	26.7		Gissing et al. 1998
	Geothlypis trichas	Common Yellowthroat	N	NT	22.0	49.0	25	Thusius et al. 2001
	Serinus canaria	Wild Canary	P	P	0	0		Voight et al. 2003
	Serinus serinus	Serin	P	P	9.4	14.9		Hoi-Leitner et al. 1999

Note: Data are from Appendix 1 in Griffith et al. (2002) and other sources. Species indicated by an asterisk were not included in Griffith et al. (2002). Also indicated are the main wintering locations of each species (Nearctic-N, Neotropical-NT, Palearctic-P, African-A). References show the sources for data on the rate of EPP, and for breeding synchrony indices (SI) that were not listed in Stutchbury and Morton (1995) or Stutchbury (1998). We also list the extra-pair paternity for populations from which the synchrony index was calculated. Four species breed in both the Nearctic and the Palearctic; data are for Palearctic populations.

ACKNOWLEDGMENTS

We thank R. Wagner and two anonymous reviewers for comments on the manuscript. T. Pitcher obtained testis mass and body size data from many museums (listed in Pitcher and Stutchbury 1998; Dunn et al. 2001). BJMS's research on migrant birds was supported by the Natural Sciences and Engineering Research Council of Canada (NSERC), a Premier's Research Excellence Award, and the Kenneth G. Molson Foundation. ESM was supported by the Christensen Fund and the Smithsonian Scholarly Studies program. TEP was supported by an NSERC scholarship.

LITERATURE CITED

Alvez , M. A. S., and D. M. Bryant. 1998. Brood parasitism in the sand martin, *Riparia riparia:* evidence for two parasitic strategies in a colonial passerine. Animal Behaviour 56:1323–1331.

Andersson, M. 1994. Sexual Selection. Princeton University Press, Princeton.

Bennett, P. M., and I. P. F. Owens. 2002. Evolutionary Ecology of Birds. Oxford University Press, Oxford.

Birkhead, T. R., and J. D. Biggins. 1987. Reproductive synchrony and extra-pair copulations in birds. Ethology 74:320–334.

Bjornstad, G., and J. T. Lifjeld. 1997. High frequency of extra-pair paternity in a dense and synchronous population of willow warblers *Phylloscopus trochilus.* Journal of Avian Biology 28:319–324.

Blakey, J. K. 1994. Genetic evidence for extra-pair fertilizations in a monogamous passerine, the great tit *Parus major.* Ibis 136:457–462.

Conrad, K. F., R. J. Robertson, and P. T. Boag. 1998. Frequency of extrapair young increases in second broods of Eastern Phoebes. Auk 115:497–502.

Cramp, S. 1977–1994. The Birds of the Western Palearctic. Oxford University Press, Oxford.

Cramp, S. (ed.). 1985, 1988, 1992. Handbook of the Birds of Europe, the Middle East and North Africa, Vols. 4–6. Oxford University Press, Oxford.

Cramp, S., and C. M. Perrins (eds.). 1993, 1994. Handbook of the Birds of Europe, the Middle East and North Africa, Vols. 7–9. Oxford University Press, Oxford.

Cramp, S., and K. E. L. Simmons (eds.). 1977, 1980, and 1983. Handbook of the Birds of Europe, the Middle East and North Africa, Vols. 1–3. Oxford University Press, Oxford.

Currie, D. R., T. Burke, R. L. Whitney, and D. B. A. Thompson. 1998. Male and female behaviour and extra-pair paternity in the wheatear. Animal Behaviour 55:689–703.

Davies, N. B., S. H. M. Butchart, T. A. Burke, N. Chaline, and I. R. K. Stewart. 2003. Reed warblers guard against cuckoos and cuckoldry. Animal Behaviour 65:285–295.

Dickinson, J. L., and J. J. Arke. 1998. Extrapair paternity, inclusive fitness, and within-group benefits of helping in western bluebirds. Molecular Ecology 7:95–105.

Dixon, A., D. Ross, S. L. C. O'Malley, and T. Burke. 1994. Paternal investment inversely related to degree of extra-pair paternity in the reed bunting. Nature 371:698–700.

Dunn, P. O., L. A. Whittingham, J. T. Lifjeld, and R. J. Robertson. 1994. Effects of breeding density, synchrony, and experience on extrapair paternity in tree swallows. Behavioral Ecology 5:123–129.

Dunn, P. O., L. A. Whittingham, and T. E. Pitcher. 2001. Mating systems, sperm competition, and the evolution of sexual dimorphism in birds. Evolution 55:161–175.

Felsenstein, J. 1985. Phylogenies and the comparative method. American Naturalist 125:1–15.

Fortsmeier, W. 2002. Factors contributing to male mating success in the polygynous dusky warbler (*Phylloscopus fuscatus*). Behaviour 139:1361–1381.

Freeman-Gallant, C. R. 1997. Extra-pair paternity in monogamous and polygynous Savannah Sparrows, *Passerculus sandwichensis.* Animal Behaviour 53:397–404.

Fridolfsson, A. K., U. B. Gyllensten, and S. Jakobsson. 1997. Microsatellite markers for paternity testing in the willow warbler *Phylloscopus trochilus:* high frequency of extra-pair young in an island population. Hereditas 126:127–132.

Gelter, H. P., and H. Tegelström. 1992. High frequency of extra-pair paternity in Swedish pied flycatchers revealed by allozyme electrophoresis and DNA fingerprinting. Behavioral Ecology and Sociobiology 31:1–7.

Gibbs, H. L., P. J. Weatherhead, P. T. Boag, B. N. White, L. M. Tabak, and D. J. Hoysak. 1990. Realized reproductive success of polygynous red-winged blackbirds revealed by DNA markers. Science 250:1394–1397.

Gissing, G. J., T. J. Crease, and A. L. A. Middleton. 1998. Extra-pair paternity associated with renesting in the American Goldfinch. Auk 115:230–234.

Griffith, S. C., I. P. F. Owens, and K. A. Thuman. 2002. Extra-pair paternity in birds: a review of interspecific variation and adaptive function. Molecular Ecology 11:2195–2212.

Gullberg, A., H. Tegelstrom, and H. P. Gelter. 1992. DNA fingerprinting reveals multiple paternity in families of great and blue tits (*Parus major* and *P. caeruleus*). Hereditas 117: 103–108.

Gwinner, E., T. Roedl, and H. Schwabl. 1994. Pair territoriality of wintering stonechats: behavior, function and hormones. Behavioral Ecology and Sociobiology 34:321–327.

Gyllensten, U. B., S. Jakobsson, and H. Temrin. 1990. No evidence for illegitimate young in monogamous and polygynous warblers. Nature 343:168–170.

Haggerty, T. M., E. S. Morton, and R. C. Fleischer. 2001. Genetic monogamy in Carolina Wrens. Auk 118: 215–219.

Hartley, I. R., M. Shepherd, T. Robson, and T. Burke. 1993. Reproductive success of polygynous male corn buntings (*Miliaria calandra*) as confirmed by DNA fingerprinting. Behavioral Ecology 4:310–317.

Harvey, P. H., and M. D. Pagel. 1991. The Comparative Method in Evolutionary Biology. Oxford University Press, Oxford.

Hasselquist, D., S. Bensch, and T. von Schantz. 1996. Correlation between male song repertoire, extra-pair fertilizations and offspring survival in the great reed warbler. Nature 381:229–232.

Henderson, I. G., P. J. B. Hart and T. Burke. 2000. Strict monogamy in a semi-colonial passerine: the Jackdaw *Corvus monedula.* Journal of Avian Biology 31:177–182.

Herremans, M. 1997. Habitat segregation and male and female Red-backed Shrikes *Lanius collurio* and Lesser Gray Shrikes *Lanius minor* in the Kalahari basin. Journal of Avian Biology 28:240–248.

Hill, G. E., R. Montgomerie, C. Roeder, and P. T. Boag. 1994. Sexual selection and cuckoldry in a monogamous bird: implications for sexual selection theory. Behavioral Ecology and Sociobiology 35:193–199.

Hoi, H., and M. Hoi-Leitner. 1997. An alternative route to coloniality in the bearded tit: females pursue extrapair fertilizations. Behavioral Ecology 8:113–119.

Hoi-Leitner, M., H. Hoi, M. Romero-Pujante, and F. Valera. 1999. Female extra-pair behaviour and environmental quality in the serin (*Serinus serinus*): a test of the "constrained female hypothesis." Proceedings of the Royal Society of London, Series B, Biological Sciences 266:1021–1026.

Johnsen, A., J. T. Lifjeld, P. A Rohde, C. R. Primmer, and H. Ellegren. 1998. Sexual conflict over fertilizations: female bluethroats escape male paternity guards. Behavioral Ecology and Sociobiology 43:401–408.

Kempenaers, B. 1993. The uses of a breeding synchrony index. Ornis Scandinavica 24:84.

Kempenaers, B., G. R. Verheyen, and A. A. Dhondt. 1995. Mate guarding and copulation behaviour in monogamous and polygynous blue tits: do males follow a best-of-a-bad-job strategy? Behavioral Ecology and Sociobiology 36:33–42.

Ketterson, E. D., and V. Nolan Jr. 1994. Male parental behavior in birds. Annual Review of Ecology and Systematics 25:601–628.

Krokene, C., K. Anthonisen, J. T. Lifjeld, and T. Amundsen. 1996. Paternity and paternity assurance behaviour in the bluethroat, *Luscinia s. svecica.* Animal Behaviour 52: 405–417.

Lack, D. 1968. Ecological adaptations for breeding in birds. Methuen, London.

Langefors, A., D. Hasselquist, and T. von Schantz. 1998. Extra-pair fertilizations in the sedge warbler. Journal of Avian Biology 29:134–144.

Leisler, B., and M. Wink. 2000. Frequencies of multiple paternity in three *Acrocephalus* species (Aves Sylviidae) with different mating systems (*A. palustris, A. arundinaceus, A. paludicola*). Ethology, Ecology and Evolution 12:237–249.

Lens, L., S. Van Dongen, M. Van den Broech, C. Van Broeckhoven, and A. A. Dhondt. 1997. Why female crested tits copulate repeatedly with the same partner: evidence for the mate assessment hypotheses. Behavioral Ecology 8:87–91.

Lifjeld, J. T., T. Slagsvold, and H. M. Lampe. 1991. Low frequency of extra-pair paternity in pied flycatchers revealed by DNA fingerprinting. Behavioral Ecology and Sociobiology 29:95–101.

Lopez Ornat, A., and R. Greenberg. 1990. Sexual segregation by habitat in migratory warblers in Quintana Roo, Mexico. Auk 107:539–543.

Lynch, J. F., E. S. Morton, and M. E. Van der Voort. 1985. Habitat segregation between the sexes of wintering Hooded Warblers (*Wilsonia citrina*). Auk 102:714–721.

Mabey, S. E., and E. S. Morton. 1992. Demography and territorial behavior of wintering Kentucky Warblers in Panama. Pages 329–336 *in* Ecology and Conservation of Neotropical Migrant Landbirds (J. M. Hagan III and D. W. Johnston, eds.). Smithsonian Institution Press, Washington, D.C.

Marra, P. P. 2000. The role of behavioral dominance in structuring patterns of habitat occupancy in a migrant bird during the nonbreeding season. Behavioral Ecology 11:299–308.

Marra, P. P., and R. T. Holmes. 2001. Consequences of dominance-mediated habitat segregation in American redstarts during the nonbreeding season. Auk 118:92–104.

Martins, E. P. 2000. Adaptation and the comparative method. Trends in Ecology and Evolution 15:296–299.

Meek, S. B., R. J. Robertson, and P. T. Boag. 1994. Extra-pair paternity and intraspecific brood parasitism in eastern bluebirds revealed by DNA fingerprinting. Auk 111: 739–744.

Morton, E. S. 1990. Habitat segregation by sex in the Hooded Warbler: experiments on proximate causation and discussion of its evolution. American Naturalist 135:319–333.

Morton, E. S. 1992. What do we know about the future of migrant landbirds? Pages 579–589 *in* Ecology and Conservation of Neotropical Migrant Landbirds (J. M. Hagan III and D. W. Johnston, eds.). Smithsonian Institution Press, Washington, D.C.

Morton, E. S., L. Forman, and M. Braun. 1990. Extrapair fertilizations and the evolution of colonial breeding in Purple Martins. Auk 107:275–283.

Morton, E. S., J. F. Lynch, K. Young, and P. Melhop. 1987. Do male Hooded Warblers exclude females from nonbreeding territories in tropical forest? Auk 104:133–135.

Morton, E. S., B. J. M. Stutchbury, J. S. Howlett, and W. H. Piper. 1998. Genetic monogamy in blue-headed vireos and a comparison with a sympatric vireo with extrapair paternity. Behavioral Ecology 5:515–524.

Møller, A. P. 1991. Sperm competition, sperm depletion, paternal care, and relative testis size in birds. American Naturalist 137:882–906.

Møller, A. P., and J. V. Briskie. 1995. Extra-pair paternity, sperm competition and the evolution of testis size in birds. Behavioral Ecology and Sociobiology 36:357–365.

Møller, A. P., and J. J. Cuervo. 2000. The evolution of paternity and paternal care in birds. Behavioral Ecology 11:472–485.

Møller, A. P., and P. Ninni. 1998. Sperm competition and sexual selection: a meta-analysis of paternity studies in birds. Behavioral Ecology and Sociobiology 43:345–358.

Møller, A. P., and H. Tegelström. 1997. Extra-pair paternity and tail ornamentation in the barn swallow *Hirundo rustica*. Behavioral Ecology and Sociobiology 41:353–360.

Murphy, M. T., A. Pierce, J. Shoen, K. L. Murphy, J. A. Campbell, and D. A. Hamilton. 2001. Population structure and habitat use by overwintering neotropical migrants on a remote oceanic island. Biological Conservation 102:333–345.

Orell, M., S. Rytkonen, V. Launonen, P. Welling, K. Koivula, K. Kumpulainen, and L. Bachmann. 1997. Low frequency of extra-pair paternity in the willow tit, *Parus montanus*, as revealed by DNA fingerprinting. Ibis 139:562–566.

Otter, K., L. Ratcliffe, D. Michaud, and P. T. Boag. 1998. Do female black-capped chickadees prefer high-ranking males as extra-pair partners? Behavioral Ecology and Sociobiology 43:25–36.

Perreault, S., R. E. Lemon, and U. Kuhnlein. 1997. Patterns and correlates of extrapair paternity in American redstarts (*Setophaga ruticilla*). Behavioral Ecology 8:612–621.

Petrie, M., C. Doums, and A. P. Møller. 1998. The degree of extra-pair paternity increases with genetic variability. Proceedings of the National Academy of Sciences USA 95:9390–9395.

Pinxten, R., O. Hanotte, M. Eens, R. F. Verheyen, A. A. Dhondt, and T. Burke. 1993. Extra-pair paternity and intraspecific brood parasitism in the European starling, *Sturnus vulgaris:* evidence from DNA fingerprinting. Animal Behaviour 45:795–809.

Pitcher, T. E., and B. J. M. Stutchbury. 1998. Latitudinal variation in testes size in six species of North American songbirds. Canadian Journal of Zoology 76:618–622.

Pitcher, T. E., and B. J. M. Stutchbury. 2000. Does parental effort limit male EPC effort in Hooded Warblers? Animal Behaviour 59:1261–1269.

Price, T. 1997. Correlated evolution and independent contrasts. Philosophical Transactions of the Royal Society of London, Series B, Biological Sciences 352:519–529.

Purvis, A., and A. Rambaut. 1995. Comparative analysis by independent contrasts (CAIC): an Apple Macintosh application for analysing comparative data. Comparative Applied Bioscience 11:247–251.

Raouf, S. A., P. G. Parker, E. D. Ketterson, V. Nolan Jr., and C. Ziegenfus. 1997. Testosterone affects reproductive success by influencing extra-pair fertilizations in male dark-eyed juncos (Aves: *Junco hyemalis*). Proceedings of the Royal Society of London, Series B, Biological Sciences 264:1599–1603.

Rappole, J. H. 1988. Intra- and intersexual competition in migratory passerine birds during the nonbreeding season. Pages 2308–2317 *in* Proceedings of the 19th International Ornithological Congress.

Rappole. J. H. 1995. The Ecology of Migrant Birds. Smithsonian Institution Press, Washington, D.C.

Rappole, J. H., and D. W. Warner. 1980. Ecological aspects of migrant bird behavior in Veracruz, Mexico. Pages 353–393 *in* Migrant Birds in the Neotropics: Ecology Behavior, Distribution, and Conservation (A. Keast and E. S. Morton, eds.). Smithsonian Institution Press, Washington, D.C.

Ratti, O., M. Hovi, A. Lundberg, H. Tegelstrom, and R. V. Alatalo. 1995. Extra-pair paternity and male characteristics in the pied flycatchers. Behavioral Ecology and Sociobiology 37:419–425.

Reyer, H. U., K. Bollman, A. R. Schlapfer, A. Schymainda, and G. Klecack. 1997. Ecological determinants of extra-pair fertilizations and egg dumping in Alpine water pipits (*Anthus spinoletta*). Behavioral Ecology 8:534–543.

Richardson, D. S., and T. Burke. 2001. Extrapair paternity and variance in reproductive success related to breeding density in Bullock's orioles. Animal Behaviour 62:519–525.

Ridgway, R. 1901–1946. The Birds of North and Middle America, Vols. 1–10. Government Printing Office, Washington, D.C.

Riley, H. T., D. M. Bryant, R. E. Carter, and D. T. Parkin. 1995. Extra-pair fertilizations and paternity defense in house martins, *Delichon urbica*. Animal Behaviour 49:495–509.

Ritchison, G., P. H. Klatt, and D. F. Westneat. 1994. Mate guarding and extra-pair paternity in northern cardinals. Condor 96:1055–1063.

Rowe, D. L., M. T. Murphy, R. C. Fleischer, and P. G. Wolf. 2001. High frequency of extra-pair paternity in Eastern Kingbirds. Condor 103:845–851.

Saino, N., C. R. Primmer, H. Ellegren, and A. P. Møller. 1997. An experimental study of paternity and tail ornamentation in the barn swallow (*Hirundo rustica*). Evolution 51:562–570.

Schmoll, T., V. Dietrich, W. Winkel, J. T. Epplen, and T. Lubjuhn. 2003. Long-term fitness consequences of female extra-pair matings in a socially monogamous species. Proceedings of the Royal Society of London, Series B, Biological Sciences 270:259–264.

Schwagmeyer, P. L., R. C. St. Clair, J. D. Moodie, T. C. Lamey, G. D. Schnell, and M. N. Moodie. 1999. Species differences in male parental care in birds: a reexamination of correlates with paternity. Auk 116:487–503.

Sheldon, B. C., and T. Burke. 1994. Copulation behavior and paternity in the chaffinch. Behavioral Ecology and Sociobiology 34:149–156.

Sheldon, B. C., and H. Ellegren. 1999. Sexual selection resulting from extra-pair paternity in collared flycatchers. Animal Behaviour 57:285–298.

Sibley, C. G., and J. E. Ahlquist. 1990. Phylogeny and classification of birds: a study in molecular evolution. Yale University Press, New Haven.

Smith, H., and T. von Schantz. 1995. Extra-pair paternity in the European starling: the effect of polygyny. Condor 95:1006–1015.

Soukop, S. S., and C. F. Thompson. 1997. Social mating system affects the frequency of extra-pair paternity in house wrens. Animal Behaviour 54:1089–1105.

Stacier, C. A. 1992. Social behavior of the Northern Parula, Cape May Warbler, and Prairie Warbler wintering in second growth forest in southwestern Puerto Rico. Pages 308–320 *in* Ecology and Conservation of Neotropical Migrant Landbirds (J. M. Hagan III and D. W. Johnston, eds.). Smithsonian Institution Press, Washington, D.C.

Stronbach, S., E. Curio, A. Bathen, J. T. Epplen, and T. Lubjuhn. 1998. Extra-pair paternity in the great tit (Parus major): a test of the "good genes" hypothesis. Behavioral Ecology 9:388–396.

Stutchbury, B. J. 1994. Competition for winter territories in a neotropical migrant: the role of age, sex and color. Auk 111:63–69.

Stutchbury, B. J. M. 1998. Female mate choice of extra-pair males: breeding synchrony is important. Behavioral Ecology and Sociobiology 43:213–215.

Stutchbury, B. J., and E. S. Morton. 1995. The effect of breeding synchrony on extra-pair mating systems in songbirds. Behaviour 132:675–690.

Stutchbury, B. J. M., and E. S. Morton. 2001. Behavioral Ecology of Tropical Birds. Academic Press, London.

Stutchbury, B. J. M., and D. Neudorf. 1998. Female control, breeding synchrony and the evolution of extra-pair mating strategies. Pages 103–122 *in* Avian Reproductive Tactics: Female and Male Perspectives (P. Parker and N. Burley, eds.). Ornithological Monographs 49.

Stutchbury, B. J. M., W. H. Piper, D. L. Neudorf, S. A. Tarof, J. M. Rhymer, G. Fuller, and R. C. Fleischer. 1997. Correlates of extra-pair fertilization success in hooded warblers. Behavioral Ecology and Sociobiology 40:119–126.

Sundberg, J., and A. Dixon. 1996. Old, colourful male yellowhammers, *Emberiza citrinella,* benefit from extra-pair copulations. Animal Behaviour 52:113–122.

Tarof, S. A., L. M. Ratcliffe, M. M. Kasumovic, and P. T. Boag. 2005. Are Least Flycatchers (*Empidonax minimus*) clusters hidden leks? Behavioral Ecology 16:207–217.

Thusius, K. J., K. A. Peterson, P. O. Dunn, and L. A. Whittingham. 2001. Male mask size is correlated with mating success in the common yellowthroat. Animal Behaviour 62:435–446.

Trivers, R. L. 1972. Parental investment and sexual selection. Pages 136–179 *in* Sexual Selection and the Descent of Man (B. Campbell, ed.). Aldine, Chicago.

Tuttle, E. 2003. Alternative reproductive strategies in the white-throated sparrow: behavioral and genetic evidence. Behavioral Ecology 14:425–432.

Valera, F., H. Hoi, and A. Kristín. 2003. Male shrikes punish unfaithful females. Behavioral Ecology 14:403–408.

Verboven, N., and A. C. Mateman. 1997. Low frequency of extra-pair fertilizations in the great tit *Parus major* revealed by DNA fingerprinting. Journal of Avian Biology 28:231–239.

Voight, C., S. Leitner, and M. Gahr. 2003. Mate fidelity in a population of island canaries (*Serinus canaria*) in the Madeiran archipelago. Journal für Ornithologie 144: 86–92.

Wagner, R. H., M. D. Schug, and E. S. Morton. 1996. Condition-dependent control of paternity by female purple martins: implications for coloniality. Behavioral Ecology and Sociobiology 38:379–389.

Webster, M. S., H. C. Chuang-Dobbs, and R. T. Holmes. 2001. Microsatellite identification of extrapair sires in a socially monogamous warbler. Behavioral Ecology 12:439–446.

Westneat, D. F. 1987. Extra-pair fertilizations in a predominantly monogamous bird: genetic evidence. Animal Behaviour 35:865–876.

Westneat, D. F. 1990. Genetic parentage in the indigo bunting: a study using DNA fingerprinting. Behavioral Ecology and Sociobiology 27:67–76.

Westneat, D. F. 1993. Polygyny and extra-pair fertilizations in eastern red-winged blackbirds (*Agelaius phoeniceus*). Behavioral Ecology 4:49–60.

Westneat, D. F., and P. W. Sherman. 1997. Density and extra-pair fertilizations in birds: a comparative analysis. Behavioral Ecology and Sociobiology 41:205–215.

Wetton, J. H., T. Burke, D. T. Parkin, and D. Walters. 1995. Single-locus DNA fingerprinting reveals that male reproductive success increases with age through extra-pair paternity in the house sparrow (*Passer domesticus*). Proceedings of the Royal Society of London, Series B, Biological Sciences 260:91–98.

Wetton, J. H., and D. T. Parkin. 1991. An association between fertility and cuckoldry in the house sparrow *Passer domesticus.* Proceedings of the Royal Society of London, Series B, Biological Sciences 245:227–233.

Whittingham, L. A., and J. T. Lifjeld. 1995. Extra-pair fertilization increases the opportunity for sexual selection in the monogamous house martin *Delichon urbica*. Journal of Avian Biology 26:283–288.

Woolfenden, B. E., B. J. M. Stutchbury, and E. S. Morton. 2005. Extra-pair fertilizations in the Acadian Flycatcher: males obtain EPFs with distant females. Animal Behavior.

Wunderle, J. M., Jr. 1992. Sexual habitat segregation in wintering Black-throated Blue Warblers in Puerto Rico. Pages 299–307 *in* Ecology and Conservation of Neotropical Migrant Landbirds (J. M. Hagan III and D. W. Johnston, eds.). Smithsonian Institution Press, Washington, D.C.

Yezerinac, S. M., P. J. Weatherhead, and P. T. Boag. 1995. Extra-pair paternity and the opportunity for sexual selection in a monogamous bird (*Dendroica petechia*). Behavioral Ecology and Sociobiology 37:179–188.

Zar, H. H. 1999. Biostatistical Analysis. Prentice-Hall, Englewood Cliffs, N.J.

DANIEL R. FROEHLICH,
SIEVERT ROHWER,
AND BRIDGET J. STUTCHBURY

Spring Molt Constraints versus Winter Territoriality

Is Conspicuous Winter Coloration Maladaptive?

AMONG TEMPERATE-BREEDING NORTH AMERICAN passerines, some males stay conspicuous during the nonbreeding season while others become dull. We used comparative data to evaluate two explanations for adult males' conspicuousness in winter—winter territoriality and constraints on spring molting. The hypothesis that conspicuous winter coloration functions in territory defense is not supported because there is no relationship between winter social behavior and color in adult males among dichromatic species. The molt constraints hypothesis predicts that species that remain conspicuous in the winter season do so because they cannot afford a complete molt of body feathers in spring, implying that conspicuous winter plumage may be maladaptive. Consistent with molt constraints, most sexually dichromatic songbirds of North America that have an extensive spring molt do not have conspicuous winter coloration in adult males. Long-distance migrants that make extensive use of plant products on the wintering grounds are more likely than obligate insectivores to undertake spring body molt. Invertebrate populations are widely depressed during the late dry season across the northern Neotropics, the period of the spring molt, whereas plant product availability is stable or somewhat elevated. For species that winter close to their temperate breeding grounds, however, diet during late winter was not a reliable predictor of spring molting. In contrast to the numerous Neotropical migrants that undertake at least some spring molt, few temperate winterers do; they are presumably under time constraints to initiate breeding as early as possible. Comparisons with molt strategies of European species suggest that both North American and European wintering species share similar constraints on spring molting, whereas molt strategies among long-distance Euro-African migrants differ from those of Nearctic migrants. Although the results of these comparative tests are, in some cases, based on incomplete quantitative data on molt patterns, winter social behavior, and diet, they nevertheless suggest that molt constraints deserve more attention in interpreting seasonal color changes in birds and in understanding winter energetics and diets.

INTRODUCTION

Three patterns of color dimorphism recur in many avian taxa (Butcher and Rohwer 1989). When color differences exist, usually (1) males are brighter than females, (2) older birds are brighter than younger birds, and (3) breeding plumages are brighter than nonbreeding plumages. These patterns are consistent with hypotheses of social competition for the evolution of conspicuous coloration (Butcher and Rohwer 1989). The first two patterns have been the subject of many empirical and theoretical studies (e.g., Rohwer et al. 1980; Hamilton and Zuk 1982; Rohwer 1982). Here we focus on seasonal color change, which has received scant attention (Hamilton and Barth 1962). For scientific names not included in the text, see table 25.2 and the Appendix to Rohwer et al. (Chap. 8, this volume).

Seasonal color changes in and the diverse winter ecology of adult birds present a unique opportunity to study the adaptive significance of color patterns. For example, adult male Summer Tanagers remain conspicuous in winter, whereas adult male Scarlet Tanagers lose their conspicuous breeding coloration altogether (Hamilton and Barth 1962). Likewise, Painted and Lazuli Buntings remain conspicuous in winter, whereas adult Indigo Buntings males do not (Rohwer 1986; Thompson 1991; Young 1991). Such contrasts among closely related species suggest that comparative analyses may be fruitful (Greenberg 1986; Holmes et al. 1989; Staicer 1992).

Seasonal Color Change

Adaptationist explanations for seasonal plumage changes focus on crypsis for dull plumages, presumably to reduce predation risk (Hamilton and Barth 1962), or status signaling for conspicuous plumages (Rohwer 1975). Conspicuous and nonvariable coloration during the nonbreeding season is viewed as an adaptation for defending winter resources, although this hypothesis has not been rigorously tested (Hamilton and Barth 1962; Morton 1980; Lynch et al. 1985; but see Butcher and Rohwer 1989). Variable winter plumages may advertise dominance status in flocking species, as confirmed experimentally (Rohwer 1977, 1978, 1985; Rohwer and Ewald 1981; Järvi et al. 1987b).

Alternatively, the unprofitable prey hypothesis (Baker and Parker 1979) suggests that conspicuous plumage signals high vigilance and agility and low probability of capture to potential predators. Under this hypothesis, conspicuous plumages should increase survivorship. Thus in areas of high predation, conspicuous plumages should be more common, not less common as under the crypsis hypothesis. Götmark (1995) provides experimental support for this idea.

Molt Constraints

Previous studies of seasonal color change in adult male passerines often ignored the costs of molt. Although a few studies demonstrated low molt costs for particular species (e.g., Brown and Bryant 1996), such cases are apparently unusual, particularly among small birds with high basal metabolic rates (Lindström et al. 1993; Klaassen 1995). More commonly, feather replacement may be constrained by costs of molting that may conflict with other important life history variables. These include physiological and energetic costs of growing new feathers (King 1981; Walsberg 1983; Murphy and King 1992; Lindström et al. 1993); aerodynamic and thermoregulatory performance costs of feather replacement (Tucker 1991; Swaddle and Witter 1997; Hedenström and Sunada 1999; Lind 2001); risks of damage to fragile developing follicles and of growing low-quality feathers (Murphy et al. 1989; Murphy and King 1991; Bortolotti et al. 2002); and time constraints imposed by slow feather growth rates that generate trade-offs in the numbers of feathers replaced (Rohwer 1999; Shugart and Rohwer 1996) or in feather quality (Nilsson and Svensson 1996; Dawson et al. 2000). Of these constraints, time and energy limitations are likely to be the most general, as they can interfere directly with the demands of other life history stages. The remaining two, performance and risk, are more strongly behavior-dependent and thus likely to vary among individuals.

Almost all North American passerines undergo a complete molt of body and flight feathers after breeding, but spring molts are highly variable in extent, ranging from absent to extensive (Rohwer and Butcher 1988; Pyle 1997). Most dramatic seasonal color changes require an extensive spring molt, allowing birds to change back into breeding coloration. Extent and timing of spring molts vary considerably even among closely related populations, suggesting either evolutionary flexibility or striking differences in molt constraints in closely related taxa (Mewaldt and King 1978; Rohwer and Manning 1990; Svensson and Hedenström 1999).

Rohwer et al. (1983) and Rohwer and Butcher (1988) first raised the possibility that spring molts may be constrained by the costs of molting. Seasonal color change could incur high energetic costs, especially if the spring molt coincides with a time of energetic stress arising from late-winter food limitation or conflicting requirements (migration or early breeding). If birds are constrained energetically from undertaking a spring molt, summer and winter plumages are no longer independent of each other. A plumage that is optimal for one season is unlikely to be optimal for the other (Rohwer et al. 1983; Rohwer and Butcher 1988). Thus, molt constraints are likely to be critical in interpreting seasonal color changes in birds.

Spring molting represents a cost-benefit trade-off: molt costs are concentrated during molt, whereas benefits from seasonal color independence and fresh plumage are distributed over the entire previous winter (plumage independence) and following summer (fresh plumage). Non-molters give up winter plumage independence (costs distributed over the entire winter) and worn feather replacement (costs distributed over the following summer), but avoid the costs of a second molt. If spring molting imposes heavy costs, then maintaining a conspicuous nonbreeding plumage may be the default condition, overcome only when selective pressures for plumage independence (predation or social domination favoring dull plumage) or feather renewal (excessive annual wear) are sufficiently high. Because molt is evolutionarily flexible, broad categorical comparisons may reveal universal pressures.

Comparisons

We evaluate the potential importance of molt constraints in determining the winter appearance of adult males by exploring seasonal color change. The molt constraint hypothesis predicts that conspicuous winter plumage is associated with a limited or absent spring molt, whereas the winter territoriality hypothesis predicts that it is associated with winter territoriality. Evidence for either association provides only weak support for the original hypothesis, as it fails to address the causative mechanism. Failure to find an association, however, offers a powerful rejection of the original hypothesis.

As mentioned, time and energy are likely the most general limitations on spring molting. If they interact, Neotropical migrants should be less constrained and more likely to have an extensive spring molt than temperate residents. This follows because food resources available to migrants in late winter and spring can be used for molting but not for breeding, since resources to support breeding are not yet available at migration destinations (Slagsvold 1976; Francis and Cooke 1986; Poulin and Lefebvre 1997). Costs of migration for all long-distance migrants are presumed to be similar and neutral with respect to spring molting because they are incurred just before migration begins. By contrast, any resources becoming available to temperate residents in spring are better invested in early breeding than in molt (Perrins 1970; Svensson 1995).

Variability in resource availability between insectivorous species and those that rely heavily on fruit, seeds, or nectar during late winter could approximate energy limitation as a molt constraint. Fruits used by birds have limited protein content, whereas seeds, grains, and nuts have intermediate and invertebrate food sources high protein content (Murphy and King 1992). Because protein demands are high during molt, insectivores might be more able to replace feathers than plant product users.

On the other hand, the widespread dry season in the northern Neotropics, though locally variable, is most severe between January and April, the spring molting period. Arthropod levels have repeatedly been shown to be significantly depressed during this time in a variety of forest habitats of the Caribbean (Parrish and Sherry 1994; Strong and Sherry 2000; Latta and Faaborg 2001), southern Central America (Buskirk and Buskirk 1976; Levings and Windsor 1982; Poulin and Lefebvre 1996, 1997), and northern South America (Lefebvre et al. 1992; Poulin and Lefebvre 1997). Several species show associated effects on body condition (Strong and Sherry 2000; Marra and Holmes 2001), diet (Morton 1980) and seasonal distribution and spacing behavior (Parrish and Sherry 1994; Johnson and Sherry 2001; Latta and Faaborg 2001; but see Karr and Brawn 1990). During the same period, fruits, seeds, Müllerian bodies, and especially flowers are often in ample supply (Morton 1980; Greenberg 1995; Poulin and Lefebvre 1996). If such differences in resource levels were sufficiently dramatic, obligate insectivores might be more likely to experience spring molt constraints, and we would expect spring molting to be more prevalent among plant product users than among insectivores.

We further reasoned that the extent of spring molt among different species could be related to the extent of breeding dichromatism. For example, the degree of breeding-season dichromatism might be correlated with the presence of distinctive subadult breeding plumage, explaining why yearling males resemble adult males in some species whereas in others they remain dull for their first potential breeding season (Rohwer et al. 1980; Rohwer and Butcher 1988). If spring molting is costly, subadult males of species in which adult males have a high percentage of their body conspicuously colored should be less likely to be able to achieve a plumage matching adult males in their first summer. Second, young males in subadult winter plumages might be under selective pressure to undertake more extensive molts than adults in order to achieve adult male breeding plumage, even though they are less efficient at securing resources (Grubb et al. 1991; Wunderle 1991; Smith and Metcalfe 1994). Third, species with partial molts should target feather replacement at those areas of the breeding plumage that are dichromatic if changing color rather than replacing worn feathers is the main selective force acting on spring molts.

Finally, comparisons of molt and migration patterns between American and European songbird species should reveal underlying similarities and differences in the overall selection regimes producing the patterns.

METHODS

We examined museum skins to determine the extent of seasonal color change in North American adult male passerines. We restricted this analysis to sexually dichromatic species (those in which almost all full adults can be correctly assigned to sex on the basis of plumage) as there are few, if any, North American species that are sexually monochromatic and seasonally dichromatic (tables 25.1, 25.2). For monochromatic species, plumage conspicuousness is difficult to evaluate without an alternate state; further, monochromatic conspicuous species in North America are generally non-territorial in winter (e.g., Kingbirds [*Tyrannus* spp.] or Blue Jay [*Cyanocitta cristata*]), and thus would reinforce our failure to find an association between conspicuousness and territoriality.

We asked three naïve independent observers to evaluate 30 of the dichromatic species for which both male and female breeding specimens and adult male specimens in fresh winter plumage were available at the Burke Museum and represented the gamut of plumage differences and genera in the full list of 101 species. Observers grouped male-female differences into four categories representing increasing degrees of difference in plumage between the sexes. To assess differences between male breeding and nonbreeding plumages, observers categorized male winter plumages as dull (little or no male-specific breeding plumage) or conspicuous (much male-specific breeding plumage). "Conspicuous" here refers to those plumage parts that distinguish adult males from the dullest females (equivalent to "colorful," sensu Butcher and Rohwer 1989). Male-specific plumage was assessed by contrasting breeding males with breeding females. In no species does either sex become more conspicuous in winter.

Interobserver agreement was high, with complete agreement for 23 of 30 species. DRF's independent categorizations agreed with at least two of three observers for 27 of these 30. Interobserver disagreement centered on the Nashville, Yellow, and Yellow-rumped Warbler and Violet-green Swallow, so we excluded these species from the seasonal color change analyses. Because of the high repeatability of categorization and the general agreement between DRF and the naïve observers, the remaining 71 species were assigned to categories by DRF.

We categorized the presence of subadult plumage in winter and summer according to Rohwer and Butcher (1988), but added species featuring less than bimodal color differences between adults and subadults on the basis of specimen examinations and descriptions in Pyle (1997) and Dunn and Garrett (1997) (table 25.1). This expands Rohwer and Butcher's list by 12 species and is desirable because the original work on delayed plumage maturation (Rohwer et al. 1980) assumed that femalelike plumages were adaptive, thus anticipating bimodality. In contrast, molt constraints do not presuppose bimodality and accommodate continuous plumage variation between subadults and adults.

We used Pyle (1997) to categorize spring body molts in Neotropical migrants as absent, partial ("limited" in Pyle 1997), and extensive. Extensive molts involve all body feathers, whereas partial molts involve only a small subset, typically on the head. In many species with partial molts, individuals may not molt at all; in any case, partial molts are unlikely to represent significant energetic costs (Pyle 1997). Thus, species with partial molts were not included in molt analyses. Species for which first winter molts are more extensive than adult molts and those with partial molts that replace only the dichromatic parts of the plumage are identified in table 25.1. Species that do not molt in spring but show limited seasonal changes in appearance by wearing away cryptic tips on the feathers of their fresh basic plumage over the winter (e.g., Lazuli Bunting and Orchard Oriole) were categorized according to their appearance in fresh fall plumage.

We attempted a thorough review of the literature on the winter spacing systems of North American songbirds and believe we provide an up-to-date assessment of current understanding despite persistently poor documentation (Keast and Morton 1980; Hagan and Johnston 1992). We classified their social organization as territorial or non-territorial (Appendix to Chap. 8, this volume), obscuring possibly important differences between home-range occupancy and agonistic territoriality (Strong and Sherry 2000; Lefebvre et al. 1992). Further, incomplete descriptive coverage across species' ranges and any failure to locate all published information may have led to some misclassifications. Even with rough categorizations, however, strong relationships can at least be suggestive and indicate where further testing may be worthwhile.

Because evidence for territoriality is often found in sources not focusing on territorial behavior, our categories are necessarily crude. For this comparison, evidence for territoriality included aggressive defense of exclusive territories from conspecifics (Lynch et al. 1985), aggressive response to playback, or, in areas where the species occurs in high density, consistent occurrence of individuals in mixed-species flocks or solo individuals (Munn and Terborgh 1979; King and Rappole 2000). In areas where they are scarce, both territorial and non-territorial species often occur singly. Territoriality was assigned liberally, even if only some individuals of a species regularly show territoriality (Staicer 1992). Non-territorial species included those that form intraspecific flocks (Greenberg 1984), do not regularly show aggression toward conspecifics (Chipley 1980), or defend ephemeral food sources from conspecifics only occasionally (Greenberg 1979).

We excluded Black-capped Vireo and Golden-cheeked Warbler from territoriality analyses for lack of clear indications of their winter social behavior. Several species treated as territorial in the analyses are identified in table 25.1 as "mixed." In these species, individual territoriality varies by region, habitat, or age and sex. Eight species for which there is circumstantial evidence that male/female pairs may occur together and in some cases even codefend winter territories are also indicated in table 25.1. For five of these (Blue-gray Gnatcatcher, Cerulean and Prothonotary Warblers, Black-headed Grosbeak, and Baltimore Oriole), consistent co-occurrence has been observed but not territorial behavior itself, so we classify these species as non-territorial.

Both migrants wintering in temperate regions and temperate permanent residents were treated as temperate winterers (table 25.2). Winter territoriality categories for these species were taken from Birds of North America accounts (Poole and Gill 1992–2003).

Table 25.1 Winter social behavior, plumage features, spring molt, and winter food sources for Neotropical migrants

Species	Spring molt	Male winter plumage	% Body dichromatic	Subadult plumage[1]	Winter territoriality	Late winter food[2]
Scissor-tailed Flycatcher	Extensive	Conspicuous	25	Winter-summer	No	I
Purple Martin	Absent	Conspicuous	100	Winter-summer	No	I
Violet-green Swallow	**Absent**		**25**	**None**	**No**	**I**
Blue-gray Gnatcatcher	Extensive	Dull	25	None	No (pairs?)	I
Black-capped Vireo	Partial*	Conspicuous	25	Winter-summer	Yes?	Mixed
Lesser Goldfinch	Extensive	Conspicuous	75	Winter	No	G
Blue-winged Warbler	Absent	Conspicuous	25	Winter-summer	Yes	?
Golden-winged Warbler	Absent	Conspicuous	25	Winter-summer	Yes	I
Tennessee Warbler	**Partial***	**Conspicuous**	**25**	**Winter-summer**	**No**	**F**
Nashville Warbler	**Partial***		**25**	**Winter+**	**No**	**?**
Virginia's Warbler	Partial*	Conspicuous	25	Winter+	No	I
Northern Parula	Partial*	Conspicuous	25	Winter-summer+	Yes (mixed)	I
Yellow Warbler	**Extensive**		**50**	**Winter**	**Yes (mixed)**	**I**
Chestnut-sided Warbler	**Extensive**	**Dull**	**50**	**Winter**	**Yes (mixed)**	**Mixed**
Magnolia Warbler	**Extensive**	**Dull**	**50**	**Winter-summer**	**Yes**	**?**
Cape May Warbler	**Extensive**	**Conspicuous**	**50**	**Winter-summer**	**Yes (mixed)**	**Mixed**
Black-thr. Blue Warbler	Absent	Conspicuous	100	Winter-summer	Yes (mxd/prs?)	I
Yellow-rumped Warbler	**Extensive**		**75**	**Winter**	**No**	**F**
Golden-cheeked Warbler	Partial*	Conspicuous	50	Winter	Yes?	?
Black-thr. Green Warbler	Partial*	Conspicuous	50	Winter-summer	Yes (mixed)	I
Townsend's Warbler	**Partial***	**Conspicuous**	**50**	**Winter**	**No**	**I**
Hermit Warbler	Partial*	Conspicuous	50	Winter	No	I
Blackburnian Warbler	**Extensive**	**Dull**	**75**	**Winter**	**No**	**I**
Prairie Warbler	Partial*	Conspicuous	50	Winter	Yes (mixed)	I
Bay-breasted Warbler	**Extensive**	**Dull**	**75**	**Winter-summer**	**No**	**F**
Blackpoll Warbler	Extensive	Dull	100	Winter	No	?
Cerulean Warbler	Extensive	Conspicuous	100	Winter+	No (pairs?)	?
Black-and-White Warbler	**Extensive**	**Conspicuous**	**50**	**Winter**	**Yes**	**I**
American Redstart	Absent	Conspicuous	100	Winter-summer+	Yes (mixed)	I
Prothonotary Warbler	**Absent**	**Dull**	**50**	**Winter-summer**	**No (pairs?)**	**I**
Kentucky Warbler	Partial*	Conspicuous	25	Winter+	Yes	I
Connecticut Warbler	Partial*	Conspicuous	25	Winter+	Yes	?
Mourning Warbler	Partial	Conspicuous	25	Winter+	Yes	Mixed
MacGillivray's Warbler	Partial	Conspicuous	25	Winter+	Yes	I
Common Yellowthroat	**Partial***	**Conspicuous**	**50**	**Winter+**	**Yes**	**?**
Hooded Warbler	Absent	Conspicuous	50	None	Yes	I
Wilson's Warbler	**Absent**	**Conspicuous**	**25**	**None**	**Yes**	**I**
Canada Warbler	Extensive	Conspicuous	25	Winter-summer	Yes (pairs?)	?
Hepatic Tanager	Partial	Conspicuous	100	Winter-summer+	No	Mixed
Scarlet Tanager	Extensive	Dull	100	Winter-summer+	Yes (mixed)	Mixed
Summer Tanager	**Extensive**	**Conspicuous**	**100**	**Winter-summer+**	**Yes (pairs?)**	**F**
Western Tanager	Extensive	Conspicuous	50	Winter-summer+	Yes	F
Lark Bunting	Extensive	Dull	100	Winter	No	G
Rose-breasted Grosbeak	**Extensive**	**Conspicuous**	**100**	**Winter-summer+**	**No (pairs?)**	**?**
Black-headed Grosbeak	Partial	Conspicuous	100	Winter-summer+	No	Mixed
Blue Grosbeak	Absent	Conspicuous	100	Winter-summer+	No	Mixed
Lazuli Bunting	Partial*	Conspicuous	75	Winter-summer+	No	Mixed
Indigo Bunting	**Extensive**	**Dull**	**100**	**Winter-summer**	**No**	**G**
Painted Bunting	**Partial**	**Conspicuous**	**100**	**Winter-summer**	**No**	**G**
Dickcissel	?	Conspicuous	25	Winter	No	G
Bobolink	**Extensive**	**Dull**	**100**	**None**	**No**	**G**
Yellow-headed Blackbird	**Absent**	**Conspicuous**	**75**	**Winter-summer+**	**No**	**G**
Hooded Oriole	?	Conspicuous	75	Winter-summer+	No	F
Baltimore Oriole	**Absent**	**Conspicuous**	**75**	**Winter-summer+**	**No (pairs?)**	**F**
Bullock's Oriole	Absent	Conspicuous	75	Winter-summer+	No	F
Orchard Oriole	Absent	Conspicuous	100	Winter-summer+	No	F
Scott's Oriole	Absent	Conspicuous	75	Winter-summer+	No	F

Note: For species in boldface, categorization of winter male plumage conspicuousness and patch size was determined through direct comparison of specimens by independent observers. See the Appendix in Rohwer et al. (Chap. 8, this volume) for territoriality references and scientific names. Information on spring molts is taken from Pyle (1997) and subadult plumage classifications are based on Rohwer and Butcher (1988), Pyle (1997), Dunn and Garrett (1997), and specimen examination (see text). Species follow Clements (2000) order. An asterisk indicates partial spring molts (i.e., not entire body plumage) that include all the dichromatic parts of the plumage. Spring molts in Dickcissel and Hooded Oriole are poorly described. A plus sign indicates species in which subadult males are currently thought to undergo more extensive prealternate molts than other age and sex classes (Pyle 1997). Superscript 1: Subadult plumage indicates the presence of distinctive subadult contour plumage during the seasons given. Superscript 2 indicates main food source in late winter, summarized from Birds of North America series (Poole and Gill 1992–2003). I = invertebrates; F = fruits and nectar; G = grains and seeds; Mixed = mix of invertebrates and plant material taken in substantial quantities; ? = no or very poor information.

Table 25.2 Spring molt, plumage features, winter social behavior, and late winter food sources for dichromatic temperate wintering species

Name	Genus	Species	Spring molt	Male winter plumage	Winter territoriality	Late winter food[1]
Vermilion Flycatcher	*Pyrocephalus*	*rubinus*	Absent	Conspicuous	No	I
Horned Lark	*Eremophila*	*alpestris*	Absent	Conspicuous	No	G
Golden-crowned Kinglet	*Regulus*	*satrapa*	Absent	Conspicuous	No	I
Ruby-crowned Kinglet	*Regulus*	*calendula*	Absent	Conspicuous	No	I
Phainopepla	*Phainopepla*	*nitens*	Absent	Conspicuous	Yes	F
Eastern Bluebird	*Sialia*	*sialis*	Absent	Conspicuous	No	F
Western Bluebird	*Sialia*	*mexicana*	Absent	Conspicuous	No	F
Mountain Bluebird	*Sialia*	*currucoides*	Absent	Conspicuous	No	F
American Robin	*Turdus*	*migratorius*	Absent	Conspicuous	No	Mixed
Varied Thrush	*Ixoreus*	*naevia*	Absent	Conspicuous	No	I
California Gnatcatcher	*Polioptila*	*californica*	Extensive	Dull	Yes	I
Black-tailed Gnatcatcher	*Polioptila*	*melanura*	Extensive	Dull	?	?
Red-breasted Nuthatch	***Sitta***	***canadensis***	**Absent**	**Conspicuous**	**?**	**Mixed**
White-breasted Nuthatch	*Sitta*	*carolinensis*	Absent	Conspicuous	Yes	Mixed
Gray-crowned Rosy-Finch	*Leucosticte*	*tephrocotis*	Absent	Conspicuous	No	G
Black Rosy-Finch	*Leucosticte*	*atrata*	Absent	Conspicuous	No	G
Brown-capped Rosy-Finch	*Leucosticte*	*australis*	Absent	Conspicuous	No	G
Pine Grosbeak	*Pinicola*	*enucleator*	Absent	Conspicuous	No	G
Cassin's Finch	*Carpodacus*	*cassinii*	Partial	Conspicuous	No	F
Purple Finch	*Carpodacus*	*purpureus*	Partial	Conspicuous	No	F
House Finch	*Carpodacus*	*mexicanus*	Absent	Conspicuous	No	G
Red Crossbill	***Loxia***	***curvirostra***	**Absent**	**Conspicuous**	**No**	**G**
White-winged Crossbill	*Loxia*	*leucoptera*	Absent	Conspicuous	No	G
Common Redpoll	***Carduelis***	***flammea***	**Absent**	**Conspicuous**	**No**	**G**
Hoary Redpoll	*Carduelis*	*hornemanni*	Absent	Conspicuous	No	G
Lawrence's Goldfinch	*Carduelis*	*lawrencei*	Extensive	Conspicuous	No	G
American Goldfinch	***Carduelis***	***tristis***	**Extensive**	**Dull**	**No**	**G**
Eastern Towhee	***Pipilo***	***erythrophthalmus***	**Absent**	**Conspicuous**	**No**	**Mixed**
Spotted Towhee	*Pipilo*	*maculatus*	Absent	Conspicuous	No	Mixed
Black-chinned Sparrow	*Spizella*	*atrogularis*	Extensive	Dull	No	G
McCown's Longspur	*Calcarius*	*mccownii*	Partial	Dull	No	G
Lapland Longspur	***Calcarius***	***lapponicus***	**Partial**	**Dull**	**No**	**G**
Smith's Longspur	*Calcarius*	*pictus*	Extensive	Dull	No	G
Chestnut-collared Longspur	*Calcarius*	*ornatus*	Partial	Dull	No	G
Snow Bunting	***Plectrophenax***	***plectrophenax***	Absent	Dull	No	G
McKay's Bunting	*Plectrophenax*	*hyperboreus*	Absent	Dull	?	?
Northern Cardinal	*Cardinalis*	*cardinalis*	Absent	Conspicuous	No	G
Red-winged Blackbird	***Agelaius***	***phoeniceus***	**Absent**	**Conspicuous**	**No**	**G**
Tricolored Blackbird	*Agelaius*	*tricolor*	Absent	Conspicuous	No	G
Brewer's Blackbird	*Euphagus*	*cyanocephalus*	Absent	Conspicuous	No	G
Boat-tailed Grackle	*Quiscalus*	*major*	Absent	Conspicuous	No	Mixed
Common Grackle	*Quiscalus*	*quiscula*	Absent	Conspicuous	No	G
Great-tailed Grackle	*Quiscalus*	*mexicanus*	Absent	Conspicuous	No	G
Brown-headed Cowbird	*Molothrus*	*ater*	Absent	Conspicuous	No	G

Note: Extent of spring molt from Pyle (1997). For species in boldface, categorization of relative winter male conspicuousness was scored through direct specimen comparison by independent observers. Winter territoriality summarized from Birds of North America series (Poole and Gill 1992–2003). Species follow Clements (2000) order. Superscript 1 indicates main food source in late winter, summarized from Birds of North America series (Poole et al. 1992-2003). I = invertebrates; F = fruits and nectar; G = grains and seeds; Mixed = mix of invertebrates and plant material taken in substantial quantities; ? = no or very poor information.

All 101 species were categorized into one of four broad trophic categories (granivores, frugivores and nectarivores, mixed-diet consumers, and invertebrate consumers) according to the primary food use in late winter, based on Birds of North America accounts (Poole and Gill 1992–2003). Insufficient information was available for 12 species and these were not included in the analyses (tables 25.1, 25.2).

Where there was variation in molt, adult winter coloration, or food resources used within genera (e.g., *Dendroica*), we did our analysis at the species level. When congeners had the same value for the variables in question (e.g., *Icterus*), the genus was treated as one data point for statistical analyses to provide for phylogenetic independence.

RESULTS

Winter Appearance and Spring Molt

Table 25.3 summarizes relationships between conspicuousness of male winter plumage color and extent of spring molts for residents and migrants. We find a strong negative relationship between conspicuousness and the presence of an extensive spring molt in all species combined (not shown; Fisher's Exact; $p < 0.001$). Most taxa with dramatic seasonal color changes (i.e., dull winter plumage) show extensive spring molts (14 out of 16 [table 25.3A,B]), and most with conspicuous winter plumage show no or partial spring body molts (32 out of 43 taxa [table 25.3A,B). This relationship holds for both Neotropical migrants and temperate winterers analyzed separately (Fisher's Exact; $p < 0.02$ and $p < 0.002$, respectively [table 25. 3A,B]). Nevertheless, 44% of taxa with extensive molts (i.e., independent winter plumage) show conspicuous winter plumage, suggesting replacement of feathers for reasons other than color change. The Prothonotary Warbler is unusual because it achieves a moderate seasonal color change without a spring molt through abrasion of "cryptic" feather tips.

Winter Social System and Winter Color

Winter social behavior was not significantly related to adult male coloration in any group tested (fig. 25.1): males of most territorial taxa are conspicuous in winter (16 of 21 taxa), but a majority of non-territorial taxa have conspicuous winter males as well (31 of 43 taxa; Fisher's Exact; $p > 0.99$ [fig. 25.1A]). Excluding temperate winterers, which include only three territorial species, does not improve the relationship much (Fisher's Exact; $p = 0.49$ [fig. 25.1A]). Considering only parulid warblers, which constitute most (20 of 23) of the Neotropical territorial species, winter social behavior is still not significantly associated with winter conspicuousness (Fisher's Exact; $p = 0.18$): although a greater proportion of territorial taxa remain conspicuous in winter (12 of 15), equal numbers of non-territorial warblers are conspicuous and dull in winter (four each) (fig. 25.1B). Excluding species with mixed strategies dropped the significance level to 0.13, suggesting a weak relationship between winter coloration and the maintenance of winter territories for a few conspicuous warblers. Even among species with extensive spring molts—rendering winter and summer appearances independent—no relationship between territoriality and winter conspicuousness could be detected, either among all migrants (Fisher's Exact; $p > 0.36$; $n = 19$) or among warblers (Fisher's Exact; $p > 0.52$; $n = 9$).

Spring Breeding and Molt Constraints

Among temperate residents, 20 taxa (pooling species with no intrageneric variation) do not molt in spring and four show extensive molts; among Neotropical migrants, 11 taxa have no spring molt and 18 molt extensively (Fisher's Exact; $p < 0.002$ [table 25.3C]). Associated with the lack of a spring molt, most temperate wintering taxa show a conspicuous winter plumage. The importance of time constraints for temperate winterers is supported by the presence of extensive spring molts in species with breeding ranges far north of their wintering ranges (Harris's Sparrow [*Zonotrichia querula*] and all four *Calcarius* longspurs), presumably be-

Table 25.3 Summary of significant findings for sexually dichromatic temperate breeders, some wintering in the Neotropics and some in North America

A. Relationship between male winter plumage conspicuousness and presence of extensive spring molt in Neotropical migrants (Fisher's Exact; $p < 0.02$)

	Molt absent	Molt extensive
Conspicuous	13	10
Dull	1	10

B. Extensive spring molt in temperate winterers (Fisher's Exact; $p < 0.002$)

	Molt absent	Molt extensive
Conspicuous	19	1
Dull	1	4

C. Extent of spring molt among all North American sexually dichromatic temperate breeders by wintering region (Fisher's Exact; $p < 0.002$)

	Molt absent	Molt extensive
Temperate winterers	20	4
Neotropical migrants	11	18

D. Relationship between the degree of sexual dichromatism among Neotropical migrants with and without a male subadult summer plumage (Fisher's Exact; $p = 0.016$)

	Dichromatism	
	High	Moderate
Absent	7	19
Present	19	12

E. Relationship between the presence of molt and diet among temperate winterers (Fisher's Exact; $p > 0.99$) and Neotropical migrants (Fisher's Exact; $p = 0.023$)

	Temperate winterers		Neotropical migrants	
	Plant products	Invertebrates	Plant products	Invertebrates
Absent	19	3	3	8
Extensive	5	1	14	5

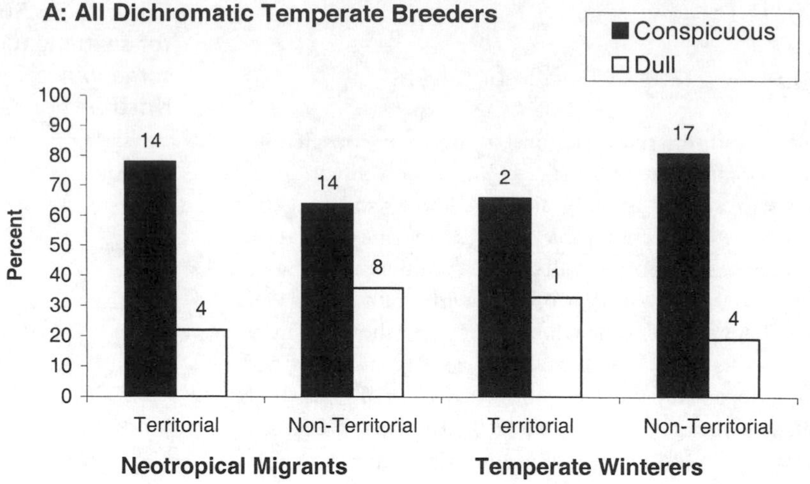

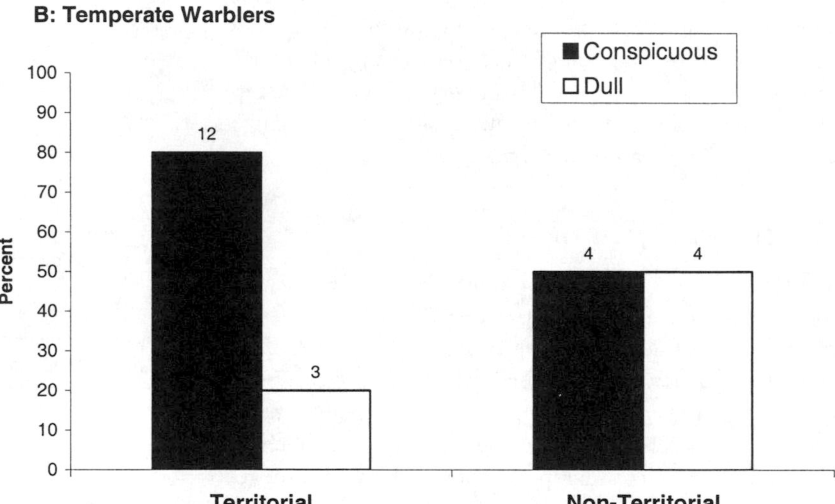

Fig. 25.1. Relationship between winter appearance of adult males compared to territorial behavior for all sexually dichromatic North American breeders (A) and for temperate-breeding Parulidae (B).

cause for these species spring molting can occur before they can breed.

Adult Badge Size and Extent of Molt

Dividing Neotropical migrant species into highly or moderately sexually dichromatic groups (having more or less than 60% of body plumage dichromatic), fully 61% of species with a summer subadult plumage ($n = 31$) are highly dichromatic, compared with only 27% of species without a summer subadult plumage ($n = 26$; Fisher's Exact; $p = 0.016$ [table 25.3D]).

Is the extent of spring molting by subadults related to the degree of plumage change required to match the conspicuousness of spring adult males? According to current reported differences in molt extent among age and sex classes (table 25.1), subadults of only nine of 31 weakly dichromatic species undertake spring molts exceeding other age and sex classes in extent. For highly dichromatic species, extensive molts are more common for subadult males (15 of 26 species; Fisher's Exact; $p = 0.04$).

Finally, 14 of 18 species with partial spring molts are thought to replace conspicuous and dichromatic body feathers, typically those around the head (table 25.1) (Pyle 1997). The other four species are scarcely contradictory. The spring molts of Mourning and McGillivray's Warblers are poorly characterized, so the absence of targeted molt is uncertain. In Black-headed Grosbeak and Painted Bunting, all of the body plumage is dichromatic but not all of it is replaced. Thus, in all of these species the main target of spring molts seems to be color change.

Food and Molt Constraints

We found no association between extensive molt in temperate wintering species that rely on plant products versus those that rely on invertebrates during winter (Fisher's Exact; $p > 0.99$ [table 25.3E]). Among Neotropical migrants, however, there was a distinct association between molts and diet: plant product users were more likely to molt than obligate insectivores (Fisher's Exact; $p = 0.023$ [table 25.3E]).

DISCUSSION

Winter Appearance and Territoriality

We found no evidence that conspicuous coloration during the nonbreeding season is an adaptation for territorial defense in Neotropical migrants. Both territorial and nonterritorial species tended to be conspicuous on the wintering grounds (table 25.3). Even for species that molt in spring, winter social behavior did not predict winter coloration. For example, Magnolia Warbler is dull in winter but shows territoriality, whereas both Scissor-tailed Flycatcher and Painted Bunting flock yet remain conspicuous in winter. Furthermore, although most temperate residents have conspicuous winter plumage, only three are territorial (table 25.2).

Can winter territoriality explain differences in winter coloration between closely related species? Summer Tanagers are conspicuous and territorial in winter but Scarlet Tanagers are dull and more likely to flock, although some Scarlet Tanagers do defend territories (table 25.1). Among *Passerina* buntings, however, all three species flock, although Indigo becomes duller and Painted remains conspicuous in winter. Even among parulid warblers, the only other group of sexually dichromatic Neotropical migrants with winter territoriality, the association between territoriality and winter conspicuousness does not have predictive value. Magnolia and Chestnut-sided Warblers have a dull winter plumage yet defend territories; four other warblers become inconspicuous in winter yet do not actively maintain winter territories (table 25.1).

Several lines of evidence suggest that conspicuous coloration is of little importance to the defense of winter territories. Many Neotropical migrant females defend exclusive individual winter territories against both males and other females (Lynch et al. 1985; Holmes et al. 1989), yet not one species features females that become more conspicuous for winter (see Morton 1989). In Hooded Warbler, male and female yearling floaters are equally likely to fill experimentally created vacancies on winter territories (Stutchbury 1994), indicating that yearling males' conspicuous coloration does not give them an advantage in territory acquisition. Finally, numerous inconspicuous and monochromatic parulid warblers are also strong defenders of winter territories, including Worm-eating, Swainson's, and the *Seiurus* Warblers (see Appendix in Rohwer et al., Chap. 8, this volume), demonstrating that the presence of territoriality is not restricted to conspicuous dichromatic wintering species.

Even for breeding birds there are too few manipulative studies to establish whether and how conspicuous coloration serves the defense of breeding territories. Such a role is clear only for the coverable badges of Red-winged Blackbirds (Peek 1972; Smith 1972; Hansen and Rohwer 1986), but coverable badges are special because the advantages of signaling subordinance maintain their reliability even when the badges have become established in the population (Hansen and Rohwer 1986; Rskaft and Rohwer 1987). Dyeing experiments manipulating the summer coloration of species without coverable badges are difficult to interpret because badges that are not variable among individuals and that evolved to advertise fighting ability should be reliable as signals only while they are becoming established in a population (Rohwer 1982). Thus, blackened adult male Yellow-headed Blackbirds (*Xanthocephalus xanthocephalus*) were both more and less successful than controls at achieving territory shifts (Rohwer and Rskaft 1989). Further manipulative experiments addressing the function of conspicuous coloration in territory defense are badly needed.

Nevertheless, there are several reasons why further experimental investigations of the relationship between dichromatic winter plumage coloration and winter territoriality may be worth pursuing. First, the number of species in which adult males and females are known to segregate by habitat (Lynch et al. 1985; Ornat and Greenberg 1990; Marra et al. 1993) or geographic areas (Cristol et al. 1999; Latta and Faaborg 2001) continues to grow. Segregation may reduce benefits in competitive ability that would accrue to females by becoming more malelike during the fall. Second, several warbler species show great individual variability in both sexes in winter coloration (e.g., Cape May and Black-throated Green Warblers), which may play a significant role in structuring social interactions in these species. Many species also show variability in white tail and wing markings (e.g., Black-throated Blue and Blackburnian Warblers) to which a number of signaling functions have been ascribed (reviewed in Butcher and Rohwer 1989). In Barn Swallows and perhaps other Palearctic species, unmelanized portions of the tail appear to have epigamic functions (Fitzpatrick 1998; Kose and Møller 1999). Tail spots should also be tested for function in nonbreeding agonistic displays, where they could serve as coverable badges. Finally, although our analysis revealed no more than a vague relationship between plumage coloration and territoriality in warblers, 12 of 15 territorial taxa were conspicuous (80%). For these taxa, conspicuousness may yet prove to play a role in establishing or maintaining territories.

Molt Constraints

Instead of serving winter territory defense, the conspicuous coloration of wintering males is more likely to result from constraints on spring molting. Most species with extensive spring molts reduce conspicuousness for winter (table 25.3A,B). Nevertheless, a number of species with conspicuous adult winter males show extensive spring molt (table 25.1), although thorough quantitative data on the extent of each of these molts are badly needed to confirm their categorizations. Alternatively, the costs of feather wear on a single annual set of feathers for these species may be sufficiently great to require semiannual replacement (see Euro-African species, below). Another possibility is that the costs of spring molting are not great enough for selection to have purged this molt from their annual cycle.

The relationship between spring molting in Neotropical

migrants and late-winter diet suggests that molt constraints could be imposed on obligate insectivores as a result of declining availability of invertebrate food resources during the dry season. Dietary flexibility on the winter grounds—in particular the ability to use plant products—may be critical for overcoming resource-based spring molt constraints faced by long-distance Neotropical migrants. Indeed, the only North American migrant to undertake its one annual complete molt on the winter grounds is the Red-eyed Vireo, an obligate winter frugivore (Mulvihill and Rimmer 1997). Similarly, several *Tyrannus* flycatchers, which are partially frugivorous in winter, regularly undertake at least part of their annual replacement of flight feathers on the winter grounds (Pyle 1997, 1998).

Time constraints rather than food habits appear to predict the presence of molt in temperate winterers, as indicated by contrasts in spring molt frequencies between temperate and Neotropical winterers. Few temperate wintering taxa have any spring molt (six of 26 taxa, including partial molters), whereas more than two-thirds (30 of 43 taxa, including partial molters) of Neotropical migrants feature at least some spring molt. Among the temperate-wintering taxa that replace body plumage in spring, goldfinches are noteworthy because their late breeding removes spring time constraints on molting. Several other temperate wintering species that have a spring molt, such as the *Calcarius* longspurs, are long-distance migrants within the temperate region. As with Neotropical migrants, spring molts in these species need not conflict with early breeding. This trade-off is further supported by geographic variation within White-crowned Sparrows (*Zonotrichia leucophrys*). Resident populations in California have very limited spring molt and yearlings of both sexes breed in a subadult plumage, whereas the highly migratory northern races have an extensive spring molt and breed in fully adult plumage (Mewaldt et al. 1968; Ralph and Pearson 1971).

Molt Constraint, Plumage Color Change, and Variability in Spring Molts

Most studies on the role of delayed plumage maturation have assumed that subadult plumage is adaptive during the breeding season (e.g., Rohwer et al. 1980; Studd and Robertson 1985; Lyon and Montgomerie 1986; Stutchbury and Robertson 1987; Hill 1988a; Enstrom 1992b). However, all North American passerines that have a subadult plumage in the breeding season also have one during the preceding winter and a partial or absent spring molt (Rohwer and Butcher 1988). This raises the possibility that dull breeding season coloration in young males results from constraints on spring molt (Rohwer and Butcher 1988). Apart from experimental work on Pied Flycatchers (*Ficedula hypoleuca* [Järvi et al. 1987a]), no studies of male subadult plumages have demonstrated unambiguous benefits of femalelike plumage for young males during the breeding season (Hill 1988b, 1989; Enstrom 1992a; Greene et al. 2000). Further, Stutchbury's (1991) dyeing experiments on Purple Martins demonstrated that subadults dyed to resemble adult males outcompeted undyed controls in competition for mates.

Species with partial spring molts typically replace those parts of the body plumage that are sexually dichromatic, suggesting that restricted spring molt strategies target plumage coloration transitions. Subadult male Painted Buntings and Hermit and Townsend's Warblers present a conundrum in that the feathers replaced in the spring molt are not always of definitive breeding coloration, retaining subadult coloration instead (Thompson 1991; Jackson et al. 1992). Similarly, first-winter Golden Orioles (*Oriolus oriolus*) from the Palearctic always undergo a complete molt in their first spring, but they grow a subadult male plumage (Jenni and Winkler 1994). This opens the possibility of summer adaptations for these species. Benefits to males of replacing the dichromatic plumage parts and costs of undertaking a spring molt may also represent a conflict between males and females, if genetic correlations encumber the evolution of sex-specific molt strategies. Species with considerable intraspecific variability in the extent of their spring molts (e.g., Summer Tanager, Black-headed Grosbeak, Indigo Bunting) should be examined for sex differences in the extent of molt.

The differences in extent of spring molt between yearling males and adults among highly dichromatic species suggest that young males in these species are balancing the costs of extensive feather replacement against the costs of sporting a subadult plumage during the breeding season (Rohwer and Butcher 1988; see Hill 1996). Further, if adult males gain in fitness by distinguishing themselves from their subadult conspecifics on the breeding grounds, they may experience selection for ever more conspicuous plumage, unattainable by their less experienced yearling conspecifics. Only molt constraints seem capable of explaining why adults, which are more experienced and more capable of securing adequate resources for molting, should, nonetheless, replace fewer feathers than subadults in spring (Grubb et al. 1991; Wunderle 1991; Smith and Metcalfe 1994).

The high variability in the extent and timing of spring molts—running from midwinter until after migration, depending on species—contrasts starkly with the relatively brief duration and cross-taxa synchronization of fall molts, suggesting adequate resources for fall molting (but see Rohwer et al., Chap. 8, this volume). Extended spring molts may reflect species-specific attempts to mitigate the costs of molting by spreading the costs over a longer period or concentrating them during periods of local resource abundance. Signaling functions may also influence the timing of spring molt. For example, juvenile male Indigo Buntings acquire the bluish feathers resembling the winter plumage of adult males during an intense molt in November and December (Rohwer 1986). Spring molt in adult Indigo Buntings is initiated as early as January and continues at the population level until spring migration. The timing of these molts may reflect an "arms race" between age-classes, with young birds benefiting socially by appearing more like adults early in winter and adult males benefiting from distinguishing themselves from young birds as soon as possible thereafter.

Euro-African Species

In contrast to North American long-distance migrants, 94% (47/50) of long-distance migrants breeding in Europe molt most or all body feathers twice a year (molt information compiled from Jenni and Winkler 1994 and Svensson 1992; migratory status and plumage characteristics summarized from Svensson 1992 and Snow and Perrins 1998 for all species breeding in Europe [not including Asia Minor]). Only three (*Phoenicurus phoenicurus, Luscinia luscinia,* and *L. megarhynchos*) among the 50 trans-Saharan migrant species fail to undertake a body molt on the wintering grounds. Flight feathers are included in the molt on the winter grounds in many species, but not on the breeding grounds. Further, 30 sexually monochromatic species with biannual body molts show negligible variation in feather color between the seasons. Euro-African migrants are clearly molting for reasons other than color change.

It is tempting to ascribe broad continental differences to phylogenetic constraints. However, species wintering in Europe, including most *Sylvia* warblers and the single short-distance migrant among European *Acrocephalus* warblers (*A. melanopogon*), show molt patterns comparable to North American winterers: complete fall molts and only occasional spring body molts (Leisler 1972; Svensson 1992; Shirihai et al. 2001; see Svensson and Hedenström 1999). Constraints on spring molts invoked for North American migrants—time constraints dictated by selection favoring early breeding—appear to hold for European residents as well. So why should long-distance migrants in the two systems show such divergent patterns?

Many Euro-African species with flight feather replacement on the winter grounds undertake this molt in spring, before leaving Africa. This is in striking contrast with their North American counterparts. North American species that replace flight feathers off the breeding grounds either interrupt fall migration to molt or molt upon arrival on their winter grounds (Rohwer et al., Chap. 8, this volume).

Several hypotheses have attempted to explain Euro-African molt strategies. Alerstam and Högstedt (1982) suggest that shifting complete molts from summer to winter may reflect post-breeding time constraints. However, this hypothesis fails to explain why a body molt in such species should, in spite of time constraints, take place on the breeding grounds before migration. Why not postpone complete molt until after migration, as do most North American winter molters?

Svensson and Hedenström (1999), noting the strong relationship between long-distance migration and winter molt in European species, hypothesize that its absence among North American *Dendroica* warblers may be due to evolutionary lag, because they were thought to be a younger group than Old World warblers.

Jones (1995) suggests that the timing of molt of Euro-African migrants tracks rainfall-related habitat suitability and food resources, taking place in early winter—when the local rainy season ends—for species wintering in northern tropical Africa and late in winter—when the local rainy season begins—for equatorial and transequatorial migrants. Such a relationship between molt and resource availability matches the food-resource constraints we suggest as limits to spring molts in Neotropical migrants.

The spring molt trade-off for Euro-African long-distance migrants, however, evidently favors incurring the costs of biannual body feather replacement over the costs of carrying worn feathers (see Holmgren and Hedenström 1995). Differences in feather wear rates between Africa and the Neotropics could account for this contrast in the respective resolutions of the spring molting trade-off. Unlike the American Tropics, the African Tropics are mostly open (Böhning-Gaese, Chap. 12, this volume). As a result, Euro-African migrants tend to spend their winters in more open habitats with greater exposure to ultraviolet radiation and airborne dust than Neotropical migrants. This should yield higher rates of feather wear (Bergman 1982; Burtt 1986) and strong selection for more frequent feather replacement. Further support for the importance of feather wear in Euro-African species is suggested by the larger number of species for which the abrasion of "cryptic tips" plays a significant role in plumage color transition. In the Neotropics, exposure and abrasion are apparently not as universal; feather wear dictates biannual molts only for species such as Bobolink that inhabit exceptionally abrasive habitats.

The scarcity of well-documented winter social strategies among dichromatic Euro-African taxa as well as lack of variability in spring molt strategy prohibits robust comparisons among Euro-African migrants to explore relationships between conspicuous coloration and winter territoriality. The Euro-African species for which information on plumage and territoriality are unambiguous suggest no pronounced relationship. Thus, although migrant male shrikes tend to be both territorial and conspicuous in winter, other males (*Monticola saxatilis, Oenanthe oenanthe, O. pleschanka,* and *O. hispanica*) are strong winter territory holders yet become cryptic in the fall molt. The frugivorous *Oriolus oriolus* remains bright, yet tends to flock. Overall, the large number of Euro-African migrants with cryptic winter plumage in comparison to Neotropical migrants suggests that predation may be a stronger selective force in unforested African wintering grounds than in the Neotropics.

FUTURE DIRECTIONS

Although our comparative analyses fail to support an association between winter territoriality and conspicuous winter plumages in Neotropical migrants, molt constraints are consistent with observed patterns, suggesting that direct tests of the molt constraints hypothesis, by manipulating specific food resource limitations, would be productive.

This chapter serves to point out serious gaps in our understanding of plumage color patterns. We need: (1) more information on how winter feeding ecology and social behavior affect the energetics of molt, (2) better quantitative data on molts to facilitate analyses of contrasts in molt patterns by age and sex, and (3) better descriptions of the winter social organization and competitive interactions among migrant songbirds. If male/female pairs truly defend winter territories in concert, what is the nature of this relationship? We especially need manipulative field experiments that address food resource effects on spring molting (no data exist), if and how plumages reinforce winter territoriality (Rohwer and Rohwer 1978; Rohwer 1985; Stutchbury 1991, 1992), and comparative predation risk (Götmark 1995). As the conflicting predictions arising from the cryptic and unprofitable prey hypotheses suggest, costs of conspicuous coloration—aside from the well-documented social costs—remain unclear. Prey selection experiments using *Accipiter* species held by falconers could help clarify predation risk associated with conspicuous plumages. Plumage comparisons based on regional differences in predation risk, for example, between the relatively low-predation Caribbean and the high-predation American mainland areas or between open savannas of Africa and cover-rich forests of Central and South America, could help assess the universality of these hypotheses.

Hypotheses emerging from our intercontinental comparisons could be tested by comparative evaluations of feather quality and relative rates of wear between different regions and groups, such as songbirds versus shorebirds or tropical residents versus migrants. For instance, feather wear rate differences between Africa and the Neotropics should be most apparent for American songbirds that benefit from greater habitat-mediated UVR protection, whereas coastal shorebirds should experience similar exposure and feather wear rates on both continents. Similarly, resident species of the two tropical regions should show differences in wear and molt patterns commensurate with those of the migrants (see Herremans 1999). Artificial sand-blasting studies such as those conducted by Burtt (1986) could be used to reveal differences in feather quality among groups and wear rate over time.

ACKNOWLEDGMENTS

Our work with specimens was conducted at the University of Washington Burke Museum. Luke Butler, Bethanne Zelano, Jennifer Hoffman, Jack DeLap, Molly Jacobs, and Melissa Keigley undertook the naïve categorizations of specimens for this project. Valuable comments by Pete Marra, two anonymous reviewers, and Jennifer Hoffman greatly improved the manuscript, as did stimulating discussions with Luke Butler and Chris Filardi. DRF received Burke Museum Eddy Fellowship support during this work.

LITERATURE CITED

Alerstam, T., and G. Högstedt. 1982. Bird migration and reproduction in relation to habitats for survival and breeding. Ornis Scandinavica 13:25–37.

Baker, R. R. and G. A. Parker. 1979. The evolution of bird coloration. Philosophical Transactions of the Royal Society of London, Series B, Biological Sciences 287:63–130.

Bergman, G. 1982. Why are the wings of *Larus f. fuscus* so dark? Ornis Fennica 59:77–83.

Bortolotti, G. R., R. D. Dawson, and G. L. Murza. 2002. Stress during feather development predicts fitness potential. Journal of Animal Ecology 71:333–342.

Brown, C. R., and D. M. Bryant. 1996. Energy expenditure during molt in Dippers (*Cinclus cinclus*): no evidence of elevated costs. Physiological Zoology 69:1036–1056.

Burtt, E. H., Jr. 1986. An analysis of physical and optical aspects of avian coloration with emphasis on wood-warblers. Ornithological Monographs 38.

Buskirk, R. E., and W. Buskirk. 1976. Changes in arthropod abundance in a highland Costa Rican Forest. American Midland Naturalist 95:288–298.

Butcher, G. S., and S. Rohwer. 1989. The evolution of conspicuous and distinctive coloration for communication in birds. Pages 51–108 *in* Current Ornithology (D. M. E. Power, ed.). Plenum Press, London.

Chipley, R. M. 1980. Nonbreeding ecology of the Blackburnian Warbler. Pages 309–317 *in* Migrant Birds in the Neotropics: Ecology, Behavior, Distribution, and Conservation (A. Keast and E. S. Morton, eds.). Smithsonian Institution Press, Washington, D.C.

Clements, J. F. 2000. Birds of the World: A Checklist (fifth ed.). Ibis, Vista, Calif.

Cristol, D. A., M. B. Baker, and C. Carbone. 1999. Differential migration revisited: latitudinal segregation by age and sex class. Pages 33–88 *in* Current Ornithology, Vol. 15 (J. V. Nolan et al., eds.). Kluwer Academic/Plenum, New York.

Dawson, A., S. A. Hinsley, P. N. Ferns, R. H. C. Bonser, and L. Eccleston. 2000. Rate of moult affects feather quality: a mechanism linking current reproductive effort to future survival. Proceedings of the Royal Society of London, Series B, Biological Sciences 267:2093–2098.

Dunn, J. L., and K. L. Garrett. 1997. Field Guide to Warblers of North America. Houghton Mifflin, New York.

Enstrom, D. A. 1992a. Breeding season communication hypotheses for delayed plumage maturation in passerines: tests in the Orchard Oriole, *Icterus spurius*. Animal Behaviour 43:463–472.

Enstrom, D. A. 1992b. Delayed plumage maturation in the Orchard Oriole (*Icterus spurius*): tests of winter adaptation hypotheses. Behavioral Ecology and Sociobiology 30:35–42.

Fitzpatrick, S. 1998. Birds' tails as signaling devices: markings, shape, length, and feather quality. American Naturalist 151:157–173.

Francis, C. M., and F. Cooke. 1986. Differential timing of spring migration in wood warblers (Parulinae). Auk 103:548–556.

Götmark, F. 1995. Black-and-white plumage in male Pied Flycatchers (*Ficedula hypoleuca*) reduces the risk of predation from sparrowhawks (*Accipiter nisus*) during the breeding season. Behavioral Ecology 6:22–26.

Greenberg, R. 1979. Body size, breeding habitat, and winter exploitation systems in *Dendroica*. Auk 96:756–766.

Greenberg, R. 1984. The winter exploitation systems of Bay-breasted and Chestnut sided Warblers in Panama. University of California Publications in Zoology 116:1–107.

Greenberg, R. 1986. Competition in migrant birds in the non-breeding season. Current Ornithology 3:281–307.

Greenberg, R. 1995. Insectivorous migratory birds in tropical ecosystems: the breeding currency hypothesis. Journal of Avian Biology 26:260–264.

Greene, E., B. E. Lyon, V. R. Muehter, L. Ratcliffe, S. J. Oliver, and P. T. Boag. 2000. Disruptive sexual selection for plumage coloration in a passerine bird. Nature 407:1000–1003.

Grubb, T. C. J., T. A. Waite, and A. J. Wiseman. 1991. Ptilochronology induced feather growth in Northern Cardinals varies with age, sex, ambient temperature and day length. Wilson Bulletin 103:435–445.

Hagan, J. M., III, and D. W. Johnston. 1992. Ecology and Conservation of Neotropical Migrant Landbirds. Smithsonian Institution Press, Washington, D.C.

Hamilton, T. H., and R. H. Barth Jr. 1962. The biological significance of season change in male plumage appearance in some new world migratory species. American Naturalist 96:129–144.

Hamilton, W. D., and M. Zuk. 1982. Heritable true fitness and bright birds: a role for parasites? Science 218:384–387.

Hansen, A. J., and S. Rohwer. 1986. Coverable badges and resource defense in birds. Animal Behaviour 34:69–76.

Hedenström, A., and S. Sunada. 1999. On the aerodynamics of moult gaps in birds. Journal of Experimental Biology 202:67–76.

Herremans, M. 1999. Biannual complete moult in the Black-chested Prinia *Prinia flavicans*. Ibis 141:115–124.

Hill, G. E. 1988a. The function of delayed plumage maturation in male Black-headed Grosbeaks. Auk 105:1–10.

Hill, G. E. 1988b. Age, plumage brightness, territory quality, and reproductive success in the Black-headed Grosbeak. Condor 90:379–388.

Hill, G. E. 1989. Late spring arrival and dull nuptial plumage: aggression avoidance by yearling males. Animal Behaviour 37:665–673.

Hill, G. E. 1996. Subadult plumage in the House Finch and tests of models for the evolution of delayed plumage maturation. Auk 113:858–874.

Holmes, R. T., T. W. Sherry, and L. Reitsma. 1989. Population structure, territoriality and overwinter survival of two migrant warbler species in Jamaica. Condor 91:545–561.

Holmgren, N., and A. Hedenström. 1995. The scheduling of molt in migratory birds. Evolutionary Ecology 9:354–368.

Jackson, W. M., C. S. Wood, and S. Rohwer. 1992. Age-specific plumage characters and annual molt schedules of Hermit Warblers and Townsend's Warblers. Condor 94:490–501.

Järvi, T., E. Rskaft, M. Bakken, and B. Zumsteg. 1987a. Evolution of variation in male secondary sexual characteristics: a test of eight hypotheses applied to Pied Flycatchers. Behavioral Ecology and Sociobiology 20:161–169.

Järvi, T., O. Walso, and M. Bakken. 1987b. Status signaling by *Parus major*: an experiment in deception. Ethology 76:334–342.

Jenni, L., and R. Winkler. 1994. Moult and Ageing of European Passerines. Academic Press, London.

Johnson, M. D., and T. W. Sherry. 2001. Effects of food availability on the distribution of migratory warblers among habitats in Jamaica. Journal of Animal Ecology 70:546–560.

Jones, P. J. 1995. Migration strategies of Palearctic Passerines in Africa. Israel Journal of Zoology 41:393–406.

Karr, J. R., and J. D. Brawn. 1990. Food resources of understory birds in central Panama: quantification and effects on avian populations. Studies in Avian Biology 13:58–64.

Keast, A., and E. S. Morton (eds.). 1980. Migrant Birds in the Neotropics: Ecology, Behavior, Distribution, and Conservation. Smithsonian Institution Press, Washington, D.C.

King, D. I., and J. H. Rappole. 2000. Winter flocking of insectivorous birds in montane pine-oak forests in Middle America. Condor 102:664–672.

King, J. R. 1981. Energetics of avian moult. Pages 312–317 *in* Proceedings of the 17th International Ornithological Conference, Vol. 1 (R. Nöhring, ed.). Verlag der Deutschen Ornithologen-Gesellschaft, Berlin.

Kose, M., and A. P. Møller. 1999. Sexual selection, feather breakage and parasites: the importance of white spots in the tail of the Barn Swallow (*Hirundo rustica*). Behavioral Ecology and Sociobiology 45:430–436.

Latta, S. C., and J. Faaborg. 2001. Winter site fidelity of Prairie Warblers in the Dominican Republic. Condor 103:455–468.

Lefebvre, G., B. Poulin, and R. McNeil. 1992. Settlement period and function of long-term territory in tropical mangrove passerines. Condor 94:83–92.

Leisler, B. 1972. Die Mauser des Mariskensängers (*Acrocephalus melanopogon*) als ökologisches Problem. Journal für Ornithologie 113:191–206.

Levings, S. C., and D. M. Windsor. 1982. Seasonal and annual variation in litter arthropod populations. Pages 355–387 *in* The Ecology of a Tropical Forest: Seasonal Rhythms and Long-term Changes (E. G. Leigh et al., eds.). Smithsonian Institution Press, Washington, D.C.

Lind, J. 2001. Escape flight in moulting Tree Sparrows (*Passer montanus*). Functional Ecology 15:29–35.

Lindström, A., G. H. Visser, and S. Daan. 1993. The energetic cost of feather synthesis is proportional to basal metabolic rate. Physiological Zoology 66:490–510.

Lynch, J. F., E. S. Morton, and M. E. van der Voort. 1985. Habitat segregation between the sexes of wintering Hooded Warblers (*Wilsonia citrina*). Auk 102:714–721.

Lyon, B. E., and R. D. Montgomerie. 1986. Delayed plumage maturation in passerine birds: reliable signaling by subordinate males? Evolution 40:605–615.

Marra, P., and R. T. Holmes. 2001. Consequences of dominance-mediated habitat segregation in American Redstarts during the nonbreeding season. Auk 118:92–104.

Marra, P. P., T. W. Sherry, and R. T. Holmes. 1993. Territorial exclusion by a long-distance migrant warbler in Jamaica: a removal experiment with American Redstarts (*Setophaga ruticilla*). Auk 110:565–572.

Mewaldt, L. R., S. S. Kibby, and M. L. Morton. 1968. Comparative biology of Pacific coastal White-crowned Sparrows. Condor 70:14–30.

Mewaldt, L. R., and J. R. King. 1978. Latitudinal variation in prenuptial molt in wintering Gambel's White-crowned Sparrows. North American Bird Bander 3:138–144.

Morton, E. S. 1980. Adaptations to seasonal changes by migrant land birds in the Panama Canal Zone. Pages 437–453 *in* Migrant Birds in the Neotropics: Ecology, Behavior, Distribution, and Conservation (A. Keast and E. S. Morton, eds.). Smithsonian Institution Press, Washington, D.C.

Morton, E. S. 1989. Female Hooded Warbler plumage does not become more male-like with age. Wilson Bulletin 101:460–462.

Mulvihill, R. S., and C. C. Rimmer. 1997. Timing and extent of the molts of adult Red-eyed Vireos on their breeding and wintering grounds. Condor 99:73–82.

Munn, C. A., and J. W. Terborgh. 1979. Multi-species territoriality in Neotropical foraging flocks. Condor 81:338–347.

Murphy, M. E., and J. R. King. 1991. Protein intake and the dynamics of the postnuptial molt in White-crowned Sparrows, *Zonotrichia leucophrys gambelii*. Canadian Journal of Zoology 69:2225–2229.

Murphy, M. E., and J. R. King. 1992. Energy and nutrient use during moult by White-crowned Sparrows *Zonotrichia leucophrys gambelii*. Ornis Scandinavica 23:304–313.

Murphy, M. E., B. T. Miller, and J. R. King. 1989. A structural comparison of fault bars with feather defects known to be nutritionally induced. Canadian Journal of Zoology 67:1311–1317.

Nilsson, J.-Å., and E. Svensson. 1996. The cost of reproduction: a new link between current reproductive effort and future reproductive success. Proceedings of the Royal Society of London, Series B, Biological Sciences 263:711–714.

Ornat, A. L., and R. Greenberg. 1990. Sexual segregation by habitat in migratory warblers in Quintana-Roo, Mexico. Auk 107:539–543.

Parrish, J. D., and T. W. Sherry. 1994. Sexual habitat segregation by American Redstarts wintering in Jamaica: importance of resource seasonality. Auk 111:38–49.

Peek, F. W. 1972. An experimental study of the territorial function of vocal and visual displays in the male Red-winged Blackbirds (*Agelaius phoeniceus*). Animal Behaviour 20:122–178.

Perrins, C. M. 1970. The timing of birds' breeding seasons. Ibis 112:242–255.

Poole, A. F., and F. B. Gill (eds.). 1992–2003. The Birds of North America: Life Histories for the 21st Century. The Birds of North America, Inc., Philadelphia.

Poulin, B., and G. Lefebvre. 1996. Dietary relationships of migrant and resident birds from a humid forest in central Panama. Auk 113:277–287.

Poulin, B., and G. Lefebvre. 1997. Estimation of arthropods available to birds: effect of trapping technique, prey distribution, and bird diet. Journal of Field Ornithology 68:426–442.

Pyle, P. 1997. Identification Guide to North American Birds. Slate Creek Press, Bolinas, Calif.

Pyle, P. 1998. Eccentric first-year molt patterns in certain tyrannid flycatchers. Western Birds 29:29–35.

Ralph, C. J., and C. A. Pearson. 1971. Correlation of age, size of territory, plumage, and breeding success in White-crowned Sparrow. Condor 73:77–80.

Rohwer, S. 1975. The social significance of avian winter plumage variability. Evolution 29:593–610.

Rohwer, S. 1977. Status signaling in Harris' Sparrows: some experiments in deception. Behaviour 61:107–129.

Rohwer, S. 1978. Passerine subadult plumages and the deceptive acquisition of resources: test of a critical assumption. Condor 80:173–179.

Rohwer, S. 1982. The evolution of reliable and unreliable badges of fighting ability. American Zoologist 22:531–546.

Rohwer, S. 1985. Dyed birds achieve higher social status than controls in Harris' Sparrows (*Zonotrichia querula*). Animal Behaviour 33:1325–1331.

Rohwer, S. 1986. A previously unknown plumage of first-year Indigo Buntings (*Passerina cyanea*) and theories of delayed plumage maturation. Auk 103:281–292.

Rohwer, S. A. 1999. Time constraints and moult-breeding tradeoffs in large birds. Pages 568–581 *in* Proceedings of the 22nd International Ornithological Congress, Durban (N. J. Adams and R. H. Slotow, eds.). BirdLife South Africa, Johannesburg.

Rohwer, S., and G. S. Butcher. 1988. Winter versus summer explanations of delayed plumage maturation in temperate passerine birds. American Naturalist 131:556–572.

Rohwer, S., and P. W. Ewald. 1981. The cost of dominance and advantage of subordination in a badge signaling system. Evolution 35:441–454.

Rohwer, S., S. D. Fretwell, and D. M. Niles. 1980. Delayed maturation in passerine plumages and the deceptive acquisition of resources. American Naturalist 115:400–437.

Rohwer, S., W. P. Klein Jr., and S. Heard. 1983. Delayed plumage maturation and the presumed prealternate molt in American redstarts (*Setophaga ruticilla*). Wilson Bulletin 95:199–208.

Rohwer, S., and J. Manning. 1990. Differences in timing and number of molts for Baltimore and Bullock's Orioles: implications to hybrid fitness and theories of delayed plumage maturation. Condor 92:125–140.

Rohwer, S., and F. C. Rohwer. 1978. Status signaling in Harris' Sparrows: experimental deceptions achieved. Animal Behaviour 26:1012–1022.

Rohwer, S., and E. Rskaft. 1989. Results of dyeing male Yellow-headed Blackbirds solid black: implications for the arbitrary identity badge hypothesis. Behavioral Ecology and Sociobiology 25:39–48.

Rskaft, E., and S. Rohwer. 1987. An experimental study of the function of the red epaulettes and the black body color of male Red-winged Blackbirds. Animal Behaviour 35:1070–1077.

Shirihai, H., G. Gargallo, and A. J. Helbig. 2001. *Sylvia* warblers. Princeton University Press, Princeton.

Shugart, G. W., and S. Rohwer. 1996. Serial descendant primary molt or Staffelmauser in Black-crowned Night-herons. Condor 98:222–233.

Slagsvold, T. 1976. Arrival of birds from spring migration in relation to vegetational development. Norwegian Journal of Zoology 24:161–173.

Smith, D. G. 1972. The role of the epaulets in the Red-winged Blackbird (*Agelaius phoeniceus*). Behaviour 41:251–268.

Smith, R. D., and N. B. Metcalfe. 1994. Success of wintering Snow Buntings. Behaviour 129:99–111.

Snow, D. W., and C. M. Perrins. 1998. The Birds of the Western Palaearctic (concise ed.), Vol. 2. Oxford University Press, Oxford.

Staicer, C. A. 1992. Social behavior of the Northern Parula, Cape May Warbler, and Prairie Warbler wintering in second-growth forest in southwestern Puerto Rico. Pages 308–321 *in* Migrant Birds in the Neotropics: Ecology, Behavior, Distribution, and Conservation (A. Keast and E. S. Morton, eds.). Smithsonian Institution Press, Washington, D.C.

Strong, A. M., and T. W. Sherry. 2000. Habitat-specific effects of food abundance on the condition of Ovenbirds wintering in Jamaica. Journal of Animal Ecology 69:883–895.

Studd, M. V., and R. J. Robertson. 1985. Life-span, competition, and delayed plumage maturation in male passerines: the breeding threshold hypothesis. American Naturalist 126: 101–115.

Stutchbury, B. J. 1991. The adaptive significance of male subadult plumage in Purple Martins: plumage dyeing experiments. Behavioral Ecology and Sociobiology 29:297–306.

Stutchbury, B. J. 1992. Experimental evidence that bright coloration is not important for territory defense in Purple Martins. Behavioral Ecology and Sociobiology 31:27–33.

Stutchbury, B. J. 1994. Competition for winter territories in a Neotropical migrant: the role of age, sex and color. Auk 111:63–69.

Stutchbury, B. J., and R. J. Robertson. 1987. Signaling subordinate and female status: two hypotheses for the adaptive significance of subadult plumage in female Tree Swallows. Auk 104:717–723.

Svensson, E. 1995. Avian reproductive timing: when should parents be prudent? Animal Behaviour 49:1569–1575.

Svensson, E., and A. Hedenström. 1999. A phylogenetic analysis of the evolution of moult strategies in Western Palearctic warblers (Aves: Sylviidae). Biological Journal of the Linnean Society 67:264–276.

Svensson, L. 1992. Identification Guide to European Passerines. Lars Svensson, Stockholm.

Swaddle, J. P., and M. S. Witter. 1997. The effects of molt on the flight performance, body mass, and behavior of European Starlings (*Sturnus vulgaris*): an experimental approach. Canadian Journal of Zoology 75:1135–1146.

Thompson, C. W. 1991. The sequence of molts and plumages in Painted Buntings and implications for theories of delayed plumage maturation. Condor 93:209–235.

Tucker, V. A. 1991. The effect of molting on the gliding performance of a Harris' Hawk (*Parabuteo unicinctus*). Auk 108: 108–113.

Walsberg, G. E. 1983. Avian ecological energetics. Avian Biology 7:161–220.

Wunderle, J. M. 1991. Age-specific foraging proficiency in birds. Pages 273–324 *in* Current Ornithology, Vol. 8. (D. M. E. Power, ed.). Plenum Press, New York.

Young, B. E. 1991. Annual molts and interruption of the fall migration for molting in Lazuli Buntings. Condor 93:235–250.

26

RUSSELL GREENBERG
AND VOLKER SALEWSKI

Ecological Correlates of Wintering Social Systems in New World and Old World Migratory Passerines

A COMPARATIVE ANALYSIS based on what is known of 194 species (56 genera and 12 subfamilies) of New World and Old World tropical migrant passerines shows that 59% of them are intraspecifically solitary during the nonbreeding season. Some type of nonbreeding territoriality is known or suspected in at least 64 species as well. Although many species and genera show tremendous plasticity in winter social behavior, major trends in the degree of gregariousness can be explained by a few ecological variables associated with higher taxonomic groups. Intraspecific gregariousness is strongly associated with omnivory, which is presumably a result of the benefits of joining conspecifics when local competition for food is at least temporarily reduced. Gregariousness is also associated with lack of long-distance aerial maneuvers to capture prey, which may relate both to the degree of specialization and the sensitivity to local competition from conspecifics. Although there are similar numbers of gregarious species in the Old World and the New World, the number of highly gregarious species (commonly occurring in flocks of more than ten individuals) is higher in the Neotropics. Omnivory is a more important correlate of gregariousness in the New World. Both of these observations may be related to the overall tendency toward omnivory in New World migrants, which in turn probably is a response to a greater availability of large crops of flowers and fruits in more humid tropical areas during the tenure of migrants.

INTRODUCTION

Ornithologists have long wondered whether there is a fundamental difference in ecological strategies between species that remain resident in tropical regions and seasonal migrants from temperate regions. However, the world is complex and migratory birds defy such simple generalizations. Detailed studies of individual species have revealed an impressive amount of inter- and intraspecific variation in resource exploitation strategies. By this, we do not mean simply that different migratory bird species feed on different foods and live in different habitats. Rather we are focusing on how birds respond to the spatial and temporal dynamics of resources (primarily food) in a particular community. At one extreme, some species depend on limiting and renewable (if sometimes declining) food resources that they sometimes invest in through active defense, whereas other species move between habitats and depend upon locating local "superabundances" of food. The extremes of these strategies have been documented, and the fact that tremendous variability exists has been long recognized. Still, the nature of resources exploited is central to our understanding of the role of resources in limiting the size and distribution of migratory bird populations. And the nature of the strategy that migratory birds use is, in almost all arguments, inexorably tied to social organization. So perhaps the fundamental question concerning the nature of a migratory strategy must first be rephrased: what factors determine the variability of winter social systems? Perhaps from this foundation we can return to the question of what factors determine the resource exploitation strategy of different migrants.

The past few decades of greatly increased field studies of avian ecology in tropical areas have provided a glimpse at the variability of social systems. A few comparative studies have focused on variation in sociality between closely related species (Greenberg 1979, 1984) or among sympatric, but not necessarily related, species (Eaton 1953; Morton 1980; Rappole and Warner 1980).

The few published studies reveal a fair degree of variability in social organization between closely related species and even within species. However, broad overviews of the patterns of variability of winter social behavior have been surprisingly few, although some aspects of winter sociality have been tabulated in various references (e.g., Levey and Stiles 1992; Rappole 1995). This is undoubtedly a result of the lack of detailed data based on focal studies of marked individuals. However, the increasing amount of fieldwork conducted in the Tropics in recent decades provides us with considerable natural historical observation of some easily detected patterns of sociality. Although these observations fall short of complete rigor, we cannot base our analysis of the correlates of social behavior solely on a few well-studied "model" species. In this chapter we therefore attempt to explore what is known about as many species of tropical migrants as possible to examine the pattern of variation in sociality and how this variation relates to key ecological attributes.

COMPONENTS OF WINTER SOCIAL SYSTEMS

To set the stage for further analysis, we present the following five major components of the winter social system, based on Greenberg (1984).

Regional Movements

Movements may consist of slow drift through a region, which apparently occurs in several species of South American and sub-Saharan African wintering migrants; actual secondary migration (itinerant migration) between two distinct wintering grounds (complete with premigratory fattening); or nomadic wanderings to track phenological changes between regions or habitats. Although they have proven difficult to study from the perspective of marked individual birds, such movements manifest themselves as dramatic fluctuations in local population size. Regional movements may manifest themselves as birds tracking particular resources (e.g., flowering and fruiting trees [see Morton 1972]) or may involve responses to asynchronous patterns of rainfall over continental and more local scales (Moreau 1972; Morton 1980; Lack 1983). In Africa, regional movements are actually composed of within-winter migrations from sites north to those south of the equator (Jones 1998).

Local Tenacity

Local tenacity refers to the degree to which individuals remain localized throughout a winter. The distinction between regional movements and low site tenacity is one of scale. Operationally, tenacity is generally discussed in reference to residency on a particular study site. With greater use of telemetry, defined search areas can be become larger but may still be defined by arbitrary limits.

Territoriality

Territoriality refers to the exclusive use (Pitelka 1959) or complete aggressive dominance (Brown 1964) of an individual of a particular area. A fine line separates the defense of a particular tree or plant because it provides a dependable and defendable source of food (site defense), and the defense of a territory. A complete understanding of a territorial system requires that individuals be identifiable and that these identifiable individuals be tracked through time. Other aspects of behavior (e.g., aggressive response to playback, presence of aggressive interactions at consistent borders) can provide some evidence that individuals are territorial (e.g., see Rappole and Warner 1980) but may not provide conclusive evidence of long-term territoriality or information on the dynamics of territory use through time. These observations also may not provide insight into the relative importance of a non-territorial component in any population. Furthermore, birds that are attempting to exclude all conspecifics from a territory may vary considerably in the efficacy of this effort. Casual observations, even of marked birds, might suggest that home ranges overlap and that a territorial system does not exist, when it actually does. Because of our belief in the need for detailed studies of populations of marked birds to adequately assess territorial systems, we have only classified species as territorial or not if such studies exist. This greatly reduces our sample size for classification over other studies that rely on more relaxed criteria.

Territorial systems themselves vary in a number of ecologically important ways. First, the territories may last days, weeks, or through entire winter periods. Increasing numbers of species have been documented to defend long-term territories (Schwartz 1965; Greenberg 1984; Holmes et al. 1989; Kelsey 1989). Stable, winter-long territorial systems are inherently variable at the individual level because territoriality implies intraspecific strife with winners and losers. In such a system, there will usually be non-territorial birds that either were unable to obtain a territory (Winker et al. 1990) or are exercising an alternative strategy based on using a different food source; there may even be birds wandering or prospecting from nearby areas. Individuals can appear to defend long-term territories, yet abandon them in the face of rapid changes in environment (usually associated with rainfall patterns in the Tropics). This pattern of territoriality has been documented in Northern Waterthrushes (*Seirurus noveboracensis* [L.Reitsma, pers. comm.]) and undoubtedly is more widespread. Another more complex phenomenon is the possible existence of "spatio-temporal" territories, as has been described for Ovenbirds (*Seiurus aurocapillus*). In this species, birds maintain an exclusive foraging range over a short period of time, but these ranges show great overlap over time (Strong 1999). This type of behavior is more difficult to discern and may help explain why so many species occur solitarily, yet without a well-defined "classic" territorial system. As we discuss below, exclusion of conspecifics from mixed-species flocks may play a role in this type of territoriality in some species.

Most migrants that defend territories do so against conspecifics, suggesting that within species is where one can expect to find the greatest potential for resource competition. The fitness consequences of defending a territory in a suboptimal habitat (Marra 2000) or of not obtaining a territory (Winker et al. 1990) have only begun to be explored through detailed studies of condition, survival, and departure schedules. Patterns of differential distributions of sex and age classes provide some insights into the advantages of successful territorial defense as well (Marra 2000).

Only a few notable species defend resources against many species (Greenberg et al. 1994). These "aggressive dominants" appear to focus on very productive and defendable sites or resources that include certain flowers (such as agaves), trees with high abundances of large arthropods (patches of riparian trees), and homopteran honeydew.

Group Size and Composition

We refer here to the number of conspecifics observed foraging in close proximity. The social group could be an aggregation of individuals at a particular resource—indicating a degree of social tolerance—or an active joining and following in a single-species flock. Species groups can vary in composition, consisting of members of the same age or sex group or related individuals. Reports are increasing for birds that occur in pairs during migration (Greenberg and Gradwohl 1980) or on the winter quarters (Gwinner et al. 1994; Jones et al. 2000). In some species it is known that the association can involve an adult male associated with a female or immature male (referred to as a satellite individual [Davies and Houston 1981, 1983; M. Foster, pers. comm.]). Furthermore, the formation of a pair may be a facultative response to temporary patterns of food abundance and distribution (Davies and Houston 1981, 1983). At least in a few of these species (Davies and Houston 1981; Gwinner et al. 1994), the pair jointly maintains a territory. Although the occurrence of such pairs during migration stopovers suggests the possibility that pair bonds may be maintained throughout the year, this has not been established empirically.

Mixed-Species Flocking

In many habitats, birds are able to obtain some of the advantages of social foraging without forming single-species flocks simply by joining mixed-species flocks. The importance of joining mixed flocks has been well described for many migratory birds (Powell 1980; Rappole and Morton 1985). However, the potential costs and limitations of joining flocks have been explored in a qualitative manner as well (Powell 1980; Hutto 1988). The costs include the need to adjust territory and home ranges away from their energetic optimum in order to have consistent access to flocks and similar adjustments in the velocity of foraging in order to keep up with flock movements. In most cases, individual migrants follow a flock nucleus comprising resident species. In a few cases, single-species flocks of migrants apparently form the nucleus of mixed-species flocks (Greenberg 1984; Hutto 1994). Mixed-species flocking can occur with a range of intraspecific social units. These are the common flocking patterns of woodland insectivorous birds, but flocks of open-country birds or granivores are often the composite grouping of single-species associations.

Sometimes in the literature mixed flocking is contrasted with territorial behavior. However, some birds have been shown to be both territorial and to join flocks (Powell 1980) and others actually defend the flock territory (Greenberg 1984). One way that observers have examined if a species is either socially intolerant or actually territorial in a mixed-flocking context is to see if the distribution of the number of individuals in a mixed-species flock is truncated compared with a Poisson distribution (Greenberg and Gradwohl 1980; Greenberg 1984; Morse 1989). The idea is that the tendency to see solitary or single pairs of individuals in a flock could represent an uncommon event, where individuals join without respect to the presence or absence of other conspecifics. However, given a high enough frequency of occurrence in a flock, the probability of always seeing single birds becomes statistically less likely. The Poisson distribution would represent a null hypothesis that flock joining is a rare but random event. Coupled with experimental presentations of playback and models, a strong inference of territoriality, or at least dominance-mediated avoidance, could be sustained. With the use of the Poisson distribution, the hypothesis of aggregating more than at random can be addressed as well.

INTRASPECIFIC VARIATION IN WINTER SOCIAL SYSTEM

Before we examine patterns of interspecific variation in social behavior, we need to relate modal or "normal" patterns (those we can use to characterize a species) with the existence or potential for intraspecific variation. Besides species that show only one type of social behavior, there are species that are both gregarious and territorial in the nonbreeding season. Most of the detailed observational and all of the experimental work has been done on species that winter in temperate or Mediterranean regions. Probably the most complete work has been done on wagtails (Motacillidae). The White Wagtail (*Motacilla alba*) has been the focus of nonbreeding socioecological studies in both Israel (Zahavi 1971) and Britain (Davies and Houston 1981, 1983). In both areas, White Wagtails were both gregarious and territorial, depending on habitat and food distribution. Davies and Houston (1981, 1983) found that birds defended territories along rivers where wave action washed ashore arthropods after they had been initially depleted. Territory owners sometimes left to forage with the flocks in nearby fields although they always returned to their territories at regular intervals even when in the short run it was more profitable to forage in the flock.

Wagtails apparently return to prevent intruders from depleting resources during their absence, so that the territory is a reliable source of food; this behavior underscores our need to understand territoriality as a long-term strategy. In Israel, Zahavi (1971) also found that some individuals formed flocks in plowed fields, whereas others defended feeding territories. Long after plowing, when the initially exposed resources were depleted, some birds established long-term territories presumably to defend the reduced, but more predictable, food supply. Hypothesizing that the distribution and local abundance of food were critical in determining whether individuals could and would defend an area, Zahavi manipulated the food supply experimentally. Distributing piles of food of different sizes, he was able to cause a shift between social foraging and territoriality. He concluded that in open fields with a patchy food supply the birds are adapted to feeding in flocks, but where the feeding conditions are more stable resources can be better exploited by territoriality.

Although this work focused on the temperate-wintering White Wagtail, observations suggest that these conclusions could also apply to the longer-distance migrating Yellow Wagtail (*Motacilla flava*). Although this species is often observed in small groups on its African wintering grounds (Elgood et al. 1966; Vande Weghe 1979; Wood 1979; Gatter and Mattes 1987), some males occupy feeding territories throughout the winter. By defending these territories, birds probably experience less of a drop in food supply with the advent of the severe dry season (Wood 1979). In support, the numbers of non-territorial birds declined more dramatically than those of territorial feeders during the course of the season, indicating that an insufficient food supply forced non-territorial birds to leave the area.

Tye (1986) performed a somewhat different experiment with the Fieldfare (*Turdus pilaris*). The species is gregarious in the nonbreeding season, and the degree of gregariousness is related to the spatial and temporal variability of food (Tye 1982), although birds might defend single apples on the ground in orchards. Tye (1986) was able to induce territorial behavior in otherwise gregarious thrushes by changing the distribution of the apples. The territory occupancy lasted until all the apples were gone, after which the birds returned to their normal gregarious behavior. According to Tye (1986) the experiments provided powerful evidence for the link between social organization and food distribution and the economic value of territorial defense. This example shows that territorial behavior can be induced in species that are normally not territorial. Fieldfares were, however, territorial on fields when the artificial apple density was not higher than that in orchards. Tye (1986) linked this difference in behavior to the longer grass in the orchard plots, which prevented a bird from seeing all available apples as well as all potential competitors. It is therefore not only the actual availability of food that plays a role in determining social behavior, but also the assessment of that availability.

Large variation in social behavior has been observed in a few species of tropical migrants but there have been no experimental manipulations of resources. For example, Tramer and Kemp (1979) noted that Tennessee Warblers (*Vermivora peregrina*) were found in large single-species flocks when visiting trees that produced a large crop of flowers and fruit. On the other hand, individuals were observed aggressively defending certain highly productive flowers. This variation in sociality in Tennessee Warblers is characteristic of this often locally abundant species and has been noted by other authors as well (Morton 1980). Yellow-rumped Warblers (*Dendroica coronata*) are often noted to be among the most intraspecifically social of all of the warblers, occurring in large flocks both at fruiting and flowering trees and while searching fields and savannas for arthropods. For certain resources, however, they have been noted to be highly territorial as well. Most interestingly, individuals of the western population ("Audubon's" Warbler) are known to defend trees that have dense populations of honeydew-producing scale insects (Margororidae [Greenberg et al. 1993]). These cases suggest that the spatial distribution and renewability of the particular resource used was critical in determining the degree to which competition mediated by aggression was a successful strategy. Where such resources are not defendable, the large local abundance and ephemerality of plant-based resources often promotes the formation of aggregations and also true flocks.

Many species show a consistent pattern of sociality and therefore seem to be obligatorily territorial (or at least solitary) or gregarious. However, when studied thoroughly, individuals of these species show alternative social strategies. Anecdotal observations of unusual behavior (e.g., see Korb and Salewski 2000) demonstrate that a certain plasticity is found in many species and that their overall consistent behavior reflects that these species face too restricted a range of ecological conditions for all possible behaviors to be manifested under natural conditions.

CORRELATES OF INTERSPECIFIC VARIATION IN WINTER SOCIALITY

Although social behavior often varies considerably within species, both the modal social behavior observed and the amount of variation around it vary considerably among species. The ecological attributes that promote consistently territorial, predominantly gregarious, or variable social behavior can be examined either through detailed comparative studies of species pairs in the same environment (where many environmental factors are controlled, except for resources used) or by broader comparative analyses involving many species. Using the first approach, Salewski et al. (2002) compared two species, the Pied Flycatcher (*Ficedula hypoleuca*) and Willow Warbler (*Phylloscopus trochilus*), that exhibit different types of sociality. They suggested that a more diverse foraging ecology enabled the Pied Flycatcher to be territorial because it could use more resources on an economically defendable patch. Alternatively (or additionally), the characteristic long attack distance may make Pied Flycatchers more sensitive to foraging interference and competition from conspecifics (see below). In contrast, Willow Warblers were more specialized feeders; thus the relative scarcity and even distribution of potential food led to foraging in either monospecific or mixed-species flocks. Although (as we noted earlier) intraspecific variation in social organization can occur in these species, much can be understood by examining the difference in modal social behavior between the two.

In this section we analyze the large patterns of variation in the sociality of migrants and examine which features of the biology of these species correlate with sociality. Although few of the hundreds of tropical migrant species have been studied in detail, we now have decades of natural history observations by seasoned observers to help us discern patterns across the Tropics.

METHODS

The methods for this analysis are similar to those of Levey and Stiles (1992). We have garnered natural historical information on as many species of tropical migrant passerines as possible (see the Appendix to this chapter). Species include passerines that breed in the Temperate Zone and winter primarily south of the Tropic of Cancer. Because of the strong ecological discontinuity they represent, swallows and martins were not included in the analysis. Inherent bias is found in any analysis where information is not available for many species. The fact that we did not restrict ourselves to the well-studied species and accepted natural history accounts of many relatively poorly known species minimizes this bias. Our analysis therefore suffers because the quality of information is far from uniform. This affects our assessment of behaviors, such as territoriality, which require the detailed study of marked populations over time.

Most of the data are either categorical or ranked, and the following are included in the data set and analysis:

- Mass (g).
- Continent: Nearctic (1) or Palearctic (2) breeding.
- Breeding Foliage Type: This is the characteristic foliage of breeding-habitat shrubs and trees (grassland or marsh birds not included); 1 = broad-leafed shrub, herbs, or grass; 2 = mixed broad leaf/coniferous; 3 = coniferous. This rather specific variable is included because of previous suggestions by the senior author (Greenberg 1986) that the dependence upon coniferous foliage during the breeding season may impose a constraint on species wintering in the Tropics, where native coniferous vegetation is limited in extent.
- Breeding Habitat: 1 = open grassland, marsh, or agricultural fields; 2 = scrub; 3 = small wooded patches or edge; 4 = woods or forest.
- Breeding Foraging Guild: 1 = ground/herb forager; 2 = shrub or tree foliage gleaner; 3 = bark gleaner; 4 = aerial; and 5 = generalist or other.
- Winter Habitat: 1 = open grassland or agricultural field; 2 = marsh or mesic scrub; 3 = scrub, edge, and small patches of trees; 4 = scrub/woodland generalist; and 5 = woods and forest.
- Dominant Maneuver: This is a ranked variable based on the degree to which a species uses long-distance aerial maneuvers to capture prey; 1 = near-surface gleaner; 2 = variable-distance gleaner and aerial striker; and 3 = long-distance aerial striker.
- Winter Foraging Guild: (see breeding foraging guild for categories).
- Omnivory: degree of use of plant products (nectar, fruit, seeds) during the "winter"; 1 = strict insectivore/carnivore; 2 = occasional use of plant prod-

ucts; 3 = regular but uncommon use, or seasonally restricted but common use of plant products; 4 common use of plant products throughout winter.

- Winter Gregariousness (intraspecific): 1 = always solitary; 2 = over 5% of observations involve small groups (2–10 individuals) either apart from or in mixed-species flocks; 3 = commonly observed in small groups and also seen in larger flocks of more than 10 individuals.
- Pairing: Observations of male-female pairing on wintering ground or during migration (M).
- Territoriality: 0 = no evidence and sufficient study to expect evidence; 1 = some individuals known to defend territories; 2 = territories a regular and dominant feature of a local population, at least in preferred habitat.
- Mixed-Species Flocks: 0 = participation in flocks rare; 1 = participation in flocks commonplace; 3 = participation in flocks nearly or completely obligatory. For this analysis, mixed flocks include any heterospecific association that moves together while foraging—even those that are a composite of several single-species flocks.

Statistical Analyses

Some might argue that the data are too subjective and of sufficiently variable quality that formal statistical testing is not warranted. Thus, the whole exercise should be considered exploratory. However, use of these tools still provides us with some objective way of assessing the importance of patterns that might emerge. Because most of the data are based on categories or ranks, we use nonparametric tests (Kruskal-Wallis and Spearman rank) to test the hypothesis that gregariousness varies with particular variables. Forward stepwise multiple regressions (Statistica, Version 6) were conducted to see which ecological variables (maneuver, omnivory, breeding habitat, winter habitat, and body mass) are most strongly associated with degree of gregariousness. These analyses were conducted using individual species and averages for genera as observations. We also ran the analyses including and excluding omnivory. We did this because omnivory itself is such a strongly explanatory variable that its inclusion may mask variables that contribute directly or indirectly to the development of gregariousness.

RESULTS

Some Overall Patterns

Most tropical migrant songbirds lead solitary lives on their wintering grounds. We categorized 59% of both the Palearctic and Nearctic species ($N = 175$) in social class 1, and only 10% of the species as highly social (class 3). However, we found documentation of territoriality in only 34% of the total species (35% New World and 32% Old World, $n = 97$ and 91)—a value that partly reflects the paucity of detailed population studies with marked birds rather than an actual rarity of the phenomenon. Participation in mixed-species flocks seems commonplace in most migrants. The regular joining of such flocks was reported for 50% of the species. Once again, further refinement of data and definitions, as well as more observations, will probably result in a higher absolute number. Interestingly, species that regularly join or form mixed-species associations were more likely to be intraspecifically gregarious as well. Of the species that do not occur in mixed flocks ($n = 67$), 80% are not themselves gregarious, and the remaining species were equally divided between categories 2 and 3 for gregariousness. Only 45% of 64 species that do regularly occur in mixed flocks were intraspecifically solitary, with the remaining species equally divided between gregariousness categories 2 and 3. The proportion of gregarious species was significantly higher for species that join mixed flocks (χ^2, $p < 0.0001$); this is primarily a result of the large portion of intraspecifically solitary species not occurring in mixed-species flocks.

Geographic Patterns in the Degree of Winter Sociality

We did not detect any overall difference in intraspecific sociality during the winter between Old World and New World species (Mann-Whitney $U_{92,82} = 3{,}740$, $p > 0.9$). However, when sociality was broken down into specific classes, there were more highly social (class 3) species in the New World than the Old World (18% vs. only 13%). Degree of omnivory showed a strong intercontinental difference, with a mean omnivory index of 1.54 for New World and only 0.84 for Old World species ($F_{1,155} = 16.01$, $p < .001$). Maneuver length class, on the other hand, showed little difference between continents ($F_{1,155} = 0.7$, $p = 0.8$).

Phylogenetic Variation in Sociality

We analyzed the intergeneric variation in gregariousness, including genera that had four or more species for which we had a ranking value. The mean value here is presented merely as a convenient index of the tendency toward sociality. Intergeneric variation was highly significant (KW = 38.6, d.f. = 16, $p < 0.001$). Highly social genera, those with a mean rank value of 2.0 or greater, include *Tyrannus* (2.5), *Vermivora* (2.1), *Emberiza* (2.8), *Passerina* (2.5), *Icterus* (2.3), *Sylvia* (2.2); less social genera include *Empidonax* (1), *Vireo* (1.2), *Wilsonia* (1.2), *Locustella* (1.3), *Acrocephalus* (1.2), *Ficedula* (1), *Luscinia* (1.3), *Lanius* (1), and *Oporornis* (1). Groups with intermediate levels of sociality include *Dendroica* (1.6), *Phylloscopus* (1.4), *Piranga* (1.4), and *Hippolais* (1.4).

We further analyzed the data by families, or where families are so subdivided, by subfamilies (as recognized by Sibley and Monroe 1990). To do so, we used the mean values for the genera constituting the higher taxonomic unit rather than the mean for the species involved. In fact, the correlation between values based on generic means and species values is very high, ranging from 0.96 to 0.99, so the choice of approach probably did not affect the outcome of any analyses based on families and subfamilies. We found considerable variation in gregariousness, with Muscicapinae, Polioptilidae, and Laniidae showing low sociality (1). Families or subfamilies showing intermediate levels of gregariousness include Vireonidae (1.2), Sylviinae (1.4), and Tyranninae (1.5), and more highly gregarious (sub)families include Emberizinae (1.7), Turdinae (1.8), Oriolidae (2), Mimidae (2), Motacillinae (2.3), and Fringillinae (3).

Association between Ecology and Sociality

We first analyzed the relationship between ecological and social variables by using univariate analysis of individual species pooled across continents. The strongest association between ecological and social variables is between degrees of omnivory and sociality ($r_s = 0.56$, $p < 0.01$ [fig. 26.1A]). We also found significant variation in sociality associated with foraging maneuver distance, with species that use a consistently long aerial attack showing a reduced tendency to flock. The Spearman rank correlation between maneuver class and sociality class was –0.39 ($p < 0.05$ [fig. 26.1B). Larger birds showed a weak but significant tendency to be more gregarious ($r_s = 0.24$ $p < 0.05$ [fig. 26.1C]) When mean values for genera were used, the correlation was somewhat stronger (and significant): –0.51 for maneuver distance, 0.63 for omnivory, and 0.34 for body mass. Finally, at the level of subfamily the correlation for standardized mean foraging class was –0.86 and for omnivory was 0.75 (fig. 26.2). The correlation with body mass was not significant (0.26).

We found no significant difference in winter sociality between species that use the different breeding habitat categories (e.g., open, scrub, patchy, and closed forest; KW = 7.3, d.f. = 4, $n = 175$, $p = 0.06$). Nor did we find a difference based on winter habitat use (KW = 4.17, d.f. = 4, $n = 173$, $p > 0.07$). We also found no significant relationship between breeding foraging guild or winter foraging guild and gregariousness. Finally, we found no relationship at any taxonomic level of analysis between use of coniferous foliage and gregariousness.

MULTIVARIATE ANALYSIS. We begin this analysis by examining the relationship of the ecological variables with the birds pooled across the hemispheres. The variables included in the stepwise regression analysis are continuous (mass) or ranked data (winter habitat, conifer use, maneuver distance, omnivory). The results (summarized in table 26.1) show that at this broad level of analysis, gregariousness is strongly related to two primary variables—the degree of omnivory and the maneuver distance—but the explanatory power of these and other variables varies with the taxonomic level and geographic scope of the analyses.

The overall R^2 of the regression of gregariousness versus ecological variables is lowest for the analysis of all species combined (0.43–0.53), higher for the analysis of mean values for genera (0.55–0.75), and highest for the standardized mean values for subfamilies and families (0.76). This suggests that the broad features of the way of life represented by the subfamilies and families play an overall defining role for sociality, but that considerable variation can be found within these major groups as well.

For the global analyses, omnivory and maneuver distance are consistently the most important variables in the regression, with more-omnivorous species and those having shorter

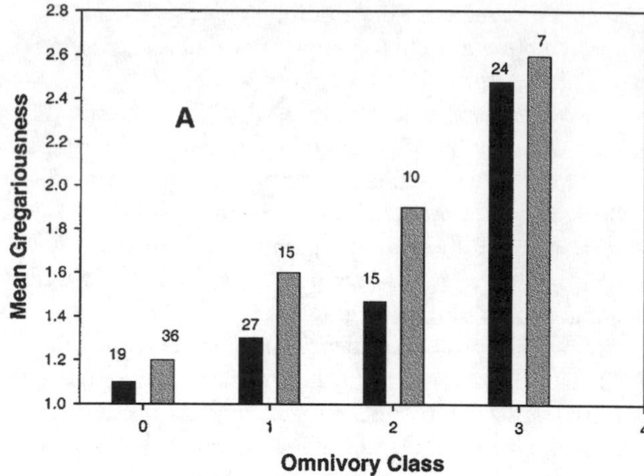

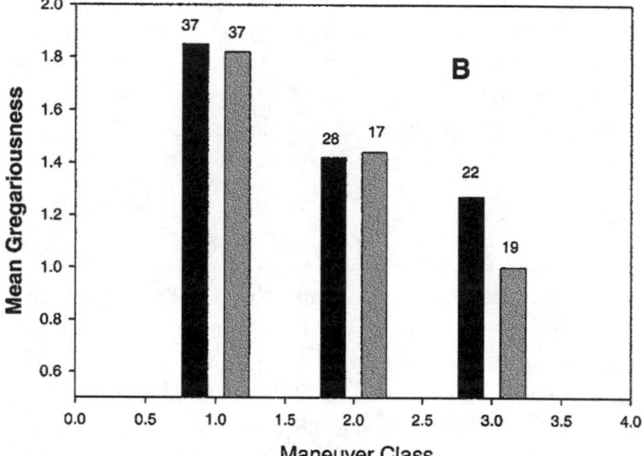

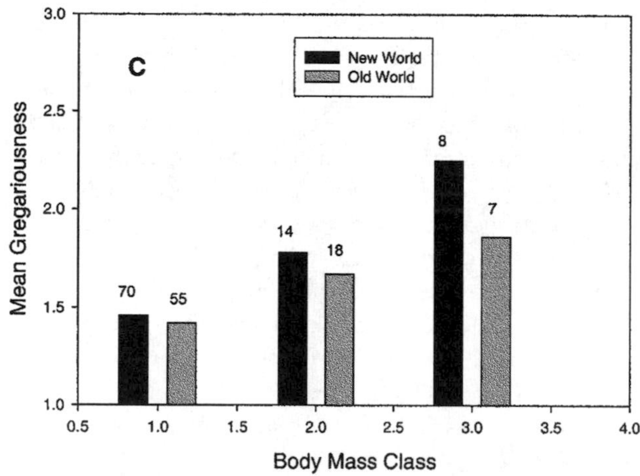

Fig. 26.1. Mean gregariousness plotted against: (A) omnivory class; (B) maneuver distance; and (C) body mass class (1 = 0–20; 2 = 20–40; 3 = >40 g) for Old World and New World migrant passerine species. Numbers over bars indicate number of species.

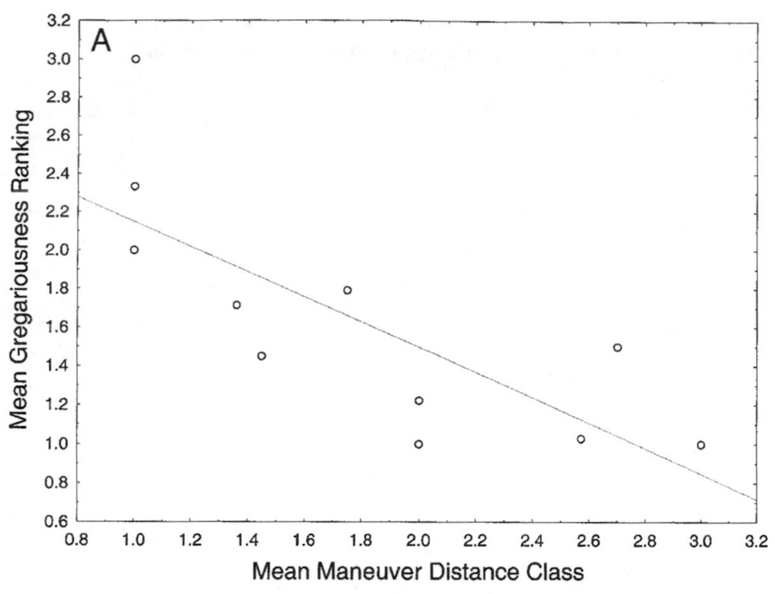

Fig. 26.2. Mean gregariousness plotted against: (A) Mean omnivory class and (B) mean maneuver distance class for subfamilies of migratory passerines.

attack distances being more gregarious than others. Omnivory is the most important variable for the New World species, and omnivory and maneuver distance are important for Old World species. The important explanatory variables differ when examined at the level of generic means, with omnivory and winter habitat showing significant betas for the New World and omnivory, maneuver, and mass for the Old World. At the level of subfamily/family, omnivory, maneuver, and winter habitat are all important. Greater degrees of gregariousness are found in subfamilies that are on the average more omnivorous, have longer maneuver distances, and/or live in open habitats.

When individual families and genera are examined, the explanatory power of the regressions declines (except for Tyrannidae, $R^2 = 0.76$) and, with one exception, only omnivory is a significant variable. The exception, *Dendroica,* shows the degree of use of coniferous foliage (see Greenberg 1979) as the only significant variable.

SOCIALITY AND OMNIVORY. The association between omnivory and sociality is well established in the literature for nonbreeding social systems of birds in general. The relationship has been noted specifically for New World migrants (e.g., see Greenberg 1979, 1984; Morton 1980; Levey and Stiles 1992). The underlying reason for the relationship relates to why social tolerance should be increased for users of fruit and flower resources. The explanation lies in the relaxation of competition for and the lower defensibility of these resources. Trees often produce large crops of fruit and flowers that provide sufficient food resources to reduce local competition and become less economically defendable. In addition, these crops often last only weeks or even

Table 26.1 Results of forward stepwise multiple regression of gregariousness and ecological variable based on total species, means for genera, means for subfamilies, and means for selected families, subfamilies, and genera

	Whole world		New World		Old World	
Comparison	R^2	Significant variables (beta)	R^2	Significant variables (beta)	R^2	Significant variables (beta)
All species	0.43	Omnivory (0.57) Maneuver (–0.19)	0.40	Omnivory (0.59)	0.53	Omnivory (0.52) Maneuver (–0.31)
Genera	0.55	Omnivory Maneuver	0.58	Omnivory Winter habitat	0.75	Omnivory Maneuver Mass
Subfamilies	0.76	Omnivory (0.79) Maneuver (–0.49) Winter habitat (–0.46)				
Tyrannidae	0.76	Omnivory 0.61				
Sylviinae	0.26	Omnivory 0.51				
Emberizinae	0.36	Omnivory 0.55				
Dendroica	0.45	Foliage type 0.21				
Phylloscopus	NS					

Note: Data from hemispheres are pooled unless otherwise noted.

days and support birds that have moved into the local area specifically to exploit this short-term resource. Greenberg (1981) showed that the preferred fruiting trees of Chestnut-sided Warblers (*Dendroica pensylvanica*) in Panamanian forest are often inaccessible to territory holders because the density of these trees is too low and their distribution is patchy. On the other hand, the non-territorial Bay-breasted Warblers (*Dendroica castanea*) had dependable access to these trees because their home range was larger and because they were socially tolerant; up to ten individuals could be found at a single tree. The most mobile and social migrant warbler in the region, Tennessee Warbler (*Vermivora peregrina*), was found regularly in the forest only when one of these preferred trees was in fruit, and then large flocks of up to 30 birds could be found visiting these trees.

The association between movement, social tolerance, and omnivory is a result of the general pattern of distribution and crop size of bird-dispersed fruit of tropical trees. Where fruit does not fit this stereotype, then, as expected, we see marked deviations from the norm in the behavior of the birds. Migrants that feed commonly on the fruits of small understory shrubs (such as the Yellow-breasted Chat [*Icteria virens*]) do not show the sociality typical of frugivorous migrants and may, in some circumstances, be territorial. Bates (1992) and Greenberg et al. (1993, 1995) documented that White-eyed (*Vireo griseus*) and Gray Vireos (*Vireo vicinior*) defend winter-long territories around *Bursera* trees and shrubs. Similar behavior has been described for the Ash-throated Flycatcher (*Myiarchus cinerascens* [Russell and Monson 1998]). *Bursera* ripen a few high-quality (high-lipid) fruits each day for the entire winter period, and so this resource becomes highly predictable and defendable. The territoriality of these species shows that it is the distribution and fruiting pattern of the plant and not the fact that it is a fruit per se that underlies the sociality of many migrants. This variation in the sociality of birds in the face of different temporal and spatial distributions of fruit and flower resources is completely consistent with a large and long-standing literature on the economics of resource defense and sociality beginning with the work of Crook (1965).

The temporal and spatial patchiness of plant resources may explain why migrants do not defend the resource or the territory around the resource; however, it does not in and of itself explain why many of these species aggregate or travel in true flocks. The participation in flocks that maintain their cohesiveness during travel between foraging sites suggests some positive advantage of associating with conspecifics that can be gained once the requirement for a competitive relationship is reduced or eliminated.

SOCIALITY AND FORAGING MANEUVER DISTANCE. The association between the tendency to use long aerial maneuvers and lack of sociality can be examined from two perspectives. The first view of this relationship focuses on the importance of local competition among individuals with similar foraging modes. If we assume that the potential for competitive disturbance is greater between conspecifics than among members of different species, then it follows that local intraspecific competition will be greatest within species whose foraging tactics are most likely to cause interference among individuals. Species that forage on rare microhabitats (such as dead-leaf clusters) or have larger effective foraging radii (sallyers to leaves and trunks), therefore, should occur in smaller intraspecific groups or solitarily. This pattern in group size has been observed in tropical resident birds (Willis 1972; Buskirk 1976; Greenberg and Gradwohl 1985). The relatively strong negative relationship between use of long foraging maneuvers and intraspecific sociality therefore suggests that local competition with conspecifics may prevent sociality. In this case, joining with other species may provide some of the benefits of flocking (antipredatory, for instance) while reducing interference competition. However, joining heterospecific flocks attaches costs associated with adjustments in optimal movement velocity, home-range, or territory to remain in the association (Powell 1980; Hutto 1988). From a strictly energetic point of view, where intraspecific interference is not an issue, one might expect birds to opt for single-species flock formation.

SPECIALIZATION, OMNIVORY, AND SOCIALITY. The second view of the relationship between foraging maneuver distance and sociality focuses on how specialization renders resources more or less predictable. Specifically, an individual that is more proficient at harvesting arthropods may not depend upon intraspecific gregariousness to locate a wider range of less predictable resources. In his brief 1986 review of migratory bird sociality, Greenberg proposed that migratory bird species can be divided into those that remain committed to insectivory throughout the winter and those that become more generalized in their use of resources. By this view, species that commonly use long-distance attack methods effectively increase the amount and quality of arthropods available and in this way are able to maintain their insectivorous specialization. In addition, Greenberg noted that larger species of migratory birds tend to be both omnivorous and more social (a conclusion that applies mainly to New World species), and he related this to the greater competition for larger arthropods, which are much rarer than smaller ones. Finally, he argued that the relationship between conifer use in the breeding grounds and sociality in the wintering grounds might also be related to the proficiency-at-insectivory hypothesis, as birds that are adapted to search for and capture arthropods in coniferous foliage may be at a distinct disadvantage when migrating to regions that have no conifers. This result also applies only to the New World, because there are very few conifer specialists among migrants of Old World (both European and Asian) temperate forests. Furthermore, such an effect of conifer foliage specialization on winter social organization was apparent only in *Dendroica*.

THE RELATIONSHIP BETWEEN SOCIALITY AND REGIONAL MOVEMENT. Although the site tenacity and stability of social groups has not been well studied for any of the tropical gregarious species, information from some of these species (particularly *Vermivora*, *Dendroica*, and *Dumetella*) indicates that local turnover of individuals tends to be high and local abundance varies with the presence of particular food resources. These observations suggest that at least some gregarious species move regionally to exploit abundant resources in the Tropics—a phenomenon that is well known among Temperate Zone birds, including certain thrushes, waxwings, and Yellow-rumped Warblers. With the development of more-sophisticated tracking technologies we should come to understand these movements better.

Thus, although confusion abounds in the middle, at the ends of the social gradient we can see species that search for locally abundant resources and move when these are depleted versus species that stand and fight for resources in a particular habitat throughout the winter. The perceived relationship between the amount of winter movement, types of resources used, and degree of social tolerance or active sociality therefore has a theoretical underpinning. These extremes define the alternative strategies available to migratory birds attempting to overwinter in tropical (as well as temperate) areas.

FUTURE DIRECTIONS

The strength of conclusions from an analysis such as that presented here depends upon having accurate information for a large number of species. In this regard, this particular analysis is very preliminary, yet it offers the glimpse of interesting patterns that will emerge as our understanding of the diversity of nonbreeding behavior becomes more complete. Certainly more-focused studies of carefully monitored, marked populations will do more to expand our understanding than anything else. However, populations of territorial and site-faithful birds lend themselves more to demographic monitoring than do species that travel in flocks. With the advent of more-sophisticated tracking technology, research should also focus on the latter species. Furthermore, with more researchers going to the field and observing migratory birds on their winter homes, some simple metrics concerning group size, participation in mixed flocks, and foraging patterns can be gathered for even the least-studied species. This may allow us to examine more subtle within-taxa correlations between ecology and sociality than were uncovered here. Guidelines for such observations and a repository of quantified natural history of tropical migrants on their winter quarters should be established so that in ten years, another more comprehensive comparative analysis can be accomplished.

CONCLUSIONS

1. Social behavior in the nonbreeding season can vary considerably within species and between closely related species. Experimental evidence suggests that intraspecific variation is directly related to the distribution of food resources in time and space. At increasingly higher taxonomic

levels, mean values of sociality become more thoroughly explained by a few ecological variables. This suggests that broad aspects of morphology and way of life are correlated with modal social behavior, but within these general constraints social behavior varies freely.

2. Most migrants are intraspecifically solitary outside of migration during the nonbreeding season, although few appear to be regularly territorial. About half of the migrant species regularly join mixed-species associations and intra- and interspecific gregariousness tend to be positively associated.

3. The New World supports a relatively large number of highly social species. These species tend to move between regions to exploit certain mass-flowering or mass-fruiting tree species.

4. Sociality is correlated with use of fruiting and flower resources, but only when large and seasonally available fruit, flower, or seed crops are visited. Omnivory is less common in Old World species and hence is a less powerful predictor of gregariousness there.

5. Sociality varies significantly between subfamilies and dominant genera, and combinations of degree of omnivory, maneuver distance, and winter habitat contribute to this variation. Variation within genera was usually related most strongly to the degree of omnivory.

6. Species that use long-distance aerial maneuvers to forage for insects tend to be less social than species that attack prey at shorter distances or merely stand and glean. This seems to be particularly true for those wintering in Africa, suggesting that intraspecific competition for arthropod resources may be more omnipresent in the dry habitats that most migrants inhabit there.

7. At least in the New World, large species tend to be more social than smaller species. This is partly because omnivory tends to increase with size, but could also be a result of the rarity of larger prey and the tendency for larger songbirds to be less specialized in the winter.

8. The prevalence of solitariness with respect to conspecifics provides prima facie evidence that intraspecific competition for resources among migrants is widespread. In species of migrants where territoriality has not been established, more research is needed on the mechanisms underlying the maintenance of solitary foraging. Further study may reveal undetected territorial systems or more subtle forms of behaviorally maintained spacing, such as spatiotemporal territories.

APPENDIX: ASPECTS OF SOCIAL SYSTEM AND ECOLOGY OF TROPICAL MIGRANT PASSERINES

Species Latin name	Mass	Contintent	Breeding foliage	Breeding habitat	Breeding guild	Winter habitat	Dominant maneuvers	Winter guild	Omnivory	Gregarious	Pairing	Territoriality	Mixed flock	References
Olive-sided Flycatcher *Contopus cooperi*	32	1	3	4	4	4	3	4	0	1	0		0	Pers. Obs.; Contr. Obs.
Western Wood-Pewee *Contopus sordidulus*	13	1	2	4	4	4	3	4	0	1	0		0	Contr. Obs.
Eastern Wood-Pewee *Contopus virens*	14	1	1	4	4	4	3	4	0	1	0	2	0	Fitzpatrick 1980
Yellow-bellied Flycatcher Empidonax flaviventris	11.6	1	3	4	4	5	3	4	1	1	0	1	0	Pers. Obs.
Acadian Flycatcher *Empidonax virescens*	12.9	1	1	4	2	5	3	2	1	1	0	2	0	Willis 1966
Alder Flycatcher *Empidonax alnorum*	13.5	1	1	2	2	3	3	2		1	1	2	0	M. Foster pers. comm.
Willow Flycatcher *Empidonax trailii*	13.5	1	1	2	2	3	3	2		1	0	2	0	Koronkiewicz, T.J. 2002.
Least Flycatcher *Empidonax mimumus*	10.3	1	1	2	2	3	3	2	1	1	0	1	0	Pers. Obs.
Hammond's Flycatcher *Empidonax hammondii*	10.1	1	3	4	2	4	3	2	1	1	0	1	0	Pers. Obs.
Gray Flycatcher *Empidonax wrightii*	10.4	1	3	4	2	3	3			1	0		0	
Dusky Flycatcher *Empidonax olberholseri*	12.5	1	3	4	2	3	3			1	0		0	
Western Flycatcher *Empidonax "difficilis"*	10.7	1	2	4	2	4	3	2		1	0		0	Hutto pers. com
Ash-throated Flycatcher *Myiarchus cinarescens*	27	1	2	3	2	3	3	2	2	1	0	1	0	Hutto pers. comm.; Russell and Monson 1998
Great-crested Flycatcher *Myiarchus crinitus*	33.5	1	1	4	2	4	3	2	1	1	0	1	0	Pers. Obs.
Cassin's Kingbird *Tyrannus vociferans*	45.6	1	1	3	4	1	1	4	3	2	0	0	0	Isabelle Bisson pers. comm.
Western Kingbird *Tyrannus verticalis*	39.6	1	1	3	4	1	3	4	3	2	0	0	0	Pers. Obs.
Scissor-tailed Flycatcher Tyrannus forficatus	43.2	1	1	2	1	1	3		3	3	0	0	0	Pers. Obs.
Eastern Kingbird *Tyrannus tyrannus*	43.6	1	1	3	4	4	3	4	3	3	0	0	0	Morton 1972; Fitzpatrick 1980
Sulphur-bellied Flycatcher *Myiodynastes luteiventris*	45.9	1	1	3	2	4	2	3	2	0	0	0		Fitzpatrick 1980
Red-backed Shrike *Lanius collario*	28.5	2		2	1	1	3	1	0	1		2	0	Bruderer 1994; Fry et al. 2000
Brown Shrike *Lanius cristatus*	29.9	2	1	2	1	1	3	1	0	1	1?	1	0	Fry et al. 2000

continued

Species Latin name	Mass	Contintent	Breeding foliage	Breeding habitat	Breeding guild	Winter habitat	Dominant maneuvers	Winter guild	Omnivory	Gregarious	Pairing	Territoriality	Mixed flock	References
Tiger Shrike *Lanius tigrinus*		2			1	4	3		0	1			0	Robson 2000
Lesser Grey Shrike *Lanius minor*	48.6	2	1	1	1	1	3	1	1	1	1	1	0	Herremans 1997; Fry et al. 2000
Woodchat Shrike *Lanius senator*	29.1	2	1	2	1	2	3	1	0	1	0	1	0	Herremans 1997; Fry et al. 2000
Bell's Vireo *Vireo bellii*	8.5	1	1	2	2	3	2	2		1	0			Contr. Obs.
Black-Capped Vireo *Vireo atricapillus*	8.5	1	1	2	2	3	2	2	1	1		0	2	
White-eyed Vireo *Vireo griseus*	11.4	1	1	2	2	4	2	2	2	1	0	2	1	Greenberg et al. 1995
Gray Vireo *Vireo vicinior*	12.8	1	2	2	2	3	2	2	2	1	1	2	0	Bates 1992
Solitary Vireo *"Vireo solitarius"*	16.6	1	2	4	2	4	2	2	1	1	0	1	1	Pers. Obs.; Contr. Obs.
Yellow-throated Vireo *Vireo flavifrons*	18	1	1	4	2	4	2	2	1	1	0		1	Pers. Obs.; Contr. Obs.
Philadelphia Vireo *Vireo philadelphicus*	12.2	1	1	4	2	4	2	2	1	1	1		1	Tramer and Kemp 1980; Pers. Obs.
Red-eyed Vireo *Vireo olivaceus*	16.7	1	1	4	2	4	2	2	3	3	0	0	1	Pers. Obs.; Contr. Obs.
Warbling Vireo *Vireo gilvus*	14.8	1	1	3	2	4	2	2	1	1			1	Pers. Obs; R. L. Hutto pers. comm.
Golden Oriole *Oriolus oriolus*	79	2	1	3	2	4	1	2	3	2	1	0	1	Baumann 2000
Rock Thrush *Monticola saxatilis*	48.5	2	1	1	1	1	3	1	2	1		1	0	Fry et al. 2000
Siberian Thrush *Zoothera sibirica*	104	2	1	4	1	4		1	2	2				Smythies 1953; Robson 2000
White's Thrush *Zoothera dauma*	75.5	2	2	4	1	5	1	1	2	1			0	Ali and Ripley 1973; Smythies 1953; Kennedy et al. 2000
Veery *Catharus fuscescens*	31.2	1	1	3	1						0		0	
Gray-cheeked Thrush *Catharus minimus*	32.8	1		4	1						0			
Bicknell's Thrush *Catharus bickneli*		1	3	2	1	4	2	2	2	1	0		1	Rimmer et al. 2001
Swainson's Thrush *Catharus ustulatus*	30.8	1	2	4	5	4	2	5	3	3	0		1	J. Sterling; Pers. Obs.
Wood Thrush *Hylocichla mustelina*	47.8	1	1	4	1	5	2	1	2	1	0	2	0	Winker et al. 1990
Grey-backed Thrush *Turdus hortulorum*	68	2				4								

Eye-Browed Thrush *Turdus obscurus*	62.6	2	2	4	1	5	1	1	2	3	0		1	Ali and Ripley 1973; Robson 2000
Pale Thrush *Turdus pallidus*	72.7	2			1		1							Kennedy et al. 2000
Dark-throated Thrush *Turdus ruficollis*	78.1	2	2	3	1	4	1	1	3	3	0		1	Ali and Ripley 1973: Smythies 1986
Spotted Flycatcher *Muscicapa striata*	14.6	2	2	3	4	4	3	4	1	1		1	0	Urban et al. 1997; Cramp 1988; Pers. Obs.
Grey-Streaked Flycatcher *Muscicapa griseisticta*	16.2	2		4	4	4	3	4	0				0	Kennedy et al. 2000
Dark-sided Flycatcher *Muscicapa sibirica*	11.7	2	2	4	4	4	3	4	0	1	0		0	Ali and Ripley 1973; Contr. Obs.
Asian Brown Flycatcher *Muscicapa daurica*	9.8	2	2	4	4	4	3	4	0	1	0		0	Ali and Ripley 1973; Grimmett et al. 1999
European Pied Flycatcher *Ficedula hypoleuca*	11.6	2	1	4	2	4	3	2	0	1		1	0	Cramp 1988; Pers. Obs.
Collared Flycatcher *Ficedula albicollis*	10.3	2	1	4	4	4	3	4		1	1M			Cramp 1988; Urban et al. 1997
Semi-collared Flycatcher *Ficedula semitorquata*	11.7	2	2	4	4	4	3		0	1	1M			Cramp 1988; Urban et al. 1997
Mugimaki Flycatcher *Ficedula mugimaki*	11.7	2	2	4	4	4	3							
Narcissus Flycatcher *Ficedula narcissina*	14	2				4	3							
Red-breasted Flycatcher *Ficedula parva*	10.7	2	2	4	2	4	3	2	0	1		1	1	Ali and Ripley 1973; Contr. Obs.
Rufous-tailed Robin *Luscinia sibilans*														
Thrush Nightingale		2	1	2	1	4	1	1		1	1	2		
Luscinia luscinia	23.8	2	1	2	1	4	1	1		1		2	0	Cramp 1988; Urban et al. 1997
Common Nightingale *Luscinia megarhynchos*	19	2	1	2	1	3	1	1	0	1		2	0	Cramp 1988; Urban et al. 1997
Siberian Rubythroat *Luscinia calliope*	18.5	2	1	2	1	3	1	1		2			0	Kennedy et al. 2000; Contr. Obs.
Bluethroat *Luscinia svecica*	18.5	2	1	1	1	3	1	1	1	1		1	0	Urban et al. 1997
Siberian Blue Robin *Luscinia cyanae*	44.8	2	1			5				1			0	Wells 1990
Red-Flanked Bluetail *Tarsiger cyanurus*	12.1	2	3	4	5	4	3			1			0	Contr. Obs.
Black Redstart *Phoenicurus ochruros*	16.5	2		2		1	3	1		1			0	Cramp 1988; Grimmett et al. 1999
Common Redstart *Phoenicurus phoenicurus*	14.6	2	2	3	2	4	3	1	1	1	1 ?		0	Cramp 1988; Urban et al. 1997
Daurian Redstart *Phoenicurus auroreus*	15.2	2		3		4	3	1		1			0	Kennedy et al. 2000
Whinchat *Saxicola rubetra*	23.1	2	1	2	1	1	3	1	0	1		1	0	Brosset 1971, 1984

continued

Species Latin name	Mass	Contintent	Breeding foliage	Breeding habitat	Breeding guild	Winter habitat	Dominant maneuvers	Winter guild	Omnivory	Gregarious	Pairing	Territoriality	Mixed flock	References
Stonechat *Saxicola torquata*	16.6	2	1	2	1	1	3	1		1	1	2	0	Gwinner et al. 1994
Northern Wheatear *Oenanthe oenanthe*	23.1	2	1	1	1	1	2	1	0	1	1 ?	1	0	Cramp 1988; Leisler 1990; Urban et al. 1997
Isabelline Wheatear *Oenanthe isabellina*	28.7	2	1		1	1		1					0	Borrow and Demey 2001
Gray Catbird *Dumetella carolinensis*	36.9	1	1	2	2	3	1		3	2	0	0	0	Pers. Obs.
Blue-gray Gnatcatcher *Polioptila caerulea*	6	1	2	3	2	4	2	2	0	1	0		1	Pers. Obs.
Chestnut-flanked White-Eye		1	1	3	2									
Zosterops erythropleurus	10.9	1	1	3	2									
Lanceolated Warbler *Locustella lanceolata*	10.6	2	1	2	1	2	1	1	0	1			0	Ali and Ripley 1973; Contr. Obs.
Grasshopper Warbler *Locustella naevia*	13.1	2	1	1.5	1	1	1	2	0	1		1	0	Cramp 1992; Urban et al. 1997; Borrow and Demey 2001
Pallas's Grasshopper Warbler *Locustella certhiola*	14.4	2	1	2	1	1	1	1	0	1			0	Ali and Ripley 1973; Contr. Obs.
Eurasian River Warbler *Locustella fluviatilis*	18.1	2	1	2	1	2	1						0	Cramp 1992; Urban et al. 1997
Savi's Warbler *Locustella luscinioides*	15	2	1	2	2/1	3	1			2			0	Cramp 1992; Urban et al. 1997; Borrow and Demey 2001
Thick-billed Warbler *Acrocephalus aedon*	26.1	2	1		2	2	2	2	0	1			0	Ali and Ripley 1973
Aquatic Warbler *Acrocephalus paludicola*	11.6	2	1	1	1		1			1			0	Cramp 1992; Urban et al. 1997; Borrow and Demey 2001
Sedge Warbler *Acrocephalus schoenobaenus*	11.2	2	1	1	1		1	2	1	2		1	0	Urban et al. 1997; Harrison et al. 1997; Borrow and Demey 2001
Paddyfield Warbler *Acrocephalus agricola*		2	1	2	1	2	2		0	1	0		0	Ali and Ripley 1973; Cramp 1992; Grimmett et al. 1999;
Eurasian Reed Warbler *Acrocephalus scirpaceus*	12.3	2	1	1	1		2		1.5	1		1	0	Cramp 1992; Urban et al. 1997; Harrison et al. 1997 Borrow and Demey 2001
Blyth's Reed Warbler *Acrocephalus dumetorum*	11.2	2	1	2	2	3	1	2	0	1.5		1	0	Ali and Ripley 1973; Contr. Obs.
Marsh Warbler *Acrocephalus palustris*	11.9	2	1	2	2	2	1	2	1	1		1	0	Kelsey 1989; Cramp 1992

Great Reed Warbler *Acrocephalus arundinaceus*	27.2	2	1	1	1		1			1		2	0	Cramp 1992; Urban et al. 1997; Harrison et al. 1997 Borrow and Demey 2001
Oriental Reed Warbler *Acrocephalus orientalis*	25.5	2	1	1	1	2				1	1	1	0	Nisbet and Medway 1972
Clamorous Reed Warbler *Acrocephalus stentoreus*	28.8	2	1	2	2	2	1	2	0	1	0	0	0	Cramp 1992; Urban et al. 1997
Booted Warbler *Hippolais caligata*	9.3	2	1	2	2	3	2	2	1	2	0	1	0	Cramp 1992; Ali and Ripley 1973; Contr. Obs.
Olivaceous Warbler *Hippolais pallida*	10.6	2	2	2	2		2	2	1	1.5		1	0	Cramp 1992; Urban et al. 1997; Borrow and Demey 2001
Upcher's Warbler *Hippolais languida*	10	2	1	2	2			2	0	1		1	0	Urban et al. 1997
Olive Tree Warbler Hippolais olivetorum	18.1	2	1	2	2			2	0	1		1	0	Urban et al. 1997
Melodious Warbler *Hippolais polygatta*	11	2	1	2	2		2	2	2	2		1	1	Urban et al. 1997; Borrow and Demey 2001
Icterine Warbler *Hippolais icterina*	14.6	2	1	3	2	4	2	2	1	1		1	1	Cramp 1992; Urban et al. 1997; Borrow and Demey 2001Pers. Obs.
Willow Warbler *Phylloscopus trochilus*	8.7	2	1	3	2	4	2	2	0	2		1	2	Pers. Obs.
Eurasian Chiffchaff *Phylloscopus collybita*	7.5	2	2	4	2	4	1	2	1	2		0		Cramp 1992; Ali and Ripley 1973
Bonelli's Warbler *Phylloscopus bonelli*	7.2	2	2	4	2	4	2	2	0	1		1	2	Cramp 1992; Urban et al. 1997; Contr. Obs.
Wood Warbler *Phylloscopus sibilitrax*	8.2	2	2	4	2	4	2.5	2	1	1.5		0	2	Pers. Obs.
Dusky Warbler Phylloscopus fuscatus	8.8	2	1	2	2	3	1	2	0	1		0	0	Cramp 1992; Grimmett et al. 1999; Contr. Obs.
Radde's Warbler *Phylloscopus schwarzi*	10	2			2	3		2	0	1		0	0	Cramp 1992; Grimmett et al. 1999; Contr. Obs.
Lemon-rumped Warbler *Phylloscopus proregulus*	7	2	3	4	2	4	2	2	0	2		0	2	Ali and Ripley 1973; Contr. Obs.
Arctic Warbler *Phylloscopus borealis*	10.6	2	2	3	2	4	2.5	2	0	1		0	2	Deignan 1945; Smythies 1953; Cramp 1992; Kennedy et al. 2000
Inornate Warbler *Phylloscopus inornatus*	7	2	2	4	2	4	2	2	0	1	0	0	2	Ali and Ripley 1973; Price Pers. Comm.
Greenish Warbler *Phylloscopus "trochiloides"*	9.5	2	2	4	2	4	2	2	0	1		2	0	Price 1981
Green Warbler *Phylloscopus nitidus*	6.8	2	2	4	2		2	2	0	1			0	Contr. Obs.

continued

Species Latin name	Mass	Contintent	Breeding foliage	Breeding habitat	Breeding guild	Winter habitat	Dominant maneuvers	Winter guild	Omnivory	Gregarious	Pairing	Territoriality	Mixed flock	References
Pale-Legged Leaf-Warbler														
Phylloscopus tenellipes	12.1	2	2	4	2	4		2		1				Cramp 1992; Grimmett et al. 1999
Large-billed Leaf-Warbler														
Phylloscopus magnirostris	11.6	2	2	4	2	5		2	0	1	0	2		Ali and Ripley 1973; Cramp 1992; Grimmett et al. 1999; Contr. Obs.
Tytler's Leaf-Warbler														
Phylloscopus tytleri	7.2	2	2	3	2		1	2	0	1	0			Ali and Ripley 1973; Contr. Obs.
Western crowned-Warbler														
Phylloscopus occipitalis	8.5	2	2	3	2	3	2	2	0	3	0		2	McDonald and Henderson 1972; Ali and Ripley 1973; Contr. Obs
Eastern Crowned Warbler														
Phylloscopus coronatus	7.4													
Yellow-faced Warbler		2		3	2									
Phylloscopus cantator	6	2			2			2						Ali and Ripley 1973
Blackcap														
Sylvia atricapilla	15.5	2	1	3	2	4	1	2	3	2			1	Cramp 1992; Urban et al. 1997; Borrow and Demey 2001
Garden Warbler														
Sylvia borin	13.9	2	1	2	2	4	1	2	2	2		0	1	Cramp 1992; Urban et al. 1997; Borrow and Demey 2001
Greater Whitethroat														
Sylvia communis	14.5	2	1	2	2	3	1	2	2	2			1	Cramp 1992; Urban et al. 1997; Borrow and Demey 2001
Lesser Whitethroat														
Sylvia curruca	10.1	2	2	3	2	3	1	2	2	2		0	1	Ali and Ripley 1973
Barred Warbler														
Sylvia nisoria	22.8	2	1	3	2	3	1	2	2	2				Cramp 1992; Urban et al. 1997
Orphean Warbler														
Sylvia hortensis	22.2	2	1	3	2	3	1	2	1	2				Ali and Ripley 1973
Subalpine Warbler														
Sylvia cantillans	10.8	2	1	2	2	3	1	2	2					Fry et al. 2000; Borrow and Demey 2001
Grey Wagtail														
Motacilla cinerea	17.7	2		1	1			1		3			1	Ali and Ripley 1973
Yellow Wagtail														
Motacilla flava	17.6	2		1	1	1	1	1	0	3		1	0	Borrow and Demey 2001
Forest Wagtail														
Motacilla indica	15.5	2	2	4	1	5	1	1	0	2			0	Ali and Ripley 1973
Tawny Pipit														
Anthus campestris	23	2		1	1	1	1	1	1	3			1	Ali and Ripley 1973
Tree Pipit														
Anthus trivialis	23.4	2	2	2	1	3	1	1	2	2		1	0	Ali and Ripley 1973;
Pechora Pipit														
Anthus gustavi		2	2	2	1	4	1	1	1	1			0	Kennedy et al. 2000
Olive-backed Pipit														
Anthus hodgsoni	19.3	2	3	4	1	4	1	1	1	2			1	Ali and Ripley 1973
Red-throated Pipit														
Anthus cervinus	20.4	2		1	1	1	1	1	1	2			2	Cramp 1988

Common Rosefinch														
Carpodacus erythrinus	24.1	2	1	3	2	3	1		3	3			1	Ali and Ripley 1973; Smythies 1986
Chestnut-eared Bunting														
Emberiza fuscata	18	2			2				3	3			1	Ali and Ripley 1973
Yellow-breasted Bunting														
Emberiza aureola	21.7	2		3	2				3	3			1	Deignan 1945; Smythies 1986;
Little Bunting														
Emberiza pusilla	13	2			2	3			3	2			1	Deignan 1945; Smythies 1986;
Black-headed Bunting														
Emberiza "melanocephala"	29.7	2			2				3	3			1	Ali and Ripley 1973; Grimmett et al. 1999
Lincoln's Sparrow														
Melospiza lincolnii	17.4	1	1	2	5	3	1	1	3	2	0	0		Pers. Obs.
Bachman's Warbler														
Vermivora bachmanii	9	1	1	4	2		1	2	3		0		1	Hamel 1986
Blue-winged Warbler														
Vermivora pinus	8.4	1	1	2	2	4	1	2	1	1	0		2	Pers. Obs.
Golden-winged Warbler														
Vermivora chrysoptera	9	1	1	2	2	4	1	2	1	1	0		2	Powell 1980
Tennessee Warbler														
Vermivora peregrina	10	1	3	4	2	4	1	2	3	3	0	1	1	Tramer and Kemp 1979; Pers. Obs.
Orange-crowned Warbler														
Vermivora celata	9	1	2	3	2	4	1	2	3	3	0		1	Hutto 1980; Pers. Obs.
Nashville Warbler														
Vermivora ruficapilla	8.7	1	3	4	2	4	1	2	3	3	0	0	1	Pers. Obs.
Virginia's Warbler														
Vermivora virginiae	7.8	1	1	3	2	3	1	2	2	1	0		1	R. L. Hutto pers. comm.
Colima Warbler														
Vermivora crissalis	9.7	1	2		2						0			
Lucy's Warbler														
Vermivora luciae	6.6	1	1	2	2	3	1	2	1	3	0		1	R. L. Hutto pers. comm.
Northern Parula														
Parula americana	8.6	1	2	4	2	4	1		1	1	0		1	Morton 1976; Staicer 1992; Eaton 1953
Yellow Warbler														
Dendroica petechia	9.5	1	1	2	2	3	2	2	1	1	0	2	0	Morton 1976; Greenberg and Salgado Ortiz 1994
Chestnut-sided Warbler														
Dendroica pensylvanica	99.5	1	1	2	2	4	2	2	2	1	0	2	1	Greenberg 1984
Magnolia Warbler														
Dendroica magnolia	88.7	1	2	3	2	4	2	2	1	1	0	1	1	Greenberg and Salgado Ortiz 1994
Cape May Warbler														
Dendroica tigrina	11	1	3	4	2	4	1	2	3	2	0	1	1	Staicer 1992; Latta and Faaborg 2002; Pers. Obs.
Black-throated Blue Warbler														
Dendroica caerulescens	10.2	1	1	4	2	4	2	2	2	1	0	2	1	Holmes et al. 1989; Wunderle 1995
Yellow-rumped Warbler														
Dendroica coronata	12.5	1	3	3	2	4	2	2	3	3		1	1	Pers. Obs.
Hermit Warbler														
Dendroica occidentalis	9.2	1	3	4	2	5	1	2	1	2	0	0	1	Greenberg et al. 2001

continued

Species Latin name	Mass	Contintent	Breeding foliage	Breeding habitat	Breeding guild	Winter habitat	Dominant maneuvers	Winter guild	Omnivory	Gregarious	Pairing	Territoriality	Mixed flock	References
Black-throated Gray Warbler *Dendroica nigrescens*	8.4	1	2	4	2	4	2	2	1	1	0		1	Hutto pers. comm.
Townsend's Warbler *Dendroica townsendi*	8.9	1	3	4	2	4	1	2	1	1	0	0	1	Greenberg et al. 2001
Black-throated Green Warbler *Dendroica virens*	8.8	1	2	4	2	4	2	2	1	2	0	0	1	Greenberg et al. 2001
Golden-cheeked Warbler *Dendroica chrysoparia*	10.2	1	2	3	2	4	2	2	1	1	0		1	Rappole et al. 1999
Blackburnian Warbler *Dendroica fusca*	9.8	1	3	4	2	4	1	2	1	2	0	0	1	Chipley 1980
Yellow-throated Warbler *Dendroica dominica*	9.4	1	2	4	2		1	2	1	1	0	1	0	Pers. Obs.
Kirtland's Warbler *Dendroica kirtlandii*	13.7	1	3	2	2	3	1		2	1	0			Wunderle pers. comm.
Prairie Warbler *Dendroica discolor*	7.7	1	2	2	2	3	1	2	1	1	0	1	1	Staicer 1992; Latta and Faaborg 2002
Palm Warbler Dendroica palmarum	10.3	1	3	2	2	3	1	2	1	2.5	0		0	Emlen 1977
Bay-breasted Warbler *Dendroica castanea*	12.5	1	3	4	2	4	1	2	2.5	3	0	1	1	Greenberg 1984
Blackpoll Warbler *Dendroica striata*	13	1	3	4	2	4	1	2		2	0		1	Contr. Obs.
Cerulean Warbler *Dendroica cerulea*	9.3	1	1	4	2	5	2	2	1	2	1		2	Greenberg 1986: Jones et al. 2000
Black-and-White Warbler *Mniotilta varia*	10.8	1	2	4	3	4	1	3	0	1	0	1	2	Pers. Obs.
American Redstart *Setophaga ruticilla*	8.3	1	1	4	2	4	3	2	0	1	0	1	1	Marra 2000
Prothonotary Warbler *Protonotaria citrea*	16.7	1	1	4	2		1	2	1	3	0	0	1	Warkentin and Morton 2000
Worm-eating Warbler *Helmitheros vermivorus*	13	1	1	4	2		1	2	0	1	0		2	Greenberg 1987
Swainson's Warbler *Limnothlypis swainsonii*	18.9	1		4	1				0	1	0	1	0	Graves 1996
Ovenbird *Seiurus aurocapillus*	19.4	1	1	4	1	4	1	1	0	1	0		0	Sherry Pers. Comm.
Northern Waterthrush *Seirurus noveboracensis*	17.8	1	2	4	1		1	1	0	1	0	1	0	Schwartz 1964 Contr. Obs.
Louisiana Waterthrush *Seirurus motacilla*	19.8	1	1	4	1	5	1	1	0	1	0	1	0	Eaton 1953
Kentucky Warbler *Oporornis formosus*	14	1	1	4	1	5	2	1	0	1	0	2	0	Mabey and Morton 1992

Connecticut Warbler *Oporornis agilis*	15.2	1		4						1	0		0	Ridgely and Tudor 1989; Contr. Obs.
Mourning Warbler *Oporornis philadelphia*	12.5	1		2		3			0	1	0		0	Ridgely and Tudor 1989; Contr. Obs.
Macgillivray's Warbler *Oporornis tolmiei*	10.4	1	1	2	2	3	1	1	0	1	0	1	0	Pers. Obs.
Common Yellowthroat *Geothlypis trichas*	10.1	1	1	2	2	3	1	2	0	1	0		0	Pers. Obs.
Hooded Warbler *Wilsonia citina*	10.5	1	1	4	5	4	3	5	0	1	0	1	1	Rappole and Warner 1980
Wilson's Warbler *Wilsonia pusilla*	7.3	1	1	2	2	4	2	2	0	1	0	1	1	Tramer and Kemp 1980; Pers. Obs.
Canada Warbler *Wilsonia canadensis*	10.4	1		4	2	4		2	1	2	1M	1	2	Greenberg and Gradwohl 1980
Red-faced Warbler *Cardellina rubrifrons*	9.8				2			2	0	1	0		2	Pers. Obs.
Yellow-breasted Chat *Icteria virens*	25.3	1	1	2	2	3	1	2	3	1	0		0	Pers. Obs.
Summer Tanager *Piranga rubra*	28.2	1	1	4		4	2	2	2	1	1		1	Pers. Obs.; Contr. Obs.
Scarlet Tanager *Piranga olivacea*	28.6	1		4	2	4		2	2	2	0		1	Pers. Obs.; Contr. Obs.
Western Tanager *Piranga ludoviciana*	28.1	1	3	4	2	4	2	2	2	2	0		1	Pers. Obs.
Dickcissel *Spiza americana*	26.9	1		1	1	1	1	1	0	3	0		0	Temple 2002
Rose-breasted Grosbeak *Pheucticus ludovicianus*	45.6	1	1	4	2	4	2	2	3	2	1		1	Pers. Obs.
Black-headed Grosbeak *Pheucticus melanocephalus*	42	1		4	2	4			2	2	0		1	Contr. Obs.
Blue Grosbeak *Passerina caerulea*	28.4	1	1	2	2	3			3	2	0		1	Pers. Obs.
Lazuli Bunting *Passerina amoena*	15.5	1		2	2	3			3	3	0		1	Pers. Obs.
Indigo Bunting *Passerina cyanea*	14.5	1	1	2	2	3	1		3	3	0		1	Pers. Obs
Painted Bunting *Passerina ciris*	15.6	1	1	2	2	3	1	2	3	2	0		1	Pers. Obs.
Baltimore Oriole *Icterus galbula*	33.8	1	1	3	2	4	1	2	3	3	1	1	1	Pers. Obs.
Bullock's Oriole *Icterus bullockii*	33.6	1	1	3	2	4	1	2	3	1	0		1	Pers. Obs.
Orchard Oriole *Icterus spurius*	19.6	1	1	3	2	3	1	2	3	3	0		0	Morton 1980; Pers. Obs.
Bobolink *Dolichonyx orysivorus*	40.7	1		1		1	1		3	3	0		2	Pettingill 1983

Note: See Methods for values or acronyms that appear in the Appendix.

ACKNOWLEDGMENTS

We received excellent information from the field notes of numerous observers with experience throughout the Tropics, particularly in South America. I thank the following people for responding to cyber-inquiries regarding poorly known species: Robert Behrstock, Isabelle Bisson, Richard Hoyer, Richard Hutto, Peter Jones, Steve Latta, Madhusudan Katti, Curtis Marantz, Trevor Price, Chris Sharpe, John Sterling, Ian Warkington, and Joe Wunderle. Scott Sillett provided valuable comments on an early draft.

LITERATURE CITED

Ali, S., and S. D. Ripley. 1973. Handbook of the Birds of India and Pakistan, Vols. 5–10. Oxford University Press, Oxford.

Bates, J. 1992. Winter territorial behavior of Gray Vireos. Wilson Bulletin 104:425–433.

Baumann, S. 2000. Vergleich von Habitatstructur und Habitatnutzung in Brutgebiet und Winterareal des Europäischen Pirols (*Oriolus o. oriolus*, L. 1758). Journal für Ornithologie 141:142–151.

Borrow, N., and D. Demey. 2001. Birds of Western Africa. Christopher Helm, London.

Brosset, A. 1971. Territorialisme et défense du territoire chez les migrateurs paléarctiques hivernant au Gabon. Alauda 39:127–131.

Brosset, A. 1984. Oiseaux migrateurs européens hivernant dans la partie guinée du Mont Nimba. Alauda 52:81–101.

Brown, J. L. 1964. The evolutionary diversity in avian territorial systems. Wilson Bulletin 76:160–169.

Bruderer, B. 1994. Habitat and niche of migrant Red-backed Shrikes *Lanius collurio* in Southern Africa. Journal für Ornithologie 135:474–475.

Buskirk, W. H. 1976. Social systems in a tropical forest avifauna. American Naturalist 110:293–310.

Chipley, R. 1980. Nonbreeding ecology of the Blackburnian Warbler. Pages 309–318 *in* Migrant Birds in the Neotropics: Ecology, Behavior, Distribution, and Conservation (A. Keast and E. S. Morton, eds.). Smithsonian Institution Press, Washington, D.C.

Cramp, S. (ed.). 1988. Handbook of the Birds of Europe, the Middle East, and North Africa: The Birds of the Western Palearctic, Vol. 5, Tyrant Flycatchers to Thrushes. Oxford University Press, Oxford.

Cramp, S. (ed.). 1992. Handbook of the Birds of Europe, the Middle East, and North Africa: The Birds of the Western Palearctic, Vol. 6, Warblers. Oxford University Press, Oxford.

Crook, J. H. 1965. The adaptive significance of avian social organization. Symposium of the Zoological Society of London 14:181–218.

Davies, N. B., and A. I. Houston. 1981. Owners and satellites: the economics of territory defense in the Pied Wagtail, *Motacilla alba*. Journal of Animal Ecology 50:157–180.

Davies, N. B., and A. I. Houston. 1983. Time allocation between territories and flocks and owner-satellite conflict in foraging Pied Wagtails, *Motacilla alba*. Journal of Animal Ecology 52:621–634.

Deignan, H. H. G. 1945. Birds of Northern Thailand. Bulletin of the U.S. National Museum 186.

Eaton, S. W. 1953. Wood warblers wintering in Cuba. Wilson Bulletin 65:169–174.

Elgood, J. H., R. E. Sharland, and P. Ward. 1966. Palaearctic migrants in Nigeria. Ibis 108: 84–116.

Emlen, J. T. 1977. Land Communities of Grand Bahama Island. A.O.U. Monograph 24.

Fitzpatrick, J. W. 1980. Wintering of North American tyrannid flycatchers in the Neotropics. Pages 67–78 *in* Migrant Birds in the Neotropics: Ecology, Behavior, Distribution, and Conservation (A. Keast and E. S. Morton, eds.). Smithsonian Institution Press, Washington D.C.

Fry, C. H., S. Keith, and E. K. Urban. 2000. The Birds of Africa, Vol. 6, Picathartes to Oxpeckers. Princeton University Press, Princeton.

Gatter, W., and H. Mattes. 1987. Anpassungen der Schafstelze Motacilla flava und afrikanischen Motacilliden an die Waldzerstörung in Liberia (Westafrika). Verhandlungen der Ornithologischen Gesellschaft in Bayern 24:467–477.

Graves, G. R. 1996. Censusing wintering populations of Swainson's warblers: surveys in the Blue Mountains of Jamaica. Wilson Bulletin 108:94–103.

Greenberg, R. 1979. Body size, breeding habitat, and winter exploitation systems in *Dendroica*. Auk 96:756–766.

Greenberg, R. 1981. Frugivory in some migrant tropical forest wood warblers. Biotropica 13:215–223.

Greenberg, R. 1984. The winter exploitation systems of Bay-breasted and Chestnut-sided Warblers in Panama. University of California Publications in Zoology 117:1–124.

Greenberg, R. 1986. The foraging ecology and social behavior of migrants during the non-breeding season. Proceedings of the 28th International Ornithological Congress, Moscow.

Greenberg, R. 1987. Seasonal foraging specialization in the Worm-eating Warbler. Condor 89:158–168.

Greenberg, R., M. Foster, and L. Marquez. 1995. The role of White-eyed Vireos in the dispersal of *Bursera simaruba* fruit. Journal of Tropical Ecology 11:619–639.

Greenberg, R, C. E. Gonzales, P. Bichier, and R. Reitsma. 2001. Non-breeding ecological differences between species in the Black-throated Green Warbler complex. Condor 103:31–37.

Greenberg, R., and J. Gradwohl. 1980. Observations of paired Canada Warblers, *Wilsonia canadensis*, during migration in Panama. Ibis 122:509–512.

Greenberg, R., and J. Gradwohl. 1985. A comparative study of the social behavior of antwrens on Barro Colorado Island, Panama. Ornithological Monographs 36:845–855.

Greenberg, R., C. Macias Caballero, and P. Bichier. 1993. Defense of homopteran honeydew by birds in the Mexican highlands and other warm temperate forests. Oikos 68:519–524.

Greenberg, R., J. Salgado, and C. Macias Caballero. 1994. Aggressive competition for critical resources among migratory birds in the Neotropics. Bird Conservation International 4:115–127.

Greenberg, R., and J. Salgado Ortiz. 1994. Interspecific defense of pasture trees by Yellow Warblers in winter. Auk 111:672–682.

Grimmett, R., L. Inskipp, and T. Inskipp. 1999. A Guide to the Birds of Pakistan, Nepal, Bangladesh, Bhutan, Sri Lanka and the Maldives. Princeton University Press, Princeton.

Gwinner, E., T. Rödl, and H. Schwabl. 1994. Pair territoriality of wintering stonechats: behavior, function, and hormones. Behavioral Ecology and Sociobiology 34:321-327.

Harrison, J. A., G. Allen, L. G. Underhill, M. Heermans, A. J. Tree, L. Parker, and C. J. Brown. 1997. The Atlas of South African Birds, Vol. 2. BirdLife South Africa, Johannesburg.

Herremans, M. 1997. Habitat segregation of male and female Redbacked Shrikes *Lanius collurio* and Lesser Grey Shrikes *Lanius minor* in the Kalahari basin, Botswana. Journal of Avian Biology 28:240–248.

Holmes, R. T., T. W. Sherry, and L. Reitsma. 1989. Population structure, territoriality, and over-winter survival of two migrant warbler species in Jamaica. Condor 91:545–561.

Hutto, R. L. 1988. Foraging behavior patterns suggest a possible cost associated with mixed species bird flocks. Oikos 51:79–83.

Hutto, R. L. 1994. The composition and social organization of mixed species flocks in tropical deciduous forest in Mexico. Condor 96:105–118.

Jones, P. J. 1998. Community dynamics of arboreal insectivorous birds in African savannas in relation to seasonal rainfall patterns and habitat change. Pages 421–447 *in* Dynamics of Tropical Communities (D. Newbery, H. H. T. Prins, and N. D. Brown, eds.). British Ecological Society Symposium no. 37. Blackwell Science, Oxford.

Jones, J., P. R. Perazzi, E. H. Carruthers, and R. J. Robertson. 2000. Sociality and foraging behavior of the Cerulean Warbler in Venezuelan shade-coffee plantations. Condor 102:958–962.

Kelsey, M. G. 1989. A comparison of song and territorial behavior of a long-distance migrant, the marsh warbler (*Acrocephalus palustris*) in summer and winter. Ibis 131:403–414.

Kennedy, R., P. C. Gonzales, E. Dickinson, and H. C. Miranda Jr. 2000. A Guide to the Birds of the Philippines. Oxford University Press, Oxford.

Korb, J., and V. Salewski. 2000. Predation on swarming termites by birds. African Journal of Ecology 38:173–174.

Koronkiewicz, T. J. 2002. Intraspecific Territoriality and Site Fidelity of Wintering Willow Flycatchers (*Empidonax traillii*) in Costa Rica. M.S. thesis, Northern Arizona University, Flagstaff.

Lack, P. C. 1983. The movement of Palearctic landbird migrants in Tsavo East National Park, Kenya. Journal of Animal Ecology 52:513–524.

Latta, S. C., and J. Faaborg. 2001. Winter fidelity of Prairie Warblers in the Dominican Republic. Condor 103:455–468.

Latta, S. C., and J. Faaborg. 2002. Demographic and population responses of Cape May Warblers wintering in multiple habitats. Ecology: 83:2502–2515.

Leisler, B. 1990. Selection and use of habitat of wintering migrants. Pages 156–174 *in* Bird Migration: Physiology and Ecophysiology (E. Gwinner, ed.). Springer-Verlag, Berlin.

Levey, D. J., and F. G. Stiles. 1992. Evolutionary precursors of long-distance migration: resource availability and movement patterns of Neotropical landbirds. American Naturalist 140:467–491.

Mabey, S. E., and E. S. Morton. 1992. Demography and territorial behavior of wintering Kentucky Warblers in Panama. Pages 329–336 *in* Ecology and Conservation of Neotropical Migrant Landbirds (J. M. Hagan III and D. W. Johnston, eds.). Smithsonian Institution Press, Washington, D.C.

Marra, P. 2000. The role of behavioral dominance in structuring patterns of habitat occupancy in a migrant bird during the non-breeding period. Behavioral Ecology 11:299–308.

McDonald, D. W., and D. G. Henderson. 1972. Aspects of the behavior and ecology of mixed species bird flocks in Kashmir. Ibis 119:481–491.

Moreau, R. E. 1972. The Palearctic-African Bird Migration System. Academic Press, London.

Morse, D. H. 1989. American Warblers. Harvard University Press, Cambridge.

Morton, E. S. 1972. Food and migration habitats of the Eastern Kingbird in Panama. Auk 88:925–926.

Morton, E. S. 1976. The adaptive significance of dull-coloration in Yellow Warblers. Condor 78:423.

Morton, E. S. 1980. Adaptations to seasonal changes in migrant land birds in the Panama Canal Zone. Pages 437–453 *in* Migrant Birds in the Neotropics: Ecology, Behavior, Distribution, and Conservation (A. Keast and E. S. Morton, eds.). Smithsonian Institution Press, Washington, D.C.

Nisbet, I. C. T., and Lord Medway. 1972. Dispersion, population ecology and migration of Eastern Great Reed Warblers *Acrocephalus orientalis* wintering in Malaysia. Ibis 114: 451–494.

Pettingill, O. S., Jr. 1983. Winter of the Bobolink. Audubon 85:102–109.

Pitelka, F. A. 1959. Numbers, breeding schedule, and territoriality in Pectoral Sandpipers of northern Alaska. Condor 61:233–264.

Powell, G. V. N. 1980. Migrant participation in mixed species flocks. Pages 477–484 *in* Migrant Birds in the Neotropics: Ecology, Behavior, Distribution, and Conservation (A. Keast and E. S. Morton, eds.). Smithsonian Institution Press, Washington, D.C.

Price, T. D. 1981. The ecology of Greenish Warbler, *Phylloscopus trochiloides,* in its winter quarters. Ibis 123:131–141.

Rappole, J. H. 1995. The Ecology of Migrant Birds: A Neotropical Perspective. Smithsonian Institution Press, Washington, D.C.

Rappole, J. H., D. I. King, and W. C. Barrow. 1999. Winter ecology of the endangered Golden-cheeked Warbler (*Dendroica chrysoparia*). Condor 101:762–770.

Rappole, J. H., and E. S. Morton. 1985. Effects of habitat alteration on a tropical avian forest community. Ornithological Monographs 36:1013–1021.

Rappole, J. H., and D. Warner. 1980. Ecological aspects of migrant bird behavior in Veracruz, Mexico. Pages 353–394 *in* Migrant Birds in the Neotropics: Ecology, Behavior, Distribution, and Conservation (A. Keast and E. S. Morton, eds.). Smithsonian Institution Press, Washington, D.C.

Ridgely, R. S., and G. Tudor. 1989. The Birds of South America, Vol. 1, The Oscine Passerines. Princeton University Press, Princeton.

Rimmer, C. C., K. P. McFarland, W. G. Ellison, and J. E. Goetz. 2001. Bicknell's Thrush (*Catharus bicknelli*). The Birds of North America, no. 592 (A. Poole and F. Gill, eds.). The Birds of North America Inc., Philadelphia.

Robson, C. 2000. A Guide to the Birds of Southeast Asia: Thailand, Peninsular Malaysia, Singapore, Myanmar, Laos, Vietnam, Cambodia. Princeton University Press, Princeton.

Russell, S. M., and G. Monson. 1998. The Birds of Sonora. University of Arizona Press, Tucson.

Salewski, V., F. Bairlein, and B. Leisler. 2002. Different wintering strategies of two Palaearctic migrants in West Africa: a consequence of foraging strategies? Ibis 144: 85–93.

Schwartz, P. 1964. The Northern Waterthrush in Venezuela. Living Bird 3:169–185.

Sibley, C., and B. L. Monroe. 1990. Distribution and Taxonomy of Birds of the World. Yale University Press, New Haven.

Smythies, B. E. 1953. The Birds of Burma (second ed.). Oliver and Boyd, Edinburgh.

Staicer, C. A. 1992. Social behavior of the Northern Parula, Cape May Warbler, and Prairie Warbler wintering in second-growth forest in southwestern Puerto Rico. Pages 308–320 *in* Ecology and Conservation of Neotropical Migrant Landbirds (J. M. Hagan III and D. W. Johnston, eds.). Smithsonian Institution Press, Washington, D.C.

Strong, A. M. 1999. Effects of food abundance on non-breeding habitat quality for two species of ground-foraging neotropical migrant warblers. Ph.D. dissertation, Tulane University, New Orleans.

Temple, S. A. 2002. Dickcissel (*Spiza americana*). Birds of North America, no. 703 (A. Poole and F. Gill, eds.). Birds of North America, Inc., Philadelphia.

Tramer, E. J., and T. R. Kemp. 1979. Diet-correlated variations in social behavior of wintering Tennessee Warblers. Auk 96:186–187.

Tramer, E. J., and T. R. Kemp. 1980. Foraging ecology of migrant and resident warblers and vireos in the highlands of Costa Rica. Pages 285–296 *in* Migrant Birds in the Neotropics: Ecology, Behavior, Distribution, and Conservation (A. Keast and E. S. Morton, eds.). Smithsonian Institution Press, Washington, D.C.

Tye, A. 1982. Social organisation and feeding in the Wheatear and Fieldfare. Ph.D. thesis, Cambridge University.

Tye, A. 1986. Economics of experimentally-induced territorial defense in a gregarious bird, the Fieldfare *Turdus pilaris*. Ornis Scandinavica 17:151–164.

Urban, E. K., H. Fry, and S. Keith. 1997. The Birds of Africa, Vol. 6, Thrushes to Puffback Flycatchers. Princeton University Press, Princeton.

Vande Weghe, J.-P. 1979. The wintering and migration of Palaearctic passerines in Rwanda. Gerfaut 69:29–43.

Warkentin, I., and E. S. Morton. 2000. Flocking and foraging behavior of wintering Prothonotary Warblers. Wilson Bulletin 112:88–98.

Wells, D. R. 1990. Migratory birds and tropical forest in the Sunda region. Pages 357–369 *in* Biogeography and Ecology of Forest Bird Communities (A. Keast, ed.). SPB Academic Publishing, The Hague.

Willis, E. O. 1966. The role of migrant birds at swarms of army ants. Living Bird 5:187–231.

Willis, E. O. 1972. The Behavior of Spotted Antbirds. Ornithological Monographs 10.

Winker, K., J. H. Rappole, and M.A. Ramos. 1990. Population dynamics of the Wood Thrush (*Hylocichla mustelina*) on its wintering grounds in southern Veracruz. Condor 92:444–460.

Wood, B. 1979. Changes in numbers of over-wintering Yellow Wagtails *Motacilla flava* and their food supplies in a West African savanna. Ibis 121:228–231.

Wunderle, J. M., Jr. 1995. Population characteristics of Black-throated Blue Warblers wintering in three sites on Puerto Rico. Auk 112:931–946.

Zahavi, A. 1971. The social behavior of the White Wagtail *Motacilla alba alba* wintering in Israel. Ibis 113:203–211.

27

TREVOR PRICE AND SEAN GROSS

Correlated Evolution of Ecological Differences among the Old World Leaf Warblers in the Breeding and Nonbreeding Seasons

THE UPLIFT OF THE HIMALAYAS created a diversity of habitats along an altitudinal gradient. Many bird species are currently confined to these habitats during the breeding season. The Himalayas are a seasonal environment and many species vacate these regions during the winter. Using as our model an assemblage of the Old World leaf warblers (*Phylloscopus*) breeding in the Himalayas, we ask how ecological diversification in the breeding season has been accompanied by and/or constrained by ecological diversification in the winter season. We study the evolution of four characteristics of each species: prey size, feeding behavior, breeding altitude, and geographical range size. For prey size and feeding method, there are similar correlates between morphology and ecology in both seasons, suggesting that these resources are partitioned in a similar way. Altitudinal differences within the breeding range evolved most recently, perhaps in association with a period of rapid uplift 8–10 Ma. In comparisons among four common species, those breeding at high altitudes forage higher in trees in winter than species breeding at lower altitudes. The causes are unclear, but in both seasons the high species fly relatively less frequently and hop more than species breeding at low altitudes. This pattern has evolved at least twice. Geographical range sizes are correlated between breeding and winter seasons. Despite strong evidence for winter limitation of population size, both the breeding season and winter season may have influenced range size on evolutionary time scales. In general, divergence may have been driven primarily by adaptation to the breeding season, the winter season, or to both, but there are sufficient similarities in resource availability for species to maintain many similarities in their habits across seasons.

INTRODUCTION

In an adaptive radiation, the origin of species is tied to ecological opportunity (Simpson 1953; Schluter 2000). As new niches become available, novel selection pressures cause divergence between populations eventually leading to speciation. Long-distance migrants live in two worlds. What is the relevance for speciation if ecological opportunity arises in one world and not the other? A priori we might expect this to depend on whether the new opportunity is a breeding-season or winter phenomenon. Reproductive isolation is manifested in the breeding season, and populations that diverge sufficiently in this season may become reproductively isolated as a correlated response (Rice and Hostert 1993; Orr and Smith 1998; Schluter 2000). In this case, ecological variation on the breeding grounds drives speciation. If populations that are geographically separated in the breeding season also remain separated in the winter, divergent selection pressures in winter could promote speciation as well. However, ecological conditions on the wintering grounds may also constrain or prevent divergence, particularly if resources are strongly limiting in that season.

In this chapter we ask how ecological differences in the breeding and winter season have evolved as new species are produced. We investigate the history of adaptive diversification among the Old World leaf warblers (genus *Phylloscopus*), using a molecular phylogeny to estimate times and patterns of ecological divergence. In the breeding season, the group reaches its maximum diversity in the Himalayas, where up to nine species may be found breeding in close proximity. Mountainous regions of the world are usually associated with high species diversity (Simpson 1964; Rahbek and Graves 2001), partly attributable to the great diversity of habitats present along elevational gradients (Martens and Eck 1995; Rahbek and Graves 2001). This is also true of the Himalayas (Martens and Eck 1995; Price et al. 2003). It appears likely that uplift of the Himalayas has driven speciation events in many taxa, including the *Phylloscopus* warblers.

Eight species of *Phylloscopus* warblers breed abundantly in the forests of the main Himalayan range in Kashmir (Price 1991) (table 27.1). We concentrate on the warblers that make up this community, because we have studied their breeding ecology extensively and are able to compare the breeding and winter ecology of several species. Although there are more than 40 species of *Phylloscopus* (Irwin et al. 2001a) much of the ecological variation across the entire genus is captured by studying this community (Richman 1996). Price et al. (1997) and Irwin et al. (2001a, 2001b) showed that six of the eight species that breed in Kashmir have large ranges extending into China and/or Siberia, where they have given rise to ecologically and morphologically similar, geographically separated "allospecies." For example, each of the three species that form the Japanese community has a close relative in Kashmir and is morphologically similar to that relative (Richman 1996). Recently it has been shown that *Phylloscopus* is not monophyletic, and some species of *Seicercus* should be included in the group (Olsson et al. 2004). This includes *Seicercus xanthoschistos,* which occurs in Kashmir, but at elevations lower than most *Phylloscopus* species, and it was not encountered at our study site. Because we have few ecological observations on this species, we do not include it here.

The Himalayas as well as the temperate regions to the north are a highly seasonal environment, and many species breeding at elevations higher than about 2,000 m, including all the *Phylloscopus* warblers, undertake altitudinal or longer-distance migrations, spending the winter in the Subtropics and Tropics of South and Southeast Asia (Ali and Ripley 1987).

Table 27.1 Morphological and ecological statistics for the eight *Phylloscopus* species breeding in Kashmir

	Morphology			Large prey (%)		Standpicks (%)		Elevation (m)
	PC1	PC2	PC3	Breeding	Winter	Breeding	Winter	Breeding
humei	−0.6	−0.3	0.7	3	5	56	45	3,350
trochiloides	1.0	−0.8	0.5	8	32	39	44	3,550
occipitalis	1.2	1.7	−0.1	8	—	55	—	2,600
pulcher	−0.5	−0.9	0.3	0	—	71	—	3,550
magnirostris	1.5	−0.1	0.2	18	—	36	56	2,500
chloronotus	−1.2	1.0	1.7	1	3	37	26	2,750
tytleri	−0.1	0.0	−1.2	0	7	82	94	2,750
affinis	−0.2	−1.4	−0.5	3	—	87	—	3,700

Note: Morphology data are from Price (1991) and give scores on the first three principal components extracted from the correlation matrix of log-transformed species means for six morphological traits. % Large prey is the fraction of times the prey item could be seen in the beak before swallowing. Breeding data are unpublished (from the study referenced in Price 1991). Winter data are from Gross and Price (2000). % Standpicks is the percentage of feeding observations in which flight was not involved in prey capture. Breeding data from Price (1991). Winter data are from Gross and Price (2000), except for *magnirostris,* which was provided by M. Katti based on 16 observations at site 10 of fig. 27.1. Elevation is the midpoint of elevational range (from Price 1991; *occipitalis* and *magnirostris* have ranges extending below the study site, and the lower boundary was taken from Ali and Ripley (1987).

METHODS

Warblers

We studied communities of *Phylloscopus* warblers in the winter at forested sites along the western half of peninsular India (fig. 27.1). Two species that breed in Kashmir were not observed: *pulcher* migrates to the east along the Himalayas and *affinis* apparently frequents more open habitats (Ali and Ripley 1987; Grimmett et al. 1999). Apart from the *Phylloscopus* species that breed at our study site in Kashmir, we also encountered *P. griseolus* at the sites in Maharashtra, but this species was never common and was little studied. We also encountered two taxa in the *trochiloides* species complex—*nitidus* (which breeds in the Caucasus) and *viridanus* (which breeds from Kashmir through Russia to Finland [Irwin et al. 2001b]). These taxa were often not distinguished and are combined here. They have different, but overlapping, winter ranges (Katti and Price 2003) and where they overlap, individuals mutually defend territories.

Morphological measurements of warblers are taken from Price (1991). We assume that geographical variation within species is small relative to interspecific variation. Studies on two species provide support for this assumption. Populations of *humei* across the whole winter range are very similar in both behavior and morphology (Gross and Price 2000). Comparisons of data in Katti and Price (1996) and Gross and Price (2000) show this to be true of *trochiloides* behavior as well. Although there is some geographical variation in *trochiloides* morphology, this is small compared with that observed between the species we are studying (Katti and Price 2003).

Winter Season Locations

We worked at five northern sites in Himachal Pradesh, with most of the data coming from Naina Devi (31° N, 76° E) and four mid-peninsula sites to the south of Mumbai (fig. 27.1). We also include data from Mundanthurai, a site near the southern tip of India (Katti and Price 1996). The five northern sites and two of the mid-peninsular sites were visited in January and February 1998 and are described in Gross and Price (2000). In January and February 2000, SG visited the two previously visited and two other mid-peninsula sites, Amboli (16°0′ N, 74°0′ E) and Molem (15°25′ N, 74°18′ E). The main purpose of visiting Molem was to

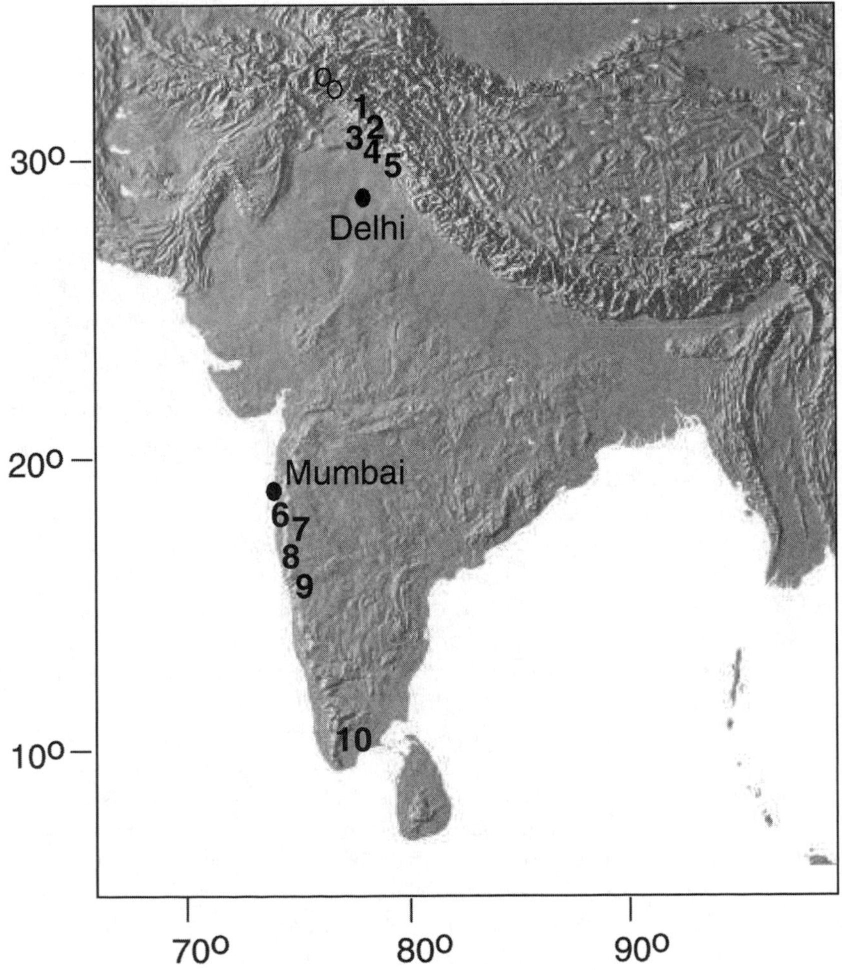

Fig. 27.1. Map of India showing approximate locations of study sites. Sites 1 (Manali, Himachal Pradesh); 2, 3 (Great Himalayan National Park, Himachal Pradesh); 4, 5 (Naina Devi, Himachal Pradesh); 6 (Karnala, Maharashtra); and 7 (Mahabaleshwar, Maharashtra) are described in Gross and Price (2000). Sites 8 (Amboli, Maharashtra) and 9 (Molem, Maharashtra) were studied in January and February 2000 (see Methods). Site 10 (Mundanthurai, Tamil Nadu) is described in Katti and Price (1996). Open circles show the breeding-season locations: the western site is in northern Pakistan (original data presented in this chapter) and the eastern site is in Kashmir (Price 1991).

study *occipitalis*, which was very rare at the locations where we worked in 1998.

Behavioral Observations

At each location we searched for individual birds and estimated the height at which they were foraging. We then recorded their next prey capture movement (whether it involved a flight, and whether the prey was large enough to be seen in the bill). These observations were collected by TP in both the breeding season (Price 1991) and winter (Gross and Price 2000). In a separate series of observations SG followed individuals for ten movements and recorded each movement as either a hop or a flight (discarding all observations that did not last ten movements). Data were collected in this way in the breeding season during a visit to the towns of Shogran and Naran (ca. 34°75′ N, 73°50′ E) in Pakistan in May and June 1999 and in the winter at sites 3–9 (fig. 27.1). Habitats studied in Pakistan were similar to those present in Kashmir, as reported in Price (1991), although willow (*Salix* spp.) was more common in Pakistan.

Range Sizes

Breeding and winter range sizes are taken from Price et al. (1997) and Katti and Price (2003), respectively. These were derived from ranges given in field guides, which were transcribed onto an equal-area map. There must be large inaccuracies in these ranges, especially in the winter, when the presence of a few stragglers could greatly augment the recorded range (see Stouffer 2001). Nevertheless variation in range size is large (more than two orders of magnitude), and systematic biases are unlikely to be introduced.

RESULTS

Historical Background

India collided with south Asia about 45–50 Ma (Harrison et al. 1992; Rowley 1998). It is thought that uplift at first proceeded slowly. By about 20 Ma there may have been elevations up to 4,000–5,000 m over a smaller land surface than at present, but sufficient to present a significant climatic barrier (Harrison et al. 1998). Elevations comparable to those of the present day were probably reached by 10 Ma (Harrison et al. 1998; Rowley et al. 2001). About 8 Ma there was a major climate shift with the onset of a modern-type monsoon (reviews in Molnar et al. 1993 and Copeland 1997). It has been suggested that this climate change may have been driven by a rapid elevation of Tibet at that time, on the order of 1,000 m (Molnar et al.1993), but given the apparent global nature of the climate shift, this is uncertain (Harrison et al. 1998). Although uplift and climate change must have affected winter habitat in peninsular India (e.g., via the monsoon), in the Himalayas uplift resulted in creation of habitats ranging from tropical to temperate. It seems likely that at least the speciation events that have been associated with altitudinally segregated breeding distributions were driven by the appearance of a diversity of breeding-season habitats.

Morphological and Ecological Diversification in the Breeding Season

In this section we review previous research on the timing and mode of adaptive divergence among the *Phylloscopus* warblers breeding in Kashmir (Richman and Price 1992). In fig. 27.2 we show an estimate of phylogenetic relationships among 27 taxa of *Phylloscopus*, using sequence data described in Price et al. (1997). Species separated across the root marking the radiation of the Himalayan species (point A in fig. 27.2) are about 23% divergent in sequence (after corrections for saturation effects). An approximate dating method (2% divergence / million years; reviewed by Klicka and Zink 1997) places the initiation of the radiation at 11–12 Ma. By this dating, the last speciation events among the warblers forming the Kashmir community were at 8 Ma (coincident with the period of major climate change and possibly rapid uplift). Since then speciation has generally produced ecologically similar allospecies in *Phylloscopus* both along the Himalayas and in Siberia and China (Price et al. 1997).

Some morphological and ecological relationships among the Kashmir warblers are summarized in table 27.1. We found that three orthogonal morphological axes (principal components) were each separately correlated with three ecological axes (Price 1991; Richman and Price 1992). PC1 (81% of the variance), a measure of body size (correlation of PC1 and body mass is $r = 0.96$), is associated with prey size. PC2 (12%), a shape measure relating tarsus length to beak size, is correlated with breeding-season altitude: species with relatively long tarsi tend to breed at high elevations. PC3 (6%), a measure of beak shape, is correlated with foraging method: species with relatively wide beaks tend to flycatch more than those with narrow beaks. The ecological associations of PC1 and PC3 are relatively well understood, by reference to other groups (e.g., flycatchers (*Ficedula* spp.) have wide beaks).

PC2 is associated with higher elevations in the breeding season (Price 1991). The reasons for this association are not clear. A large contributor to PC2 is tarsus length (Price 1991). Species with relatively longer tarsi tend to hop more and fly less during foraging according to a study of five species (table 27.2, fig. 27.3). The correlation between mean number of hops (from table 27.2) and PC2 (from table 27.1) is $r = -0.81$, $N = 5$, $p = 0.1$, in the breeding season and $r = -0.66$, $N = 5$, $p = 0.2$ in the winter. The reason why species breeding at tops of mountains hop more is not well understood but is considered further in the discussion.

Historical reconstruction suggests that initial adaptive diversification was primarily along PC1, the body size axis (Richman and Price 1992; Richman 1996). There are three well-supported clades in Kashmir, whose ancestors are indicated by the numbers 1, 2, and 3 in fig. 27.2. The first clade

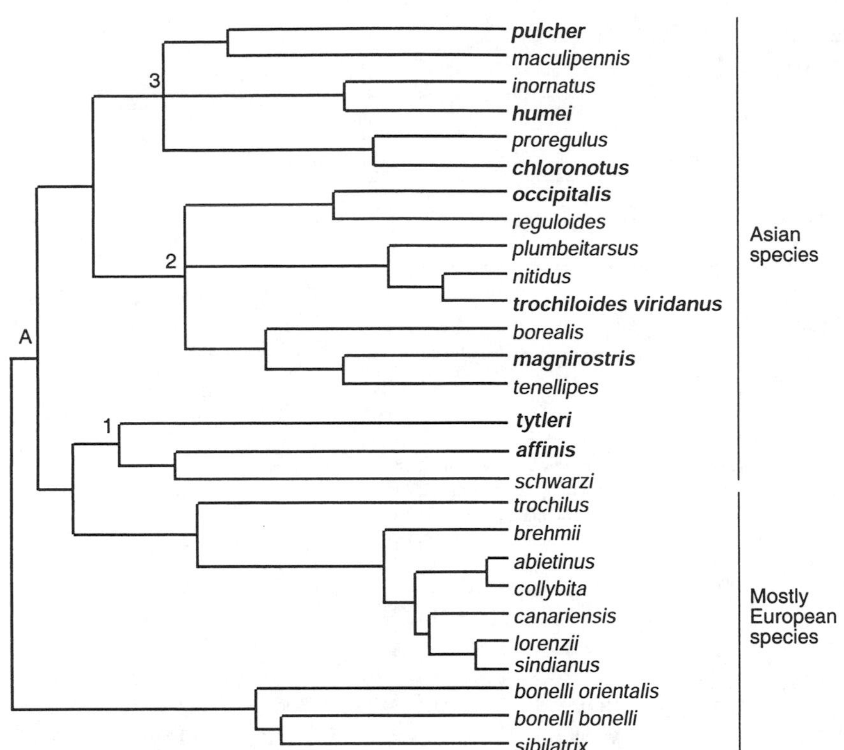

Fig. 27.2. Phylogenetic relationships for 27 taxa of *Phylloscopus* warblers. A maximum likelihood phylogeny based on the mtDNA cytochrome b gene was generated in PUZZLE (Strimmer and von Haeseler 1996), using sequence data referenced in Price et al. (1997). A molecular clock was enforced with the HKY model of substitution and gamma distributed rates. Scale bar indicates 10% sequence evolution along a branch. The species breeding in Kashmir are in bold. The node A marks the root of the clade that includes all the Kashmir species, and nodes 1, 2, and 3 mark the speciation events leading to three clades within which species are similar in body size (Richman and Price 1992). In general, taxa on branches of less than 4% sequence evolution are allospecies or subspecies. Note that divergence between species is double the amount of evolution along a branch, so that 5% sequence evolution is equivalent to 10% divergence and is dated at ca. 5 million years (reviewed by Klicka and Zink 1997).

consists of two intermediate-sized species (*affinis, tytleri*), the second clade consists of the three largest species (*magnirostris, occipitalis, trochiloides*), and the third clade consists of the three smallest species *(pulcher, humei, chloronotus)*. The ancestors to these three clades are thus inferred to have differed in body size (Richman and Price 1992). In addition, the first clade is distinct from the other two in that its members feed more than 80% of the time by hopping and picking ("standpicks") rather than catching prey in flight. Members of the other clades standpick less frequently (table 27.1).

Speciation events within the three clades occurred more or less simultaneously to produce species that differ in altitude. For example, among species in the small-bodied clade, *humei* breeds at high elevations in birch whereas the similar-sized *chloronotus* breeds at lower elevations in conifer. The inference is that, perhaps in association with Himalayan uplift or climate change, ancestral species of different body size gave rise to one or more altitudinal replacements. These altitudinal replacements showed relatively little evolution along the body size dimension; hence the orthogonal relationship of altitudinal distribution (associated with PC2) to body size (PC1).

BREEDING RANGE SIZE. Geographical range sizes in the breeding season vary by almost two orders of magnitude (fig. 27.4). Kashmir species with small ranges are confined to the Himalayas, whereas other species are present in both the Himalayas and farther north. Price et al. (1997) showed that the species whose ranges extend north tend to breed at higher elevations and/or occupy habitats in the Himalayas that are also found in the north. For example, only two species have spread west from Kashmir around the Tibetan plateau to the Altay region of Siberia, *humei* and *trochiloides*. These are relatively high-elevation species. This suggests that differences in breeding ranges, and by inference in the abundance and rarity of different species, are de-

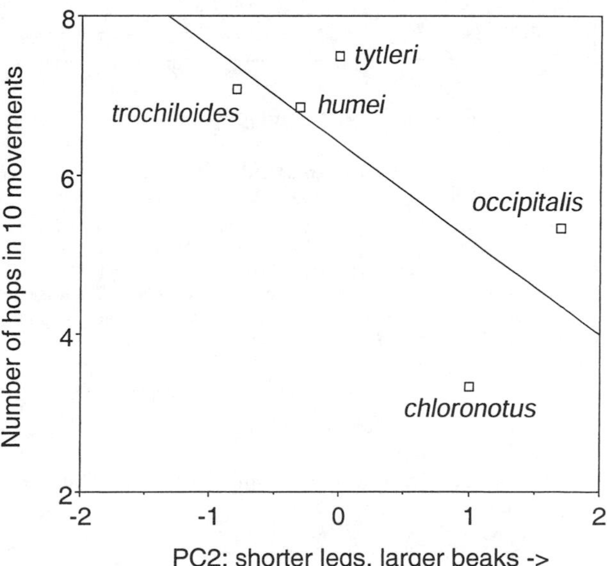

Fig. 27.3. Correlation between PC2, a morphological measure largely influenced by tarsus length, and the number of hops individuals took in observations of ten consecutive movements (breeding and winter season combined, $r = -0.71$). Standard errors are less than the width of the symbols in both dimensions.

Table 27.2 Movements made in ten sequential locomotory movements

	Number of hops		% Observations with a long flight	
	Breeding Mean ± s.e. (*N*)	Winter Mean ± s.e. (*N*)	Breeding	Winter
trochiloides	6.8 ± 0.03 (54)	7.3 ± 0.03 (69)	50	6
humei	7.2 ± 0.03 (47)	6.3 ± 0.07 (33)	26	9
tytleri	6.4 ± 0.22 (5)	7.6 ± 0.03 (69)	40	7
chloronotus	3.6 ± 0.05 (40)	3.1 ± 0.03 (58)	40	60
occipitalis	5.0 ± 0.04 (58)	5.5 ± 0.02 (112)	47	27

Note: Species are arranged in order of morphology based on a shape axis (PC2; see table 27.1 and fig. 27.3): *trochiloides* has a relatively long tarsus and small beak; *occipitalis* has a relatively short tarsus and large beak. % Observations with a long flight is the percentage of individuals that had at least one flight between a bush or tree as part of the ten movements. *N* is the number of individuals observed. The correlation between mean number of hops in the breeding and winter seasons is $r = 0.88$, $p < 0.05$. Breeding-season observations were made in northern Pakistan in 1999. Winter-season observations were made in India (across sites 3–9 of fig. 27.1) in 1998 and 2000.

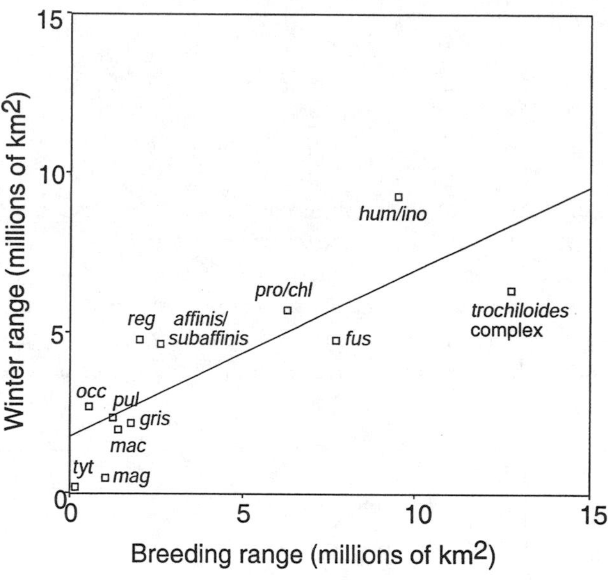

Fig. 27.4. Correlation between breeding and winter range area for all species that have at least part of their breeding or winter range in India ($N = 12$, regression equation: $Y = 1.7 + 0.5X$, $r = 0.81$, $p < 0.01$). Allospecies (closely related species that are allopatric to each other and ecologically very similar) in superspecies complexes are grouped together. For species abbreviations see fig. 27.2; except *fus* = *fuscatus; gris* = *griseolus*. Breeding ranges are from Price et al. (1997). Winter ranges are from Katti and Price (2003).

termined at least in part by availability of breeding-season habitats (Price et al. 1997).

Winter Ecology

We investigate how the winter ecology of a species is related to its breeding-season ecology, considering in turn each of the three morphological axes and range size.

BODY SIZE / PREY SIZE. Body size (PC1) is correlated with prey size in both the breeding and winter seasons (Price 1991; Gross and Price 2000) (see table 27.1). Diversification along this axis apparently proceeded with little conflict between the seasons.

In midwinter, south India has more and larger prey available than north India (Gross and Price 2000; Katti and Price 2003) and this difference is correlated with the range positions of the wintering *Phylloscopus:* larger species occur farther south (Katti and Price 2003). Southern wintering communities consist of two or three common large species (*trochiloides* with *magnirostris* and/or *occipitalis*). Northern wintering communities consist of one or two common small species (*chloronotus* and/or *humei*). Sites in central India consist of a small species (*humei*) and a large species (*trochiloides*) and in some places an intermediate species (*tytleri*). A parsimonious reconstruction of range position suggests that ancestrally a single small species wintered in northern India and a single large species wintered in the south. Both body size and winter range positions have been relatively conserved through later speciation events.

BEAK SHAPE / FEEDING METHOD. Some species tend to flycatch and hover (i.e., capture prey while in flight), whereas others tend to standpick (i.e., capture prey without flight). In the breeding season, standpicking is strongly correlated with beak shape (PC3) (Price 1991): species with long pointed beaks standpick more. Corresponding data for five species in the winter season are given in table 27.1. There is a very similar and strong correlation of percentage standpicks with the same beak shape measure in the winter ($r = -0.99$, $p < 0.01$, $N = 5$). Thus the association of morphology with feeding method is conserved across seasons, implying that prey exploitable by all feeding methods are available in both seasons and ecological adaptation to one season is adaptive, or at least accommodated, in the other.

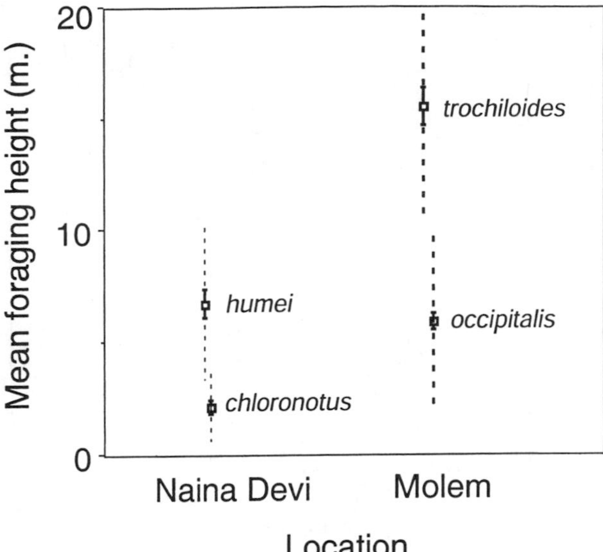

Fig. 27.5. Foraging heights (mean, standard error, standard deviation) at locations where two species from the same body size clade co-occur in winter (Naina Devi, site 5; and Molem, site 9; see fig. 27.1). Note that canopy height is higher at Molem than at Naina Devi: at both sites the tree crowns are fully occupied. Based on 30–90 observations per species.

TARSUS LENGTH / ELEVATION. We summarize results for morphological/ecological associations along the PC2 axis based on the four species we have been able to study in detail, which demonstrate remarkable patterns of parallel evolution. The two species occupying high elevations in the breeding season (*humei* and *trochiloides*) occupy tree crowns in the winter. The two species occupying low elevations in the breeding season (*chloronotus* and *occipitalis*) tend to occupy the subcanopy in the winter (fig. 27.5). *P. humei* is related to *chloronotus* and *trochiloides* to *occipitalis*. There has been parallel evolution of altitudinal distributions in the breeding season, therefore, accompanied by parallel evolution of foraging height in the winter season (fig. 27.6). These ecological differences have been accompanied by parallel evolution of hopping rather than flying in search of prey as well as the relative length of the tarsus (fig. 27.3). We suggest that microhabitat similarity between high-altitude breeding sites and tree canopies in winter sites favors the use of hopping, but we have no direct evidence. We consider possible influences of habitat on locomotion in more detail in the discussion.

In summary, species of different body sizes are relatively distantly related. At each altitude in the breeding season there is both a large and a small species, but the larger species winters to the south of the small species. Species of similar body size are closer relatives. They segregate by altitude in the breeding season and by foraging height in the winter. These relationships are summarized in fig. 27.6.

WINTER RANGE SIZE. Winter range sizes are correlated with breeding-season range sizes (fig. 27.4). Winter range position is a relatively conserved trait. However, neither breeding range position nor breeding range size are conserved, and they can differ substantially among close relatives (Price et al. 1997). Because range sizes are correlated between seasons, winter range size differs among relatives, even though approximate position does not (e.g., fig. 27.6).

Resource Limitation

In winter, correlations of food abundance and warbler abundance are very strong across both time and space (fig. 27.7). If the high correlations ($r \sim 1$) are taken literally, the associations across time in particular imply that the number of warblers returning from the winter to the breeding grounds is set by winter conditions and is largely indepen-

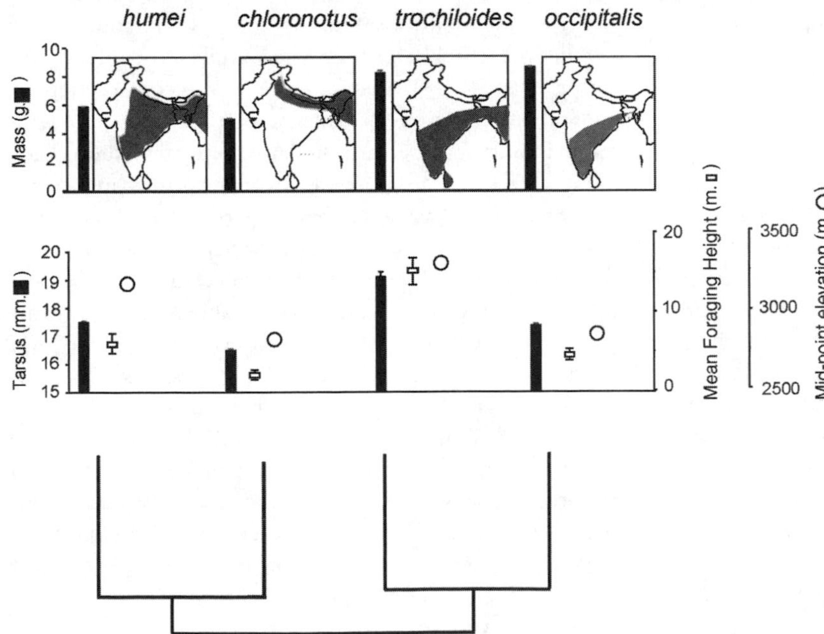

Fig. 27.6. Conserved traits (top row) and parallel evolution (second row) among the four common species of *Phylloscopus* warblers considered in this chapter. Relationships and branch lengths among the four relevant species are redrawn from fig. 27.2. Bootstrap support for the depicted relationship among these four species is 98%. Morphology (± s.e.) and median elevations in the breeding season are from Price (1991). Foraging height in the winter is from this chapter (fig. 27.5). Winter season range maps are from Katti and Price (2003) and are very approximate. Note that the larger species, especially *trochiloides*, are found in low densities farther north than the limit drawn here (e.g., *trochiloides* occurs rarely in Delhi in midwinter), and *occipitalis* is apparently rare in the middle of southern India (Grimmett et al. 1999).

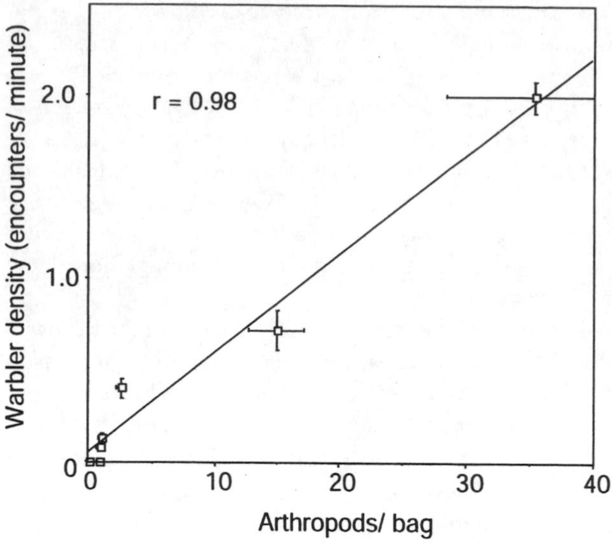

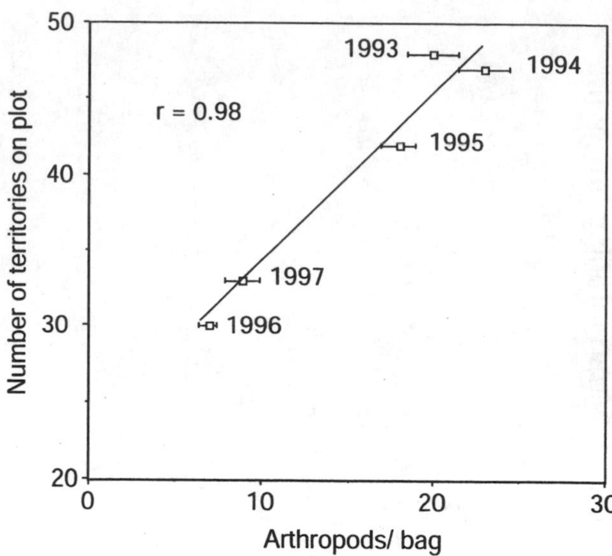

Fig. 27.7. Associations of bird density with arthropod density in winter. Arthropods were collected by placing garbage bags over branches and sorting the contents (Katti and Price 1996). Left: The *Y*-axis is total warbler density (including several non-*Phylloscopus* species not considered in this chapter) across sites 1–7 of fig. 27.1 in January/February 1998 (from Gross and Price 2000:table 1); there is a perfect ranking of latitude with both arthropod and bird abundance. Bird censuses are numbers heard on 1-h transects and standard errors are based on replicates across different routes on different mornings. At each site, 20–50 bags of arthropods were collected, and standard error is based on the mean per bag. Similar results are obtained if only *Phylloscopus* warblers are included and if analyses are restricted to different size ranges of arthropods (Gross and Price 2000). Right: Estimates of density of *Phylloscopus trochiloides* (number of territories on a 15-ha plot) and arthropods in January of 5 years at Mundanthurai (site 10 of fig. 27.1). Fifty bags were collected each year; standard error is based on the mean per bag. From Katti and Price (1996) with data for 1997 provided by M. Katti (pers. comm.).

dent of breeding productivity in the previous year. Information in Irwin (2000:fig. 5) for *trochiloides* shows the opposite relationship in the breeding season, at least across space. In this season, the amount of available food (caterpillars) is far higher in Siberia than in the Himalayas, even though densities of *trochiloides* are much lower in Siberia (Irwin 2000), and eventually the range limit is reached. Other *Phylloscopus* species also have low densities in Siberia (Forstmeier et al. 2001). Density may decrease toward the north in the breeding season because of the long distance from the wintering grounds, rather than through any direct influence of food.

DISCUSSION

The Himalayan region is species-rich. Price et al. (2003) used published information on altitudinal and geographical breeding distributions (e.g., Ali and Ripley 1987; Grimmett et al. 1999) to show that of the 976 land and freshwater bird species breeding in the Indian subcontinent, 154 (16%) are restricted to altitudes above 2,000 m in the Himalayas. This implies that uplift of the Himalayas stimulated production of many new species. Several speciation events within the genus *Phylloscopus* apparently correspond to a period of major climate change at about 8 Ma, which may itself be related to rapid Himalayan uplift (Molnar et al. 1993).

At this time, a parsimonious reconstruction implies that three ecologically different *Phylloscopus* species each gave rise to one or more altitudinal replacements in the breeding season, with relatively little alteration in body size. The parallel production of these altitudinal replacements, particularly within the clades of large and small body size, has led to parallel evolution of many other features, so that apart from body size, species at similar altitudes appear more similar to each other than they are to their closer relatives at another altitude (fig. 27.6, table 27.3). Features that have evolved in parallel include morphology, locomotory habit, and plumage patterns (table 27.3).

Parallel evolution is quite common when adaptive radiations are founded by similar ancestors in different places (Schluter and McPhail 1993). For example, Losos (1992) and Losos et al. (1998) used phylogenetic reconstruction to show that buildup of *Anolis* lizard communities across different islands occurred in a similar manner, with divergence along different ecological axes happening in a predictable sequence. The *Phylloscopus* example differs in three ways: first, the parallelism occurs across clades of substantially different body size, which are a priori expected to be subject to different ecologies; second, a remarkable number of morphological and behavioral traits have evolved in parallel; and third, the parallel evolution took place between pairs of species that all currently occur in the same locality.

We suggest that the parallel ecological differences between species were likely initially driven by adaptations to the breeding season. Nonetheless, these adaptations have to be accommodated in the winter season, when conditions seem to be quite different. Breeding in high-altitude habitats apparently selected for adaptations similar to those selected for by canopy foraging in winter.

Table 27.3 Parallel evolution within the small and large body-sized clades, based on the Kashmir breeding community

	humei (small) *trochiloides* (large)	*chloronotus* (small) *occipitalis* (large)
Breeding-season habitat	High altitudes (willow, birch, rhododendron[a])	Low altitudes (conifer)
Winter habitat	Tree canopy	Subcanopy, bushes
Morphology	Long legs, small beaks	Short legs, large beaks
Plumage pattern	No crown stripe	Crown stripe
Locomotion	Many hops	Fewer hops
Social behavior in winter	Solitary	Mixed foraging flocks

[a]*Humei* is replaced by the morphologically similar *pulcher* in rhododendron.

Perhaps the most ecologically relevant morphological difference between the top-of-mountain/top-of-tree ("top") species and the base-of-the-mountain/subcanopy ("base") species is in relative tarsus length. Top species have relatively long tarsi and hop more. Although we have few quantitative data, our impression is that in both seasons top species occur in habitats of moderate and relatively evenly distributed vegetative density (willow/birch/rhododendron during the breeding season [Price 1991] and broad-leaf tree crowns in winter [Katti and Price 1996; Gross and Price 2000]) and they travel through this matrix primarily by hopping. It appears to us that base species occur in habitats of more variable density. In the winter, the subcanopy consists of small trees and shrubs. Small trees may be less dense than canopy habitat, whereas shrubs are more dense. Further, the distance between these subcanopy microhabitats is more variable than the distance between the tree crowns. Base species navigate the subcanopy often by flying (table 27.2), particularly when shrubs and small trees are sparse. After landing on a low-density microhabitat (i.e., a small tree), birds often forage by flying from one twig to another, as perches are too widely spaced to hop. After landing on a high-density microhabitat (i.e., shrub), birds typically forage on the outside of the shrub, rather than traveling through its center. In effect, they are using the surface of the shrub (rather than the volume), which would be most effectively traversed by a hopping specialist. This also occurs in the breeding season, during which base species (particularly *chloronotus*) hover and flit about the surface of dense fir trees.

The top species—*humei* and *trochiloides*—are solitary throughout the winter and defend individual territories against conspecifics (Price 1981; Katti and Price 1996; Gross and Price 2000). Price (1981) suggested that they were able to maintain territories because they are often the only small insectivorous species in the habitats in which they are found (semideciduous tropical forest). By contrast, the base species encounter several other insectivorous species in their habitat and form the nucleus of mixed foraging flocks. We might expect foraging birds holding territories to travel shorter distances between perches than birds that are associated with flocks, and indeed flocking species do fly between trees more than territorial species (table 27.2). Thus social behavior and vegetative structure both appear to favor relatively more flying among the base species.

We have argued that the changes in feeding behavior associated with high-altitude habitats and tree canopies were initially driven by adaptation to the breeding season, primarily because of the inferred association of Himalayan uplift with the creation of novel habitats. For many other traits the relative roles of the breeding season, the winter, and their interaction are unclear. For example, the base species—*chloronotus* and *occipitalis*—have bright crown stripes whereas the top species—*humei* and *trochiloides*—do not. Marchetti (1993) inferred that these stripes were a result of selection to be visible in their dark coniferous breeding habitat. However, the base species are the ones that form flocks in the winter and it is possible that the stripes also have a role in communication among flock members (see Brooke 1998 and Beauchamp and Heeb 2001).

The relative roles of the two seasons are particularly unclear for range size. Range sizes in the breeding season and winter are correlated. Price et al. (1997) argued that in some cases range size is primarily driven by breeding-season conditions. Thus some species and superspecies occupying higher-altitude habitats have breeding-season ranges extending to higher latitudes: this may ultimately be attributed to climatic similarities between high latitudes and high altitudes. The suggestion that range size is determined by breeding-season conditions seems to conflict with the finding that population sizes are strongly limited by winter food (fig. 27.6). The paradox can be resolved if the different species compete so strongly for food in the winter season that limitation is taking place at the level of a group of species. If this is the case, increased breeding productivity in one species results in more arriving at the beginning of the winter and lower survival of individuals of another species.

Studies of patterns of local recruitment within species in North America show that breeding productivity does influence population size in the following year. Sherry and Holmes (1992) and Sillett et al. (2000), studying the American Redstart (*Setophaga ruticilla*) and Black-throated Blue Warbler (*Dendroica caerulescens*), respectively, found evidence for breeding-season productivity on a subsequent year's recruitment, despite apparent food limitation in the winter in Jamaica (Johnson and Sherry 2001; Sillett et al. 2000). These results could be explained if high breeding productivity in one location coupled with strong intraspecific competition in the winter results in lower returns to other populations. In fact Sillett et al. (2000) found breeding returns to be positively correlated across localities, and they suggested that winter limitation was less strong elsewhere in the winter quarters. An alternative is that competition in the winter extends across species, and increased breeding productivity in one species affects population sizes of another.

In *Phylloscopus,* the influence of breeding productivity on population size in the following year has not been studied. It is possible that there is only a weak correlation, so that food abundance in the winter is the prime determinant of population size of any given species and consequently the breeding-season range sizes. In this case, in winter different species have the competitive edge on resources with different abundances and distributions, resulting in variable ranges. It may be that in winter the tree canopy is a more abundant habitat than the subcanopy, leading to higher abundances of the "top" species, and this drives the large breeding-season ranges of these species. We are unable to separate these breeding and winter effects, which almost certainly interact. For example, range expansions in the breeding season may lead to high population sizes, resulting in adaptation to a broader niche in the winter.

Newton (1995) noted a pattern of relatively small winter ranges and large breeding ranges among Palearctic-African migrants. He attributed the relatively large size of breeding-season ranges to intraspecific competition for resources in that season. Our studies lead us to suggest that the small ranges in winter may be best attributed to intense interspecific competition for food, so that small environmental differences can tilt the advantage to one or another species. Relatively large ranges in the breeding season may more reflect a much smaller fitness penalty for increased dispersal, given the apparent availability of food and relatively low density of birds at high latitudes.

COMPARISONS WITH THE NEW WORLD WARBLERS. In the New World, high correlations of arthropod availability and warbler abundance demonstrate winter food limitation (Johnson and Sherry 2001; Sherry et al., Chap. 31, this volume), but there are few comparative studies of winter ecology among closely related species. Greenberg (1984) argued that in Panama in winter foraging differences among *Dendroica* species of different body sizes were initially driven by adaptation to coniferous or broadleaf habitat in the breeding season. In the breeding season larger species are found in conifer, where they often glean from the tops of branches; in the winter in tropical forest these species forage from leaf tops. Diversification in foraging and diet among species of *Dendroica* is quite different from that in the *Phylloscopus* (Price et al. 2000), but both may have been initially driven by adaptations to breeding-season habitats.

In a second comparison Greenberg at al. (2001) studied three species of warbler within the *Dendroica virens* superspecies complex that winter in the same locality in Mexico. These three species are geographically separated in the breeding season in North America and ecologically similar to each other. They are much younger than *Phylloscopus* (Price et al. 2000) and speciation has probably been driven by cycles of breeding-range expansion followed by geographical isolation in the Pleistocene (Mengel 1964; Greenberg et al. 2001). The two most closely related species segregate by habitat in winter. In the breeding season one species is dominant over the other, and Greenberg et al. (2001) suggest that this competitive dominance drives habitat segregation in the winter. The most comparable situation to these species in our study is the co-occurrence of two members of the *Phylloscopus trochiloides* superspecies. Unlike the *Dendroica* species studied by Greenberg et al. (2001), which forage in flocks in winter, individuals in these taxa hold individual territories throughout the winter, which they defend against other individuals in both taxa (Katti and Price 1996). The larger taxon (*nitidus*) generally occurs farther south, in what are inferred to be the more productive areas (Katti and Price 2003), and restriction of the smaller taxon (*viridanus*) farther north may also reflect competitive dominance.

Our results imply that conditions in both the breeding season and winter can affect the population size of a species, through adaptation to resources in one or the other season, as well as current availability of habitat. To the extent that different species exploit different resources, availability of those resources in winter can strongly affect relative population sizes. Production of new species as a result of changes in the breeding season has been accompanied by parallel partitioning of resources in both seasons, with a major role attributed to ecological interactions in the winter season.

ACKNOWLEDGMENTS

We particularly thank Madhusudan Katti for laying the foundations for this study with his original work at Mundanthurai tiger reserve (site 10) and for providing unpublished data from that site. We also thank Adam Richman, who pioneered the study of ecological differences by using phylogenies and applied that to the breeding-season warbler communities discussed here. We thank Peter Marra for persuading us to take part in the symposium, and Russell Greenberg and anonymous reviewers for comments on the manuscript. For field assistance we thank Karen Marchetti and Heather Tinsman. P. Alström provided a useful unpublished manuscript. The new field research reported here was supported by the National Geographic Society, the National Science Foundation, the Chapman fund of the American Museum of Natural History, and the David Marc Belkin fund for undergraduates at the University of California, San Diego.

LITERATURE CITED

Ali, S., and S. D. Ripley. 1987. Compact Handbook of the Birds of India and Pakistan Together with Those of Bangladesh, Nepal, Bhutan and Sri Lanka. Oxford University Press, Delhi and New York.

Beauchamp, G., and P. Heeb. 2001. Social foraging and the evolution of white plumage. Evolutionary Ecology Research 3:703–720.

Brooke, M. D. 1998. Ecological factors influencing the occurrence of "flash marks" in wading birds. Functional Ecology 12:339–346.

Copeland, P. 1997. The when and where of the growth of the Himalaya and the Tibetan plateau. Pages 19–40 *in* Tectonic Uplift and Climate Change (W. F. Ruddiman, ed.). Plenum Press, New York and London.

Forstmeier, W., O. V. Bourski, and B. Leisler. 2001. Habitat choice in *Phylloscopus* warblers: the role of morphology, phylogeny and competition. Oecologia 128:566–576.

Greenberg, R. 1984. The winter exploitation systems of bay-breasted and chestnut-sided warblers in Panama. University of California Publications in Zoology 116. University of California Press, Berkeley.

Greenberg R., C. E. Gonzales, P. Bichier, and R. Reitsma. 2001. Non-breeding habitat selection and foraging behavior of the Black-throated Green Warbler complex in southeastern Mexico. Condor 103:31–37.

Grimmett, R., C. Inskipp, and T. Inskipp. 1999. A guide to the birds of India, Pakistan, Nepal, Bangladesh, Bhutan, Sri Lanka, and the Maldives. Princeton University Press, Princeton.

Gross, S. J., and T. D. Price. 2000. Determinants of the northern and southern range limits of a warbler. Journal of Biogeography 27:869–878.

Harrison, T. M., P. Copeland, W. S. Kidd, and A. Yin. 1992. Raising Tibet. Science 255:1663–1670.

Harrison, T. M., A. Yin and F. J. Ryerson. 1998. Orographic evolution of the Himalaya and Tibetan plateau. Pages 39–72 *in* Tectonic Boundary Conditions for Climate Reconstructions (T. J. Crowley and K. C. Burke, eds.). Oxford University Press, New York and Oxford.

Irwin, D. E. 2000. Song variation in an avian ring species. Evolution 54:998–1010.

Irwin, D. E., P. Alstrom, U. Olsson, and Z. M. Benowitz-Fredericks. 2001a. Cryptic species in the genus *Phylloscopus* (Old World leaf warblers). Ibis 143:233–247.

Irwin, D. E., S. Bensch, and T. D. Price. 2001b. Speciation in a ring. Nature 409:333–337.

Johnson, M. D., and T. W. Sherry. 2001. Effects of food availability on the distribution of migratory warblers among habitats in Jamaica. Journal of Animal Ecology 70:546–560.

Katti, M., and T. Price. 1996. Effects of climate on Palaearctic warblers over-wintering in India. Journal of the Bombay Natural History Society 93:411–427.

Katti, M., and T. Price. 2003. Latitudinal trends in body size among over-wintering leaf warblers (genus *Phylloscopus*). Ecography 26:69–79.

Klicka, J., and R. M. Zink. 1997. The importance of recent ice ages in speciation: a failed paradigm. Science 277:1666–1669.

Losos, J. B. 1992. The evolution of convergent structure in Caribbean *Anolis* communities. Systematic Biology 41:403–420.

Losos, J. B., T. R. Jackman, A. Larson, K. de Queiroz, and L. Rodriguez-Schettino. 1998. Contingency and determinism in replicated adaptive radiations of island lizards. Science 279:2115–2118.

Marchetti, K. 1993. Dark habitats and bright birds illustrate the role of the environment in species divergence. Nature 362:149–152.

Martens, J., and S. Eck. 1995. Towards an Ornithology of the Himalayas: Systematics, Ecology and Vocalizations of Nepal Birds. Bonner Zoologische Monographien. Zoologisches Forschungsinstitut und Museum Alexander Koenig, Bonn.

Mengel, R. 1964. The probable history of species formation in some northern wood warblers (Parulidae). Living Bird 3:9–43.

Molnar, P., P. England, and J. Martinod. 1993. Mantle dynamics, uplift of the Tibetan Plateau, and the Indian monsoon. Reviews of Geophysics 31:357–396.

Newton, I. 1995. Relationship between breeding and wintering ranges in Palaearctic-African migrants. Ibis 137:241–249.

Olsson, U., P. Alström and P. Sundberg. 2004. Non-monophyly of the avian genus *Seicercus* (Aves: Sylviidae) revealed by mitochondrial DNA. Zoologica Scripta 33:501–510.

Orr, M. R., and T. B. Smith. 1998. Ecology and speciation. Trends in Ecology and Evolution 13:502–506.

Price, T. D. 1981. The ecology of the greenish warbler, *Phylloscopus trochiloides,* in its winter quarters. Ibis 123:131–144.

Price, T. D. 1991. Morphology and ecology of breeding warblers along an altitudinal gradient in Kashmir, India. Journal of Animal Ecology 60:643–664.

Price, T. D., A. J. Helbig, and A. D. Richman. 1997. Evolution of breeding distributions in the Old World leaf warblers (genus *Phylloscopus*). Evolution 51:552–561.

Price, T., I. J. Lovette, E. Bermingham, H. L. Gibbs, and A. D. Richman. 2000. The imprint of history on communities of North American and Asian warblers. American Naturalist 156:354–367.

Price T, J. Zee J, K. Jamdar K, N. Jamdar 2003: Bird species diversity along the Himalaya: a comparison of Himachal Pradesh with Kashmir. Journal of the Bombay Natural History Society 100: 394-410

Rahbek, C., and G. R. Graves. 2001. Multiscale assessment of patterns of avian species richness. Proceedings of the National Academy of Sciences USA 98:4534–4539.

Rice, W. R., and E. E. Hostert. 1993. Laboratory experiments on speciation: what have we learned in 40 years? Evolution 47:1637–1653.

Richman, A. D. 1996. Ecological diversification and community structure in the Old World leaf warblers (genus *Phylloscopus*): a phylogenetic perspective. Evolution 50:2461–2470.

Richman, A. D., and T. Price. 1992. Evolution of ecological differences in the Old World leaf warblers. Nature 355: 817–821.

Rowley, D. B. 1998. Minimum age of initiation of collision between India and Asia north of Everest based on the subsidence history of the Zhepure Mountain section. Journal of Geology 106:229–235.

Rowley, D. B., R. T. Pierrehumbert, and B. S. Currie. 2001. A new approach to stable isotope-based paleoaltimetry: implications for paleoaltimetry and paleohypsometry of the High Himalaya since the Late Miocene. Earth and Planetary Science Letters 188:253–268.

Schluter, D. 2000. The Ecology of Adaptive Radiation. Oxford Series in Ecology and Evolution. Oxford University Press, Oxford.

Schluter, D., and J. D. McPhail. 1993. Character displacement and replicate adaptive radiation. Trends in Ecology and Evolution 8:197–200.

Sherry, T. W., and R. T. Holmes. 1992. Population fluctuations in a long-distance Neotropical migrant: demographic evidence for the importance of breeding season events in the American redstart. Pages 431–442 *in* Ecology and Conservation of Neotropical Migrant Landbirds (J. M. Hagan and D. W. Johnston, eds.). Smithsonian Institution Press, Washington, D.C., and London.

Sillett, S. K., R. T. Holmes, and T. W. Sherry. 2000. Impacts of a global climate cycle on population dynamics of a migratory songbird. Science 288:2040–2042.

Simpson, G. G. 1953. The Major Features of Evolution. Columbia University Press, New York.

Simpson, G. G. 1964. Species density of North American recent mammals. Systematic Zoology 13:57–73.

Stouffer, P. C. 2001. Do we know what we think we know about winter ranges of migrants to South America? The case of the veery *(Catharus fuscescens)*. Auk 118:832–837.

Strimmer, K., and A. von Haeseler. 1996 Quartet puzzling: a quartet maximum-likelihood method for reconstructing tree topologies. Molecular Biology and Evolution 13:964–969.

Part 7 / POPULATION ECOLOGY

RICHARD T. HOLMES

Overview

THE NEED TO UNDERSTAND the processes and factors that determine the distribution and abundance of organisms has long been a central theme of ecology. This requires knowledge of population dynamics, including reproductive output, mortality, dispersal, and habitat use, and how these vary over time and space. Such information is also essential for predicting the impacts of both natural and human-mediated environmental changes on populations, and hence for the development of sound conservation and management plans.

Migratory birds provide a particular challenge for population ecologists for several reasons. First, they are distributed widely, often at global scales. Second, individuals that constitute a "local population" in one season of the year may be widely dispersed and co-occur with a mix of other individuals from other "populations" at other times of the annual cycle (see Part 4 of this volume and references therein), making the definition of a "population" difficult. Third, the importance and relative impact of limiting factors may differ at different seasons of the year, as may the strength and importance of density dependence. Thus, to develop a complete understanding of the population ecology of a migratory species requires a knowledge of the year-round population biology of that species. To date, this has not yet been achieved for any bird species, although as chapters in this volume attest, progress is being made in that direction.

The chapters in this section are diverse, examining very different aspects of the population ecology of migratory birds. In the first chapter, Michael Runge and Peter Marra examine the hypothesis that events in different parts of the annual cycle interact to influence overall population dynamics. They develop a population model for the annual cycle of a migratory species that includes seasonal interactions, such as carry-over effects, by incorporating the effects of behavioral dominance on habitat-specific rates of both survival and fecundity. They conclude that habitat availability and behavioral dominance during both breeding and nonbreeding phases of the annual cycle interact to regulate local sex ratios and population dynamics. They emphasize that understanding and quantifying density dependence in migratory species year-round is central to advancing our knowledge of population dynamics and our ability to conserve these species.

Tibor Szép and Anders Pape Møller present a novel approach for assessing the influence of environmental factors on the distribution and survival of migratory swallows (Barn Swallow [*Hirundo rustica*] and Bank Swallow [*Riparia riparia*]) during migratory and nonbreeding periods. Assuming that mortality occurs mainly away from the breeding grounds, they related survival between breeding seasons in Europe to satellite-determined data on the condition of rainfall-dependent vegetation across the African continent (the wintering area for populations of their study species). In so doing, they could examine their data for correlations between bird survival and environmental conditions in different wintering regions. Interestingly, one region of high importance for survival of *H. rustica* was identified in Algeria, which was also the site of all 13 spring recoveries of banded birds that had been previously marked in Denmark. Although considerable refinement and validation for this method remain to be done, it does appear promising as a way of identifying areas of potential importance for survival of migratory birds.

David Winkler takes a different approach and examines why birds disperse, as well as the relationship between dispersal and migration. He focuses most directly on natal dispersal—how to measure it, why it happens, what selective forces are operating, how it affects fitness, and how it might relate to migration. His study organism is another hirundine (Tree Swallow [*Tachycineta bicolor*]), for which he has obtained estimates of male and female natal dispersal in a 400-km radius around Ithaca, New York, based on an intensive mark-recapture program. The surprising result is that the mean distance dispersed by juveniles from their hatching locality to their first site of breeding, that is, natal dispersal, is only 2.5 km for males and 6.9 km for females. These are probably the best quantitative estimates of natal dispersal currently available for a migratory passerine. Winkler also discusses the possible relationship between migration and natal dispersal, as well possible mechanisms by which swallows may prospect for breeding sites in their first summer and other mechanisms of site selection.

Next, Thomas Sherry, Matthew Johnson, and Allan Strong review evidence concerning how and whether food is limiting in the winter period. They consider the relationship between food and habitat quality, bird dispersion patterns, body condition, timing of annual cycle events (e.g., migratory departures), and survival. Although finding general support for the winter food limitation hypothesis, they point out that few studies have tested directly whether food affects survival or local population size and that there have been few studies in which the effects of food availability on migrant birds have been manipulated and tested experimentally. Most evidence for food limitation, at least for passerines, relates to habitat-specific, climate-caused changes in insect abundance over the winter season that affect body condition and perhaps survival. They conclude by proposing a number of new approaches for further work to test for the importance of winter food limitation.

The last chapter of this section concerns limiting factors, population regulation, and seasonal interactions in migratory birds. Scott Sillett and Richard Holmes examine the breeding-season processes that limit and regulate local abundance of a migratory passerine species. They analyze long-term data on reproductive ecology of a parulid warbler (Black-throated Blue Warbler [*Dendroica caerulescens*]) and identify which abiotic and biotic processes limit population size during the breeding period and which function in a density-dependent manner. They then use a stochastic simulation model to assess whether the density dependence measured can regulate local abundance within the range observed locally during the14 years of their study. Their results indicate that the effects of predators, climate, and food availability on warbler fecundity were independent of population density and can best be considered as limiting factors. Annual survival rates of adults were also density independent. Fecundity of the Black-throated Blue Warbler population, however, was negatively correlated with population density. Furthermore, recruitment of new breeders into the population was related to fecundity and warbler density in the preceding summer. Modeling results showed that density-dependent fecundity alone was sufficient to regulate population size within observed levels. Abundance of this population can thus be explained in part by a complex interaction of abiotic and biotic processes operating during the breeding period.

Each of the chapters in this section makes major contributions to our understanding of different aspects of the population dynamics of migratory species. Nonetheless, the importance of winter mortality, dispersal, and the operation of limiting factors and regulatory mechanisms in both breeding and nonbreeding periods is still not understood for any one migratory species. Further research is needed to develop ways of integrating the effects of these different approaches and processes into single-species populations to assess what really drives the population dynamics of migratory birds.

MICHAEL C. RUNGE
AND PETER P. MARRA

Modeling Seasonal Interactions in the Population Dynamics of Migratory Birds

UNDERSTANDING THE POPULATION DYNAMICS of migratory birds requires understanding the relevant biological events that occur during breeding, migratory, and overwintering periods. The few available population models for passerine birds focus on breeding-season events, disregard or oversimplify events during nonbreeding periods, and ignore interactions that occur between periods of the annual cycle. Identifying and explicitly incorporating seasonal interactions into population models for migratory birds could provide important insights about when population limitation actually occurs in the annual cycle. We present a population model for the annual cycle of a migratory bird, based on the American Redstart (*Setophaga ruticilla*) but more generally applicable, that examines the importance of seasonal interactions by incorporating: (1) density dependence during the breeding and winter seasons, (2) a carry-over effect of winter habitat on breeding-season productivity, and (3) the effects of behavioral dominance on seasonal and habitat-specific demographic rates. First, we show that habitat availability on both the wintering and breeding grounds can strongly affect equilibrium population size and sex ratio. Second, sex ratio dynamics, as mediated by behavioral dominance, can affect all other aspects of population dynamics. Third, carry-over effects can be strong, especially when winter events are limiting. These results suggest that understanding the population dynamics of migratory birds may require more consideration of the seasonal interactions induced by carry-over effects and density dependence in multiple seasons. This model provides a framework in which to explore more fully these seasonal dynamics and a context for estimation of life history parameters.

INTRODUCTION

The factors that limit and the mechanisms that regulate the dynamics of bird populations are often poorly understood (Murdoch 1994; Sutherland 1996; Rodenhouse et al. 1997; Rodenhouse et al. 2003; Sillett and Holmes, Chap. 32, this volume), especially for migratory birds. Identifying the factors driving the population dynamics of migratory birds requires understanding the relevant biological events that occur during breeding, migration and overwintering periods. Available population models for migratory passerine birds tend to focus on breeding-season events, oversimplify events during nonbreeding periods, and ignore interactions that occur between periods of the annual cycle. Models that identify relevant events in the nonbreeding season and explicitly incorporate how different periods of the annual cycle interact could provide a more accurate picture of population dynamics than models that omit these dynamics. In addition, such models may provide important insights about when in the annual cycle limitation occurs.

Despite considerable effort to study the factors that limit migratory bird populations, our understanding of when and how these populations are limited remains poor (Marra and Holmes 2001; Sillett and Holmes 2002; Rodenhouse et al. 2003). Until recently, the prevailing view was that migratory bird populations were limited primarily by events on their wintering grounds (e.g., Fretwell 1972; Alerstam and Högstedt 1982; Robbins et al. 1989; Baillie and Peach 1992; Rappole and MacDonald 1994). Support for this hypothesis comes mainly from evidence such as population declines associated with weather extremes in winter and declines correlated with winter-habitat loss. Alternative hypotheses include: (1) summer limitation, perhaps due to high nest predation and parasitism at high densities (e.g., Holmes et al. 1986; Sherry and Holmes 1992; Böhning-Gaese et al. 1993), with evidence seen in correlations between local reproductive success and changes in local breeding populations; and (2) simultaneous summer and winter limitation (Sherry and Holmes 1995).

Results of recent research suggest that while the summer-limitation, winter-limitation, and summer-and-winter limitation hypotheses have some empirical support, they are likely overly simplistic (Marra et al. 1998; Marra and Holmes 2001; Sillett and Holmes 2002; Webster and Marra, Chap. 16, this volume). Periods of the annual cycle appear to be linked inextricably, such that ecological circumstances within one season subsequently influence reproductive success and/or survival in a subsequent season, effects we term "seasonal interactions" (fig. 28.1). We propose two general mechanisms by which there might be ecological interactions between the seasons of the annual cycle: at the *individual* or *population* levels.

The essence of seasonal interactions at the individual level is that individuals carry over effects, such as poor physical condition or late arrival, from one season to the next, and that these residual effects explain ecologically relevant variation in demographic rates in a later season. In contrast, seasonal interactions at the population level occur when the size of the population "carries over" the seasonal effect and is driven by density-dependent processes in each season. For example, an increase in population size leaving the winter grounds leads to higher densities, hence lower reproductive success, in the following breeding season. Theoretically, carry-over effects can influence density-dependent effects, and both individual- and population-level seasonal interactions simultaneously influence population dynamics.

Evidence is mounting for the importance of seasonal interactions at the individual level via carry-over effects. For example, poor physical condition during one season may ex-

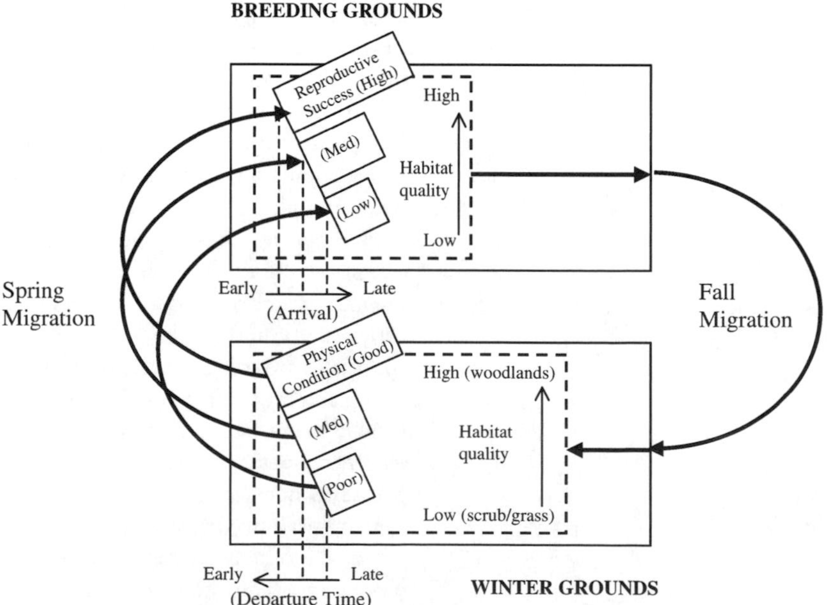

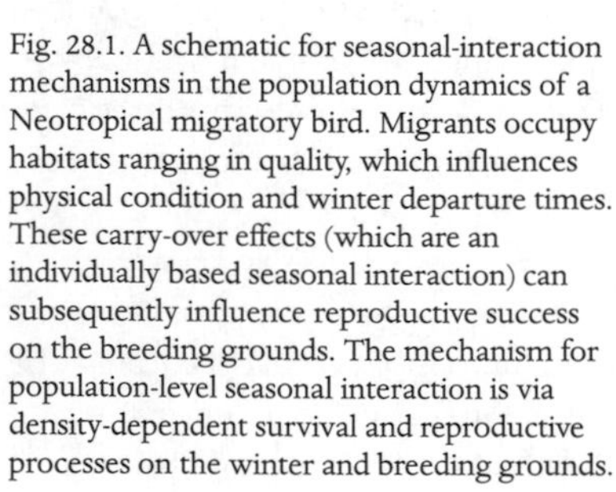

Fig. 28.1. A schematic for seasonal-interaction mechanisms in the population dynamics of a Neotropical migratory bird. Migrants occupy habitats ranging in quality, which influences physical condition and winter departure times. These carry-over effects (which are an individually based seasonal interaction) can subsequently influence reproductive success on the breeding grounds. The mechanism for population-level seasonal interaction is via density-dependent survival and reproductive processes on the winter and breeding grounds.

plain variation in reproductive success or survival in a subsequent season. Research on waterfowl initially supported this idea when it was discovered that good environmental conditions on the wintering grounds correlated with higher recruitment the following summer (Heitmeyer and Fredrickson 1981; Kaminski and Gluesing 1987). More recently, Marra et al. (1998) have shown that winter habitat influences the timing of spring migration and the physical condition of American Redstarts (*Setophaga ruticilla*) at time of departure, which in turn influences arrival time and body condition on breeding grounds and ultimately reproductive success (Norris et al. 2004). By affecting the timing of arrival on breeding grounds, carry-over effects could influence access to high-quality territories and the number of possible breeding attempts. Changes in the timing of and physical condition upon arrival are both mechanisms by which effects from the wintering grounds and/or migration period can persist into the breeding season.

Little evidence supports the importance of seasonal interactions at the population level, but this is likely a demonstration of the difficulty of obtaining such data, rather than testimony against such interactions. Direct evidence that seasonal interactions affect population processes would be provided, for instance, by observing a smaller decline in breeding density than expected from a known loss of winter habitat, which would suggest that density-dependent processes in a later season compensated for the loss of habitat. Such evidence is extremely difficult to obtain, because the corresponding winter and breeding ranges and demography are very rarely known and even less often successfully monitored. The oft-debated evidence for compensatory harvest mortality in ducks (Anderson and Burnham 1976) and the oystercatcher work of Goss-Custard et al. (1995c, 1995d) are both cases where such interactions are suggested. There is also some tantalizing evidence for another mode of seasonal interaction, namely, a single external process acting on more than one season: Sillett et al. (2000) found that both breeding- and winter-season dynamics of Black-throated Blue Warblers (*Dendroica caerulescens*) were affected by climatic variation associated with the El Niño Southern Oscillation.

Few attempts have been made to develop a year-round population model for a migratory bird that explicitly incorporates winter and breeding-season events. Sutherland and Dolman (1994) present a population model for a migratory bird that demonstrates how equilibrium population size depends on how individuals interact and compete year-round. Such interactions result in density-dependent interference competition, resource depletion, and, ultimately, mortality. Further work by Sutherland (1996, 1998) has generally concluded that equilibrium population size for a migratory bird species is determined by the relative strengths of density dependence operating during both the breeding and nonbreeding seasons. These models incorporate population-level seasonal interactions but do not consider individual-level carry-over effects or distinguish between the sexes.

To understand better the significance of individual- and population-level seasonal interactions, we developed an empirical model for a migratory songbird that experiences density dependence on the breeding and winter grounds, shows behavioral dominance, and can carry individual effects over from winter to summer. The development of this model relies heavily on insights from our studies of the American Redstart. On their breeding grounds, redstarts show strong age-specific habitat segregation (Ficken and Ficken 1967; Sherry and Holmes 1997) driven by dominance behavior of older males (Sherry and Holmes 1989). In addition, redstarts exhibit strong territorial behavior (Holmes et al. 1989; Ornat and Greenberg 1990; Marra et al. 1993) and sexual habitat segregation on their wintering grounds (Ornat and Greenberg 1990; Sliwa 1991; Marra and Holberton 1998), also the result of behavioral dominance by older males (Marra 2000). Winter territoriality probably functions primarily to secure a dependable source of food over the winter period (Price 1981; Greenberg 1986), and secondarily to provide safe haven from predators and inclement weather. Regardless of cause, redstarts relegated to poor winter habitat (largely females) lose mass over winter, depart later on spring migration, and have lower annual survival (Marra et al. 1998; Marra and Holmes 2001). Thus, redstarts present a convenient case study with which to investigate broader issues of seasonal interactions; indeed, their dynamics motivated our thoughts on the subject.

In this chapter, we develop a population model that is motivated by the dynamics of American Redstarts but is applicable to many species with similar patterns of sexual habitat segregation on the wintering grounds (Marra and Holmes 2001). Through a series of simulations with this model, we investigate: (1) how the amounts of breeding and nonbreeding habitat interact to determine equilibrium population size; (2) the extent to which sexual habitat segregation in winter influences equilibrium population size and sex ratio; and (3) the importance of carry-over effects from the nonbreeding to the breeding season. The focus of this modeling work is not to make specific predictions about a particular population or species, but rather, to understand patterns of population dynamics driven by seasonal interactions.

METHODS

Model Description

We developed a matrix population model to describe a migratory bird species that experiences habitat limitation and segregation on both the breeding and nonbreeding grounds, with the potential for carry-over effects between seasons (fig. 28.2). Upon arrival at the wintering grounds, birds compete for territories in "good" habitat, which is limited and has a carrying capacity of K_{wg}; those that lose this competition must occupy territories in "poor" habitat, which is unlimited (if not actually unlimited, then practi-

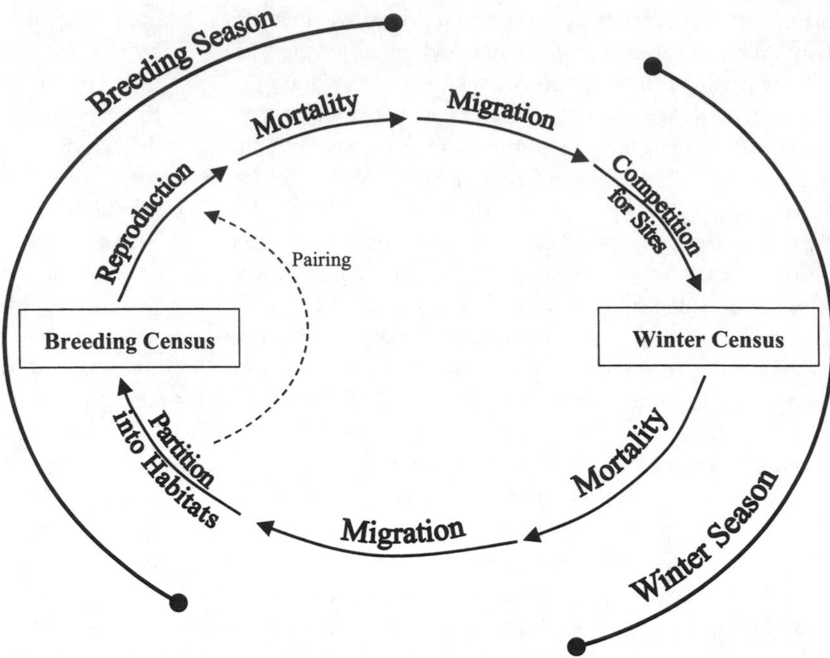

Fig. 28.2. Diagram of the stages in the annual cycle of a migratory songbird, as captured by the population model described in this chapter. The winter season encompasses two processes: competition for high-quality habitat upon arrival on the wintering grounds and winter mortality. The breeding season encompasses three processes: partitioning of breeding habitat by arrival time, reproduction, and summer mortality. The winter and breeding seasons are linked by migratory periods, which have mortality associated with them. See "Model Description" in the text for more details about this annual cycle.

cally so). After all the birds have arrived and settled onto territories in year *t*, the population can be described as a vector

$$\mathbf{W}(t) = \begin{bmatrix} W_{mg}(t) \\ W_{mp}(t) \\ W_{fg}(t) \\ W_{fp}(t) \end{bmatrix} \quad (1)$$

where W_{mg} is the number of males in "good" habitat, W_{fp} is the number of females in "poor" habitat, and so on. (In the description that follows, we drop the year-specific notation for simplicity of expression, thus referring to $\mathbf{W}(t)$ as $\mathbf{W}$, and similarly for other quantities. We bring back the year-specific notation at the very end.) These birds experience mortality over the winter season that is both sex- and habitat-specific. At the end of the winter season, the population structure is

$$\mathbf{W}' = \begin{bmatrix} s_{wmg} & 0 & 0 & 0 \\ 0 & s_{wmp} & 0 & 0 \\ 0 & 0 & s_{wfg} & 0 \\ 0 & 0 & 0 & s_{wfp} \end{bmatrix} \cdot \mathbf{W} \quad (2)$$

where, for example, s_{wmg} is the winter survival rate for males in good habitat, and the prime-notation ($\mathbf{W}'$) is used to indicate the population structure post-winter. At the end of the winter season, birds migrate north to breeding grounds. Mortality during migration depends upon the sex of the bird and upon the winter habitat from which it leaves. Birds that spend the winter in poor habitat leave the wintering grounds later and in poorer condition (fig. 28.1). After spring migration, the structure of the population is

$$\mathbf{W}'' = \begin{bmatrix} s_{smg} & 0 & 0 & 0 \\ 0 & s_{smp} & 0 & 0 \\ 0 & 0 & s_{sfg} & 0 \\ 0 & 0 & 0 & s_{sfp} \end{bmatrix} \cdot \mathbf{W}' \quad (3)$$

where s_{smg} is the survival rate over spring migration for males that came from good habitat, and the double-prime-notation ($\mathbf{W}''$) is used to indicate the population structure after spring migration.

The breeding ground contains two types of habitat, both of which are limited: the "source" habitat has a carrying capacity of K_{bc} (breeding ground, source habitat) pairs; the "sink" habitat, which is of lesser quality, has a carrying capacity of K_{bk} (sink) pairs. Because birds that wintered in good habitat arrive on the breeding grounds earlier, they fill the source habitat first. The number of females in source habitat is given by

$$B_{fc} = \begin{cases} W''_{fg} + W''_{fp} & \text{if } W''_{fg} + W''_{fp} < K_{bc} \\ K_{bc} & \text{otherwise} \end{cases} \quad (4)$$

and the number of females in sink habitat is given by

$$B_{fk} = \begin{cases} 0 & \text{if } W''_{fg} + W''_{fp} < K_{bc} \\ W''_{fg} + W''_{fp} - K_{bc} & \text{if } K_{bc} \le W''_{fg} + W''_{fp} < K_{bc} + K_{bk}. \\ K_{bk} & \text{if } W''_{fg} + W''_{fp} \ge K_{bc} + K_{bk} \end{cases} \quad (5)$$

Note that density dependence is implicit in equation (5): females displaced from sink habitat are presumed to die. This is a "ceiling" form of density dependence—no effect is evident until the number of arriving females exceeds the combined carrying capacity of the source and sink habitats. The number of males in source habitat is described similarly,

$$B_{mc} = \begin{cases} W''_{mg} + W''_{mp} & \text{if } W''_{mg} + W''_{mp} < K_{bc} \\ K_{bc} & \text{otherwise} \end{cases} \tag{6}$$

but the number of males in sink habitat is limited by being able to find a mate, hence by B_{fk} rather than K_{bk}:

$$B_{mk} = \begin{cases} 0 & \text{if } W''_{mg} + W''_{mp} < K_{bc} \\ W''_{mg} + W''_{mp} - K_{bc} & \text{if } K_{bc} \le W''_{mg} + W''_{mp} < K_{bc} + B_{fk} \\ B_{fk} & \text{if } W''_{mg} + W''_{mp} \ge K_{bc} + B_{fk} \end{cases}. \tag{7}$$

In contrast to females, if the number of arriving males exceeds the carrying capacity or the number of females, the additional males do not die, but become "drain" males—non-territorial males that move around looking for a vacated territory or the potential for extra-pair copulation. Thus, the number of "drain" males is given by

$$B_{md} = \max(0, W''_{mg} + W''_{mp} - K_{bc} - B_{fk}). \tag{8}$$

Pairs formed from arriving males and females can be placed in eight classes, depending on whether the pair is in source or sink habitat, whether the male spent the previous winter in good or poor habitat, and whether the female spent the winter in good or poor habitat. Because birds coming from good winter habitat arrive on breeding grounds first, as many "good-good" pairs as possible form. Of all pairs in source habitat, the proportion composed of both a male and female from good habitat is

$$P_{cgg} = \begin{cases} 1 & \text{if } W''_{mg} > K_{bc} \text{ and } W''_{fg} > K_{bc} \\ \dfrac{\min(W''_{mg}, W''_{fg})}{\min(B_{mc}, B_{fc})} & \text{otherwise} \end{cases}. \tag{9}$$

The proportion composed of a male from good winter habitat and a female from poor winter habitat is

$$P_{cgp} = \begin{cases} \dfrac{\min(W''_{mg}, B_{fc}) - W''_{fg}}{\min(B_{mc}, B_{fc})} & \text{if } W''_{mg} > W''_{fg} \text{ and } W''_{fg} < K_{bc} \\ 0 & \text{otherwise} \end{cases}. \tag{10}$$

The proportion composed of a "poor" male and a "good" female is

$$P_{cpg} = \begin{cases} \dfrac{\min(W''_{fg}, B_{mc}) - W''_{mg}}{\min(B_{mc}, B_{fc})} & \text{if } W''_{mg} > W''_{fg} \text{ and } W''_{mg} < K_{bc} \\ 0 & \text{otherwise} \end{cases}. \tag{11}$$

Finally, the proportion composed of a male and female both from poor habitat can be found by subtraction,

$$P_{cpp} = 1 - P_{cgg} - P_{cgp} - P_{cpg}. \tag{12}$$

A similar logic is needed to determine the proportion of pairs, by class, in the sink habitat, but the number of cases is somewhat larger. Of all the pairs in sink habitat, the proportion composed of a male and female both from good winter habitat is

$$P_{kgg} = \begin{cases} 1 & \text{if } W''_{mg} \ge K_{bc} + K_{bk} \text{ and } W''_{fg} \ge K_{bc} + K_{bk} \\ 0 & \text{if } W''_{mg} \le K_{bc} \text{ or } W''_{fg} \le K_{bc} \\ \dfrac{\min(W''_{mg}, W''_{fg}) - K_{bc}}{\min(B_{mk}, B_{fk})} & \text{otherwise} \end{cases}. \tag{13}$$

The proportion composed of a "good" male and a "poor" female is

$$P_{kgp} = \begin{cases} 1 & \text{if } W''_{fg} < K_{bc} \text{ and } W''_{mg} \ge K_{bc} + K_{bk} \\ \dfrac{W''_{mg} - K_{bc}}{\min(B_{mk}, B_{fk})} & \text{if } W''_{fg} < K_{bc} \text{ and } K_{bc} < W''_{mg} < K_{bc} + K_{bk} \\ \dfrac{\min(W''_{mg}, B_{fk} + K_{bc}) - W''_{fg}}{\min(B_{mk}, B_{fk})} & \text{if } K_{bc} < W''_{fg} < K_{bc} + K_{bk} \text{ and } W''_{mg} > W''_{fg} \\ 0 & \text{otherwise} \end{cases}. \tag{14}$$

The proportion composed of a "poor" male and a "good" female is

$$P_{kpg} = \begin{cases} 1 & \text{if } W''_{mg} < K_{bc} \text{ and } W''_{fg} \ge K_{bc} + K_{bk} \\ \dfrac{W''_{fg} - K_{bc}}{\min(B_{mk}, B_{fk})} & \text{if } W''_{mg} < K_{bc} \text{ and } K_{bc} < W''_{fg} < K_{bc} + K_{bk} \\ \dfrac{\min(W''_{fg}, B_{mk} + K_{bc}) - W''_{mg}}{\min(B_{mk}, B_{fk})} & \text{if } K_{bc} < W''_{mg} < K_{bc} + K_{bk} \text{ and } W''_{fg} > W''_{mg} \\ 0 & \text{otherwise} \end{cases}. \tag{15}$$

Finally, the proportion composed of a male and female both from poor habitat is

$$P_{kpp} = \begin{cases} 1 - P_{kgg} - P_{kgp} - P_{kpg} & \text{if } B_{mk} > 0 \text{ and } B_{fk} > 0 \\ 0 & \text{otherwise} \end{cases}. \tag{16}$$

Thus, after the birds have settled into pairs, the population structure on the breeding ground (corresponding to the "breeding census" in fig. 28.2) is

$$\mathbf{B} = \begin{bmatrix} B_{mc} \\ B_{mk} \\ B_{md} \\ B_{fc} \\ B_{fk} \end{bmatrix}. \tag{17}$$

Fecundity rates for a pair depend on whether the pair is in source or sink habitat, and also on the composition of the pair with regard to habitat in the previous winter. This is the mechanism for a "carry-over effect"—individuals from good winter habitat arrive earlier and in better condition, and translate these advantages into increased production. The average fecundity for pairs in source and sink habitat is given by

$$\begin{bmatrix} R_{source} \\ R_{sink} \end{bmatrix} = \begin{bmatrix} P_{cgg} & P_{cgp} & P_{cpg} & P_{cpp} & 0 & 0 & 0 & 0 \\ 0 & 0 & 0 & 0 & P_{kgg} & P_{kgp} & P_{kpg} & P_{kpp} \end{bmatrix} \begin{bmatrix} R_{cgg} \\ R_{cgp} \\ R_{cpg} \\ R_{cpp} \\ R_{kgg} \\ R_{kgp} \\ R_{kpg} \\ R_{kpp} \end{bmatrix} \tag{18}$$

where the R_i values are the habitat- and class-specific fecundities. The number of young produced is given by

$$\mathbf{Y} = \begin{bmatrix} Y_{mc} \\ Y_{mk} \\ Y_{fc} \\ Y_{fk} \end{bmatrix} = \begin{bmatrix} 1-f & 0 \\ 0 & 1-f \\ f & 0 \\ 0 & f \end{bmatrix} \begin{bmatrix} \min(B_{mc}, B_{fc}) & 0 \\ 0 & \min(B_{mk}, B_{fk}) \end{bmatrix} \begin{bmatrix} R_{source} \\ R_{sink} \end{bmatrix} \tag{19}$$

where f is the fraction of young that are female, and Y_{mc}, for instance, is the number of male young produced from source habitat.

Adult birds experience both sex- and habitat-specific mortality over the breeding season. At the end of the breeding season, the population structure is

$$\mathbf{B}' = \begin{bmatrix} s_{bmc} & 0 & 0 & 0 & 0 \\ 0 & s_{bmk} & 0 & 0 & 0 \\ 0 & 0 & s_{bmd} & 0 & 0 \\ 0 & 0 & 0 & s_{bfc} & 0 \\ 0 & 0 & 0 & 0 & s_{bfk} \end{bmatrix} \cdot \mathbf{B} \tag{20}$$

where, for example, s_{bmc} is the breeding season survival rate for males in source habitat. Following breeding, birds migrate south to the wintering grounds. Mortality during migration depends upon the sex of the bird and the breeding habitat it used. This structure can be used to portray any pattern in habitat-specific effects; for example, sink adults could have lower survival rates than source adults (perhaps because their food resources were poorer) or they could have higher survival rates than source adults (perhaps because they chose to expend less energy on producing young); in the simulations in this chapter, we assumed the former (see "Parameter Values" below) but the model is flexible in this regard. After fall migration, the structure of the adult population is

$$\mathbf{B}'' = \begin{bmatrix} s_{fmc} & 0 & 0 & 0 & 0 \\ 0 & s_{fmk} & 0 & 0 & 0 \\ 0 & 0 & s_{fmd} & 0 & 0 \\ 0 & 0 & 0 & s_{ffc} & 0 \\ 0 & 0 & 0 & 0 & s_{ffk} \end{bmatrix} \cdot \mathbf{B}' \tag{21}$$

where s_{fmc} is the survival rate over fall migration for males that came from source habitat. The young also experience mortality during fall migration that is sex- and habitat-specific, so that after fall migration, the young that arrive on the wintering grounds are described by

$$\mathbf{Y}'' = \begin{bmatrix} s_{ymc} & 0 & 0 & 0 \\ 0 & s_{ymk} & 0 & 0 \\ 0 & 0 & s_{yfc} & 0 \\ 0 & 0 & 0 & s_{yfk} \end{bmatrix} \cdot \mathbf{Y} \tag{22}$$

where s_{ymc} is the survival rate over fall migration for young males from source habitat (note that there is no vector $\mathbf{Y}'$ needed in this model). The population that arrives on the wintering grounds is composed of both adult and young

$$\mathbf{A} = \begin{bmatrix} \mathbf{B}''_{5\times1} \\ \mathbf{Y}''_{4\times1} \end{bmatrix}. \tag{23}$$

These birds compete for territories in "good" winter habitat, which is limited to K_{wg} individuals (note that this carrying capacity is measured in individuals, not pairs). Their ability to compete depends upon an intrinsic age-, sex-, and condition (habitat)-specific competitive factor (γ) and the number of birds in each class. Note that there is also a carry-over effect implicit in this competition if there is a difference in γ's based on breeding habitat. The number of birds in each class that successfully compete for good habitat is given by

$$A_i^G = \frac{A_i \gamma_i}{\sum_i A_i \gamma_i} K_{wg} \tag{24}$$

where i indexes the nine classes found in equation (23), with provisions made so that

$$A_i^G \leq A_i \text{ for all } i. \tag{25}$$

The number of birds in each class that are relegated to poor habitat is found by subtraction,

$$A_i^P = A_i - A_i^G. \tag{26}$$

After the competition, the distinctions between young and adult, source and sink, are lost, so

$$\begin{aligned} W_{mg} &= A_{mc}^G + A_{mk}^G + A_{md}^G + A_{ymc}^G + A_{ymk}^G \\ W_{mp} &= A_{mc}^P + A_{mk}^P + A_{md}^P + A_{ymc}^P + A_{ymk}^P \\ W_{fg} &= A_{fc}^G + A_{fk}^G + A_{yfc}^G + A_{yfk}^G \\ W_{fp} &= A_{fc}^P + A_{fk}^P + A_{yfc}^P + A_{yfk}^P \end{aligned} \tag{27}$$

and finally,

$$\mathbf{W}(t+1) = \begin{bmatrix} W_{mg}(t+1) \\ W_{mp}(t+1) \\ W_{fg}(t+1) \\ W_{fp}(t+1) \end{bmatrix} = \begin{bmatrix} W_{mg} \\ W_{mp} \\ W_{fg} \\ W_{fp} \end{bmatrix} \quad (28)$$

where we now bring back the year-specific notation to indicate that a year has passed.

Parameter Values

As described above, this model requires 43 parameters: 22 survival rates, eight fecundity rates, nine competition parameters, the fraction of young that are female, and three carrying capacities. If our purpose were prediction or assessment for a particular species, careful attention to formal parameter estimation would be critical; but for the purposes of understanding the patterns in the dynamics of the model, point estimates for parameters are less important than the patterns among them. Thus, while the American Redstart provided a guidepost for the articulation of model parameters, the following section should not be viewed as a formal exercise in estimation.

We assumed that winter survival rates did not differ by sex or habitat (Marra and Holmes 2001), but that spring migration survival rates were lower for birds from poor habitat than good habitat (table 28.1). During the breeding season, we assumed that survival did not differ by sex, but did depend on habitat, with birds in sink habitat having lower survival rates than birds in source habitat, and drain males having the lowest survival rates (table 28.2). A similar pattern was used for fall migration survival rates, except we assumed that drain males were more similar to source males because drain males do not incur any costs of reproduction (table 28.2). We assumed that the survival rates of young during fall migration were the same as adults from the same habitat (table 28.2). Note that young are not kept separate in the model once they arrive and settle on wintering habitat.

We simplified the eight fecundity rates into three parameters: a base rate, a habitat effect, and a carry-over effect (table 28.3). Note that these fecundity rates are the number of young per pair that are alive at the end of the breeding season; thus, they incorporate all components of productivity, including fledgling survival. We set the base rate at 1.8 young per pair in source habitat, and assumed that fecundity was half that in sink habitat (Sherry and Holmes 1997). To simplify the carry-over effect of winter habitat on subsequent productivity, we assumed that for each member of a pair that spent the previous winter in poor habitat, the fecundity of the pair was reduced by a factor c (table 28.3). Thus, with a carry-over effect $c = 2$, a pair composed of a male from good habitat and a female from poor habitat would have a fecundity half that of a "good-good" pair, and a "poor-poor" pair would have a fecundity half that again. We assumed that the sex ratio of young was 1:1 (thus, $f = 0.5$).

We simplified the nine competitive factors into three effects: the competitive ability for good winter habitat of birds that bred in or fledged from source habitat relative to those from sink habitat; the relative competitive ability of young compared with adults; and the relative competitive ability of males compared with females (table 28.4). Birds from sink habitat were assumed to have one-tenth the competitive ability of birds from source habitat, owing possibly to a later arrival date. Likewise, young were assumed to have one-fifth the competitive ability of adults. To explore the effect of male-biased competition for good winter habitat, we defined a male dominance parameter, γ, that we allowed to vary from 1 (equal competitive ability between males and females) to 5 (males five times more competitive than females). Marra and Holmes (2001) found a male-to-female ratio of 6:4 in good winter habitat for American Redstarts. This is not

Table 28.1 Winter survival rates and spring migration survival rates used in the model, by sex and habitat type

	Winter survival		Spring migration survival	
Males on good habitat	s_{wmg}	0.80	s_{smg}	0.90
Males on poor habitat	s_{wmp}	0.80	s_{smp}	0.80
Females on good habitat	s_{wfg}	0.80	s_{sfg}	0.90
Females on poor habitat	s_{wfp}	0.80	s_{sfp}	0.80

Table 28.2 Breeding season survival rates and fall migration survival rates used in the model, by sex, age, and habitat type

	Breeding season survival		Fall migration (adults)		Fall migration (young)	
Males on source habitat	s_{bmc}	0.95	s_{fmc}	0.80	s_{ymc}	0.80
Males on sink habitat	s_{bmk}	0.85	s_{fmk}	0.75	s_{ymk}	0.75
Drain males	s_{bmd}	0.80	s_{fmd}	0.80	—	—
Females on good habitat	s_{bfc}	0.95	s_{ffc}	0.80	s_{yfc}	0.80
Females on poor habitat	s_{bfk}	0.85	s_{ffk}	0.75	s_{yfk}	0.75

Table 28.3 Fecundity rates by habitat and pair-class

Habitat	Pair-class	Parameter	Formula	$c = 1$	$c = 2$
Source	Good-Good	R_{cgg}	1.8	1.8	1.8
	Good-Poor	R_{cgp}	$1.8/c$	1.8	0.9
	Poor-Good	R_{cpg}	$1.8/c$	1.8	0.9
	Poor-Poor	R_{cpp}	$1.8/c^2$	1.8	0.45
Sink	Good-Good	R_{kgg}	0.9	0.9	0.9
	Good-Poor	R_{kgp}	$0.9/c$	0.9	0.45
	Poor-Good	R_{kpg}	$0.9/c$	0.9	0.45
	Poor-Poor	R_{kpp}	$0.9/c^2$	0.9	0.225

Note: In the simulations described in this chapter, for each member of a pair that spends the winter on poor habitat, the fecundity is reduced by a factor c. This factor represents the strength of the carry-over effect. The fecundities are shown for two values of the carry-over effect: no effect ($c = 1$) and a strong effect ($c = 2$).

a direct measure of competitive ability because the sex ratio depends on the overall sex ratio in the population and the differential survival rates between the sexes, as well as the relative competitive abilities, but it is nevertheless a rough indication of the level of competition. We chose a range approximately twice as large as this observation to capture a potential range for the male dominance parameter. Note that we did not investigate a range of values for the other two competitive effects (source vs. sink individuals, young vs. adults), but such an investigation is warranted.

In the simulations with the model, a wide range of values was explored for two of the three carrying capacities (K_{bc} and K_{wg}). The third carrying capacity (K_{bk}, the capacity of sink breeding habitat) was set at a level (10,000) much higher than the other carrying capacities, so that it was effectively unlimited.

Simulations

We performed three sets of simulations, all designed to look at properties of the model at equilibrium. To calculate equilibrium results, we iterated the model through time with a fixed set of parameters, until population vectors during the breeding and winter seasons stabilized. Typically, equilibrium was reached within 50 simulated annual cycles, but we always ran the model for 300 annual cycles to be certain.

The three sets of simulations were designed to examine, in turn, the effects of three dynamics: the relative amounts of breeding and winter carrying capacity; sexual habitat segregation; and winter-to-summer carry-over effects. In the first simulation, we varied carrying capacities of source breeding and good winter habitat and examined the resulting equilibrium population size on the breeding ground. For this simulation, we held the carry-over effect at $c = 1$, and the male dominance parameter at $\gamma = 5$. From the results of this simulation, we chose three combinations of carrying capacities, all of which produced an equilibrium breeding-ground population size of 500 birds: a summer-limited case ($K_{bc} = 205$ pairs, $K_{wg} = 900$ individuals); an intermediate case ($K_{bc} = 224$ pairs, $K_{wg} = 580$ individuals); and a winter-limited case ($K_{bc} = 800$ pairs, $K_{wg} = 485$ individuals). We used these three cases in the second and third simulations. In the second simulation, we varied the male dominance parameter, held the carry-over effect constant ($c = 1$), and looked at the resulting equilibrium sex ratio (male:female) during the breeding season. In the third simulation, we varied the

Table 28.4 Competitive factors by habitat, sex, and age

Sex and age	Habitat	Parameter	Formula	$\gamma = 1$	$\gamma = 5$
Adult male	Source	γ_1	1	1	1
	Sink	γ_2	0.1	0.1	0.1
	Drain	γ_3	0.01	0.01	0.01
Adult female	Source	γ_4	$1/\gamma$	1	0.2
	Sink	γ_5	$0.1/\gamma$	0.1	0.02
Young male	Source	γ_6	0.2	0.2	0.2
	Sink	γ_7	0.01	0.01	0.01
Young female	Source	γ_8	$0.2/\gamma$	0.2	0.04
	Sink	γ_9	0.01	0.01	0.01

Note: For the simulations in this chapter, these competitive factors are assumed to be governed by a male dominance parameter γ—the stronger this factor, the greater the competitive edge males have over females for good winter habitat. With $\gamma = 1$, there is no difference between the sexes in competition for good winter habitat; with $\gamma = 5$, the odds of a single male outcompeting a single female are 5:1.

strength of the carry-over effect, held the male dominance parameter constant ($\gamma = 5$), and looked at the resulting equilibrium breeding population size.

RESULTS

Equilibrium population size on breeding grounds ($\Sigma B = B_{mc} + B_{mk} + B_{md} + B_{fc} + B_{fk}$) varied as a function of the carrying capacities of source breeding habitat (K_{bc}, measured in pairs) and good wintering habitat (K_{wg}, measured in individuals) (fig. 28.3). This variation indicates different conditions for winter and breeding limitation (fig. 28.3). The equilibrium surface shows two pronounced areas: one (marked "W") where the population is winter limited, and one (marked "B") where the population is breeding limited, with a transition zone ("T") between them. In the area marked "W," equilibrium population size increases (or decreases) as the amount of good winter habitat (hence the carrying capacity K_{wg}) increases (or decreases), regardless of changes in the amount of source breeding habitat (hence carrying capacity K_{bc}). In this area, equilibrium population size does not change with a change in the amount of source habitat, provided the amount of good winter habitat is held constant. Thus, the population must be winter limited, because only changes in winter habitat can change the equilibrium population size. A similar argument shows why the population is breeding limited at the point marked "B." In either case, sufficient increase in suitable habitat in the limiting season eventually causes the limitation to switch to the other season. For instance, as the amount of source breeding habitat increases from point "B," while holding winter habitat constant, equilibrium population size increases, *up to a point.* Beyond that, the population becomes winter limited, and further increases in breeding habitat will not change the equilibrium population size. The results for equilibrium *winter* population size (not shown) are qualitatively similar to those for breeding population size (fig. 28.3).

The dynamics in the transition zone ("T" in fig. 28.3) between the winter- and breeding-limited regions are due to subtle interactions between the seasons. In this region, the population is largely breeding limited but is affected slightly by the amount of winter habitat. The seasonal interaction can be understood by considering the bold line in fig. 28.3, along which source breeding habitat increases while good winter habitat is held constant (at K_{wg} = 900 individuals). This slice through the three-dimensional surface in fig. 28.3 is shown in two dimensions in fig. 28.4, as the top line in the upper panel. The remaining lines in the upper panel divide the total population into source, sink, and drain individuals, and the lower panel shows the corresponding equilibrium *winter* population size, divided into individuals in good and poor habitats. Initially, as the capacity of source breeding habitat (hence carrying capacity K_{bc}) increases, the equilibrium summer (upper panel) and winter (lower panel) population sizes increase as well, because under these circumstances, the population is breeding-season limited. Left of reference line a, where the population is breeding limited, all individuals are able to occupy good winter habitat (see lower panel), and enough birds return to the breeding grounds that some must spill over into sink habitat (see upper panel). At a certain point (reference line a), there is enough source habitat that the good winter habitat fills (its carrying capacity is 900). With continued increases in source habitat, birds must compete for good winter habitat and some must spill over into poor winter habitat (lower panel, between lines a and b). Two things begin to happen on the breeding grounds (upper panel): first, because survival during spring migration is lower for birds from poor winter habitat, proportionally fewer birds return to the breeding

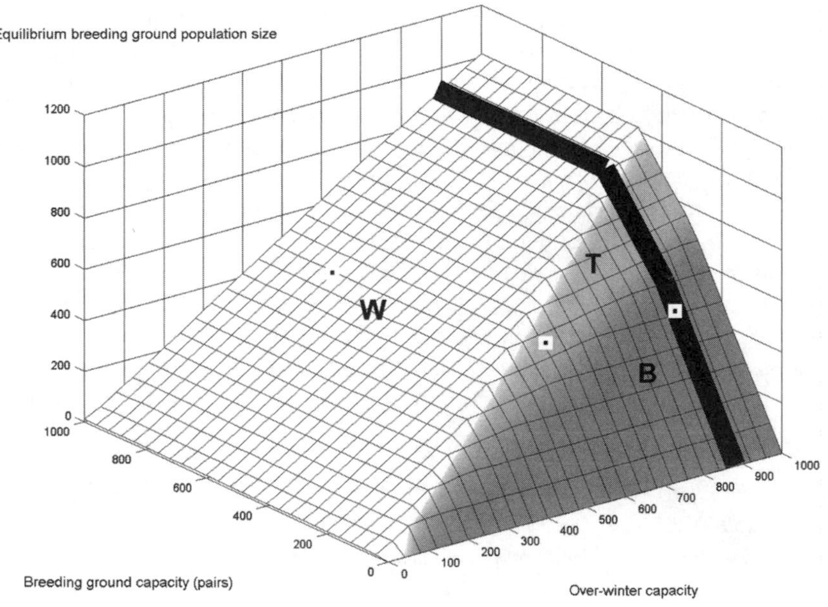

Fig. 28.3. Equilibrium population size on the breeding grounds (ΣB, equation [17]) as a function of the carrying capacities of source breeding habitat (K_{bc}, in pairs) and good winter habitat (K_{wg}, in individuals). "W" refers to the conditions under which the population is winter limited, "B" to conditions of breeding-season limitation, and "T" to a transition zone between the two. The bold line shows a slice through this curve when K_{wg} is held constant at 900; this is shown in more detail in fig. 28.4. The three squares represent pairs of carrying capacities for which the equilibrium breeding population size is 500, but which differ in being winter limited, summer limited, or intermediate; these are the three cases shown in figs. 28.5 and 28.6. The equilibrium population sizes in this figure were generated with $c = 1$ and $\gamma = 5$.

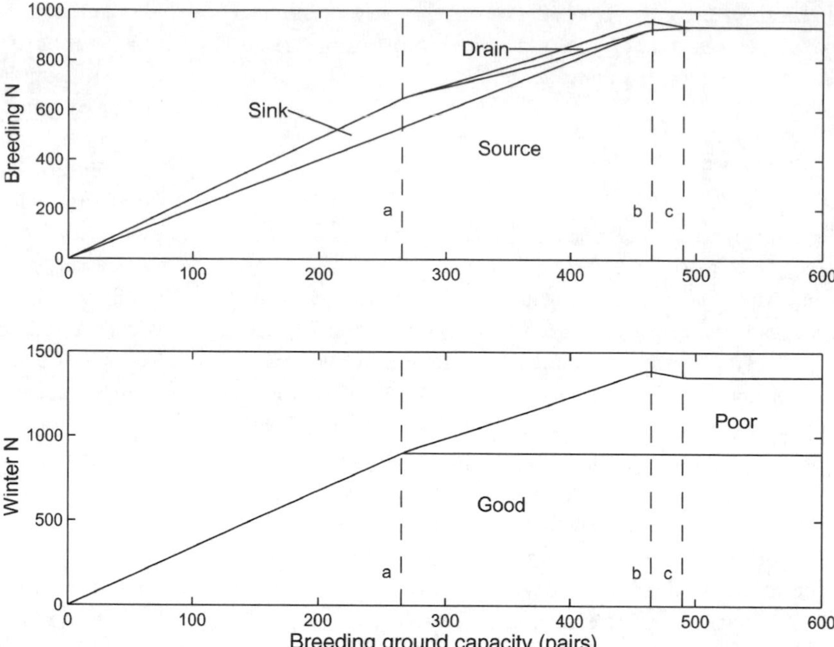

Fig. 28.4. Equilibrium breeding (ΣB, equation [17]) and winter (ΣW, equation [1]) population sizes as a function of the carrying capacity of source breeding habitat (K_{bc}), with the carrying capacity of good winter habitat (K_{wg}) held constant at 900. The regions between the curves in the first graph show the equilibrium population sizes in source, sink, and "drain" breeding habitats, males and females combined. The regions between the curves in the second graph show the equilibrium population sizes in good and poor winter habitats, males and females combined. The reference lines (a, b, and c) mark important transition points in the dynamics, generated by seasonal interactions (see "Results").

grounds and the use of sink habitat decreases; second, because males compete more effectively for the good winter habitat, the sex ratio shifts, and the number of "drain" (unpaired) males increases. At reference line b, the average survival rate of females has decreased enough that all the females that return to the breeding grounds can find territories in source habitat; thus there are no pairs in sink habitat (upper panel). Up to this point, the sex ratio in the source breeding habitat is 1:1—all males in source habitat find a mate; but the overall sex ratio is male biased, due to the "drain" males. But beyond line b, as the average survival rate of females continues to drop (because a greater proportion of them are in poor winter habitat), there are not enough females to mate with the males, even in source habitat. With continued increases in source habitat, all the males can find a territory in source habitat, even though not all find mates. At reference line c, the average winter survival rates have decreased to the point that continued increases in source breeding habitat produce no further increases in equilibrium population size (both upper and lower panels). Beyond this, the population is entirely winter limited. The conditions between reference lines a and c in fig. 28.4 correspond to the transition zone in fig. 28.3.

Equilibrium sex ratio on the breeding grounds is influenced by the male dominance parameter on the winter grounds (fig. 28.5). If a population is winter limited, and the male dominance parameter is greater than 1, a greater proportion of females will be forced into the poor winter habitat. This lowers the average spring migration survival rates of females relative to males, resulting in a male-biased sex ratio on the breeding grounds and, in turn, a male-biased sex ratio on the wintering grounds the following year. As the strength of the male dominance parameter increases, so does the bias in the sex ratio. On the other hand, if a population is summer limited, then competition for good winter habitat becomes unimportant, and the male dominance parameter has no effect on the sex ratio. Note that the 1:1 sex ratio seen in fig. 28.5 is a consequence of the base survival rates being equal between the sexes; this need not be the case. If sex-specific survival rates are caused by factors other than competition for good winter habitat, the base sex ratio would still depart from 1:1; however, the sex ratio would not be affected by the dominance parameter in a summer-limited population. Populations in the transition zone ("T") show an intermediate effect of the male dominance parameter on sex ratio.

Equilibrium breeding population size ΣB can be influenced by the strength of the carry-over effect (c), particularly when the population is winter limited (fig. 28.6). The carry-over effect is the ratio of productivity of individuals from good versus poor winter habitat. For instance, with a carry-over effect of 2, a "good-good" pair has twice the productivity of a "good-poor" or "poor-good" pair, and four times the productivity of a "poor-poor" pair. In a winter-limited population, as the strength of this carry-over effect increases, the equilibrium population size decreases, because the net productivity of the population decreases. In a summer-limited population, no such effect is observed, because there are no birds spending the winter in poor habitat (see fig. 28.4, lower panel, left of line a). For populations in the transition zone ("T" in fig. 28.3, and between lines a and c in fig. 28.4), the effect is intermediate, because a smaller portion of the population spends the winter in poor habitat.

DISCUSSION

Modeling provides an indispensable tool for identifying critical aspects of an organism's annual cycle and understand-

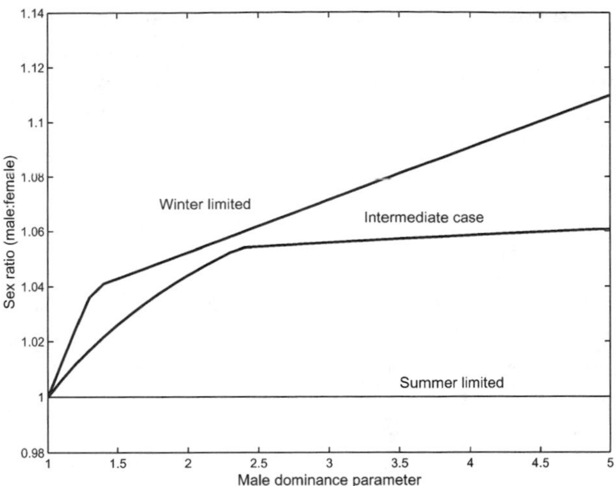

Fig. 28.5. Equilibrium sex ratio (male:female) on the breeding grounds as a function of the male dominance parameter on the wintering grounds (γ). The carry-over effect (c) is held at 1. The three cases correspond to the squares in fig. 28.3: a winter-limited case (K_{bc} = 800 pairs, K_{wg} = 485 individuals); an intermediate case (K_{bc} = 224 pairs, K_{wg} = 580 individuals); and a summer-limited case (K_{bc} = 205 pairs, K_{wg} = 900 individuals).

ing the interactions among them. To date, few attempts have been made to develop either theoretical or empirical population models for migratory birds that explicitly incorporate seasonal dynamics; as a result, we have a limited set of tools available for understanding the factors that drive the dynamics of such populations (but see Goss-Custard et al. 1995a, 1995b, 1995c, 1995d). Undertaking such model development is daunting, however, because of the practical difficulties associated with parameter estimation. Migratory birds move over large geographic areas, often thousands of miles between breeding, stopover, and winter sites. Such behavior makes acquiring demographic information from the entire annual cycle extremely difficult (but see Sillett and Holmes 2002). Despite these issues, developing theoretical models remains a valuable exercise for generating research hypotheses, setting conservation priorities, assessing management options, and identifying which parameters have the greatest potential impact on population dynamics.

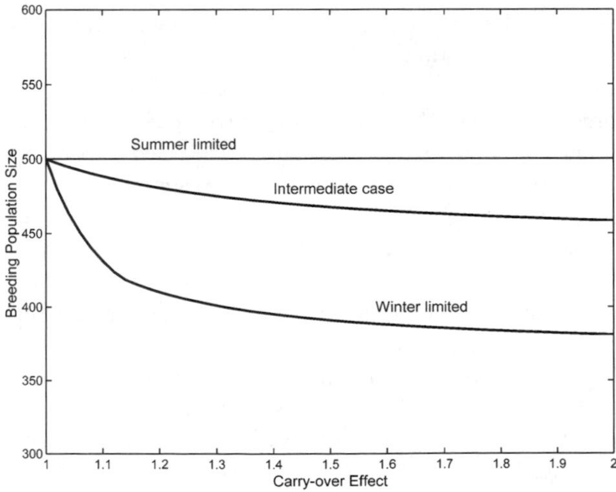

Fig. 28.6. Equilibrium breeding population size (ΣB, equation [17]) as a function of the strength of the carry-over effect (c), with the male dominance parameter held at $\gamma = 5$. The three cases are as described in the legend of fig. 28.5 (and depicted in fig. 28.3).

In this chapter, we developed a theoretical population model to help understand the significance of density dependence, behavioral dominance, and carry-over effects in the annual cycle of a long-distance migratory bird. Although much of our model development and results were based on information collected from American Redstarts, there are several similarities between this and other species of migratory birds allowing for broader generalization. One of the primary mechanisms driving the dynamics of this model is the dominance behavior in winter and summer. To date, 16 species of migratory songbirds have been shown to exhibit sexual habitat segregation on their nonbreeding grounds (e.g., Nisbet and Medway 1972; Ornat and Greenberg 1990; Greenberg et al. 1997; P. P. Marra, unpubl. data), a spacing pattern probably caused by dominance (Marra 2000). Furthermore, most species of songbirds also exhibit some form of dominance-mediated spacing pattern during the breeding season. Because dominance-mediated spacing systems and their associated consequences for physical condition and survival appear to be relatively common, the model we present here may well be applicable to other migratory bird species.

Seasonal Population Limitation

To address the issue of seasonal population limitation, we varied the capacities of source breeding and good winter habitat, and looked at the resulting equilibrium population sizes throughout the annual cycle. We found large sets of conditions under which the population was entirely limited by either winter or breeding habitat (fig. 28.3). A similar question was investigated by Dolman and Sutherland (1994) and applied specifically to Oystercatchers (*Haematopus ostralegus*) (see also Sutherland and Dolman 1994; Goss-Custard et al. 1995a, 1995b, 1995c, 1995d; Sutherland 1996, 1998). Our results (fig. 28.3) differ from those of Dolman and Sutherland (1994:S41, their fig. 2a): region "M" of their figure corresponds to region "B" or ours; and their region "P" corresponds to our region "W"; but between those regions, the results of Dolman and Sutherland (1994) show a much more gradual transition, such that equilibrium population size changes with change in the amount of either winter or breeding habitat. These differences are due to the nature of the density dependence implicit within each model. In the Oystercatcher model, productivity and survival decreased gradually with increases in density. This form of density dependence might imply a crowding mechanism, such that average mortality and productivity decrease with each additional individual (Fretwell 1972). In our model, we used a site-dependent form of density dependence (Rodenhouse et al. 1997). Under this mechanism, once the source breeding or good winter habitat is filled, each additional bird that attempts to settle will be forced into the next best available habitat, resulting in a sharp change in the consequences for those birds forced into suboptimal habitat. We have no crowding mechanism per se implicit in our model. A more detailed site-dependent model could allow for a continuous range of habitat quality; such a model might produce results more like those of Dolman and Sutherland (1994), since there would be a gradual change in the consequences for each additional bird in the population. In reality, multiple mechanisms probably interact to regulate population size in migratory birds (Rodenhouse et al. 2003), a dynamic that might further change the nature of density dependence. The critical point here is that the dominant mechanism for density dependence can strongly affect the nature of seasonal limitation. The importance of the functional form of density dependence, as an expression of the mechanism, has been demonstrated for other applications of population biology (Runge and Johnson 2002).

The potential importance and subtlety of population-level seasonal interactions are illustrated in fig. 28.4. Especially in the "transition" zone, an understanding of dynamics in any one season requires knowledge about how it interacts with processes in other seasons. For instance, as in Sutherland (1996), loss of winter habitat does not result in as severe a decline in equilibrium population size as might otherwise be expected, because increased production (due to density dependence on the breeding grounds) partially offsets the impact of the habitat loss. Understanding seasonal compensation thus requires understanding how the seasons interact with one another through population-level effects, especially as mediated by seasonal density-dependent processes.

Direct study of the form and strength of density dependence during a particular season would require formal estimates of seasonal survival (through, say, mark-recapture or radio-telemetry methods) in conjunction with estimates of density, over a long-enough period to observe a range of densities. Several such studies of seasonal survival have been conducted for waterfowl (e.g., Blohm et al. 1987; Reinecke et al. 1987), but density has not been measured, and the survival rates have typically varied so little that a relationship with density would have been undetectable. Sillett and Holmes (2002) estimated seasonal survival rates for Black-throated Blue Warblers, but found little evidence for density-dependent survival (Sillett and Holmes, Chap. 32, this volume). Goss-Custard et al. (1995d) have used some clever indirect methods to develop an estimate for nonbreeding-season density dependence, based on game theory models combined with data from measures of individual variation in competitive abilities (Dolman and Sutherland 1994; Goss-Custard et al. 1995a, 1995b; Sutherland 1996, 1998). This estimate, however, was generated from one research site and little is known about how this relationship varies geographically. Our modeling work, as well as that of Dolman and Sutherland (1994), suggests that understanding density dependence during the nonbreeding season could be critical to understanding the population dynamics of long-distance migratory birds. This, in turn, suggests that new research is needed that quantifies the density dependence of nonbreeding-season survival (during the stationary and migratory periods) for a variety of avian taxa over large geographic spatial scales.

Sex Ratio Dynamics

The sex ratio dynamics throughout the annual cycle also need to be understood in the context of the interactions between seasons. Our results show that the equilibrium sex ratio on the breeding grounds can be influenced by male dominance on winter grounds. The sex ratio, however, is sensitive to the relative competition parameters only in a winter-limited population, because only then will there be consequences to female survival. These results may be particularly important for understanding the dynamics of redstarts and other species that show sexual habitat segregation. American Redstarts exhibit age-specific habitat segregation on their breeding grounds, and sex and age-specific habitat segregation on their wintering grounds, all known to be induced by the dominance behavior of males (Marra 2000). Such year-round intraspecific competition may be the primary behavioral mechanism driving the distribution of redstarts across the landscape. Ultimately, the consequences of this distribution depend on the relative amounts of suitable breeding and wintering habitats. Because male redstarts are behaviorally dominant over females during winter and exclude them into poor habitat, where their fitness may be lower, dominance behavior may play an integral role in determining the population sex ratio. Thus, when winter is the limiting season, limitation may act primarily through females, influencing population dynamics by reducing the number of females available for breeding (Marra and Holmes 1997, 2001).

It is important to note that simply observing a biased sex ratio on the breeding grounds is not evidence for intraspecific competition in the winter. Male-biased sex ratios could also be due to higher female mortality during the breeding season (Trivers 1972; Breitwisch 1989) or at any other point during the annual cycle. Intraspecific competition on the wintering grounds would be suggested by a shift in sex ratio following a change in the availability of good winter habitat. For example, if the sex ratio became male biased as good winter habitat was lost, that would suggest not only that the population was winter limited, but also that there was competition between the sexes on the winter grounds. Finally, note that the ratio of males to females in good winter habitat or in poor winter habitat is not simply a function of the relative competition parameters; it is a complex function of the primary sex ratio, the survival rates in good versus poor winter habitat, the relative survival rates of males and females in the rest of the annual cycle, and the relative amounts of habitat available on the breeding and wintering grounds.

Carry-Over Effects

Seasonal population dynamics can also interact through carry-over mechanisms that involve the fitness of individual birds. Our results show that carry-over effects can be substantial, but only if the population is limited in the season when the carry-over originates. Note that this carry-over effect can interact with the seasonal compensation effect: in the "transition" zone (fig. 28.6, intermediate case), the impact of the carry-over effect is moderated by the density dependence on the breeding grounds. A stronger carry-over effect leads to lower production, hence, lower equilibrium population size; but a lower population size also leads to increased production. Thus, the carry-over effect is partially offset. Again, to understand the carry-over dynamics or to be able to predict their effects, it is critical to understand them in the context of the annual cycle and all the other seasonal interactions. This implies that assigning causation for particular phenomena may require understanding events in the prior period, and for migratory birds, this could mean thousands of kilometers away from the point of measurement. Although this discussion implies that carry-over effects can interact with density dependence to affect population regulation, it does not follow that carry-over effects are a necessary condition for regulation. For instance, in the Black-throated Blue Warbler, which does not exhibit dramatic sexual habitat segregation and does not seem to be strongly limited by events on winter quarters, Sillett and Homes (Chap. 32, this volume) show that a population can be regulated solely by density dependence operating within the breeding season. Our point is just that carry-over effects *can* influence regulation, and we have only begun to scratch the surface of how to measure and understand the significance of carry-over effects for migratory birds (Marra et al. 1998; Norris et al. 2004; Webster and Marra, Chap. 16, this volume; Szép and Møller, Chap. 29, this volume). We believe it is important to frame future research efforts on carry-over effects in the context of seasonal interactions.

We have focused our attention on a winter-to-summer carry-over effect, but the model also contains a summer-to-winter carry-over effect, the dynamics of which we have not yet explored. Because the model allows sink individuals to differ from source individuals in being able to compete for good winter habitat (through differential competitive factors, i.e., the γ_i's in equation 24), the habitat conditions of birds on the breeding grounds carry over to affect survival during winter. The strength of this summer-to-winter carry-over effect could be expressed as the ratio of competitive factors for source individuals to those for sink individuals. In the simulations shown herein, that ratio was fixed at 10:1; an interesting extension would be to explore the sensitivity of the equilibrium population sizes to this ratio.

FUTURE DIRECTIONS

We have shown that: (1) the relative amounts of breeding and wintering habitat strongly affect both equilibrium population size and sex ratio; (2) sex ratio dynamics, as mediated by behavioral dominance, can affect all other aspects of population dynamics; and (3) carry-over effects can play a critical role in driving population dynamics. The implications of these results are profound. To understand how populations are regulated we must consider carry-over effects as well as density-dependent mechanisms acting year-round. Otherwise, conclusions about the mechanisms driving population dynamics and predictions about future changes may be misleading. Moreover, for conservation of migratory species to be effective in the event of a population decline, intervention measures will depend on understanding where and how the population is limited.

But, we have much work to do if we hope to understand the nature of population limitation for migratory birds. First, we need to undertake the challenging work of obtaining direct estimates for several population parameters. We need more estimates for habitat-specific survival during the nonbreeding period, and these estimates should be measured throughout the winter ranges of particular species. We need estimates of the strength and functional form of density dependence during the breeding and wintering periods, either through long-term observational studies that quantify productivity and survival in relation to density; or through experimental studies that manipulate density and measure the impacts on productivity and survival. Second, we need a better understanding of how events during migration and stopover affect individual and population-level processes, and how migration events connect breeding and nonbreeding periods (DeSante 1995; Sherry and Holmes 1995; Sillett and Holmes 2002). It is generally accepted that density-independent factors are more significant than density-dependent processes for migratory birds during migration, but little is known definitively about this period (Sillett and Holmes 2002, Chap. 32, this volume). Direct estimation of survival during migration would be valuable, especially if it could be tied to habitat in the preceding stationary period to elucidate any carry-over effects. Better estimates of population connectivity among the periods of the annual cycle will ultimately bring us far in our pursuit of understanding carry-over effects and population dynamics. The challenge, of course, with the migration and connectivity questions, is how to follow individuals across vast geographic distances.

Regarding the model we have presented in this chapter, our analysis of its properties has been admittedly cursory. Greater understanding of the population dynamics implied by this model may help to motivate and focus additional field research. Several details invite further exploration: (1) In the transition zone ("T" in fig. 28.3), where seasonal compensation is acting, how is the carry-over effect moderated? (2) Likewise, how does seasonal compensation interact with the sex ratio dynamics induced by behavioral dominance? (3) What hypotheses about the effects of habitat loss, management, or enhancement can be generated by this model and tested in the field? (4) Which general results in this chapter are not sensitive to the particular mechanisms of density dependence or carry-over?

In addition to further theoretical exploration of the model presented in this chapter, there is much value in developing an empirical application of it for a particular species. Such development may elucidate the challenges of parameter estimation, motivate development of new field studies for estimation, and allow us to understand the conservation needs of that species in greater detail.

ACKNOWLEDGMENTS

Much of the biological information used in the model was collected at PPM's study sites in Jamaica. Thanks are due to several groups and individuals who made research at these sites possible. Petroleum Corporation of Jamaica allowed us to conduct this research at the Font Hill Nature Preserve, Yvette Strong and the National Environment & Planning Agency provided cooperation and assistance with our research, Robert Sutton, Anne Haynes-Sutton, Pam (Williams) Becsy, Steve and Sue Callaghan, and Peter Williams kindly provided enormous amounts of support and hospitality during our long stays in Jamaica. Funding for this research was provided to PPM by the National Science Foundation. The manuscript was improved greatly by the comments of Scott Sillett. Finally, we dedicate this chapter to the memory of Robert Sutton.

LITERATURE CITED

Alerstam, T., and G. Högstedt. 1982. Bird migration and reproduction in relation to habitats for survival and breeding. Ornis Scandinavica 13:25–37.

Anderson, D. R., and K. P. Burnham. 1976. Population ecology of the mallard. Pt. 6. The effect of exploitation on survival. U.S. Fish and Wildlife Service Resource Publication 128. U.S. Department of Interior, Washington, D.C.

Baillie, S. R., and W. J. Peach. 1992. Population limitation in Palearctic-African migrant passerines. Ibis 134 (Supplement):120–132.

Blohm, R. J., R. E. Reynolds, J. P. Bladen, J. D. Nichols, J. E. Hines, K. P. Pollock, and R. T. Eberhardt. 1987. Mallard mortality rates on key breeding and wintering areas. Transactions of the North American Wildlife and Natural Resources Conference 52:246–263.

Böhning-Gaese, K., M. L. Taper, and J. H. Brown. 1993. Are declines in North American insectivorous songbirds due to causes on the breeding range? Conservation Biology 7:76–86.

Breitwisch, R. 1989. Mortality patterns, sex ratios, and parental investment in monogamous birds. Current Ornithology 6:1–50.

DeSante, D. F. 1995. Suggestions for future directions for studies of marked migratory landbirds from the perspective of a practitioner in population management and conservation. Journal of Applied Statistics 22:949–965.

Dolman, P. M., and W. J. Sutherland. 1994. The response of bird populations to habitat loss. Ibis 137:S38–S46.

Ficken, M. S., and R. W. Ficken. 1967. Age specific differences in the breeding behavior and ecology of the American redstart. Wilson Bulletin 79:188–199.

Fretwell, S. D. 1972. Populations in a seasonal environment. Monographs in Population Biology. Princeton University Press, Princeton.

Goss-Custard, J. D., R. W. G. Caldow, R. T. Clarke, S. E. A. Le V. dit Durell, and W. J. Sutherland. 1995a. Deriving population parameters from individual variations in foraging behaviour: Pt.1. Empirical game theory distribution model of oystercatchers *Haematopus ostralegus* feeding on mussels *Mytilus edulis*. Journal of Animal Ecology 64:265–276.

Goss-Custard, J. D., R. W. G. Caldow, R. T. Clarke, and A. D. West. 1995b. Deriving population parameters from individual variations in foraging behaviour: Pt. 1.Model tests and population parameters. Journal of Animal Ecology 64:277–289.

Goss-Custard, J. D., R. T. Clarke, K. B. Briggs, B. J. Ens, K.-M. Exo, C. Smit, A. J. Beintema, R. W. G. Caldow, D. C. Catt, N. A. Clark, S. E. A. Le V. dit Durell, M. P. Harris, J. B. Hulscher, P. L. Meininger, N. Picozzi, R. Prŷs-Jones, U. N. Safriel, and A. D. West. 1995c. Populations consequences of winter habitat loss in a migratory shorebird: Pt. 1. Estimating model parameters. Journal of Applied Ecology 32:320–336.

Goss-Custard, J. D., R. T. Clarke, S. E. A. Le V. dit Durell, R. W. G. Caldow, and B. J. Ens. 1995d. Populations consequences of winter habitat loss in a migratory shorebird: Pt. 2. Model Predictions. Journal of Applied Ecology 32:337–351.

Greenberg, R. 1986. Competition in migrant birds in the nonbreeding season. Current Ornithology 3:281–307.

Greenberg, R., P. Bichier, and J. Sterling. 1997. Acacia, cattle and migratory birds in southeastern Mexico. Biological Conservation 80:235–247.

Heitmeyer, M. E., and L. H. Fredrickson. 1981. Do wetland conditions in the Mississippi Delta hardwoods influence Mallard recruitment? Transactions of the North American Wildlife and Natural Resources Conference 46:44–57.

Holmes, R. T., T. W. Sherry, and L. Reitsma. 1989. Population structure, territoriality and overwinter survival of two migrant warbler species in Jamaica. Condor 91:545–561.

Holmes, R. T., T. W. Sherry, and F. W. Sturges. 1986. Bird community dynamics in a temperate deciduous forest: long-term trends at Hubbard Brook. Ecological Monographs 56:201–220.

Kaminski, R. M., and E. A. Gluesing. 1987. Density- and habitat-related recruitment in Mallards. Journal of Wildlife Management 51:141–148.

Marra, P. P. 2000. The role of behavioral dominance in structuring patterns of habitat occupancy in a migrant bird during the non-breeding period. Behavioral Ecology 11:299–308.

Marra, P. P., K. A. Hobson, and R. T. Holmes. 1998. Linking winter and summer events in a migratory bird by using stable-carbon isotopes. Science 282:1884–1886.

Marra, P. P., and R. L. Holberton. 1998. Corticosterone levels as indicators of habitat quality in a migratory bird on its non-breeding grounds. Oecologia 116:284–292.

Marra, P. P., and R. T. Holmes. 1997. Breeding season removal experiments: do they test for habitat saturation or female availability? Ecology 78:947–952.

Marra, P. P., and R. T. Holmes. 2001. Consequences of dominance-mediated habitat segregation in American Redstarts during the nonbreeding season. Auk 118:92–104.

Marra, P. P., T. W. Sherry, and R. T. Holmes. 1993. Territorial exclusion by older males in a Neotropical migrant bird in winter: removal experiments in American Redstarts (*Setophaga ruticlla*). Auk 110:565–572.

Murdoch, W. W. 1994. Population regulation in theory and practice. Ecology 75:271–287.

Nisbet, I. C. T., and L. Medway. 1972. Dispersion, population ecology, and migration of Eastern Great Reed Warblers (*Acrocephalus orientalis*) wintering in Malaysia. Ibis 114: 451–494.

Norris, D. R., P. P. Marra, T. K. Kyser, T. W. Sherry, and L. M. Ratcliffe. 2004. Tropical winter habitat limits reproductive success on the temperate breeding grounds in a migratory bird. Proceedings of the Royal Society of London, Series B, Biological Sciences 271:59–64.

Ornat, A. L., and R. Greenberg. 1990. Sexual segregation by habitat in migratory warblers in Quintana Roo, Mexico. Auk 107: 539–543.

Price, T. 1981. The ecology of the greenish warbler (*Phylloscopus trochiloides*) in its wintering quarters. Ibis 123:131–144.

Rappole, J. H., and M. V. McDonald. 1994. Cause and effect in population declines of migratory birds. Auk 111:652–660.

Reinecke, K. J., C. W. Shiffer, and D. Delnicki. 1987. Winter survival of female mallards in the Lower Mississippi Valley. Transactions of the North American Wildlife and Natural Resources Conference 52:258–263.

Robbins, C. S., J. R. Sauer, R. S. Greenberg, and S. Droege. 1989. Population declines in North American birds that migrate to the Neotropics. Proceedings of the National Academy of Sciences USA 86:7658–7662.

Rodenhouse, N. L., T. W. Sherry, and R. T. Holmes. 1997. Site-dependent regulation of population size: a new synthesis. Ecology 78:2025–2042.

Rodenhouse, N. L., T. S. Sillett, P. J. Doran, and R. T. Holmes. 2003. Multiple density dependence mechanisms regulate a migratory bird population during the breeding season. Proceedings of the Royal Society of London, Series B, Biological Sciences 270:2105–2110.

Runge, M. C., and F. A. Johnson. 2002. The importance of functional form in optimal control solutions of problems in population dynamics. Ecology 83:1357–1371.

Sherry, T. W., and R. T. Holmes. 1989. Age-specific social dominance affects habitat use by breeding American redstarts (*Setophaga ruticilla*): a removal experiment. Behavior, Ecology, and Sociobiology 25:327–333.

Sherry, T. W., and R. T. Holmes. 1992. Population fluctuations in a long-distance Neotropical migrant: demographic evidence for the importance of breeding season events in the American Redstart. Pages 431–442 *in* Ecology and Conservation of Neotropical Migrant Landbirds (J. M. Hagan III and D. W. Johnston, eds.). Smithsonian Institution Press, Washington, D.C.

Sherry, T. W., and R. T. Holmes. 1995. Summer versus winter limitation of populations: what are the issues and what is the evidence? Pages 85–120 *in* Ecology and Management of Neotropical Migratory Birds (T. E. Martin and D. M. Finch, eds.). Oxford University Press, New York.

Sherry, T. W., and R. T. Holmes. 1997. American Redstart (*Setophaga ruticilla*). The Birds of North America, no. 277 (A. Poole and F. Gill, eds.). The Birds of North America, Inc., Philadelphia.

Sillett, T. S., and R. T. Holmes. 2002. Variation in survivorship of a migratory songbird throughout its annual cycle. Journal of Animal Ecology 71:296–308.

Sillett, T. S., R. T. Holmes, and T. W. Sherry. 2000. Impacts of a global climate cycle on population dynamics of a migratory songbird. Science 288:2040–2042.

Sliwa, A. 1991. Age- and sex-specific habitat and geographic segregation patterns of two New World wood warblers (Parulinae) wintering in Jamaica. Master's thesis, Freie Universität Berlin, Berlin.

Sutherland, W. J. 1996. Predicting the consequences of habitat loss for migratory populations. Proceedings of the Royal Society of London, Series B, Biological Sciences 263:1325–1327.

Sutherland, W. J. 1998. The effect of local change in habitat quality on populations of migratory species. Journal of Applied Ecology 35:418–421.

Sutherland, W. J., and P. M. Dolman. 1994. Combining behaviour and population dynamics with applications for predicting consequences of habitat loss. Proceedings of the Royal Society of London, Series B, Biological Sciences 255:133–138.

Trivers, R. L. 1972. Parental investment and sexual selection. Pages 136–179 *in* Sexual Selection and the Descent of Man (B. Campbell, ed.). Aldine Press, Chicago.

29

TIBOR SZÉP
AND ANDERS PAPE MØLLER

Using Remote Sensing Data to Identify Migration and Wintering Areas and to Analyze Effects of Environmental Conditions on Migratory Birds

MIGRATORY BIRDS EXPERIENCE MOST MORTALITY during the nonbreeding season, but the relative effect of environmental conditions during migration and on the wintering grounds remains unknown. In addition, these environmental conditions may influence morphological and other phenotypic variables as well as reproduction in migratory birds. We developed a novel method using information on the condition of the vegetation on migratory routes and in the winter quarters, based on satellite images, to assess the relative importance of environmental conditions during these parts of the year for survival of migratory birds. Using time series analyses of the relationship between survival and the Normalized Difference Vegetation Index (NDVI), we identified areas of potential importance for survival. We selected among areas of importance for survival and their interactions, using mark-recapture analyses of group covariates for long-term population studies of birds. Once areas of importance for survival had been identified, we used information on environmental conditions in these areas to predict the phenotype of returning adults at the breeding grounds and their reproductive performance. We used a variety of methods of cross-validation to verify the importance of the chosen areas for survival. We illustrate the use of this method with our own long-term data on Bank Swallows (*Riparia riparia*) and Barn Swallows (*Hirundo rustica*). Finally, we compare this method with other methods used to link events in the different parts of the annual range of migratory birds.

INTRODUCTION

Migratory birds divide their year among the breeding grounds, migration stopovers, and the winter quarters, which may be thousands of kilometers apart in different hemispheres. Attempts to link events in these separate parts of the distribution range of migratory birds are still few and isolated. Because of the current limitations of satellite telemetry for studying the migration of small birds directly, indirect methods are needed to link populations during breeding, migration, and wintering. The main problem with studies of the ecology of migrants is that we do not know where the migrants go, or where they come from once they return to the breeding grounds. Many researchers have hypothesized that declines in population sizes of migratory birds in the Palearctic and Nearctic are due partly to environmental changes in the winter quarters (Berthold et al. 1986, 1998; Robbins et al. 1989; Terborgh 1989; Askins et al. 1990; Marchant 1992; James et al. 1996). Deterioration of winter habitat can reduce populations by affecting adult survival (Møller 1989; Kanyamibwa et al. 1990; Peach et al. 1991; Szép 1995). Similarly, habitat changes along migratory routes can also influence population trends. Mass mortality during migration is common (Alerstam 1991; Berthold 1993), although the relative importance of mortality during migration for overall annual survival has not yet been assessed. Despite the potential importance of winter and migration survival to populations, the extent to which conditions on migration and wintering sites affect survival rates remains largely unresolved (Marra et al. 1998; Sillett et al. 2000; Booth and Visser 2001). Furthermore, interactions between environmental conditions during migration and winter are completely unexplored. Such effects may be important because changes in environmental conditions in one part of the world may affect survival during migration in another part of the world.

The phenotype of many migratory birds changes during winter because of a partial or complete molt. Environmental conditions during that period may affect feather growth and hence the phenotype of returning migrants in the breeding grounds the subsequent summer (Butcher and Rohwer 1989). In addition, environmental conditions during winter and migration may, through their effects on phenotypic selection, affect the mean and the variance in phenotypic values among birds returning to the breeding grounds. We are aware of two studies showing such "carry-over" effects (Jones 1987; Marra et al 1998). Whether such effects are common remain to be determined.

The performance of migratory birds once they have returned to the breeding grounds may thus reflect not only local environmental conditions at the breeding grounds, but also conditions during migration and the previous winter (Marra et al. 1998). Information on connectivity among breeding, staging, and wintering grounds is essential for understanding these problems (Marra et al. 1998). In addition, it is crucial to understand whether such carry-over effects arise from differences in the intensity of selection or from phenotypic responses to environmental conditions in these disparate parts of the annual range. We are unaware of any study testing for or showing such carry-over effects, let alone distinguishing between the different mechanisms giving rise to such effects.

Assessing the importance of environmental conditions in the staging areas during migration and in the winter quarters for annual survival is prevented by the difficulties of (1) determining the locations of the staging areas during migration and the winter quarters, and (2) recording meaningful measures of environmental conditions for a given species in these areas. First, although many millions of birds have been banded over the years, resulting in numerous recoveries, we still have but a sketchy knowledge of the migration routes and wintering grounds of most migratory species. Even large numbers of recoveries may not reliably represent the distribution of a species, because the location and the number of recoveries depend on factors such as human population density, literacy, and socioeconomic status. Because these factors vary among and within countries and with time, the distribution of recoveries is most likely biased. Second, once the areas of potential importance for survival have been identified, we still need to obtain measures of environmental conditions that are meaningful for a given species. This is not simple, particularly if many areas need to be sampled.

Here we provide a novel possible solution to these two obstacles. We do that by relying on long-term survival data between breeding seasons and information on the Normalized Difference Vegetation Index (NDVI), based on satellite images that indicate the condition of rainfall-dependent vegetation in space and time (Prince and Justice 1991). NDVI provides a simple measure of the amount and vigor of vegetation at the land surface related to the level of photosynthetic activity (Prince and Justice 1991); the higher the value, the greater the amount of green vegetation. Such vegetation indices have been employed for both qualitative and quantitative studies (Tucker et al. 1999). This new method can: (1) identify which areas in the wintering grounds and along migration are important for the survival of a given population of a migratory bird species; (2) quantify how environmental conditions in these areas, and interactions between conditions in these areas, affect survival; and (3) quantify how environmental conditions in the breeding and nonbreeding areas, and interactions between these conditions, affect phenotype, including phenotypic traits affected by molt and subsequent reproduction.

METHOD

Our method is based on the assumption that survival, phenotype and reproduction of migratory birds are partly determined by conditions outside the breeding grounds. This assumption is justified by estimates that less than 5% of the annual adult mortality takes place during the breeding season (Marra and Holmes 2001; Sillett and Holmes 2002), and by previous studies indicating that adult survival rate is greatly affected by environmental conditions in the winter quarters on (Møller1989; Kanyamibwa et al. 1990; Peach et al. 1991; Szép 1995; Marra and Holmes 2001). Our method is an explanatory tool that allows identification of areas of importance for survival, and the selection of these areas must be validated by independent methods such as recoveries of banded birds, stable isotope analyses, or molecular markers.

In the following sections we first describe the steps of computation, including ways in which the method and its output can be cross-validated, and, finally, we discuss the implications of this method for the study of migratory birds and their ecology.

The approach is exemplified by using long-term capture-recapture data from the breeding season for two species of hirundines, the Barn Swallow (*Hirundo rustica*) and the Bank Swallow (*Riparia riparia*). The Barn Swallow breeds solitarily or in colonies of up to more than 100 pairs. The Danish study population arrives from spring migration in April–June, with fall departure taking place in August–October. Wintering takes place south of the Sahara, with the Danish study population wintering in South Africa. Barn Swallows roost communally in winter in large roosts that may number up to 5 million individuals. During daytime, Barn Swallows disperse to the surroundings, with birds covering distances of more than 50 km. Wintering birds are nomadic, following seasonal rains that give rise to the appearance of swarming termites and other insects (Cramp 1988). Body mass may increase by 25% during several days following rains (A. P. Møller, unpubl. data). The Bank Swallow breeds in larger colonies than the Barn Swallow, with colonies reaching more than 1,000 pairs. The Hungarian study population arrives in April–May, with fall departure in August. Wintering takes place mainly in the Sahel zone and other arid parts of sub-Saharan Africa. Bank Swallow winter roosts may number more than 1 million birds. During daytime, Bank Swallows disperse to feed on flying insects (Cramp 1988).

STEPS OF COMPUTATION

STEP 1: ESTIMATE SURVIVAL BETWEEN BREEDING SEASONS FROM TIME SERIES OF MARK-RECAPTURE DATA. We estimate survival rates by standard analysis of capture-recapture data (Lebreton et al. 1992). Survival and recapture rates can be estimated by analysis of capture-recapture data, using the program MARK (White and Burnham 1999). The goodness-of-fit of the data to the general Cormack-Jolly-Seber model (Clobert and Lebreton 1987) can be determined by the standard program RELEASE (Burnham et al. 1987). Akaike's Information Criterion (AIC) is then used to choose a model with the lowest AIC values (Lebreton et al. 1992). This provides the opportunity to find the statistically most relevant model, one that estimates the survival and recapture parameters with a high accuracy and low bias (Lebreton et al. 1992; Anderson and Burnham 1999). Selected models can be compared with the nested ones by using likelihood ratio tests to investigate main effects and interactions of factors on survival and recapture rates, as in analysis of variance (Burnham et al. 1987).

In our analysis, we used long-term capture-recapture data sets of a Danish Barn Swallow population (Møller and Szép 2002) and a Hungarian Bank Swallow population (Szép 1995). Data for adults captured and recaptured at the following breeding colonies were used. Barn Swallow: 1984–2001, Kraghede, Denmark (57°12′ N, 10°00′ E); Bank Swallow: 1986–2001, Tiszatelek, Hungary (48°12′ N, 21°47′ E). In the case of the Bank Swallow, we used 12 survival rates for calculating correlation coefficients between survival rate and environmental conditions during migration and in winter. This was necessary because flooding during the breeding season in 1998 prevented the capture of birds in adequate numbers that year, and so the survival rate estimate for 1997–1998 was poor. Moreover, estimates of survival and recapture rate for 2000–2001 could not be determined (Burnham et al. 1987).

For the two data sets we found that the general Cormack-Jolly-Seber capture-recapture model fitted the data (Lebreton et al. 1992). The best models with the lowest AIC values, following the procedure of model selection, showed that survival of adult males and females varied in parallel (without interaction). We only used the survival rates of males for studying the relationship between annual changes in survival rate and NDVI values, but future analyses could readily investigate differences in responses of the two sexes.

STEP 2: DIVISION OF RANGE INTO APPROPRIATE UNITS FOR ANALYSIS. In this step we have divided the migration and wintering range into squares of 0.25 by 0.25° (633–781 km^2, depending on latitude). This level of subdivision must make a compromise between what is biologically relevant and what is computationally feasible. We chose this square size on the assumption that this is within the range of diurnal movements of aerially foraging hirundines, which frequently move 25 km or more per day (Møller and Aarestrup 1980). Different dimensions may be appropriate for species with different range sizes, although that remains to be investigated.

STEP 3: DIVIDE THE YEAR INTO BIOLOGICALLY MEANINGFUL PERIODS. Using information on the annual cycle of the birds (Turner and Rose 1989), we divided the annual cycle of the two species into four periods: (i) Breeding season (June–August), (ii) fall migration (September–November), (iii) winter (December–February), and (iv) spring migration (March–May). Although the duration of these periods generally reflects the duration of the different parts of the annual cycle, we chose durations that were multiples of ten because NDVI data are available for each 10-day period.

STEP 4: MAKE CORRELATION MATRIX BETWEEN REMOTE SENSING DATA AND SURVIVAL TIME SERIES. We estimated environmental conditions during migration and winter from the NDVI by using satellite images indicating the condition of rainfall-dependent vegetation in space and time (Prince and Justice 1991). This index has been calculated for each 10-day period since 1982 for the entire African continent. Because NDVI is directly related to the level of photosynthetic activity, it provides a measure of the amount and vigor of vegetation at the land surface (Prince and Justice 1991). Many migratory birds rely directly on insects for food, which in turn depends on plant productivity, so the NDVI index is likely to reflect the abundance of insect food. This is particularly likely to be the case for species that rely on swarming insects such as termites, which respond directly to precipitation. NDVI is derived from data collected by National Oceanic and Atmospheric Administration (NOAA) satellites, and processed by the Global Inventory Monitoring and Modeling Studies at the National Aeronautics and Space Administration (NASA) (Prince and Justice 1991). Vegetation indices derived from the NOAA Advanced Very High Resolution Radiometer sensor have been used for both qualitative and quantitative studies of numerous phenomena related to the condition of the vegetation (Tucker et al. 1999). NDVI can be used as an indicator of relative biomass and greenness of the vegetation (Paurelo et al. 1997; Chen and Brutsaert 1998; Boone et al. 2000), precipitation (Schmidt and Karnieli 2000), and quantity and quality of bird habitats (Wallin et al. 1992). Maurer (1994) indicated a correlation between abundance of birds and NDVI, and NDVI has the potential to be used for mapping bird distributions at large spatial scales (Osborne et al. 2001). The NDVI data are available at NASA for the entire African continent at [http://edcintl.cr.usgs.gov/adds/adds.html]. These images are calibrated for intersensor and intrasensor degradation for the period 1982–2001.

Using WinDisp v4.0 software [www.fao.org/WAICENT/faoinfo/economic/giews/english/windisp/windisp.htm]), we averaged the NDVI data for each pixel (8 × 8 km) available at the website to obtain a mean value for each square and for each period. NDVI values below 2 indicate barren surfaces or cloud cover, so we excluded these pixels from our analysis. We calculated NDVI data only for squares in which at least 30% of the pixels had valid data with NDVI values from 2 to 255, and only for years for which we had survival estimates. This excluded 15% of the roughly 40,000 squares in Africa. The effects of exclusion of certain squares are likely to be minor, given the procedures adopted in step 5 below.

Subsequently, we calculated the Spearman rank correlation between the survival estimate and NDVI for a given period for each of the squares. An example of a map showing the geographical distribution of these correlations for the Barn Swallow can be found in fig. 29.1A.

STEP 5: SELECT CONTIGUOUS AREAS THAT EXCEED A CHOSEN THRESHOLD OF CORRELATION. For each of the three periods of the nonbreeding season we identified squares that had a positive correlation between annual survival estimate and NDVI exceeding a critical threshold. We identified areas of importance for survival using five different thresholds of the rank correlation between NDVI and survival rate, $r_S \geq 0.20$, 0.30, 0.40, 0.50, and 0.60, to determine the robustness of the conclusions to the choice of threshold. The results for the threshold of 0.60 provided the best fit, because a value of 0.80 yielded insufficient areas, and values less than 0.60 yielded many more areas with an overall poorer fit. An area was defined as the aggregation of all neighboring squares that had a positive correlation exceeding the threshold. In a second series of analyses we defined an area as the aggregation of squares that were maximally separated by one square without data or not exceeding the threshold. This procedure allowed us to investigate the robustness of our conclusions to the exclusion of squares that did not have NDVI data for the entire period. Use of this second criterion did not provide quantitatively different results.

For each of the three periods, we then identified squares whose correlation between annual survival and NDVI exceeded this threshold. The NDVI data were averaged across all squares constituting an area for the subsequent analyses in the following steps.

We used the mean NDVI values of these areas as independent variables in a stepwise linear regression with annual survival estimates as the dependent variable. We chose the areas that explained the largest amount of variation across the three time periods of the annual cycle provided they explained at least 10% of the variance in a partial correlation analysis.

STEP 6: USE NDVI FROM SELECTED AREAS IN GROUP COVARIATE SURVIVAL MODEL. Using the program MARK (White and Burnham 1999), we used Analysis of Deviance to test the significance of group covariates on survival (Skalski et al. 1993). We used NDVI values from the chosen areas (see above) as group covariates in generalized linear models. We did that by comparing the global model with the covariate model and the constant model. The use of two- and three-way interactions in this framework allowed us to test whether conditions during one period had a differential influence on the effects of conditions during later periods.

We tested the fit of the NDVI values of the selected areas as group covariates by using capture-recapture regression methods (Lebreton et al. 1992). Only area selection based on the most restricted threshold values ($r_S > 0.6$ and size of the area > 4 squares) produced a model that significantly fit the data. This covariates model had the best fit using the AIC criterion (Anderson and Burnham 1999). Selected and nonselected areas for the Barn Swallow and the Bank Swallow are shown in figs. 29.1B and 29.1C, respectively.

Figure 29.2 illustrates examples of relationships between adult survival rates of male Barn Swallows and NDVI. Here we use NDVI from an area in Algeria that was found to be particularly important for survival during spring migration.

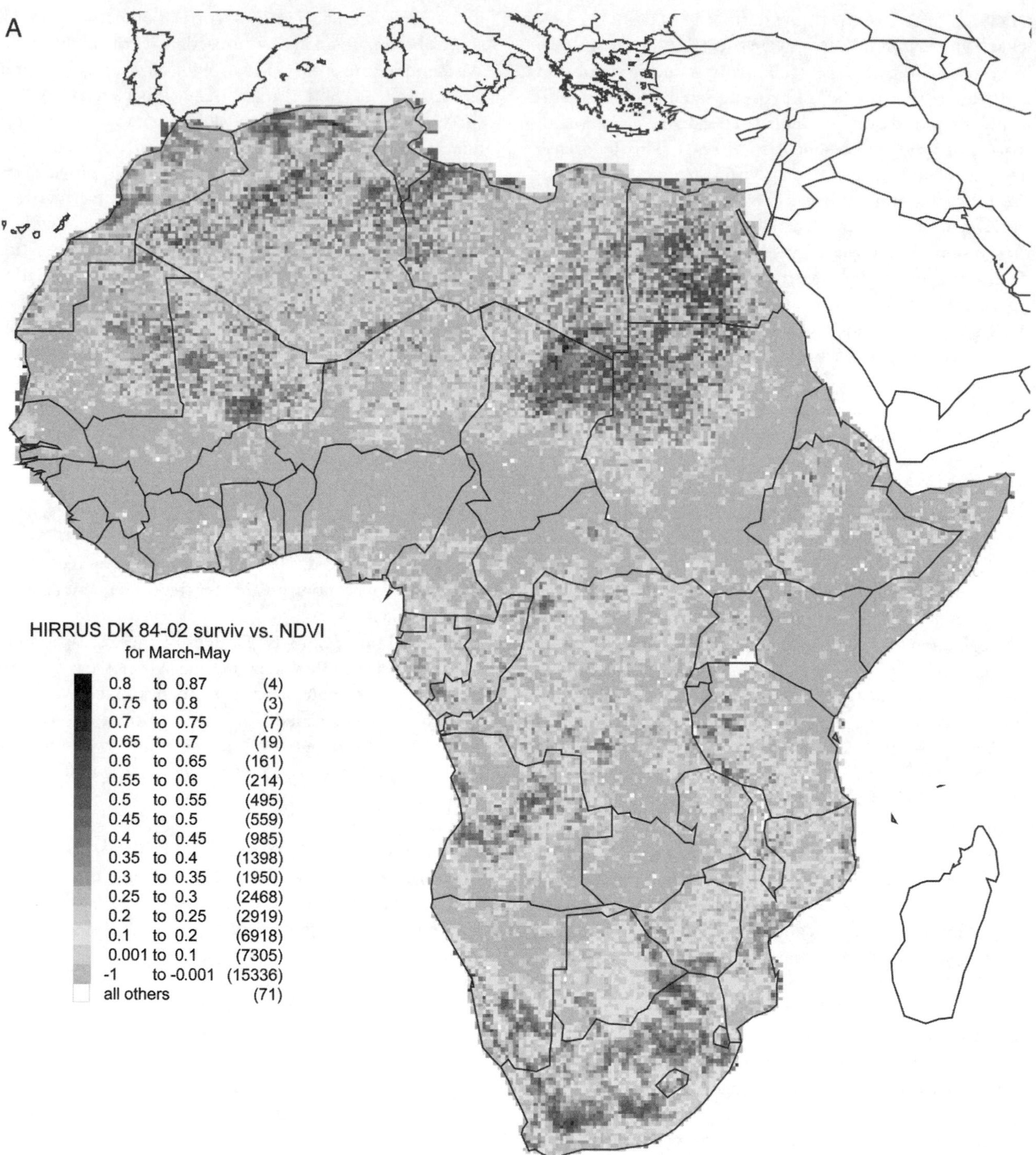

Fig. 29.1A. Spearman rank correlation coefficients between annual adult survival of adult male Barn Swallows from a Danish population during the years 1984–2001 and NDVI during spring migration (March–May). Darker color reflects more strongly positive correlations. The number of 0.25 × 0.25° squares where the correlations were in the given ranges are shown in brackets.

This area also held all 13 spring recoveries in Africa north of the Sahara of Barn Swallows banded in Denmark. Figure 29.2A shows the positive relationship between survival rate and NDVI in Algeria during spring migration (linear regression: $F = 11.63$, d.f. $= 1{,}15$, $r^2 = 0.399$, $p = 0.004$, slope [SE] = 1.85 [0.54]).

Environmental conditions in different parts of the range may interact to influence survival rate because the effects of adverse conditions in one area may become exacerbated by adverse conditions in another area. Figure 29.2B shows an example of such an interaction, in this case between NDVI in Algeria during spring migration and the intensity of the

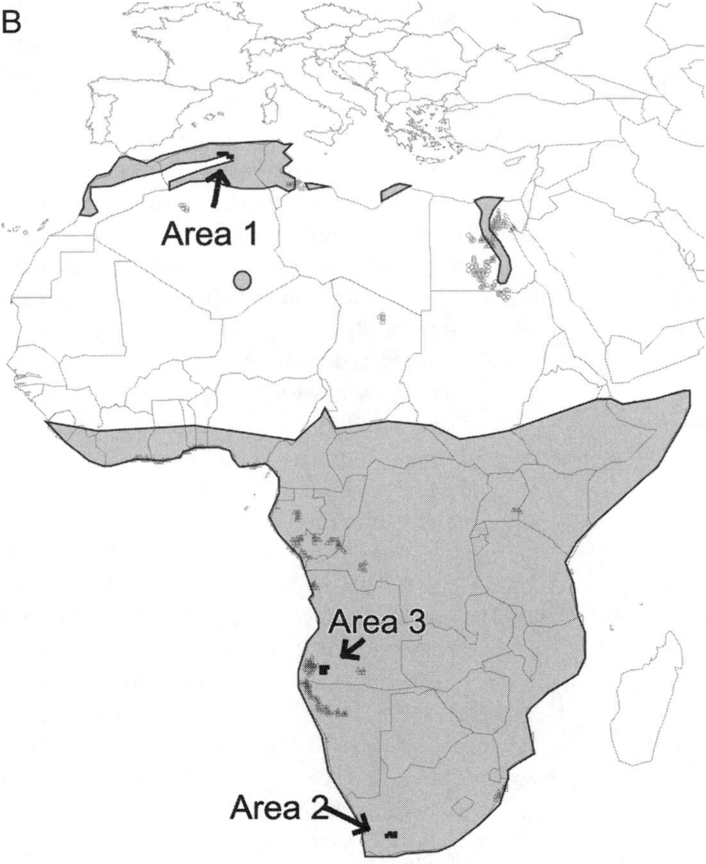

Fig. 29.1B. Areas in Africa during fall (September–November), winter (December–February), and spring (March–May) that were important (correlation coefficient ≥ 0.6 and number of squares ≥ 4) for annual survival rate of adult Barn Swallows from a Danish population. Area 1: Algeria in spring (6 squares, 3,757 km²); Area 2: South Africa in spring (5 squares, 3,254 km²); and Area 3: Angola in fall (7 squares, 5,197 km²). Gray shaded areas show the distribution of the Barn Swallow in Africa (based on Keith et al. 1992).

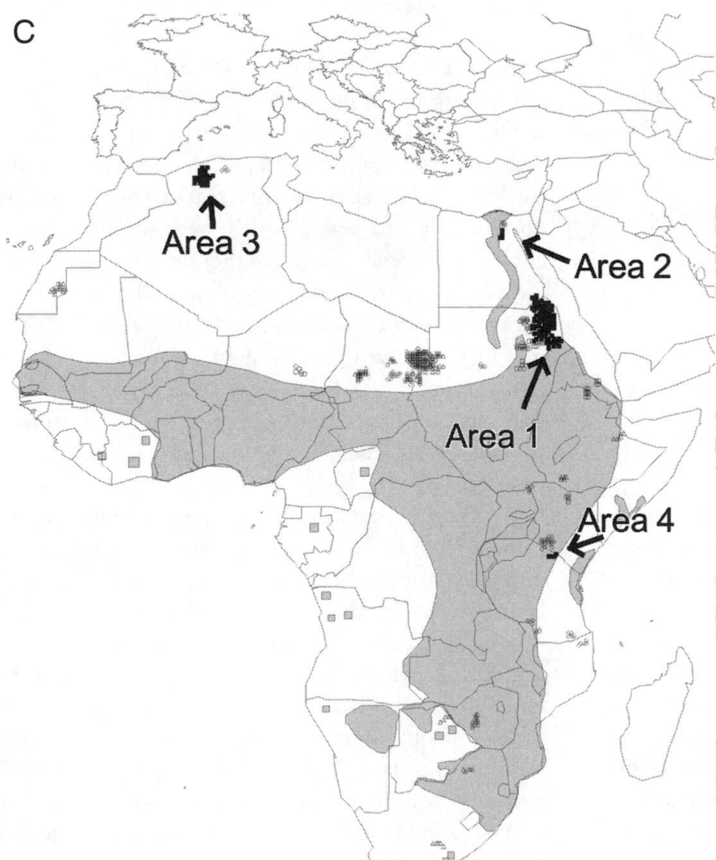

Fig. 29.1C. Areas in Africa during fall (September–November), winter (December–February), and spring (March–May) that were important (correlation coefficient ≥ 0.6 and number of squares ≥ 4) for annual survival rate of adult Bank Swallows from a Hungarian population (hatched areas) and areas that were selected (black areas) to be included as predictors of annual survival rate in a group covariate model. Area 1: Northern part of Nubian Desert (NE Sudan) in winter (123 squares, 88,560 km²); Area 2: along the Nile near Cairo in spring (5 squares, 3,600 km²); Area 3: N Algeria in winter (34 squares, 24,480 km²); and Area 4: SW Kenya and NE Tanzania in winter (4 squares, 2,880 km²). Gray shaded areas show the distribution of the Sand Martin in Africa (based on Keith et al. 1992).

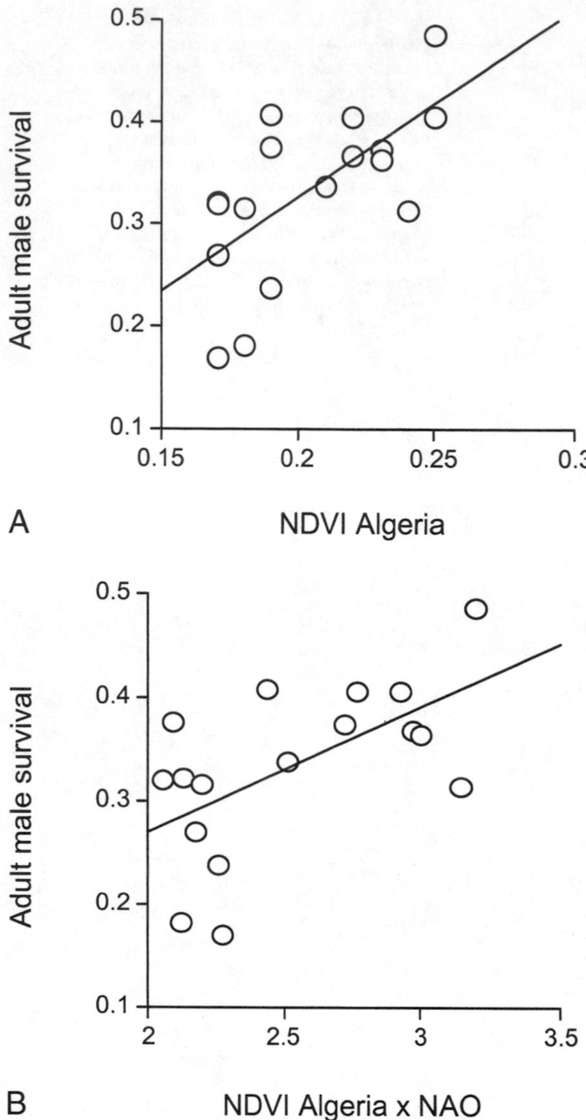

Fig. 29.2. Male adult survival rate of adult Barn Swallows from a Danish population versus (A) NDVI from Algeria in spring and (B) the interaction between NDVI from Algeria in spring and North Atlantic Oscillation from the breeding grounds the previous winter.

North Atlantic Oscillation (NAO) on adult survival rate (linear regression: $F = 8.26$, d.f. $= 1,15$, $r^2 = 0.312$, $p = 0.021$, slope [SE]) $= 0.12$ [0.04]). The NAO is a general index of climatic conditions in the breeding areas in Europe, affecting the Danish population of Barn Swallows (Møller 2002), with high values being associated with warm and wet spring and summer weather in northern Europe (Hurrell 1995).

STEP 7: USE NDVI VALUES FROM SELECTED AREAS TO PREDICT MORPHOLOGY AND REPRODUCTION. The selected areas have been identified and verified to be within the known distribution of the species, and to coincide with the distribution of recoveries of banded birds. NDVI data from these chosen areas can then be used as predictors of morphology and other measures of phenotype of returning adults, but also as predictors of reproductive variables. Again, environmental conditions in the breeding areas, migration ranges, and winter quarters, and their interactions, can be used as predictors of these phenotypic traits.

Annual differences in phenotype and reproductive variables will partly reflect differences in composition of the cohort due to differences in intensity of selection and partly reflect phenotypic responses to environmental conditions. The intensity of selection will clearly be linked directly to annual survival rate, and this effect of selection can be controlled statistically by including annual survival rate as a variable first forced into the statistical model. As an example of this approach, we show arrival date of Barn Swallows to the Danish study site in relation to environmental conditions during winter, on migration, and at the breeding grounds. Arrival date is delayed in years with large NDVI during spring migration in Algeria (fig. 29.3A; linear regression: $F = 6.59$, d.f. $= 1,16$, $r^2 = 0.247$, $p = 0.021$, slope [SE] $=$ 46.11 [17.97]). This may seem counterintuitive, but a simple explanation is that in years with high NDVI in Algeria, survival is better (fig. 29.2A), resulting in more individuals of poor phenotypic quality arriving late to the breeding grounds (Møller 1994a, 1994b). In fact, residual arrival date (after controlling statistically for survival rate) was no longer significantly related to NDVI in Algeria (fig. 29.3B; linear regression: $F = 0.88$, d.f. $= 1.14$, $r^2 = 0.000$, $p = 0.364$).

Finally, when analyzing time-series it is important to account for temporal autocorrelation. Test statistics can be adjusted for such correlation to achieve appropriate probabilities, as already emphasized by Bartlett (1952).

STEP 8: VALIDATION OF THE METHOD. We emphasize that the method described here is not based on arguments of circularity. Obviously, the areas included in the final model by definition exceed the threshold level that was used to choose areas in the first place. However, the method does not a priori define which areas were or were not included in the final model. Neither does the method by itself include any interaction terms in the final model. Initial areas chosen for test of suitability as predictors of survivorship may include a number of biologically meaningful areas, but also a number of false positives. Adoption of cross-validation using the four different approaches listed below should allow distinction between biologically meaningful and false positives.

1. *The low probability of finding large continuous areas of migration and wintering distribution with a positive Spearman rank correlation between NDVI and survival.* Spearman rank correlations exceeding the threshold value of 0.60 occur only in 250, 444, and 236 of the squares in the case of the Bank Swallow in fall, winter, and spring, respectively. Because the four chosen areas had 123, 5, 34, and 4 squares, respectively, the probability of finding so many squares exceeding the threshold can be calculated. It is simply the probability of having a correlation coefficient exceeding the threshold of 0.60 raised to the power of the number of squares in an area, that is, 0.00735 raised to the power of 123, 5, 34, and

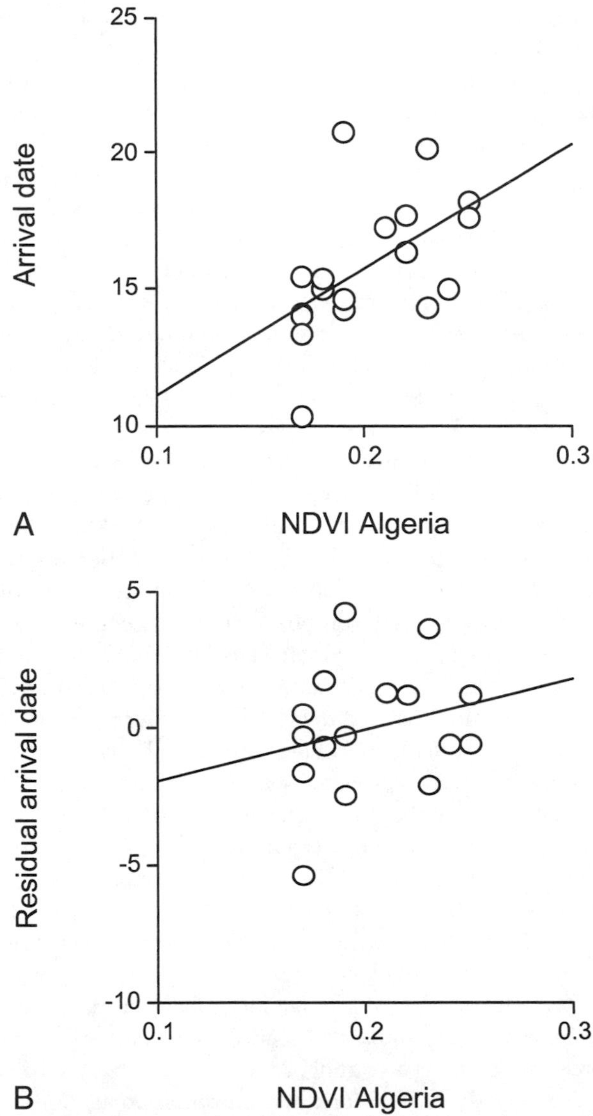

Fig. 29.3. (A) Mean annual arrival date of Barn Swallows from a Danish population and NDVI during spring migration in Algeria. (B) Mean annual arrival date (± SE) of Barn Swallows from a Danish population versus NDVI during spring migration in Algeria after controlling arrival date for differences in annual survival rate by using residuals from a regression of mean annual arrival date on annual survival. The lines are the linear regression lines.

4, respectively. These probabilities are exceedingly small, although they may to some extent be inflated by a certain degree of spatial autocorrelation. Preliminary analyses of patterns of spatial autocorrelation in southern Africa revealed significant correlations at a scale of up to a couple of hundred kilometers (S. Selmi et al., unpubl. data). Even when taking this autocorrelation into account the calculated probabilities are still exceedingly small compared with random expectations.

2. *Coincidence between areas of distribution and areas of importance for survival.* For many migratory birds we know the geographical distribution during winter, and we can calculate that fraction of the entire continent and compare it with the distribution of selected areas with respect to the known distribution. For the Bank Swallow in Africa, almost 54% of the selected areas were within the known distribution (Keith et al. 1992) and supposed migration areas. The Nubian Desert and the Algerian areas are not indicated on the maps in Keith et al. (1992), although it is very possible that these areas lie along the migration routes of the species. In the present study we have considered only half the number of squares of these areas to be situated within the supposed migration areas. The predicted proportion within the distribution of the species would have been 39.4% for randomly distributed areas. Similarly, for the Barn Swallow in Africa, 100% of the selected areas were within the known distribution (Keith et al. 1992), whereas the predicted proportion would have been 51% for randomly distributed areas. For the southern part of Africa, we found 80% of the selected areas to be within the known distribution (Harison et al. 1997), while the predicted proportion would have been 74%.

3. *Distribution of banding recoveries and selected areas.* If selected areas were within the true ranges, we would expect coincidence between the distribution of recoveries and selected areas. For Danish Barn Swallows we found that 73% of the 15 winter recoveries were from Angola, Namibia, and South Africa near the area of importance for survival, and all 13 recoveries from North Africa during spring migration were from Morocco, Algeria, and Tunisia. This shows a high degree of coincidence between the identified areas and the known distribution of recoveries. We have been equally successful in predicting the winter range important for survival of Spanish Barn Swallows, which, contrary to the Danish Barn Swallows, is in West Africa (F. de Lope et al., unpubl. data).

4. *Coincidence of DNA and isotope profile of birds from selected areas with those of birds from the known breeding areas.* Once selected areas have been identified, it is immediately feasible to go to these areas at the right time of the year and catch individuals of a given species. Profiles of molecular markers such as RAPDs, microsatellites or others, and chemical markers such as isotopes should then match profiles of birds from the known breeding grounds. Juveniles of many species do not molt completely until they reach the winter quarters. Hence, old feathers from juveniles should match the isotope profile of nestlings from the breeding grounds, and new feathers grown in the winter quarters should match the isotope profile of adults that have returned to the breeding grounds. Such tests are currently underway for different Barn Swallow populations wintering in Africa.

Thus, three of the four methods of validation have confirmed the choice of selected areas of importance for survival of Barn Swallows and Bank Swallows. We are currently investigating the fourth method of validation.

Finally, we note that other methods linking breeding population to wintering populations are equally in need of cross-validation. Winter quarters identified by stable isotope analysis will also require validation because several areas may have similar isotope composition. Likewise, molecular identification of wintering populations may not be reliable if nearby populations have similar molecular profiles. Even recoveries may provide limited information about the distribution in winter because their distribution will reflect human population density, prosperity, efficiency of local administration and postal services, and many other factors.

FUTURE DIRECTIONS

We have presented a novel method to identify areas of importance during migration and in winter for survival of migratory birds, based on remote sensing information. For numerous species we still have only the slightest evidence of where populations are wintering, let alone information on the effects of environmental conditions in these areas on survival rate. Our method can be used to identify such candidate areas, which in turn can be cross-validated by field studies to obtain samples for stable isotope analysis or molecular analysis. We have carefully described the procedures used for adopting this method, and we have listed a number of ways in which it can be cross-validated. Using this method, we can build predictive models based on group covariate analyses of mark-recapture data that can be used to determine the population consequences of climatic change in different parts of the annual range of migratory birds. The method has a number of additional uses that include assessment of carry-over effects on adult phenotype and reproduction from migration and winter to the breeding grounds (Marra et al. 1998). The method can also be used to assess the relative importance of selection and phenotypic change for adult phenotype and reproduction.

The main problems with this method are as follows: First, it can be used only when long-term mark-recapture data are available. We estimate that for Palearctic and Nearctic migratory passerines such data are available for at least 50 different populations of migratory birds (A. P. Møller, unpubl. data). Knowing migration and wintering areas of importance for these populations would be an almost unimaginable leap forward in our knowledge.

We can only think of molecular and chemical marking methods as alternative ways of addressing the problem of linking events during breeding, migration, and in winter. Such efforts have been limited to linking winter and breeding events (Webster et al. 2002). Thus, we are unaware of any other method that also includes the importance of migration, and we are similarly unaware of any method that allows tests for interactions between environmental conditions in these three parts of the annual cycle of migratory birds.

Density dependence of events during migration and in winter for migratory birds remains largely unknown. As a notable exception, a study of Barn Swallows from Denmark showed that survival from one breeding season to the next was density-independent, based on key-factor analysis of the number of birds present at the end of the breeding season and at the beginning of the next season (Møller 1989). Breeding studies of migrants can readily quantify the size of the population at the end of the breeding season. Similarly, the size of the breeding population the following year can be directly estimated from a census of the breeding population. We are currently attempting to use estimates of population size in early fall at the end of the breeding season as a group covariate in mark-recapture analyses of survival. These analyses are based on the untested assumption that emigration equals immigration. A main effect of population size on survival in such models would be consistent with density dependence. Interactions between NDVI in selected areas and population size would indicate density dependence interacting with environmental conditions during that particular part of the year.

Age-dependent survival and its relationship with environmental conditions during migration and in the winter quarters also remain largely unknown, with Marra and Holmes (2001) providing the single exception. Because we can estimate the proportion of juveniles at the end of the breeding season and the proportion of yearlings the following year, we can obtain an indirect estimate of juvenile survival rate. Studies of recruitment in the Danish Barn Swallow population have shown that local recruitment of banded birds, calculated as a proportion of all nestlings, was strongly positively correlated with the proportion of local recruits of all yearlings the following year (A. P. Møller, unpubl. data). Likewise, the number of nestlings produced in a given year can predict the number of yearlings the subsequent year (Møller 2002). Finally, local population changes in the Danish Barn Swallow population are a good predictor of national population changes in Denmark (Møller 1994a). The estimate of juvenile survival rate proposed here assumes that any unbanded individual at the start of the breeding season is a yearling, an assumption that has been independently verified for three European populations of Barn Swallows (Møller 1994a; F. de Lope et al., unpubl. data). These estimates of juvenile survival rate can be investigated using the general method developed here to identify areas of importance for survival. This approach would allow tests of whether the same areas are of importance for survival of both juveniles and adults. Likewise, we could test whether phenotype and reproduction of yearlings and adults are affected similarly by environmental conditions in the same or in different areas during migration and in winter.

In the previous sections we described and discussed ways in which remote sensing data can be exploited to further our understanding of the ecology, demography, and evolution of migratory birds. However, it is very important to remember that this approach can be coupled with studies of isotope profiles, trace elements, and DNA to provide additional insights. We have already suggested how time-series analyses of survival rate in combination with remote-sensing data can be used to pinpoint populations during winter and migration that subsequently can be sampled for chemical or molecular analyses. We expect that many more such possibilities exist.

What are the future prospects of studies of migrants using remote sensing? We have shown a variety of applications of this approach, and we have listed other potential uses. It is our belief that efforts to link the use of remote-sensing analyses with other approaches to the study of bird migration will considerably advance our understanding.

ACKNOWLEDGMENTS

We thank the National Aeronautics and Space Administration and the U.S. Geological Survey/FEWS for providing NDVI satellite pictures and E. S. Pfirman for advice on using WinDisp. Recoveries of banded birds were provided by the Zoological Museum, Copenhagen, Denmark. This work was supported by the Hungarian Scientific Research Fund with grants OTKA F17709, T29853, and KKA 3203/98, and a Magyary Zoltán postdoctorate grant to TSz.

LITERATURE CITED

Alerstam, T. 1991. Bird Migration. Cambridge University Press, Cambridge.

Anderson, D. R., and K. P. Burnham. 1999. Understanding information criteria for selection among capture-recapture or ring recovery models. Bird Study 46:S14–S21.

Askins, R. A., J. F. Lynch, and R. Greenberg. 1990. Population declines in migratory birds in eastern North America. Current Ornithology 7:1–57.

Bartlett, M. S. 1952. On the theoretical specification of sampling properties of autocorrelated time series. Journal of the Royal Statistical Society (Supplement 8):27–411.

Berthold, P. 1993. Bird Migration. Oxford University Press, Oxford.

Berthold, P., W. Fiedler, R. Schlenker, and U. Querner. 1998. 25-year study of the population development of Central European songbirds: a general decline, most evident in long-distance migrants. Naturwissenschaften 85:350–353.

Berthold, P., G. Fliege, U. Querner, and H. Winkler. 1986. Die Bestandsentwicklung von Kleinvögeln in Mitteleuropa: Analyse von Fangzahlen. Journal für Ornithologie 127:397–437.

Boone, R. B., K. A. Galvin, N. M. Smith, and S. J. Lynn. 2000. Generalizing El Niño effects upon Maasai livestock using hierarchical clusters of vegetation patterns. Photogrammetric Engineering and Remote Sensing 66:737–744.

Booth, C., and M. E. Visser. 2001. Adjustment to climate change is constrained by arrival date in a long-distance migrant bird. Nature 411:296–298.

Burnham, K. P., D. R. Anderson, G. C. White, C. Brownie, and K. H. Pollock. 1987. Design and analysis methods for fish survival experiments based on release-recapture. American Fisheries Society Monograph 5.

Butcher, G. S., and S. Rohwer. 1989. The evolution of conspicuous and distinctive coloration for communication in birds. Current Ornithology 6:51–108.

Chen, D., and W. Brutsaert. 1998. Satellite-sensed distribution and spatial patterns of vegetation parameters over a tallgrass prairie. Journal of the Atmospheric Sciences 55:1225–1238.

Clobert, J., and J.-D. Lebreton. 1987. Recent models for mark-recapture and mark resighting data: a response to C. Brownie. Biometrics 43:1019–1022.

Cramp, S. (ed.). 1988. The Birds of the Western Palearctic, Vol. 5. Oxford University Press, Oxford.

Harison, J. A., D. G. Allan, L. G. Underhill, M. Herremans, A. J. Tree, V. Parker, and C. J. Crown (eds.). 1997. The Atlas of Southern African Birds. BirdLife South Africa, Johannesburg.

Hurrell, J. W. 1995. Decadal trends in the North Atlantic Oscillation: regional temperatures and precipitation. Science 269:676–679.

James, F. C., E. McCulloch, and D. A. Wiedenfeld. 1996. New approaches to the analysis of population trends in land birds. Ecology 77:13–27.

Jones, G. 1987. Selection against large size in the sand martin (*Riparia riparia*) during a dramatic population crash. Ibis 129:274–280.

Kanyamibwa, S., A. Schierer, R. Pradel, and J. D. Lebreton. 1990. Changes in adult annual survival rates in a western European population of the White Stork (*Ciconia ciconia*). Ibis 132: 27–35.

Keith, S., E. Urban, and H. Fry. 1992. The Birds of Africa, Vol. 4. Academic Press, London.

Lebreton, J.-D., K. P. Burnham, J. Clobert, and D. R. Anderson. 1992. Modeling survival and testing biological hypotheses using marked animals: a unified approach with case studies. Ecological Monographs 62:67–118.

Marchant, J. H. 1992. Recent trends in breeding populations of some common trans-Saharan migrant birds in northern Europe. Ibis 134 (Supplement 1):113–119.

Marra, P. P., K. A. Hobson, and R. T. Holmes. 1998. Linking winter and summer events in a migratory bird by using stable-carbon isotopes. Science 282:1884–1886.

Marra, P. P., and R. T. Holmes. 2001. Consequences of dominance-mediated habitat segregation in American Redstarts during the nonbreeding season. Auk 107:96–106.

Maurer, B. A. 1994. Geographical population analysis tools for the analysis of biodiversity. Blackwell Scientific, Oxford.

Møller, A. P. 1989. Population dynamics of a declining swallow *Hirundo rustica* L. population. Journal of Animal Ecology 58:1051–1063.

Møller, A. P. 1994a. Sexual Selection and the Barn Swallow. Oxford University Press, Oxford.

Møller, A. P. 1994b. Phenotype-dependent arrival time and its consequences in a migratory bird. Behavioral Ecology and Sociobiology 35:115–122.

Møller, A. P. 2002. North Atlantic Oscillation (NAO) effects of climate on the relative importance of first and second clutches in a migratory passerine bird. Journal of Animal Ecology 71:201–210.

Møller, A. P., and W. C. Aarestrup. 1980. Swallows *Hirundo rustica* at a communal roost in North Jutland. Dansk Ornitologisk Forenings Tidsskrift 74:149–152. [In Danish with English summary.]

Møller, A. P., and T. Szép. 2002. Survival rate of adult barn swallows *Hirundo rustica* in relation to sexual selection and reproduction. Ecology 83:2220–2228.

Osborne, P. E., J. C. Alonso, and R. G. Bryant. 2001. Modelling landscape-scale habitat use using GIS and remote sensing: a case study with great bustards. Journal of Applied Ecology 38:458–471.

Paruelo, J. M., H. E. Epstein, W. K. Lauenroth, and I. C. Burke. 1997. ANPP estimates from NDVI for the Central Grassland Region of the United States. Ecology 78:953–958.

Peach, W. J., S. R. Baillie, and L. Underhill. 1991. Survival of British Sedge Warblers (*Acrocephalus schoenobaenus*) in relation to west African rainfall. Ibis 133:300–305.

Prince, S. D., and C. O. Justice (eds.). 1991. Coarse Resolution Remote Sensing of the Sahelian Environment. Special Issue, International Journal of Remote Sensing 12:1133–1421.

Robbins, C. S., J. R. Sauer, R. S. Greenberg, and S. Droege. 1989. Population declines in North American birds that migrate to the Neotropics. Proceedings of the National Academy of Sciences USA 86:7658–7662.

Schmidt, H., and A. Karnieli. 2000. Remote sensing of the seasonal variability of vegetation in a semi-arid environment. Journal of Arid Environments 45:43–60.

Sillett, T. S., and R. T. Holmes. 2002. Variation in survivorship of a migratory songbird throughout its annual cycle. Journal of Animal Ecology 71:296–308.

Sillett, T. S., R. T. Holmes, and T. W. Sherry. 2000. Impacts of global climate cycle on population dynamics of a migratory songbird. Science 288:2040–2042.

Skalski, J. R., A. Hoffmann, and S. G. Smith. 1993. Testing the significance of individual- and cohort-level covariates in animal survival studies. Pages 9–28 *in* Marked Individuals in the Study of Bird Population (J.-D. Lebreton and P. M. North, eds.). Birkhäuser Verlag, Basel, Switzerland.

Szép, T. 1995. Relationship between West African rainfall and the survival of central European Bank Swallows *Riparia riparia*. Ibis 137:162–168.

Terborgh, J. W. 1989. Where Have All the Birds Gone? Princeton University Press, Princeton.

Tucker, C. J., H. E. Dregne, and W. W. Newcomb. 1991. Expansion and contraction of the Sahara Desert between 1980 and 1990. Science 253:299–301.

Turner, A., and C. Rose. 1989. Handbook of the Swallows and Martins of the World. Christopher Helm, London.

Wallin, D. O., C. C. H. Elliott, H .H. Shugart, C. J. Tucker, and F. Wilhelmi.1992. Satellite remote sensing of breeding habitat for an African weaver-bird. Landscape Ecology 7:87–99.

Webster, M. S., P. P. Marra, S. M. Haig, S. Bensch, and R. T. Holmes. 2002. Links between worlds: unraveling migratory connectivity. Trends in Ecology and Evolution 17:76–83.

White, G. C., and K. P. Burnham. 1999. Program MARK: survival estimation from populations of marked animals. Bird Study 46:S120–S138.

30

DAVID W. WINKLER

How Do Migration and Dispersal Interact?

DISPERSAL AND MIGRATION are the two largest types of movements in the lives of most birds, and a better understanding of these movements and how they relate to each other would illuminate many aspects of their ecology and evolution. Although a causal linkage between migratory and dispersal distances is intuitively appealing, there is actually very little direct evidence of such a link; future direct (mark-recapture) and indirect (molecular biological) measures of dispersal distances in pairs of species carefully selected to differ in migratory distances, and very little else, have the greatest chance of shedding light on the biological link between how far birds go in migration and dispersal. Other connections between migration and dispersal are possible, and some of the most interesting comparisons are between ways that species with different movement patterns acquire and process information about their destinations. Analyses from an ongoing study of Tree Swallow (*Tachycineta bicolor*) dispersal suggest that information gathered during migration may sometimes influence the direction of natal dispersal, and further work exploring the cognitive and sensory ecology of birds as they move throughout their annual cycles seems sure to be rewarding and important.

INTRODUCTION

Like most ornithologists, I find the migrations of birds fascinating: the miraculous feats performed by organisms light enough to post with a single first-class stamp are a constant inspiration. In my research, however, I have always focused on the breeding biology and life histories of birds, and one of the life history traits that I have been actively investigating for the past 10 years has been the dispersal biology of Tree Swallows (*Tachycineta bicolor*), a Neotropical migrant passerine. In studying Tree Swallow dispersal, I have begun to ask questions about the relation between dispersal and migration, and this chapter explores this connection in Tree Swallows and in birds more generally. I begin with some comments on both migration and dispersal separately: this is how they have been treated in the past, and one of the goals of this chapter is to see whether this separate treatment should continue.

Migration

Throughout this chapter I use terms for movement as they are generally applied by ornithologists. Thus, "migration" is a regular annual movement away from and back to breeding grounds, not a displacement of breeding sites. Unlike its use in other literatures, "migration" here does not imply any gene flow. Avian annual movements can be thought of as occurring in at least four modes (fig. 30.1). The null pattern of movement is that displayed by residents, which do not undergo any large-scale annual movements: their breeding and wintering grounds are the same. By contrast, migrants can travel great distances per year, but they are distinguished by the very high probability that, given they survive, they will return to their breeding grounds in succeeding breeding seasons.

Two modes of avian migration are generally recognized. Facultative (or "partial" or "environmental" [Howard 1960]) migrants are those that move away from the breeding grounds a variable distance from year to year, depending on

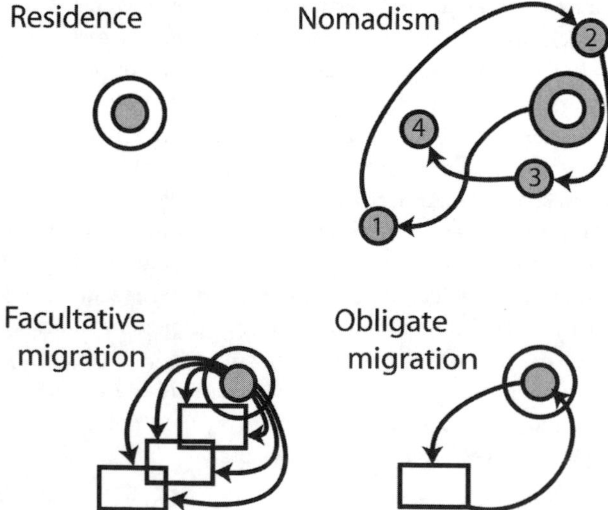

Fig. 30.1. The four principal modes of avian annual movement. Breeding localities (the shaded circles) are in the same general breeding range for residents and facultative and obligate migrants, but they can be in broadly disparate localities in succeeding years in nomads. For further details see text.

environmental conditions (Berthold 1984; Chan 2001). The facultative migrants are very interesting, as it is they (and the nomads, below) that have the greatest chance of showing organismal influences that affect the expression of their movement patterns over ecological time. There is, for instance, good reason to expect that different classes (defined, e.g., by age, sex, or phenotypic quality) of individuals may covary in their phenology and extent of migration (Kokko 1999). Variation across individuals in the extent of migration in a given year produces several well-studied examples of sex bias in winter distributions (Myers 1981; Ketterson and Nolan 1983), patterns that can change remarkably rapidly when populations colonize new areas (Belthoff and Gauthreaux 1991).

Obligate (or "full" or "innate" [Howard 1960]) migrants, on the other hand, engage in a preprogrammed itinerary of flights, often with very little influence from local environmental variations other than photoperiod, and it is species of this type that have the potential to be site-faithful to both their breeding and wintering grounds (Ioale and Benvenuti 1981; Gauthreaux 1982; Robertson and Cooke 1999). Even in obligate migrants it is well known that juveniles have greater variance in their migratory paths than adults, and it may be tempting to ascribe this to a phenotypic response to their unsettled ecological conditions, much as the facultative migrants might display. Very few, if any, studies have studied populations longitudinally, however, to be sure that reduced variance in movement patterns with age is not the simple result of the selective elimination throughout ontogeny of the most deviant individuals.

In addition to these two migratory types are the nomads, species that can, if conditions in one site deteriorate, change their breeding locations by a substantial distance the following year. Note that I restrict the term "nomads" to species that shift their breeding localities, regardless of their patterns of movement outside the breeding season. These shifts in breeding localities seem to result from the search for suitable conditions for breeding, and the shifts in breeding distribution are very often associated with very large-scale movements. Nomads appear to be adapted to environments in which breeding conditions are sometimes very rich but are variably unpredictable. There is probably a spectrum of birds, from those that have a fairly predictable route, with breeding occurring somewhere along that route when conditions are good, to those in which the distribution of breeding sites is much more unpredictable. It is often difficult to distinguish apparently random observations of a nomadic species from those expected from more regular movements accompanied by large-scale fluctuations in population size (e.g., Ford 1978). But what does distinguish nomads from facultative migrants is the probability that birds will breed in very different places in succeeding years. This has resulted in a life history that must include the gathering and processing of information over a very wide spatial extent, and the nomads likely have a very different cognitive map of their world than do more "typical" residents or migrants.

For all their fascination, nomads are by their nature very hard to study. For many nomadic species, the idea of direct mark-recapture studies is not feasible, as the probability of recapturing *any* birds at one site after their initial capture is remote indeed. Among the more than 9,000 birds of the world, probably not more than a few hundred species are nomadic, and, though some nomads (e.g., crossbills [Newton 1972]) occasionally breed in the vicinity of large concentrations of ornithologists, most are confined to the relatively little-studied Arctic (e.g., lemming-reliant raptors [Galushin 1974]) or arid (e.g., *Quelea quelea* [Jones et al. 2002], *Taeniopygia guttata* [Zann and Runciman 1994], Australian honeyeaters [Keast 1968; Ford 1978, 1989]) regions of the world. The nomadic lifestyle is especially interesting in the present context as it represents a melding of what in most birds are two separate life history stages: migration and dispersal. It is as though nomads retain the possibility throughout their lives of undertaking a new natal dispersal event (potentially of extraordinarily long spatial extent) whenever breeding conditions deteriorate.

Most breeding birds (except the nomads), for reasons that are not entirely clear (see, e.g., Pärt 1994, 1995), are generally quite faithful to their breeding sites: they are "philopatric" once they have begun breeding. This breeding philopatry results in "breeding dispersal" displacements between successive breeding sites that are generally much smaller than the displacement associated with the initial natal dispersal event. Given the relative rarity of nomadism, and the prevalence of breeding philopatry, for most birds only natal dispersal generally has the potential to result in a significant net displacement of breeding localities and a net shift in the mean distribution of genetic variants. Thus, for most ornithologists, and for the purposes of this chapter, natal dispersal is *the* life history stage that has the potential to result in large net displacements of breeding birds, and for the rest of this chapter I limit my attention to natal dispersal.

Dispersal

Many of the chapters in this volume serve as eloquent testimony to our long-standing fascination with migration and the pervasiveness of the influence of migration on other aspects of the life history. By contrast, the importance of dispersal to studies of migrant birds has become especially interesting only of late. Beginning in the late 1980s (e.g., Ball et al. 1988) it began to emerge that the general morphological homogeneity of many North Temperate migrant populations over large expanses of their continental ranges is mirrored in the lack of genetic differences. This genetic homogeneity may be due to: (1) recent post-Pleistocene range expansions from small founding populations, (2) selective sweeps or evolution of a generalized genotype that is adaptable to conditions throughout the current broad ranges, and/or (3) pervasive and extensive gene flow (= dispersal) across large parts of the ranges. Thus, to understand the evolution of temperate migrant birds better, population biologists need to learn more about how far, and how frequently, these birds disperse.

Dispersal is also important more generally. Of all the fundamental life history processes that affect population biology, dispersal is probably the most pervasive and least understood (Clobert et al. 2001). Population geneticists have long known that dispersal can act both as a source of genetic variation for evolutionary change (e.g., Wright 1982; Bohonak 1999) and a limit to local adaptation (e.g., Dhondt et al. 1990; Hendry et al. 2001; Lenormand 2002). Birds with shorter dispersal distances may be more prone to local differentiation (Belliure et al. 2000). Population ecologists have also recently re-discovered (Andrewartha and Birch 1954 vs., e.g., Durrett and Levin 1994; Tilman and Kareiva 1997) the importance of dispersal in metapopulation and range dynamics. Dispersal is the glue that binds together the components of a metapopulation, causing the demographic interconnection that is essential to metapopulation dynamics. And without dispersal, the dramatic range extensions that we see today (Shaw 1995; Veit and Lewis 1996) and that we infer in the recent past (Mila et al. 2000) would not have been possible. Finally, demographers know dispersal as the principal confounding factor in estimating survival rates (e.g., Lindberg et al. 2001; Blums et al. 2002): a bird that disappears from a marked open population can only be known with certainty to have dispersed or died if it is found again after leaving.

Why Disperse?

For all its importance for population-level process, dispersal is fundamentally an organismal life history trait, and a great deal of theory has been developed to explain the proportion of offspring that disperse (e.g., Hamilton and May 1977; Johnson and Gaines 1990). Just why fledging birds disperse is a little unclear. In their review of the fitness consequences of dispersal and philopatry, Bélichon et al. (1996) cited only the study by Pärt (1991) on *Ficedula albicollis* as providing direct evidence on the fitness consequences of natal dispersal, and in these Collared Flycatchers, individuals that dispersed actually had lower reproductive success than those that stayed near their natal site. By contrast, Altwegg et al. (2000) more recently reported that dispersing juvenile House Sparrows (*Passer domesticus*) survived better than those that stayed home at either the natal or the colonized sites, and they suggested that natal dispersal was an adaptation to escape poor conditions at the natal site. Forero et al. (2002) added to the confusion by showing that male Black Kites (*Milvus migrans*) that dispersed shorter distances had higher lifetime reproductive success, whereas females experienced no change in reproduction with dispersal distance, though those that dispersed the farthest obtained the most-experienced mates.

It may not be surprising, then, that many of the putative advantages of dispersal (e.g., discovery of new habitat, avoidance of overcrowding and inbreeding) can be countered by apparent disadvantages (e.g., leaving a familiar area, risk of never encountering good habitat). There is ample room for novel hypotheses regarding the costs and benefits of dispersal, and some of these suggest a causal connection between dispersal and migration. For instance, given that the bulk of breeder mortality in most migrants takes place away from the breeding grounds, perhaps natal dispersal provides first-time breeders with the best chances of locating a breeding vacancy during the intense period of territorial and mate selection that is wedged in between migratory return and the start of breeding (R. Greenberg, pers. comm. 2002). O'Connor (1985) proposed that European migrant birds more commonly occupy intermediate successional habitats than residents, and the more rapid turnover of breeding opportunities may thus favor stronger natal dispersal in migrants as a result (Paradis et al. 1998). Given the dramatic post-Pleistocene expansions of most migrant bird populations (e.g., Mila et al. 2000), extensive dispersal must have been strongly favored in the recent geological past, and, in the absence of strong selection to reduce dispersal, its preponderance in the current fauna may be an artifact of historical conditions. On a global scale, strong dispersal abilities, in both migrants and residents, may now be favored as Earth once again experiences rapid changes in climate and attendant habitat distributions (Parmesan and Yohe 2002).

Although it is probably true that birds that disperse more often tend to disperse farther on average (cf. Pulido et al. 1996 for migration), there is relatively little understanding of the selective forces affecting variation in how far they disperse (e.g., Pärt 1990 review in Colbert and Lebreton 1991; Payne 1991; McCarthy 1996; Ezoe 1998; Rousset and Gandon 2002). For many of dispersal's population functions, this latter understanding is more critical, as we often are as curious about *how far* they disperse as about *whether* they do so.

Migration's Relation to Dispersal

Given that migration and dispersal are both significant movements of birds, how are they related? Migration involves finding the ecological conditions that make surviving the annual cycle possible, whereas dispersal locates a place to breed. These are different functions, and thus the source of different selective pressures. Although much of this chapter explores the connections between them, it remains a viable possibility that the connection between dispersal and migration is weak or nonexistent. Migrants are known to have remarkable powers of orientation and navigation (e.g., Berthold et al. 2003) that suggest that in some sense *they know where they are going.* The mechanisms that allow them to reach their goals may be very different from those that allow a dispersing juvenile to *explore and find a place* to breed. Nomads probably use distinctive methods for finding their way, and residents clearly disperse with no migration. The connection between dispersal and migration, then, definitely needs justification.

As suggested above, dispersal and migration may be connected through historical associations in the selective factors affecting the population biology of migrants and their dispersal biology. Another potential historical connection is that migration in many cases seems likely to have evolved from the gradual extension of dispersal or food-finding movements (Levey and Stiles 1992; Berthold 1999). It is clearly possible, however, to see the action of very different selection pressures in producing long-distance annual movements; some groups (e.g., warblers, tanagers, flycatchers) can be viewed as tropical birds being selected to venture poleward to breed (Levey and Stiles 1992; Chesser and Levey 1998), whereas other groups (e.g., shorebirds) can be seen as Arctic birds selected to escape equatorward for the winter. But it is clear that, even for shorebirds, these simple scenarios for the evolution of migration cannot stand detailed scrutiny: "the time of unidimensional explanatory hypotheses is behind us" (Piersma 1996:479; see also Joseph et al. 1999).

To the extent that long-distance migrants have distinctive physiology and morphology (e.g., Holberton and Dufty, Chap. 23, this volume, and Winkler and Leisler, Chap. 7, this volume), it would not be at all surprising if their increased capacity for movement affected their dispersal distances and methods. It would also not be surprising if the neurological and cognitive equipment possessed by exploratory dispersers influenced the ways and places that they migrated. In sum, though, it must be granted that there is very little hard empirical evidence for a relationship between dispersal and migration, and I now, after a methodological detour, attempt to assess directly the correlations between them.

METHODS FOR STUDYING DISPERSAL

The many methods for studying the migrations of birds rely heavily on the facts that most birds migrate in flocks or aggregate at stopover sites during their migrations and that they occur in some seasons in ranges largely or totally disjunct from their breeding distributions. If birds generally migrated solitarily and only within the bounds of their overall species breeding range, imagine how difficult a study of migration would be! This is exactly the situation faced by those wishing to study dispersal. Dispersal is generally a solitary activity, and we are forced to pursue its study through the direct method of banding pre-fledglings and recapturing them on their breeding sites or through a series of indirect methods.

Genetic Methods

Because direct mark-recapture methods are notoriously demanding of researcher time (and thus expense), several indirect methods of estimating dispersal have been developed. The most commonly available of these is to infer rates of dispersal between populations by using F_{ST}, the ratio of within- and between-population variance in frequencies of neutral genetic markers. This method has the advantages that it is fairly easy to accomplish and that the lab work to characterize a large number of neutral loci is becoming increasingly routine and inexpensive. Unfortunately, the most common methods to infer dispersal from estimates of F_{ST} for the set of populations under consideration entails a large number of assumptions (e.g., that there is no selection, drift, or mutation; that populations are of the same size, connected by equiprobable dispersal; that the entire system is at equilibrium) that likely are never met in the majority of organisms (e.g., Whitlock and McCauley 1999), including birds (Crochet 1996). Even if one is willing to make these assumptions, one is left with the problem that the F_{ST} approach can only measure the product of the effective population size (*N*) and the "migration rate" (m = dispersal rate for ornithologists). Thus, to accurately estimate dispersal rates, one must have some way of estimating effective population sizes over the period of time during which the markers being studied have been changing in frequency. This is a major hurdle, and this need to account for the history of population size is joined by the other concern, that patterns of genetic variation may be the effect of the "ghost of dispersal past." Patterns of genetic variation observed today are the result of a complicated mix of factors acting at present with those acting up to millions of years ago. Thus F_{ST} approaches yield *very* indirect estimates of dispersal rates between populations.

Recent coalescent models (Beerli and Felsenstein 1999, 2001; Nielsen and Wakeley 2001) may eventually replace F_{ST} approaches, as they make use of all the genealogical information contained in molecular data. These approaches can be very flexible as to the sorts of dispersal and population growth that are to be considered, but they are computationally intensive.

A less apparent problem with both the F_{ST} and coalescent approaches as they are usually practiced is that they require the a priori identification of the geographic limits of the populations that are sampled. In practice, this is usually done by designating different sampling localities as different populations, but it is important to realize that this implicit assumption about population definition may vary dramatically from the actual amount of genetic independence that exists on the ground and that different designations of population membership affect the calculations of Nm. Methods have been developed for the estimation of Nm in more complicated continuous spatial structures (e.g., Epperson 1995, 1999; Epperson and Li 1996), but these have apparently not been used in any empirical avian studies. More substantively for studies of dispersal, these methods only produce an estimate of the rate of movement between populations, and any estimate of dispersal distance or direction must rest on calculations based on the mean distances and directions between populations. This would be fine if populations were small and discretely defined, but, at least for most temperate bird distributions, populations are probably very widely distributed and their members generally very sparsely sampled.

A more direct approach to using molecular methods to estimate dispersal distance was developed by Neigel et al. (1991, 1997). Assuming a random-walk model of genetic change, one can work from the phylogeny and geographic distribution of genetic variants to estimate mean dispersal distance per generation. This method has now been applied in two migrant temperate passerines (Zink and Dittmann 1993a, 1993b), three passerines of successional habitats in Panama (Brawn et al. 1996), and five Amazonian forest resident passerines (Bates 2002), the last of which evidence estimated dispersal distances that are roughly an order of magnitude smaller than those of the temperate and successional tropical species. Thus, the similarity of distance estimates from Zink and Dittmann's migrants and Brawn et al.'s residents suggests that habitat may be more important than migratory movements per se in affecting dispersal distances. Clearly, more applications of these techniques to larger numbers of (preferably closely related) species would be very informative.

Neigel's method has been developed for mitochondrial DNA variation, and given the explosion in studies of nuclear microsatellite variation, it will be interesting to see if phylogenetic methods for this DNA can be developed, and the method of Neigel et al. applied in future. Neigel (pers. comm. 2002) is updating his method to a maximum-likelihood approach, and further developments of the technique and applications to avian data sets seem very desirable.

In any event, if large numbers of indirect estimates of dispersal rates or distances are obtained, comparisons of residents and migrants might seem a straightforward way to investigate the connection between migration and dispersal. To control for the variable effects of different population genetic and genomic histories in different lineages, we would probably want to make our comparisons as tightly constrained phylogenetically as possible, and with similar range sizes, comparing, for example, the dispersal behavior of the Tree Swallow with that of its resident (or very low movement) congener the Mangrove Swallow (*T. albilinea*). There are many other such comparisons possible, especially in groups such as the Tyrannidae, Parulidae, and Sylviidae, but when one considers such comparisons (e.g., Buerkle 1999), one very often encounters the problem that resident forms tend to have more restricted (and more equatorial) ranges than their migrant relatives. Smaller ranges and smaller population sizes go hand in hand, and variation in population size has a tremendous effect on measures of genetic variability, thus confounding any attempt to measure dispersal in a comparable way in populations of differing migratory status.

Stable Isotopes to the Rescue?

Another potential indirect approach to studying dispersal rates and distances is to use geographic changes in the distribution of stable nuclear isotopes to determine the rates of movement between populations. Although the method requires a mass spectrometer, individual sample costs can be quite reasonable, and spatial variation in isotope distributions sometimes seems well suited to estimating variation in dispersal. Graves et al. (2002) recently interpreted differences in isotope variability of two age-classes of a warbler population to be due to the effects of dispersal. But so far, no one appears to have used isotope distributions to estimate dispersal rates, directions, or distances. Given their potentially large range of movement and the difficulty of recapturing them, nomadic species may be especially attractive subjects for isotopic research.

The Direct Mark-Recapture Method

Even when these indirect methods are refined further, it will be difficult to do more than estimate the rates of movement between populations or arbitrarily defined samples of individuals or habitats, because all of these methods (except the phylogeographic methods of Neigel et al.) rely on the comparisons of mean characteristics of collections of individuals from a population or sampling locality. To do more than estimate rates, and to actually estimate the distributions of distances that individuals disperse, will still require the marking and recapturing of large numbers of dispersing young. Such direct mark-recapture studies that have been done (Sutherland et al. 2000) have often included histograms of the frequencies of individuals dispersing different distances from the natal site. Such histograms have been influential in guiding our thought about how far the two sexes disperse (Greenwood 1980) and the role that such sex differences in dispersal distance can play in such things as the avoidance (Moore and Ali 1984) and optimization (Shields 1983) of inbreeding. But several authors (e.g., Barrowclough 1978; van Noordwijk 1995) have also pointed out how dependent these estimates of dispersal distance distributions are on artifacts induced by the limited study areas in which most studies of dispersal have been conducted: across a range of distances from the natal site, birds often are recaptured in proportion to the recapture opportunity that the study site provides at each distance. And no matter what corrections for this bias are applied (e.g., Baker et al. 1995; Koenig et al. 1996; Thomson et al. 2003), no study of dispersal distance distributions can extend to potential distances greater than the study area allows.

Insights from Ithaca Studies of Tree Swallows

The direct way to deal with this problem is to increase the size of the study area until the spatial scale of the study area does not artificially truncate the distribution of dispersal distances. For the past 10 years I have been endeavoring to do just that in our studies of dispersal in Tree Swallows centered on Ithaca, New York (42°30′ N, 76°28′ W). By recruiting amateur banders, we have extended the size of our study area to a circle of 400-km radius around Ithaca. The details of this study are described elsewhere (Winkler, unpubl. data), but at this spatial extent, we are finally to the point where the decay in potential dispersal recaptures with distance is much slower than the decay with distance in actual recaptures. Thus, we believe we have measured the dispersal distance distribution relatively free of study-area artifacts. For the present purposes, there are two relevant conclusions from this work. First, the great majority of dispersal events that we have detected have been less than 50 km in extent, and the mean distance dispersed by Tree Swallows is only 6.92 km for females and 2.46 km for males. Thus, it is clear that swallows, despite the fact that they migrate to the Tropics and back every year, do not disperse far.

Why don't migrants disperse farther than they do, when distance traveled is clearly no great obstacle? Is it possible that they have been selected not to leave behind local adaptations (to parasites, mate compatibility, and so on) or knowledge of familiar predators, competitors, and feeding and breeding sites? Does the fact that they fledged there indicate that at least some reproduction at the natal site is clearly possible, and that it is thus preferable to a suite of unknown alternatives? Is it possible that birds are optimizing their inbreeding (Shields 1982, 1983) by not moving farther from their natal site? Both the costs and benefits of dispersal (e.g., Bélichon et al. 1996) must be known better before we can hope to understand variations in the means and distributions of dispersal distances.

No matter what the ultimate value of dispersal distance variation, there is another conclusion from the Tree Swallow study. Like other observed dispersal distance distributions (Greenwood et al. 1979; Greenwood 1980; Payne 1991), the distribution of dispersal distances for Tree Swallows drops off steeply with increasing distance from the natal site (fig. 30.2). There is, however, a very heavy long-distance tail in the distribution of dispersal distances, at least in females, raising the possibility of a bimodality in dispersal behavior that has only been hinted at before (e.g., Haas 1990). This possible bimodality, together with the fact that most study areas cannot extend to the scales of even the mean dispersal distance in this study, suggests considerable caution in the interpretation of dispersal distance distributions from less-broad-scaled studies.

Armed with this caveat, we can have a look, however, at several recent compendia of dispersal distance distributions. Sutherland et al. (2000) reviewed 77 studies that obtained at least some information, most from studies in a single area, with capture-recapture methods. Of these, only 39 had sufficient data to estimate a dispersal distance distribution. To combine the dispersal distances of different taxa into a single data set, Sutherland et al. divided the distance axis for each distance distribution by the median dispersal distance for each species. This resulted in an aggregate distance distribution that, like other individual distributions that have been published in the past, declines steeply with distance (fig. 30.2). Sutherland et al., citing the work of Turchin (1998), point out that such distributions are, relative to a normal distribution, both left skewed and platykurtotic:

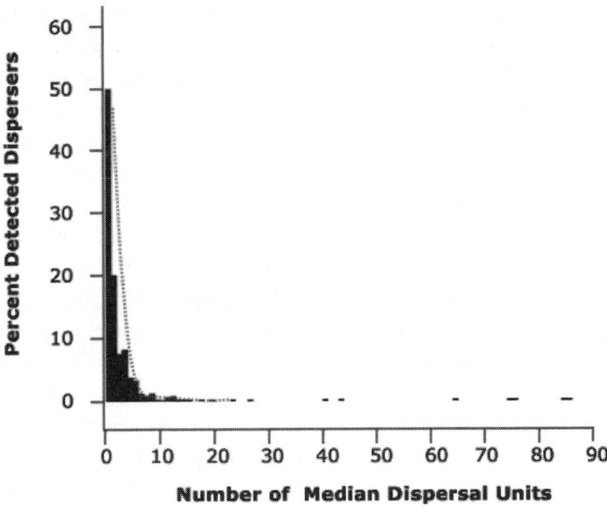

Fig. 30.2. The distribution of 791 natal dispersal events detected in a study area of 400 km radius centered on Ithaca, New York, in the years 1985–1998. The dashed lined is the distribution reported by Sutherland et al. (2000). Note the considerable number of recaptures at distances beyond the limit of about 23 median dispersal units inherent in the data reviewed by Sutherland et al. Also, note that the observed numbers of dispersers in the range reported by Sutherland et al. is almost uniformly smaller than those predicted by the Sutherland et al. function. For further details see text.

there are relatively many individuals that stay close to home and there are "heavy tails" to the distribution made up of individuals that disperse a very long way from the natal site (see also Paradis et al. 2002). The interesting conclusion that emerges from superimposing the curve of Sutherland et al. on the New York Tree Swallow data (fig. 30.2) is that the heaviness of the tails may likely be even greater than Sutherland et al. might have imagined, suggesting to me the possibility of a second long-distance mode of dispersal. If we had limited ourselves to a little over 20 median dispersal units as did Sutherland et al., the distributions of distances in the two data sources would be very similar. But the addition of the much longer distance dispersal events has the effect of further emphasizing the tail of the distribution and thus depressing the percentage of recaptures nearer the origin. It is dangerous to interpret too much from a comparison of one species with a distribution derived from many, but one clear possibility to explain this discrepancy is that all or most of the studies cited by Sutherland et al. have been rather severely distance-censored by limited study areas.

An alternative approach to the essentially single-site-based studies reviewed by Sutherland et al. is to use data from national banding schemes to detect displacement of birds from on or near their natal ground to recoveries during the succeeding breeding season. This approach overcomes study-area artifacts by analyzing data from anywhere that the species has been banded throughout its range. But this broad perspective comes at the cost of detailed knowledge. Detail on short-distance movements, which may constitute a large proportion of the dispersal-distance distribution (fig. 30.2), can be lost, especially in countries (such as the United States) where the national banding schemes do not store data at a scale finer than a 10′ lati-long block (which corresponds to a rectangle of about 6.5 × 9 km in the Temperate Zone). Most challenging, the approach requires troublesome assumptions about natal sites and breeding status on recovery, as the researcher is forced to make judgments about whether the initial capture site was indeed the natal site and whether the subsequent capture site was indeed a site where the bird was actually breeding. Nevertheless, this approach has been used in North American birds (Moore and Dolbeer 1989; C. M. Francis, pers. comm. 2002), and in its most comprehensive application in the study of 75 British terrestrial birds by Paradis et al. (1998).

What Do These Data Sets Tell Us about Migration and Dispersal?

Both of these approaches can be used to test for a correlation between migratory status and dispersal distance. Sutherland et al. (2000) tested such a correlation, after controlling for body size and diet, but comparisons of median dispersal distance yielded no suggestion of an effect of migration ($p = 0.93$). The larger data set, in which only an estimate of maximum dispersal distance was possible, yielded a mean maximum dispersal distance of 28.4 km for migrants (both obligate and facultative, $n = 39$) and 24.3 km for residents ($n = 34$). These differences in means were not significant ($p = 0.41$) in a direct multivariate test. There was, however, a significant ($p = 0.046$) interaction (not diagnosed further in the chapter), between social system and migratory status in their effect on maximum dispersal distance. Maximum dispersal distance is a very unreliable indicator of dispersal distance because it is so heavily influenced by sample size and detectability. All told, this is very scant evidence of any relation between migration and dispersal distances, though it is conceivable that further analysis of the interaction with social system could yield a stronger conclusion.

Paradis et al. reported a mean natal dispersal distance of 22.8 km for migrants ($n = 16$) and 15.6 km for residents ($n = 59$), a difference that was highly significant ($p < 0.001$) after controlling for body size. Despite the strength of this relationship, Paradis et al. (1998:531) stressed that this result needed "to be interpreted cautiously," as they were well aware of the problems of interpreting the initial banding location as the natal site. They conducted special analyses and corrections in their data to get as accurate a measure as possible, but they still considered their result as "provisional."

Interpretations of Correlations between Migration and Dispersal Distances

Clearly there are big problems in interpreting too strongly the summaries of dispersal distances provided by Sutherland et al. and Paradis et al. as evidence of the correlation between migration and dispersal distances. But there seem to be good reasons to expect such a correlation, and these results, especially those of Paradis et al., are not likely to be considered "provisional" by many (e.g., Belliure et al. 2000). More direct mark-recapture studies of the relation between migration and dispersal distance are clearly needed. Given that most temperate migrant passerines appear to have originated ultimately from tropical ancestors, a more phylogenetically controlled comparison for the effect of migration on dispersal would be, as suggested above for indirect methods, to compare the dispersal of temperate migrants with that of tropical residents. Tropical residents are notoriously reluctant to cross habitat barriers (e.g., Capparella 1992), and what little is known of dispersal in these species suggests that fewer of these birds disperse, and at far smaller distances, than do temperate migrants (Greenberg and Gradwohl 1997). Clearly, parallel direct studies of movement using the same methods on closely related species are badly needed, as all we have from direct studies is the tantalizing possibility that tropical residents may indeed have far smaller ranges of movement than migrants.

Despite the slim empirical evidence for a connection, the biological properties of migrants and residents persistently suggest one. Aren't migrants just more prone to movement and thus more likely to move longer distances during dispersal? It has long been known that morphological characters like wing length correlate with migratory distance, and migrants have the endocrine and metabolic machinery that fits them for long flights (e.g., Holberton and Dufty, Chap. 23, this volume). Perhaps possession of this greater flight distance potential carries over directly into different dispersal distances. Isn't it also possible that migrants are more exploratory / less neophobic than their sedentary relatives (Greenberg 1990; Grünberger and Leisler 1993; Greenberg and Mettke-Hofmann 2001; Mettke-Hofmann and Greenberg, Chap. 10, this volume), a trait that presumably would aid in the exploration of new habitats (see Why Disperse? above). Could such psychological differences between migrants and nonmigrants lead to differences in how they disperse and how far they go when they do so? And how are the cognitive abilities of migrants and residents likely to differ? Do migrants have the ability to store and process much more spatially structured data than residents? These and many other fascinating questions remain to be explored, but taken together they suggest a basis for expecting a connection between migration and dispersal. The nomads also serve as a constant reminder of the likely diversity in avian modes of movement and the physiological and psychological means to achieve them.

What about Dispersal Directions?

In migration studies, much attention has been paid to orientation, and a similar focus in dispersal studies may suggest another tangible way that migration and dispersal may connect. The directions taken from the home nest for natal dispersal are especially interesting because the process of migration may have an effect on where dispersers end up. Interest in the behavioral mechanisms underlying dispersal has surged, especially in those mechanisms involved with the selection of the first breeding site (e.g., Brewer and Harrison 1975; Reed et al. 1999); most attention has focused on the possibility that birds gather information on available breeding sites, perhaps even ranking them according to the density (Muller at al 1997; Serrano et al. 2001) or mean breeding success (Danchin et al. 1998; Brown et al. 2000; Doligez et al. 2002) of breeders during the summer that they fledge. Under this hypothesis, birds in their first breeding summers are returning to sites that they "prospected" the previous summer immediately after fledging. Such an interpretation was suggested early on by Lombardo (1987) for the cavity-visiting behavior of juvenile Tree Swallows, and our fragmentary observations from upstate New York are consistent with the possibility that juveniles do indeed prospect in their first summer for available sites. This prospecting at a relatively fine spatial scale presumably interacts with the larger-scale information opportunities and challenges faced by a long-distance migrant. If one considers a northbound migrant in spring, on the way to its first breeding site, it is easy to imagine that it is collecting information on its way about potential breeding sites. It is also easy to see that, either for adaptive reasons or because of mistakes in orientation or navigation or the vagaries of hostile weather conditions, variance in the length and bearing of the return migration could arise. Such variance in where they return may impart variance on where the returnees could potentially settle to breed: if variance in migration length were most prevalent, then we might expect to see a north-south bias in dispersal directions from the natal site, and if variance in return bearing were more common, then dispersal directions could be biased in an east-west direction.

Our Tree Swallow data set allows us to ask whether the dispersal directions of the birds have been biased in direction. For every natal dispersal event in our data set, we divided the landscape surrounding each natal site into four "pie slices" centered on each of the primary compass directions. We further divided each of these slices into concentric ring segments of 0–10 km, 10–20 km, and 20–400 km radii. Within each of these segments, we tabulated for each dispersing bird the number of captures we made in the year of settlement, and we used these totals to construct the expected distribution of dispersal directions. (To explicitly consider the possibility of prospecting-derived information, we also ran these analyses with expected distributions based on captures in the fledging year (rather than the year of settlement), but the results were not qualitatively different from those reported below.)

Comparing these expected distributions with those observed (table 30.1) provides some interesting results. Those birds that disperse to breeding localities near (<10 km from) the natal site are less likely to show up north, and more likely east, of the natal site. Birds that settle in the range of 10–20 km from the natal site show no departure from a random dispersion, and those that settle far (>20 km) from the

Table 30.1 The observed distributions of recaptures of Tree Swallows in "pie slices" around each compass direction from their natal site

	0–10 km		10–20 km		20–400 km	
	Observed	% Deviation	Observed	% Deviation	Observed	% Deviation
North	236	–4.99	18	–3.48	7	–8.57
South	176	1.73	15	3.63	8	–29.43
East	135	3.63	19	5.60	11	21.71
West	133	–0.38	14	5.76	9	16.29
	$\chi^2 = 10.53$	$p = 0.015$	$\chi^2 = 2.42$	$p = 0.490$	$\chi^2 = 25.78$	$p < 0.001$

natal site show a significant biasing toward east-west, and away from north-south, dispersal directions.

This pattern might be interpreted in may ways, but I suggest the following. It may be that short-distance dispersal movements are informed by prospecting, with a slight bias to the southeast because that is the direction from which the birds are returning to the Finger Lakes region (they migrate north in spring along the Atlantic Coast before heading inland). And it is possible that birds dispersing more than 10 km from their natal site end up settling beyond the range that is well known from the previous season's prospecting and only after a randomizing search. Finally, the birds that disperse the farthest (>20 km) from their natal sites may indeed have gotten lost or displaced by weather in their northward journey and settled out in the vicinity of where they returned without revisiting the vicinity of their natal site at all.

Interpreting these results further will require more-detailed investigations of the biology of dispersing birds, but it may also be worthwhile to note that the indirect methods of Neigel (1997) can produce separate estimates of movement in both the north-south ("latitudinal") and east-west ("longitudinal") directions. In the only such separate estimates for a temperate bird, Zink and Dittmann (1993a) reported greater estimated dispersal distance in the longitudinal direction (6.1 km) than in the latitudinal direction (3.6 km) in their study of molecular variation in Song Sparrows (*Melospiza melodia*). By contrast, Bates (2002) reported smaller longitudinal estimates in all five species of Amazonian forest passerines that he studied.

Other Benefits of Post-Fledging Movements

Besides prospecting for future breeding sites, several other functions for post-fledging movements have been proposed. R. R. Baker (1993) suggested that birds may adaptively bias the direction of their post-fledging movements in order to build a cognitive map of the region that they intended to return to in the following breeding season. Møller and Erritzøe (2001) recently suggested that juveniles, by prospecting, may be exposing themselves to novel antigens, thus vaccinating themselves against what may be local parasites in future breeding seasons. Neither of these ideas is exclusive of the interpretation of post-fledging movements as prospecting, and clearly, we have a great deal to learn about post-fledging movements and the roles they play in the lives of birds.

Movements and the Gathering of Information

All I am suggesting here is that natal dispersal patterns in migrant species are probably a mix of the effects of where migration settles the dispersers and how much they know about the site in advance. Do nocturnal migrants have different sources of information, different reliability of their orientation, and thus different patterns of dispersal directions than diurnal ones? Does foraging mode have an effect? It seems possible, for instance, that aerial insectivores and other planktivores may process and store environmental information at a different scale than foliage gleaners. What about breeding sociality / territoriality / coloniality? Do some taxa have group migration and group dispersal, and are solitary migrants more prone to migration mistakes? Are migrants likely to make more wide-ranging prospecting movements, either before departure or immediately upon their return (Baker 1993; Reed et al. 1999)? Migration and prospecting clearly rely on alternative means for gathering information. The mysteries of how birds acquire information about where to go, during both migration and dispersal, are likely to fascinate us for a very long time.

Variation in dispersal distances may also depend on migration in another way. Returning first-time breeders may know just where they are and where they want to breed on the basis of past prospecting, but they may be precluded from doing so because another bird got there first. Thus, migration, by posing a challenge that sorts among birds of different individual quality (e.g., Kokko 1999), could easily result in greater dispersal distances or dispersal to less preferred sites in those birds that are later in returning to the breeding grounds (Forstmeier 2002). And, of course, one of the ways in which individuals may differ in individual quality is in how they acquire, store, and process information on the availability and defensibility of breeding opportunities.

CONCLUSIONS AND PROSPECTS

Migration and dispersal are the two most important movements in birds' lives, and, despite the paucity of direct empirical confirmation, the balance of evidence suggests that they are not independent phenomena. Future research will profitably explore how the sensory and cognitive machinery of migrants is harnessed during the process of dispersal and how physical capacities for movement affect both migration and dispersal. Population biologists also have much to uncover about the effects of movements across landscapes over geological time. In every aspect on which I have touched in this chapter, further creative research is sorely needed, and the connection between migration and dispersal is worthy of much more continued study.

Specific studies that seem especially pressing are comparative studies of dispersal biology, in both tropical and temperate latitudes, comparing migrants and residents. These studies must be done in a well-resolved phylogenetic and phylogeographic context, so that differences observed can be confidently attributed to differences in movements rather than differences in habitats, population sizes, or their varied history. This is an extremely tall order, but promising pairs of taxa seem available in warblers (of both the Old and New Worlds), flycatchers, and swallows, and both mark-recapture and the indirect methods of Neigel should probably be applied.

As we learn more about how brain structure affects behavior, it will be very interesting to compare neurological structure and function of migrants and residents, dispersers, and settled breeders. Already, it would be interesting to compare hippocampal size in migrants versus residents versus nomads, and in juveniles while prospecting versus young breeders in their first breeding season. As the spectrum of available neurological tests increases, the possibility that we may be able to assess the influence of such factors as neophobia on the exploration of space during dispersal and migration is an exciting prospect indeed.

ACKNOWLEDGMENTS

Many thanks to my lab group at Cornell for several profitable discussions of these ideas, especially to Peter Wrege, Laura Stenzler, Becca Saffran, Dave Hiebeler, and Gary Langham. My colleagues Rasmus Nielsen (Department of Biological Statistics and Computational Biology) and Irby Lovette at Cornell provided very valuable advice on indirect methods and their applications to dispersal. David Paton (University of Adelaide, Australia) first taught me of the nomadic birds of Australia, and he and his wife, Penny, together with Mari Kimura, Irby Lovette, Claudia Mettke-Hofmann, Theunis Piersma, Trevor Price, and Laura Stenzler helped with some key references. David Serrano, an anonymous reviewer, and especially Russ Greenberg provided valuable comments on the manuscript.

Dozens of Swallow/Bluebird Dispersal Study participants from around New York and surrounding states and provinces made our extended studies of Tree Swallow dispersal possible, an effort that benefited greatly in its formative years from a joint National Science Foundation-Informal Science Education grant with Rick Bonney, André Dhondt, and John Fitzpatrick that supported the efforts of Paul Allen, Tracey Kast, and Pixie Senesac. Peter Wrege performed the analyses of the Tree Swallow data, and John Tautin and his staff at the Bird Banding Laboratory of U.S. Geological Survey have been patient and supportive.

And finally, I thank Peter Marra and Russell Greenberg for inviting me to think with them about the meeting from which this chapter grew.

LITERATURE CITED

Altwegg, R., T. H. Ringsby, and B. E. Saether. 2000. Phenotypic correlates and consequences of dispersal in a metapopulation of house sparrows *Passer domesticus*. Journal of Animal Ecology 69:762–770.

Andrewartha, H. G., and L. C. Birch. 1954. The Distribution and Abundance of Animals. University of Chicago Press, Chicago.

Baker, M., N. Nur, and G. R. Geupel. 1995. Correcting biased estimates of dispersal and survival due to limited study area: theory and an application using Wren Tits. Condor 97:663–674.

Baker, R. R. 1993. The function of postfledging exploration: a pilot study of three species of passerines ringed in Britain. Ornis Scandinavica 24:71–79.

Ball, R. M., S. Freeman, F. C. James, E. Bermingham, and J. C. Avise. 1988. Phylogeographic population structure of Red-winged Blackbirds assessed by mitochondrial DNA. Proceedings of the National Academy of Sciences USA 85:1558–1562.

Barrowclough, G. F. 1978. Sampling bias in dispersal studies based on finite area. Bird-Banding 49:333–341.

Bates, J. M. 2002. The genetic effects of forest fragmentation on five species of Amazonian birds. Journal of Avian Biology 33:276–294.

Beerli, P., and J. Felsenstein. 1999. Maximum-likelihood estimation of migration rates and effective population numbers in two populations using a coalescent approach. Genetics 152:763–773.

Beerli, P., and J. Felsenstein. 2001. Maximum likelihood estimation of a migration matrix and effective population sizes in *n* subpopulations by using a coalescent approach. Proceedings of the National Academy of Sciences USA 98:4563–4568.

Bélichon, S., J. Clobert, and M. Massot. 1996. Are there differences in fitness components between philopatric and dispersing individuals? Acta Oecologica 17:503–517.

Belliure, J., G. Sorci, A. P. Møller, and J. Clobert. 2000. Dispersal distances predict subspecies richness in birds. Journal of Evolutionary Biology 13:480–487.

Belthoff, J. R., and S. A. Gauthreaux, Jr. 1991. Partial migration and differential winter distribution of house finches in the eastern United States. Condor 93:374–382.

Berthold, P. 1984. The control of partial migration in birds: a review. Ring 10:253–265.

Berthold, P. 1999. A comprehensive theory for the evolution, control and adaptability of avian migration. Ostrich 70:1–11.

Berthold, P., E. Gwinner, and E. Sonnenschein (eds.). 2003. Avian Migration. Springer-Verlag, Heidelberg and New York.

Blums, P., J. D. Nichols, J. E. Hines, and A. Mednis. 2002. Sources of variation in survival and breeding site fidelity in three species of European ducks. Journal of Animal Ecology 71: 438–450.

Bohonak, A. 1999. Dispersal, gene flow, and population structure. Quarterly Review of Biology 74:21–45.

Brawn, J. D., T. M. Collins, M. Medina, and E. Bermingham. 1996. Associations between physical isolation and geographical variation within three species of Neotropical birds. Molecular Ecology 5:33–46.

Brewer, R., and K. G. Harrison. 1975. The time of habitat selection by birds. Ibis 117:521–522.

Brown, C. R, M. B. Brown, and E. Danchin. 2000. Breeding habitat selection in cliff swallows: the effect of conspecific reproductive success on colony choice. Journal of Animal Ecology 69:133–142.

Buerkle, C. A. 1999. The historical pattern of gene flow among migratory and nonmigratory populations of prairie warblers (Aves: Parulinae). Evolution 53:1915–1924.

Capparella, A. P. 1992. Neotropical avian diversity and riverine barriers. Pages 307–316 *in* Proceedings of the 20th International Ornithological Congress, Christchurch.

Chan, K. 2001. Partial migration in Australian landbirds: a review. Emu 101:281–292.

Chesser R. T., and D. J. Levey. 1998. Austral migrants and the evolution of migration in new world birds: diet, habitat, and migration revisited. American Naturalist 152:311–319.

Clobert, J., E. Danchin, A. A. Dhondt, and J. D. Nichols (eds.). 2001. Dispersal. Oxford University Press, Oxford.

Clobert, J., and J. D. Lebreton. 1991. Estimation of demographic parameters in bird populations. Pages 75–104 *in* Bird Population Studies: Relevance to Conservation and Management (C. M. Perrins, J. D. Lebreton, and G. Hirons, eds.). Oxford University Press, Oxford.

Crochet, P. A. 1996. Can measures of gene flow help to evaluate bird dispersal? Acta Oecologica 17:459–474.

Danchin, D., T. Boulinier, and M. Massot. 1998. Conspecific reproductive success and breeding habitat selection: implications for the evolution of coloniality. Ecology 79:2415–2428.

Dhondt, A. A., F. Adriaensen, E. Matthysen, and B. Kempenaers. 1990. Nonadaptive clutch sizes in tits. Nature 348:723–725.

Doligez, B., E. Danchin, and J. Clobert. 2002. Public information and breeding habitat selection in a wild bird population. Science 297:1168–1170.

Durrett, R., and S. A. Levin. 1994. Stochastic spatial models: a user's guide to ecological applications. Proceedings of the Royal Society of London, Series B, Biological Sciences 343: 329–350.

Epperson, B. K. 1995. Spatial distribution of genotypes under isolation by distance. Genetics 140:1431–1440.

Epperson, B. K. 1999. Gene genealogies in geographically structured populations. Genetics 152:797–806.

Epperson, B. K., and T. Q. Li. 1996. Measurement of genetic structure within populations using Moran's spatial autocorrelation statistics. Proceedings of the National Academy of Sciences USA 93:10528–10532.

Ezoe, H. 1998. Optimal dispersal range and seed size in a stable environment. Journal of Theoretical Biology 190:287–293.

Ford, H. A. 1978. The Black Honeyeater: nomad or migrant? South Australian Ornithologist 27:263–269.

Ford, H. A. 1989. Ecology of Birds: An Australian Perspective. Surrey Beatty and Sons Ltd., Chipping Norton, Australia.

Forero, M. G., J. A. Donazar, and F. Hiraldo. 2002. Causes and fitness consequences of natal dispersal in a population of black kites. Ecology 83:858–872.

Forstmeier, W. 2002. Benefits of early arrival at breeding grounds vary between males. Journal of Animal Ecology 71:1–9.

Galushin, V. M. 1974. Synchronous fluctuations in populations of some raptors and their prey. Ibis 116:127–134.

Gauthreaux, S. A., Jr. 1982. The ecology and evolution of avian migration systems. Pages 93–167 *in* Avian Biology, Vol. 6 (D. S. Farner and J. R. King, eds.). Academic Press, New York.

Graves, G. R., C. S. Romanek, and A. R. Navarro. 2002. Stable isotope signature of philopatry and dispersal in a migratory songbird. Proceedings of the National Academy of Sciences USA 99:8096–8100.

Greenberg, R. 1990. Feeding neophobia and ecological plasticity: a test of the hypothesis with captive sparrows. Animal Behaviour 39:375–379.

Greenberg, R., and J. Gradwohl. 1997. Territoriality, adult survival, and dispersal in the Checker-throated Antwren in Panama. Journal of Avian Biology 28:103–110.

Greenberg, R., and C. Mettke-Hofmannn. 2001. Ecological aspects of neophobia and neophilia in birds. Current Ornithology 16:119–178.

Greenwood, P. J. 1980. Mating systems, philopatry and dispersal in birds and mammals. Animal Behavior 28:1140–1162.

Greenwood, P. J., P. H. Harvey, and C. M. Perrins. 1979. The role of dispersal in the Great Tit (*Parus major*): the causes, consequences and heritability of natal dispersal. Journal of Animal Ecology 48:123–142.

Grünberger, S., and B. Leisler. 1993. Die Ausbildung von Habitatpräferenzen bei der Tannenmeise (*Parus ater*): genetische Prädisposition und Einfluß der Jugenderfahrung. Journal für Ornithologie 134:355-358. [Abstract in English.]

Haas, C. A. 1990. Breeding ecology and site fidelity of American Robins, Brown Thrashers and Loggerhead Shrikes in shelterbelts in North Dakota. Ph.D. dissertation, Cornell University, Ithaca.

Hamilton, W. D., and R. M. May. 1977. Dispersal in stable habitats. Nature 269:578–581.

Hendry A. P., T. Day, and E. B. Taylor. 2001. Population mixing and the adaptive divergence of quantitative traits in discrete populations: a theoretical framework for empirical tests. Evolution 55:459–466.

Howard, W. E. 1960. Innate and environmental dispersal of individual vertebrates. American Midland Naturalist 63:152–161.

Ioale, P., and S. Benvenuti. 1981. Winter philopatry and homing performance in some passerine birds. Italian Journal of Zoology 15:315–316.

Johnson, C. N., and M. S. Gaines. 1990. Evolution of dispersal: theoretical models and empirical tests using birds and mammals. Annual Review of Ecology and Systematics 21: 449–480.

Jones, P. J, C. C. H. Elliott, and R. A. Cheke. 2002. Methods for ageing juvenile Red-billed Queleas, *Quelea quelea,* and their potential for the detection of juvenile dispersal patterns. Ostrich 73:43–48.

Joseph, L., E. P. Lessa , and L. Christidis. 1999. Phylogeny and biogeography in the evolution of migration: shorebirds of the *Charadrius* complex. Journal of Biogeography 26: 329–342.

Keast, J. A. 1968. Seasonal movements in the Australian Honeyeaters (Meliphagidae) and their ecological significance. Emu 67:159–209.

Ketterson, E. D., and V. Nolan, Jr. 1983. The evolution of differential bird migration. Pages 357–402 *in* Current Ornithology, Vol. 1 (R. F. Johnston, ed.). Plenum Press, New York.

Koenig, W. D., D. Van Vuren, and P. N. Hooge. 1996. Detectability, philopatry, and the distribution of dispersal distances in vertebrates. Trends in Ecology and Evolution 11:514–517.

Kokko, H. 1999. Competition for early arrival in migratory birds. Journal of Animal Ecology 68:940–950.

Lenormand, T. 2002. Gene flow and the limits to natural selection. Trends in Ecology and Evolution 17:183–189.

Levey, D. J., and F. G. Stiles. 1992. Evolutionary precursors of long-distance migration-resource availability and movement patterns in neotropical landbirds. American Naturalist 140:447–476.

Lindberg, M. S, W. L. Kendall, J. E. Hines, and M. G. Anderson. 2001. Combining band recovery data and Pollock's robust design to model temporary and permanent emigration. Biometrics 57:273–281.

Lombardo, M. P. 1987. Attendants at Tree Swallow nests: the exploratory dispersal hypothesis. Condor 89:138–149.

McCarthy, M. A. 1996. Natal dispersal distance under the influence of competition. Pages 163–171 *in* Frontiers of Population Ecology (R. B. Floyd, A. W. Sheppard, and P. J. De Barro, eds.). CSIRO Publishing, Melbourne.

Mila, B., D. J. Girman, M. Kimura, and T. B. Smith. 2000. Genetic evidence for the effect of a postglacial population expansion on the phylogeography of a North American songbird. Proceedings of the Royal Society of London, Series B, Biological Sciences 267:1033–1040.

Møller, A. P., and J. Erritzøe. 2001. Dispersal, vaccination and regression of immune defense organs. Ecology Letters 4:484–490.

Moore, J., and R. Ali. 1984. Are dispersal and inbreeding avoidance related? Animal Behaviour 32:94–112.

Moore, W. S., and R. A. Dolbeer. 1989. The use of banding recovery data to estimate dispersal rates and gene flow in avian species: case studies in the red-winged blackbird and common grackle. Condor 91:242–253.

Muller, K. L., J. A. Stamps, V. V. Krishnan, and N. H. Willits. 1997. The effects of conspecific attraction and habitat quality on habitat selection in territorial birds (*Troglodytes aedon*). American Naturalist 150:650–661.

Myers, J. P. 1981. A test of three hypotheses for latitudinal segregation of the sexes in wintering birds. Canadian Journal of Zoology 59:1527–1534.

Neigel, J. E. 1997. A comparison of alternative strategies for estimating gene flow from genetic markers. Annual Review of Ecology and Systematics 28:105–128.

Neigel, J. E., R. M. Ball, Jr., and J. C. Avise. 1991. Estimation of single generation migration distances from geographic variation in animal mitochondrial DNA. Evolution 45:423–432.

Newton, I. 1972. Finches. William Collins Sons & Co. Ltd., Glasgow.

Nielsen, R., and J. W. Wakeley. 2001. Distinguishing migration from isolation: an MCMC approach. Genetics 158:885–896.

O'Connor, R. J. 1985. Behavioural regulation of bird populations: a review of habitat use in relation to migration and residency. Pages 105–142 *in* Behavioral Ecology: Ecological Consequences of Adaptive Behaviour (R. M. Sibly and R. H. Smith, eds.). Blackwell Scientific, Oxford.

Paradis, E, S. R. Baillie, and W. J. Sutherland. 2002. Modeling large-scale dispersal distances. Ecological Modeling 151: 279–292.

Paradis, E., S. R. Baillie, W. J. Sutherland, and R. D. Gregory. 1998. Patterns of natal and breeding dispersal in birds. Journal of Animal Ecology 67:518–536.

Parmesan, C., and G. Yohe. 2002. A globally coherent fingerprint of climate change impacts across natural systems. Nature 421:37–42.

Pärt, T. 1990. Natal dispersal in the collared flycatcher: possible causes and reproductive consequences. Ornis Scandinavica 21:83–88.

Pärt, T. 1991. Philopatry pays: a comparison between collared flycatcher sisters. American Naturalist 138:790–796.

Pärt, T. 1994. Male philopatry confers a mating advantage in the migratory collared flycatcher, *Ficedula albicollis*. Animal Behaviour 48:401–409.

Pärt, T. 1995. The importance of local familiarity and search costs for age- and sex-biased philopatry in the collared flycatcher. Animal Behaviour 49:1029–1038.

Payne, R. B. 1991. Natal dispersal and population structure in a migratory songbird, the Indigo Bunting. Evolution 45:49–62.

Piersma, T. 1996. Family Scolopacidae (sandpipers, snipes and phalaropes). Pages 444–533 *in* Handbook of Birds of the World, Vol. 3, Hoatzin to Auks (J. del Hoyo, A. Elliott, and J. Sargatal, eds.). Lynx Ediciones, Barcelona.

Pulido F., P. Berthold, and A. J. van Noordwijk. 1996. Frequency of migrants and migratory activity are genetically correlated in a bird population: evolutionary implications. Proceedings of the National Academy of Sciences USA 93:14642–14647.

Reed, J. M., T. Boulinier, E. Danchin, and L. W. Oring. 1999. Informed dispersal: prospecting by birds for breeding sites. Current Ornithology 15:189–259.

Robertson G., and F. Cooke. 1999. Winter philopatry in migratory waterfowl. Auk 116:20–34.

Rousset, F., and S. Gandon. 2002. Evolution of the distribution of dispersal distance under distance-dependent cost of dispersal. Evolutionary Biology 15:515–523.

Serrano, D., J. L. Tella, M. G. Forero, and J. A. Donazar. 2001. Factors affecting breeding dispersal in the facultatively colonial lesser kestrel: individual experience vs. conspecific cues. Journal of Animal Ecology 70:568–578.

Shaw, M. W. 1995. Simulation of population expansion and spatial pattern when individual dispersal distributions do not decline exponentially with distance. Proceedings of the Royal Society of London, Series B, Biological Sciences 259:243–248.

Shields, W. M. 1982. Philopatry, inbreeding and the evolution of sex. State University of New York Press, Albany.

Shields, W. M. 1983. Optimal inbreeding and the evolution of philopatry. Pages 132–159 *in* The Ecology of Animal Movement (I. R. Swingland and P. J. Greenwood, eds.). Clarendon Press, Oxford.

Sutherland, G. D., A. S. Harestad, K. Price, and K. P. Lertzman. 2000. Scaling of natal dispersal distances in terrestrial birds and mammals. Conservation Ecology 4:16. [Online] URL: [www.consecol.org/vol4/iss1/art16].

Thomson, D. L., A. Van Noordwijk, and W. Hagemeijer. 2003. Estimating avian dispersal distances from data on ringed birds. Journal of Applied Statistics 30:1003–1008.

Tilman, D., and P. Kareiva (eds.). 1997. Spatial Ecology: The Role of Space in Population Dynamics and Interspecific Interactions. Monographs in Population Biology 30. Princeton University Press, Princeton.

Turchin, P. 1998. Quantitative analysis of movement: measuring and modeling population redistribution in animals and plants. Sinauer Associates, Sunderland, Mass.

van Noordwijk, A. J. 1995. On bias due to observer distribution in the analysis of data on natal dispersal in birds. Journal of Applied Statistics 22:683–694.

Veit, R. R., and M. A. Lewis. 1996. Dispersal, population growth, and the Allee effect: dynamics of the House Finch invasion of Eastern North America. American Naturalist 148:255–274.

Whitlock, M. C., and D. E. McCauley. 1999. Indirect measures of gene flow and migration: F_{ST} not equal 1/(4Nm+1). Heredity 82:117–125.

Wright, S. 1982. Character change, speciation, and the higher taxa. Evolution 36:427–443.

Zann, R., and D. Runciman. 1994. Survivorship, dispersal and sex ratios of Zebra Finches *Taeniopygia guttata* in southeast Australia. Ibis 136:136–146.

Zink, R. M., and D. L. Dittmann. 1993a. Gene flow, refugia, and evolution of geographic variation in the Song Sparrow (*Melospiza melodia*). Evolution 47:717–729.

Zink, R. M., and D. L. Dittmann. 1993b. Population structure and gene flow in the chipping sparrow and a hypothesis for evolution in the genus *Spizella*. Wilson Bulletin 105:399–413.

31

THOMAS W. SHERRY,
MATTHEW D. JOHNSON,
AND ALLAN M. STRONG

Does Winter Food Limit Populations of Migratory Birds?

IN THEORY, WINTER HABITAT QUALITY, particularly food abundance, contributes to year-round population limitation in migratory birds, but this hypothesis has rarely been tested in the field. We hypothesize here that food availability in winter limits population size via winter habitat quality. From this Winter Food-Limitation Hypothesis (WFLH) we deduce and assess three testable predictions concerning winter populations in time and in space as well as delayed effects during and following migration. We find widespread support for the WFLH, involving diverse avian migratory taxa and geographic regions, although few studies have tested the strongest-inference prediction, namely that change in winter food affects survival and immediate winter population size. The lack of stronger support for the WFLH to date is primarily due to the expansive spatial scale of migration, the difficulty of distinguishing mortality from emigration, the paucity of dietary and available food data, and the multiple mechanisms by which food impacts populations. Further, the nearly complete lack of manipulative experiments makes it difficult to eliminate predators, parasites, diseases, and competitors as confounding factors that all potentially limit migratory bird populations. An important generalization emerging from many terrestrial studies is that prey abundance tends to track a late-winter dry season, causing decreased bird body condition, especially in some habitats and winters. This general response to decreased prey availability suggests that seasonal changes in food availability may have contributed to the evolution of migration. Although the frequently opportunistic response by migratory birds to variable prey availability challenges our ability to document population responses, the mobility of these birds makes them useful indicators for conservation efforts.

INTRODUCTION

Population biologists are increasingly addressing the challenges presented by migratory animals, whose distribution, local abundance, and life history traits are influenced by ecological factors operating at different times and places. In the case of long-distance migrant birds, the ecological factors influencing reproductive success operate in different locations and times from nonbreeding-season factors that influence survival. These birds' life histories play out over spatial scales of continents and even different biogeographic regions. This scale of movement challenges scientists' ability to follow and study individuals throughout their lifetimes to address population questions, particularly in small-bodied species (Webster et al. 2002).

Understanding population dynamics of migratory animals is also daunting because of the range of possible influences operating at different temporal and spatial scales. Populations of migrant birds can in theory be limited during the reproductive period, the nonbreeding and nonmigratory phases of the life cycle ("winter"), or migration (Sherry and Holmes 1995). Moreover, ecological events in one season may carry over into, and interact with, demographic processes in a subsequent season (e.g., Marra et al. 1998; Sillett et al. 2000). Thus the question of when and how migratory bird populations are limited (or regulated), and by what resources, is still a challenge.

Abundant research supports the idea that breeding success in summer has an important impact on migrants' population sizes and dynamics (e.g., Robinson et al. 1995; Askins 2000). Although less studied, the winter period has also been postulated to limit migrant bird populations (e.g., Baillie and Peach 1992; Rappole and McDonald 1994; Sherry and Holmes 1995, 1996; Sillett et al. 2000); however, the causal mechanisms behind winter population limitation are largely untested. Assuming that aggressive (e.g., territorial) behavior is adaptive under conditions of limited resources, the widespread occurrence of such behavior in winter supports the idea that resources are limiting in this season (Morse 1980; Rappole and Warner 1980; Marra 2000). Both intraspecific and interspecific aggressive behavior are frequent in winter and often associated with local food resources (Greenberg 1986; Greenberg and Salgado-Ortiz 1994; Greenberg et al. 1994), implicating food as an important limiting factor. Furthermore, numerous authors have noted the association of migratory birds with particular kinds or concentrations of food in winter (e.g., Willis 1966; Sherry 1984; Lack 1986; Leisler 1990; Greenberg et al. 1993; Greenberg and Salgado Ortiz 1994). These observations suggest that food in winter is the most likely resource limiting populations—the "Winter Food-Limitation Hypothesis" (WFLH). Despite the likelihood that winter food contributes to habitat quality, and ultimately to population limitation, the WFLH has rarely been tested, and when it has the evidence has been correlative and rarely considered in relation to factors other than food.

The purpose of the present review was to assess food in the nonbreeding season as a factor limiting migratory bird populations both in winter and in subsequent seasons. Here we deduce testable predictions from this WFLH, rank them in decreasing strength of inference, and assess them with available data. Considerable support for the predictions of the WFLH encourages discussion of several of its implications: How do food and other factors such as predators and parasites interact to limit migratory bird populations year-round? How does winter food limitation influence life history evolution? What are the major ecological factors that affect winter food availability? What are the conservation implications of the WFLH?

PREDICTIONS AND TESTS OF THE WINTER FOOD-LIMITATION HYPOTHESIS

The idea that winter habitat quality and availability limits bird populations on their wintering grounds is widely accepted (e.g., Baillie and Peach 1992; Rappole and McDonald 1994; Goss-Custard et al. 1995; Sherry and Holmes 1995, 1996; Newton 1998), but a mechanistic link between food and habitat use is not well established. By habitat quality we mean the survival or reproductive benefit a habitat confers upon its occupants (Hall et al. 1997). If food is the most important factor controlling winter habitat quality, then food controls the population size surviving the winter, because increases in food would increase habitat quality and thereby boost spring population size. Food can also cause delayed responses by a population, for example, influencing survival during migration or subsequent reproductive success. Unless an increase in winter survival is completely compensated by decreased migratory survival or breeding productivity, then winter food supply will help regulate migratory birds' overall population size (Sherry and Holmes 1995). With this background, we deduce three predictions from the WFLH, and we review evidence for these predictions primarily among long-distance migrants.

Prediction 1. Temporal change in winter food availability should affect winter population size proportionately. The simplest and most direct prediction of the WFLH is that a decrease in winter food supply should reduce the wintering population size and an increase in food should increase the population, up to the point permitted by the next most limiting factor. Support for this prediction provides the strongest evidence in favor of the hypothesis because it postulates a direct and immediate cause-effect link between food and wintering population size, N (fig. 31.1). Support of this prediction is both necessary and sufficient to demonstrate winter food limitation.

Evidence to support this prediction could be direct, via

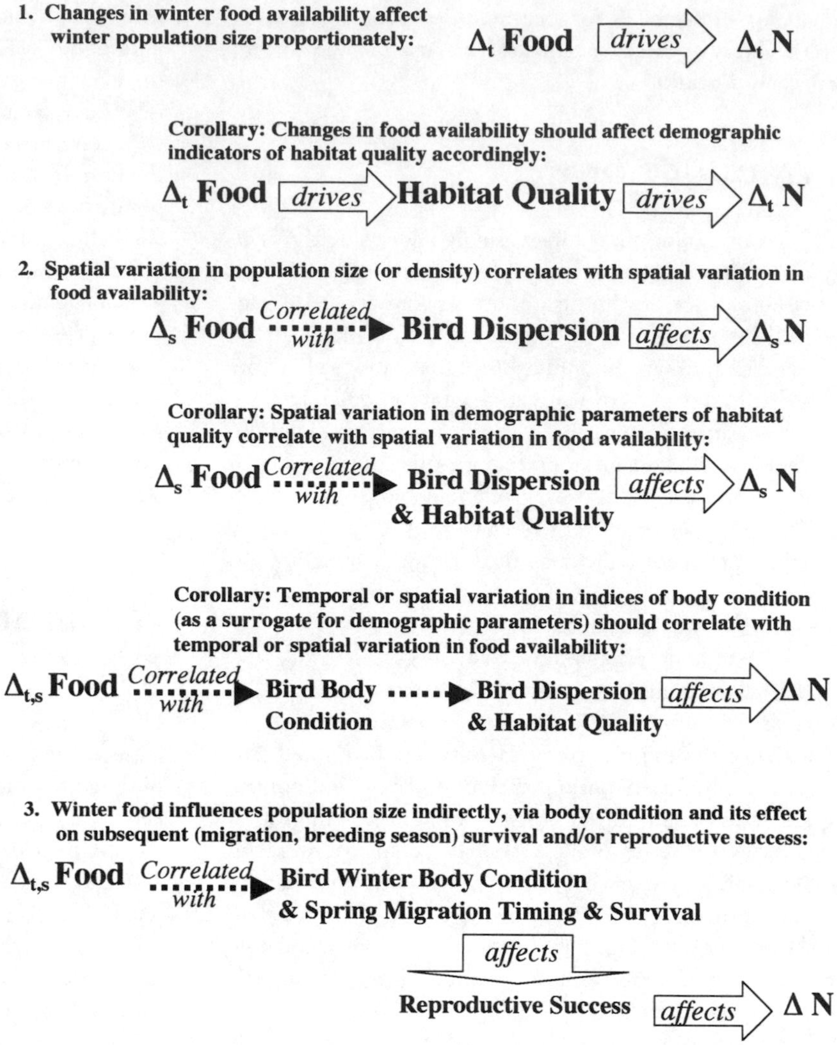

Fig. 31.1. Schematic diagram of causation pathways by which food or food surrogates affect population size according to the Winter Food Limitation Hypothesis. Subscripts "t" and "s" represent time and space, respectively.

1. Changes in winter food availability affect winter population size proportionately:

Corollary: Changes in food availability should affect demographic indicators of habitat quality accordingly:

Δ_t Food *drives* **Habitat Quality** *drives* **Δ_t N**

2. Spatial variation in population size (or density) correlates with spatial variation in food availability:

Δ_s Food *Correlated with* **Bird Dispersion** *affects* **Δ_s N**

Corollary: Spatial variation in demographic parameters of habitat quality correlate with spatial variation in food availability:

Δ_s Food *Correlated with* **Bird Dispersion & Habitat Quality** *affects* **Δ_s N**

Corollary: Temporal or spatial variation in indices of body condition (as a surrogate for demographic parameters) should correlate with temporal or spatial variation in food availability:

$\Delta_{t,s}$ Food *Correlated with* **Bird Body Condition** → **Bird Dispersion & Habitat Quality** *affects* **Δ N**

3. Winter food influences population size indirectly, via body condition and its effect on subsequent (migration, breeding season) survival and/or reproductive success:

$\Delta_{t,s}$ Food *Correlated with* **Bird Winter Body Condition & Spring Migration Timing & Survival**

affects

Reproductive Success *affects* **Δ N**

experimental food manipulation, or correlational, via population response to extrinsic changes in food supply. Direct food manipulations have been conducted for wintering *resident* birds, and have generally been followed by enhanced survival, consistent with the hypothesis of food limitation (Berndt and Frantzen 1964; Watson and O'Hare 1979; Smith et al. 1980; Jansson et al. 1981). Strong (1999) reduced ant abundance in territories of the migratory Ovenbird (*Seiurus aurocapillus*) wintering in Jamaica, causing increased territory size. However, what effect this may have had on individual ovenbirds or their population is not known, and the experiment is being repeated on a larger spatial scale (D. R. Brown and T. W. Sherry, unpubl. data). We know of no other study that has manipulated winter food so as to assess its effect on a wintering migratory bird population.

Studies of diverse migratory taxa nonetheless provide correlational evidence for the first prediction of the WFLH by documenting changes in local density or population size immediately following natural changes in food availability (table 31.1). These taxa include perching birds, hummingbirds, ducks, shorebirds, and raptors. Although correlational, the evidence reported by these studies provides some of the strongest support for the WFLH.

The interpretation of such changes in migrant numbers coincident with changing food is not always straightforward. First, unmeasured ecological conditions that change concomitantly with food supply could be responsible for any observed change in bird numbers. For example, drought depresses food supply and is correlated with dramatic declines in Paleotropical migrants (Peach et al. 1991; Baillie and Peach 1992), but high mortality can also result from changes in temperature or other physiological stressors, not to mention predators or parasites. Nonetheless, without controlled experiments we can only speculate on the relative importance of factors other than food in governing population changes coincident with changing food supply. A second complication in interpreting temporal correlations between food and wintering migrant numbers is the temporal scale of observation. Short-term changes in migrant abundance over days or weeks may result simply from redistribution of birds without consequence to overall population size (Johnson and Sherry 2001). Thus, the strongest evidence to date for the WFLH comes from the longer-term studies (see temporal scale in table 31.1).

If we assume that migrants' body condition and, ultimately, survival influence their overall population size, then a corollary of prediction 1 is that changes in food should affect these demographic parameters. However, support of this corollary alone is insufficient to support the WFLH because the assumption is invalidated when a change in one demographic parameter following a change in food availability is compensated by a change in another parameter, so that overall abundance remains stable. For example, a decrease in winter food availability could lower winter survival, consistent with this prediction, but the resulting smaller population could experience greater per capita reproduction, yielding an unchanged (regulated) population. Although partial compensation for winter events could occur during spring migration, complete compensation is unlikely because some migration mortality is likely independent of population size, as suggested by the effects of storms during migration (e.g., Zumeta and Holmes 1978; Sillett and Holmes 2002), and thus not regulatory. Food-mediated changes in demographic indicators are therefore likely to exert some influence on overall breeding population size.

Habitat-specific demography has received less attention in the winter than in the breeding season, perhaps because of the long-term data required to quantify survival, the most relevant nonbreeding demographic trait. Of the few studies that have investigated nonbreeding season, habitat-specific survival for migrants (e.g., Conway et al. 1995; Marra and Holmes 2001; Brown et al. 2002), most have not simultaneously quantified food, so this corollary of the first prediction remains inadequately tested. Nonetheless, the studies that have measured food show a general tendency for survival to fluctuate in correspondence with variation in winter food availability. Latta and Faaborg (2001) and Johnson and Sherry (2001), working with wintering warblers, found that persistence over the winter (an indirect measure of survival, because survival is confounded with emigration) decreased following decreases in arthropod availability.

Prediction 1 involves changes in abundance that result from temporal changes in food supply. Where those changes are controlled and randomly assigned by experimentation, we can rule out most alternative factors; where those changes are natural, the inference that observed bird responses are due to food alone is weakened. With or without experimentation, however, prediction 1 provides some control of alternative factors because changes in population size follow changes in food supply in particular locations. In contrast, the following prediction is more poorly controlled because it relies on variation in food availability across space; that is, confounding factors in ecology are often bet-

Table 31.1 Evidence for the winter food limitation hypothesis based on temporal changes in bird abundance or density in correspondence with changes in food availability (tests of Prediction 1)

Species	Food	Temporal scale	Region	Reference
Various sparrows	Seeds	8–21 years	Arizona	Pulliam and Parker 1979 Dunning and Brown 1982
Rufous Hummingbird *Selasphorus rufus*	Nectar	Days–weeks	California	Gass et al. 1976
Various ducks	Mussels	Tens of years	Lake Erie	Wormington and Leach 1992
Various seaducks	Invertebrates	6 winters	Scotland	Campbell 1984
Dunlin *Calidris alpina*	Invertebrates (inferred, not measured)	>3 years	Britain	Goss-Custard and Moser 1988
Brant *Branta bernicla hrota*	Eel grass	~10 years	Svalbard	Clausen et al. 1998
Prairie Warblers Dendroica discolor	Insects	3 months	Dominican Republic	Latta and Faaborg 2001
Various frugivores-insectivores	Fruits and flowers	Monthly variation	Argentina	Malizia 2001
Northern Pintail *Anas acuta*	Seeds and invertebrates	2 winters	California	Miller and Newton 1999
Merlin *Falco columbarius*	Rodents	16 years	Sweden	Wiklund 2001
Various warblers	Insects	Weeks	Jamaica	Johnson and Sherry 2001
Various insectivores	Insects	3 years data but correlations within 1 year	Tanzania-Kenya	Sinclair 1978
Yellow Wagtail *Motacilla flava*	Arthropods	5 months	West Africa	Wood 1979

ter controlled over time than across space (Eberhardt and Thomas 1991; Manly 1992).

Prediction 2. Winter food availability should correlate spatially (e.g., by habitat) with winter population size. This prediction rests explicitly on the links among bird dispersion, habitat quality, and population limitation (e.g., Sherry and Holmes 1995). As explained earlier, winter food limits migrant populations if it is the primary determinant of habitat quality, and animals should in theory disperse among habitats of different quality so as to match population size to food resources (Moreau 1972). The effect of food could be restricted to a simple redistribution of birds among habitats as food supply fluctuates, without any population consequences. This prediction is thus weaker than the previous one because it invokes a relatively indirect link between winter food and a migrant's population size (fig. 31.1), and support for this prediction is necessary, but not sufficient, to support the WFLH.

This prediction has received the most attention, with the majority of investigations showing a general spatial correspondence between bird distribution and food availability. The evidence is strongest for shorebirds, for which spatial correlation is documented between bird numbers and benthic macro-invertebrate density under diverse circumstances and scales (Goss-Custard 1970; Wolff and Smit 1990; Hockey et al. 1992; Yates et al. 1993). Although less well documented, migratory waterfowl and seabird abundance has also been found to vary in correspondence to their respective foods, including invertebrates (Yates et al. 1993), krill (Heinemann et al. 1989; Veit et al. 1993), and eel grass (Clausen et al. 1998).

Prediction 2 has received less attention for migratory landbirds, perhaps because of the comparative difficulty in quantifying resource availability in structurally complex habitats. Hutto (1980) and Johnson and Sherry (2001) found a positive correlation between migrant and insect numbers in the Neotropics, but Folse (1982) working in Africa did not. Price and Gross (Chap. 27, this volume) found correlations between food and *Phylloscopus* warbler abundance wintering in India. Studies with too few sites for correlation analysis corroborate the idea that wintering migrant landbirds tend to become more abundant in sites or habitats with more food (Sinclair 1978; Poulin et al. 1992; Lefebvre et al. 1994; Parrish and Sherry 1994; Katti and Price 1999; Latta and Faaborg 2001; 2002; Perez-Tris and Telleria 2002).

This second prediction invokes food as a proximate limiting factor via its effect on survival, a demographic indicator of habitat quality, which in turn drives population size (fig. 31.1). Like the corollary of prediction 1, this deduction therefore relies on the assumption that food's influence on survival is not compensated elsewhere in the birds' annual cycle. Few studies have examined a corollary of prediction 2, namely that the mechanism of food's effect on a population is via survival or other indicators of body condition, again because quantifying survival is difficult. Marra and Holmes (2001) found higher survival for American Redstarts (*Setophaga ruticilla*) in mangrove swamp than in adjacent scrub forest, and Parrish and Sherry (1994) found correspondingly greater abundance of at least some insect taxa in mangroves, although this tendency may be restricted to old-growth mangrove swamps (Johnson and Sherry 2001). Working in three habitats, Johnson et al. (unpubl. data) found that apparent annual survival was highest in the food-rich and lowest in the food-poor habitat. Similarly, overwinter persistence was highest (among three habitats) in food-rich sites for both the insectivorous Prairie Warbler (*Dendroica striata* [Latta and Faaborg 2001]) and the omnivorous Cape May Warbler (*Dendroica tigrina* [Latta and Faaborg 2002]).

Thus far, we have reviewed evidence that links food to population size either directly or indirectly via demography, chiefly survival. If we assume that measures of individual birds' body condition determine or indicate survival, then a corollary of prediction 2 is that surrogate variables for survival should vary spatially with variation in food availability. Marra and Holmes (2001) showed that this assumption was valid for wintering American Redstarts, but because the assumption has been inadequately tested for other taxa, we have ranked evidence involving variability in body condition lower in strength of inference than variability in survival.

Several studies of terrestrial migrants in winter have documented temporal or spatial variation in body condition, but as before, few have done so concurrently with measures of food availability. Those studies that have examined body condition have focused on surrogates for fitness, including change in body mass (Marra et al. 1998; Johnson et al., unpubl. data), rate of feather regrowth (e.g., Strong and Sherry 2000), stress hormones (Marra and Holberton 1998), subcutaneous fat (Katti and Price 1999), and pectoral muscle mass (Latta and Faaborg 2002). Insectivorous Ovenbirds varied in maintenance of body mass and feather regrowth rate both spatially (among territories) and temporally with ant biomass, their primary winter food (Strong and Sherry 2000).

Measures of body mass have generally varied as predicted with measures of food availability. Copious data obtained from winter-harvested waterfowl indicate a clear correlation between food availability and body mass (e.g., Hobaugh 1985; Miller and Newton 1999), and nonharvested species appear to exhibit similar patterns (e.g., Nariko et al. 1999). Studies of shorebirds also suggest a positive correlation between variation in food across time and/or space and variation in body mass (Tsipoura and Burger 1999; Mitchell et al. 2000). An important caveat here is that body mass is composed of separate components (e.g., body fat, overall body size in relation to skeletal characters, and pectoral muscle mass) that may respond independently over space and time to various ecological circumstances; moreover, predation risk and food predictability likely affect body lipids and pectoral body mass (Rogers, Chap. 9, this volume). Thus body mass likely does not depend simply on overall food abundance.

In a variety of studies foraging behavior has been used as a surrogate for measurements of food (e.g., Lovette and Holmes 1995; Wunderle and Latta 1998). When behavior is used to estimate food availability, the linkage between the predictor variable (e.g., foraging) and the population response is even more indirect than the pathways in fig. 31.1, and thus more tenuous. Better understanding of the link between foraging behavior and food availability is necessary to justify the use of foraging behavior as a surrogate for food availability.

Prediction 3. Winter food influences population size indirectly, via body condition and its effect on subsequent (migration, breeding season) survival and/or reproductive success. Because food abundance influences body condition in winter as reviewed above, winter food can therefore exert delayed, indirect effects on populations that ecologists have only begun to suspect (e.g., Marra et al. 1998; Sillett et al. 2000). Support of this third prediction is sufficient, but not necessary, to support the WFLH, because a particular species could in theory respond immediately in winter to concurrent changes in food abundance (in support of predictions 1 and 2), but show no delayed responses (prediction 3). We also rank evidence in support of this third prediction of WFLH as weaker than that for prediction 1 because the delay between winter food abundance and subsequent population responses allows the possibility that ecological conditions subsequent to winter can intervene and modify the population's response to winter food (fig. 31.1). The "carry-over effects" implicit in prediction 3 are poorly understood, in part because of the challenge of linking migratory populations spatially during their annual travels (Webster et al. 2002). Nonetheless, these delayed population effects may be widespread (e.g., Sillett et al. 2000; Møller and Hobson 2004; Norris et al. 2004; Saino et al. 2004, and they certainly increase the range of population responses to winter food (Marra et al. 1998), and thus the challenge of testing such food impacts.

Our third prediction of the WFLH converges with a deduction from the idea of seasonal interactions in migratory birds (Webster et al. 2002), which in general predicts delayed effects of ecological conditions in one season on migrants' population and individual responses in subsequent phases of the annual cycle (winter to summer, summer to winter, migration to summer, and so on). The WFLH differs from seasonal interaction predictions in specifying food as the dominant ecological factor and in focusing on impacts of ecological conditions arising solely in winter.

Our review of data relevant to the three predictions of the WFLH suggests that food can be an important factor limiting migratory bird populations, affecting their distribution, year-round abundance, survival, and even reproduction following a migration away from the winter range. These data come from a variety of species on most continents, and they represent different temporal and spatial scales of study. Nonetheless, strong direct support of the hypothesis (prediction 1) is infrequent.

ALTERNATIVE MECHANISMS OF POPULATION LIMITATION IN WINTER

Despite the foregoing evidence supporting the WFLH, we still understand only sketchily the mechanisms by which food limits populations, for a variety of reasons. First, food abundance influences populations in diverse and often indirect ways, including socially constrained dispersion of individuals among habitats (Marra and Holmes 2001), interaction with predators and parasites, and time delays. For example, resource distribution patterns can influence bird dispersion among habitats, which can in turn differentially influence survival and body condition via predation, parasitism, and other factors that vary by habitat. Another consideration is that food availability is influenced not only by prey abundance (i.e., "standing crop"), but also by resource productivity (i.e., turnover rate), and standing crop alone may be a poor estimate of food supply for highly renewable foods. The abundance of migratory shorebirds supported by high tropical marine productivity in the Banc d'Arguin (Zwarts et al. 1990) exemplifies how low arthropod standing crop of small-bodied prey may not predict abundance as well as prey population productivity—a circumstance that may apply to "terrestrial" habitats such as mangroves.

The relationship between food and population size can also be obscured by competitors, including other bird species (Greenberg 1986), other vertebrates such as lizards and frogs, and potentially even invertebrates such as ants and spiders (Van Bael et al. 2003). A variety of recent studies document effects of migratory species on the feeding success and habitat choice of other migrants (Greenberg et al. 1993; Greenberg and Salgado-Ortiz 1994; Greenberg et al. 1994; Latta and Faaborg 2002). Resident tropical birds can also be important competitors (e.g., Lack 1976; Greenberg 1986). Greenberg's (1995) Breeding Currency Hypothesis (BCH) formalizes a general mechanism of competition between migrant and resident insectivorous birds. Johnson et al. (in press) found support for the primary prediction of the BCH, namely that large arthropods ("breeding currency") most suitable for the reproductive success of resident birds tended to be proportionately more abundant in Jamaican resident-dominated habitats, whereas smaller arthropods in winter tended to be most abundant in the habitats with proportionately more wintering migrants. The implication of this relationship is that wintering terrestrial migrant birds compete diffusely with resident birds for subsistence or survival food, and resident species displace migrants from the best local breeding habitats. This example of migrant-resident avian niche partitioning is a special case of resource partitioning among consumers in the Tropics (e.g., Lack 1976), an appealing, if still inadequately understood phenomenon. An important implication of our review for studies of competition is that the diverse pathways by which winter food can influence a population (fig. 31.1) should lead to similarly diverse ways that competitors can influence a population.

Predators and parasites can also obscure the relationship of the winter population response to food both directly through mortality and indirectly by influencing feeding behavior and habitat choice. For example, animals in poor body condition, which can be the result of food shortages, may be more susceptible to predators, as illustrated by young Dunlin (*Calidris alpina*) wintering on the Banc D'Arguin, Mauritania (Bijlsma 1990). Expanded home range size and dispersal movements of animals, which can be precipitated by low food abundance, should also put individuals at greater risk of both predation (Hubbard et al. 1999) and parasitism. Both predators and parasites cause mortality in wintering migrants. Scaley-leg mite infests migrant and resident birds wintering in the Caribbean region and it affects both body condition and survival (e.g., Latta and Faaborg 2001; Latta 2003), but the quantitative impact on populations remains poorly understood. Rogers (Chap. 9, this volume) emphasizes the importance of predators to wintering migrants by illustrating the complex ways in which predators can shape adaptive patterns of body mass variation.

Our ability to test the relationship between available food and population responses also depends on methodological challenges, some of which have been mentioned already, including problems of establishing population connectivity (Webster et al. 2002) and quantifying resource abundance. In addition, food resources are filtered by consumers—depending on their searching, handling, and digestive adaptations—which makes it challenging to quantify what foods are effectively available to a particular population (Hutto 1990; Strong 2000). Predation and disease can also be difficult to measure because of their infrequent occurrence and indirect effects on survival and subsequent reproduction.

IMPLICATIONS OF WINTER FOOD LIMITATION FOR MIGRANT LIFE HISTORY EVOLUTION

One unifying pattern for both Old and New World terrestrial migratory systems is a decline in food resources in the late dry season (Katti and Price 1996; Rödl 1999; Strong and Sherry 2000; Johnson and Sherry 2001; Latta and Faaborg 2001, 2002), a simultaneous deterioration of body condition (Sherry and Holmes 1996; Marra and Holberton 1998; Marra et al. 1998; Strong and Sherry 2000; Latta and Faaborg 2002), and decreased survival or persistence both in drought-stressed habitats (Katti and Price 1996; Sherry and Holmes 1996; Marra et al. 1998; Rödl 1999; Strong and Sherry 2000; Latta and Faaborg 2001; Marra and Holmes 2001) and in relatively dry winters (Peach et al. 1991; Baillie and Peach 1992; Sillett et al. 2000). If ancestral, resident species/populations were constrained to occupy similarly seasonal, unbuffered habitats just before spring migration, then intra- or interspecific competition combined with emigration could be an intermediate step in the evolution of a long-distance migratory strategy.

Consider a hypothetical resident individual's annual cycle in an early successional habitat. If the lowest level of resource availability occurs at the end of the dry season, despotic interactions over the winter (e.g., Greenberg 1986; Greenberg and Salgado-Ortiz 1994; Marra 2000) may constrain individuals facing a negative energy balance either to starve or to emigrate just before the breeding season begins. If emigration leads to the location of increased food resources, then breeding may take place in areas with more favorable food resources. Although this hypothesis ignores many of the intermediate steps (e.g., physiological, anatomical, neurological) that are necessary for the evolution of a completely migratory life history, it is consistent with several of the evolutionary precursors that may have been responsible for migration: dry-season food limitation, despotic interactions, and increased emigration rates.

Studies of the migratory origins of parulid warblers support this hypothesis. Cox (1985) modeled the evolution of the Neotropical migration system in parulids and concluded that the Mexican Plateau, characterized by strong seasonality and interannual variation in rainfall, was the likely point of origin for many migratory taxa. Similar environmental conditions and subsequent forest fragmentation throughout North America during the Miocene-Pliocene boundary may have contributed to the rapid adaptive radiation of *Dendroica* (Lovette and Bermingham 1999). Thus, xeric conditions appear to be conducive to the proliferation of migratory species. Entomological studies corroborate the idea that drought reduces arthropod abundance (e.g., Wolda 1978; Levings 1983; Pearson and Derr 1986; Frith and Frith 1990), lending credence to dry-season declines in food availability as a driving mechanism for migration.

Chesser and Levey (1998) argued that habitats with poorly buffered resources in the nonbreeding season, rather than diet per se, are most likely to favor migratory behavior. They argued that although diet, specifically frugivory, was not supported as a precursor to migration, the use of ephemeral food resources might be a consequence of selection for poorly buffered habitats. There is strong empirical evidence that many (but not all) migrants use such disturbed, early successional habitats (Hutto 1980; Hagan and Johnston 1992; Petit et al. 1995; Johnson and Sherry 2001), and these species are generally "replaced" spatially by resident species in more buffered habitats (Keast and Morton 1980; Douglas 2002). This may occur because resident species are not as effective as migrants in tracking resources in space because of the constraint of defending suitable home ranges year round with sufficient breeding currency (sensu Greenberg 1995; Johnson et al. 2004). Many migrants, by contrast, appear to track ephemeral food resources spatially over the course of the nonbreeding season, as reviewed under hypothesis 2 above. Thus, ecological conditions typical of large parts of the Subtropics may have favored evolution of latitudinal migration, via patterns of food resource distribution and abundance.

CONSERVATION IMPLICATIONS OF WINTER FOOD LIMITATION

The mobility of migrant birds, their often opportunistic food habits as described above, and the lack of habitat-specific demographic data for most species make the conservation of migratory birds during the nonbreeding season challenging. The fact that individuals use multiple habitats outside the breeding season (e.g., Rappole et al. 1989; Wunderle and Latta 2000; Johnson and Sherry 2001) underscores the necessity of a landscape or regional approach in winter, just as in summer (Askins 2000). However, critical to any conservation strategy is the need to address issues of both habitat quantity and quality. Habitat quantity has long concerned avian conservationists (e.g., Keast and Morton 1980; Terborgh 1989). What we emphasize here is that conservation efforts need to increase consideration of habitat quality, including factors that maintain or degrade prey abundance. Increasing use of demographic data (birth and death rates) in assessing avian habitat quality is a positive development, but we nonetheless recommend greater effort to address the ecological determinants of habitat quality, and our review suggests that food availability merits more attention.

Humans can influence habitat quality to birds most obviously by influencing the kinds and abundance of resources (Greenberg 1992; Petit et al. 1995). For example, Johnson (2000) demonstrated that the species of shade trees used in coffee cultivation influenced the abundance of birds via foliage arthropods. Similarly, plant species influence the distributions of migrant and resident Jamaican birds via fruit and nectar resources (Douglas 2002). Anthropogenic habitat conversion and landscape changes (e.g., increased edge vegetation) will in general alter habitat quality for birds and other kinds of wildlife. Humans can also influence habitat quality for birds indirectly via weather patterns (Sillett et al. 2000), and potentially via effects on predators and pathogens (such as West Nile Virus).

Migratory birds may be useful indicator species in habitat restoration efforts because of their exceptional ability to discover and use resources opportunistically. Additionally, the ability of many migratory birds to track ephemeral resources (e.g., Sherry 1984; Johnson and Sherry 2001) makes them particularly effective consumers of some insect pests (Greenberg et al. 2000; Van Bael et al. 2003), because an increase in arthropods of the sizes and types consumed by these birds will result in a local increase in migratory bird populations. Such functional and numerical responses, coupled with birds' high metabolic rate, will increase consumption of these arthropods. Thus, our studies of the feeding behavior and food limitation underscore the potential of wintering migratory bird populations for arthropod control in tropical wintering areas.

FUTURE DIRECTIONS

Our inability to link winter food availability more strongly to changes in migratory bird population size, that is, to test the WFLH more rigorously, can be ascribed to several factors that indicate future research needs: paucity of dietary data for most species, prevalence of feeding opportunism and thus diet variability in space and time, large spatial scale necessary for manipulative experiments, unclear relationship between climate and food availability, difficulty in distinguishing between mortality and emigration as responses to food shortage, lack of understanding of population connectivity between breeding and wintering phases of the life cycle, the need to examine alternative hypotheses, and the need for a better understanding of the relationship between body condition and fitness. This review of winter food emphasizes diet as a logical link between the population dynamics, habitat-selection, life history evolution, and conservation needs of migratory birds. Future diet studies promise to advance our understanding of migratory birds in diverse ways, and such studies are feasible, if challenging, as reviewed above.

Dietary opportunism can both complicate and simplify our understanding of how food availability limits migratory bird populations in winter. The use of small, poorly digestible, abundant prey types—subsistence resources that are inadequate as breeding currency—have been documented as prey for arboreal species, ground foragers, and wetland species (Morton 1980; Zwarts et al. 1990; Poulin and Lefebvre 1996; Medori 1998; Strong 2000). Such breadth of prey types within species potentially complicates measurement of food availability, but recent studies have demonstrated the tendency for migrants to take prey in direct proportion to their abundance, an observation also consistent with dietary opportunism (Sherry 1984; Strong 2000). An opportunistic diet has the potential to simplify the daunting task of measuring food availability by using generalized sampling techniques. Although species-tailored food-sampling methods certainly provide better data for particular populations (e.g., Rödl 1999; Strong and Sherry 2000, 2001; Johnson and Sherry 2001), a sampling scheme with standard methodology used by collaborating researchers across a broad geographic scale should also help unravel the effects of climate, soils, and other factors influencing food availability. Moreover, more studies need to test simultaneously the impacts of food limitation in winter versus alternative factors (e.g., Johnson and Sherry 2001) and interacting factors.

Measuring a population's demographic response to resources is just as challenging as measuring the resources themselves, and this topic deserves a major review of its own. For example, distinguishing mortality from emigration remains one of the greatest challenges to studying these far-ranging animals (Webster et al. 2002). We need studies of the congruence and redundancy of different measures to assess body condition conducted under diverse ecological circumstances. A related need is the examination of delayed responses of migrants to food conditions in winter, that is, seasonal interactions via body condition and demographic parameters. Such studies will be most feasible in species with limited geographical ranges or in those whose breeding and wintering population connectivity are known.

Two important conservation needs are to learn how to improve habitat quality for wintering migratory birds and to identify both the ecological and economic value of these birds. This review indicates that food abundance is an important determinant of habitat quality, which suggests that manipulating food abundance will be a fruitful experimental approach. Possible ways to do this include changing plant species (Johnson 2000) and more generally manipulating vegetation cover and habitat type. Future research should emphasize agricultural habitats such as irrigated and forest-based crops, because these are habitats in which humans and migratory birds potentially can coexist. Conversely, we need to quantify the value of the ecological and economic services provided by migratory birds, including their potential indirect beneficial effects on plants via the birds' consumption of insect herbivores (Marquis and Whelan 1994; Greenberg et al. 2000; Strong et al. 2000; Van Bael et al. 2003).

ACKNOWLEDGMENTS

We dedicate this chapter to the memory of Robert Sutton, passionate bird scientist, author, conservationist, and friend, who was instrumental in all phases of our research in Jamaica. Many other Jamaicans both facilitated our work and made it far more pleasurable: Anne Haynes-Sutton, Leo Douglas, Catherine Levy, the late Pamela Williams, Peter Vogel, Errol Ziadie of the Portland Ridge Hunting Club, and numerous field assistants.

We thank the National Science Foundation for continued financial support of our research in the Caribbean, and Dick Holmes and Pete Marra for constructive suggestions on the manuscript.

LITERATURE CITED

Askins, R. 2000. Restoring North America's birds: lessons from landscape ecology. Yale University Press, New Haven.

Baillie, S. R., and W. J. Peach. 1992. Population limitation in Palaearctic-African migrant passerines. Ibis 134:120–132.

Berndt, R., and M. Franzen. 1964. The influence of the severe winter of 1962/63 on the breeding status of hole nesting birds in Braunschweig. Ornithologische Mitteilungen 16: 126–130.

Bijlsma, R. G. 1990. Predation by large falcons on wintering waders on the Banc D'Arguin, Mauritania. Pages 75–82 *in* Homeward Bound: Problems Waders Face When Migrating from the Banc D'Arguin, Mauritania, to Their Northern Breeding Grounds in Spring (B. J. Ens, T. Piersma, W. J. Wolff, and L. Zwarts, eds.). Special edition of Ardea 78(1/2).

Brown, D. R., C. M. Strong, and P. C. Stouffer. 2002. Demographic effects of habitat selection by hermit thrushes wintering in a pine plantation landscape. Journal of Wildlife Management 66:407–416.

Campbell, L. H. 1984. The impact of changes in sewage treatment on seaducks wintering in the Firth of Forth, Scotland. Biological Conservation 28:173–180.

Chesser, R. T., and D. J. Levey. 1998. Austral migrants and the evolution of migration in New World birds: diet, habitat, and migration revisited. American Naturalist 152:311–319.

Clausen, P., J. Madsen, S. M. Percival, D. O'Connor, and G. Q. Anderson. 1998. Population development and changes in winter site use by the Svalbard light-bellied brant goose, *Branta bernicla hrota* 1980–1994. Biological Conservation 84:157–165.

Conway, C. J., G. V. N. Powell, and J. D. Nichols. 1995. Overwinter survival of Neotropical migratory birds in early-successional and mature tropical forests. Conservation Biology 9:855–864.

Cox, G. W. 1985. The evolution of avian migration systems between temperate and tropical regions of the New World. American Naturalist 126:451–474.

Douglas, L. 2002. Impact of human habitat degradation on resident and Neotropical migratory birds occupying the Tropical Dry Forest Life Zone of southern Jamaica. M.S. thesis, University of the West Indies, Mona Campus, Jamaica.

Dunning, J. B., and J. H. Brown. 1982. Summer rainfall and winter sparrow densities—a test of the food limitation hypothesis. Auk 99:123–129.

Eberhardt, L. L., and J. M. Thomas. 1991. Designing environmental field studies. Ecological Monographs 61:53–73.

Folse, L. J., Jr. 1982. An analysis of avifauna-resource relationships on the Serengeti Plains. Ecological Monographs 52: 111–127.

Frith, D., and C. Frith. 1990. Seasonality of litter invertebrate populations in an Australian upland tropical rain forest. Biotropica 22:181–190.

Gass, C. L., G. Angehr, and J. Centa. 1976. Regulation of food supply by breeding territoriality in the Rufous Hummingbird. Canadian Journal of Zoology 54:2046–2054.

Goss-Custard, J. D. 1970. The response of Redshank (*Tringa totanus* (L.)) to spatial variations in the density of their prey. Journal of Animal Ecology 39:91–113.

Goss-Custard, J. D., R. T. Clarke, K. B. Briggs, B. J. Ens, K.-M. Exo, C. Smit, A. J. Beintema, R. W. G. Caldow, D. C. Catt, N. A. Clark, S. E. A. Le V. dit. Durell, M. P. Harris, J. B. Hulscher, P. L. Meininger, N. Picozzi, R. Prys-Jones, U. N. Safriel, and A. D. West. 1995. Population consequences of winter habitat loss in a migrating shorebird: Pt.. 1. Estimating model parameters. Journal of Applied Ecology 32:320–336.

Goss-Custard, J. D., and M. E. Moser. 1988. Rates of change in the numbers of Dunlin wintering in British estuaries in relation to the spread of *Spartina anglica*. Journal of Applied Ecology 25:95–109.

Greenberg, R. 1986. Competition in migrant birds in the nonbreeding season. Current Ornithology 3:281–307.

Greenberg, R. 1992. Forest migrants in non-forest habitats on the Yucatan Peninsula. Pages 273–286 *in* Ecology and Conservation of Neotropical Migrant Landbirds (J. M. Hagan III and D. W. Johnston, eds.). Smithsonian Institution Press, Washington, D.C.

Greenberg, R. 1995. Insectivorous migratory birds in tropical ecosystems: the breeding currency hypothesis. Journal of Avian Biology 26:260–264.

Greenberg, R., P. Bichier, A. Cruz Angon, C. MacVean, R. Perez, and E. Cano. 2000. The impact of avian insectivory on arthropods and leaf damage in some Guatemalan coffee plantations. Ecology 81:1750–1755.

Greenberg, R., C. Macias Caballero, and P. Bichier. 1993. Defense of homopteran honeydew by birds in the Mexican highlands and other warm temperate forests. Oikos 68:519–524.

Greenberg, R., and J. Salgado-Ortiz. 1994. Interspecific defense of pasture trees by wintering Yellow Warblers. Auk 111: 672–682.

Greenberg, R., J. Salgado Ortiz, and C. Macias Caballero. 1994. Aggressive competition for critical resources among migratory birds in the Neotropics. Bird Conservation International 4:115–127.

Hagan, J. M., III, and D. W. Johnston (eds.). 1992. Ecology and Conservation of Neotropical Migrant Landbirds. Smithsonian Institution Press, Washington, D.C.

Hall, L. S., P. R. Krausman, and M. L. Morrison. 1997. The habitat concept and a plea for standard terminology. Wildlife Society Bulletin 25:171–182.

Heinemann, D., G. Hunt, and I. Eversson. 1989. The distribution of marine avian predators and their prey, *Euphausia superba,* in Bransfield Strait, Southern Drake Passage, Antarctica. Marine Ecology Progress Series 58:3–16.

Hobaugh, W. C. 1985. Body condition and nutrition of snow geese wintering in southeastern Texas. Journal of Wildlife Management 49:1028–1037.

Hockey, P. A. R., R. A. Navarro, B. Kalejta, and C. R. Velasquez. 1992. The riddle of the sands: why are shorebird densities so high in southern estuaries? American Naturalist 140: 961–979.

Hubbard, M. W., D. L. Garner, and E. E. Klaas. 1999. Factors influencing Wild Turkey hen survival in south-central Iowa. Journal Wildlife Management 63:731–738.

Hutto, R. L. 1980. Winter habitat distribution of migratory land birds in western Mexico with special reference to small foliage gleaning insectivores. Pages 181–203 *in* Migrant Birds in the Neotropics: Ecology, Behavior, Distribution, and Conservation (A. Keast and E. S. Morton, eds.). Smithsonian Institution Press, Washington, D.C.

Hutto, R. L. 1990. Measuring the availability of food resources. Studies in Avian Biology 13:20–28.

Jansson, C., J. Ekman, and A. von Brömssen. 1981. Winter mortality and food supply in tits (*Parus* spp.). Oikos 37: 313–322.

Johnson, M. D. 2000. Effects of shade-tree species and crop structure on the winter arthropod and bird communities in a Jamaican shade coffee plantation. Biotropica 32:133–145.

Johnson, M. D., and T. W. Sherry. 2001. Effects of food availability on the distribution of migratory warblers among habitats in Jamaica. Journal of Animal Ecology 70:546–560.

Johnson, M. D., T. W. Sherry, and R. T. Holmes. Manuscript. Measuring habitat quality for a migratory songbird wintering in natural and agricultural habitats. Submitted to Conservation Biology.

Johnson, M. D., T. W. Sherry, A. M. Strong, and A. Medori. In Press. Migrants in Neotropical bird communities: an assessment of the breeding currency hypothesis. Journal of Animal Ecology.

Katti, M., and T. Price. 1996. Effects of climate on Palaearctic warblers over-wintering in India. Journal of the Bombay Natural History Society 93:411–427.

Katti, M., and T. Price. 1999. Annual variation in fat storage by a migrant warbler overwintering in the Indian tropics. Journal of Animal Ecology 68:815–823.

Keast, A., and E. S. Morton, eds. 1980. Migrant Birds in the Neotropics: Ecology, Behavior, Distribution, and Conservation. Smithsonian Institution Press, Washington, D.C.

Lack, D. 1976. Island Biology: Illustrated by the Land Birds of Jamaica. University of California Press, Berkeley and Los Angeles.

Lack, P. R. 1986. Ecological correlates of migrants and residents in a tropical African savanna. Ardea 74:111–119.

Latta, S. C. 2003. Effects of scaley-leg mite infestations on body condition and site fidelity of migratory warblers in the Dominican Republic. Auk 120:730–743.

Latta, S. C., and J. Faaborg. 2001. Winter site fidelity of Prairie Warblers in the Dominican Republic. Condor 103:455–468.

Latta, S. C., and J. Faaborg. 2002. Demographic and population responses of Cape May Warblers wintering in multiple habitats. Ecology 83:2502–2515.

Lefebvre, G., B. Poulin, and R. McNeil. 1994. Temporal dynamics of mangrove bird communities in Venezuela with special reference to migrant warblers. Auk 111:405–415.

Leisler, B. 1990. Selection and use of habitat of wintering migrants. Pages 156–174 *in* Bird Migration (E. Gwinner, ed.). Springer-Verlag, Berlin.

Levings, S. C. 1983. Seasonal, annual, and among-site variation in the ground ant community of a deciduous tropical forest: some causes of patchy species distributions. Ecological Monographs 53:435–455.

Lovette, I. J., and E. Bermingham. 1999. Explosive speciation in the New World *Dendroica* warblers. Proceedings of the Royal Society of London, Series B, Biological Sciences 266:1629–1637.

Lovette, I. J., and R. T. Holmes. 1995. Foraging behavior of American redstarts in breeding and wintering habitats: implications for relative food availability. Condor 97:782–791.

Malizia, L. R. 2001. Seasonal fluctuations of birds, fruits, and flowers in a subtropical forest of Argentina. Condor 103: 45–61.

Manly, B. F. J. 1992. The Design and Analysis of Research Studies. Cambridge University Press, Cambridge.

Marquis, R. J., and C. J. Whelan. 1994. Insectivorous birds increase growth of white oak through consumption of leaf-chewing insects. Ecology 75:2007–2014.

Marra, P. P. 2000. The role of behavioral dominance in structuring patterns of habitat occupancy in a migrant bird during the nonbreeding season. Behavioral Ecology 11:299–308.

Marra, P. P., K. A. Hobson, and R. T. Holmes. 1998. Linking winter and summer events in a migratory bird by using stable-carbon isotopes. Science 282:1884–1886.

Marra, P. P., and R. L. Holberton. 1998. Corticosterone levels as indicators of habitat quality: effects of habitat segregation in a migratory bird during the non-breeding season. Oecologia 116:284–292.

Marra, P. P., and R. T. Holmes. 2001. Consequences of dominance-mediated habitat segregation in American Redstarts during the nonbreeding season. Auk 118:92–104.

Medori, A. 1998. Seasonal and habitat changes in the diet of a Neotropical migrant warbler, with special emphasis on the conservation value of shade coffee plantations. Unpublished undergraduate thesis, Tulane University, New Orleans.

Miller, M. R., and W. E. Newton. 1999. Population energetics of northern pintails wintering in the Sacramento Valley, California. Journal of Wildlife Management 63:1222–1238.

Mitchell, P. I., I. Scott, and P. R. Evans. 2000. Vulnerability to severe weather and regulation of body mass of Icelandic and British redshank *Tringa totanus*. Journal of Avian Biology 31:511–521.

Møller, A. P., and K. A. Hobson. 2004. Heterogeneity in stable isotope profiles predicts coexistence of two populations of barn swallows *Hirundo rustica* differing in morphology and reproductive performance. Proceedings of the Royal Society of London, Series B, Biological Sciences 271: 1355–1362.

Moreau, R. E. 1972. The Palaearctic-African Bird Migration Systems. Academic Press, San Diego.

Morse, D. H. 1980. Population limitation: breeding or wintering grounds? Pages 505–516 *in* Migrant Birds in the Neotropics: Ecology, Behavior, Distribution, and Conservation (A. Keast and E. S. Morton, eds.). Smithsonian Institution Press, Washington, D.C.

Morton, E. S. 1980. Adaptations to seasonal changes by migrant land birds in the Panama Canal zone. Pages 437–453 *in* Migrant Birds in the Neotropics: Ecology, Behavior, Distribution, and Conservation (A. Keast and E. S. Morton, eds.). Smithsonian Institution Press, Washington, D.C.

Nariko, O., M. Yamamuro, J. Hiratsuka, and H. Satoh. 1999. Habitat selection by wintering tufted ducks with special reference to their digestive organ and to possible segregation between neighboring populations. Ecological Research 14:303–315.

Newton, I. 1998. Population Limitation in Birds. Academic Press, San Diego.

Norris, D. R., P. P. Marra, T. K. Kyser, T. W. Sherry, and L. M. Ratcliffe. 2004. Tropical winter habitat limits reproductive success on the temperate breeding grounds in a migratory bird. Proceedings of the Royal Society of London, Series B, Biological Sciences 271:59–64.

Parrish, J. D., and T. W. Sherry. 1994. Sexual habitat segregation by American Redstarts wintering in Jamaica: importance of resource seasonality. Auk 111:38–49.

Peach, W. J., S. R. Baillie, and L. Underhill. 1991. Survival of sedge Warblers *Acrocephalus schoenobaenus* in relation to west African rainfall. Ibis 133:300–305.

Pearson, D. L., and J. A. Derr. 1986. Seasonal patterns of lowland forest floor arthropod abundance in southeastern Peru. Biotropica 18:244–256.

Perez-Tris, J., and J. L. Telleria. 2002. Migratory and sedentary blackcaps in sympatric non-breeding grounds: implications for the evolution of avian migration. Journal of Animal Ecology 71:211–224.

Petit, D. R., J. F. Lynch, R. L. Hutto, J. G. Blake, and R. B. Waide. 1995. Habitat use and conservation in the Neotropics. Pages 145–197 *in* Ecology and Management of Neotropical Migrant Birds: A Synthesis and Review of Critical Issues (T. E. Martin and D. M. Finch, eds.). Oxford University Press, New York.

Poulin, B., and G. Lefebvre. 1996. Dietary relationships of migrant and resident birds from a humid forest in central Panama. Auk 113:277–287.

Poulin, B., G. Lefebvre, and R. McNeil. 1992. Tropical avian phenology in relation to abundance and exploitation of food resources. Ecology 73:2295–2309.

Pulliam, H. R., and T. A. Parker. 1979. Population regulation of sparrows. Fortschritte der Zoologie 25:137–147.

Rappole, J. H., and M. V. McDonald. 1994. Cause and effect in population declines of migratory birds. Auk 111:652–660.

Rappole, J. H., M. A. Ramos, and K. Winker. 1989. Wintering Wood Thrush movements and mortality in southern Veracruz. Auk 106:402–410.

Rappole, J. H., and D. W. Warner. 1980. Ecological aspects of migrant bird behavior in Veracruz, Mexico. Pages 353–393 *in* Migrant Birds in the Neotropics: Ecology, Behavior, Distribution, and Conservation (A. Keast and E. S. Morton, eds.). Smithsonian Institution Press, Washington, D.C.

Robinson, S. K., F. R. Thompson III, T. M. Donovan, D. R. Whitehead, and J. Faaborg. 1995. Regional forest fragmentation and the nesting success of migratory songbirds. Science 267:1987–1990.

Rödl, T. 1999. Environmental factors determine numbers of overwintering European stonechats *Saxicola rubicola*—a long-term study. Ardea 87:247–259.

Saino, N., T. Szép, R. Ambrosini, M. Romano, and A. P. Møller. 2004. Ecological conditions during winter affect sexual selection and breeding in a migratory bird. Proceedings of the Royal Society of London, Series B, Biological Sciences 271:681–686.

Sherry, T. W. 1984. Comparative dietary ecology of sympatric, insectivorous Neotropical flycatchers (Tyrannidae). Ecological Monographs 54:313–338.

Sherry, T. W., and R. T. Holmes. 1995. Summer versus winter limitation of populations: issues and the evidence. Pages 85–120 *in* Ecology and Management of Neotropical Migratory Birds (T. M. Martin and D. Finch, eds.). Oxford University Press, Oxford.

Sherry, T. W., and R. T. Holmes. 1996. Winter habitat limitation in Neotropical-Nearctic migrant birds: implications for population dynamics and conservation. Ecology 77:36–48.

Sillett, T. S., and R. T. Holmes. 2002. Variation in survivorship of a migratory songbird throughout its annual cycle. Journal of Animal Ecology 71:296–308.

Sillett, T. S., R. T. Holmes, and T. W. Sherry. 2000. The El Niño Southern Oscillation impacts population dynamics of a migratory songbird throughout its annual cycle. Science 288:2040–2042.

Sinclair, A. R. E. 1978. Factors affecting the food supply and breeding season of resident birds and movements of Palearctic migrants in a tropical African savannah. Ibis 120:480–497.

Smith, J. N. M., R. D. Montgomerie, M. J. Taitt, and Y. Yom-Tov. 1980. A winter feeding experiment on an island Song Sparrow population. Oecologia 47:164–170.

Strong, A. M. 1999. Effects of prey abundance on winter habitat quality for two species of ground-foraging Neotropical migrant warblers. Ph.D. dissertation, Tulane University, New Orleans.

Strong, A. M. 2000. Divergent foraging strategies of two Neotropical migrant warblers: implications for winter habitat use. Auk 117:381–392.

Strong, A. M., and T. W. Sherry. 2000. Habitat-specific effects of food abundance on the condition of ovenbirds wintering in Jamaica. Journal of Animal Ecology 69:883–895.

Strong, A. M., and T. W. Sherry. 2001. Body condition of Swainson's Warblers wintering in Jamaica, with emphasis on the conservation value of Caribbean dry forests. Wilson Bulletin 113:410–418.

Strong, A. M., T. W. Sherry, and R. T. Holmes. 2000. Bird predation on herbivorous insects: indirect effects on sugar maple saplings. Oecologia 125:370–379.

Terborgh, J. 1989. Where Have All the Birds Gone? Princeton University Press, Princeton,.

Tsipoura, N., and J. Burger. 1999. Shorebird diet during spring migration stopover on Delaware Bay. Condor 101:635–644.

Van Bael, S. A., J. D Brawn, and S. K. Robinson. 2003. Birds defend trees from herbivores in a Neotropical forest canopy. Proceedings of the National Academy of Sciences USA 100:8304–8307.

Veit, R. R., E. D. Silverman, and I. Everson. 1993. Aggregation patterns of pelagic predators and their principal prey, Antarctic krill, near South Georgia. Journal of Animal Ecology 62:551–564.

Watson, A., and P. J. O'Hare. 1979. Red grouse populations on experimentally treated and untreated Irish bog. Journal of Applied Ecology 16:433–452.

Webster, M. S., P. P. Marra, S. M. Haig, S. Bensch, and R. T. Holmes. 2002. Links between worlds: unraveling migratory connectivity. Trends in Ecology and Evolution 17:76–83.

Wiklund, C. G. 2001. Food as a mechanism of density-dependent regulation of breeding numbers in the merlin, *Falco columbarius*. Ecology 82:860–867.

Willis, E. O. 1966. The role of migrant birds at swarms of army ants. Living Bird 5:187–231.

Wolda, H. 1978. Seasonal fluctuations in rainfall, food and abundance of tropical insects. Journal of Animal Ecology 47:369–381.

Wolf, W. J., and C. J. Smit. 1990. The Banc d'Arguin, Mauritania, as an environment for coastal birds. Pages 17–37 *in* Homeward Bound: Problems Waders Face When Migrating from the Banc D'Arguin, Mauritania, to Their Northern Breeding Grounds in Spring (B. J. Ens, T. Piersma, W. J. Wolff, and L. Zwarts, eds.). Special edition of Ardea 78(1/2).

Wood, B. 1979. Changes in numbers of overwintering Yellow Wagtails and their food supplies in a west Africa savannah. Ibis 121:228–231.

Wormington, A., and J. H. Leach. 1992. Concentrations of migrant Diving Ducks at Point Pelee National Park, Ontario, in response to invasion of Zebra Mussels, *Dreissena polymorpha*. Canadian Field Naturalist 106:376–380.

Wunderle, J. M., Jr., and S. C. Latta. 1998. Avian resource use in Dominican shade coffee plantations. Wilson Bulletin 110:255–265.

Wunderle, J. M., Jr., and S. C. Latta. 2000. Winter site fidelity of Nearctic migrants in shade coffee plantations of different sizes in the Dominican Republic. Auk 117:596–614.

Yates, M. G., J. D. Goss-Custard, S. McGrorty, K. H. Lakhani, S. E. A. Le V. dit Durell, R. T. Clarke, W. E. Rispin, I. Moy, T. Yates, R. A. Plant, and A. J. Frost. 1993. Sediment characteristics, invertebrate densities and shorebird densities on the inner banks of the Wash. Journal of Applied Ecology 30:599–614.

Zumeta, D., and R. T. Holmes. 1978. Habitat shift and roadside mortality of Scarlet Tanagers during a cold, wet New England spring. Wilson Bulletin 90:575–586.

Zwarts, L., A.-M. Blomert, B. J. Ens, R. Hupkes, and T. M. van Spanje. 1990. Why do waders reach high feeding densities on the intertidal flats of the Banc d'Arguin, Mauritania? Pages 39–52 *in* Homeward Bound: Problems Waders Face When Migrating from the Banc D'Arguin, Mauritania, to Their Northern Breeding Grounds in Spring (B. J. Ens, T. Piersma, W. J. Wolff, and L. Zwarts, eds.). Special edition of Ardea 78(1/2).

T. SCOTT SILLETT
AND RICHARD T. HOLMES

Long-Term Demographic Trends, Limiting Factors, and the Strength of Density Dependence in a Breeding Population of a Migratory Songbird

EFFECTIVE CONSERVATION AND MANAGEMENT of migratory bird species requires an understanding of when and how their populations are limited and regulated. From 1986 to 1999, we studied the processes that determine the abundance of a temperate-tropical migrant passerine, the Black-throated Blue Warbler (*Dendroica caerulescens*), during the breeding portion of its annual cycle. We found that young fledged per territory, fledgling mass, and yearling recruitment were negatively correlated with adult density, whereas the proportion of nests depredated per year was not correlated with density. Annual fecundity was also negatively correlated with density-independent nest predation and fluctuated with variation in weather and food supply associated with the El Niño Southern Oscillation. Using matrix population models, we discovered that local abundance of Black-throated Blue Warblers could be regulated by the negative feedback on annual fecundity generated by the density of breeding adults. Taken together, these results indicate that warbler abundance is determined in part by a complex interaction between biotic and abiotic factors during the breeding season and provide the first assessment of the strength of density dependence in a temperate-tropical migrant passerine.

INTRODUCTION

The processes that limit and regulate bird abundance remain poorly known (Sinclair 1989; Murdoch 1994). Limiting factors affect the average fecundity, survival, and thus the size of a population (Sinclair 1989). The effects of limiting factors, such as weather and food supply, are not related to population size; that is, they are density independent. Regulatory processes, however, involve density-dependent mechanisms. These mediate local interactions among individuals, their natural enemies, and food supply to cause demographic rates (e.g., fecundity, survival) to decline when population density increases, and vice versa (Begon et al. 1996). Regulation, which requires density dependence, implies that there is a mean population size, or an equilibrium, around which the population fluctuates and that the variance of population size is bounded (Royama 1977; Turchin 1995; Hixon et al. 2002). The influence of a density-dependent process, however, could be masked in some years by a density-independent factor like climate variation. Thus, simply because a process operates in a density-dependent manner does not mean that it is consistently regulatory. The regulatory nature of a process can be assessed more precisely through the use of simulation models that can verify whether a density-dependent process can regulate abundance within the bounds actually observed in nature (Pennycuick 1969; Klomp 1980; Ekman 1984; Arcese et al. 1992; McCallum et al. 2000). The identification of limiting and regulatory processes is a key element for understanding population dynamics and is likewise necessary for the parameterization of models that can aid management and conservation efforts (Lebreton and Clobert 1991; Murdoch 1994; Runge and Johnson 2002).

The importance of knowing what factors affect the size of bird populations is underscored by recent declines in abundances of many species, especially temperate-tropical migratory passerines (Robbins et al. 1989; Terborgh 1989; Askins et al. 1990; Baillie and Peach 1992; Peterjohn et al. 1995; Peach et al. 1998). Documenting when and how populations of migratory songbirds are limited and regulated is complicated by the fact that these birds inhabit a variety of temperate and tropical habitats during their annual cycle. Furthermore, despite mixing of breeding populations on winter quarters (Chamberlain et al. 1997; Kimura et al. 2002; Rubenstein et al. 2002), events in one season can affect population dynamics in subsequent seasons (Baillie and Peach 1992; Marra et al. 1998; Sillett et al. 2000). However, the long-term demographic data required to determine the relative contribution of the breeding, migratory, and overwinter seasons to year-round population dynamics are lacking for most migratory bird species (Rappole and McDonald 1994; Sherry and Holmes 1995, 1996; Latta and Baltz 1997; Askins 2000).

Since 1986, we have been studying the year-round ecology of a migratory songbird, the Black-throated Blue Warbler (*Dendroica caerulescens*) in both its breeding and wintering areas (Holmes et al. 1989; Rodenhouse and Holmes 1992; Holmes et al. 1992, 1996; Sillett et al. 2000; Sillett and Holmes 2002). Recently, we have shown that multiple density-dependent mechanisms operate during the breeding period (Rodenhouse et al. 2003) at our study site in New Hampshire, where the warbler population has been stable for at least the past 30 years (Holmes and Sherry 2001). In this chapter, we examine in detail the demography of this breeding population from 1986 to 1999 to elucidate the factors that determine its local abundance. Our objectives are twofold. First, we analyze long-term data on reproductive ecology to identify which abiotic and biotic processes limit the warbler population during the breeding period and which function in a density-dependent manner. Second, we use a stochastic simulation model, parameterized with field data, to assess whether the density dependence measured in our study population can regulate local abundance within the range observed from 1986 to 1999. We use these results to consider the relative importance of breeding-season events to songbird population dynamics.

METHODS

Study Site and Species

We conducted our field research in the 3100-ha Hubbard Brook Experimental Forest in Woodstock and Thornton, New Hampshire, from May to August, 1986–1999. The Hubbard Brook valley, which was extensively logged in the early 1900s, is contiguous with the much larger White Mountain National Forest. Vegetation on our study plot was dominated by hardwoods 20–25 m tall, primarily American beech (*Fagus grandifolia*), sugar maple (*Acer saccharum*), and yellow birch (*Betula alleghaniensis*). The dense understory consisted mainly of hobblebush (*Viburnum alnifolium*), as well as striped maple (*A. pensylvanicum*) and seedlings and saplings of canopy trees. Further details on habitat can be found in Bormann and Likens (1979), Holmes and Sturges (1975), and Holmes (1990).

Our study site at Hubbard Brook represents high-quality habitat for the Black-throated Blue Warbler (Holmes et al. 1992, 1996), one of the most common breeding birds in hardwood forests in northern New England (Holmes 1994). Black-throated Blue Warblers are sexually dichromatic and insectivorous, and they breed in the understory of mature forests in northeastern North America (Holmes 1994). This open-cup-nesting species is territorial during the breeding season and adults have strong fidelity to territory sites from year to year. At Hubbard Brook, Black-throated Blue Warblers are typically socially monogamous breeders, although a few males (5–10%) are bigamous each year (Holmes et al. 1992; authors' unpubl. data). Clutch size is relatively invariant, ranging from three to five for first broods and from two to four for second broods, with a mean and modal clutch size of four (Holmes 1994). Pairs can successfully raise two broods of young per summer. Both parents share in feeding nestlings and fledglings, but only females build nests and incubate eggs. Natal dispersal for our population is high: none of over 2,500 young fledged from our study plot have returned to the plot as adults, although a very small number (<10) have been resighted as breeders within the larger Hubbard Brook valley. New breeders, or yearlings, however, disperse from other areas and settle on the study plot each year. The number of these yearling recruits is strongly and positively correlated with annual fecundity on the study site in the previous year (Sillett et al. 2000), indicating that our study population reflects Black-throated Blue Warbler population dynamics occurring at a broader, regional scale.

Field Methods

Reproductive ecology of Black-throated Blue Warblers was studied on a 64-ha plot at Hubbard Brook. This study plot was gridded into 25-m quadrats to facilitate daily territory mapping and locating nests. Territories were delineated by using the minimum convex polygon method (Ford and Myers 1981). Annual warbler density on the 64-ha plot (44.6 ± 2.4 adults) was determined by counting the number of whole or fractional territories within the plot, taking into account the pairing status of each territorial male (e.g., unpaired, monogamous, bigamous). A small number of territorial males (0–2) were unmated each year, and no unmated females were detected. Birds were captured in mist-nets and individually marked with a unique combination of one U.S. Fish and Wildlife Service aluminum leg band and two colored plastic leg bands. Individuals were aged as yearlings or older, by using plumage characters (Pyle et al. 1987).

We searched for nests of all Black-throated Blue Warblers breeding within the 64-ha plot, and found most nests during nest building. Nests were checked every 2 days until fledging or failure. In nearly all cases we determined cause of failure from the condition of failed nests (e.g., egg fragments or missing eggs indicated nest predation). Nestlings were weighed and banded on day 6 of the 8-day nestling period (hatching = day 0), the last day to safely handle young without causing premature fledging. Nestling mass on day 6 was used as an estimate of mass at fledging. Identity of males belonging to nests was verified by observing males and recording their leg band combinations when they came to nests to feed nestlings. Successful fledging was confirmed by noting parents feeding fledglings. Adults were monitored closely after a first brood fledged to ensure that we discovered if pairs attempted a second brood. In total, we collected data from 386 nests, not counting those abandoned during nest building.

Availability of the Black-throated Blue Warbler's primary food in summer, lepidopteran larvae, was quantified each year on four biweekly surveys beginning in late May. A survey entailed visually inspecting understory sugar maple and American beech leaves on transects through the study plot and recording the length of each caterpillar present (Rodenhouse and Holmes 1992). A total of 4,000 leaves per tree species were examined on each survey. Caterpillar lengths were converted to biomass using length-mass regressions (Rodenhouse 1986). Black-throated Blue Warblers usually forage in the understory (Robinson and Holmes 1982), and caterpillar biomass is not strongly stratified by height at our study site (Holmes and Schultz 1988).

Analysis of Long-Term Demographic Data

Linear regression was used to analyze the correlations between annual warbler density and several variables related to annual fecundity: young fledged, mean clutch size, proportion of females attempting second broods, mean fledgling mass, proportion of bigamous males, yearling recruitment, and annual variation in fledging success. Young fledged equaled the mean number of young fledged per territory per year. Mean fledgling mass was computed by first averaging mass on a per nest basis, then averaging means for all nests per territory per year. Yearling recruitment was defined as the number of whole and fractional territories of yearling males on the 64-ha plot in year_{i+1}. Annual variation in fledging success was represented by the standardized coefficient of variation (Sokal and Rohlf 1995) in number of young fledged per territory per year. The Durbin-Watson test for independence was used to test for first-order temporal autocorrelations in regression model residuals. All statistical analyses were performed with the JMP computer program (SAS Institute 2000).

We also used linear regression to examine the correlations between annual warbler density and two external variables that affect reproductive success: food availability and nest predation rate. Food availability was defined as total biomass of all lepidopteran larvae recorded in each of the four surveys described above. Nest predation rate was defined as the mean proportion of nests that were depredated per territory per year.

To test for an effect of warbler density on annual adult survival (ϕ) and recapture (p) probabilities, we used Cormack-Jolly-Seber (CJS) models (Pollock et al. 1990; Lebreton et al. 1992) and the MARK computer program (White and Burnham 1999). The data set for this analysis came from the color-banded population breeding on our 64-ha study plot and was composed of capture histories of 171 females and 165 males. Black-throated Blue Warbler survival and recapture probabilities at Hubbard Brook differed between females and males (model notation: [$\phi_{\text{sex}}\ p_{\text{sex}}$]), but not by bird age (Sillett and Holmes 2002). We therefore chose a set of candidate models (see Results) in which ϕ and p were always functions of sex. Adult density was included in models as a covariate; additive models were not considered because annual variation in ϕ differed by sex (Sillett and Holmes 2002). Model selection methods based on Akaike's Information Criterion (AIC) were used to assess the statistical evidence for an effect of adult density on ϕ and p (Burnham and Anderson 1998). Following Burnham and Anderson (1998), models were ranked by second-order AIC differences (AIC_c) and the relative likelihood of each model was estimated with AIC weights (w_i).

Multiple regression was used to examine the relative effects of biotic and abiotic limiting factors, as well as adult density, on two measures of annual warbler fecundity: number of young fledged and fledgling mass. The latter has been shown to be a predictor of fledgling survival in many bird species (Newton 1998). Because fecundity of Black-throated Blue Warblers can be strongly affected by nest predation (Holmes et al. 1992), nest predation rate was included in the two multiple regression models. Fecundity of this species is also limited by food availability and climatic variation (Rodenhouse 1992; Rodenhouse and Holmes 1992; Sillett et al. 2000). The El Niño Southern Oscillation, or ENSO, is one source of climatic variation in New England (Ropelewski and Halpert 1986; Montroy et al. 1998). At Hubbard Brook, warbler fecundity and the availability of the birds' primary food source, lepidopteran larvae, were significantly correlated with mean monthly values of the Southern Oscillation Index or SOI (Sillett et al. 2000), a standardized measure of ENSO phase (Kiladis and Diaz 1989). Strongly negative values of SOI indicate El Niño conditions, and strongly positive values indicate La Niña conditions. Between 1986 and 1998, warbler fecundity and caterpillar abundance were low in El Niño years and high in La Niña years (Sillett et al. 2000). SOI was not correlated with adult density ($r = 0.19$, $p = 0.51$) or with nest predation rate ($r = 0.16$, $p = 0.59$) from 1986 to 1999. We thus used SOI to represent variation in both climate and warbler food supply in the multiple regression models. Significance tests for model effects were depicted graphically with partial regression leverage plots (Rawlings 1988; Sall 1990).

Simulation Model

We constructed a stochastic Leslie matrix model (Caswell 2001) to test whether density-dependent fecundity (see Results) was sufficient to regulate local warbler abundance within the levels observed on our study area. Model output was compared with the actual number of males recorded on the 64-ha plot from 1986 to 1999. The two stages in the matrix were "yearling" and "adult" males. Matrix elements were yearlings per yearling (i.e., the number of yearlings recruited in year_{i+1} per yearling breeder in year_i), yearlings per adult, yearling survival, and adult survival.

We parameterized the model with estimates of survivorship and fecundity from the Black-throated Blue Warbler population at Hubbard Brook (table 32.1). Stage-specific survival estimates were taken from Sillett and Holmes (2002). Data on survival of Black-throated Blue Warblers from fledgling until their first breeding season were not available, but survival probabilities of small passerines during this interval typically ranges from 0.1 to 0.3 (Anders et al. 1997). We used 0.3 as our estimate of fledgling survival because values less than this resulted in a population growth rate (ϕ) of less than 1 in a deterministic model without density-dependent fecundity. Stage-specific fecundity values in the stochastic model were linear functions of population size produced by regressing annual male density on the 64-ha plot against (mean number of fledglings per year/2) for both yearlings and adults. Survivorship and the intercepts of the fecundity functions in the stochastic model were set as random variables, where survival probability was drawn from a beta distribution and fecundity was drawn from a log-normal distribution (Hilborn and Mangel 1997). Estimates of process variance (i.e., environmental stochasticity) in survival rates were taken from Sillett and Holmes (2002). For males, this process variance was close to 0. We chose a value of 0.01 for the model. The estimate of process variance in female survival rate (0.04 [see Sillett and Holmes 2002]) was used for fledglings. Initial population size for both models was one male. Matrices were projected for 50 years; this process was repeated 1,000 times for the stochastic model. All model calculations were performed in Excel (Microsoft Corporation 1999) with PopTools (Hood 2000).

RESULTS

Long-Term Demographic Patterns

Several variables related to warbler reproductive success were correlated with adult density. Mean number of young fledged per year, mean annual fledgling mass, and recruitment of yearlings in year_{i+1} were all significantly negatively correlated with density (fig. 32.1). Likewise, the proportion of females attempting second broods each year was

Table 32.1 Parameter values for a projection matrix model of the Black-throated Blue Warbler population breeding at Hubbard Brook Experimental Forest, New Hampshire

Parameter	Value
Adult, yearling survival	0.51 ± 0.01[a]
Fledgling survival	0.30 ± 0.04[b]
Yearling fecundity	[(1.47 ± 0.65)[c] – 0.01*(no. males)]*fledgling survival
Adult fecundity	[(4.04 ± 0.66)[c] – 0.09*(no. males)]*fledgling survival

Note: Estimates of adult survival are from Sillett and Holmes (2002). See Methods for parameter details.

[a]Mean ± 1 SD; drawn from a beta distribution.

[b]Drawn from a beta distribution; covaried with adult and yearling survival.

[c]Drawn from a log-normal distribution.

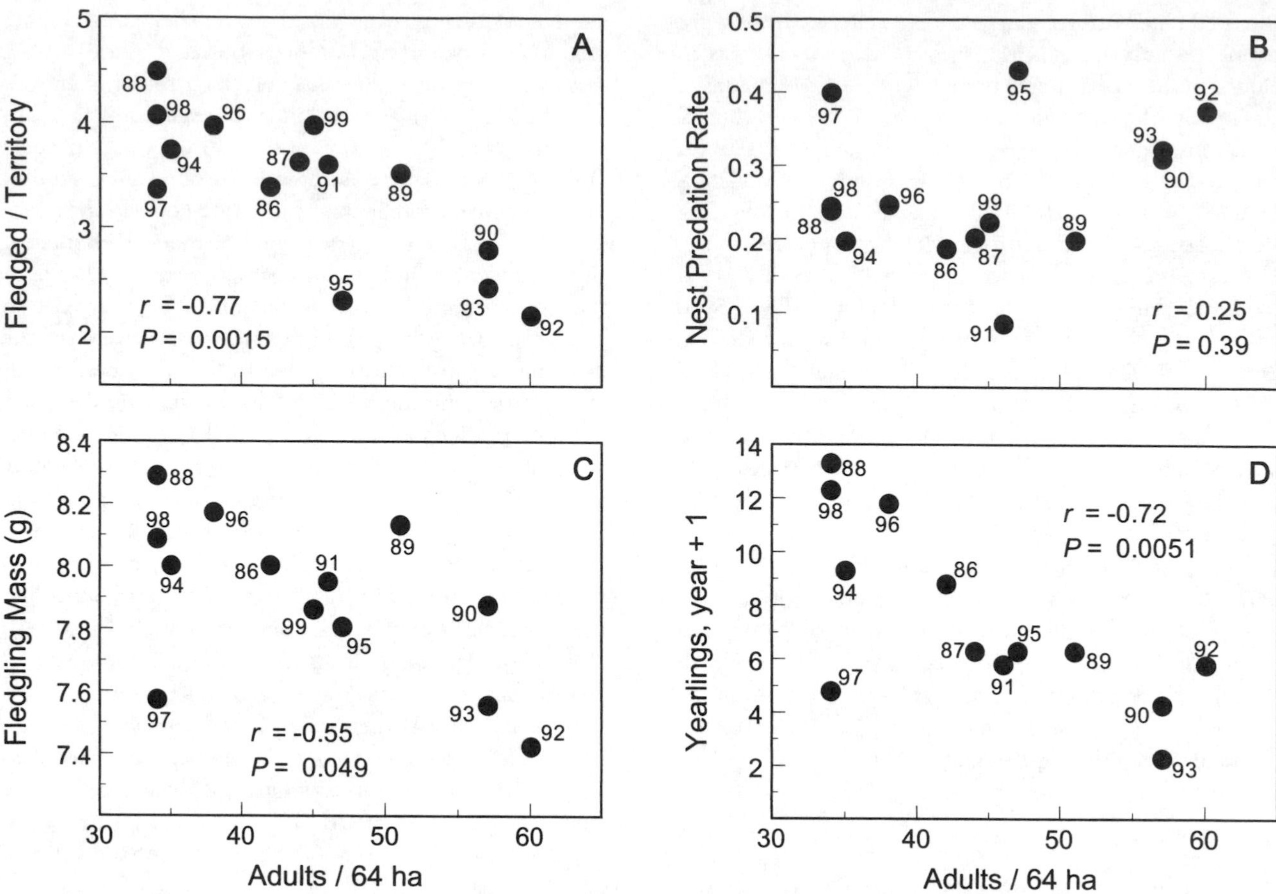

Fig. 32.1. Density of Black-throated Blue Warblers on a 64-ha plot at Hubbard Brook Experimental Forest, New Hampshire, 1986–1999, versus four variables related to reproductive success: mean number of young fledged annually per territory (A), mean nest predation rate per territory per year (B), mean annual fledgling mass (C), and number of yearling male recruits in year$_{i+1}$ (D). Numbers by points indicate year. Fledgling mass data were lacking for 1987. First-order temporal autocorrelation in model residuals was not significant in any of the four regressions (Durbin-Watson $d > 1.61$, $p > 0.19$).

strongly density-dependent ($r = -0.78$, $p = 0.001$). In contrast, mean clutch size for both first and second brood nests was not density-dependent (first broods: $r = -0.39$, $p = 0.17$; second broods: $r = -0.38$, $p = 0.19$), nor was the proportion of bigamous males per year ($r = -0.14$, $p = 0.63$). Similarly, variance in fledging success of Black-throated Blue Warblers did not increase with warbler density (fig. 32.2). Adult density was also not significantly correlated with either annual nest predation rate (fig. 32.1B) or with caterpillar biomass (for each of four counts, $|r| < 0.15$, $p > 0.61$).

Population density did not have a strong effect on CJS estimates of annual survival and recapture probabilities (table 32.2). According to Σw_i, simple sex-specific survival (table 32.2: models 1, 3) was 4.7 times more likely to be the best fit to our data than density-dependent survival (table 33.2: models 2, 4). Similarly, sex-specific recapture probabilities (table 32.2: models 1, 2) were 6.5 times more likely than density-dependent recapture probabilities (table 32.2: models 3, 4), given our data.

Multiple regression models (table 32.3) indicate that adult density, nest predation rate, and ENSO phase explained the

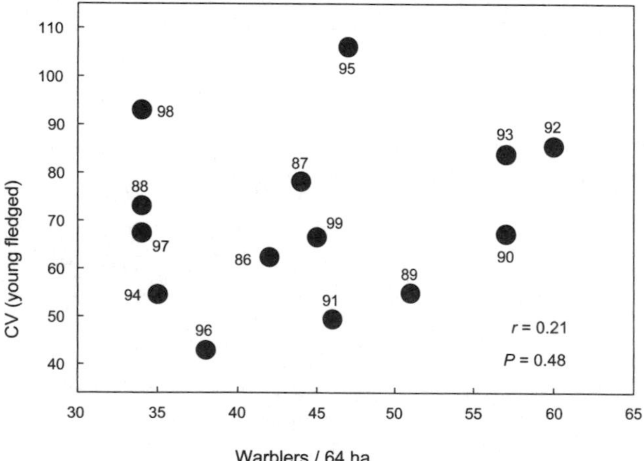

Fig. 32.2. Density of Black-throated Blue Warblers was not correlated with the unbiased coefficient of variation (CV) in mean number of young fledged per territory. Data are from a 64-ha plot at Hubbard Brook Experimental Forest, New Hampshire, 1986–1999. The unbiased CV was calculated as (standard deviation$_{\text{young fledged}}$*100/mean$_{\text{young fledged}}$)*(1 + 1/4n), where n = sample size (Sokal and Rohlf 1995).

Table 32.2 Models of annual survival (ϕ) and recapture (p) probabilities for Black-throated Blue Warblers at Hubbard Brook Experimental Forest, New Hampshire

Model	K	AIC$_c$	Δi	w_i
1. ϕ_{sex}, p_{sex}	4	768.84	0.00	0.71
2. $\phi_{sex*density}, p_{sex}$	6	771.88	3.05	0.16
3. $\phi_{sex}, p_{sex*density}$	6	772.55	3.72	0.11
4. $\phi_{sex*density}, p_{sex*density}$	8	775.85	7.02	0.02
5. $\phi_{sex*year}, p_{sex*year}$	39	811.02	42.18	0.00

Note: Columns give model notation, number of estimable parameters (K), second-order Akaike's information criterion values (AIC$_c$), AIC$_c$ differences (Δ_i), and AIC$_c$ weights (w_i). Subscripts describe parameterizations of ϕ and p: "sex" = parameters different between females and males; "density" = parameter a function of adult density; "year" = parameter varies annually. Subscripts joined by a * indicate a factorial model. All models were fit by using a logit link function. The program RELEASE (Burnham et al. 1987) indicates that the global model ($\phi_{sex*year}, p_{sex*year}$) provided a good fit to the data ($\chi^2_{38} = 19.37, p = 0.99$).

great majority of variation in both mean number of young fledged per year ($R^2 = 0.91$, $p < 0.0001$) and mean annual fledgling mass ($R^2 = 0.80$, $p = 0.002$). Number fledged decreased as both adult density and nest predation rate increased and increased with increasing values of SOI, indicating that more young were fledged in La Niña years (fig. 32.3A–C). Mean nestling mass was also negatively affected by increasing adult density and nest predation rate (fig. 32.3D–E), and was low in El Niño years and high in La Niña years (fig. 32.3F). Number fledged was most strongly correlated with adult density, followed by nest predation rate, then ENSO phase (table 32.3). Fledgling mass was most strongly correlated with ENSO phase, then with nest predation rate, and was only marginally correlated with adult density (table 32.3). Thus, two biotic factors—density-independent nest predation and density-dependent resource competition—the latter in combination with climatically affected food abundance, appear to be the most critical variables influencing warbler fecundity.

Simulation Model

The Leslie matrix model indicated that local warbler abundance can be tightly regulated by density-dependent fecundity observed in the field study (fig. 32.4). Size of simulated populations typically reached an asymptote after 20 years. Mean population size from the 1,000 model iterations was similar to the mean number of males recorded on the study plot from 1969 to 1999 (fig. 32.4). Furthermore, the 95% confidence interval from the 1,000 model iterations closely bracketed observed variation in warbler abundance (fig. 32.4).

DISCUSSION

Our results demonstrate that local abundance of Black-throated Blue Warblers at Hubbard Brook can be explained by a complex interaction of abiotic and biotic factors operating on breeding grounds. Fledging success, and to a lesser extent, fledgling mass, were correlated with nest predation rates and with fluctuations in climate and food supply associated with ENSO. However, nest predation, climatic variation, and food availability must be considered limiting fac-

Table 32.3 Analysis of variance and parameter estimates from two multiple regression models testing the effects of population density, nest predation rate, and ENSO phase on mean number of young fledged per year and fledgling mass from 1986 to 1999

Dependent variable	Source	d.f.	Estimate ± 1 SE	Mean square	F	p
Young fledged[a]						
	Adult density	1	−0.04 ± 0.008	1.96	34.14	0.0002
	Nest predation rate	1	−3.34 ± 0.73	1.22	21.19	0.001
	SOI	1	0.28 ± 0.09	0.59	10.35	0.009
	Error[b]	10		0.06		
Fledgling mass[c]						
	Adult density	1	−0.009 ± 0.004	0.09	4.86	0.06
	Nest predation rate	1	−1.13 ± 0.41	0.13	7.39	0.02
	SOI	1	0.17 ± 0.05	0.19	10.75	0.009
	Error[d]	9		0.02		

Note: Fledgling mass data were lacking for 1987. See Methods for description of independent variables.

[a]Intercept = 6.24 ± 0.35.

[b]Durbin-Watson $d = 2.14$, $p = 0.57$.

[c]Intercept = 8.64 ± 0.19.

[d]Durbin-Watson $d = 1.57$, $p = 0.17$.

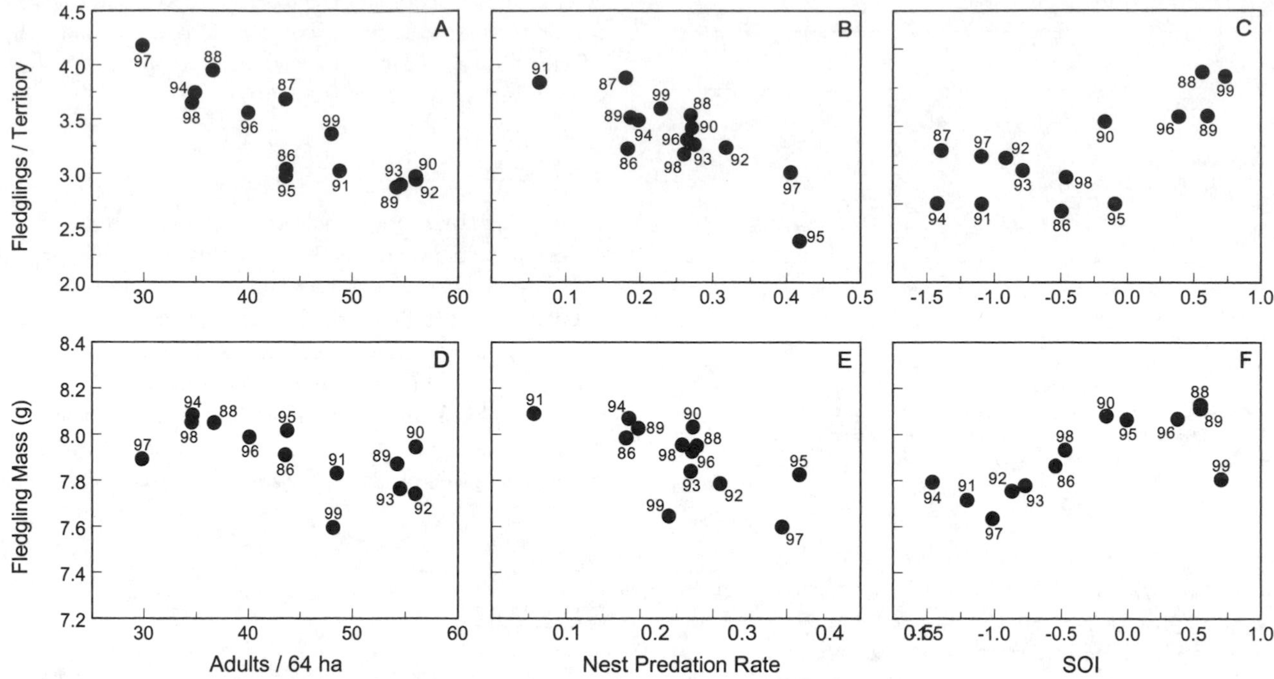

Fig. 32.3. Warbler density, nest predation rate, and ENSO phase explained 91% of the variation in mean number of Black-throated Blue Warbler young fledged per territory from 1986 to 1999 (partial regression leverage plots: A–C) and 80% of the variation in mean annual fledgling mass (partial regression leverage plots: D–F). ENSO phase for each year is represented by annual mean monthly values of the Southern Oscillation Index, or SOI (see Methods). Numbers by points indicate year. Fledgling mass data were lacking for 1987. Test statistics and parameter estimates for the two multiple regression models are given in table 33.3.

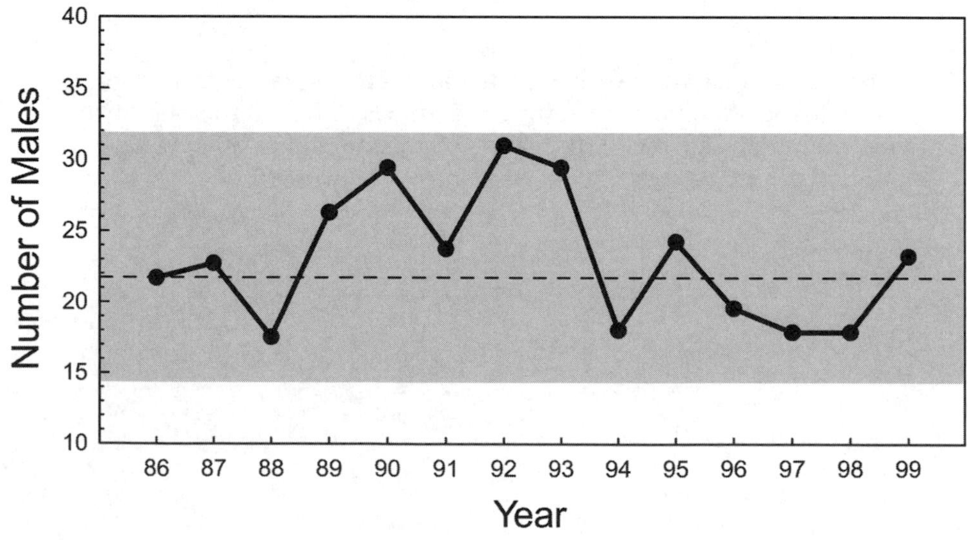

Fig. 32.4. Number of male Black-throated Blue Warblers recorded on a 64-ha plot at Hubbard Brook Experimental Forest, New Hampshire from 1986 to 1999 (solid line with • symbols) compared with output generated by a single sex; stochastic Leslie matrix model parameterized with estimates in table 33.1. The gray region indicates the 95% confidence interval for the model population at year 50. The dashed line gives mean population size from 1,000 model iterations. See Results for further details.

tors because their impact on fecundity was not related to warbler density. Similarly, the factors that determine annual survival rates of adults on our study plot were not density-dependent. Warbler abundance was regulated by the effect of breeding season population density on fledgling success and possibly by the effect of population density on fledgling mass. Recruitment of yearling breeders each May, and hence local population size, was in turn correlated with both fecundity (Sillett et al. 2000) and warbler density in the preceding summer (fig. 32.1D). Multiple regression indicates that the fecundity of this population was strongly affected by density-dependent and density-independent processes. Adult density appeared to have the greatest effect on warbler fledging success, whereas fledgling mass was most influenced by density-independent factors.

Density-dependent fecundity is a result of multiple demographic processes that determine the quantity and quality of offspring produced. Several factors could result in more young being produced at low density, such as larger clutch sizes (reviewed by Newton 1998), increased levels of bigamy, and higher frequency of double brooding (Kluyver 1971; Both 1998a). Fledgling quality (e.g., as indicated by fledgling mass) is thought to be negatively associated with density because of competition for limited food (Gaston 1985; Both 1998a; Both et al. 1999). For Black-throated Blue Warblers, density-dependent fledgling production was probably related to a declining probability of double brooding as density increased (see Results). Double brooding has a major effect on the seasonal fecundity of this warbler population (Nagy 2002). Fledgling mass in Black-throated Blue Warblers may be limited in part by food (Rodenhouse and Holmes 1992), but we found no direct correlation between warbler density and caterpillar abundance in this study.

Although many studies have documented density dependence in passerine birds (Sinclair 1989; Newton 1998; Paradis et al. 2002), few have examined whether the density-dependent processes measured can actually regulate abundance within meaningful boundaries. Model results given here show that density-dependent fledging success can regulate local abundance of Black-throated Blue Warblers within observed levels of variation observed in our study area. Density-dependent fecundity has also been shown to regulate populations of Great Tits (*Parus major*) and Song Sparrows (*Melospiza melodia*), although density-dependent juvenile survival and recruitment were of equal or greater importance for these populations (Pennycuick 1969; Klomp 1980; Arcese et al. 1992). For Willow Tits (*Parus montanus*) breeding in Sweden, density-dependent survival of yearlings was the only feedback mechanism capable of regulating population size (Ekman 1984). McCallum et al. (2000) used a simulation model to show that abundance of a tropical resident species, the Gray-breasted Silvereye (*Zosterops lateralis*) on Heron Island in Australia, is regulated by a density-dependent process, possibly juvenile survival. Estimating juvenile survival for migratory passerines is difficult and has only been done in a few studies (e.g., Anders et al. 1997). Although we lack such data for Black-throated Blue Warblers, we suspect that the Hubbard Brook population could also be weakly regulated by density-dependent juvenile survival, because fledgling mass is density-dependent (fig. 32.3D, table 32.2), and because fledgling mass is often correlated with first-year survival (Newton 1998).

Variance in fledging success of Black-throated Blue Warblers did not vary with density (see fig. 32.2) on our study plot. This implies that all individuals on the plot were affected equally by population density (Ferrer and Donazar 1996; Both 1998b). Furthermore, this result suggests that a crowding mechanism (Newton 1998; Rodenhouse et al. 1999) based on intraspecific resource competition may be the primary factor regulating fecundity, and thus local abundance, in areas of relatively homogeneous habitat.

Natural enemies are another factor that can limit bird populations, and many studies have demonstrated density-dependent rates of nest predation (Martin 1996; Newton 1998), brood parasitism (Arcese et al. 1992; Soler et al. 1998), and disease (Moss et al. 1990; Hochachka and Dhondt 2000). Our analysis of long-term demographic data indicate that nest predation is not density-dependent, and is thus not likely to be an important regulatory mechanism for the Black-throated Blue Warbler population at Hubbard Brook. This conclusion is further supported by Reitsma (1992), who found no evidence for density-dependent depredation of artificial nests in the shrub layer at Hubbard Brook. Brown-headed Cowbird (*Molothrus ater*) eggs have never been recorded in nests at Hubbard Brook, indicating that brood parasitism is not a factor affecting our study population. We have no data on parasitism by invertebrates such as *Protocalliphora* flies or on infection by disease organisms.

The results from this study demonstrate that the abundance of Black-throated Blue Warblers can be regulated by breeding-ground events. However, we do not contend that events during other phases of the annual cycle are unimportant to population dynamics. Abundance of migratory birds is certainly determined in part by density-dependent processes during the nonbreeding season. In particular, physical condition and thus survival of overwintering songbirds are likely to be affected by an interaction between population density and habitat availability because: (1) high-quality habitat on winter quarters appears to be limiting (Rappole and McDonald 1994; Sherry and Holmes 1995, 1996; Latta and Baltz 1997; Marra et al. 1998), (2) many species are strongly territorial, and (3) dominant individuals acquire the best territory sites (Marra 2000; Marra and Holmes 2001). Competitor density could also influence physical condition of migrating birds at stopover sites (Moore and Young 1991; Kelly et al. 2002).

Density-independent limiting factors, especially climatic fluctuations, also have important effects on adult and juvenile survival in the nonbreeding season (Baillie and Peach 1992; Sillett et al. 2000) and could overwhelm density-dependent processes. For example, matrix elasticity analyses (Caswell 2001) indicate that population growth rate and equilibrium population size in Black-throated Blue Warblers are most strongly affected by adult survival (Sillett 2000), which we have shown here to be density-independent. A factor that dramatically limits future adult survival could therefore have a greater influence on warbler population dynamics than density-dependent fecundity. This scenario is likely to have the strongest effect on population size if the rate of winter habitat destruction increases (see Sutherland 1996, 1998).

Because of well-documented population declines, temperate-tropical migratory birds have been the focus of much research on reproductive ecology and habitat selection. Nevertheless, basic demographic information is lacking for most species, data which are needed to parameterize models to assist in species management and conservation (Sherry and Holmes 1995). This study provides evidence that fecundity of at least one species of temperate-tropical migrant is influenced by conspecific density within breeding habitat and that this density dependence can regulate abundance within boundaries observed in the field. Knowledge of density dependence, and of the functional form of density-dependent relationships, is required to make accurate predictions regarding population viability and the response of populations to habitat change (May 1986; Sæther et al. 1998; Runge and Johnson 2002). As a caveat to the above, it is important to note that the relationships between density and demographic rates that we present here come from a source population (Holmes et al. 1996) breeding in fairly uniform, high-quality habitat. The strength of density-dependent processes is unknown for populations breeding in more heterogeneous environments or for sink populations breeding in suboptimal habitat.

ACKNOWLEDGMENTS

This research was funded by the U.S. National Science Foundation, Sigma Xi, and the Cramer Fund of Dartmouth College. We are grateful to the dozens of people who helped with fieldwork, especially J. Barg, P. Marra, N. Rodenhouse, T. Sherry, and M. Webster. This chapter benefited from the advice and comments of M. Ayres, D. Bolger, A. Dhondt, P. Doran, B. Dunning, G. Hood, P. Marra, M. McPeek, S. Morrison, L. Nagy, N. Rodenhouse, T. Sherry, A. Strong, and S. Zens. We thank the Hubbard Brook Experimental Forest of the U.S. Department of Agriculture Northeastern Research Station for their cooperation.

LITERATURE CITED

Anders, A. D., D. C. Dearborn, J. Faaborg, and F. R. Thompson. 1997. Juvenile survival in a population of Neotropical migrant birds. Conservation Biology 11:698–707.

Arcese, P., J. N. M. Smith, W. M. Hochachka, C. M. Rogers, and D. Ludwig. 1992. Stability, regulation, and the determination of abundance in an insular Song Sparrow population. Ecology 73:805–822.

Askins, R. A. 2000. Restoring North America's Birds: Lessons from Landscape Ecology. Yale University Press, New Haven.

Askins, R. A., J. F. Lynch, and R. Greenberg. 1990. Population declines in migratory birds in eastern North America. Current Ornithology 7:1–57.

Baillie, S. R., and W. J. Peach. 1992. Population limitation in Palearctic-African migrant passerines. Ibis 134:120–132.

Begon, M., J. L. Harper, and C. R. Townsend. 1996. Ecology. Blackwell Scientific, Boston.

Bormann, F. H., and G. E. Likens. 1979. Pattern and Process in a Forested Ecosystem. Springer-Verlag, New York.

Both, C. 1998a. Experimental evidence for density dependence of reproduction in Great Tits. Journal of Animal Ecology 67:667–674.

Both, C. 1998b. Density dependence of clutch size: habitat heterogeneity or individual adjustment? Journal of Animal Ecology 67:659–666.

Both, C., M. E. Visser, and N. Verboven. 1999. Density-dependent recruitment rates in Great Tits: the importance of being heavier. Proceedings of the Royal Society of London, Series B, Biological Sciences 266:465–469.

Burnham, K. P., and D. R. Anderson. 1998. Model Selection and Inference: A Practical Information-Theoretic Approach. Springer, New York.

Burnham, K. P., D. R. Anderson, G. C. White, C. Brownie, and K. H. Pollock. 1987. Design and analysis methods for fish survival experiments based on release-recapture. American Fisheries Society Monograph 5.

Caswell, H. 2001. Matrix Population Models: Construction, Analysis, and Interpretation. Sinauer Associates, Sunderland, Mass.

Chamberlain, C. P., J. D. Blum, R. T. Holmes, X. H. Feng, T. W. Sherry, and G. R. Graves. 1997. The use of isotope tracers for identifying populations of migratory birds. Oecologia 109:132–141.

Ekman, J. 1984. Stability and persistence of an age-structured avian population in a seasonal environment. Journal of Animal Ecology 53:135–146.

Ferrer, M., and J. A. Donazar. 1996. Density-dependent fecundity by habitat heterogeneity in an increasing population of Spanish Imperial Eagles. Ecology 77:69–74.

Ford, R. G., and J. P. Myers. 1981. An evaluation and comparison of techniques for estimating home range and territory size. Studies in Avian Biology 6:461–465.

Gaston, A. J. 1985. Development of young in the Atlantic Alcidae. Pages 319–354 *in* The Atlantic Alcidae (D. N. Nettleship and T. R. Birkhead, eds.). Academic Press, London.

Hilborn, R., and M. Mangel. 1997. The Ecological Detective: Confronting Models with Data. Princeton University Press, Princeton.

Hixon, M. A., S. W. Pacala, and S. A. Sandin. 2002. Population regulation: historical context and contemporary challenges of open vs. closed systems. Ecology 83:1490–1508.

Hochachka, W. M., and A. A. Dhondt. 2000. Density-dependent decline of host abundance resulting from a new infectious disease. Proceedings of the National Academy of Sciences USA 97:5303–5306.

Holmes, R. T. 1990. The structure of a temperate deciduous forest bird community: variability in time and space. Pages 121–139 *in* Biogeography and Ecology of Forest Bird Communities (A. Keast, ed.). Academic Publishing, The Hague, The Netherlands.

Holmes, R. T. 1994. Black-throated Blue Warbler (*Dendroica caerulescens*). Pages 1–24 *in* The Birds of North America, no. 87 (A. Poole and F. Gill, eds.). Academy of Natural Sciences, Philadelphia.

Holmes, R. T., P. P. Marra, and T. W. Sherry. 1996. Habitat-specific demography of breeding Black-throated Blue Warblers (*Dendroica caerulescens*): implications for population dynamics. Journal of Animal Ecology 65:183–195.

Holmes, R. T., and J. C. Schultz. 1988. Food availability for forest birds: effects of prey distribution and abundance on bird foraging. Canadian Journal of Zoology 66:720–728.

Holmes, R. T., and T. W. Sherry. 2001. Thirty-year bird population trends in an unfragmented temperate deciduous forest: importance of habitat change. Auk 118:589–610.

Holmes, R. T., T. W. Sherry, P. P. Marra, and K. E. Petit. 1992. Multiple brooding and productivity of a Neotropical migrant, the Black-throated Blue Warbler (*Dendroica caerulescens*), in an unfragmented temperate forest. Auk 109:321–333.

Holmes, R. T., T. W. Sherry, and L. Reitsma. 1989. Population structure, territoriality and overwinter survival of two migrant warbler species in Jamaica. Condor 91:545–561.

Holmes, R. T., and F. W. Sturges. 1975. Bird community dynamics and energetics in a northern hardwoods ecosystem. Journal of Animal Ecology 44:175–200.

Hood, G. 2000. PopTools, Version 2.06. www.dwe.csiro.au/vbc/poptools/.

Kelly, J. F., L. S. DeLay, and D. M. Finch. 2002. Density-dependent mass gain by Wilson's Warblers during stopover. Auk 119: 210–213.

Kiladis, G. N., and H. F. Diaz. 1989. Global climate anomalies associated with the Southern Oscillation. Journal of Climate 2:1069–1090.

Kimura, M., S. M. Clegg, I. J. Lovette, K. R. Holder, D. J. Girman, B. Mila, P. Wade, and T. B. Smith. 2002. Phylogeographical approaches to assessing demographic connectivity between breeding and overwintering regions in a Nearctic-Neotropical warbler (*Wilsonia pusilla*). Molecular Ecology 11:1605–1616.

Klomp, H. 1980. Fluctuations and stability in Great Tit populations. Ardea 68:205–224.

Kluyver, H. N. 1971. Regulation of numbers in populations of Great Tits (*Parus m. major* L.). Pages 507–524 *in* Dynamics of Populations (P. J. den Boer and G. R. Gradwell, eds.). Center for Agricultural Publishing and Documentation, Wageningen.

Latta, S. C., and M. E. Baltz. 1997. Population limitation in neotropical migratory birds: comments. Auk 114:754–762.

Lebreton, J.-D., K. P. Burnham, J. Clobert, and D. R. Anderson. 1992. Modeling survival and testing biological hypotheses using marked animals: a unified approach with case studies. Ecological Monographs 62:67–118.

Lebreton, J.-D., and J. Clobert. 1991. Bird population dynamics, management, and conservation: the role of mathematical modeling. Pages 105–125 *in* Bird Population Studies: Relevance to Conservation and Management (C. M. Perrins, J.-D. Lebreton, and G. J. N. Hirens, eds.). Oxford University Press, Oxford.

Marra, P. P. 2000. The role of behavioral dominance in structuring patterns of habitat occupancy in a migrant bird during the nonbreeding season. Behavioral Ecology 11:299–308.

Marra, P. P., K. A. Hobson, and R. T. Holmes. 1998. Linking winter and summer events in a migratory bird by using stable-carbon isotopes. Science 282:1884–1886.

Marra, P. P., and R. T. Holmes. 2001. Consequences of dominance-mediated habitat segregation in American Redstarts during the nonbreeding season. Auk 107:96–106.

Martin, T. E. 1996. Fitness costs of resource overlap among coexisting bird species. Nature 380:338–340.

May, R. M. 1986. When two and two do not make four: nonlinear phenomena in ecology. Proceedings of the Royal Society of London, Series B, Biological Sciences 228:241–266.

McCallum, H., J. Kikkawa, and C. Catterall. 2000. Density dependence in an island population of silvereyes. Ecology Letters 3:95–100.

Microsoft Corporation. 1999. Excel 2000. Microsoft Corporation, Redmond, Wash.

Montroy, D. L., M. B. Richman, and P. J. Lamb. 1998. Observed nonlinearities of monthly teleconnections between tropical Pacific sea surface temperature anomalies and central and eastern North American precipitation. Journal of Climate 11:1812–1835.

Moore, F. R., and W. Young. 1991. Evidence of food-based competition among passerine migrants during stopover. Behavioral Ecology and Sociobiology 28:85–90.

Moss, R., I. B. Trenholm, A. Watson, and R. Parr. 1990. Parasitism, predation and survival of hen Red Grouse *Lagopus lagopus scoticus* in spring. Journal of Animal Ecology 59: 631–642.

Murdoch, W. W. 1994. Population regulation in theory and practice. Ecology 75:271–287.

Nagy, L. R. 2002. Causes and consequences of individual variation in reproductive output in a forest-dwelling Neotropical migrant songbird. Ph.D. dissertation, Dartmouth College, Hanover, N.H.

Newton, I. 1998. Population Limitation in Birds. Academic Press, London.

Paradis, E., S. R. Baillie, W. J. Sutherland, and R. D. Gregory. 2002. Exploring density-dependent relationships in demographic parameters in populations of birds at a large spatial scale. Oikos 97:293–307.

Peach, W. J., S. R. Baillie, and D. E. Balmer. 1998. Long-term changes in the abundance of passerines in Britain and Ireland as measured by constant effort mist-netting. Bird Study 45:257–275.

Pennycuick, L. 1969. A computer model of the Oxford Great Tit population. Journal of Theoretical Biology 22:381–400.

Peterjohn, B. G., J. S. Sauer, and C. S. Robbins. 1995. Population trends from the North American Breeding Bird Survey. Pages 3–39 *in* Ecology and Management of Neotropical Migratory Birds: A Synthesis and Review of Critical Issues (T. E. Martin and D. M. Finch, eds.). Oxford University Press, New York.

Pollock, K. H., J. D. Nichols, C. Brownie, and J. E. Hines. 1990. Statistical inference for capture-recapture experiments. Wildlife Monographs 107:1–97.

Pyle, P., S. N. G. Howell, R. P. Yunick, and D. F. DeSante. 1987. Identification Guide to North American Passerines. Slate Creek Press, Bolinas, Calif.

Rappole, J. H., and M. V. McDonald. 1994. Cause and effect in population declines of migratory birds. Auk 111:652–660.

Rawlings, J. O. 1988. Applied Regression Analysis. Wadsworth, Pacific Grove, Calif.

Reitsma, L. 1992. Is nest predation density dependent? A test using artificial nests. Canadian Journal of Zoology 70: 2498–2500.

Robbins, C. S., J. R. Sauer, R. S. Greenberg, and S. Droege. 1989. Population declines in North American birds that migrate to the Neotropics. Proceedings of the National Academy of Sciences USA 86:7658–7662.

Robinson, S. K., and R. T. Holmes. 1982. Foraging behavior of forest birds: the relationships among search tactics, diet, and habitat structure. Ecology 63:1918–1931.

Rodenhouse, N. L. 1986. Food limitation for forest passerines: effects of natural and experimental food reductions. Ph.D. dissertation, Dartmouth College, Hanover, N.H.

Rodenhouse, N. L. 1992. Potential effects of climatic change on a Neotropical migrant landbird. Conservation Biology 6:263–272.

Rodenhouse, N. L., and R. T. Holmes. 1992. Results of experimental and natural food reductions for breeding Black-throated Blue Warblers. Ecology 73:357–372.

Rodenhouse, N. L., T. W. Sherry, and R. T. Holmes. 1999. Multiple mechanisms of population regulation: contributions of site dependence, crowding, and age structure. Pages 2939–2952 *in* Proceedings of the 22nd International Ornithological Congress (N. J. Adams and R. H. Slotow, eds.).

Rodenhouse, N. L., T. S. Sillett, P. J. Doran, and R. T. Holmes. 2003. Multiple density-dependence mechanisms regulate a migratory bird population during the breeding season. Proceedings of the Royal Society of London, Series B, Biological Sciences 270:2105–2110.

Ropelewski, C. F., and M. S. Halpert. 1986. North American precipitation and temperature patterns associated with the El Niño Southern Oscillation (ENSO). Monthly Weather Review 114:2352–2362.

Royama, T. 1977. Population persistence and density dependence. Ecological Monographs 47:1–35.

Rubenstein, D. R., C. P. Chamberlain, R. T. Holmes, M. P. Ayres, J. R. Waldbauer, G. R. Graves, and N. C. Tuross. 2002. Linking breeding and wintering ranges of a migratory songbird using stable isotopes. Science 295:1062–1065.

Runge, M. C., and F. A. Johnson. 2002. The importance of functional form in optimal control solutions of problems in population dynamics. Ecology 83:1357–1371.

Sall, J. P. 1990. Leverage plots for general linear hypotheses. American Statistician 44:303–315.

SAS Institute. 2000. JMP, Version 4. SAS Institute, Cary, N.C.

Sæther, B.-E., S. Engen, A. Islam, R. McCleery, and C. Perrins. 1998. Environmental stochasticity and extinction risk in a population of a small songbird, the Great Tit. American Naturalist 151:441–450.

Sherry, T. W., and R. T. Holmes. 1995. Summer versus winter limitation of populations: what are the issues and what is the evidence. Pages 85–120 *in* Ecology and Management of Neotropical Migratory Birds: A Synthesis and Review of Critical Issues (T. E. Martin and D. M. Finch, eds.). Oxford University Press, New York.

Sherry, T. W., and R. T. Holmes. 1996. Winter habitat quality, population limitation, and conservation of Neotropical-Nearctic migrant birds. Ecology 77:36–48.

Sillett, T. S. 2000. Long-term demographic trends, population limitation, and the regulation of abundance in a migratory songbird. Ph.D. dissertation, Dartmouth College, Hanover, N. H.

Sillett, T. S., and R. T. Holmes. 2002. Variation in survivorship of a migratory songbird throughout its annual cycle. Journal of Animal Ecology 71:296–308.

Sillett, T. S., R. T. Holmes, and T. W. Sherry. 2000. Impacts of a global climate cycle on population dynamics of a migratory songbird. Science 288:2040–2042.

Sinclair, A. R. E. 1989. Population regulation in animals. Pages 197–241 *in* Ecological Concepts: The Contribution of Ecology to an Understanding of the Natural World (J. Cherrett, ed.). 29th Symposium of the British Ecological Society. Blackwell, Oxford.

Sokal R. R., and F. J. Rohlf. 1995. Biometry: The Principles and Practice of Statistics in Biological Research (third ed.). W. H. Freeman and Co., San Francisco.

Soler, M., J. J. Soler, J. G. Martinez, T. Perez-Contreras, and A. P. Møller. 1998. Micro-evolutionary change and population dynamics of a brood parasite and its primary host: the intermittent arms race hypothesis. Oecologia 117:381–390.

Sutherland, W. J. 1996. Predicting the consequences of habitat loss for migratory populations. Proceedings of the Royal Society of London, Series B, Biological Sciences 263:1325–1327.

Sutherland, W. J. 1998. The effect of local change in habitat quality on populations of migratory species. Journal of Applied Ecology 35:418–421.

Terborgh, J. 1989. Where Have All the Birds Gone? Princeton University Press, Princeton.

Turchin, P. 1995. Population regulation: old arguments and a new synthesis. Pages 19–40 *in* Population Dynamics: New Approaches and Synthesis (N. Cappuccino and P. W. Price, eds.). Academic Press, San Diego.

White, G. C., and K. P. Burnham. 1999. Program MARK: survival estimation from populations of marked animals. Bird Study 46:120–139.

RUSSELL GREENBERG
AND PETER P. MARRA

The Renaissance of Migratory Bird Biology

A Synthesis

NEW APPROACHES TO OLD QUESTIONS

The study of bird migration has fascinated scientists since the early days of ornithological research. Our aim in assembling this book was to provide readers with a sense of renewal in all aspects of the study of bird migration from evolution and ecology to physiology and morphology. The major questions that drive research on migration systems continue to dangle out there like brass rings eluding our grasp: How did long-distance migration originate? How do different events and seasons shape the life history and adaptations of migratory species? How do migratory species interact with resident species? Although we await final answers to these and other questions that may never come, new approaches have been brought to bear on these time-honored issues. As witnessed within this volume, researchers are increasingly integrating studies of migratory birds with other seemingly unrelated disciplines of biology such as biogeochemistry and remote sensing.

The Evolution of Long-Distance Migration

Probably the single most significant improvement in our understanding of the evolution of migration has been the use of molecular genetics to determine the phylogenetic relationships between migratory and resident taxa. Several studies have been published to date (see Joseph, Chap. 2 [All such references herein are to chapters in this volume.]) and it is clear that birds have shown surprising evolutionary lability in migratory behavior. Although migration has deep roots in the major taxa involved, migratory patterns appear to be quite responsive to environmental change. The development of direct (e.g., fossil evidence) and indirect (e.g., isotopes analyses of climate change) approaches to establish what the paleoecosystems might have been in the regions where migration might have evolved provides a powerful backdrop for understanding these evolutionary studies (Steadman, Chap. 1). The use of new phylogenetic techniques, combined with traditional approaches within historical biogeography, provides a fresh breath for testing specific hypotheses about when and where migration evolved (Outlaw et al. 2003). These same techniques also provide better resolution than morphological analysis for how migratory and nonmigratory taxa are related (Joseph, Chap. 2).

The Diversity of Migration Systems

The diversity of avian migration systems is just becoming apparent. There are six major migration systems around the world: Palearctic-Afrotropical, Nearctic-Neotropical, Palearctic-Asian, Austral-African, Austral-Neotropical, and Austral-Asian (Hayes 1995). Only recently has the extent of migrations from Southern Temperate regions into the tropics (austral migration) been appreciated. Probably because of the limited landmass at high latitudes and the less continental climate of these regions, the diversity of austral migrants has been thought to be low. Also, the state of biogeography of Southern Hemisphere avifaunas still lags behind their boreal equivalents. However, upon close examination, we now know that some of these austral systems are quite diverse. For example, over 220 species of birds from the Southern Cone of South America are now classified as complete or partial migrants (Chesser, Chap. 14). Furthermore, we now have a much better understanding of some of the basic ecological and biogeographical features of at least the South American austral migration systems (Chesser, Chap. 14). We can begin to perceive finer patterns within austral migration systems, similar to those made for Northern Temperate Zone birds. Joseph (Chap. 2) makes a compelling argument that what we have considered to be the austral migration systems involves two very distinct strategies with respect to response to climate. Furthermore, the diversity of movements within tropical regions is receiving increased recognition. Such tropical movements include north/south movements confined to tropical areas—some of which involve crossing the equator to take advantage of different seasonal regimes. Intratropical movements can also involve altitudinal movements of birds (Loiselle and Blake 1992), particularly omnivorous species, and migrations between wet and dry habitats. Given that migration may have originally developed in multiple lineages under more tropical conditions during the early Tertiary (Steadman, Chap. 1), the study of present-day tropical migration systems may provide insight into the ecological prerequisites of migration (Chesser and Levey 1998).

Among landbirds, Old and New World migration systems are dominated by distantly related taxa. Hemispheric comparisons of migratory traits can be quite robust because a shared phylogenetic history is minimal. Such comparisons of unrelated taxa allow for the determination of which features are general and which are responsive to present-day ecological and historical differences between temperate and tropical ecosystems. For example, current conditions in northern Africa are considerably more arid than those in the northern Neotropics, although both regions showed periods of distinct aridity through the Pleistocene. We also know that the vegetation of Europe was more extremely affected during the glaciations than that of North America. The geography of barriers to migration differs as well. How these large-scale and the myriad more subtle differences shaped migration in the two hemispheres is just beginning to be explored in a comprehensive way.

Comparisons between Old and New World patterns is a pervasive theme in many of the chapters in the book. Patterns of distribution of migratory birds across the Holarctic are compared by Mönkkonen and Forsman (Chap. 11). The difference in habitat distribution of migrants in the Western Palearctic and Nearctic is strongly related to differences in habitats in Africa and the Neotropics (Böhning-Gaese, Chap. 11). However, differences in the response of vegetation to Pleistocene events need to be factored into future comparisons between the Old and New Worlds. Greenberg and Salewski (Chap. 26) find similarities in the degree of sociality in migrants in the two hemispheres, with telling differences (such as the lack of highly social and omnivorous Old World tropical migrants). Some of the differences between the New and Old Worlds are unexpected and demand further research to develop explanatory hypotheses. An example of this is the lower frequency of extra-pair fertilization in Old World migratory passerines (Stutchbury et al., Chap. 24). It should be noted that most comparisons focus on Western Palearctic / African migration versus North America (often just eastern North America). In many ways, the Eastern Palearctic is more similar to North America in its present-day habitat and Pleistocene vegetation history. We hope the next synthesis will involve more comparative work with this region as well.

The Nonbreeding Season

Thirty years ago, the Tropics formed a mysterious realm into which migratory birds disappeared for long periods. Perhaps nothing has changed more dramatically than the accessibility of the Tropics and the increase in our knowledge of the natural history of tropical avifaunas, both migratory and nonmigratory. Basic information on the natural history of migratory species on their tropical wintering grounds has increased but this period of the annual cycle is still seriously understudied (Rohwer, Chap. 8; Stutchbury et al., Chap. 24; Froehlich et al., Chap. 25; Greenberg and Salewski, Chap. 26; Runge and Marra, Chap. 28). For some tropical regions around the world, we have information on the geographic distributions of winter populations of migratory birds and some rudimentary habitat use data for many species, along with anecdotal observations of social and foraging behavior. We are just now beginning to see signs of greater attention to individual species and specific demographic classes within species during the nonbreeding season. However, quantified data have been garnered for only a few well-studied species. Information on the ecology of migrants wintering in East Asia, in particular, is still almost nonexistent.

The Macro-Distribution of Migratory Birds

MacArthur's (1959) focus on breeding landbirds of North America was the first attempt to find patterns in the relative abundance and diversity of migratory and resident birds. As more information has become available, researchers have looked across the different continents to establish general rules governing the distribution of migrant and resident taxa (Newton and Dale 1996). These approaches have generally been applied in accounting for patterns of distribution during the breeding season in the North Temperate Zone, but several authors in this volume have synthesized what we know about the relative importance of climate, habitat structure, and food resources across most of the major migration systems and periods of the annual cycle. Other chapters in this volume introduce innovative approaches to the question of the relationship between the distribution of long-distance migrant and resident birds in both the Temperate Zone and the Tropics. Various studies (Hockey, Chap. 5; Mönkkönen and Forsman, Chap. 11; Böhning-Gaese, Chap. 12; Gauthreaux et al., Chap. 15) each integrate, in their own way, the broad distributional patterns with the ecological processes that might mediate the relationships between migrants and their environment. This new integrated approach to the distribution of migrants is based on an array of concepts, some old and some new: evolutionary origins, global patterns of climate (temperature and wind patterns in the Temperate Zones, rainfall in the Tropics), the properties of ecosystems that contribute to climatic buffering, the seasonality of particular resource types, and the competitive and cooperative interactions between migrant and resident species (and how these change between climatic regimes). These concepts go a long way toward explaining the relative abundance and diversity of migratory and resident species, particularly in the Temperate Zone.

Studies of ecological interactions between migrants and residents in tropical areas have become less popular in recent years. It is increasingly clear from such analyses that our ability to investigate large-scale questions seeking to explain the distribution and abundance of species depends upon a more complete knowledge than we now have of the population ecology of individual species and specific guilds of species.

The Micro-Distribution of Migratory Birds

Avian ecology has traditionally focused on the role of resources (particularly food) in determining the distribution and abundance of species. Evaluating the relationship between migratory bird populations and their food supplies requires sound information on the abundance and seasonality of food and its variation between years and habitats. Furthermore, the data are more interpretable if data on diet are used to help select a resource-sampling scheme that focuses on the relevant prey items. With new attention to empirical evidence, Sherry et al. (Chap. 31) have established a prima facie case that food supply is often the most important limiting factor in the distribution and abundance of nonbreeding migratory birds in the Tropics. On the breeding ends of things, annual fluctuations in food supply also underlie variation in reproductive success and this variation appears to have important consequences for population size in subsequent years (Sillett et al. 2000).

Our understanding of population dynamics also requires detailed knowledge of the effect of different predators and parasites (avian brood parasites, ectoparasites, disease-causing microbial agents). The role of nest predators and brood parasites has long been incorporated into breeding season studies of migratory birds. However the effect of predation and disease on adult birds has been more elusive. Recent studies of incidence of human disease (e.g., West Nile virus) in migratory birds underscore their potential as long-distance vectors (Ricklefs et al., Chap. 17) and the topic is certain to receive much more research attention.

The analysis of morphological and behavioral adaptation has become increasingly sophisticated, allowing us to use the traditional comparative approaches with increasing rigor. The multivariate models necessary to handle complex and interrelated traits are far more robust than they were two decades ago, enabling researchers to simultaneously examine the effect of many different types of variables. More importantly, comparative analysis has seen the advent of techniques to tease apart correlated characters that are a result of shared phylogenetic history from those that are adaptive responses to similar environments. These statistical tools have resulted in our ability to draw out emergent patterns of behavior and morphology that distinguish migratory taxa from ecologically similar or related resident species. For example, adaptations that promote long-distance flight clearly provide constraints on trophic adaptations and related bill and hind-limb morphology (Winkler and Leisler, Chap. 7). The patterns that emerge from these broad comparative approaches should certainly inform the development of more focused experimental tests of function.

Demography and Life History

Demographic data, in particular, are crucial to answering key questions such as when in a migratory species' life cycle do population limitation and regulation occur and what are the life history trade-offs involved in the evolution of migration (Runge and Marra, Chap. 28, and Sillett and Holmes, Chap. 32). Life history data are also becoming increasingly important to address the conservation needs of migratory species. Our understanding of the population biology of migratory birds has grown from an increase in studies of marked populations as much as—perhaps more than—from any particularly theoretical development. These studies, particularly those carried out over many years, provide necessary fodder for the development of theory in the field. For example, the intriguing question of why birds migrate as far as they do (to the point that northern populations "leap-frog" over the winter range of more southerly breeding populations) has inspired some sophisticated models (Bell, Chap. 4) that require much more detailed demographic data to test.

We are beginning to see studies of populations of migratory species with uniquely marked individuals during the nonbreeding season and a few studies (too few) of species that encompass populations at both ends. The estimation of survival and reproductive output is increasingly gaining prominence in studies of migratory bird ecology. New statistical techniques for estimating survivorship and reproductive success have moved the field forward in the past decade. Estimating and untangling juvenile survivorship and natal dispersal remain the greatest methodological challenge to developing a complete picture of the demography of migratory species. Winkler (Chap. 30) reviews a number of promising approaches to the study of dispersal in migratory birds; most evidence still appears to support the notion that migratory birds tend to have longer dispersal distances and that this results in more open and less differentiated local populations.

NEW PERSPECTIVES

Cross-Seasonal Approaches Are Necessary

We believe that the single most profound epiphany in the study of migratory birds is that their life histories and adaptations simply cannot be understood through increasingly detailed studies in a single season. Although this idea is not completely new, the degree to which it has permeated the thinking of ornithologists studying migratory species represents a clear paradigm shift that is reflected in all of the contributions to this volume. The contributions on morphology, cognition, foraging behavior, population biology, macroecology, and evolutionary processes (such as speciation) all underscore the fact that we need to integrate the effects of the different phases in the life of migratory birds to achieve a holistic understanding of each of these topics.

For example, early research examining the role of natural selection in shaping life histories and adaptations in migrants analyzed it as two competing hypotheses: either the breeding season is critical or the winter is. It should be clear that for a migratory bird to reproduce successfully—the hallmark of evolutionary success—it must possess a phenotype that functions in all seasons. This will almost invariably result in a convex fitness set, where traits are not specialized for one particular season or habitat, but instead function successfully at all seasons. This generally requires adaptive compromises for any one season that can only be understood by examining how things work at all seasons. Price and Gross (Chap. 27) provides an intriguing example of how the ecological niches of *Phylloscopus* warblers are constrained across seasons.

Adaptive compromise is a critical theme in understanding various specific adaptations in migratory birds. For example, studies in this volume of external morphology (Winkler and Leisler, Chap. 32), physiology (McWilliams and Karasov, Chap. 6; Holberton and Dufty, Chap. 23), energy storage and management (Rogers, Chap. 9) are beginning to show how we can integrate selection through the annual cycle to understand adaptation in migratory species. Of course, the need to be successful in the face of changing conditions across the calendar is not unique to migratory birds, but can be found in all birds facing seasonal environments. Migratory birds are, therefore, a model system for studying adaptation in seasonal environments—an issue of general biological significance. Webster and Marra (Chap. 16) provide clear evidence that it is critical to understand the contribution of selection in different seasons and that events in one season can carry over to another. The importance of carry-over effects is modeled by Runge and Marra (Chap. 28) and show that competition for resources during the winter expresses itself in differential reproductive success on the breeding grounds thousands of kilometers away.

Ecological decisions made in one season are integrated into the selective factors shaping other necessary activities at other times in the annual cycle. The tools of cognitive ethology and comparative psychology are just beginning to be applied to learning how individuals acquire and use information during and between migratory excursions (Mettke-Hoffman and Greenberg, Chap. 10). A more holistic approach to the annual calendar must go beyond considerations of winter and summer events (two worlds) to include all time-sensitive activities. Constraints on where and when birds must migrate to optimize resources and flying conditions along the way certainly affect the choices birds make during the more stationary periods of their annual cycle. Data are beginning to confirm that migration is a time of year that accounts for much of the mortality (Sillett and Holmes, Chap. 32) and influences the condition of individual birds as they face the breeding season (Moore et al., Chap. 20, and Piersma et al., Chap. 21). As a consequence, we are beginning better integration of migration studies with our understanding of life history and demography as a whole. However, because individuals are hard to follow for extended periods, migration research requires additional innovative approaches to make major advances. As little as we know about how the period of migration affects the overall life history of migratory birds, the transition periods—such as the period immediately following breeding—

are even more mysterious. Rohwer et al. (Chap. 8) reveal the richness of strategies influencing where and when to complete molt, arguably the most critical of the post-breeding activities. A thorough understanding of when and where birds molt is not only important in its own right, but is critical for using isotopes in feathers (see below) to study migratory connectiveness.

Migration, Speciation, and Population Connectivity

The role of migration in influencing evolutionary divergence of populations has long been an area of speculation (Salomenson 1955). Generally, this thinking has focused on the role of migration in increasing dispersal distances and gene flow between breeding populations—thus inhibiting genetic divergence. Additionally, selection during the nonbreeding season may counter any selection for local adaptation that occurs when populations segregate in the breeding season. Such are the classic areas of inquiry into the evolutionary significance of migration. However, recent research has revealed other ways in which migration can influence the fundamental process of speciation. The study of migratory divides, which is exemplified in chapters by Irwin and Irwin (Chap. 3) and by Smith et al. (Chap. 18), suggest that geographically imposed migratory divides may select against hybrids and hence hasten the rate at which species form on either side of the divide. Migration may also contribute to speciation through a hypothesized "migration dosing," as is proposed by Bildstein and Zalles (Chap. 13). They develop the argument that migrants blown off course (i.e., vagrants) may form the nucleus of populations that colonize islands and other isolated landforms. In the rare circumstance that these birds form stable reproductive populations, speciation may eventually occur. In summary, migration appears, at least theoretically, able to either impede or facilitate speciation depending on the situation.

Processes such as speciation, acting primarily on breeding areas, have consequences on the distribution of migratory birds during the nonbreeding season. Webster and Marra (Chap. 16) outline a broad and important research agenda on population connectivity that needs to be addressed as we learn more about how breeding populations map onto winter distributions and vice versa. Because such mapping will allow researchers to pursue many different kinds of questions, we spent considerable space in this volume discussing the various tools for the job. These tools include the use of genetic markers, stable isotopes, and trace elements in bird tissues (Smith et al., Chap. 18, and Hobson, Chap. 19). With the tools available to characterize populations, we may soon be able to link performance between particular areas of the breeding and wintering grounds (given that such structure exists). Szép and Møller (Chap. 29) take this bold step by relating the survival of two swallow species between breeding seasons in Europe to satellite-determined data on the condition of rainfall-dependent vegetation across the African continent. Clearly, understanding both within- and between-season population connectivity will provide a more accurate picture of what a "population" truly is for a migratory species.

Expanding Research in Time and Space

Ornithologists have come a long way in the past two decades in recognizing the importance of long-term climatic cycles and rare disturbance events in shaping the life histories of birds. This increasing emphasis on long-term events has been fueled by the inadequacy of equilibrium-based models for explaining real populations in natural ecosystems, as well as by the need to place human-caused change in its proper context. We have made real progress in demonstrating the role of known climatic events (such as the El Niño Southern Oscillation) in shaping the life history of migratory and nonmigratory birds (Sillett et al. 2000). And we now understand that disturbance factors, such as severe weather and fire, have probably affected migratory birds as much as periods of stasis have (Rotenberry et al. 1995). Taking an even broader view of environmental instability, Winkler (Chap. 30) makes a tantalizing suggestion that dispersal strategies of temperate zone birds may have been shaped by the historical loss and gain of large areas of appropriate habitat during the Pleistocene.The unit of study has been changing as well. Avian ecologists now recognize that "communities" the size of study plots do not necessarily represent the dynamics of populations that interact across vast areas of natural and human-influenced landscapes (Pulliam 1988). We have a long way to go before we can understand the dynamics of populations over the relevant landscape mosaic during both the breeding and nonbreeding seasons, but this should remain a research priority.

A Focus on Individuals

Recent advances in behavioral ecology have brought us a major shift in emphasis over the past two decades. The individual bird is now taking center stage in our efforts to understand the ecology and evolution of migratory birds. Accompanying this greater interest in the performance of individuals is an increase in our ability to track their fates. With radiotelemetry we are better able to examine the decision-making strategies of individuals at the local scale, and thanks to advances in technology in satellite telemetry capabilities for larger species, we can now track birds through entire migrations. Cochran and Wikelski (Chap. 22) elaborate on these advances and provide an exciting vision for our ability to track individuals from space over even greater distances.

Ornithologists often focus on measuring the abundance of individuals over space. However, aware that interpreting presence / absence data is fraught with problems, ecologists and conservation biologists are increasingly concerned with the actual fitness consequences of different ecological decisions. Although fitness itself is the measure of ultimate importance, for most organisms—particularly highly mobile ones—fitness is almost impossible to measure. It is even more difficult to assess how any one aspect of the life history of a bird affects its fitness. So ornithologists are employing, and in some cases developing, new techniques to measure the condition of individuals that in all likelihood correlate with fitness itself. Fitness indices, such as hormone levels, mass changes, and immune responses, do indeed represent an advance from the past but must still be compared with more direct measures of fitness so that we can be certain of their utility.

The traditional view is that most migratory birds form pair bonds that facilitate cooperation between the sexes to raise young. Recently, the nature of the pair bond has been challenged in several different ways (Stutchbury et al., Chap. 24). First, genetic analyses have shown that in migratory songbirds, many pairs are not truly monogamous. The male that participates in the cooperative activities associated with the pair bond does not necessarily sire all of the young. This discovery has led to a rich line of research on what factors favor the occurrence of extra-pair fertilizations and which pair member is in control of the reproductive decisions. Second, several interesting facets of male-female dynamics during the nonbreeding season have been discovered. For example, the number of taxa where it is known that pair bonds are developed and maintained during the nonbreeding season has expanded. Classically, extra-seasonal pair bonds were thought to be restricted to a few groups of long-lived migratory species—such as waterfowl. Now we know that a scattering of songbirds also develop such bonds. However, the possibility that pair bonds are maintained between seasons needs further exploration. Also interesting is the recent evidence that males and females compete for resources during the nonbreeding season. Evidence thus far suggests that males are behaviorally dominant over females and that this has consequences for female survival. Not understood, however, is how this can act as an evolutionary stable strategy.

MISSING PERSPECTIVES

As comprehensive as we believe this book is, we cannot help thinking about the important topics that did not receive coverage. Some of these areas may prove to be critical in any future synthesis about the ecology and evolution of bird migration. We would therefore like to touch briefly upon some of the holes in the tapestry of this volume with the hope that mentioning these gaps will still inspire future research.

Plasticity in Migratory Programs

The role of genetic control versus developmental plasticity and environmental influence of the migratory program, and their interaction, is a subject that will loom large in any future discussion of the evolution of migration patterns. Research in the past two decades has established for some species the importance of genetically controlled endogenous rhythms in migratory behavior that play a key role in establishing the timing and at least general features of the geography of migration. However, the facultative response of Zugunruhe and other components of the migration program have been confirmed only in short-distance migrants and one or two Palearctic migratory species. How individuals make their own unique migration decisions and respond to variation in the environment is an area for physiologist and cognitive ethologists to explore, but it has profound implications for how migration can be coupled with or decoupled from environmentally based selection. Furthermore, the responsiveness of hard-wired behaviors to selection is an area that is just beginning to be explored.

Such plasticity in avian migratory programs is evident at both the individual and population levels. It is likely that even the most spectacular long-distance journeys evolved through gradual stages from residency, to partial migration where only some individuals in a population migrate, to complete short-distance movements, and finally to long-distance migrations. Detailed study of populations displaying partial migration systems (Cristol et al. 1998) may provide insights into the trade-offs underlying all migration systems. Partial migration systems are not a focus of any of the contributions to this volume, but their study represents a continuing research opportunity.

The Role of Pioneers and Vagrants

When it comes to normal dispersal, we are just beginning to frame the right questions and develop the methodologies to approach such questions for migratory species (Winkler, Chap. 30). However, the role of individuals that disperse well outside of the "normal" range of a species—so-called vagrant birds—has not been investigated from an evolutionary or ecological perspective (but see Bildstein and Zalles, Chap. 13). While these individuals provide much of the sport for birders, the numbers in which they occur suggest that they may also play an important role in the long-term success of lineages of migratory birds. In the highly unpredictable and rapidly changing Quaternary and Holocene climates and biomes, birds that produce pioneering phenotypes may be ecological winners. The existence of extreme vagrants may be a highly visible endpoint to a gradient of hyperdispersal in Temperate Zone migratory birds. The ecological and taxonomic correlates of avian vagrancy may be an area of high-risk research with a large potential pay-off, but the possible role of vagrants in speciation (migration-dosing [Bildstein and Zalles, Chap. 13]) remains an important area of research and discovery.

From Orientation to Habitat Selection

Orientation mechanisms have inspired some of the most elegant experimental work in animal behavior and physiology. After decades of this research, however, we are still far from understanding how birds locate and choose particular sites at either end of their migratory route and how they are able to retrace their trips so precisely. The prospect of tracking individuals opens up the possibility of developing greater understanding for the mechanisms by which particular habitat patches are selected.

FUTURE DIRECTIONS

The hallmark of success of any scientific field is the degree to which it attracts the best and brightest of the next generation of investigators. We believe that the understanding of how migratory behavior shapes the adaptations of birds is inherently fascinating. More generally, migration systems provide an excellent opportunity for examining how environmental challenges at different stages of a life cycle shape avian adaptations. New technologies, better natural history information, and new perspectives have reinvigorated the field of ornithology and, more specifically, migratory bird biology. Thus we believe that this volume, in each of its chapters, provides any young investigator a wealth of important and interesting questions to pursue. We offer the following additional questions as our take on what some of the most interesting challenges of the coming decades might be. Our hope is that when another symposium is organized in 10 years, we have a few answers and a new list of even more compelling questions.

1. How can we integrate phylogenetic hypotheses for the evolution of different clades with our increasing understanding of paleoenvironments in different regions? How much were the current migration systems shaped by deep (Tertiary) versus recent (Quaternary) events?
2. What are the likely ecological attributes of taxa that gave rise to migratory populations?
3. To what degree do density-dependent versus density-independent events during migration contribute to the population dynamics of migratory birds?
4. What factors determine seasonal population limitation and at what time of year are populations of migratory birds most limited?
5. What are the primary factors that regulate populations of migratory birds?
6. What are the trade-offs in foraging ecology and trophic adaptations in species that migrate between temperate and tropical habitats?
7. How do social interactions and dominance relationships in one season affect events in subsequent seasons?
8. What cognitive abilities distinguish migratory and non-migratory species?
9. How do breeding populations map onto winter distributions across migratory species and vice versa? What are the social and ecological factors that underlie these patterns?
10. To what degree do competitive interactions between migrant species and between migrant and resident species shape the distribution and habitat use of migrant species year round?
11. What ecological and climatic factors shape patterns of migration and determine the travel and stopover strategies?
12. How does migration contribute to evolutionary processes within species and to the formation of new species?
13. To what degree are the annual time activity budget and migration strategies genetically constrained and what is the importance of behavioral plasticity?

LITERATURE CITED

Chesser, T., and D. J. Levey. 1998. Austral migrants and the evolution of migration in New World birds: diet, habitat, and migration revisited. American Naturalist 152:311–319.

Cristol, D. A., M. B. Baker, and C. Carbone. 1999. Differential migration: latitudinal segregation by age and sex class. Current Ornithology 15:33–88.

Hayes, F. E. 1995. Definitions for migrant birds: what is a Neotropical migrant? Auk 112:521–523.

Loiselle, B. A., and J. G. Blake. 1992. Population variation in a tropical bird community: implications for conservation. BioScience 42:838–845.

MacArthur, R. H. 1959. On the breeding distribution patterns of North American migrant birds. Auk 76:318–325.

Newton, I., and L. Dale. 1996. Relationship between migration and latitude among west European birds. Journal of Animal Ecology 65:137–146.

Outlaw, D. C., G. Voelker, B. Milá, and D. Girman. 2003. The evolution of long-distance migration and historical biogeography of the *Catharus* thrushes: a molecular phylogenetic approach. Auk 120(2):299–310.

Pulliam, H. R. 1988. Sources, sinks, and population regulation. American Naturalist 132:652–661.

Rotenberry, J. T., R. J. Cooper, J. M. Wunderle, and K. G. Smith. 1995. When and how are populations limited? The roles of insect outbreaks, fire, and other natural perturbations. Pages 55–84 *in* Ecology and Management of Neotropical Migratory Birds (T. E. Martin and D. M. Finch, eds.). Oxford University Press, Oxford.

Salomenson, F. 1955. The evolutionary significance of bird migration. Biologiske Meddelelser 22:1–62.

Sillett T. S., R. T. Holmes, and T. W. Sherry. 2000. Impacts of a global climate cycle on population dynamics of a migratory songbird. Science 288:2040–2042.

Index